| Distribution | Probability Density Functions in One Variable | | |
|---|---|---|---|
| | Probability Density Function | Mean | Variance |
| Binomial, $B(n, p)$ | $f(x) = \binom{n}{x} p^x q^{n-x}, x = 0, 1, \ldots, n;$ $0 < p < 1, q = 1 - p$ | $np$ | $npq$ |
| Bernoulli, $B(1, p)$ | $f(x) = p^x q^{1-x}, x = 0, 1$ | $p$ | $pq$ |
| Geometric | $f(x) = pq^{x-1}, x = 1, 2, \ldots;$ $0 < p < 1, q = 1 - p$ | $\frac{1}{p}$ | $\frac{q}{p^2}$ |
| Poisson, $P(\lambda)$ | $f(x) = e^{-\lambda} \frac{\lambda^x}{x!}, x = 0, 1, \ldots; \lambda > 0$ | $\lambda$ | $\lambda$ |
| Hypergeometric | $f(x) = \dfrac{\dbinom{m}{x}\dbinom{n}{r-x}}{\dbinom{m+n}{r}},$ where $x = 0, 1, \ldots, r \left(\binom{m}{r} = 0, r > m\right)$ | $\frac{mr}{m+n}$ | $\frac{mnr(m+n-r)}{(m+n)^2(m+n-1)}$ |
| Gamma | $f(x) = \frac{1}{\Gamma(\alpha)\beta^\alpha} x^{\alpha-1} \exp\left(-\frac{x}{\beta}\right), x > 0;$ $\alpha, \beta > 0$ | $\alpha\beta$ | $\alpha\beta^2$ |
| Negative Exponential | $f(x) = \lambda \exp(-\lambda x), x > 0; \lambda > 0;$ or $f(x) = \frac{1}{\mu} e^{-x/\mu}, x > 0; \mu > 0$ | $\frac{1}{\lambda}$ $\mu$ | $\frac{1}{\lambda^2}$ $\mu^2$ |
| Chi-square | $f(x) = \frac{1}{\Gamma(\frac{r}{2})2^{r/2}} x^{\frac{r}{2}-1} \exp\left(-\frac{x}{2}\right), x > 0;$ $r > 0$ integer | $r$ | $2r$ |
| Normal, $N(\mu, \sigma^2)$ | $f(x) = \frac{1}{\sqrt{2\pi}\sigma} \exp\left[-\frac{(x-\mu)^2}{2\sigma^2}\right],$ $x \in \Re; \mu \in \Re, \sigma > 0$ | $\mu$ | $\sigma^2$ |
| Standard Normal, $N(0, 1)$ | $f(x) = \frac{1}{\sqrt{2\pi}} \exp\left(-\frac{x^2}{2}\right), x \in \Re$ | $0$ | $1$ |
| Uniform, $U(\alpha, \beta)$ | $f(x) = \frac{1}{\beta-\alpha}, \alpha \le x \le \beta;$ $-\infty < \alpha < \beta < \infty$ | $\frac{\alpha+\beta}{2}$ | $\frac{(\alpha-\beta)^2}{12}$ |

| Probability Density Functions in Many Variables | | | |
|---|---|---|---|
| **Distribution** | **Probability Density Function** | **Means** | **Variances** |
| **Multinomial** | $f(x_1, \ldots, x_k) = \frac{n!}{x_1! x_2! \ldots x_k!} \times$ $p_1^{x_1} p_2^{x_2} \ldots p_k^{x_k}, x_i \geq 0$ integers, $x_1 + x_2 + \cdots + x_k = n; p_j > 0, j = 1,$ $2, \ldots, k, p_1 + p_2 + \cdots + p_k = 1$ | $np_1, \ldots, np_k$ | $np_1 q_1, \ldots, np_k q_k.$ $q_i = 1 - p_i, j =$ $1, \ldots, k$ |
| **Bivariate Normal** | $f(x_1, x_2) = \frac{1}{2\pi \sigma_1 \sigma_2 \sqrt{1-\rho^2}} \exp\left(-\frac{q}{2}\right),$ $q = \frac{1}{1-\rho^2}\left[\left(\frac{x_1-\mu_1}{\sigma_1}\right)^2 - 2\rho\left(\frac{x_1-\mu_1}{\sigma_1}\right)\right.$ $\left. \times \left(\frac{x_2-\mu_2}{\sigma_2}\right) + \left(\frac{x_2-\mu_2}{\sigma_2}\right)^2\right],$ $x_1, x_2, \in \Re; \mu_1, \mu_2 \in \Re, \sigma_1, \sigma_2 > 0,$ $-1 \leq \rho \leq 1, \rho = $ correlation coefficient | $\mu_1, \mu_2$ | $\sigma_1^2, \sigma_2^2$ |
| **$k$-Variate Normal, $N(\mu, \Sigma)$** | $f(\mathbf{x}) = (2\pi)^{-k/2} |\Sigma|^{-1/2} \times$ $\exp\left[-\frac{1}{2}(\mathbf{x} - \mu)' \Sigma^{-1}(\mathbf{x} - \mu)\right],$ $\mathbf{x} \in \Re^k; \mu \in \Re^k, \Sigma : k \times k$ nonsingular symmetric matrix | $\mu_1, \ldots, \mu_k$ | Covariance matrix: $\Sigma$ |

# An Introduction
# to Probability and
# Statistical Inference
## Second Edition

# An Introduction to Probability and Statistical Inference

## Second Edition

**George G. Roussas**

Department of Statistics
University of California, Davis

AMSTERDAM • BOSTON • HEIDELBERG • LONDON
NEW YORK • OXFORD • PARIS • SAN DIEGO
SAN FRANCISCO • SINGAPORE • SYDNEY • TOKYO
Academic Press is an imprint of Elsevier

Academic Press is an imprint of Elsevier
32 Jamestown Road, London NW1 7BY, UK
525 B Street, Suite 1800, San Diego, CA 92101-4495, USA
225 Wyman Street, Waltham, MA 02451, USA
The Boulevard, Langford Lane, Kidlington, Oxford OX5 1GB, UK

Second edition 2015

**Library of Congress Cataloging-in-Publication Data**
A catalog record for this book is available from the Library of Congress

**British Library Cataloguing in Publication Data**
A catalogue record for this book is available from the British Library

For information on all Academic Press publications
visit our web site at store.elsevier.com

This book has been manufactured using Print On Demand technology. Each copy is produced
to order and is limited to black ink. The online version of this book will show color figures
where appropriate.

ISBN: 978-0-12-800114-1

TO: SOPHIA, NIKO, ANGELO

**ARGOS**

# Contents

# Preface

## OVERVIEW

This book is an introductory textbook in probability and statistical inference. No prior knowledge of either probability or statistics is required, although prior exposure to an elementary precalculus course would prove beneficial in the sense that the student would not see the basic concepts discussed here for the first time.

The mathematical prerequisite is a year of calculus and familiarity with the basic concepts and some results of linear algebra. Elementary differential and integral calculus will suffice for the majority of the book. In some parts, such as Chapters 4–6, the concept of a multiple integral is used. Also, in Chapter 6, the student is expected to be at least vaguely familiar with the basic techniques of changing variables in a single or a multiple integral.

## CHAPTER DESCRIPTIONS

The material discussed in this book is enough for a 1-year course in introductory probability and statistical inference. It consists of a total of 15 chapters. Chapters 1 through 7 are devoted to probability, distributional theory, and related topics. Chapter 8 presents an overview of statistical inference. Chapters 9 through 14 discuss the standard topics of parametric statistical inference, namely, point estimation, interval estimation, and testing hypotheses. This is done first in a general setting and then in the special models of linear regression and analysis of variance. Chapter 15 is devoted to discussing selected topics from nonparametric inference.

## FEATURES

This book has a number of features that differentiate it from existing books. First, the material is arranged in such a manner that Chapters 1 through 7 can be used independently for an introductory course in probability. The desirable duration for such a course would be a semester, although a quarter would also be long enough if some of the proofs were omitted. Chapters 1 through 7.1–7.2 would suffice for this purpose. The centrally placed Chapter 8 plays a twofold role. First, it serves as a window into what statistical inference is all about for those taking only the probability part of the course. Second, it paints a fairly broad picture of the material discussed in considerable detail in the subsequent chapters. Accordingly and purposely, no specific results are stated, no examples are discussed, no exercises are included. All these things are done in the chapters following it. As already mentioned, the sole objective here is to take the reader through a brief orientation trip to statistical inference; to indicate why statistical inference is needed in the first place, how the relevant main problems are formulated, and how we go about resolving them.

The second differentiating feature of this book is the relatively large number of examples discussed in detail. There are more than 220 such examples, not including scores of numerical examples and applications. The first chapter alone is replete with 44 examples selected from a variety of applications. Their purpose is to impress upon the student the breadth of applications of probability and statistics, to draw attention to the wide range of applications, where probabilistic and statistical questions are pertinent. At this stage, one could not possibly provide answers to the questions posed without the methodology developed in the subsequent chapters. Answers to these questions are given in the form of examples and exercises throughout the remaining chapters.

This book contains more than 780 exercises placed strategically at the ends of sections. The exercises are closely related to the material discussed in the respective sections, and they vary in the degree of difficulty. Detailed solutions to all of them are available in the form of a *Solutions Manual* for the instructors of the course, when this textbook is used. Brief answers to even-numbered exercises are provided at the end of the book. Also included in the textbook are approximately 60 figures that help illustrate some concepts and operations.

Still another desirable feature of this textbook is the effort made to minimize the so-called arm waving. This is done by providing a substantial number of proofs, without ever exceeding the mathematical prerequisites set. This also helps ameliorate the not so unusual phenomenon of insulting students' intelligence by holding them incapable of following basic reasoning.

Regardless of the effort made by the author of an introductory book in probability and statistics to cover the largest possible number of areas, where probability and statistics apply, such a goal is unlikely to be attained. Consequently, no such textbook will ever satisfy students who focus exclusively on their own area of interest. It is also expected that this book will come as a disappointment to students who are oriented more toward vocational training rather than college or university education. This book is not meant to codify answers to questions in the form of framed formulas and prescription recipes. Rather, its purpose is to introduce the student to a thinking process and guide her or him toward the answer sought to a posed question. To paraphrase a Chinese saying, if you are taught how to fish, you eat all the time, whereas if you are given a fish, you eat only once.

On several occasions, the reader is referred for proofs and more comprehensive treatment of some topics to the book *A Course in Mathematical Statistics*, 2nd edition *(1997)*, Academic Press, by G.G. Roussas. This reference book was originally written for the same audience as that of the present book. However, circumstances dictated the adjustment of the level of the reference book to match the mathematical preparation of the anticipated audience.

On the practical side, a number of points of information are given here. Thus, $\log x$ (logarithm of $x$), whenever it occurs, is always the natural logarithm of $x$ (the logarithm of $x$ with base e), whether it is explicitly stated or not.

The rule followed in the use of decimal numbers is that we retain three decimal digits, the last of which is rounded up to the next higher number, if the fourth omitted decimal is greater or equal to 5. An exemption to this rule is made when the division is exact, and also when the numbers are read out of tables. This book is supplied with an appendix consisting of excerpts of tables: *Binomial tables, Poisson tables, Normal tables, t-tables, Chi-square tables,* and *F-tables*. Table 7 consists of a list of certain often-occurring distributions along with some of their characteristics. The last table, Table 8, is a handy reference to some formulas used in the text. The appendix is followed by a list of some notation and abbreviations extensively used throughout this book, and the body of the book is concluded with brief answers to the even-numbered exercises.

In closing, a concerted effort has been made to minimize the number of inevitable misprints and oversights in this book. We have no illusion, however, that the book is free of them. This author would greatly appreciate being informed of any errors; such errors will be corrected in a subsequent printing of the book.

## BRIEF PREFACE OF THE REVISED VERSION

This is a revised version of the book copyrighted in 2003. The revision consists of correcting misprints and oversights, which occurred in the original book. Also, in providing references to exercises, as well as numerous hints associated with exercises. Most importantly, some of the material is rearranged (in particular, in Chapter 11) to facilitate the exposition and ensure coherence among several results. Some new material is added (see, e.g., Chapter 12), including some exercises and a table, Table 8, in the appendix. In all other respects, the basic character of the book remains the same.

The revision was skillfully implemented by my associate Randy Lai, to whom I express here my sincere appreciation.

**George G. Roussas**
Davis, California
November 2013

## ACKNOWLEDGMENTS AND CREDITS

I would like to thank Subhash Bagui, University of West Florida; Matthew Carlton, Cal Polytechnic State University; Tapas K. Das, University of South Florida; Jay Devore, Cal Polytechnic State University; Charles Donaghey, University of Houston; Pat Goeters, Auburn University; Xuming He, University of Illinois, and Krzysztof M. Ostaszewski, Illinois State University, Champaign-Urbana for their many helpful comments.

Some of the examples discussed in this book have been taken and/or adapted from material included in the book *Statistics: Principles and Methods*, 2nd edition (1992), [ISBN: 0471548421], by R. A. Johnson, G. K. Bhattacharyya, Copyright ©1987, 1992, by John Wiley & Sons, Inc., and are reprinted by permission of John Wiley & Sons, Inc. They are Table 4 on page 74, Examples 8, 1, 2, 4, 12, 4, 2, 1, and 7 on pages 170, 295, 296, 353, 408, 439, 510, 544, and 562, respectively, and Exercises 4.18, 3.19, 4.21, 5.22, 5.34, 8.16, 4.14, 6.34, 3.16, 6.6 and 3.8 on pages 123, 199, 217, 222, 225, 265, 323, 340, 356, 462, and 525, respectively. The reprinting permission is kindly acknowledged herewith.

# Some motivating examples and some fundamental concepts

# 1

This chapter consists of three sections. The first section is devoted to presenting a number of examples (25 to be precise), drawn from a broad spectrum of human activities. Their purpose is to demonstrate the wide applicability of probability (and statistics). In the formulation of these examples, certain terms, such as at random, average, data fitted by a line, event, probability (estimated probability, probability model), rate of success, sample, and sampling (sample size), are used. These terms are presently to be understood in their everyday sense and will be defined precisely later on.

In the second section, some basic terminology and fundamental quantities are introduced and are illustrated by means of examples. In the closing section, the concept of a random variable is defined and is clarified through a number of examples.

## 1.1 SOME MOTIVATING EXAMPLES

**Example 1.** In a certain state of the Union, $n$ landfills are classified according to their concentration of three hazardous chemicals: Arsenic, barium, and mercury. Suppose that the concentration of each one of the three chemicals is characterized as either high or low. Then some of the questions which can be posed are as follows: (i) If a landfill is chosen at random from among the $n$, what is the probability it is of a specific configuration? In particular, what is the probability that it has: (a) High concentration of barium? (b) High concentration of mercury and low concentration of both arsenic and barium? (c) High concentration of any two of the chemicals and low concentration of the third? (d) High concentration of any one of the chemicals and low concentration of the other two? (ii) How can one check whether the proportions of the landfills, falling into each one of the eight possible configurations (regarding the levels of concentration), agree with *a priori* stipulated numbers? (See Exercise 1.5 in Chapter 12.)

**Example 2.** Suppose a disease is present in $100p_1\%$ $(0 < p_1 < 1)$ of a population. A diagnostic test is available but is yet to be perfected. The test shows $100p_2\%$

false positives ($0 < p_2 < 1$) and $100p_3\%$ false negatives ($0 < p_3 < 1$). That is, for a patient not having the disease, the test shows positive (+) with probability $p_2$ and negative (−) with probability $1 - p_2$. For a patient having the disease, the test shows "−" with probability $p_3$ and "+" with probability $1 - p_3$. A person is chosen at random from the target population, and let $D$ be the event that the person is diseased and $N$ be the event that the person is not diseased. Then, it is clear that some important questions are as follows: In terms of $p_1, p_2$, and $p_3$: (i) Determine the probabilities of the following configurations: $D$ and +, $D$ and −, $N$ and +, $N$ and −. (ii) Also, determine the probability that a person will test +, or the probability the person will test −. (iii) If the person chosen tests +, what is the probability that he/she is diseased? What is the probability that he/she is diseased, if the person tests −?

**Example 3.** In the circuit drawn below (Figure 1.1), suppose that switch $i = 1, \ldots, 5$ turns on with probability $p_i$ and independently of the remaining switches. What is the probability of having current transferred from point A to point B?

**Example 4.** A travel insurance policy pays $1000 to a customer in case of a loss due to theft or damage of luggage on a 5-day trip. If the risk of such a loss is assessed to be 1 in 200, what is a fair premium for this policy?

**Example 5.** Jones claims to have extrasensory perception (ESP). In order to test the claim, a psychologist shows Jones five cards that carry different pictures. Then Jones is blindfolded and the psychologist selects one card and asks Jones to identify the picture. This process is repeated $n$ times. Suppose, in reality, that Jones has no ESP but responds by sheer guesses.

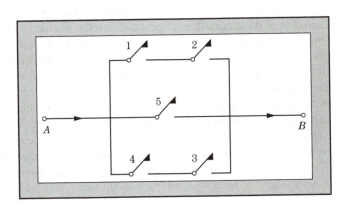

**FIGURE 1.1**

Circuit referred to in Example 3

(i) Decide on a suitable probability model describing the number of correct responses. (ii) What is the probability that at most $n/5$ responses are correct? (iii) What is the probability that at least $n/2$ responses are correct?

**Example 6.** A government agency wishes to assess the prevailing rate of unemployment in a particular county. It is felt that this assessment can be done quickly and effectively by sampling a small fraction $n$, say, of the labor force in the county. The obvious questions to be considered here are: (i) What is a suitable probability model describing the number of unemployed? (ii) What is an estimate of the rate of unemployment?

**Example 7.** Suppose that, for a particular cancer, chemotherapy provides a 5-year survival rate of 75% if the disease could be detected at an early stage. Suppose further that $n$ patients, diagnosed to have this form of cancer at an early stage, are just starting the chemotherapy. Finally, let $X$ be the number of patients among the $n$ who survive 5 years.

Then the following are some of the relevant questions which can be asked: (i) What are the possible values of $X$, and what are the probabilities that each one of these values are taken on? (ii) What is the probability that $X$ takes values between two specified numbers $a$ and $b$, say? (iii) What is the average number of patients to survive 5 years, and what is the variation around this average?

**Example 8.** An advertisement manager for a radio station claims that over $100p\%$ $(0 < p < 1)$ of all young adults in the city listen to a weekend music program. To establish this conjecture, a random sample of size $n$ is taken from among the target population and those who listen to the weekend music program are counted. (See Exercise 3.1 in Chapter 11.)

(i) Decide on a suitable probability model describing the number of young adults who listen to the weekend music program. (ii) On the basis of the collected data, check whether the claim made is supported or not. (iii) How large a sample size $n$ should be taken to ensure that the estimated average and the true proportion do not differ in absolute value by more than a specified number with prescribed (high) probability?

**Example 9.** When the output of a production process is stable at an acceptable standard, it is said to be "in control." Suppose that a production process has been in control for some time and that the proportion of defectives has been $p$. As a means of monitoring the process, the production staff will sample $n$ items. Occurrence of $k$ or more defectives will be considered strong evidence for "out of control."

(i) Decide on a suitable probability model describing the number $X$ of defectives; what are the possible values of $X$, and what is the probability that each of these values is taken on? (ii) On the basis of the data collected, check whether or not the process is out of control. (iii) How large a sample size $n$ should be taken to ensure that the estimated proportion of defectives will not differ in absolute value from the true proportion of defectives by more than a specified quantity with prescribed (high) probability? (See Exercise 3.2 in Chapter 11.)

**Example 10.**  An electronic scanner is believed to be more efficient in determining flaws in a material than a mechanical testing method which detects $100p\%$ $(0 < p < 1)$ of the flawed specimens. To determine its success rate, $n$ specimens with flaws are tested by the electronic scanner.

   (i) Decide on a suitable probability model describing the number $X$ of the flawed specimens correctly detected by the electronic scanner; what are the possible values of $X$, and what is the probability that each one of these values is taken on? (ii) Suppose that the electronic scanner detects correctly $k$ out of $n$ flawed specimens. Check whether or not the rate of success of the electronic scanner is higher than that of the mechanical device. (See Exercise 3.3 in Chapter 11.)

**Example 11.**  At a given road intersection, suppose that $X$ is the number of cars passing by until an observer spots a particular make of a car (e.g., a Mercedes).

   Then some of the questions one may ask are as follows: (i) What are the possible values of $X$? (ii) What is the probability that each one of these values is taken on? (iii) How many cars would the observer expect to observe until the first Mercedes appears?

**Example 12.**  A city health department wishes to determine whether the mean bacteria count per unit volume of water at a lake beach is within the safety level of 200. A researcher collected $n$ water samples of unit volume and recorded the bacteria counts.

   Relevant questions here are: (i) What is the appropriate probability model describing the number $X$ of bacteria in a unit volume of water; what are the possible values of $X$, and what is the probability that each one of these values is taken on? (ii) Do the data collected indicate that there is no cause for concern? (See Exercise 3.7 in Chapter 11.)

**Example 13.**  Consider an aptitude test administered to aircraft pilot trainees, which requires a series of operations to be performed in quick succession.

   Relevant questions here are: (i) What is the appropriate probability model for the time required to complete the test? (ii) What is the probability that the test is completed in no less than $t_1$ minutes, say? (iii) What is the percentage of candidates passing the test, if the test is to be completed within $t_2$ minutes, say?

**Example 14.**  Measurements of the acidity (pH) of rain samples were recorded at $n$ sites in an industrial region.

(i) Decide on a suitable probability model describing the number $X$ of the acidity of rain measured. (ii) On the basis of the measurements taken, provide an estimate of the average acidity of rain in that region.

**Example 15.**   To study the growth of pine trees at an early state, a nursery worker records $n$ measurements of the heights of 1-year-old red pine seedlings.
(i) Decide on a suitable probability model describing the heights $X$ of the pine seedlings. (ii) On the basis of the $n$ measurements taken, determine average height of the pine seedlings. (iii) Also, check whether these measurements support the stipulation that the average height is a specified number. (See Exercise 4.3 in Chapter 11.)

**Example 16.**   It is claimed that a new treatment is more effective than the standard treatment for prolonging the lives of terminal cancer patients. The standard treatment has been in use for a long time, and from records in medical journals the mean survival period is known to have a certain numerical value (in years). The new treatment is administered to $n$ patients, and their duration of survival is recorded.
(i) Decide on suitable probability models describing the survival times $X$ and $Y$ under the old and the new treatments, respectively. (ii) On the basis of the existing journal information and the data gathered, check whether or not the claim made is supported. (See Exercise 3.9 in Chapter 11.)

**Example 17.**   A medical researcher wishes to determine whether a pill has the undesirable side effect of reducing the blood pressure of the user. The study requires recording the initial blood pressures of $n$ college-age women. After the use of the pill regularly for 6 months, their blood pressures are again recorded.
(i) Decide on suitable probability models describing the blood pressures, initially and after the 6-month period. (ii) Do the observed data support the claim that the use of the pill reduces blood pressure? (See Exercise 4.4 in Chapter 11.)

**Example 18.**   It is known that human blood is classified into four types denoted by A, B, AB, and O. Suppose that the blood of $n$ persons who have volunteered to donate blood at a plasma center has been classified in these four categories. Then a number of questions can be posed; some of them are:
(i) What is the appropriate probability model to describe the distribution of the blood types of the $n$ persons into the four types? (ii) What is the estimated probability that a person, chosen at random from among the $n$, has a specified blood type (e.g., O)? (iii) What are the proportions of the $n$ persons falling into each one of the four categories? (iv) How can one check whether the observed proportions are in agreement with *a priori* stipulated numbers? (See Exercise 1.1 in Chapter 12.)

**Example 19.**   The following record shows a classification of 41,208 births in Wisconsin (courtesy of Professor Jerome Klotz). Set up a suitable probability model and check whether or not the births are Uniformly distributed over all 12 months of the year. (See Exercise 1.2 in Chapter 12.)

| Jan. | 3478 | July | 3476 |
|------|------|-------|--------|
| Feb. | 3333 | Aug. | 3495 |
| March | 3771 | Sept. | 3490 |
| April | 3542 | Oct. | 3331 |
| May | 3479 | Nov. | 3188 |
| June | 3304 | Dec. | 3321 |
| | | Total | 41,208 |

**Example 20.** To compare the effectiveness of two diets $A$ and $B$, 150 infants were included in a study. Diet $A$ was given to 80 randomly selected infants and diet $B$ was given to the other 70 infants. At a later time, the health of each infant was observed and classified into one of the three categories: "excellent," "average," and "poor." The frequency counts are tabulated as follows:

| | Health Under Two Different Diets | | | |
|--------|-----------|---------|------|-------------|
| | **Excellent** | **Average** | **Poor** | **Sample Size** |
| Diet A | 37 | 24 | 19 | 80 |
| Diet B | 17 | 33 | 20 | 70 |
| Total | 54 | 57 | 39 | 150 |

Set up a suitable probability model for this situation and, on the basis of the observed data, compare the effectiveness of the two diets. (See Exercise 1.3 in Chapter 12.)

**Example 21.** Osteoporosis (loss of bone minerals) is a common cause of broken bones in the elderly. A researcher on aging conjectures that bone mineral loss can be reduced by regular physical therapy or by certain kinds of physical activity. A study is conducted on $n$ elderly subjects of approximately the same age divided into control, physical therapy, and physical activity groups. After a suitable period of time, the nature of change in bone mineral content is observed.

Set up a suitable probability model for the situation under consideration, and check whether or not the observed data indicate that the change in bone mineral varies for different groups. (See Exercise 1.4 in Chapter 12.)

| | Change in Bone Mineral | | | |
|---------|--------------------|--------------|------------------------|-------|
| | **Appreciable Loss** | **Little Change** | **Appreciable Increase** | **Total** |
| Control | 38 | 15 | 7 | 60 |
| Therapy | 22 | 32 | 16 | 70 |
| Activity | 15 | 30 | 25 | 70 |
| Total | 75 | 77 | 48 | 200 |

**Example 22.**   In the following table, the data $x$ = undergraduate GPA and $y$ = score in the Graduate Management Aptitude Test (GMAT) are recorded.

| Data of Undergraduate GPA ($x$) and GMAT Score ($y$) | | | | | |
|---|---|---|---|---|---|
| $x$ | $y$ | $x$ | $y$ | $x$ | $y$ |
| 3.63 | 447 | 2.36 | 399 | 2.80 | 444 |
| 3.59 | 588 | 2.36 | 482 | 3.13 | 416 |
| 3.30 | 563 | 2.66 | 420 | 3.01 | 471 |
| 3.40 | 553 | 2.68 | 414 | 2.79 | 490 |
| 3.50 | 572 | 2.48 | 533 | 2.89 | 431 |
| 3.78 | 591 | 2.46 | 509 | 2.91 | 446 |
| 3.44 | 692 | 2.63 | 504 | 2.75 | 546 |
| 3.48 | 528 | 2.44 | 336 | 2.73 | 467 |
| 3.47 | 552 | 2.13 | 408 | 3.12 | 463 |
| 3.35 | 520 | 2.41 | 469 | 3.08 | 440 |
| 3.39 | 543 | 2.55 | 538 | 3.03 | 419 |
|  |  |  |  | 3.00 | 509 |

(i) Draw a scatter plot of the pairs $(x, y)$. (ii) On the basis of part (i), set up a reasonable model for the representation of the pairs $(x, y)$. (iii) Indicate roughly how this model can be used to predict a GMAT score on the basis of the corresponding GPA score. (See the beginning of Section 13.1, Examples 1–3, 6, and Exercises 3.2–3.5, 4.1, 4.2, 5.1, 5.2, all in Chapter 13.)

**Example 23.**   In an experiment designed to determine the relationship between the doses of a compost fertilizer $x$ and the yield $y$ of a crop, $n$ values of $x$ and $y$ are observed. On the basis of prior experience, it is reasonable to assume that the pairs $(x, y)$ are fitted by a straight line, which can be determined by certain summary values of the data. Later on, it will be seen how this is specifically done and also how this model can be used for various purposes, including that of predicting a value of $y$ on the basis of a given value of $x$. (See the beginning of Section 13.1, Examples 4, 7, and Exercise 5.3, all in Chapter 13.)

**Example 24.**   In an effort to improve the quality of recording tapes, the effects of four kinds of coatings A, B, C, and D on the reproducing quality of sound are compared. Twenty-two measurements of sound distortions are given in the following table:

| Sound Distortions Obtained with Four Types of Coating | |
|---|---|
| Coating | Observations |
| A | 10, 15, 8, 12, 15 |
| B | 14, 18, 21, 15 |
| C | 17, 16, 14, 15, 17, 15, 18 |
| D | 12, 15, 17, 15, 16, 15 |

In connection with these data, several questions may be posed (and will be posed later on; see Example 2 in Chapter 14). The most immediate of them all is the question of whether or not the data support the existence of any significant difference among the average distortions obtained using the four coatings.

**Example 25.** Charles Darwin performed an experiment to determine whether self-fertilized and cross-fertilized plants have different growth rates. Pairs of *Zea mays* plants, one self- and the other cross-fertilized, were planted in pots, and their heights were measured after a specified period of time. The data Darwin obtained were tabulated as follows:

| Plant Height (in 1/8 in.) | | | | | |
|---|---|---|---|---|---|
| Pair | Cross- | Self- | Pair | Cross- | Self- |
| 1 | 188 | 139 | 9 | 146 | 132 |
| 2 | 96 | 163 | 10 | 173 | 144 |
| 3 | 168 | 160 | 11 | 186 | 130 |
| 4 | 176 | 160 | 12 | 168 | 144 |
| 5 | 153 | 147 | 13 | 177 | 102 |
| 6 | 172 | 149 | 14 | 184 | 124 |
| 7 | 177 | 149 | 15 | 96 | 144 |
| 8 | 163 | 122 | | | |

Source: Darwin, C., *The Effects of Cross- and Self-Fertilization in the Vegetable Kingdom*, D. Appleton and Co., New York, 1902.

These data lead to many questions, the most immediate being whether cross-fertilized plants have a higher growth rate than self-fertilized plants. This example will be revisited later on. (See Exercise 5.14 in Chapter 9 and Example 4 in Chapter 15.)

## 1.2 SOME FUNDAMENTAL CONCEPTS

One of the most basic concepts in probability and statistics is that of a *random experiment*. Although a more precise definition is possible, we will restrict ourselves here to understanding a random experiment as a procedure which is carried out under a certain set of conditions; it can be repeated any number of times under the same set of conditions; and upon the completion of the procedure certain results are observed. The results obtained are denoted by $s$ and are called *sample points*. The set of all possible sample points is denoted by $S$ and is called a *sample space*. Subsets of $S$ are called *events* and are denoted by capital letters A, B, C, etc. An event consisting of one sample point only, $\{s\}$, is called a *simple* event and *composite* otherwise. An event A *occurs* (or *happens*) if the outcome of the random experiment (that is, the sample point $s$) belongs in A, $s \in A$; A *does not occur* (or *does not happen*)

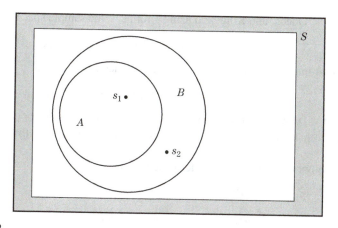

**FIGURE 1.2**

$A \subseteq B$; in fact, $A \subset B$, because $s_2 \in B$, but $s_2 \notin A$.

if $s \notin A$. The event $S$ always occurs and is called the *sure* or *certain* event. On the other hand, the event $\emptyset$ never happens and is called the *impossible* event. Of course, the relation $A \subseteq B$ between two events $A$ and $B$ means that the event $B$ *occurs whenever A does*, but not necessarily the opposite. (See Figure 1.2 for the *Venn diagram* depicting the relation $A \subseteq B$.) The events $A$ and $B$ are *equal* if both $A \subseteq B$ and $B \subseteq A$.

Some random experiments are given in the following along with corresponding sample spaces and some events.

**Example 26.**   Tossing three distinct coins once.

Then, with $H$ and $T$ standing for "heads" and "tails," respectively, a sample space is

$$S = \{HHH, HHT, HTH, THH, HTT, THT, TTH, TTT\}.$$

The event $A =$ "no more than 1 $H$ occurs" is given by:

$$A = \{TTT, HTT, THT, TTH\}.$$

**Example 27.**   Rolling once two distinct dice.

Then a sample space is:

$$S = \{(1, 1), (1, 2), \ldots, (1, 6), \ldots, (6, 1), (6, 2), \ldots, (6, 6)\},$$

and the event $B =$ "the sum of numbers on the upper faces is $\leq 5$" is

$$B = \{(1, 1), (1, 2), (1, 3), (1, 4), (2, 1), (2, 2), (2, 3), (3, 1), (3, 2), (4, 1)\}.$$

**Example 28.**   Drawing a card from a well-shuffled standard deck of 52 playing cards. Denoting by $C, D, H,$ and $S$ clubs, diamonds, hearts, and spades, respectively, by $J, Q, K$ Jack, Queen, and King, and using 1 for aces, the sample space is given by:

$$S = \{1_C, \ldots, 1_S, \ldots, 10_C, \ldots, 10_S, \ldots, K_C, \ldots, K_S\}.$$

An event $A$ may be described by: $A =$ "red and face card," so that

$$A = \{J_D, J_H, Q_D, Q_H, K_D, K_H\}.$$

**Example 29.**   Drawing (without replacement) two balls from an urn containing $m$ numbered (from 1 through $m$) black balls and $n$ numbered (from 1 through $n$) red balls.

Then, in obvious notation, a sample space here is

$$S = \{b_1 b_2, \ldots, b_1 b_m, \ldots, b_m b_1, \ldots, b_m b_{m-1},$$

$$b_1 r_1, \ldots, b_1 r_n, \ldots, b_m r_1, \ldots, b_m r_n,$$

$$r_1 b_1, \ldots, r_1 b_m, \ldots, r_n b_1, \ldots, r_n b_m,$$

$$r_1 r_2, \ldots, r_1 r_n, \ldots, r_n r_1, \ldots, r_n r_{n-1}\}.$$

An event $A$ may be the following: $A =$ "the sum of the numbers on the balls does not exceed 4." Then

$$A = \{b_1 b_2, b_1 b_3, b_2 b_1, b_3 b_1, b_1 r_1, b_1 r_2, b_1 r_3,$$

$$b_2 r_1, b_2 r_2, b_3 r_1, r_1 b_1, r_1 b_2, r_1 b_3, r_2 b_1,$$

$$r_2 b_2, r_3 b_1, r_1 r_2, r_1 r_3, r_2 r_1, r_3 r_1\} \quad \text{(assuming that } m, n \geq 3).$$

**Example 30.**   Recording the gender of children of two-children families.

With $b$ and $g$ standing for boy and girl, and with the first letter on the left denoting the older child, a sample space is: $S = \{bb, bg, gb, gg\}$. An event $B$ may be: $B =$ "children of both genders." Then $B = \{bg, gb\}$.

**Example 31.**   Ranking five horses in a horse race.

Then the suitable sample space $S$ consists of 120 sample points, corresponding to the 120 permutations of the numbers 1, 2, 3, 4, 5. (We exclude ties.) The event $A =$ "horse #3 comes second" consists of the 24 sample points, where 3 always occurs in the second place.

**Example 32.**   Tossing a coin repeatedly until $H$ appears for the first time.

The suitable sample space here is

$$S = \{H, TH, TTH, \ldots, TT, \ldots, TH, \ldots\}.$$

Then the event $A =$ "the 1st $H$ does not occur before the 10th tossing" is given by:

$$A = \left\{ \underbrace{T \ldots T}_{9} H, \; \underbrace{T \ldots T}_{10} H, \ldots \right\}.$$

**Example 33.**   Recording the number of telephone calls served by a certain telephone exchange center within a specified period of time.

   Clearly, the sample space here is: $S = \{0, 1, \ldots, C\}$, where $C$ is a suitably large number associated with the capacity of the center. For mathematical convenience, we often take $S$ to consist of all nonnegative integers; that is, $S = \{0, 1, \ldots\}$.

**Example 34.**   Recording the number of traffic accidents which occurred in a specified location within a certain period of time.

   As in the previous example, $S = \{0, 1, \ldots, M\}$ for a suitable number $M$. If $M$ is sufficiently large, then $S$ is taken to be $S = \{0, 1, \ldots\}$.

**Example 35.**   Recording the number of particles emitted by a certain radioactive source within a specified period of time.

   As in the previous two examples, $S$ is taken to be $S = \{0, 1, \ldots, M\}$, where $M$ is often a large number, and then as before $S$ is modified to be $S = \{0, 1, \ldots\}$.

**Example 36.**   Recording the lifetime of an electronic device, or of an electrical appliance, etc.

   Here, $S$ is the interval $(0, T]$ for some reasonable value of $T$; that is, $S = (0, T]$. Sometimes, for justifiable reasons, we take $S = (0, \infty)$.

**Example 37.**   Recording the distance from the bull's eye of the point where a dart, aiming at the bull's eye, actually hits the plane. Here, it is clear that $S = [0, \infty)$.

**Example 38.**   Measuring the dosage of a certain medication, administered to a patient, until a positive reaction is observed.

   Here, $S = (0, D]$ for some suitable $D$ (not rendering the medication lethal!).

**Example 39.**   Recording the yearly income of a target population.

   If the incomes are measured in \$ and cents, the outcomes are fractional numbers in an interval $[0, M]$ for some reasonable $M$. Again, for reasons similar to those cited in Example 36, $S$ is often taken to be $S = [0, \infty)$.

**Example 40.**   Waiting until the time the Dow-Jones Industrial Average index reaches or surpasses a specified level.

   Here, with reasonable qualifications, we may chose to take $S = (0, \infty)$.

   Examples 1–25, suitably interpreted, may also serve as further illustrations of random experiments. All examples described previously will be revisited on various occasions.

   For instance, in Example 1 and in self-explanatory notation, a suitable sample space is

$$S = \{A_h B_h M_h, \; A_h B_h M_\ell, \; A_h B_\ell M_h, \; A_\ell B_h M_h, \; A_h B_\ell M_\ell,$$

$$A_\ell B_h M_\ell, \; A_\ell B_\ell M_h, \; A_\ell B_\ell M_\ell\}.$$

Then the events $A = $ "no chemical occurs at high level" and $B = $ "at least two chemicals occur at high levels" are given by:

$$A = \{A_\ell B_\ell M_\ell\}, \qquad B = \{A_h B_h M_\ell,\ A_h B_\ell M_h,\ A_\ell B_h M_h,\ A_h B_h M_h\}.$$

In Example 2, a patient is classified according to the result of the test, giving rise to the following sample space:

$$S = \{D+, D-, N+, N-\},$$

where $D$ and $N$ stand for the events "patient has the disease" and "patient does not have the disease," respectively. Then the event $A = $ "false diagnosis of test" is given by: $A = \{D-, N+\}$.

In Example 5, the suitable probability model is the so-called Binomial model. The sample space $S$ is the set of $2^n$ points, each point consisting of a sequence of $n$ $S$'s and $F$'s, $S$ standing for success (on behalf of Jones) and $F$ standing for failure. Then the questions posed can be answered easily.

Examples 6 through 10 can be discussed in the same framework as that of Example 5 with obvious modifications in notation.

In Example 11, a suitable sample space is

$$S = \{M, M^c M, M^c M^c M, \ldots, M^c \ldots M^c M, \ldots\},$$

where $M$ stands for the passing by of a Mercedes car. Then the events $A$ and $B$, where $A = $ "Mercedes was the fifth car passed by" and $B = $ "Mercedes was spotted after the first three cars passed by" are given by:

$$A = \{M^c M^c M^c M^c M\} \quad \text{and} \quad B = \{M^c M^c M^c M,\ M^c M^c M^c M^c M, \ldots\}.$$

In Example 12, a suitable sample space is $S = \{0, 1, \ldots, M\}$ for an appropriately large (integer) $M$; for mathematical convenience, $S$ is often taken to be $S = \{0, 1, 2, \ldots\}$.

In Example 13, a suitable sample space is $S = (0, T)$ for some reasonable value of $T$. In such cases, if $T$ is very large, mathematical convenience dictates replacement of the previous sample space by $S = (0, \infty)$.

Examples 14 and 15 can be treated in the same framework as Example 13 with obvious modifications in notation.

In Example 18, a suitable sample space $S$ is the set of $4^n$ points, each point consisting of a sequence of $n$ symbols $A, B, AB$, and $O$. The underlying probability model is the so-called Multinomial model, and the questions posed can be discussed by available methodology. Actually, there is no need even to refer to the sample space $S$. All one has to do is to consider the outcomes in the $n$ trials and then classify the $n$ outcomes into four categories $A, B, AB$, and $O$.

Example 19 fits into the same framework as that of Example 18. Here, the suitable $S$ consists of $12^{41,208}$ points, each point being a sequence of symbols representing the 12 months. As in the previous example, there is no need, however, even to refer to this sample space. Example 20 is also of the same type.

In many cases, questions posed can be discussed without reference to any explicit sample space. This is the case, for instance, in Examples 16–17 and 21–25.

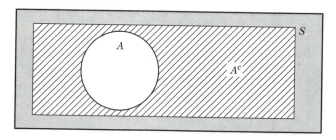

**FIGURE 1.3**

$A^c$ is the shaded region.

In the examples discussed previously, we have seen sample spaces consisting of finitely many sample points (Examples 26–31), sample spaces consisting of countably infinite many points (e.g., as many as the positive integers) (Example 32 and also Examples 33–35 if we replace $C$ and $M$ by $\infty$ for mathematical convenience), and sample spaces consisting of as many sample points as there are numbers in a nondegenerate finite or infinite interval in the real line, which interval may also be the entire real line (Examples 36–40). Sample spaces with countably many points (i.e., either finitely many or countably infinitely many) are referred to as *discrete* sample spaces. Sample spaces with sample points as many as the numbers in a nondegenerate finite or infinite interval in the real line $\Re = (-\infty, \infty)$ are referred to as *continuous* sample spaces.

Returning now to events, when one is dealing with them, one may perform the same operations as those with sets. Thus, the *complement* of the event $A$, denoted by $A^c$, is the event defined by: $A^c = \{s \in \mathcal{S}; s \notin A\}$. The event $A^c$ is presented by the Venn diagram in Figure 1.3. So $A^c$ occurs whenever $A$ does not, and vice versa.

The *union* of the events $A_1, \cdots, A_n$, denoted by $A_1 \cup \cdots \cup A_n$ or $\bigcup_{j=1}^{n} A_j$, is the event defined by $\bigcup_{j=1}^{n} A_j = \{s \in \mathcal{S}; s \in A_j, \text{ for } at \ least \ one \ j = 1, \ldots, n\}$. So the event $\bigcup_{j=1}^{n} A_j$, occurs whenever at least one of $A_j, j = 1, \ldots, n$ occurs. For $n = 2, A_1 \cup A_2$ is presented in Figure 1.4. The definition extends to an infinite number of events. Thus, for countably infinite many events $A_j, j = 1, 2, \ldots$, one has $\bigcup_{j=1}^{\infty} A_j = \{s \in \mathcal{S}; s \in A_j, for \ at \ least \ one \ j = 1, 2, \ldots\}$.

The *intersection* of the events $A_j, j = 1, \ldots, n$, is the event denoted by $A_1 \cap \cdots \cap A_n$ or $\bigcap_{j=1}^{n} A_j$ and is defined by $\bigcap_{j=1}^{n} A_j = \{s \in \mathcal{S}; s \in A_j, for \ all \ j = 1, \ldots, n\}$. Thus, $\bigcap_{j=1}^{n} A_j$ occurs whenever all $A_j, J = 1, \ldots, n$ occur simultaneously. For $n = 2, A_1 \cap A_2$ is presented in Figure 1.5. This definition extends to an infinite number of events. Thus, for countably infinite many events $A_j, j = 1, 2, \ldots$, one has $\bigcap_{j=1}^{\infty} A_j = \{s \in \mathcal{S}; s \in A_j, for \ all \ j = 1, 2, \ldots\}$.

If $A_1 \cap A_2 = \emptyset$, the events $A_1$ and $A_2$ are called *disjoint* (see Figure 1.6). The events $A_j, j = 1, 2, \ldots$, are said to be *mutually* or *pairwise disjoint*, if $A_i \cap A_j = \emptyset$ whenever $i \neq j$. The *differences* $A_1 - A_2$ and $A_2 - A_1$ are the events defined by $A_1 - A_2 = \{s \in \mathcal{S}; s \in A_1, s \notin A_2\}$, $A_2 - A_1 = \{s \in \mathcal{S}; s \in A_2, s \notin A_1\}$ (see Figure 1.7).

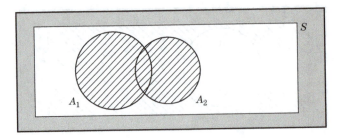

**FIGURE 1.4**

$A_1 \cup A_2$ is the shaded region.

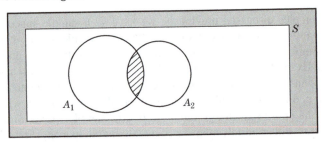

**FIGURE 1.5**

$A_1 \cap A_2$ is the shaded region.

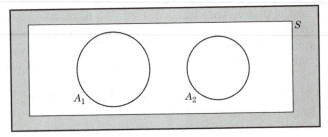

**FIGURE 1.6**

$A_1$ and $A_2$ are disjoint; that is, $A_i \cap A_j = \varnothing$.

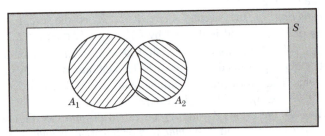

**FIGURE 1.7**

$A_1 - A_2$ is ////, $A_2 - A_1$ is \\\\.

From the definition of the preceding operations, the following properties follow immediately, and they are listed here for reference.

**Proposition 1.**

(i) $S^c = \emptyset, \emptyset^c = S, (A^c)^c = A$.

(ii) $S \cup A = S, \emptyset \cup A = A, A \cup A^c = S, A \cup A = A$.

(iii) $S \cap A = A, \emptyset \cap A = \emptyset, A \cap A^c = \emptyset, A \cap A = A$.

The previous statements are all obvious, as is the following: $\emptyset \subseteq A$ for every event $A$ in $S$. Also,

**Proposition 2.**

(i) $\left.\begin{array}{l} A_1 \cup (A_2 \cup A_3) = (A_1 \cup A_2) \cup A_3 \\ A_1 \cap (A_2 \cap A_3) = (A_1 \cap A_2) \cap A_3 \end{array}\right\}$  *(associative laws)*

(ii) $\left.\begin{array}{l} A_1 \cup A_2 = A_2 \cup A_1 \\ A_1 \cap A_2 = A_2 \cap A_1 \end{array}\right\}$  *(commutative laws)*

(iii) $\left.\begin{array}{l} A \cap (\cup_j A_j) = \cup_j(A \cap A_j) \\ A \cup (\cap_j A_j) = \cap_j(A \cup A_j) \end{array}\right\}$  *(distributive laws)*

In the last relations, as well as elsewhere, when the range of the index $j$ is not indicated explicitly, it is assumed to be a finite set, such as $\{1, \ldots, n\}$, or a countably infinite set, such as $\{1, 2, \ldots\}$.

For the purpose of demonstrating some of the set-theoretic operations just defined, let us consider some concrete examples.

**Example 41.**    Consider the sample space $S = \{s_1, s_2, s_3, s_4, s_5, s_6, s_7, s_8\}$ and define the events $A_1, A_2$, and $A_3$ as follows: $A_1 = \{s_1, s_2, s_3\}, A_2 = \{s_2, s_3, s_4, s_5\}, A_3 = \{s_3, s_4, s_5, s_8\}$. Then observe that:

$A_1^c = \{s_4, s_5, s_6, s_7, s_8\}$,    $A_2^c = \{s_1, s_6, s_7, s_8\}$,    $A_3^c = \{s_1, s_2, s_6, s_7\}$;

$A_1 \cup A_2 = \{s_1, s_2, s_3, s_4, s_5\}$,    $A_1 \cup A_3 = \{s_1, s_2, s_3, s_4, s_5, s_8\}$,

$A_2 \cup A_3 = \{s_2, s_3, s_4, s_5, s_8\}$,    $A_1 \cup A_2 \cup A_3 = \{s_1, s_2, s_3, s_4, s_5, s_8\}$;

$A_1 \cap A_2 = \{s_2, s_3\}$,    $A_1 \cap A_3 = \{s_3\}$,    $A_2 \cap A_3 = \{s_3, s_4, s_5\}$,    $A_1 \cap A_2 \cap A_3 = \{s_3\}$;

$A_1 - A_2 = \{s_1\}$,    $A_2 - A_1 = \{s_4, s_5\}$,    $A_1 - A_3 = \{s_1, s_2\}$,

$A_3 - A_1 = \{s_4, s_5, s_8\}$,    $A_2 - A_3 = \{s_2\}$,    $A_3 - A_2 = \{s_8\}$;

$\left(A_1^c\right)^c = \{s_1, s_2, s_3\}(=A_1)$,    $\left(A_2^c\right)^c = \{s_2, s_3, s_4, s_5\}(=A_2)$,

$\left(A_3^c\right)^c = \{s_3, s_4, s_5, s_8\}(=A_3)$.

An identity and DeMorgan's laws stated subsequently are of significant importance. Their justifications are left as exercises (see Exercises 2.14 and 2.15).

**Proposition 3** (An Identity).

$$\cup_j A_j = A_1 \cup \left(A_1^c \cap A_2\right) \cup \left(A_1^c \cap A_2^c \cap A_3\right) \cup \cdots$$

**Example 42.**   From Example 41, we have:

$$A_1 = \{s_1, s_2, s_3\}, \qquad A_1^c \cap A_2 = \{s_4, s_5\}, \qquad A_1^c \cap A_2^c \cap A_3 = \{s_8\}.$$

Note that $A_1, A_1^c \cap A_2, A_1^c \cap A_2^c \cap A_3$ are pairwise disjoint. Now $A_1 \cup (A_1^c \cap A_2) \cup (A_1^c \cap A_2^c \cap A_3) = \{s_1, s_2, s_3, s_4, s_5, s_8\}$, which is equal to $A_1 \cup A_2 \cup A_3$; that is,

$$A_1 \cup A_2 \cup A_3 = A_1 \cup \left(A_1^c \cap A_2\right) \cup \left(A_1^c \cap A_2^c \cap A_3\right)$$

as the preceding identity states.

The significance of the identity is that the events on the right-hand side are pairwise disjoint, whereas the original events $A_j, j \geq 1$, need not be so.

**Proposition 4** (DeMorgan's Laws).

$$(\cup_j A_j)^c = \cap_j A_j^c, \qquad (\cap_j A_j)^c = \cup_j A_j^c.$$

**Example 43.**   Again from Example 41, one has:

$$(A_1 \cup A_2)^c = \{s_6, s_7, s_8\}, \quad A_1^c \cap A_2^c = \{s_6, s_7, s_8\};$$
$$(A_1 \cup A_2 \cup A_3)^c = \{s_6, s_7\}, \quad A_1^c \cap A_2^c \cap A_3^c = \{s_6, s_7\};$$
$$(A_1 \cap A_2)^c = \{s_1, s_4, s_5, s_6, s_7, s_8\}, \quad A_1^c \cup A_2^c = \{s_1, s_4, s_5, s_6, s_7, s_8\};$$
$$(A_1 \cap A_2 \cap A_3)^c = \{s_1, s_2, s_4, s_5, s_6, s_7, s_8\},$$
$$A_1^c \cup A_2^c \cup A_3^c = \{s_1, s_2, s_4, s_5, s_6, s_7, s_8\},$$

so that

$$(A_1 \cup A_2)^c = A_1^c \cap A_2^c, \quad (A_1 \cup A_2 \cup A_3)^c = A_1^c \cap A_2^c \cap A_3^c, \quad \text{as DeMorgan's}$$
$$(A_1 \cap A_2)^c = A_1^c \cup A_2^c, \quad (A_1 \cap A_2 \cap A_3)^c = A_1^c \cup A_2^c \cup A_3^c, \quad \text{laws state.}$$

As a further demonstration of how complements, unions, and intersections of events are used for the expression of new events, consider the following example.

**Example 44.**   In terms of the events $A_1, A_2$, and $A_3$ (in some sample space $S$) and, perhaps, their complements, unions, and intersections, express the following events:

$D_i = $ "$A_i$ does not occur," $i = 1, 2, 3$, so that $D_1 = A_1^c, D_2 = A_2^c, D_3 = A_3^c$;

$E = $ "all $A_1, A_2, A_3$ occur," so that $E = A_1 \cap A_2 \cap A_3$;

$F = $ "none of $A_1, A_2, A_3$ occurs," so that $F = A_1^c \cap A_2^c \cap A_3^c$;

$G = $ "at least one of $A_1, A_2, A_3$ occurs," so that $G = A_1 \cup A_2 \cup A_3$;

$H = $ "exactly two of $A_1, A_2, A_3$ occur," so that $H = \left(A_1 \cap A_2 \cap A_3^c\right) \cup \left(A_1 \cap A_2^c \cap A_3\right) \cup \left(A_1^c \cap A_2 \cap A_3\right)$;

$I = $ "exactly one of $A_1, A_2, A_3$ occurs," so that $I = \left(A_1 \cap A_2^c \cap A_3^c\right) \cup \left(A_1^c \cap A_2 \cap A_3^c\right) \cup \left(A_1^c \cap A_2^c \cap A_3\right)$.

It also follows that:

$$G = \text{"exactly one of } A_1, A_2, A_3 \text{ occurs"} \cup \text{"exactly two of } A_1, A_2, A_3 \text{ occur"} \cup$$
$$\text{"all } A_1, A_2, A_3 \text{ occur"}$$
$$= I \cup H \cup E.$$

This section is concluded with the concept of a monotone sequence of events; namely, the sequence of events $\{A_n\}, n \geq 1$, is said to be *monotone*, if either $A_1 \subseteq A_2 \subseteq \cdots$ (*increasing*) or $A_1 \supseteq A_2 \supseteq \cdots$ (*decreasing*). In case of an increasing sequence, the union $\bigcup_{j=1}^{\infty} A_j$ is called the *limit* of the sequence, and in case of a decreasing sequence, the intersection $\bigcap_{j=1}^{\infty} A_j$ is called its *limit*.

The concept of the limit is also defined, under certain conditions, for nonmonotone sequences of events, but we are not going to enter into it here. The interested reader is referred to Definition 1, page 5, of the book *A Course in Mathematical Statistics*, 2nd edition (1997), Academic Press, by G. G. Roussas.

---

## EXERCISES

**2.1**  An airport limousine departs from a certain airport with three passengers to be delivered in any one of the three hotels denoted by $H_1, H_2, H_3$. Let $(x_1, x_2, x_3)$ denote the number of passengers (not which ones!) left at hotels $H_1, H_2$, and $H_3$, respectively.

    **(i)** Write out the sample space $S$ of all possible deliveries.

    **(ii)** Consider the events $A, B, C$, and $D$, defined as follows, and express them in terms of sample points.

        $A = $ "one passenger in each hotel,"

        $B = $ "all passengers in $H_1$,"

        $C = $ "all passengers in one hotel,"

        $D = $ "at least two passengers in $H_1$,"

        $E = $ "fewer passengers in $H_1$ than in each one of $H_2$ and $H_3$."

**2.2**  A machine dispenses balls that are either red or black or green. Suppose we operate the machine three successive times and record the color of the balls dispensed, to be denoted by $r, b$, and $g$ for the respective colors.

    **(i)** Write out an appropriate sample space $S$ for this experiment.

    **(ii)** Consider the events $A, B$, and $C$, defined as follows, and express them by means of sample points.

        $A = $ "all three colors appear,"

        $B = $ "only two colors appear,"

        $C = $ "at least two colors appear."

**2.3**  A university library has five copies of a textbook to be used in a certain class. Of these copies, numbers 1 through 3 are of the 1st edition, and numbers 4 and 5 are of the 2nd edition. Two of these copies are chosen at random to be placed on a 2-hour reserve.

(i) Write out an appropriate sample space $S$.

(ii) Consider the events $A, B, C$, and $D$, defined as follows, and express them in terms of sample points.

$A = $ "both books are of the 1st edition,"
$B = $ "both books are of the 2nd edition,"
$C = $ "one book of each edition,"
$D = $ "no book is of the 2nd edition."

**2.4**   A large automobile dealership sells three brands of American cars, denoted by $a_1, a_2, a_3$; two brands of Asian cars, denoted by $b_1, b_2$; and one brand of a European car, denoted by $c$. We observe the cars sold in two consecutive sales. Then:

(i) Write out an appropriate sample space for this experiment.

(ii) Express the events defined as follows in terms of sample points:

$A = $ "American brands in both sales,"
$B = $ "American brand in the first sale and Asian brand in the second sale,"
$C = $ "American brand in one sale and Asian brand in the other sale,"
$D = $ "European brand in one sale and Asian brand in the other sale."

**Hint.** For part (i), denote by $(x_1, x_2)$ the typical sample point, where $x_1, x_2$ stand for one of $a_1, a_2, a_3$; $b_1, b_2$; $c$.

**2.5**   Of two gas stations I and II located at a certain intersection, I has five gas pumps and II has six gas pumps. On a given time of a day, observe the numbers $x$ and $y$ of pumps in use in stations I and II, respectively.

(i) Write out the sample space $S$ for this experiment.

(ii) Consider the events $A, B, C$, and $D$, defined as follows and express them in terms of sample points.

$A = $ "only three pumps are in use in station I,"
$B = $ "the number of pumps in use in both stations is the same,"
$C = $ "the number of pumps in use in station II is larger than that in station I,"
$D = $ "the total number of pumps in use in both stations is not greater than 4."

**2.6**   At a certain busy airport, denote by $A, B, C$, and $D$ the events defined as follows:

$A = $ "at least five planes are waiting to land,"
$B = $ "at most three planes are waiting to land,"
$C = $ "at most two planes are waiting to land,"
$D = $ "exactly two planes are waiting to land."

In terms of the events $A, B, C$, and $D$ and, perhaps, their complements, express the following events:

$E = $ "at most four planes are waiting to land,"
$F = $ "at most one plane is waiting to land,"

$G =$ "exactly three planes are waiting to land,"
$H =$ "exactly four planes are waiting to land,"
$I =$ "at least four planes are waiting to land."

**2.7**  Let $S = \{(x, y) \in \Re^2 = \Re \times \Re;\ -3 \leq x \leq 3, 0 \leq y \leq 4, x$ and $y$ integers$\}$,
and define the events $A, B, C,$ and $D$ as follows:

$$A = \{(x, y) \in S; x = y\}, \quad B = \{(x, y) \in S; x = -y\},$$
$$C = \{(x, y) \in S; x^2 = y^2\}, \quad D = \{(x, y) \in S; x^2 + y^2 \leq 5\}.$$

  **(i)** List explicitly the members of $S$.
  **(ii)** List the members of the events just defined.

**2.8**  In terms of the events $A_1, A_2, A_3$ in a sample space $S$ and, perhaps, their
complements, express the following events:
  **(i)** $B_0 = \{s \in S; s$ belongs to none of $A_1, A_2, A_3\}$,
  **(ii)** $B_1 = \{s \in S; s$ belongs to exactly one of $A_1, A_2, A_3\}$,
  **(iii)** $B_2 = \{s \in S; s$ belongs to exactly two of $A_1, A_2, A_3\}$,
  **(iv)** $B_3 = \{s \in S; s$ belongs to all of $A_1, A_2, A_3\}$,
  **(v)** $C = \{s \in S; s$ belongs to at most two of $A_1, A_2, A_3\}$,
  **(vi)** $D = \{s \in S; s$ belongs to at least one of $A_1, A_2, A_3\}$.

**2.9**  If for three events $A, B,$ and $C$ it happens that either $A \cup B \cup C = A$ or
$A \cap B \cap C = A$, what conclusions can you draw? What about the case that
both $A \cup B \cup C = A$ and $A \cap B \cap C = A$ hold simultaneously?

**2.10**  Show that $A$ is the impossible event (i.e., $A = \emptyset$), if and only if
$(A \cap B^c) \cup (A^c \cap B) = B$ for every event $B$.

**2.11**  Let $A, B,$ and $C$ be arbitrary events in $S$. Determine whether each of the
following statements is correct or incorrect:
  **(i)** $(A - B) \cup B = (A \cap B^c) \cup B = B$,
  **(ii)** $(A \cup B) - A = (A \cup B) \cap A^c = B$,
  **(iii)** $(A \cap B) \cap (A - B) = (A \cap B) \cap (A \cap B^c) = \emptyset$,
  **(iv)** $(A \cup B) \cap (B \cup C) \cap (C \cup A) = (A \cap B) \cup (B \cap C) \cup (C \cap A)$.
  **Hint.** For part (iv), you may wish to use Proposition 2(iii).

**2.12**  For any three events $A, B,$ and $C$ in a sample space $S$ show that the transitive
property relative to $\subseteq$ holds (i.e., $A \subseteq B$ and $B \subseteq C$ imply that $A \subseteq C$).

**2.13**  Establish the distributive laws; namely, $A \cap (\cup_j A_j) = \cup_j (A \cap A_j)$ and
$A \cup (\cap_j A_j) = \cap_j (A \cup A_j)$.
  **Hint.** Show that the event of either side is contained in the event of the
other side.

**2.14**  Establish the identity:

$$\cup_j A_j = A_1 \cup \left(A_1^c \cap A_2\right) \cup \left(A_1^c \cap A_2^c \cap A_3\right) \cup \cdots$$

  **Hint.** As in Exercise 2.13.

**2.15** Establish DeMorgan's laws, namely,

$$(\cup_j A_j)^c = \cap_j A_j^c \quad \text{and} \quad (\cap_j A_j)^c = \cup_j A_j^c.$$

**Hint.** As in Exercise 2.13.

**2.16** Let $S = \Re$ and, for $n = 1, 2, \ldots$, define the events $A_n$ and $B_n$ by:

$$A_n = \left\{ x \in \Re; \ -5 + \frac{1}{n} < x < 20 - \frac{1}{n} \right\}, \qquad B_n = \left\{ x \in \Re; 0 < x < 7 + \frac{3}{n} \right\}.$$

(i) Show that the sequence $\{A_n\}$ is increasing and the sequence $\{B_n\}$ is decreasing.

(ii) Identify the limits, $\lim_{n \to \infty} A_n = \bigcup_{n=1}^{\infty} A_n$ and $\lim_{n \to \infty} B_n = \bigcap_{n=1}^{\infty} B_n$.

**Remark.**  See discussion following Example 44.

## 1.3 RANDOM VARIABLES

For every random experiment, there is at least one sample space appropriate for the random experiment under consideration. In many cases, however, much of the work can be done without reference to an explicit sample space. Instead, what are used extensively are random variables and their distributions. Those quantities will be studied extensively in subsequent chapters. What is done in this section is the introduction of the concept of a random variable.

Formally, a *random variable*, to be shortened to r.v., is simply a function defined on a sample space $S$ and taking values in the real line $\Re = (-\infty, \infty)$. Random variables are denoted by capital letters, such as $X, Y, Z$, with or without subscripts. Thus, the value of the r.v. $X$ at the sample point $s$ is $X(s)$, and the set of all values of $X$, that is, the *range* of $X$, is usually denoted by $X(S)$. The only difference between a r.v. and a function in the usual calculus sense is that the domain of a r.v. is a sample space $S$, which may be an abstract set, unlike the usual concept of a function, whose domain is a subset of $\Re$ or of a Euclidean space of higher dimension. The usage of the term "random variable" employed here rather than that of a function may be explained by the fact that a r.v. is associated with the outcomes of a random experiment. Thus, one may argue that $X(s)$ is not known until the random experiment is actually carried out and $s$ becomes available. Of course, on the same sample space, one may define many distinct r.v.'s.

In reference to Example 26, instead of the sample space $S$ exhibited there, one may be interested in the number of heads appearing each time the experiment is carried out. This leads to the definition of the r.v. $X$ by: $X(s) = \#$ of $H$'s in $s$. Thus, $X(HHH) = 3, X(HHT) = X(HTH) = X(THH) = 2, X(HTT) = X(THT) = X(TTH) = 1$, and $X(TTT) = 0$, so that $X(S) = \{0, 1, 2, 3\}$. The notation $(X \leq 1)$ stands for the event $\{s \in S; X(s) \leq 1\} = \{TTT, HTT, THT, TTH\}$. In the general case and for $B \subseteq \Re$, the notation $(X \in B)$ stands for the event $A$ in the sample space $S$ defined by $A = \{s \in S; X(s) \in B\}$. It is also denoted by $X^{-1}(B)$.

In reference to Example 27, a r.v. $X$ of interest may be defined by $X(s) = $ sum of the numbers in the pair $s$. Thus, $X((1,1)) = 2, X((1,2)) = X((2,1)) = 3, \ldots, X((6,6)) = 12$, and $X(\mathcal{S}) = \{2, 3, \ldots, 12\}$. Also, $X^{-1}(\{7\}) = \{s \in \mathcal{S}; X(s) = 7\} = \{(1,6), (2,5), (3,4), (4,3), (5,2), (6,1)\}$. Similarly for Examples 28–31.

In reference to Example 32, a natural r.v. $X$ is defined to denote the number of tosses needed until the first head occurs. Thus, $X(H) = 1, X(TH) = 2, \ldots, X(\underbrace{T \ldots T}_{n-1} H) = n, \ldots$, so that $X(\mathcal{S}) = \{1, 2, \ldots\}$. Also, $(X > 5) = (X \geq 6) = \{TTTTTH, TTTTTTH, \ldots\}$.

In reference to Example 33, an obvious r.v. $X$ is: $X(s) = s, s = 0, 1, \ldots$, and similarly for Examples 34–35.

In reference to Example 36, a r.v. $X$ of interest is $X(s) = s, s \in \mathcal{S}$, and similarly for Examples 37–40.

Also, in reference to Example 5, an obvious r.v. $X$ may be defined as follows: $X(s) = \#$ of $S$'s in $s$. Then, clearly, $X(\mathcal{S}) = \{0, 1, \ldots, n\}$. Similarly for Examples 6–10.

In reference to Example 11, a r.v. $X$ may be defined thus: $X(s) = $ the position of $M$ in $s$. Then, clearly, $X(\mathcal{S}) = \{1, 2, \ldots\}$.

In reference to Example 18, the r.v.'s of obvious interests are: $X_A = \#$ of those persons, out of $n$, having blood type A, and similarly for $X_B, X_{AB}, X_O$. Similarly for Examples 19 and 20.

From the preceding examples, two kinds of r.v.'s emerge: Random variables that take on countably many values, such as those defined in conjunction with Examples 26–35, and r.v.'s that take on all values in a nondegenerate (finite or not) interval in $\mathfrak{R}$. Such r.v.'s are defined in conjunction with Examples 36–40. Random variables of the former kind are called discrete r.v.'s (or r.v.'s of the discrete type), and r.v.'s of the latter type are called continuous r.v.'s (or r.v.'s of the continuous type).

More generally, a r.v. $X$ is called *discrete* (or *of the discrete type*), if $X$ takes on countably many values; i.e., either finitely many values such as $x_1, \ldots, x_n$, or countably infinitely many values such as $x_0, x_1, \ldots$ or $x_1, x_2, \ldots$. On the other hand, $X$ is called *continuous* (or *of the continuous type*) if $X$ takes all values in a nondegenerate interval $I \subseteq \mathfrak{R}$. Although there are other kinds of r.v.'s, in this book we will habitually restrict ourselves to discrete and continuous r.v.'s as just defined.

The study of r.v.'s is one of the main objectives of this book.

## EXERCISES

**3.1**  In reference to Exercise 2.1, define the r.v.'s $X_i, i = 1, 2, 3$ as follows: $X_i = \#$ of passengers delivered to hotel $H_i$.

Determine the values of each $X_i, i = 1, 2, 3$, and specify the values of the sum $X_1 + X_2 + X_3$.

**3.2**  In reference to Exercise 2.2, define the r.v.'s $X$ and $Y$ as follows: $X = \#$ of red balls dispensed, $Y = \#$ of balls other than red dispensed.

Determine the values of $X$ and $Y$ and specify the values of the sum $X + Y$.

**3.3** In reference to Exercise 2.5, define the r.v.'s $X$ and $Y$ as follows: $X = \#$ of pumps in use in station I, $Y = \#$ of pumps in use in station II. Determine the values of $X$ and $Y$, and also of the sum $X + Y$.

**3.4** In reference to Exercise 2.7, define the r.v. $X$ by: $X((x, y)) = x + y$. Determine the values of $X$, as well as the following events: $(X \leq 2), (3 < X \leq 5), (X > 6)$.

**3.5** Consider a year with 365 days, which are numbered serially from 1 to 365. Ten of those numbers are chosen at random and without repetition, and let $X$ be the r.v. denoting the largest number drawn. Determine the values of $X$.

**3.6** A four-sided die has the numbers 1 through 4 written on its sides, one on each side. If the die is rolled twice:
  **(i)** Write out a suitable sample space $S$.
  **(ii)** If $X$ is the r.v. denoting the sum of numbers appearing, determine the values of $X$.
  **(iii)** Determine the events: $(X \leq 3), (2 \leq X < 5), (X > 8)$.
  **Hint.** For part (i), the typical sample point is a pair $(x, y)$, where $x$ and $y$ run through the values $1, 2, 3, 4$.

**3.7** From a certain target population, $n$ individuals are chosen at random and their blood types are determined. Let $X_1, X_2, X_3$, and $X_4$ be the r.v.'s denoting the number of individuals having blood types A, B, AB, and O, respectively. Determine the values of each one of these r.v.'s, as well as the values of the sum $X_1 + X_2 + X_3 + X_4$.

**3.8** A bus is expected to arrive at a specified bus stop any time between 8:00 and 8:15 a.m., and let $X$ be the r.v. denoting the actual time of arrival of the bus.
  **(i)** Determine the suitable sample space $S$ for the experiment of observing the arrival of the bus.
  **(ii)** What are the values of the r.v. $X$?
  **(iii)** Determine the event: "The bus arrives within 5 min before the expiration of the expected time of arrival."

# The concept of probability and basic results

This chapter consists of five sections. The first section is devoted to the definition of the concept of probability. We start with the simplest case, where complete symmetry occurs, proceed with the definition by means of relative frequency, and conclude with the axiomatic definition of probability. The defining properties of probability are illustrated by way of examples. Also, a number of basic properties, resulting from the definition, are stated and justified. Some of them are illustrated by means of examples. This section is concluded with two theorems, which are stated but not proved.

In the second section, the probability distribution function of a r.v. is introduced. Also, the distribution function and the probability density function of a r.v. are defined, and we explain how they determine the distribution of the r.v.

The concept of the conditional probability of an event, given another event, is taken up in the following section. Its definition is given and its significance is demonstrated through a number of examples. This section is concluded with three theorems, formulated in terms of conditional probabilities. Through these theorems, conditional probabilities greatly simplify calculation of otherwise complicated probabilities.

In the fourth section, the independence of two events is defined, and we also indicate how it carries over to any finite number of events. A result (Theorem 6) is stated which is often used by many authors without its use even being acknowledged. The section is concluded with an indication of how independence extends to random experiments. The definition of independence of r.v.'s is deferred to Chapter 5.

In the final section of the chapter, the so-called fundamental principle of counting is discussed; combinations and permutations are then obtained as applications of this principle. Several illustrative examples are also provided.

## 2.1 DEFINITION OF PROBABILITY AND SOME BASIC RESULTS

When a random experiment is entertained, one of the first questions which arise is, what is the probability that a certain event occurs? For instance, in reference to Example 26 in Chapter 1, one may ask: What is the probability that exactly one head occurs; in other words, what is the probability of the event $B = \{HTT, THT, TTH\}$? The answer to this question is almost automatic and is 3/8. The relevant reasoning goes like this: Assuming that the three coins are balanced, the probability of each one of the 8 outcomes, considered as simple events, must be 1/8. Since the event $B$ consists of 3 sample points, it can occur in 3 different ways, and hence its probability must be 3/8.

This is exactly the intuitive reasoning employed in defining the concept of probability when two requirements are met: First, the sample space $S$ has finitely many outcomes, $S = \{s_1, \ldots, s_n\}$, say, and second, each one of these outcomes is "equally likely" to occur, has the same chance of appearing, whenever the relevant random experiment is carried out. This reasoning is based on the underlying symmetry. Thus, one is led to stipulating that each one of the (simple) events $\{s_i\}$, $i = 1, \ldots, n$, has probability $1/n$. Then the next step, that of defining the probability of a composite event $A$, is simple; if $A$ consists of $m$ sample points, $A = \{s_{i_1}, \ldots, s_{i_m}\}$, say $(1 \le m \le n)$ (or none at all, in which case $m = 0$), then the probability of $A$ must be $m/n$. The notation used is: $P(\{s_1\}) = \cdots = P(\{s_n\}) = \frac{1}{n}$ and $P(A) = \frac{m}{n}$. Actually, this is the so-called *classical* definition of probability. That is,

***Classical Definition of Probability*** Let $S$ be a sample space, associated with a certain random experiment and consisting of finitely many sample points $n$, say, each of which is equally likely to occur whenever the random experiment is carried out. Then the probability of any event $A$, consisting of $m$ sample points $(0 \le m \le n)$, is given by $P(A) = \frac{m}{n}$.

In reference to Example 26 in Chapter 1, $P(A) = \frac{4}{8} = \frac{1}{2} = 0.5$. In Example 27 (when the two dice are unbiased), $P(X = 7) = \frac{6}{36} = \frac{1}{6} \simeq 0.167$, where the r.v. $X$ and the event $(X = 7)$ are defined in Section 1.3 of Chapter 1. In Example 29, when the balls in the urn are thoroughly mixed, we may assume that all of the $(m + n)(m + n - 1)$ pairs are equally likely to be selected. Then, since the event $A$ occurs in 20 different ways, $P(A) = \frac{20}{(m+n)(m+n-1)}$. For $m = 3$ and $n = 5$, this probability is $P(A) = \frac{20}{56} = \frac{5}{14} \simeq 0.357$.

From the preceding (classical) definition of probability, the following simple properties are immediate: For any event $A$, $P(A) \ge 0$; $P(S) = 1$; if two events $A_1$ and $A_2$ are disjoint $(A_1 \cap A_2 = \emptyset)$, then $P(A_1 \cup A_2) = P(A_1) + P(A_2)$. This is so because, if $A_1 = \{s_{i_1}, \ldots, s_{i_k}\}$, $A_2 = \{s_{j_1}, \ldots, s_{j_\ell}\}$, where all $s_{i_1}, \ldots, s_{i_k}$ are distinct from all $s_{j_1}, \ldots, s_{j_\ell}$, then $A_1 \cup A_2 = \{s_{i_1}, \ldots, s_{i_k} s_{j_1}, \ldots, s_{j_\ell}\}$ and $P(A_1 \cup A_2) = \frac{k+\ell}{n} = \frac{k}{n} + \frac{\ell}{n} = P(A_1) + P(A_2)$.

In many cases, the stipulations made in defining the probability as above are not met, either because $S$ has not finitely many points (as is the case in Examples 32, 33–35 (by replacing $C$ and $M$ by $\infty$), and 36–40 in Chapter 1), or because the (finitely many) outcomes are not equally likely. This happens, for instance, in Example 26 when the coins are not balanced and in Example 27 when the dice are biased. Strictly speaking, it also happens in Example 30. In situations like this, the way out is provided by the so-called *relative frequency* definition of probability. Specifically, suppose a random experiment is carried out a large number of times $N$, and let $N(A)$ be the *frequency* of an event $A$, the number of times $A$ occurs (out of $N$). Then the *relative frequency* of $A$ is $\frac{N(A)}{N}$. Next, suppose that, as $N \to \infty$, the relative frequencies $\frac{N(A)}{N}$

oscillate around some number (necessarily between 0 and 1). More precisely, suppose that $\frac{N(A)}{N}$ converges, as $N \to \infty$, to some number. Then this number is called the *probability* of $A$ and is denoted by $P(A)$. That is, $P(A) = \lim_{N \to \infty} \frac{N(A)}{N}$. (It will be seen later in this book that the assumption of convergence of the relative frequencies $N(A)/N$ is justified subject to some qualifications.) To summarize,

**Relative Frequency Definition of Probability** Let $N(A)$ be the number of times an event $A$ occurs in $N$ (identical) repetitions of a random experiment, and assume that the relative frequency of $A$, $\frac{N(A)}{N}$, converges to a limit as $N \to \infty$. This limit is denoted by $P(A)$ and is called the probability of $A$.

At this point, it is to be observed that the empirical data show that the relative frequency definition of probability and the classical definition of probability agree in the framework in which the classical definition applies.

From the relative frequency definition of probability and the usual properties of limits, it is immediate that: $P(A) \geq 0$ for every event $A$; $P(S) = 1$; and for $A_1, A_2$ with $A_1 \cap A_2 = \emptyset$,

$$P(A_1 \cup A_2) = \lim_{N \to \infty} \frac{N(A_1 \cup A_2)}{N} = \lim_{N \to \infty} \left( \frac{N(A_1)}{N} + \frac{N(A_2)}{N} \right)$$

$$= \lim_{N \to \infty} \frac{N(A_1)}{N} + \lim_{N \to \infty} \frac{N(A_2)}{N} = P(A_1) + P(A_2);$$

that is, $P(A_1 \cup A_2) = P(A_1) + P(A_2)$, provided $A_1 \cap A_2 = \emptyset$. These three properties were also seen to be true in the classical definition of probability. Furthermore, it is immediate that under either definition of probability, $P(A_1 \cup \cdots \cup A_k) = P(A_1) + \cdots + P(A_k)$, provided the events are pairwise disjoint; $A_i \cap A_j = \emptyset, i \neq j$.

The above two definitions of probability certainly give substance to the concept of probability in a way consonant with our intuition about what probability should be. However, for the purpose of cultivating the concept and deriving deep probabilistic results, one must define the concept of probability in terms of some basic properties, which would not contradict what we have seen so far. This line of thought leads to the so-called axiomatic definition of probability due to Kolmogorov.

**Axiomatic Definition of Probability** Probability is a function, denoted by $P$, defined for each event of a sample space $S$, taking on values in the real line $\Re$, and satisfying the following three properties:

**P(1)**  $P(A) \geq 0$ for every event $A$ (nonnegativity of $P$).
**P(2)**  $P(S) = 1$ ($P$ is normed).
**P(3)**  For countably infinite many pairwise disjoint events $A_i, i = 1, 2, \ldots$, $A_i \cap A_j = \emptyset, i \neq j$, it holds

$$P(A_1 \cup A_2 \cup \cdots) = P(A_1) + P(A_2) + \cdots; \quad \text{or} \quad P\left( \bigcup_{i=1}^{\infty} A_i \right) = \sum_{i=1}^{\infty} P(A_i)$$

(sigma-additivity ($\sigma$-additivity) of $P$).

### Comments on the Axiomatic Definition

(1) Properties (P1) and (P2) are the same as the ones we have seen earlier, whereas property (P3) is new. What we have seen above was its so-called *finitely additive* version; that is, $P(\bigcup_{i=1}^{n} A_i) = \sum_{i=1}^{n} P(A_i)$, provided $A_i \cap A_j = \emptyset, i \neq j$. It will be seen below that finite additivity is implied by $\sigma$-additivity but not the other way around. Thus, if we are to talk about the probability of the union of countably infinite many pairwise disjoint events, property (P3) must be stipulated. Furthermore, the need for such a union of events is illustrated as follows: In reference to Example 32, calculate the probability that the first head does not occur before the $n$th tossing. By setting $A_i = \{\underbrace{T \ldots T}_{i-1} H\}, i = n, n + 1, \ldots,$ what we are actually after here is $P(A_n \cup A_{n+1} \cup \cdots)$ with $A_i \cap A_j = \emptyset, i \neq j, i$ and $j \geq n$.

(2) Property (P3) is superfluous (reduced to finite-additivity) when the sample space $S$ is finite, which implies that the total number of events is finite.

(3) Finite-additivity is implied by additivity for two events, $P(A_1 \cup A_2) = P(A_1) + P(A_2), A_1 \cap A_2 = \emptyset$, by way of mathematical induction.

Here are two examples in calculating probabilities.

**Example 1.** In reference to Example 1 in Chapter 1, take $n = 58$, and suppose we have the following configuration:

| | Barium | | | |
| --- | --- | --- | --- | --- |
| | High Mercury | | Low Mercury | |
| **Arsenic** | **High** | **Low** | **High** | **Low** |
| High | 1 | 3 | 5 | 9 |
| Low | 4 | 8 | 10 | 18 |

Calculate the probabilities mentioned in (i) (a)-(d).

**Discussion.** For simplicity, denote by $B_h$ the event that the site selected has a high barium concentration, and likewise for other events figuring below. Then:

(i)(a)  $B_h = (A_h \cap B_h \cap M_h) \cup (A_h \cap B_h \cap M_\ell) \cup (A_\ell \cap B_h \cap M_h) \cup (A_\ell \cap B_h \cap M_\ell)$ and the events on the right-hand side are pairwise disjoint. Therefore (by the basic property #2 in Section 2.1.1),

$$P(B_h) = P(A_h \cap B_h \cap M_h) + P(A_h \cap B_h \cap M_\ell)$$
$$+ P(A_\ell \cap B_h \cap M_h) + P(A_\ell \cap B_h \cap M_\ell)$$
$$= \frac{1}{58} + \frac{3}{58} + \frac{4}{58} + \frac{8}{58} = \frac{16}{58} = \frac{8}{29} \simeq 0.276.$$

**(i)(b)** Here, $P(M_h \cap A_\ell \cap B_\ell) = P(A_\ell \cap B_\ell \cap M_h) = \frac{10}{58} = \frac{5}{29} \simeq 0.172$.

**(i)(c)** Here, the required probability is as in (a):

$$P(A_h \cap B_h \cap M_\ell) + P(A_h \cap B_\ell \cap M_h) + P(A_\ell \cap B_h \cap M_h) = \frac{12}{58} = \frac{6}{29} \simeq 0.207.$$

**(i)(d)** As above,

$$P(A_h \cap B_\ell \cap M_\ell) + P(A_\ell \cap B_h \cap M_\ell) + P(A_\ell \cap B_\ell \cap M_h) = \frac{27}{58} \simeq 0.466.$$

**Example 2.** In ranking five horses in a horse race (Example 31 in Chapter 1), calculate the probability that horse #3 terminates at least second.

**Discussion.** Let $A_i$ be the event that horse #3 terminates at the $i$th position, $i = 1, \ldots, 5$. Then the required event is $A_1 \cup A_2$, where $A_1, A_2$ are disjoint. Thus,

$$P(A_1 \cup A_2) = P(A_1) + P(A_2) = \frac{24}{120} + \frac{24}{120} = \frac{2}{5} = 0.4.$$

**Example 3.** In tossing a coin repeatedly until $H$ appears for the first time (Example 32 in Chapter 1), suppose that $P\{\underbrace{T \ldots T}_{i-1} H\} = P(A_i) = q^{i-1} p$ for some $0 < p < 1$ and $q = 1 - p$ (in anticipation of Definition 3 in Section 2.4). Then

$$P\left(\bigcup_{i=n}^{\infty} A_i\right) = \sum_{i=n}^{\infty} P(A_i) = \sum_{i=n}^{\infty} q^{i-1} p = p \sum_{i=n}^{\infty} q^{i-1} = p \frac{q^{n-1}}{1-q} = p \frac{q^{n-1}}{p} = q^{n-1}.$$

For instance, for $p = 1/2$ and $n = 3$, this probability is $\frac{1}{4} = 0.25$. That is, when tossing a fair coin, the probability that the first head does not appear either the first or the second time (and therefore it appears either the third time or the fourth time, etc.) is 0.25. For $n = 10$, this probability is approximately $0.00195 \simeq 0.002$.

Next, we present some basic results following immediately from the defining properties of the probability. First, we proceed with their listing and then with their justification.

## 2.1.1 SOME BASIC PROPERTIES OF A PROBABILITY FUNCTION

1. $P(\emptyset) = 0$.
2. For any pairwise disjoint events $A_1, \ldots, A_n$, $P(\bigcup_{i=1}^{n} A_i) = \sum_{i=1}^{n} P(A_i)$.
3. For any event $A$, $P(A^c) = 1 - P(A)$.
4. $A_1 \subseteq A_2$ implies $P(A_1) \leq P(A_2)$ and $P(A_2 - A_1) = P(A_2) - P(A_1)$.
5. $0 \leq P(A) \leq 1$ for every event $A$.
6. **(i)** For any two events $A_1$ and $A_2$:

$$P(A_1 \cup A_2) = P(A_1) + P(A_2) - P(A_1 \cap A_2).$$

**(ii)** For any three events $A_1, A_2$, and $A_3$:

$$P(A_1 \cup A_2 \cup A_3) = P(A_1) + P(A_2) + P(A_3) - [P(A_1 \cap A_2)$$
$$+ P(A_1 \cap A_3) + P(A_2 \cap A_3)] + P(A_1 \cap A_2 \cap A_3).$$

**7.** For any events $A_1, A_2, \ldots, P(\bigcup_{i=1}^{\infty} A_i) \leq \sum_{i=1}^{\infty} P(A_i)$ ($\sigma$-sub-additivity), and $P(\bigcup_{i=1}^{n} A_i) \leq \sum_{i=1}^{n} P(A_i)$ (finite-sub-additivity).

### 2.1.2 JUSTIFICATION

**1.** From the obvious fact that $S = S \cup \emptyset \cup \emptyset \cup \cdots$ and property (P3),

$$P(S) = P(S \cup \emptyset \cup \emptyset \cup \cdots) = P(S) + P(\emptyset) + P(\emptyset) + \cdots$$

or $P(\emptyset) + P(\emptyset) + \cdots = 0$. By (P1), this can only happen when $P(\emptyset) = 0$.

Of course, that the impossible event has probability 0 does not come as a surprise. Any reasonable definition of probability should imply it.

**2.** Take $A_i = \emptyset$ for $i \geq n + 1$, consider the following obvious relation, and use (P3) and #1 to obtain:

$$P\left(\bigcup_{i=1}^{n} A_i\right) = P\left(\bigcup_{i=1}^{\infty} A_i\right) = \sum_{i=1}^{\infty} P(A_i) = \sum_{i=1}^{n} P(A_i).$$

**3.** From (P2) and #2, $P(A \cup A^c) = P(S) = 1$ or $P(A) + P(A^c) = 1$, so that $P(A^c) = 1 - P(A)$.

**4.** The relation $A_1 \subseteq A_2$, clearly, implies $A_2 = A_1 \cup (A_2 - A_1)$, so that, by #2, $P(A_2) = P(A_1) + P(A_2 - A_1)$. Solving for $P(A_2 - A_1)$, we obtain $P(A_2 - A_1) = P(A_2) - P(A_1)$, so that, by (P1), $P(A_1) \leq P(A_2)$.

At this point it must be pointed out that $P(A_2 - A_1)$ *need* not be $P(A_2) - P(A_1)$, if $A_1$ is *not* contained in $A_2$.

**5.** Clearly, $\emptyset \subseteq A \subseteq S$ for any event $A$. Then (P1), #1 and #4 give: $0 = P(\emptyset) \leq P(A) \leq P(S) = 1$.

**6.** **(i)** It is clear (e.g., by means of a Venn diagram) that

$$A_1 \cup A_2 = A_1 \cup (A_2 \cap A_1^c) = A_1 \cup (A_2 - A_1 \cap A_2).$$

Then, by means of #2 and #4:

$$P(A_1 \cup A_2) = P(A_1) + P(A_2 - A_1 \cap A_2) = P(A_1) + P(A_2) - P(A_1 \cap A_2).$$

**(ii)** Apply part (i) to obtain:

$$P(A_1 \cup A_2 \cup A_3) = P[(A_1 \cup A_2) \cup A_3] = P(A_1 \cup A_2) + P(A_3)$$
$$- P[(A_1 \cup A_2) \cap A_3]$$
$$= P(A_1) + P(A_2) - P(A_1 \cap A_2) + P(A_3)$$
$$- P[(A_1 \cap A_3) \cup (A_2 \cap A_3)]$$

$$= P(A_1) + P(A_2) + P(A_3) - P(A_1 \cap A_2)$$
$$- [P(A_1 \cap A_3) + P(A_2 \cap A_3) - P(A_1 \cap A_2 \cap A_3)]$$
$$= P(A_1) + P(A_2) + P(A_3) - P(A_1 \cap A_2) - P(A_1 \cap A_3)$$
$$- P(A_2 \cap A_3) + P(A_1 \cap A_2 \cap A_3).$$

**7.** By the identity in Section 1.2 and (P3):

$$P\left(\bigcup_{i=1}^{\infty} A_i\right) = P[A_1 \cup (A_1^c \cap A_2) \cup \cdots \cup (A_1^c \cap \cdots \cap A_{n-1}^c \cap A_n) \cup \cdots]$$

$$= P(A_1) + P(A_1^c \cap A_2) + \cdots + P(A_1^c \cap \cdots \cap A_{n-1}^c \cap A_n) + \cdots$$

$$\leq P(A_1) + P(A_2) + \cdots + P(A_n) + \cdots \text{ (by #4).}$$

For the finite case:

$$P\left(\bigcup_{i=1}^{n} A_i\right) = P[A_1 \cup (A_1^c \cap A_2) \cup \cdots \cup (A_1^c \cap \cdots \cap A_{n-1}^c \cap A_n)]$$

$$= P(A_1) + P(A_1^c \cap A_2) + \cdots + P(A_1^c \cap \cdots \cap A_{n-1}^c \cap A_n)$$

$$\leq P(A_1) + P(A_2) + \cdots + P(A_n).$$

Next, some examples are presented to illustrate some of the properties #1–#7.

**Example 4.**

(i) For two events $A$ and $B$, suppose that $P(A) = 0.3, P(B) = 0.5$, and $P(A \cup B) = 0.6$. Calculate $P(A \cap B)$.

(ii) If $P(A) = 0.6, P(B) = 0.3, P(A \cap B^c) = 0.4$, and $B \subset C$, calculate $P(A \cup B^c \cup C^c)$.

**Discussion.**

(i) From $P(A \cup B) = P(A) + P(B) - P(A \cap B)$, we get $P(A \cap B) = P(A) + P(B) - P(A \cup B) = 0.3 + 0.5 - 0.6 = 0.2$.

(ii) The relation $B \subset C$ implies $C^c \subset B^c$ and hence $A \cup B^c \cup C^c = A \cup B^c$. Then $P(A \cup B^c \cup C^c) = P(A \cup B^c) = P(A) + P(B^c) - P(A \cap B^c) = 0.6 + (1 - 0.3) - 0.4 = 0.9$.

**Example 5.** Let $A$ and $B$ be the respective events that two contracts I and II, say, are completed by certain deadlines, and suppose that: $P$(at least one contract is completed by its deadline) $= 0.9$ and $P$(both contracts are completed by their deadlines) $= 0.5$. Calculate the probability: $P$(exactly one contract is completed by its deadline).

**Discussion.** The assumptions made are translated as follows: $P(A \cup B) = 0.9$ and $P(A \cap B) = 0.5$. What we wish to calculate is: $P[(A \cap B^c) \cup (A^c \cap B)] = P[(A \cup B) - (A \cap B)] = P(A \cup B) - P(A \cap B) = 0.9 - 0.5 = 0.4$.

**Example 6.**

(i) For three events $A, B$, and $C$, suppose that $P(A \cap B) = P(A \cap C)$ and $P(B \cap C) = 0$. Then show that $P(A \cup B \cup C) = P(A) + P(B) + P(C) - 2P(A \cap B)$.

(ii) For any two events $A$ and $B$, show that $P(A^c \cap B^c) = 1 - P(A) - P(B) + P(A \cap B)$.

**Discussion.**

(i) We have $P(A \cup B \cup C) = P(A) + P(B) + P(C) - P(A \cap B) - P(A \cap C) - P(B \cap C) + P(A \cap B \cap C)$. But $A \cap B \cap C \subset B \cap C$, so that $P(A \cap B \cap C) \leq P(B \cap C) = 0$, and therefore $P(A \cup B \cup C) = P(A) + P(B) + P(C) - 2P(A \cap B)$.

(ii) Indeed, $P(A^c \cap B^c) = P((A \cup B)^c) = 1 - P(A \cup B) = 1 - P(A) - P(B) + P(A \cap B)$.

**Example 7.** In ranking five horses in a horse race (Example 31 in Chapter 1), what is the probability that horse #3 will terminate either first or second or third?

**Discussion.** Denote by $B$ the required event and let $A_i =$ "horse #3 terminates in the $i$th place," $i = 1, 2, 3$. Then the events $A_1, A_2, A_3$ are pairwise disjoint, and therefore

$$P(B) = P(A_1 \cup A_2 \cup A_3) = P(A_1) + P(A_2) + P(A_3).$$

But $P(A_1) = P(A_2) = P(A_3) = \frac{24}{120} = 0.2$, so that $P(B) = 0.6$.

**Example 8.** Consider a well-shuffled deck of 52 playing cards (Example 28 in Chapter 1), and suppose we draw at random three cards. What is the probability that at least one is an ace?

**Discussion.** Let $A$ be the required event, and let $A_i$ be defined by $A_i =$ "exactly $i$ cards are aces," $i = 0, 1, 2, 3$. Then, clearly, $P(A) = P(A_1 \cup A_2 \cup A_3)$. Instead, we may choose to calculate $P(A)$ through $P(A^c) = 1 - P(A_0)$, where

$$P(A_0) = \frac{\binom{48}{3}}{\binom{52}{3}} = \frac{48 \times 47 \times 46}{52 \times 51 \times 50} = \frac{4324}{5525}, \quad \text{so that } P(A) = \frac{1201}{5525} \simeq 0.217.$$

**Example 9.** Refer to Example 3 in Chapter 1 and let $C_1, C_2, C_3$ be defined by $C_1 =$ "both $S_1$ and $S_2$ work," $C_2 =$ "$S_5$ works," $C_3 =$ "both $S_3$ and $S_4$ work," and let $C =$ "current is transferred from point $A$ to point $B$." Then $P(C) = P(C_1 \cup C_2 \cup C_3)$. At this point (in anticipation of Definition 3 in Section 2.4; see also Exercise 4.14 in this chapter), suppose that:

$$P(C_1) = p_1 p_2, \quad P(C_2) = p_5, \quad P(C_3) = p_3 p_4,$$

$$P(C_1 \cap C_2) = p_1 p_2 p_5, \quad P(C_1 \cap C_3) = p_1 p_2 p_3 p_4,$$

$$P(C_2 \cap C_3) = p_3 p_4 p_5, \quad P(C_1 \cap C_2 \cap C_3) = p_1 p_2 p_3 p_4 p_5.$$

Then:

$$P(C) = p_1 p_2 + p_5 + p_3 p_4 - p_1 p_2 p_5 - p_1 p_2 p_3 p_4 - p_3 p_4 p_5 + p_1 p_2 p_3 p_4 p_5.$$

For example, for $p_1 = p_2 = p_3 = p_4 = p_5 = 0.9$, we obtain

$$P(C) = 0.9 + 2(0.9)^2 - 2(0.9)^3 - (0.9)^4 + (0.9)^5 \simeq 0.996.$$

This section is concluded with two very useful results stated as theorems. The first is a generalization of property #6 to more than three events, and the second is akin to the concept of continuity of a function as it applies to a probability function.

**Theorem 1.**  *The probability of the union of any n events, $A_1, \ldots, A_n$, is given by:*

$$P\left(\bigcup_{j=1}^{n} A_j\right) = \sum_{j=1}^{n} P(A_j) - \sum_{1 \le j_1 < j_2 \le n} P(A_{j_1} \cap A_{j_2})$$

$$+ \sum_{1 \le j_1 < j_2 < j_3 \le n} P(A_{j_1} \cap A_{j_2} \cap A_{j_3})$$

$$- \cdots + (-1)^{n+1} P(A_1 \cap \cdots \cap A_n).$$

Although its proof (which is by mathematical induction) will not be presented, the pattern of the right-hand side above follows that of property #6(i) and it is clear. First, sum up the probabilities of the individual events, then subtract the probabilities of the intersections of the events, taken two at a time (in the ascending order of indices), then add the probabilities of the intersections of the events, taken three at a time as before, and continue like this until you add or subtract (depending on $n$) the probability of the intersection of all $n$ events.

Recall that, if $A_1 \subseteq A_2 \subseteq \cdots$, then $\lim_{n \to \infty} A_n = \bigcup_{n=1}^{\infty} A_n$, and if $A_1 \supseteq A_2 \supseteq \cdots$, then $\lim_{n \to \infty} A_n = \bigcap_{n=1}^{\infty} A_n$.

**Theorem 2.**  *For any monotone sequence of events $\{A_n\}$, $n \ge 1$, it holds $P(\lim_{n \to \infty} A_n) = \lim_{n \to \infty} P(A_n)$.*

This theorem will be employed in many instances, and its use will be then pointed out.

# EXERCISES

**1.1**   If $P(A) = 0.4$, $P(B) = 0.6$, and $P(A \cup B) = 0.7$, calculate $P(A \cap B)$.

**1.2**   If for two events $A$ and $B$, it so happens that $P(A) = \frac{3}{4}$ and $P(B) = \frac{3}{8}$, show that:

$$P(A \cup B) \ge \frac{3}{4} \quad \text{and} \quad \frac{1}{8} \le P(A \cap B) \le \frac{3}{8}.$$

**1.3** If for the events $A, B$, and $C$, it so happens that $P(A) = P(B) = P(C) = 1$, then show that:

$$P(A \cap B) = P(A \cap C) = P(B \cap C) = P(A \cap B \cap C) = 1.$$

**Hint.** Use Properties 2.1.1 (4), 6(i), (ii).

**1.4** If the events $A, B$, and $C$ are related as follows: $A \subset B \subset C$ and $P(A) = \frac{1}{4}, P(B) = \frac{5}{12}$, and $P(C) = \frac{7}{12}$, compute the probabilities of the following events:

$$A^c \cap B, \qquad A^c \cap C, \qquad B^c \cap C, \qquad A \cap B^c \cap C^c, \qquad A^c \cap B^c \cap C^c.$$

**Hint.** Use Properties 2.1.1 (3), (4) here, and Proposition 4 in Chapter 1.

**1.5** Let $S$ be the set of all outcomes when flipping a fair coin four times, so that all 16 outcomes are equally likely. Define the events $A$ and $B$ by:

$$A = \{s \in S; s \text{ contains more } T's \text{ than } H's\},$$

$$B = \{s \in S; \text{ there are both } H's \text{ and } T's \text{ in } s \text{ and every } T \text{ in } s \text{ precedes}$$

$$\text{every } H \text{ in } s\}.$$

Compute the probabilities $P(A), P(B)$.

**1.6** Let $S = \{x \text{ integer}; 1 \leq x \leq 200\}$, and define the events $A, B$, and $C$ as follows:

$$A = \{x \in S; x \text{ is divisible by } 7\},$$
$$B = \{x \in S; x = 3n + 10, \text{ for some positive integer } n\},$$
$$C = \{x \in S; x^2 + 1 \leq 375\}.$$

Calculate the probabilities $P(A), P(B)$, and $P(C)$.

**1.7** If two fair dice are rolled once, what is the probability that the total number of spots shown is:
 (i) Equal to 5?
 (ii) Divisible by 3?

**1.8** Students in a certain college subscribe to three news magazines $A, B$, and $C$ according to the following proportions:

$$A : 20\%, \qquad B : 15\%, \qquad C : 10\%,$$

both $A$ and $B$: 5%, both $A$ and $C$: 4%, both $B$ and $C$: 3%, all three $A, B$, and $C$: 2%.
If a student is chosen at random, what is the probability he/she subscribes to none of the news magazines?
**Hint.** Use Proposition 4 in Chapter 1, and Property 2.1.1 (6) (ii) here.

**1.9** A high-school senior applies for admissions to two colleges $A$ and $B$, and suppose that: $P(\text{admitted at } A) = p_1$, $P(\text{rejected by } B) = p_2$, and $P(\text{rejected by at least one, } A \text{ or } B) = p_3$.

(i) Calculate the probability that the student is admitted by at least one college.
(ii) Find the numerical value of the probability in part (i), if $p_1 = 0.6, p_2 = 0.2$, and $p_3 = 0.3$.

**1.10** An airport limousine service has two vans, the smaller of which can carry six passengers and the larger nine passengers. Let $x$ and $y$ be the respective numbers of passengers carried by the smaller and the larger van in a given trip, so that a suitable sample space $S$ is given by:

$$S = \{(x, y); x = 0, \ldots, 6 \text{ and } y = 0, 1, \ldots, 9\}.$$

Also, suppose that, for all values of $x$ and $y$, the probabilities $P(\{(x, y)\})$ are equal. Finally, define the events $A, B$, and $C$ as follows:

$A$ = "the two vans together carry either 4 or 6 or 10 passengers,"
$B$ = "the larger van carries twice as many passengers as the smaller van,"
$C$ = "the two vans carry different numbers of passengers."

Calculate the probabilities: $P(A), P(B)$, and $P(C)$.

**1.11** In the sample space $S = (0, \infty)$, consider the events $A_n = (0, 1 - \frac{2}{n}), n = 1, 2, \ldots, A = (0, 1)$, and suppose that $P(A_n) = \frac{2n-1}{4n}$.
(i) Show that the sequence $\{A_n\}$ is increasing and that $\lim_{n \to \infty} A_n = \bigcup_{n=1}^{\infty} A_n = A$.
(ii) Use the appropriate theorem (cite it!) in order to calculate the probability $P(A)$.

**1.12** Show that for any $n$ events $A_1, \ldots, A_n$, it holds:

$$P\left(\bigcap_{i=1}^{n} A_i^c\right) \geq 1 - \sum_{i=1}^{n} P(A_i).$$

**1.13** Show that, if $P(A_i) = 1, i = 1, 2, \ldots$, then $P\left(\bigcap_{i=1}^{\infty} A_i\right) = 1$.
**Hint.** Use DeMorgan's laws and Theorem 2 in Chapter 2.

**1.14** In a small community 15 families have a number of children as indicated below.

| Number of family | 4 | 5 | 3 | 2 | 1 | |
|---|---|---|---|---|---|---|
| Number of children | 1 | 2 | 3 | 4 | 5 | |
| Total number of children | 4 | 10 | 9 | 8 | 5 | 36 |

(i) If a family is chosen at random, what is the probability $P(C_i)$, where $C_i$ = "the family bas $i$ children," $i = 1, 2, 3, 4, 5$?
(ii) If a child is chosen at random (from among the 36 children in the community), what is the probability $P(F_i)$, where $F_i$ = "child belongs in a family with $i$ children," $i = 1, 2, 3, 4, 5$?

## 2.2 DISTRIBUTION OF A RANDOM VARIABLE

For a r.v. $X$, define the set function $P_X(B) = P(X \in B)$. Then $P_X$ is a probability function because: $P_X(B) \geq 0$ for all $B$, $P_X(\Re) = P(X \in \Re) = 1$, and, if $B_j, j = 1, 2, \ldots$ are pairwise disjoint then, clearly, $(X \in B_j), j \geq 1$, are also pairwise disjoint and $X \in (\bigcup_{j=1}^{\infty} B_j) = \bigcup_{j=1}^{\infty} (X \in B_j)$. Therefore,

$$P_X\left(\bigcup_{j=1}^{\infty} B_j\right) = P\left[X \in \left(\bigcup_{j=1}^{\infty} B_j\right)\right] = P\left[\bigcup_{j=1}^{\infty} (X \in B_j)\right]$$

$$= \sum_{j=1}^{\infty} P(X \in B_j) = \sum_{j=1}^{\infty} P_X(B_j).$$

The probability function $P_X$ is called the *probability distribution function of the r.v. X*. Its significance is extremely important because it tells us the probability that $X$ takes values in any given set $B$. Indeed, much of probability and statistics revolves around the distribution of r.v.'s in which we have an interest.

By selecting $B$ to be $(-\infty, x], x \in \Re$, we have $P_X(B) = P(X \in (-\infty, x]) = P(X \leq x)$. In effect, we define a point function which we denote by $F_X$; that is, $F_X(x) = P(X \leq x), x \in \Re$. The function $F_X$ is called the *distribution function* (d.f.) of $X$. Clearly, if we know $P_X$, then we certainly know $F_X$. Somewhat unexpectedly, the converse is also true. Namely, if we know the (relatively "few") probabilities $F_X(x), x \in \Re$, then we can determine precisely all probabilities $P_X(B)$ for $B$ subset of $\Re$. This converse is a deep theorem in probability that we cannot deal with here. It is, nevertheless, the reason for which it is the d.f. $F_X$ we deal with, a familiar point function for which so many calculus results hold, rather than the unfamiliar set function $P_X$.

Clearly, the expressions $F_X(+\infty)$ and $F_X(-\infty)$ have no meaning because $+\infty$ and $-\infty$ are not real numbers. They are defined as follows:

$$F_X(+\infty) = \lim_{n \to \infty} F_X(x_n), \quad x_n \uparrow \infty \quad \text{and} \quad F_X(-\infty) = \lim_{n \to \infty} F_X(y_n), \quad y_n \downarrow -\infty.$$

These limits exist because $x < y$ implies $(-\infty, x] \subset (-\infty, y]$ and hence

$$P_X((-\infty, x]) = F_X(x) \leq F_X(y) = P_X((-\infty, y]).$$

The d.f. of a r.v. $X$ has the following basic properties:

1. $0 \leq F_X(x) \leq 1$ for all $x \in \Re$;
2. $F_X$ is a nondecreasing function;
3. $F_X$ is continuous from the right;
4. $F_X(+\infty) = 1, F_X(-\infty) = 0$.

The first and the second properties are immediate from the definition of the d.f.; the third follows by Theorem 2, by taking $x_n \downarrow x$; so does the fourth, by taking $x_n \uparrow +\infty$, which implies $(-\infty, x_n] \uparrow \Re$, and $y_n \downarrow -\infty$, which implies $(-\infty, y_n] \downarrow \emptyset$. Figures 2.1 and 2.2 show the graphs of the d.f.'s of some typical cases.

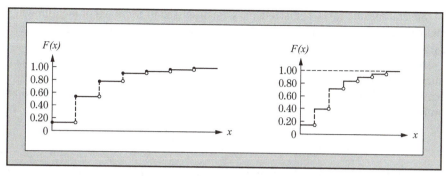

**FIGURE 2.1**

Examples of graphs of d.f.'s.

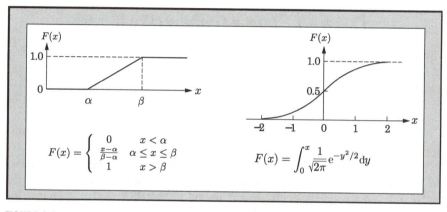

**FIGURE 2.2**

Examples of graphs of d.f.'s.

Now, suppose that the r.v. $X$ is discrete and takes on the values $x_j, j = 1, 2, \ldots, n$. Take $b = \{x_j\}$ and on the set $\{x_1, x_2, \ldots, x_n\}$ define the function $f_X$ as follows: $f_X(x_j) = P_X(\{x_j\})$. Next, extend $f_X$ over the entire $\Re$ by setting $f_X(x) = 0$ for $x \neq x_j, j = 1, 2, \ldots, n$. Then $f_X(x) \geq 0$ for all $x$, and it is clear that $P(X \in B) = \sum_{x_j \in B} f_X(x_j)$ for $B \subseteq \Re$. In particular, $\sum_{j=1}^{n} f_X(x_j) = \sum_{x_j \in \Re} f_X(x_j) = P(X \in \Re) = 1$. The function $f_X$ just defined is called the *probability density function* (p.d.f.) of the r.v. $X$. By selecting $B = (-\infty, x]$ for some $x \in \Re$, we have $F_X(x) = \sum_{x_j \leq x} f_X(x_j)$. Furthermore, if we assume at this point that $x_1 < x_2 < \cdots < x_n$, it is clear that

$$f_X(x_j) = F_X(x_j) - F_X(x_{j-1}), j = 2, 3, \ldots, n \quad \text{and} \quad f_X(x_1) = F_X(x_1);$$

we may also allow $j$ to take the value 1 above by setting $F_X(x_0) = 0$. Likewise if $X$ takes the values $x_j, j = 1, 2, \ldots$ These two relations state that, in the case that $X$ is a discrete r.v. as above, either one of the $F_X$ or $f_X$ specifies uniquely the other.

Several specific cases of discrete r.v.'s are discussed in Section 3 of the following chapter. Also, Examples 4–12, 18–20, 26, 27, and 32–35 in Chapter 1 lead to discrete r.v.'s.

Now, suppose that $X$ is a continuous r.v., one which takes on all values in a proper interval $I$ (finite or not) in the real line $\Re$, so that $I = (a, b)$ with $-\infty \leq a < b \leq \infty$. Suppose further that there exists a function $f : I \to [0, \infty)$ having the following property: $F_X(x) = \int_a^x f(t)\, dt, x \in I$. In particular,

$$\int_a^b f(t)\, dt = F_X(b) = P(X \leq b) = P(a \leq X \leq b) = 1.$$

If $I$ is not all of $\Re$, extend $f$ off $I$ by setting $f(x) = 0$ for $x \notin I$. Thus, for all $x : f(x) \geq 0$ and $F_X(x) = \int_{-\infty}^x f(t)\, dt$. As has already been pointed out elsewhere, $F_X$ uniquely determines $P_X$. The implication of it is that $P(X \in B) = P_X(B) = \int_B f(t)\, dt, B \subseteq \Re$, and, in particular,

$$\int_\Re f(t)\, dt = \int_{-\infty}^\infty f(t)\, dt = F_X(+\infty) = P(X \in \Re) = 1.$$

The function $f$ with the properties: $f(x) \geq 0$ all $x$ and $P(X \in B) = \int_B f(t)\, dt, B \subseteq \Re, (\int_{-\infty}^\infty f(t)\, dt = 1)$, is the p.d.f. of the r.v. $X$. In order to emphasize its association with the r.v. $X$, we often write $f_X$.

Most of the continuous r.v.'s we are dealing with in this book do have p.d.f.'s. In Section 3.3, a number of such r.v.'s will be presented explicitly.

Also, Examples 13–17, 21–25, and 36–40 in Chapter 1, under reasonable assumptions, lead to continuous r.v.'s, as will be seen on various occasions later.

It is to be observed that for a continuous r.v. $X$, $P(X = x) = 0$ for all $x \in \Re$. That is, the probability that $X$ takes on any specific value $x$ is 0; $X$ takes on values with positive probabilities in a nondegenerate interval around $x$. That $P(X = x) = 0$ follows, of course, from the definition of the p.d.f. of a continuous r.v., as

$$P(X = x) = \int_{\{x\}} f(t)\, dt = 0.$$

For a case of a continuous r.v., refer to Example 37 in Chapter 1 and let $X$ and $Y$ be r.v.'s denoting the cartesian coordinates of the point $P$ of impact. Then the distance of $P$ from the origin is the r.v. $R = \sqrt{X^2 + Y^2}$, which (reasonably enough) may be assumed to take every value in $[0, \infty)$. As will be seen, it is reasonable to assume that $X$ and $Y$ are independently Normally distributed with mean 0 and variance $\sigma^2$. This leads to the fact that $R^2$ is a multiple of a Chi-square distributed r.v., so that the p.d.f. of $R$ is precisely determined. (See Exercise 2.14 in Chapter 5).

If $X$ is a continuous r.v. with p.d.f. $f_X$, then its d.f. $F_X$ is given by $F_X(x) = \int_{-\infty}^x f_X(t)\, dt, x \in \Re$, so that $f_X$ uniquely determines $F_X$. It is also true that $\frac{dF_X(x)}{dx} = f_X(x)$ (for continuity points $x$ of $f_X$). Thus, $F_X$ also determines $f_X$.

In summary then, with a r.v. $X$, we have associated three quantities:

**(i)** The probability distribution function (or just distribution) of $X$, denoted by $P_X$, which is a set function and gives the probabilities $P(X \in B), B \subseteq \Re$.

(ii) The d.f. of $X$, denoted by $F_X$, which is a point function and gives the probabilities $P(X \in (-\infty, x]) = P(X \le x) = F_X(x), x \in \mathfrak{R}$.
(iii) The p.d.f. $f_X$ which is a nonnegative (point) function and gives all probabilities we may be interested in, either through a summation (for the discrete case) or through an integration (for the continuous case). Thus, for every $x \in \mathfrak{R}$:

$$F_X(x) = P(X \le x) = \begin{cases} \sum_{x_i \le x} f_X(x_i) & \text{for the discrete case,} \\ \int_{-\infty}^{x} f_X(t)\, dt & \text{for the continuous case,} \end{cases}$$

or, more generally:

$$P_X(B) = P(X \in B) = \begin{cases} \sum_{x_i \in B} f_X(x_i) & \text{for the discrete case,} \\ \int_{B} f_X(t)\, dt & \text{for the continuous case.} \end{cases}$$

In the discrete case, $f_X(x_i) = P(X = x_i) = P_X(\{x_i\})$, whereas in the continuous case, $f_X(x) = 0$ for every $x$. The p.d.f. $f_X$, clearly, determines the d.f. $F_X$, and the converse is also true. Of course, the p.d.f. $f_X$ also determines the probability distribution $P_X$.

At this point, notice that it makes sense to ask whether a given function $f$ is the p.d.f. of a r.v. $X$. The required conditions for this to be the case are: $f(x) \ge 0$ for all $x$, and either

$$f(x_j) > 0, \quad j = 1, 2, \ldots, \quad \text{with } \sum_j f(x_j) = 1, \text{ and } f(x) = 0 \text{ for all } x \ne x_j, j \ge 1;$$

$$\text{or} \quad \int_{-\infty}^{\infty} f(x)\, dx = 1.$$

Let us conclude this section with the following concrete examples.

**Example 10.**    The number of light switch turn-ons at which the first failure occurs is a r.v. $X$ whose p.d.f. is given by: $f(x) = c(\frac{9}{10})^{x-1}, x = 1, 2, \ldots$ (and $0$ otherwise).

(i) Determine the constant $c$.
(ii) Calculate the probability that the first failure will not occur until after the 10th turn-on.
(iii) Determine the corresponding d.f. $F$.

**Hint.** Refer to #4 in Table 8 in the Appendix.

**Discussion.**

(i) The constant $c$ is determined through the relationship: $\sum_{x=1}^{\infty} f(x) = 1$ or $\sum_{x=1}^{\infty} c(\frac{9}{10})^{x-1} = 1$. However, $\sum_{x=1}^{\infty} c(\frac{9}{10})^{x-1} = c \sum_{x=1}^{\infty} (\frac{9}{10})^{x-1} = c[1 + (\frac{9}{10}) + (\frac{9}{10})^2 + \cdots] = c \frac{1}{1 - \frac{9}{10}} = 10c$, so that $c = \frac{1}{10}$.

**(ii)** Here $P(X > 10) = P(X \geq 11) = c \sum_{x=11}^{\infty} (\frac{9}{10})^{x-1} = c[(\frac{9}{10})^{10} + (\frac{9}{10})^{11} +$

$\cdots] = c\frac{(\frac{9}{10})^{10}}{1-\frac{9}{10}} = c \times 10(\frac{9}{10})^{10} = \frac{1}{10} \times 10(\frac{9}{10})^{10} = (0.9)^{10} \simeq 0.349.$

**(iii)** First, for $x < 1, F(x) = 0$. Next, for $x \geq 1, F(x) = \sum_{t=1}^{x} c(\frac{9}{10})^{t-1} = 1 -$

$\sum_{t=x+1}^{\infty} c \times (\frac{9}{10})^{t-1} = 1 - c \sum_{t=x+1}^{\infty} (\frac{9}{10})^{t-1} = 1 - \frac{1}{10} \times \frac{(\frac{9}{10})^x}{1-\frac{9}{10}} = 1 - (\frac{9}{10})^x.$

Thus, $F(x) = 0$ for $x < 1$, and $F(x) = 1 - (\frac{9}{10})^x$ for $x \geq 1$.

**Example 11.** The recorded temperature in an engine is a r.v. $X$ whose p.d.f. is given by: $f(x) = n(1 - x)^{n-1}, 0 \leq x \leq 1$ (and 0 otherwise), where $n \geq 1$ is a known integer.

**(i)** Show that $f$ is, indeed, a p.d.f.
**(ii)** Determine the corresponding d.f. $F$.

**Discussion.**

**(i)** Because $f(x) \geq 0$ for all $x$, we simply have to check that $\int_0^1 f(x)\,dx = 1$. To this end, $\int_0^1 f(x)\,dx = \int_0^1 n(1 - x)^{n-1}\,dx = -n\frac{(1-x)^n}{n}|_0^1 = -(1 - x)^n|_0^1 = 1.$

**(ii)** First, $F(x) = 0$ for $x < 0$, whereas for $0 \leq x \leq 1, F(x) = \int_0^x n(1 - t)^{n-1}\,dt = -(1 - t)^n|_0^x$ (from part (i)), and this is equal to: $-(1 - x)^n + 1 = 1 - (1 - x)^n$. Thus,

$$F(x) = \begin{cases} 0, & x < 0, \\ 1 - (1 - x)^n, & 0 \leq x \leq 1, \\ 1, & x > 1. \end{cases}$$

# EXERCISES

**2.1** A sample space describing a three-children family is as follows: $S = \{bbb, bbg, bgb, gbb, bgg, gbg, ggb, ggg\}$, and assume that all eight outcomes are equally likely to occur. Next, let $X$ be the r.v. denoting the number of girls in such a family. Then:
   **(i)** Determine the set of all possible values of $X$.
   **(ii)** Determine the p.d.f. of $X$.
   **(iii)** Calculate the probabilities: $P(X \geq 2), P(X \leq 2)$.

**2.2** A r.v. $X$ has d.f. $F$ given by:

$$F(x) = \begin{cases} 0, & x \leq 0, \\ 2c(x^2 - \frac{1}{3}x^3), & 0 < x \leq 2, \\ 1, & x > 2. \end{cases}$$

   **(i)** Determine the corresponding p.d.f. $f$.
   **(ii)** Determine the constant $c$.

**2.3** The r.v. $X$ has d.f. $F$ given by:

$$F(x) = \begin{cases} 0, & x \le 0, \\ x^3 - x^2 + x, & 0 < x \le 1, \\ 1, & x > 1. \end{cases}$$

(i) Determine the corresponding p.d.f. $f$.
(ii) Calculate the probability $P(X > \frac{1}{2})$.

**2.4** The r.v. $X$ has d.f. $F$ given by:

$$F(x) = \begin{cases} 0, & x < 4, \\ 0.1, & 4 \le x < 5, \\ 0.4, & 5 \le x < 6, \\ 0.7, & 6 \le x < 8, \\ 0.9, & 8 \le x < 9, \\ 1, & x \ge 9. \end{cases}$$

(i) Draw the graph of $F$.
(ii) Calculate the probabilities

$$P(X \le 6.5), \quad P(X > 8.1), \quad P(5 < x < 8).$$

**2.5** Let $X$ be a r.v. with p.d.f. $f(x) = cx^{-(c+1)}$, for $x \ge 1$, where $c$ is a positive constant.
(i) Determine the constant $c$, so that $f$ is, indeed, a p.d.f.
(ii) Determine the corresponding d.f. $F$.

**2.6** Let $X$ be a r.v. with p.d.f. $f(x) = cx + d$, for $0 \le x \le 1$, and suppose that $P(X > \frac{1}{2}) = \frac{1}{3}$. Then:
(i) Determine the constants $c$ and $d$.
(ii) Find the d.f. $F$ of $X$.

**2.7** Show that the function $f(x) = (\frac{1}{2})^x, x = 1, 2, \ldots$ is a p.d.f.
**Hint.** See #4 in Table 8 in Appendix.

**2.8** For what value of $c$ is the function $f(x) = c\alpha^x, x = 0, 1, \ldots$ a p.d.f.? The quantity $\alpha$ is a number such that $0 < \alpha < 1$.
**Hint.** As in Exercise 2.7.

**2.9** For what value of the positive constant $c$ is the function
$f(x) = c^x, x = 1, 2, \ldots$ a p.d.f.?
**Hint.** As in Exercise 2.7.

**2.10** The p.d.f. of a r.v. $X$ is $f(x) = c(\frac{1}{3})^x$, for $x = 0, 1, \ldots$, where $c$ is a positive constant.
(i) [etermine the value of $c$.
(ii) [alculate the probability $P(X \ge 3)$.
**Hint.** As in Exercise 2.7.

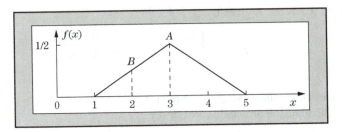

**FIGURE 2.3**

The graph of the p.d.f. referred to in Exercise 2.15.

**2.11** The r.v. $X$ has p.d.f. $f$ given by: $f(x) = c(1 - x^2)$, $-1 \leq x \leq 1$.
  **(i)** Determine the constant $c$.
  **(ii)** Calculate the probability $P(-0.9 < X < 0.9)$.

**2.12** Let $X$ be a r.v. denoting the lifetime of an electrical equipment, and suppose that the p.d.f. of $X$ is: $f(x) = ce^{-cx}$, for $x > 0$ (for some constant $c > 0$).
  **(i)** Determine the constant $c$.
  **(ii)** Calculate the probability that $X$ is at least equal to 10 (time units).
  **(iii)** If the probability in part (ii) is 0.5, what is the value of $c$?

**2.13** The r.v. $X$ has the so-called Pareto p.d.f. given by: $f(x) = \frac{1+\alpha}{x^{2+\alpha}}$, for $x > 1$, where $\alpha$ is a positive constant.
  **(i)** Verify that $f$ is, indeed, a p.d.f.
  **(ii)** Calculate the probability $P(X > c)$ (in terms of $c$ and $\alpha$), for some $c > 1$.

**2.14** Suppose that the r.v. $X$ takes on the values $0, 1, \ldots$ with the respective probabilities $P(X = j) = f(j) = \frac{c}{3^j}, j = 0, 1, \ldots$ . Then:
  **(i)** Determine the constant $c$. Compute the probabilities:
  **(ii)** $P(X \geq 3)$.
  **(iii)** $P(X = 2k + 1, \ k = 0, 1, \ldots)$.
  **(iv)** $P(X = 3k + 1, \ k = 0, 1, \ldots)$.
  **Hint.** As in Exercise 2.7.

**2.15** Let $X$ be a r.v. with p.d.f. $f$ whose graph is given is Figure 2.3. Without calculating $f$ and by using Geometric arguments, compute the following probabilities:

$$P(X \leq 3), \qquad P(1 \leq X \leq 2), \qquad P(X > 2), \qquad P(X > 5).$$

**2.16** Let $X$ be the r.v. denoting the number of a certain item sold by a merchant in a given day, and suppose that its p.d.f. is given by:

$$f(x) = \left(\frac{1}{2}\right)^{x+1}, \quad x = 0, 1, \ldots$$

Calculate the following probabilities:

(i)  No items are sold.
(ii)  More than three items are sold.
(iii)  An odd number of items is sold.
**Hint.** As in Exercise 2.7.

**2.17**  Suppose a r.v. $X$ has p.d.f. given by: $f(x) = \lambda e^{-\lambda x}, x > 0, (\lambda > 0)$, and you are
invited to bet whether $X$ would be $\geq c$ or $< c$ for some positive
constant $c$.
  (i)  An terms of the probabilities: $P(X \geq c)$ and $P(X < c)$, decide for what $c$
      (expressed in terms of $\lambda$) would you bet in favor of $X \geq c$?
  (ii)  That is the answer in part (i) if $\lambda = 4 \log 2$? (log is the natural logarithm.)

**2.18**  The lifetime in hours of electric tubes is a r.v. $X$ with p.d.f. $f(x) = c^2 x e^{-cx}$, for
$x \geq 0$, where $c$ is a positive constant.
  (i)  Determine the constant $c$ for which $f$ is, indeed, a p.d.f.
  (ii)  Calculate the probability that the lifetime will be at least $t$ hours.
  (iii)  Find the numerical value in part (ii) for $c = 0.2$ and $t = 10$.

**2.19**  Let $X$ be the r.v. denoting the number of forms required to be filled out by a
contractor for participation in contract bids, where the values of $X$ are 1, 2, 3,
4, and 5, and suppose that the respective probabilities are proportional to $x$;
that is, $P(X = x) = f(x) = cx, x = 1, \ldots, 5$.
  (i)  Determine the constant $c$.
  (ii)  Calculate the probabilities:

$$P(X \leq 3), \qquad P(2 \leq X \leq 4).$$

**2.20**  The recorded temperature in an engine is a r.v. $X$ whose p.d.f. is given by:
$f(x) = n(1 - x)^{n-1}, 0 < x < 1 (n \geq 1$, known integer). The engine is
equipped with a thermostat which is activated when the temperature exceeds a
specified level $x_0$. If the probability of the thermostat being activated is
$1/10^{2n}$, determine $x_0$.

**2.21**  A fair coin is tossed independently and repeatedly until the pair $HT$ appears
(in this order) for the first time. Let $X$ be the r.v. denoting the number of
trials needed for this outcome to occur. Then, clearly, $X$ takes on the
values: $2, 3, \ldots$.
  (i)  Show that $P(X = x) = f(x) = \frac{x-1}{2^x}, x = 2, 3, \ldots$
  (ii)  Verify that $f(x), x = 2, 3, \ldots$ is, indeed, a p.d.f.
  **Hint.** Any $x$ outcomes consist of $k$ $H$'s and $x - k$ $T$'s and their (common)
  probability is $(1/2)^k \times (1/2)^{x-k} = 1/2^x$. Furthermore, there is only one such
  outcome if the first result is $H$, and there are $x - 2$ such outcomes if the first
  result is $T$. For part (ii), use formula #5 in Table 8.

**2.22**  Let the r.v. $X$ have p.d.f. $f_X(x) > 0$ for $(-\infty \leq)a < x < b(\leq \infty)$, and 0
otherwise, and set $Y = |X|$. Derive the p.d.f. $f_Y$ of $Y$ in terms of $f_X$.

**2.23** Let $X$ be a (continuous) r.v. with p.d.f. $f$ given by $f(x) = 2c(2x - x^2)$ for $0 < x < 2$, and 0 otherwise.
  **(i)** For what value of $c$ is $f$, indeed, a p.d.f.?
  **(ii)** Compute the probability $P(X < 1/2)$.
  **(iii)** Determine the corresponding d.f. $F$.

**2.24** The r.v. $X$ is said to have the Lognormal distribution with parameters $\alpha$ and $\beta$ $(\alpha > 0, \beta > 0)$ if its p.d.f. is given by:

$$f(x) = \frac{1}{x\beta\sqrt{2\pi}}e^{-(\log x - \log \alpha)^2/2\beta^2}, \quad x > 0 \text{ (and 0 for } x \leq 0).$$

  **(i)** Show that $f$ is, indeed, a p.d.f.
  **(ii)** Set $Y = \log X$ and show that $Y \sim N(\log \alpha, \beta^2)$ by computing first its d.f. $F_Y$ and then differentiating to obtain $f_Y$.

## 2.3 CONDITIONAL PROBABILITY AND RELATED RESULTS

Conditional probability is a probability in its own right, as will be seen, and it is an extremely useful tool in calculating probabilities. Essentially, it amounts to suitably modifying a sample space $\mathcal{S}$, associated with a random experiment, on the evidence that a certain event has occurred. Consider the following examples, by way of motivation, before a formal definition is given.

**Example 12.** In tossing three distinct coins once (Example 26 in Chapter 1), consider the events $B =$ "exactly 2 heads occur" $= \{HHT, HTH, THH\}$, $A =$ "2 specified coins (e.g., coins #1 and #2) show heads" $= \{HHH, HHT\}$. Then $P(B) = \frac{3}{8}$ and $P(A) = \frac{2}{8} = \frac{1}{4}$. Now, suppose we are told that event $B$ has occurred and we are asked to evaluate the probability of $A$ on the basis of this evidence. Clearly, what really matters here is the event $B$, and, given that $B$ has occurred, the event $A$ occurs only if the sample point $HHT$ appeared; that is, the event $\{HHT\} = A \cap B$ occurred. The required probability is then $\frac{1}{3} = \frac{1/8}{3/8} = \frac{P(A \cap B)}{P(B)}$, and the notation employed is $P(A \mid B)$ (probability of $A$, given that $B$ has occurred or, just, given $B$). Thus, $P(A \mid B) = \frac{P(A \cap B)}{P(B)}$. Observe that $P(A \mid B) = \frac{1}{3} > \frac{1}{4} = P(A)$.

**Example 13.** In rolling two distinct dice once (Example 27 in Chapter 1), consider the event $B$ defined by: $B =$ "the sum of numbers on the upper face is $\leq 5$", so that $B = \{(1, 1), (1, 2), (1, 3), (1, 4), (2, 1), (2, 2), (2, 3), (3, 1), (3, 2), (4, 1)\}$, and let $A =$ "the sum of numbers on the upper faces is $\geq 4$." Then $A^c =$ "the sum of numbers on the upper faces is $\leq 3$" $= \{(1, 1), (1, 2), (2, 1)\}$, so that $P(B) = \frac{10}{36} = \frac{5}{18}$ and $P(A) = 1 - P(A^c) = 1 - \frac{3}{36} = \frac{33}{36} = \frac{11}{12}$. Next, if we are told that $B$ has occurred, then the only way that $A$ occurs is if $A \cap B$ occurs, where $A \cap B =$ "the sum of numbers on the upper faces is both $\geq 4$ and $\leq 5$ (i.e., either

4 or 5)" $= \{(1,3), (1,4), (2,2), (2,3), (3,1), (3,2), (4,1)\}$. Thus, $P(A \mid B) = \frac{7}{10} = \frac{7/36}{10/36} = \frac{P(A \cap B)}{P(B)}$, and observe that $P(A \mid B) = \frac{7}{10} < \frac{11}{12} = P(A)$.

**Example 14.**   In recording the gender of children in a two-children family (Example 30 in Chapter 1), let $B = \{bg, gb\}$ and let $A = $ "older child is a boy" $= \{bb, bg\}$, so that $A \cap B = \{bg\}$. Then $P(B) = \frac{1}{2} = P(A), P(A \mid B) = \frac{1}{2} = \frac{1/4}{1/2} = \frac{P(A \cap B)}{P(B)}$.

These examples motivate the following definition of conditional probability.

**Definition 1.**   The *conditional* probability of an event $A$, given the event $B$ with $P(B) > 0$, is denoted by $P(A \mid B)$ and is defined by: $P(A \mid B) = P(A \cap B)/P(B)$.

Replacing $B$ by the entire sample space $\mathcal{S}$, we are led back to the (*unconditional*) probability of $A$, as $\frac{P(A \cap \mathcal{S})}{P(\mathcal{S})} = \frac{P(A)}{1} = P(A)$. Thus, the conditional probability is a generalization of the concept of probability where $\mathcal{S}$ is restricted to an event $B$.

That the conditional probability is, indeed, a probability is seen formally as follows: $P(A \mid B) \geq 0$ for every $A$ by definition;

$$P(\mathcal{S} \mid B) = \frac{P(\mathcal{S} \cap B)}{P(B)} = \frac{P(B)}{P(B)} = 1;$$

and if $A_1, A_2, \ldots$ are pairwise disjoint, then

$$P\left(\bigcup_{j=1}^{\infty} A_j \mid B\right) = \frac{P[(\bigcup_{j=1}^{\infty} A_j) \cap B]}{P(B)} = \frac{P[\bigcup_{j=1}^{\infty}(A_j \cap B)]}{P(B)}$$

$$= \frac{\sum_{j=1}^{\infty} P(A_j \cap B)}{P(B)} = \sum_{j=1}^{\infty} \frac{P(A_j \cap B)}{P(B)} = \sum_{j=1}^{\infty} P(A_j \mid B).$$

It is to be noticed, furthermore, that the $P(A \mid B)$ can be smaller or larger than the $P(A)$, or equal to the $P(A)$. The case that $P(A \mid B) = P(A)$ is of special interest and will be discussed more extensively in the next section. This point is made by Examples 12–14.

Here are another three examples pertaining to conditional probabilities.

**Example 15.**   When we are recording the number of particles emitted by a certain radioactive source within a specified period of time (Example 35 in Chapter 1), we are going to see that, if $X$ is the number of particles emitted, then $X$ is a r.v. taking on the values $0,1,\ldots$ and that a suitable p.d.f. for it is $f_X(x) = e^{-\lambda} \frac{\lambda^x}{x!}, x = 0, 1, \ldots$, for some constant $\lambda > 0$. Next, let $B$ and $A$ be the events defined by: $B = (X \geq 10), A = (X \leq 11)$, so that $A \cap B = (10 \leq X \leq 11) = (X = 10 \text{ or } X = 11)$. Then

$$P(B) = \sum_{x=10}^{\infty} e^{-\lambda} \frac{\lambda^x}{x!} = e^{-\lambda} \sum_{x=10}^{\infty} \frac{\lambda^x}{x!},$$

$$P(A) = \sum_{x=0}^{11} e^{-\lambda} \frac{\lambda^x}{x!} = e^{-\lambda} \sum_{x=0}^{11} \frac{\lambda^x}{x!}, \quad \text{and}$$

$$P(A \mid B) = \left( e^{-\lambda} \frac{\lambda^{10}}{10!} + e^{-\lambda} \frac{\lambda^{11}}{11!} \right) \bigg/ e^{-\lambda} \sum_{x=10}^{\infty} \frac{\lambda^x}{x!}.$$

For a numerical example, take $\lambda = 10$. Then we have (by means of Poisson tables)

$$P(B) \simeq 0.5421, \quad P(A) \simeq 0.6968, \quad \text{and} \quad P(A \mid B) \simeq 0.441.$$

**Example 16.**   When recording the lifetime of an electronic device or of an electrical appliance, etc. (Example 36 in Chapter 1), if $X$ is the lifetime under consideration, then $X$ is a r.v. taking values in $(0, \infty)$, and a suitable p.d.f. for it is seen to be the function $f(x) = \lambda e^{-\lambda x}, x \geq 0$, for some constant $\lambda > 0$. Let $B$ and $A$ be the events: $B =$ "at the end of 5 time units, the equipment was still operating" $= (X \geq 5)$, $A =$ "the equipment lasts for no more than 2 additional time units" $= (X \leq 7)$. Then $A \cap B = (5 \leq X \leq 7)$, and:

$$P(B) = \int_5^{\infty} \lambda e^{-\lambda x} \, dx = e^{-5\lambda}, \quad P(A) = \int_0^7 \lambda e^{-\lambda x} \, dx = 1 - e^{-7\lambda},$$

$$P(A \cap B) = \int_5^7 \lambda e^{-\lambda x} \, dx = e^{-5\lambda} - e^{-7\lambda}, \quad \text{so that}$$

$$P(A \mid B) = \frac{P(A \cap B)}{P(B)} = \frac{e^{-5\lambda} - e^{-7\lambda}}{e^{-5\lambda}} = 1 - e^{-2\lambda}.$$

Take, for instance, $\lambda = \frac{1}{10}$. Then, given that $e^{-1} \simeq 0.36788$, the preceding probabilities are

$$P(B) \simeq 0.607, \quad P(A) \simeq 0.503, \quad \text{and} \quad P(A \mid B) \simeq 0.181.$$

**Example 17.**   If for the events $A$ and $B$, $P(A)P(B) > 0$, then show that: $P(A \mid B) > P(A)$ if and only if $P(B \mid A) > P(B)$. Likewise, $P(A \mid B) < P(A)$ if and only if $P(B \mid A) < P(B)$.

**Discussion.**   Indeed, $P(A \mid B) > P(A)$ is equivalent to $\frac{P(A \cap B)}{P(B)} > P(A)$ or $\frac{P(A \cap B)}{P(A)} > P(B)$ or $P(B \mid A) > P(B)$. Likewise, $P(A \mid B) < P(A)$ is equivalent to $\frac{P(A \cap B)}{P(B)} < P(A)$ or $\frac{P(A \cap B)}{P(A)} < P(B)$ or $P(B \mid A) < P(B)$.

This section is concluded with three simple but very useful results. They are the so-called multiplicative theorem, the total probability theorem, and the Bayes formula.

**Theorem 3** (Multiplicative Theorem). *For any n events $A_1, \ldots, A_n$ with* $P(\bigcap_{j=1}^{n-1} A_j) > 0$, *it holds:*

$$P\left(\bigcap_{j=1}^{n} A_j\right) = P(A_n \mid A_1 \cap \cdots \cap A_{n-1})P(A_{n-1} \mid A_1 \cap \cdots \cap A_{n-2})$$
$$\ldots P(A_2 \mid A_1)P(A_1).$$

Its justification is simple, is done by mathematical induction, and is left as an exercise (see Exercise 3.8). Its significance is that we can calculate the probability of the intersection of $n$ events, step by step, by means of conditional probabilities. The calculation of these conditional probabilities is far easier. Here is a simple example that amply illustrates the point.

**Example 18.**     An urn contains 10 identical balls of which 5 are black, 3 are red, and 2 are white. Four balls are drawn one at a time and without replacement. Find the probability that the first ball is black, the second red, the third white, and the fourth black.

**Discussion.**     Denoting by $B_1$ the event that the first ball is black, and likewise for $R_2, W_3$, and $B_4$, the required probability is

$$P(B_1 \cap R_2 \cap W_3 \cap B_4) = P(B_4 \mid B_1 \cap R_2 \cap W_3)P(W_3 \mid B_1 \cap R_2)P(R_2 \mid B_1)P(B_1).$$

Assuming equally likely outcomes at each step, we have

$$P(B_1) = \frac{5}{10}, \quad P(R_2 \mid B_1) = \frac{3}{9}, \quad P(W_3 \mid B_1 \cap R_2) = \frac{2}{8},$$
$$P(B_4 \mid B_1 \cap R_2 \cap W_3) = \frac{4}{7}.$$

Therefore,

$$P(B_1 \cap R_2 \cap W_3 \cap B_4) = \frac{4}{7} \times \frac{2}{8} \times \frac{3}{9} \times \frac{5}{10} = \frac{1}{42} \simeq 0.024.$$

For the formulation of the next result, the concept of a partition of $\mathcal{S}$ is required. The events $\{A_1, A_2, \ldots, A_n\}$ form a *partition* of $\mathcal{S}$, if these events are pairwise disjoint, $A_i \cap A_j = \emptyset, i \neq j$, and their union is $\mathcal{S}$, $\bigcup_{j=1}^{n} A_j = \mathcal{S}$. Then it is obvious that any event $B$ in $\mathcal{S}$ may be expressed as follows, in terms of a partition of $\mathcal{S}$; namely, $B = \bigcup_{j=1}^{n}(A_j \cap B)$. Furthermore,

$$P(B) = \sum_{j=1}^{n} P(A_j \cap B) = \sum_{j=1}^{n} P(B \mid A_j)P(A_j), \quad \text{provided } P(A_j) > 0 \text{ for all } j.$$

The concept of partition is defined similarly for countably infinite many events, and the probability $P(B)$ is expressed likewise. In the sequel, by writing $j = 1, 2, \ldots$ and $\sum_j$ we mean to include both cases: Finitely many indices and countably infinitely many indices.

Thus, we have the following result.

**Theorem 4** (Total Probability Theorem). *Let $\{A_1, A_2, \ldots\}$ be a partition of $S$, and let $P(A_j) > 0$ for all $j$. Then, for any event $B$,*

$$P(B) = \sum_j P(B \mid A_j)P(A_j).$$

The significance of this result is that, if it happens that we know the probabilities of the partitioning events, $P(A_j)$, as well as the conditional probabilities of $B$, given $A_j$, then these quantities may be combined, according to the preceding formula, to produce the probability $P(B)$. The probabilities $P(A_j), j = 1, 2, \ldots$, are referred to as *a priori* or *prior* probabilities. The following examples illustrate the theorem and also demonstrate its usefulness.

**Example 19.** In reference to Example 2 in Chapter 1, calculate the probability $P(+)$.

**Discussion.** Without having to refer specifically to a sample space, it is clear that the events $D$ and $N$ form a partition. Then,

$$P(+) = P(+ \text{ and } D) + P(+ \text{ and } N) = P(+ \mid D)P(D) + P(+ \mid N)P(N).$$

Here, the *a priori* probabilities are $P(D) = p_1, P(N) = 1 - p_1$, and

$$P(+ \mid D) = 1 - P(- \mid D) = 1 - p_3, \quad P(+ \mid N) = p_2.$$

Therefore, $P(+) = (1 - p_3)p_1 + p_2(1 - p_1)$. For a numerical application, take $p_1 = 0.02$ and $p_2 = p_3 = 0.01$. Then $P(+) = 0.0296$. So, on the basis of this testing procedure, about 2.96% of the population would test positive.

**Example 20.** The proportions of motorists in a given gas station using regular unleaded gasoline, extra unleaded, and premium unleaded over a specified period of time are 40%, 35%, and 25%, respectively. The respective proportions of filling their tanks are 30%, 50%, and 60%. What is the probability that a motorist selected at random from among the patrons of the gas station under consideration and for the specified period of time will fill his/her tank?

**Discussion.** Denote by $R, E$, and $P$ the events of a motorist using unleaded gasoline which is regular, extra unleaded, and premium, respectively, and by $F$ the event of having the tank filled. Then the translation into terms of probabilities of the proportions given above is

$$P(R) = 0.40, \qquad P(E) = 0.35, \qquad P(P) = 0.25,$$
$$P(F \mid R) = 0.30, \qquad P(F \mid E) = 0.50, \qquad P(F \mid P) = 0.60.$$

Then the required probability is

$$P(F) = P((F \cap R) \cup (F \cap E) \cup (F \cap P))$$
$$= P(F \cap R) + P(F \cap E) + P(F \cap P)$$

$$= P(F \mid R)P(R) + P(F \mid E)P(E) + P(F \mid P)P(P)$$

$$= 0.30 \times 0.40 + 0.50 \times 0.35 + 0.60 \times 0.25$$

$$= 0.445.$$

In reference to Theorem 4, stipulating the prior probabilities $P(B \mid A_j), j = 1, 2, \ldots,$ is often a precarious thing and guesswork. This being the case, the question then arises as to whether experimentation may lead to reevaluation of the prior probabilities on the basis of new evidence. To put it more formally, is it possible to use $P(A_j)$ and $P(B \mid A_j), j = 1, 2, \ldots$ in order to calculate $P(A_j \mid B)$? The answer to this question is in the affirmative, is quite simple, and is the content of the next result.

**Theorem 5** (Bayes' Formula).   *Let $\{A_1, A_2, \ldots\}$ and $B$ be as in the previous theorem. Then, for any $j = 1, 2, \ldots$:*

$$P(A_j \mid B) = \frac{P(B \mid A_j)P(A_j)}{\sum_i P(B \mid A_i)P(A_i)}.$$

*Proof.* Indeed, $P(A_j \mid B) = P(A_j \cap B)/P(B) = P(B \mid A_j)P(A_j)/P(B)$, and then the previous theorem completes the proof. ∎

The probabilities $P(A_j \mid B)$, $j = 1, 2, \ldots$, are referred to as *posterior* probabilities in that they are reevaluations of the respective prior $P(A_j)$ after the event $B$ has occurred.

**Example 21.**   Referring to Example 19, a question of much importance is this: Given that the test shows positive, what is the probability that the patient actually has the disease? In terms of the notation adopted, this question becomes: $P(D \mid +) = ?$ Bayes' formula gives

$$P(D \mid +) = \frac{P(+ \mid D)P(D)}{P(+ \mid D)P(D) + P(+ \mid N)P(N)} = \frac{p_1(1 - p_3)}{p_1(1 - p_3) + p_2(1 - p_1)}.$$

For the numerical values used above, we get

$$P(D \mid +) = \frac{0.02 \times 0.99}{0.0296} = \frac{0.0198}{0.0296} = \frac{198}{296} \simeq 0.669.$$

So $P(D \mid +) \simeq 66.9\%$. This result is both reassuring and surprising. Reassuring, in that only 66.9% of those testing positive actually have the disease. Surprising, in that this proportion looks rather low, given that the test is quite good: It identifies correctly 99% of those having the disease. A reconciliation between these two seemingly contradictory aspects is as follows: The fact that $P(D) = 0.02$ means that, on the average, 2 out of 100 persons have the disease. So, in 100 persons, 2 will have the disease and 98 will not. When 100 such persons are tested, $2 \times 0.99 = 1.98$ will be correctly confirmed as positive (because 0.99 is the probability of a correct positive), and $98 \times 0.01 \doteq 0.98$ will be incorrectly diagnosed as positive (because 0.01 is the probability of an incorrect positive). Thus, the proportion of correct positives is equal to:

$$\text{(correct positives)/(correct positives + incorrect positives)}$$

$$= 1.98/(1.98 + 0.98) = 1.98/2.96 = 198/296 \simeq 0.669.$$

**Remark 1.** The fact that the probability $P(D \mid +)$ is less than 1 simply reflects the fact that the test, no matter how good, is imperfect. Should the test be perfect $(P(+ \mid D) = P(- \mid D^c) = 1)$, then $P(D \mid +) = 1$, as follows from the preceding calculations, no matter what $P(D)$ is. The same, of course, is true for $P(D^c \mid -)$.

**Example 22.** Refer to Example 20 and calculate the probabilities: $P(R \mid F)$, $P(E \mid F)$, and $P(P \mid F)$.

**Discussion.** By Bayes' formula and Example 20,

$$P(R \mid F) = \frac{P(R \cap F)}{P(F)} = \frac{P(F \mid R)P(R)}{P(F)} = \frac{0.30 \times 0.40}{0.445} \simeq 0.270,$$

and likewise,

$$P(E \mid F) = \frac{0.50 \times 0.35}{0.445} \simeq 0.393, \qquad P(P \mid F) = \frac{0.60 \times 0.25}{0.445} \simeq 0.337.$$

## EXERCISES

**3.1**  If $P(A \mid B) > P(A)$, then show that $P(B \mid A) > P(B)$, by assuming that both $P(A)$ and $P(B)$ are positive.

**3.2**  If $A \cap B = \emptyset$ and $P(A \cup B) > 0$, express the probabilities $P(A \mid A \cup B)$ and $P(B \mid A \cup B)$ in terms of $P(A)$ and $P(B)$.

**3.3**  A girls' club has in its membership rolls the names of 50 girls with the following descriptions:
20 blondes, 15 with blue eyes, and 5 with brown eyes;
25 brunettes, 5 with blue eyes, and 20 with brown eyes;
5 redheads, 1 with blue eyes, and 4 with green eyes.
If one arranges a blind date with a club member, what is the probability that:
(i)  The girl is blonde?
(ii)  The girl is blonde, if it was revealed only that she has blue eyes?

**3.4**   Suppose that the probability that both of a pair of twins are boys is 0.30 and
that the probability that they both are girls is 0.26. Given that the probability
of the first child being a boy is 0.52, what is the probability that:
  **(i)**   The second twin is a boy, given that the first is a boy?
  **(ii)**   The second twin is a girl, given that the first is a girl?
  **(iii)**   The second twin is a boy?
  **(iv)**   The first is a boy and the second is a girl?
  **Hint.** Denote by $b_i$ and $g_i$ the events that the $i$th child is a boy or a girl,
  respectively, $i = 1, 2$.

**3.5**   A shipment of 20 TV tubes contains 16 good tubes and 4 defective tubes.
Three tubes are chosen successively and at random each time and are also
tested successively. What is the probability that:
  **(i)**   The third tube is good if the first two were found to be good?
  **(ii)**   The third tube is defective if the first was found to be good and the
  second defective?
  **(iii)**   The third tube is defective if the first was found to be defective and the
  second was found to be good?
  **(iv)**   The third tube is defective if one of the other two was found to be good
  and the other was found to be defective?
  **Hint.** Denote by $D_i$ and $G_i$ the events that the $i$th tube is defective or good,
  respectively, $i = 1, 2, 3$.

**3.6**   For any three events $A, B$, and $C$ with $P(A)P(B)P(C) > 0$, show that:
  **(i)**   $P(A^c \mid B) = 1 - P(A \mid B)$.
  **(ii)**   $P(A \cup B \mid C) = P(A \mid C) + P(B \mid C) - P(A \cap B \mid C)$.
  Also, by means of counterexamples, show that the following equations
  need not be true:
  **(iii)**   $P(A \mid B^c) = 1 - P(A \mid B)$.
  **(iv)**   $P(C \mid A \cup B) = P(C \mid A) + P(C \mid B)$, where $A \cap B = \varnothing$.

**3.7**   If $A, B$, and $C$ are any events in the sample space $\mathcal{S}$, show that
  $\{A, A^c \cap B, A^c \cap B^c \cap C, (A \cup B \cup C)^c\}$ is a partition of $\mathcal{S}$.
  **Hint.** Show that $A \cup (A^c \cap B) \cup (A^c \cap B^c \cap C) = A \cup B \cup C$.

**3.8**   Use mathematical induction to prove Theorem 3.

**3.9**   Let $\{A_j, j = 1, \ldots, 5\}$ be a partition of the sample space $\mathcal{S}$ and suppose that:

$$P(A_j) = \frac{j}{15} \quad \text{and} \quad P(A \mid A_j) = \frac{5 - j}{15}, \quad j = 1, \ldots, 5.$$

  Compute the probabilities $P(A_j \mid A), j = 1, \ldots, 5$.

**3.10**   A box contains 15 identical balls except that 10 are white and 5 are black.
Four balls are drawn successively and without replacement.
Calculate the probability that the first and the fourth balls are white.
  **Hint.** Denote by $W_i$ and $B_i$ the events that the $i$th ball is white or black,
  respectively, $i = 1, \ldots, 4$.

**3.11** A box contains $m + n$ identical balls except that $m$ of them are red and $n$ are black. A ball is drawn at random, its color is noticed, and then the ball is returned to the box along with $r$ balls of the same color. Finally, a ball is drawn also at random.
    (i) That is the probability that the first ball is red?
    (ii) That is the probability that the second ball is red?
    (iii) Compare the probabilities in parts (i) and (ii) and comment on them.
    (iv) That is the probability that the first ball is black if the second is red?
    (v) Find the numerical values in parts (i), (ii), and (iv) if $m = 9, n = 6$, and $r = 5$.

**Hint.** Denote $R_i$ and $B_i$ the events that the $i$th ball is red or black, respectively, $i = 1, 2$.

**3.12** A test correctly identifies a disease $D$ with probability 0.95 and wrongly diagnoses $D$ with probability 0.01. From past experience, it is known that disease $D$ occurs in a targeted population with frequency 0.2%. An individual is chosen at random from said population and is given the test.
Calculate the probability that:
    (i) The test is $+$, $P(+)$.
    (ii) The individual actually suffers from disease $D$ if the test turns out to be positive, $P(D \mid +)$.

**Hint.** Use Theorems 4 and 5.

**3.13** Suppose that the probability of correct diagnosis (either positive or negative) of cervical cancer in the Pap test is 0.95 and that the proportion of women in a given population suffering from this disease is 0.01%. A woman is chosen at random from the target population and the test is administered.
What is the probability that:
    (i) The test is positive?
    (ii) The subject actually has the disease, given that the diagnosis is positive?

**Hint.** Use Theorems 4 and 5.

**3.14** A signal $S$ is sent from point $A$ to point $B$ and is received at $B$ if both switches I and II are closed (Figure 2.4). It is assumed that the probabilities of I and II being closed are 0.8 and 0.6, respectively, and that
$P(\text{II is closed}|\text{I is closed}) = P(\text{II is closed})$. Calculate the following probabilities:

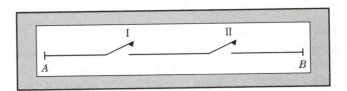

**FIGURE 2.4**

The circuit referred to in Exercise 3.14.

(i) The signal is received at $B$.

(ii) The (conditional) probability that switch I was open, given that the signal was not received at $B$.

(iii) The (conditional) probability that switch II was open, given that the signal was not received at $B$.

**Hint.** Use Theorems 4 and 5.

**3.15** The student body in a certain college consists of 55% women and 45% men. Women and men smoke cigarettes in the proportions of 20% and 25%, respectively. If a student is chosen at random, calculate the probability that:

(i) The student is a smoker.

(ii) The student is a man, given that he/she is a smoker.

**Hint.** Use Theorems 4 and 5.

**3.16** From a population consisting of 52% females and 48% males, an individual, drawn at random, is found to be color blind. If we assume that the proportions of color-blind females and males are 25% and 5%, respectively, what is the probability that the individual drawn is a male?

**Hint.** Use Theorems 4 and 5.

**3.17** Drawers I and II contain black and red pencils as follows:

Drawer I: $b_1$ black pencils and $r_1$ red pencils,

Drawer II: $b_2$ black pencils and $r_2$ red pencils.

A drawer is chosen at random and then a pencil is also chosen at random from that drawer.

(i) What is the probability that the pencil is black?

(ii) If it is announced that the pencil is black, what is the probability it was chosen from drawer I?

(iii) Give numerical values in parts (i) and (ii) for:

$b_1 = 36, r_1 = 12, b_2 = 60, r_2 = 24$.

**Hint.** For parts (i) and (ii), use Theorem 4 and 5. In parts (i) and (ii), probabilities are to be expressed in terms of $b_1, b_2, r_1$, and $r_2$.

**3.18** Three machines I, II, and III manufacture 30%, 30%, and 40%, respectively, of the total output of certain items. Of them, 4%, 3%, and 2%, respectively, are defective. One item is drawn at random from the total output and is tested.

(i) That is the probability that the item is defective?

(ii) If it is found to be defective, what is the probability the item was produced by machine I?

(iii) Same question as in part (ii) for each one of the machines II and III.

**Hint.** Use Theorems 4 and 5.

**3.19** Suppose that a multiple choice test lists $n$ alternative answers of which only one is correct. If a student has done the homework, he/she is certain to identify the correct answer; otherwise the student chooses an answer at random. Denote by $A$ the event that the student does the homework, set $p = P(A)$, and let $B$ be the event that he/she answers the question correctly.

(i) Express the probability $P(A \mid B)$ in terms of $p$ and $n$.

(ii) If $0 < p < 1$ and fixed, show that the probability $P(A \mid B)$, as a function of $n$, is increasing.

(iii) Does the result in part (ii) seem reasonable?

**Hint.** Use Theorem 4 for the computation of the $P(B)$.

**3.20** If the p.d.f. of the r.v. $X$ is: $f(x) = \lambda e^{-\lambda x}$, for $x > 0$ ($\lambda > 0$), calculate:

(i) $P(X > t)$ (for some $t > 0$).

(ii) $P(X > s + t \mid X > s)$ (for some $s, t > 0$).

(iii) Compare the probabilities in parts (i) and (ii), and draw your conclusion.

**3.21** Let $A$ and $B$ be two events with $P(A) > 0$, $P(B) > 0$. Then, if $P(A|B) > P(A)$, show that $P(B|A) > P(B)$.

**3.22** Suppose that 10% of a target population has committed a traffic violation (within a certain period of time). A person chosen at random from this population is subjected to a lie detector, which is known to indicate that a nonviolator is guilty with a probability 0.08, and that a violator is innocent with probability 0.15.

(i) Denote by $V$ the event that the person chosen committed a traffic violation, and write down the numerical values of the probabilities $P(V)$ and $P(V^c)$.

(ii) Denote by $I$ and $G$ the events that the lie detector indicates innocence and guilt, respectively, for the person tested. Give the numerical value for:

  (a) The probability that the test reports "guilty," given that the person is a nonviolator.

  (b) The probability that the test reports "innocent," given that the person is a violator.

(iii) Compute the probability of guilt, $P(G)$.

(iv) Compute the probability $P(V|G)$.

**3.23** At a certain gas station, 40% of the customers use regular unleaded gas (event $A_1$), 35% use extra unleaded gas (event $A_2$), and 25% use premium unleaded gas (event $A_3$). Of those customers using regular unleaded gas, only 30% fill their tanks (event $B$). Of those customers using extra unleaded gas, 60% fill their tanks, while of those using premium, 50% fill their tanks.

(i) What is the probability that the next customer fills the tank?

(ii) If the next customer fills the tank, what is the probability that regular unleaded gas is pumped? Extra unleaded gas? Premium gas?

**3.24** A certain target population has a disease $D$ at the rate of 0.5%. At test correctly detects the presence of the disease at the rate of 95%, and also provides correct negative diagnosis at the same rate. For an individual chosen at random from the target population, calculate the probability that:

(i) The result of the test is positive.

**(ii)** The individual actually has the disease, given that the test results is a positive diagnosis.

**(iii)** Provide an interpretation of the probability calculated in part (ii).

**3.25** For any two events $A$ and $B$ with $P(B) > 0$, show that:

**(i)** $P(A|A \cup B) = \frac{P(A)}{P(A \cup B)}$.

**(ii)** $P(A|A \cup B) = \frac{P(A)}{P(A) + P(B)}$ if $A \cap B = \emptyset$.

**3.26** An appellate panel of three judges reviews a criminal case, and each judge votes independently of the others. If the defendant whose case is reviewed is, indeed, guilty, each judge casts a "guilty" vote with probability 0.8, and if it is innocent each judge casts a "guilty" vote with probability 0.1. It is assumed that 70% of the defendants are actually guilty.

Denote by $G$ and $G_i$, the events: $G =$ "the defendant is guilty," $G_i =$ "the $i$th judge casts a "guilty" vote," $i = 1, 2, 3$. Then:

**(i)** Compute the probability $P(G_i)$, $i = 1, 2, 3$.

**(ii)** Compute the (conditional) probability $P(G|G_i)$, $i = 1, 2, 3$.

**(iii)** If the defendant is declared guilty when at least two judges cast "guilty" votes, compute the probability $P($defendant is declared guilty$)$.

**3.27** Let $R_i$ be the event that it rains the $i$th day, $i = 1, 2, 3$, and suppose that:
$$P(R_2|R_1) = P(R_3|R_2) = p_1, \ P(R_2|R_1^c) = P(R_3|R_2^c) = p_2,$$
$$P(R_3|R_1 \cap R_2) = P(R_3|R_1^c \cap R_2) = P(R_3|R_2) = P(R_2|R_1) \ (= p_1),$$
$$P(R_3|R_1 \cap R_2^c) = P(R_3|R_1^c \cap R_2^c) = P(R_3|R_2^c) = P(R_2|R_1^c) \ (= p_2).$$
(That is, the probability that it will rain tomorrow depends only on what is happening today and not yesterday.)

**(i)** Calculate the conditional probability $P(R_3|R_1)$ in terms of $p_1$ and $p_2$.

**(ii)** What is the numerical value of $P(R_3|R_1)$ if $p_1 = 0.4$ and $p_2 = 0.2$?

**3.28** Let $A$ and $B$ be events with $P(B) > 0$. Then:

**(i)** Show that: $P(A^c|B) = 1 - P(A|B)$ (as it should be, since $P(\cdot|B)$ is a probability measure).

**(ii)** The following examples shows that:
$$P(A|B^c) \neq 1 - P(A|B) \quad (0 < P(B) < 1).$$

Let $S = \{1, 2, 3, 4, 5\}$, accompanied with the Uniform probability, and let $A = \{1, 2\}, B = \{2, 3, 4\}$.

**3.29** **(i)** For any three events $A, B, C$ with $P(C) > 0$, show that:
$$P(A \cup B|C) = P(A|C) + P(B|C) - P(A \cup B|C)$$

(as it should be, since $P(\cdot|C)$ is a probability measure).

**(ii)** The following example shows that:
$$P(C|A \cup B) \neq P(C|A) + P(C|B) \quad \text{even if } A \cap B = \emptyset.$$

Let $S = \{1, 2, 3, 4, 5\}$ accompanied with the Uniform probability measure, and let $A = \{1, 2\}$ and $B = \{3, 4\}, C = \{2, 3\}$.

**3.30** Each one of $k$ urns $U_j, j = 1, \ldots, k$ contains $m + n$ balls of which $m$ are white and $n$ are black. A ball, chosen at random, is transferred from urn $U_1$ to urn $U_2$. Next, a ball, also chosen at random, is transferred from urn $U_2$ to urn $U_3$, and continue like this until a ball, chosen at random, is transferred from urn $U_{k-1}$ to urn $U_k$. Finally, a ball is drawn at random from urn $U_k$. Express the probability, in terms of $m$ and $n$, that the ball drawn from urn $U_k$ is black.

**3.31** Consider three urns $U_j, j = 1, 2, 3$ such that urn $U_j$ contains $m_j$ white balls and $n_j$ black balls. One urn is chosen at random, and one ball is drawn from it also at random. If the ball drawn was white:
  **(i)** What is the probability that the urn chosen was urn $U_1$ or $U_2$?
  **(ii)** Find the numerical value of the probability in part (i) for: $m_1 = n_1$; $m_2 = n_2$; $m_3 = n_3$.

**3.32** From an urn containing $r$ red ball and $b$ black balls, three balls are drawn without replacement.
  **(i)** If the first two balls were of the same color, what is the probability that they were red?
  **(ii)** What is the probability that the second red ball appears the third time?
  **(iii)** Give the numerical values in parts (i) and (ii) for $r = 4$ and $b = 6$.

**3.33** It has been observed that, when people dine out in restaurants, 40% start out will soup and 60% start out with salad. Of those starting out with soup, 25% continue with a meatless dish, while of those starting out with salad, 35% continue with a meatless dish.
  If a diner had a dish with meat, what is the probability that he/she:
  **(i)** Had soup as a first serving?
  **(ii)** Had salad as a first serving?

**3.34** Two cards are drawn at random from a well-shuffled standard deck of playing cards. What is the probability that the second card is a club when the sampling is done:
  **(i)** With replacement?
  **(ii)** Without replacement?
  **(iii)** Compare the answers and elaborate on them. Also, consider some obvious conditional probabilities.

**3.35** Consider two urns $U_1$ and $U_2$ such that urn $U_j$ contains $m_j$ white balls and $n_j$ black balls, $j = 1, 2$. A ball is drawn at random and independently from each one of the two urns and is placed into a third urn $U_3$, and then a ball is drawn at random from the third urn.
  **(i)** Express the probability that the ball is black in terms of $m_j, n_j, j = 1, 2$.
  **(ii)** Find the numerical value of the probability in part (i) for $m_1 = 6$, $n_1 = 4$; $m_2 = 10, n_2 = 16$.

**3.36** From an urn containing five red balls, three blue balls, and two white balls, three balls are drawn at random and without replacement. If the third ball is not white, what is the probability of the first two balls being of the same color?

**3.37** For $i = 1, 2, 3$, urn $U_i$ contains $r_i$ red balls and $w_i$ white balls. A ball chosen at random from urn $U_1$ is transferred to urn $U_2$; then a ball also chosen at random from urn $U_2$ is transferred to urn $U_3$; finally, all ball is chosen at random from urn $U_3$. Denote by $R_i$ and $W_i$ the event that the balls chosen from the $i$th urn are red and white, respectively, $i = 1, 2, 3$.
  - **(i)** Then find the expressions, in terms of $r_i, w_i, i = 1, 2, 3$, for the probabilities:
    - **(a)** $P(R_3)$;
    - **(b)** $P(R_3|R_1)$;
    - **(c)** $P(R_3|W_1)$.
  - **(ii)** Show that $P(R_3|W_1) < P(R_3) < P(R_3|R_1)$.

**3.38** Consider two urns $U_j, j = 1, 2$ such that urn $U_j$ contains $m_j$ white balls and $n_j$ black balls. A balanced die is rolled, and if an even number appears, a ball, chosen at random from urn $U_1$, is transferred to urn $U_2$. If an odd number appears, a ball, chosen at random from urn $U_2$, is transferred to urn $U_1$.
  - **(i)** What is the probability that, after the above experiment is performed twice, the number of white balls in urn $U_2$ remains the same?
  - **(ii)** Find the numerical value of the probability in part (i) for: $m_1 = 5$, $n_1 = 3$, $m_2 = 6$, $n_2 = 4$.

## 2.4 INDEPENDENT EVENTS AND RELATED RESULTS

In Example 14, it was seen that $P(A \mid B) = P(A)$. Thus, the fact that the event $B$ occurred provides no information in reevaluating the probability of $A$. Under such a circumstance, it is only fitting to say that $A$ is independent of $B$. For any two events $A$ and $B$ with $P(B) > 0$, we say that $A$ is *independent* of $B$, if $P(A \mid B) = P(A)$. If, in addition, $P(A) > 0$, then $B$ is also *independent* of $A$ because

$$P(B \mid A) = \frac{P(B \cap A)}{P(A)} = \frac{P(A \cap B)}{P(A)} = \frac{P(A \mid B)P(B)}{P(A)} = \frac{P(A)P(B)}{P(A)} = P(B).$$

Because of this symmetry, we then say that $A$ and $B$ are independent. From the definition of either $P(A \mid B)$ or $P(B \mid A)$, it follows then that $P(A \cap B) = P(A)P(B)$. We further observe that this relation is true even if one or both of $P(A), P(B)$ are equal to 0. We take this relation as the defining relation of independence.

**Definition 2.** Two events $A_1$ and $A_2$ are said to be *independent (statistically or stochastically or in the probability sense)*, if $P(A_1 \cap A_2) = P(A_1)P(A_2)$. When $P(A_1 \cap A_2) \neq P(A_1)P(A_2)$ they are said to be *dependent*.

**Remark 2.** At this point, it should be emphasized that disjointness and independence of two events are two distinct concepts; the former does not even require the concept of probability. Nevertheless, they are related in that, if $A_1 \cap A_2 = \emptyset$, then they are independent if and only if at least one of $P(A_1), P(A_2)$ is equal to 0. Thus (subject to $A_1 \cap A_2 = \emptyset$), $P(A_1)P(A_2) > 0$ implies that $A_1$ and $A_2$ are definitely dependent.

The definition of independence extends to three events $A_1, A_2, A_3$, as well as to any number $n$ of events $A_1, \ldots, A_n$. Thus, three events $A_1, A_2, A_3$ for which $P(A_1 \cap A_2 \cap A_3) > 0$ are said to be independent, if all conditional probabilities coincide with the respective (unconditional) probabilities:

$$
\left.
\begin{aligned}
P(A_1 \mid A_2) = P(A_1 \mid A_3) = P(A_1 \mid A_2 \cap A_3) = P(A_1), \\
P(A_2 \mid A_1) = P(A_2 \mid A_3) = P(A_2 \mid A_1 \cap A_3) = P(A_2), \\
P(A_3 \mid A_1) = P(A_3 \mid A_2) = P(A_3 \mid A_1 \cap A_2) = P(A_3), \\
P(A_1 \cap A_2 \mid A_3) = P(A_1 \cap A_2), P(A_1 \cap A_3 \mid A_2) \\
= P(A_1 \cap A_3), P(A_2 \cap A_3 \mid A_1) = P(A_2 \cap A_3).
\end{aligned}
\right\} \quad (1)
$$

From the definition of conditional probability, relations (1) are equivalent to:

$$
\left.
\begin{aligned}
P(A_1 \cap A_2) = P(A_1)P(A_2), P(A_1 \cap A_3) = P(A_1)P(A_3), \\
P(A_2 \cap A_3) = P(A_2)P(A_3), P(A_1 \cap A_2 \cap A_3) = P(A_1)P(A_2)P(A_3).
\end{aligned}
\right\} \quad (2)
$$

Furthermore, it is to be observed that relations (2) hold even if any of $P(A_1), P(A_2)$, $P(A_3)$ are equal to 0. These relations are taken as defining relations of independence of three events $A_1, A_2, A_3$.

As one would expect, all four relations (2) are needed for independence (i.e., in order for them to imply relations (1)). That this is, indeed, the case is illustrated by the following examples.

**Example 23.** Let $S = \{1, 2, 3, 4\}$ and let $P(\{1\}) = P(\{2\}) = P(\{3\}) = P(\{4\}) = 1/4$. Define the events $A_1, A_2, A_3$ by: $A_1 = \{1, 2\}, A_2 = \{1, 3\}, A_3 = \{1, 4\}$. Then it is easily verified that: $P(A_1 \cap A_2) = P(A_1)P(A_2), P(A_1 \cap A_3) = P(A_1)P(A_3), P(A_2 \cap A_3) = P(A_2)P(A_3)$. However, $P(A_1 \cap A_2 \cap A_3) \neq P(A_1)P(A_2)P(A_3)$.

**Example 24.** Let $S = \{1, 2, 3, 4, 5\}$ and let $P(\{1\}) = \frac{2}{16}, P(\{2\}) = P(\{3\}) = P(\{4\}) = \frac{3}{16}, P(\{5\}) = \frac{5}{16}$. Define the events $A_1, A_2, A_3$ by: $A_1 = \{1, 2, 3\}, A_2 = \{1, 2, 4\}, A_3 = \{1, 3, 4\}$. Then it is easily verified that: $P(A_1 \cap A_2 \cap A_3) = P(A_1)P(A_2) \times P(A_3)$ but none of the other three relations in (2) is satisfied.

Relations (2) provide the pattern of the definition of independence of $n$ events. Thus:

**Definition 3.**   The events $A_1, \ldots, A_n$ are said to be *independent (statistically or stochastically or in the probability sense)* if, for all possible choices of $k$ out of $n$ events $(2 \leq k \leq n)$, the probability of their intersection equals the product of their probabilities. More formally, for any $k$ with $2 \leq k \leq n$ and any integers $j_1, \ldots, j_k$ with $1 \leq j_1 < \cdots < j_k \leq n$, we have:

$$P\left(\bigcap_{i=1}^{k} A_{j_i}\right) = \prod_{i=1}^{k} P(A_{j_i}). \tag{3}$$

If at least one of the relations in (3) is violated, the events are said to be *dependent*. The number of relations of the form (3) required to express independence of $n$ events is:

$$\binom{n}{2} + \binom{n}{3} + \cdots + \binom{n}{n} = 2^n - \binom{n}{1} - \binom{n}{0} = 2^n - n - 1.$$

For example, for $n = 2, 3$, these relations are: $2^2 - 2 - 1 = 1$ and $2^3 - 3 - 1 = 4$, respectively.

Typical cases where independent events occur are whenever we are sampling with replacement from finite populations, such as selecting successively and with replacement balls from an urn containing balls of several colors, pulling successively and with replacement playing cards out of a standard deck of such cards, and the like.

The following property of independence of events is often used without even being acknowledged; it is stated here as a theorem.

**Theorem 6.**

(i) *If the events $A_1, A_2$ are independent, then so are all three sets of events: $A_1, A_2^c; A_1^c, A_2; A_1^c, A_2^c$.*

(ii) *More generally, if the events $A_1, \ldots, A_n$ are independent, then so are the events $A_1', \ldots, A_n'$, where $A_i'$ stands either for $A_i$ or $A_i^c, i = 1, \ldots, n$.*

For illustrative purposes, we present the proof of part (i) only.

*Proof.*   Clearly, $A_1 \cap A_2^c = A_1 - A_1 \cap A_2$. Thus,

$$P(A_1 \cap A_2^c) = P(A_1 - A_1 \cap A_2) = P(A_1) - P(A_1 \cap A_2) \text{ (since } A_1 \cap A_2 \subseteq A_1)$$

$$= P(A_1) - P(A_1)P(A_2) \quad \text{(by independence of } A_1, A_2)$$

$$= P(A_1)[1 - P(A_2)] = P(A_1)P(A_2^c).$$

The proof of $P(A_1^c \cap A_2) = P(A_1^c)P(A_2)$ is entirely symmetric. Finally,

$$P(A_1^c \cap A_2^c) = P((A_1 \cup A_2)^c) \quad \text{(by DeMorgan's laws)}$$

$$= 1 - P(A_1 \cup A_2)$$

$$= 1 - P(A_1) - P(A_2) + P(A_1 \cap A_2)$$

$$= 1 - P(A_1) - P(A_2) + P(A_1)P(A_2) \quad \text{(by independence of } A_1, A_2)$$

$$= [1 - P(A_1)] - P(A_2)[1 - P(A_1)]$$

$$= P(A_1^c)P(A_2^c). \qquad \blacksquare$$

The following examples will help illustrate concepts and results discussed in this section.

**Example 25.**   Suppose that $P(B)P(B^c) > 0$. Then the events $A$ and $B$ are independent if and only if $P(A \mid B) = P(A \mid B^c)$.

**Discussion.**   First, if $A$ and $B$ are independent, then $A$ and $B^c$ are also independent, by Theorem 6. Thus, $P(A \mid B^c) = \frac{P(A \cap B^c)}{P(B^c)} = \frac{P(A)P(B^c)}{P(B^c)} = P(A)$. Since also $P(A \mid B) = P(A)$, the equality $P(A \mid B) = P(A \mid B^c)$ holds. Next, $P(A \mid B) = P(A \mid B^c)$ is equivalent to $\frac{P(A \cap B)}{P(B)} = \frac{P(A \cap B^c)}{P(B^c)}$ or $P(A \cap B)P(B^c) = P(A \cap B^c)P(B)$ or $P(A \cap B)[1 - P(B)] = P(A \cap B^c)P(B)$ or $P(A \cap B) - P(A \cap B)P(B) = P(A \cap B^c)P(B)$ or $P(A \cap B) = [P(A \cap B) + P(A \cap B^c)]P(B) = P(A)P(B)$, since $(A \cap B) \cup (A \cap B^c) = A$. Thus, $A$ and $B$ are independent.

**Remark 3.**   It is to be pointed out that the condition $P(A \mid B) = P(A \mid B^c)$ for independence of the events $A$ and $B$ is quite natural, intuitively. It says that the (conditional) probability of $A$ remains the same no matter which one of $B$ or $B^c$ is given.

**Example 26.**   Let $P(C)P(C^c) > 0$. Then the inequalities $P(A \mid C) > P(B \mid C)$ and $P(A \mid C^c) > P(B \mid C^c)$ imply $P(A) > P(B)$.

**Discussion.**   The inequalities $P(A \mid C) > P(B \mid C)$ and $P(A \mid C^c) > P(B \mid C^c)$ are equivalent to $P(A \cap C) > P(B \cap C)$ and $P(A \cap C^c) > P(B \cap C^c)$. Adding up these inequalities, we obtain $P(A \cap C) + P(A \cap C^c) > P(B \cap C) + P(B \cap C^c)$ or $P(A) > P(B)$, since $A = (A \cap C) \cup (A \cap C^c)$ and $B = (B \cap C) \cup (B \cap C^c)$.

**Remark 4.**   Once again, that the inequalities of the two conditional probabilities should imply the same inequality for the unconditional probabilities is quite obvious on intuitive grounds. The justification given above simply makes it rigorous.

**Example 27.**   If the events $A, B$, and $C$ are independent, then $P(A \cup B \cup C) = 1 - [1 - P(A)][1 - P(B)][1 - P(C)]$.

**Discussion.**   Clearly,

$$P(A \cup B \cup C) = P[(A^c \cap B^c \cap C^c)^c] \qquad \text{(by DeMorgan's laws)}$$

$$= 1 - P(A^c \cap B^c \cap C^c) \qquad \text{(by property \#3 in Section 2.1.1)}$$

$$= 1 - P(A^c)P(B^c)P(C^c) \qquad \text{(by Theorem 6(ii) applied with } n = 3)$$

$$= 1 - [1 - P(A)][1 - P(B)][1 - P(C)].$$

**Example 28.** A mouse caught in a maze has to maneuver through three successive escape hatches in order to escape. If the hatches operate independently and the probabilities for the mouse to maneuver successfully through them are 0.6, 0.4, and 0.2, respectively, calculate the probabilities that the mouse: (i) will be able to escape, (ii) will not be able to escape.

**Discussion.** Denote by $H_1, H_2$, and $H_3$ the events that the mouse successfully maneuvers through the three hatches, and by $E$ the event that the mouse is able to escape. We have that $H_1, H_2$, and $H_3$ are independent, $P(H_1) = 0.6, P(H_2) = 0.4$, and $P(H_3) = 0.2$, and $E = H_1 \cap H_2 \cap H_3$. Then: (i) $P(E) = P(H_1 \cap H_2 \cap H_3) = P(H_1)P(H_2)P(H_3) = 0.6 \times 0.4 \times 0.2 = 0.048$, and (ii) $P(E^c) = 1 - P(E) = 1 - 0.048 = 0.952$.

The concept of independence carries over to random experiments. Although a technical definition of independence of random experiments is available, we are not going to indulge in it. The concept of independence of random experiments will be taken in its intuitive sense, and somewhat more technically, in the sense that random experiments are independent if they give rise to independent events associated with them.

Finally, independence is also defined for r.v.'s. This topic will be taken up in Chapter 5 (see Definition 1 there). Actually, independence of r.v.'s is one of the founding blocks of most discussions taking place in this book.

## EXERCISES

**4.1** If $P(A) = 0.4, P(B) = 0.2$, and $P(C) = 0.3$, calculate the probability $P(A \cup B \cup C)$, if the events $A, B$, and $C$ are:
   **(i)** Pairwise disjoint.
   **(ii)** Independent.

**4.2** Show that the event $A$ is independent of itself if and only if $P(A) = 0$ or $P(A) = 1$.

**4.3**   **(i)** For any two events $A$ and $B$, show that $P(A \cap B) \geq P(A) + P(B) - 1$.
   **(ii)** If $A$ and $B$ are disjoint, then show that they are independent if and only if at least one of $P(A)$ and $P(B)$ is zero.
   **(iii)** If the events $A, B$, and $C$ are pairwise disjoint, under what conditions are they independent?

**4.4** Suppose that the events $A_1, A_2$, and $B_1$ are independent, the events $A_1, A_2$, and $B_2$ are independent, and that $B_1 \cap B_2 = \emptyset$. Then show that the events $A_1, A_2, B_1 \cup B_2$ are independent.

**4.5** (i) If for the events $A, B,$ and $C$, it so happens that:
$P(A) = P(B) = P(C) = \frac{1}{2}, P(A \cap B) = P(A \cap C) = P(B \cap C) = \frac{1}{4}$, and
$P(A \cap B \cap C) = \frac{1}{6}$, determine whether or not these events are independent. Justify your answer.

(ii) For the values given in part (i), calculate the probabilities:
$P(A^c), P(A \cup B), P(A^c \cap B^c), P(A \cup B \cup C),$ and $P(A^c \cap B^c \cap C^c)$.

**4.6** For the events $A, B, C$ and their complements, suppose that:

$$P(A \cap B \cap C) = \frac{1}{16}, \quad P(A \cap B^c \cap C) = \frac{5}{16}, \quad P(A \cap B \cap C^c) = \frac{3}{16},$$

$$P(A \cap B^c \cap C^c) = \frac{2}{16}, \quad P(A^c \cap B \cap C) = \frac{2}{16}, \quad P(A^c \cap B \cap C^c) = \frac{1}{16},$$

$$P(A^c \cap B^c \cap C) = \frac{1}{16}, \quad \text{and} \quad P(A^c \cap B^c \cap C^c) = \frac{1}{16}.$$

(i) Calculate the probabilities: $P(A), P(B), P(C)$.
(ii) Determine whether or not the events $A, B,$ and $C$ are independent.
(iii) Calculate the (conditional) probability $P(A \mid B)$.
(iv) Determine whether or not the events $A$ and $B$ are independent.
**Hint.** Use Theorem 4 for part (i) and for the calculations of the $P(A \cap B)$ in part (iii).

**4.7** If the events $A_1, \ldots, A_n$ are independent, show that

$$P\left(\bigcup_{j=1}^{n} A_j\right) = 1 - \prod_{j=1}^{n} P(A_j^c) = 1 - \prod_{j=1}^{n}[1 - P(A_j)].$$

**Hint.** Use DeMorgan's laws, property #3 in Section 2.1.1, and Theorem 6(ii).

**4.8** (i) Three coins, with probability of falling heads being $p$, are tossed once and you win, if all three coins show the same face (either all $H$ or all $T$). What is the probability of winning?
(ii) What are the numerical answers in part (i) for $p = 0.5$ and $p = 0.4$?

**4.9** Suppose that men and women are distributed in the freshman and sophomore classes of a college according to the proportions listed in the following table.

| Class\\Gender | M | W | Totals |
|---|---|---|---|
| F | 4 | 6 | 10 |
| S | 6 | $x$ | $6+x$ |
| Totals | 10 | $6+x$ | $16+x$ |

A student is chosen at random and let $M, W, F,$ and $S$ be the events, respectively, that the student is a man, a woman, a freshman, or a sophomore.

Then, being a man or a woman and being a freshman or sophomore are independent, if:

$$P(M \cap F) = P(M)P(F), \qquad P(W \cap F) = P(W)P(F),$$
$$P(M \cap S) = P(M)P(S), \qquad P(W \cap S) = P(W)P(S).$$

Determine the number $x$, so that the preceding independence relations hold. **Hint.** Determine $x$ by using only one of the above four relations (and check that this value of $x$ also satisfies the remaining three relations).

**4.10** The r.v. $X$ has p.d.f. given by:

$$f(x) = \begin{cases} cx, & 0 \le x < 5, \\ c(10 - x), & 5 \le x < 10, \\ 0, & \text{elsewhere.} \end{cases}$$

  **(i)** Determine the constant $c$.
  **(ii)** Draw the graph of $f$.
    Define the events $A$ and $B$ by: $A = (X > 5), B = (5 < X < 7.5)$.
  **(iii)** Calculate the probabilities $P(A)$ and $P(B)$.
  **(iv)** Calculate the conditional probability $P(B \mid A)$.
  **(v)** Are the events $A$ and $B$ independent or not? Justify your answer.

**4.11** Three players *I, II, III* throw simultaneously three coins with respective probabilities of falling heads ($H$) $p_1, p_2$, and $p_3$. A sample space describing this experiment is

$$S = \{HHH, HHT, HTH, THH, HTT, THT, TTH, TTT\}.$$

Define the events $A_i, i = 1, 2, 3$ and $B$ by:

$$A_1 = \{HTT, THH\} \quad A_2 = \{THT, HTH\}, \quad A_3 = \{TTH, HHT\}$$

(i.e., the outcome for the *i*th player, $i = 1, 2, 3$, is different from those for the other two players),

$$B = \{HHH, TTT\}.$$

If any one of the events $A_i, i = 1, 2, 3$, occurs, the *i*th player wins and the game ends. If event $B$ occurs, the game is repeated independently as many times as needed until one of the events $A_1, A_2, A_3$ occurs.
  **(i)** Calculate the probabilities: $P(A_i), i = 1, 2, 3$.
  **(ii)** How do these probabilities become for $p_1 = p_2 = p_3 = p$?
  **(iii)** That is the numerical value in part (ii) if $p = 0.5$?
  **Hint.** By symmetry, it suffices to calculate $P(A_1)$. Let $A_{1j} = $ "event $A_1$ occurs the *j*th time," $B_j = $ "event $B$ occurs the *j*th time." Then (with slight abuse of notation)

$$A_1 = A_{11} \cup (B_1 \cap A_{12}) \cup (B_1 \cap B_2 \cap A_{13}) \cup \cdots$$

At this point, use #4 in Table 8 in Appendix.

**4.12** Jim takes the written and road driver's license tests repeatedly until he passes them. It is given that the probability that he passes the written test is 0.9, that he passes the road test is 0.6, and that the tests are independent of each other. Furthermore, it is assumed that the road test cannot be taken unless he passes the written test, and that once he passes the written test, he does not have to take it again ever, no matter whether he passes or fails his road tests. Also, it is assumed that the written and the road tests are distinct attempts.

  **(i)**  That is the probability that he will pass the road test on his $n$th attempt?

  **(ii)**  That is the numerical value in part (i) for $n = 5$?

**Hint.** Denote by $W_i$ and $R_j$ the events that Jim passes the written test and the road test the $i$th and $j$th time, respectively. Then the required event is expressed as follows:

$$\left(W_1 \cap R_1^c \cap \cdots \cap R_{n-2}^c \cap R_{n-1}\right) \cup \left(W_1^c \cap W_2 \cap R_1^c \cap \cdots \cap R_{n-3}^c \cap R_{n-2}\right)$$
$$\cup \cdots \cup \left(W_1^c \cap \cdots \cap W_{n-2}^c \cap W_{n-1} \cap R_n\right).$$

**4.13** The probability that a missile fired against a target is not intercepted by an antimissile missile is $\frac{2}{3}$. If the missile is not intercepted, then the probability of a successful hit is $\frac{3}{4}$.

If four missiles are fired independently, what is the probability that:

  **(i)**  The $i$th missile ($i = 1, \ldots, 4$) will successfully hit the target?

  **(ii)**  All four will successfully hit the target?

  **(iii)**  It least one will do so?

  **(iv)**  That is the minimum number of missiles to be fired so that at least one is not intercepted with probability at least 0.95?

  **(v)**  That is the minimum number of missiles to be fired so that at least one hits the target with probability at least 0.99?

**4.14** Electric current is transmitted from point A to point B, provided at least one of the circuits #1 through #$n$ here is closed (Figure 2.5). It is assumed that the $n$ circuits close independently of each other and with respective probabilities $p_1, \ldots, p_n$.

Determine the following probabilities:

  **(i)**  To circuit is closed.

  **(ii)**  It least one circuit is closed.

  **(iii)**  Exactly one circuit is closed.

  **(iv)**  How do the expressions in parts (i)-(iii) simplify if $p_1 = \cdots = p_n = p$?

  **(v)**  That are the numerical values in part (iv) for $n = 5$ and $p = 0.6$?

**Hint.** Use Theorem 6(ii). Probabilities in parts (i)-(iii) are to be expressed in terms of $p_1, \ldots, p_n$ and $p$, respectively.

**4.15** Consider two urns $U_1$ and $U_2$ such that urn $U_1$ contains $m_1$ white balls and $n_1$ black balls, and urn $U_2$ contains $m_2$ white balls and $n_2$ black balls. All balls are identical except for color. One ball is drawn at random from each of the urns $U_1$ and $U_2$ and is placed into a third urn. Then a ball is drawn at random from the third urn. Compute the probability that the ball is

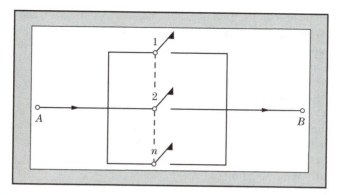

**FIGURE 2.5**

The circuit referred to in Exercise 4.14.

  **(i)** Black;
  **(ii)** White.
  **(iii)** Give numerical answers to parts (i) and (ii) for: $m_1 = 10, n_1 = 15$;
      $m_2 = 35, n_2 = 25$.
  **Hint.** For parts (i) and (ii), denote by $B_i$ and $W_i$ the events that the ball drawn
  from the $i$th urn, $i = 1, 2$, is black and white, respectively, and by $B$ and $W$ the
  events that the ball drawn from the third urn is black or white, respectively,
  and then use Theorem 4.

**4.16** If $p_1$ is the probability that $A$ has a headache, and $p_2$ is the probability that $B$
  has a headache, and the two events are independent:
  **(i)** Compute the probability $P(A$ or $B$ has a headache, but not both) as a
      function of $p_1$ and $p_2$.
  **(ii)** What is the numerical value in part (i) for $p_1 = 0.1, p_2 = 0.09$?

**4.17** The events $A$, $B$, $C$ are independent and $P(A) = 0.3$, $P(B) = 0.2$, $P(C) = 0.6$.
  Compute: $P(A^c \cap B \cap C)$; $P(A^c \cap B^c \cap C)$; $P(A^c \cap B^c \cap C^c)$, and justify your
  reasoning.

**4.18** For any three events $A$, $B$, and $C$, let $D$ be defined by: $D = $ "exactly 2 of the
  events $A$, $B$, and $C$ occur."
  **(i)** Express the event $D$ in terms of $A$, $B$, and $C$, and, perhaps,
      their complements.
  **(ii)** Show that:

$$P(D) = P(A \cap B) + P(A \cap C) + P(B \cap C) - 3P(A \cap B \cap C).$$

  **(iii)** If $A$, $B$, and $C$ are independent, with $P(A) = 0.2$, $P(B) = 0.7$,
      $P(C) = 0.4$, compute the $P(D)$.
  **Hint.** For part (ii), you may want to use the fact that, for any two events $E$ and
  $F$, say, it holds: $E \cap F^c = E - (E \cap F)$, so that $P(E \cap F^c) = P(E) - P(E \cap F)$.

**4.19** In the game of roulette, each one of the numbers 1 through 36 and the symbols 0 and 00 have equal probability of occurring in each spin of the wheel.

  **(i)** Suppose that in the first spin, one puts bets on two distinct numbers, e.g., 6 and 36. What is the probability that one has:
  
  **(a)** Two wins?
  
  **(b)** At least one win?
  
  **(c)** Exactly one win?
  
  **(ii)** Suppose that in two (independent) spins, one puts a bet on number 6 the first time and number 36 the second time. What is the probability that one:
  
  **(a)** Wins both times?
  
  **(b)** At least one time?
  
  **(c)** Exactly one time?
  
  **(iii)** Compare the results in (i)(a) and (ii)(a); (i)(b) and (ii)(b); (i)(c) and (ii)(c), and explain the difference.

## 2.5 BASIC CONCEPTS AND RESULTS IN COUNTING

In this brief section, some basic concepts and results are discussed regarding the way of counting the total number of outcomes of an experiment or the total number of different ways we can carry out a task. Although many readers will, undoubtedly, be familiar with parts of or the entire material in this section, it would be advisable, nevertheless, to invest some time here in introducing and adopting some notation, establishing some basic results, and then using them in computing probabilities in the classical probability framework.

Problems of counting arise in a great number of different situations. Here are some of them. In each one of these situations, we are asked to compute the number of different ways that something or other can be done. Here are a few illustrative cases.

**Example 29.**

  **(i)** Attire yourself by selecting a T-shirt, a pair of trousers, a pair of shoes, and a cap out of $n_1$ T-shirts, $n_2$ pairs of trousers, $n_3$ pairs of shoes, and $n_4$ caps (e.g., $n_1 = 4, n_2 = 3, n_3 = n_4 = 2$).

  **(ii)** Form all $k$-digit numbers by selecting the $k$ digits out of $n$ available numbers (e.g., $k = 2, n = 4$ such as $\{1, 3, 5, 7\}$).

  **(iii)** Form all California automobile license plates by using one number, three letters and then three numbers in the prescribed order.

  **(iv)** Form all possible codes by using a given set of symbols (e.g., form all "words" of length 10 by using the digits 0 and 1).

  **(v)** Place $k$ books on the shelf of a bookcase in all possible ways.

  **(vi)** Place the birthdays of $k$ individuals in the 365 days of a year in all possible ways.

  **(vii)** Place $k$ letters into $k$ addressed envelopes (one letter to each envelope).

  **(viii)** Count all possible outcomes when tossing $k$ distinct dice.

**(ix)** Select $k$ cards out of a standard deck of playing cards (e.g., for $k = 5$, each selection is a poker hand).

**(x)** Form all possible $k$-member committees out of $n$ available individuals.

The calculation of the numbers asked for in situations (i) through (x) just outlined is in actuality a simple application of the so-called *fundamental principle of counting*, stated next in the form of a theorem.

**Theorem 7** (Fundamental Principle of Counting).   *Suppose a task is completed in $k$ stages by carrying out a number of subtasks in each one of the $k$ stages. If the numbers of these subtasks are $n_1, \ldots, n_k$ for the $k$ stages, respectively, then the total number of different ways the overall task is completed is: $n_1 \times \cdots \times n_k$.*

Thus, in (i) above the number of different attires is $4 \times 3 \times 2 \times 2 = 48$.

In (ii), the number of all 2-digit numbers formed by using $1, 3, 5, 7$ is $4 \times 4 = 16$ ($11, 13, 15, 17$; $31, 33, 35, 37$; $51, 53, 55, 57$; $71, 73, 75, 77$).

In (iii), the number of all possible license plates (by using indiscriminately all 10 digits from 0 through 9 and all 26 letters of the English alphabet, although this is not the case in practice) is $10 \times (26 \times 26 \times 26) \times (10 \times 10 \times 10) = 175{,}760{,}000$.

In (iv), the number of all possible "words" is found by taking $k = 10$ and $n_1 = \cdots = n_{10} = 2$ to obtain: $2^{10} = 1024$.

In (v), all possible arrangements are obtained by taking $n_1 = k, n_2 = k - 1, \ldots, n_k = k - (k - 1) = 1$ to get: $k(k - 1) \ldots 1 = 1 \ldots (k - 1)k$. For example, for $k = 10$, the number of arrangements is: $3{,}628{,}800$.

In (vi), the required number is obtained by taking $n_1 = \cdots = n_k = 365$ to get: $365^k$. For example, for $k = 3$, we have $365^3 = 48{,}627{,}125$.

In (vii), the required number is: $k(k - 1) \ldots 1 = 1 \ldots (k - k)k$ obtained by taking $n_1 = k, n_2 = k - 1, \ldots, n_k = k - (k - 1) = 1$.

In (viii), the required number is: $6^k$ obtained by taking $n_1 = \cdots = n_k = 6$. For example, for $k = 3$, we have $6^3 = 216$, and for $k = 10$, we have $6^{10} = 60{,}466{,}176$.

In (ix), the number of poker hands is: $\frac{52 \times 51 \times 50 \times 49 \times 48}{120} = 2{,}598{,}960$. The numerator is obtained by taking $n_1 = 52, n_2 = 51, n_3 = 50, n_4 = 49, n_5 = 48$. The division by 120 accounts for elimination of hands consisting of the same cards but drawn in different order.

Finally, in (x), the required number is: $\frac{n(n-1) \ldots (n-k+1)}{1 \times 2 \times \cdots \times k}$, by arguing as in (ix). For example, for $n = 10$ and $k = 3$, we have: $\frac{10 \times 9 \times 8}{1 \times 2 \times 3} = 120$.

In all of the situations (i) through (x), the required numbers were calculated by the appropriate application of Theorem 7. Furthermore, in many cases, as clearly exemplified by cases (ii), (iii), (v), (vii), (ix), and (x), the task performed consisted of selecting and arranging a number of objects out of a set of available objects. In so doing, the order in which the objects appear in the arrangement may be of significance, as is, indeed, the case in situations (ii), (iii), (iv), (v), (vi), and (vii), or it may be just irrelevant, as happens, for example, in cases (ix) and (x). This observation leads us to the concepts of permutations and combinations. More precisely, we have:

**Definition 4.** An *ordered* arrangement of $k$ objects taken from a set of $n$ objects ($1 \leq k \leq n$) is a *permutation* of the $n$ objects taken $k$ at time. An *unordered* arrangement of $k$ objects taken from a set of $n$ objects is a *combination* of the $n$ objects taken $k$ at a time.

The question then arises as to how many permutations and how many combinations can be formed. The answer to this question is given next.

**Corollary** (to Theorem 7).

(i) *The number of ordered arrangements of a set of $n$ objects taken $k$ at a time ($1 \leq k \leq n$) is $n^k$ when repetitions are allowed. When no repetitions are allowed, this number becomes the permutations of $n$ objects taken $k$ at a time, is denoted by $P_{n,k}$, and is given by:*

$$P_{n,k} = n(n-1)\ldots(n-k+1). \tag{4}$$

*In particular, for $k = n$,*

$$P_{n,n} = n(n-1)\ldots 1 = 1 \ldots (n-1)n = n!,$$

*where the notation $n!$ is read "$n$ factorial."*

(ii) *The number of combinations (i.e., the number of unordered and without repetition arrangements) of $n$ objects taken $k$ at a time ($1 \leq k \leq n$) is denoted by $\binom{n}{k}$ and is given by:*

$$\binom{n}{k} = \frac{P_{n,k}}{k!} = \frac{n!}{k!(n-k)!}. \tag{5}$$

**Remark 5.** Whether permutations or combinations are appropriate in a given problem follows from the nature of the problem. For instance, in (ii), permutations rather than combinations are appropriate as, e.g., 13 and 31 are distinct entities. The same is true of cases (iii)–(viii), whereas combinations are appropriate for cases (ix) and (x).

As an example, in part (ii), $P_{4,2} = 4 \times 3 = 12$ (leave out the numbers with identical digits 11, 22, 33, and 44), and in part (ix), $\binom{52}{5} = \frac{52!}{5!47!} = 2,598,960$, after cancellations and by carrying out the arithmetic.

**Remark 6.** In (5), set $k = n$. Then the left-hand side is, clearly, 1, and the right-hand side is $\frac{n!}{n!0!} = \frac{1}{0!}$. In order for this to be 1, we *define* $0! = 1$. From formula (5), it also follows that $\binom{n}{0} = 1$.

This section is concluded with the justification of Theorem 7 and its corollary and some applications of these results in calculating certain probabilities.

*Proof of Theorem 7.* It is done by mathematical induction. For $k = 2$, all one has to do is to pair out each one of the $n_1$ ways of carrying out the subtask at stage 1 with each one of the $n_2$ ways of carrying out the subtask at stage 2 in order to obtain $n_1 \times n_2$ for the number of ways of completing the task. Next, make the induction hypothesis that the conclusion is true for $k = m$ and establish it for $k = m + 1$. So, in the first $m$

stages, the total number of ways of doing the job is $n_1 \times \cdots \times n_m$, and there is still the final $(m+1)$st stage for completing the task. Clearly, all we have to do here is to combine each one of the $n_1 \times \cdots \times n_m$ ways of doing the job in the first $m$ stages with each one of the $n_{m+1}$ ways of carrying out the subtask in the $(m+1)$st stage to obtain the number $n_1 \times \cdots \times n_m \times n_{m+1}$ of completing the task. ∎

*Proof of the Corollary.*

(i) Here, we are forming an ordered arrangement of objects in $k$ stages by selecting one object at each stage from among the $n$ available objects (because repetitions are allowed). Thus, the theorem applies with $n_1 = \cdots = n_k = n$ and gives the result $n^k$. When repetitions are not allowed, the only thing that changes from the case just considered is that: $n_1 = n, n_2 = n - 1, \ldots, n_k = n - (k-1) = n - k + 1$, and formula (4) follows.

(ii) Let $\binom{n}{k}$ be the number of combinations (unordered without repetition arrangements) of the $n$ objects taken $k$ at a time. From each one of these unordered arrangements, we obtain $k!$ ordered arrangements by permutation of the $k$ objects. Then $k! \times \binom{n}{k}$ is the total number of ordered arrangements of the $n$ objects taken $k$ at a time, which is $P_{n,k}$, by part (i). Solving for $\binom{n}{k}$, we obtain the first expression in (5). The second expression follows immediately by multiplying by $(n-k)\ldots 1$ and dividing by $1 \ldots (n-k) = (n-k)!$. ∎

There are many interesting variations and deeper results based on Theorem 7 and its corollary. Some of them may be found in Sections 2.4 and 2.6 of Chapter 2 of the book *A Course in Mathematical Statistics*, 2nd edition (1997), Academic Press, by G.G. Roussas.

We only state and prove here one result regarding permutations (factorials) when the objects involved are not all distinct. More precisely, we have the following result.

**Proposition 1.** *Consider $n$ objects that are divided into $k$ groups $(1 \le k \le n)$ with the property that the $m_i$ members of the $i$th group are identical and distinct from the members of the remaining groups, $m_1 + \ldots + m_k = n$. Then the number of distinct arrangements of the $n$ objects is $n!/(m_1! \times \ldots \times m_k!)$*

*Proof.* One way of generating all distinct arrangements of the $n$ objects is to select $m_i$ positions out of $n$ available in $\begin{pmatrix} n \\ m_i \end{pmatrix}$ possible ways and place there the $m_i$ identical objects, $i = 1, \ldots, k$. Then, by Theorem 1, the total number of arrangements is:

$$\begin{pmatrix} n \\ m_1 \end{pmatrix} \times \begin{pmatrix} n - m_1 \\ m_2 \end{pmatrix} \times \ldots \times \begin{pmatrix} n - m_1 - \ldots - m_{k-1} \\ m_k \end{pmatrix} =$$

$$\frac{n!}{m_1!(n-m_1)!} \times \frac{(n-m_1)!}{m_2!(n-m_1-m_2)!} \times \ldots \times \frac{(n-m_1-\ldots-m_{k-1})!}{m_k!(n-m_1-\ldots-m_{k-1}-m_k)!} =$$

$$\frac{n!}{m_1! \times m_2! \times \ldots \times m_k!}, \text{ since } (n-m_1-\ldots-m_{k-1}-m_k)! = 0! = 1.$$

An alternative way to look at this problem would be to consider the $n!$ arrangements of the $n$ objects, and then make the $m_i!$ arrangements within the $i$th group, $i = 1, \ldots, k$, which leave the overall arrangement unchanged. Thus, the number of distinct arrangements of the $n$ objects is $n!/(m_1! \times m_2! \times \ldots \times m_k!)$.   ∎

**Example 30.**   It happens that four hotels in a certain large city have the same name, e.g., Grand Hotel. Four persons make an appointment to meet at the Grand Hotel. If each one of the four persons chooses the hotel at random, calculate the following probabilities:

(i)  All four choose the same hotel.
(ii)  All four choose different hotels.

**Discussion.**

(i)  If $A =$ "all 4 choose the same hotel," then $P(A) = \frac{n(A)}{n(S)}$, where $n(A)$ is the number of sample points in $A$. One explicit way of computing the numbers in $n(S)$ and $n(A)$ is the following self-explanatory:

$$n(S) = \binom{4}{1} \times 4 \times \binom{3}{1} \times 4 \times \binom{2}{1} \times 4 \times \binom{1}{1} \times 4 = 4! \times 4^4,$$

by Theorem 7 applied with $k = 8$ and $n_1 = \binom{4}{1}$, $n_3 = \binom{3}{1}$, $n_5 = \binom{2}{1}$, $n_7 = \binom{1}{1}$, and $n_2 = n_4 = n_6 = n_8 = 4$, and

$$n(A) = \binom{4}{1} \times 4 \times \binom{3}{1} \times 1 \times \binom{2}{1} \times 1 \times \binom{1}{1} \times 1 = 4! \times 4,$$

by Theorem 7 again applied with $k = 8$ and $n_1 = \binom{4}{1}$, $n_2 = 4$, $n_3 = \binom{3}{1}$, $n_5 = \binom{2}{1}$, $n_7 = \binom{1}{1}$, and $n_4 = n_6 = n_8 = 1$. Thus, $P(A) = \frac{4! \times 4}{4! \times 4^4} = \frac{4}{4^4} = \frac{1}{4^3} = \frac{1}{64} \simeq 0.015625 \simeq 0.016$.

(ii)  If $B =$ "all four choose different hotels", then $n(B) = \binom{4}{1} \times 4 \times \binom{3}{1} \times 3 \times \binom{2}{1} \times 2 \times \binom{1}{1} \times 1 = 4! \times 4!$, by Theorem 7 applied with $k = 8$ and $n_1 = n_2 = 4$, $n_3 = n_4 = 3$, $n_5 = n_6 = 2$, $n_7 = n_8 = 1$, so that

$$P(B) = \frac{4! \times 4!}{4! \times 4^4} = \frac{4!}{4^4} = \frac{6}{64} = \frac{3}{32} \simeq 0.09375 \simeq 0.094. y$$

(Alternative approaches are also possible.)

**Example 31.**   Out of a set of three keys, only one opens a certain door. Someone tries the keys successively and let $A_k$ be the event that the right key appears the $k$th time. Calculate the probability $P(A_k)$:

(i)  If the keys tried are not replaced, $k = 1, 2, 3$.
(ii)  If the keys tried are replaced, $k = 1, 2, \ldots$.

**Discussion.**

(i) $P(A_1) = \frac{1}{3}; P(A_2) = \frac{2 \times 1}{3 \times 2} = \frac{1}{3}; P(A_3) = \frac{2 \times 1 \times 1}{3 \times 2 \times 1} = \frac{1}{3}$. So, $P(A_1) = P(A_2) = P(A_3) = \frac{1}{3} \simeq 0.333$.

(ii) Clearly, with notation explained in Remark 7 below, $P(A_k) = P(W_1 \cap \cdots \cap W_{k-1} \cap R_k) = \left(\frac{2}{3}\right)^{k-1} \times \frac{1}{3}$ for all $k = 1, 2, \ldots$

**Remark 7.** To calculate the probabilities in part (i) in terms of conditional probabilities, set: $R_k = $ "the right key appears the $k$th time," $W_k = $ "a wrong key appears the $k$th time," $k = 1, 2, 3$. Then: $P(A_1) = P(R_1) = \frac{1}{3}, P(A_2) = P(W_1 \cap R_2) = P(R_2 \mid W_1)P(W_1) = \frac{1}{2} \times \frac{2}{3} = \frac{1}{3}$, and $P(A_3) = P(W_1 \cap W_2 \cap R_3) = P(R_3 \mid W_1 \cap W_2)P(W_2 \mid W_1)P(W_1) = \frac{1}{1} \times \frac{1}{2} \times \frac{2}{3} = \frac{1}{3}$.

**Example 32.** The faculty in an academic department in UC-Davis consists of four assistant professors, six associate professors, and five full professors. Also, it has 30 graduate students. An *ad hoc* committee of five is to be formed to study a certain curricular matter.

(i) What is the number of all possible committees consisting of faculty alone?

(ii) How many committees can be formed if two graduate students are to be included and all academic ranks are to be represented?

(iii) If the committee is to be formed at random, what is the probability that the faculty will not be represented?

**Discussion.** It is clear that combinations are the appropriate tool here. Then we have:

(i) This number is $\binom{15}{5} = \frac{15!}{5!10!} = \frac{11 \times 12 \times 13 \times 14 \times 15}{1 \times 2 \times 3 \times 4 \times 5} = 3003$.

(ii) Here the number is $\binom{30}{2}\binom{4}{1}\binom{6}{1}\binom{5}{1} = \frac{30!}{2!28!} \times 4 \times 6 \times 5 = \frac{29 \times 30}{2} \times 120 = 52{,}200$.

(iii) The required probability is

$$\frac{\binom{30}{5}\binom{15}{0}}{\binom{45}{5}} = \frac{\binom{30}{5}}{\binom{45}{5}} = \frac{30!/5!25!}{45!/5!40!} = \frac{26 \times 27 \times 28 \times 29 \times 30}{41 \times 42 \times 43 \times 44 \times 45} = \frac{2262}{19{,}393} \simeq 0.117.$$

**Example 33.** What is the probability that a poker hand contains four pictures, including at least two Jacks? It is recalled here that there are 12 pictures consisting of 4 Jacks, 4 Queens, and 4 Kings.

**Discussion.** A poker hand can be selected in $\binom{52}{5}$ ways. The event described, call it $A$, consists of the following number of sample points: $n(A) = n(J_2) + n(J_3) + n(J_4)$, where $J_i = $ "the poker hand contains exactly $i$ Jacks," $i = 2, 3, 4$. But

$$n(J_2) = \binom{4}{2}\binom{8}{2}\binom{40}{1}, \quad n(J_3) = \binom{4}{3}\binom{8}{1}\binom{40}{1}, \quad n(J_4) = \binom{4}{4}\binom{8}{0}\binom{40}{1},$$

so that

$$P(A) = \frac{\binom{4}{2}\binom{8}{2} + \binom{4}{3}\binom{8}{1} + \binom{4}{4}\binom{8}{0}\binom{40}{1}}{\binom{52}{5}} = \frac{8040}{2{,}598{,}960} \simeq 0.003.$$

(For the calculation of $\binom{52}{5}$ see Example 29(ix).)

**Example 34.**  Each of the $2n$ members of a committee flips a fair coin in deciding whether or not to attend a meeting of the committee; a committee member attends the meeting if an $H$ appears. What is the probability that a majority will show up for the meeting?

**Discussion.**  There will be majority if there are at least $n + 1$ committee members present, which amounts to having at least $n + 1$ $H$'s in $2n$ independent throws of a fair coin. If $X$ is the r.v. denoting the number of $H$'s in the $2n$ throws, then the required probability is: $P(X \geq n + 1) = \sum_{x=n+1}^{2n} P(X = x)$. However,

$$P(X = x) = \binom{2n}{x}\left(\frac{1}{2}\right)^x \left(\frac{1}{2}\right)^{2n-x} = \frac{1}{2^{2n}}\binom{2n}{x},$$

since there are $\binom{2n}{x}$ ways of having $x$ $H$'s in $2n$ throws. Therefore,

$$P(X \geq n + 1) = \frac{1}{2^{2n}} \sum_{x=n+1}^{2n} \binom{2n}{x} = \frac{1}{2^{2n}}\left[ \sum_{x=0}^{2n} \binom{2n}{x} - \sum_{x=0}^{n} \binom{2n}{x} \right]$$

$$= \frac{1}{2^{2n}}\left[ 2^{2n} - \sum_{x=0}^{n} \binom{2n}{x} \right] = 1 - \frac{1}{2^{2n}} \sum_{x=0}^{n} \binom{2n}{x}.$$

For example, for $2n = 10, P(X \geq 6) = 1 - 0.6230 = 0.377$ (from the Binomial tables).

---

## EXERCISES

**5.1**  Telephone numbers at UC-Davis consist of seven-digit numbers the first 3 of which are 752. It is estimated that about 15,000 different telephone numbers are needed to serve the university's needs.

Are there enough telephone numbers available for this purpose? Justify your answer.

**5.2**  An experimenter is studying the effects of temperature, pressure, and a catalyst on the yield of a certain chemical reaction. Three different temperatures, four different pressures, and five different catalysts are under consideration.

    (i) If any particular experimental run involves the use of a single temperature, pressure, and catalyst, how many experimental runs are possible?

    (ii) How many experimental runs are there that involve use of the lowest temperature and two lowest pressures?

    (iii) How many experimental runs are possible if a specified catalyst is to be used?

**5.3**   (i) Given that a zip code consists of a five-digit number, where the digits are selected from among the numbers $0, 1, \ldots, 9$, calculate the number of all different zip codes.

    (ii) If $X$ is the r.v. defined by: $X(\text{zip code}) = \#$ of nonzero digits in the zip code, which are the possible values of $X$?

    (iii) Give three zip codes and the respective values of $X$.

**5.4** How many five-digit numbers can be formed by using the numbers 1, 2, 3, 4, and 5, so that odd positions are occupied by odd numbers and even positions are occupied by even numbers, if:

    (i) Repetitions are allowed.

    (ii) Repetitions are not allowed.

**5.5** How many three-digit numbers can be formed by using the numbers: 0, 1, 2, 3, 4, 5, 6, 7, 8, and 9, and satisfying one of the following requirements?

    (i) No restrictions are imposed.

    (ii) All three digits are distinct.

    (iii) All three-digit numbers start with 1 and end with 0.

       If the three-digit numbers are formed at random, calculate the probability that such a number will be:

    (iv) As described in (ii).

    (v) As described in (iii).

**5.6** On a straight line, there are $n$ spots to be filled in by either a dot or a dash. What is the number of the distinct groups of resulting symbols? What is this number if $n = 5, 10, 15, 20$, and 25?

**5.7** A child's set of blocks consists of two red, four blue, and five yellow blocks. The blocks can be distinguished only by color. If the child lines up the blocks in a row at random, calculate the following probabilities:

    (i) Red blocks appear at both ends.

    (ii) All yellow blocks are adjacent.

    (iii) Blue blocks appear at both ends.

**Hint.** Use Proposition 1.

**5.8** Suppose that the letters $C, E, F, F, I$, and $O$ are written on six chips and placed into a box. Then the six chips are mixed and drawn one by one without replacement. What is the probability that the word "OFFICE" is formed?

**Hint.** Use Proposition 1.

**5.9** For any integers $m$ and $n$ with $0 \leq m \leq n$, show that $\binom{n}{m} = \binom{n}{n-m}$ either by calculation or by using a suitable argument without writing out anything.

**5.10** Show that $\binom{n+1}{m+1} / \binom{n}{m} = \frac{n+1}{m+1}$.
**Hint.** Write out each expression in terms of factorials.

**5.11** If $M$ and $m$ are positive integers with $m \leq M$, show that:

$$\binom{M}{m} = \binom{M-1}{m} + \binom{M-1}{m-1},$$

by recalling that $\binom{k}{x} = 0$ for $x > k$.
**Hint.** As in Exercise 5.10, starting with the right-hand side.

**5.12** Without any calculations and by recalling that $\binom{k}{x} = 0$ for $x > k$, show that:

$$\sum_{x=0}^{r} \binom{m}{x} \binom{n}{r-x} = \binom{m+n}{r}.$$

**5.13** The Binomial expansion formula states that, for any $x$ and $y$ real and $n$ a positive integer:

$$(x+y)^n = \sum_{k=0}^{n} \binom{n}{k} x^k y^{n-k}.$$

  **(i)** Justify this formula by using relation (5).
  **(ii)** Use this formula in order to show that:

$$\sum_{k=0}^{n} \binom{n}{k} = 2^n \quad \text{and} \quad \sum_{k=0}^{n} (-1)^k \binom{n}{k} = 0.$$

**5.14** In the plane, there are $n$ points such that no three of them lie on a straight line. How many triangles can be formed? What is this number for $n = 10$?

**5.15** Beethoven wrote 9 symphonies, Mozart wrote 27 piano concertos, and Schubert wrote 15 string quartets.
  **(i)** If a university radio station announcer wishes to play first a Beethoven symphony, then a Mozart concerto, and then a Schubert string quartet, in how many ways can this be done?
  **(ii)** What is the number in part (i) if all possible orderings are considered?

**5.16** A course in English composition is taken by 10 freshmen, 15 sophomores, 30 juniors, and 5 seniors. If 10 students are chosen at random, calculate the probability that this group will consist of 2 freshman, 3 sophomores, 4 juniors, and 1 senior.

**5.17** If $n$ countries exchange ambassadors, how many ambassadors are involved? What is this number for $n = 10, 50, 100$?

**5.18** From among $n$ eligible draftees, $m$ are to be drafted in such a way that all possible combinations are equally likely to occur. What is the probability that a specified man is not drafted?

**5.19** From 10 positive and 6 negative numbers, 3 numbers are chosen at random and without repetitions. What is the probability that their product is a negative number?

**5.20** Two people toss independently $n$ times each a coin whose probability of falling heads is $p$. What is the probability that they have the same number of heads? What does this probability become for $p = \frac{1}{2}$ and any $n$? Also, for $p = \frac{1}{2}$ and $n = 5$?

**5.21** A shipment of 2000 light bulbs contains 200 defective items and 1800 good items. Five hundred bulbs are chosen at random and are tested, and the entire shipment is rejected if more than 25 bulbs from among those tested are found to be defective. What is the probability that the shipment will be accepted?

**5.22** A student is given a test consisting of 30 questions. For each question, 5 different answers (of which only one is correct) are supplied. The student is required to answer correctly at least 25 questions in order to pass the test. If he/she knows the right answers to the first 20 questions and chooses an answer to the remaining questions at random and independently of each other, what is the probability that the student will pass the test?

**5.23** Three cards are drawn at random and without replacement from a standard deck of 52 playing cards. Compute the probabilities $P(A_i), i = 1, \ldots, 4$, where the events $A_i, i = 1, \ldots, 4$ are defined as follows:
$A_1 = $ "all three cards are black," $A_2 = $ "exactly one card is an ace,"
$A_3 = $ "one card is a diamond, one card is a heart, and one card is a club."
$A_4 = $ "at least two cards are red."

**5.24** From an urn containing $n_R$ red balls, $n_B$ black balls, and $n_W$ white balls (all identical except for color), three balls are drawn at random. Calculate the following probabilities:
  **(i)** All three balls are red.
  **(ii)** It least one ball is red.
  **(iii)** One ball is red, one is black, and one is white.
    Do this when the balls are drawn:
    **(a)** Successively and with replacement.
    **(b)** Without replacement.

**5.25** A student committee of 12 people is to be formed from among 100 freshmen (40 male + 60 female), 80 sophomores (30 male and 50 female), 70 juniors (24 male and 46 female), and 40 seniors (12 male and 28 female). Calculate the following probabilities:

(i)   Seven students are female and five are male.

(ii)   The committee consists of the same number of students from each class.

(iii)   The committee consists of two female students and one male student from each class.

(iv)   The committee includes at least 1 senior (one of whom will serve as the chairperson of the committee).

The following tabular form of the data facilitates the calculations:

| Class\Gender | Male | Female | Totals |
|---|---|---|---|
| Freshman | 40 | 60 | 100 |
| Sophomore | 30 | 50 | 80 |
| Junior | 24 | 46 | 70 |
| Senior | 12 | 28 | 40 |
| Totals | 106 | 184 | 290 |

**5.26**   Of six marbles, one is red, one is blue, two are yellow, one is black, and one is green. In how many distinct ways can they be arranged next to each other?
**Hint.** Use Proposition 1.

**5.27**   From 10 positive and 6 negative numbers, 3 numbers are chosen at random and without repetition. What is the probability that their product is a negative number?

**5.28**   From six identical balls, numbered from 1 through 6, three balls are drawn at random. What is the probability that the largest number drawn is five, when:

(i)   The balls are drawn without regard to the order.

(ii)   The order does matter.

**5.29**   A meal is prepared by using meat, a vegetable, and a starch. If there are available four kinds of meat, six kinds of vegetables, and three kinds of starches, how many different meals can be prepared?

**5.30**   At the beginning of a school year, the school district superintendent is to assign six new teachers to three schools. How many assignments are possible, if:

(i)   Each school is to be given two teachers?

(ii)   Schools $S_1$ and $S_2$, say, are to be given three and two teachers, respectively?

**5.31**   Consider five line segments of length 1, 3, 5, 7, and 9, and choose three of them at random. What is the probability that a triangle can be formed by using these three chosen line segments?
**Hint.** In any triangle, each side is smaller than the sum of the other two.

**5.32**   A committee consists of $2n + 1$ people, one of whom (a specific one) is to chair the committee. In how many different ways can these $2n + 1$ people by seated, if the chairperson is to be in the middle?

**5.33** A container contains six balls numbered 1 through 6. Three balls are drawn at random and without replacement. What is the probability that the largest number drawn is 5?

**5.34** (i) One fair die is rolled independently 6 times. Compute the probability of obtaining at least one 6.
   (ii) Two fair dice are rolled independently 24 times. Compute the probability of obtaining at least one double 6; i.e., (6, 6).
   (iii) Compare the probabilities in parts (i) and (ii).

**5.35** An elevator in a building makes three stops, and each one of three passengers picks an exit at random and independently of any one else. What is the probability that the three passengers:
   (i) Take the same exit?
   (ii) Take three different exists?

**5.36** For any integers $n$ and $k$ with $1 \leq k \leq n$, show that

$$\binom{n}{k} = \binom{n-1}{k} + \binom{n-1}{k-1}$$

by using:
   (i) A combinatorial argument.
   (ii) An algebraic argument.

**5.37** A student committee consist of eight students, two from each class (Fr., So., Jr., Sr.). A subcommittee of 4 is to be formed. How many such subcommittees can be formed:
   (i) Without any restrictions?
   (ii) All classes are to be represented?
   (iii) Only two classes are represented?
   (iv) If the subcommittee members are chosen at random, what are the probabilities that the situations in part (ii) and (iii) will occur?

**5.38** For any integers $n$ and $k$ with $1 \leq k \leq n$, show that:

$$\binom{n}{k} = \binom{k-1}{k-1} + \binom{k}{k-1} + \cdots + \binom{n-1}{k-1} = \sum_{i=k}^{n} \binom{i-1}{k-1}.$$

**Hint.** Use Exercise 5.36.

**5.39** An urn contains $r$ red balls and $w$ white balls. The $r + w$ balls are all drawn at random one at a time and without replacement. Show that the probability that all $r$ red balls are drawn before all $w$ white balls are drawn is equal to $\frac{w}{r+w}$.

**5.40** An urn contains $2n$ balls of which $n$ are red and $n$ are white, and suppose that the red and the white balls are numbered from 1 through $n$. Next, for an integer $k$ with $1 \leq k \leq n$, $k$ balls are drawn at random and without replacement. When a red ball and a white ball bearing the same number are shown, we say that a "match occurred."

Find expressions for the following probabilities:

  **(i)** No match occurs.

  **(ii)** At least an match occurs.

  **(iii)** Compute the numerical values in part (i) and (ii) for: $2n = 20$ and $k = 3$?

**5.41** A sociologist draws a random sample of 6 individuals from a group consisting of 8 males and 12 females. How many samples are there:

  **(i)** Altogether?

  **(ii)** Consisting only of males?

  **(iii)** Consisting of two males and four females?

  **(iv)** Consisting of persons of the same gender?

  **(v)** What is the probability that the sample consists of two males and four females?

**5.42** Three (fair six-sided) dice, colored red, blue, and yellow, are rolled once. Denote by $R$, $B$, and $Y$ the numbers appearing on the upper side of the three dice, respectively, and compute the following probabilities:

  **(i)** $P(R \neq B)$.

  **(ii)** $P(R < B)$ by using part (i) and symmetry.

  **(iii)** $P(R \neq B \neq Y)$.

  **(iv)** $P(R < B < Y)$ by using part (iii) and symmetry.

  **(v)** $P(R < B < Y | R \neq B)$.

# Numerical characteristics of a random variable, some special random variables

In this chapter, we discuss the following material. In Section 3.1, the concepts of expectation and variance of a r.v. are introduced and interpretations are provided. Higher order moments are also defined and their significance is pointed out. Also, the moment generating function of a r.v. is defined and its usefulness as a mathematical tool is commented upon. In Section 3.2, the Markov and Tchebichev inequalities are introduced and their role in estimating probabilities is explained. Section 3.3 is devoted to discussing some of the most commonly occurring distributions: They are the Binomial, Geometric, Poisson, Hypergeometric, Gamma (Negative Exponential and Chi-square), Normal, and Uniform distributions. In all cases, the mathematical expectation, variance, and the moment generating function involved are presented. The chapter is concluded with a discussion of the concepts of median and mode, which are illustrated by concrete examples.

## 3.1 EXPECTATION, VARIANCE, AND MOMENT GENERATING FUNCTION OF A RANDOM VARIABLE

The ideal situation in life would be to know with certainty what is going to happen next. This being almost never the case, the element of chance enters in all aspects of our life. A r.v. is a mathematical formulation of a random environment. Given that we have to deal with a r.v. $X$, the best thing to expect is to know the values of $X$ and the probabilities with which these values are taken on, for the case that $X$ is discrete, or the probabilities with which $X$ takes values in various subsets of the real line $\Re$ when $X$ is of the continuous type. That is, we would like to know the *probability distribution function* of $X$. In real life, often, even this is not feasible. Instead, we are forced to settle for some numerical characteristics of the distribution of $X$. This line of arguments leads us to the concepts of the mathematical expectation and variance of a r.v., as well as to moments of higher order.

**Definition 1.** Let $X$ be a (discrete) r.v. taking on the values $x_i$ with corresponding probabilities $f(x_i), i = 1, \ldots, n$. Then the *mathematical expectation* of $X$ (or just *expectation* or *mean value* of $X$ or just *mean* of $X$) is denoted by $EX$ and is defined by:

$$EX = \sum_{i=1}^{n} x_i f(x_i). \tag{1}$$

If the r.v. $X$ takes on (countably) infinite many values $x_i$ with corresponding probabilities $f(x_i)$, $i = 1, 2, \ldots$, then the expectation of $X$ is defined by:

$$EX = \sum_{i=1}^{\infty} x_i f(x_i), \quad \text{provided} \sum_{i=1}^{\infty} |x_i| f(x_i) < \infty. \tag{2}$$

Finally, if the r.v. $X$ is continuous with p.d.f. $f$, its expectation is defined by:

$$EX = \int_{-\infty}^{\infty} x f(x) dx, \quad \text{provided this integral exists.} \tag{3}$$

The alternative notations $\mu(X)$ or $\mu_X$ are also often used.

**Remark 1.**

(i) The condition $\sum_{i=1}^{\infty} |x_i| f(x_i) < \infty$ is needed because, if it is violated, it is known
that $\sum_{i=1}^{\infty} x_i f(x_i)$ may take on different values, depending on the order in which
the terms involved are summed up. This, of course, would render the definition
of $EX$ meaningless.

(ii) An example will be presented later on (see Exercise 1.16) where the integral
$\int_{-\infty}^{\infty} x f(x) dx = \infty - \infty$, so that it does not exist.

The expectation has several interpretations, some of which are illustrated by the
following Examples 1 and 2. One basic interpretation, however, is that of center of
gravity. Namely, if one considers the material system where mass $f(x_i)$ is placed at
the point $x_i$, $i = 1, \ldots, n$, then $EX$ is the *center of gravity* (point of equilibrium) of this
system. In this sense, $EX$ is referred to as a *measure of location of the center* of the
distribution of $X$. The same interpretation holds when $X$ takes on (countably) infinite
many values or is of the continuous type.

**Example 1.**    Suppose an insurance company pays the amount of $1000 for lost
luggage on an airplane trip. From past experience, it is known that the company pays
this amount in 1 out of 200 policies it sells. What premium should the company
charge?

**Discussion.**    Define the r.v. $X$ as follows: $X = 0$ if no loss occurs, which happens with probability $1 - (1/200) = 0.995$, and $X = -1000$ with probability $\frac{1}{200} =
0.005$. Then the expected loss to the company is: $EX = -1000 \times 0.005 = -5$. Thus,
the company must charge $5 to break even. To this, it will Normally add a reasonable
amount for administrative expenses and a profit.

Or, if $P$ is the premium required for the insurance company to break even, then the r.v. $X$ takes on the value $-1000 + P$ with probability 0.005, and the value $P$ with probability 0.995. Then $EX = (-1000 + P) \times 0.005 + P \times 0.995 = 0$ gives $P = \$5$.

Even in this simple example, but most certainly so in more complicated cases, it is convenient to present the values of a (discrete) r.v. and the corresponding probabilities in a tabular form as follows:

| $x$ | 0 | $-1000$ | Total |
|------|---|---------|-------|
| $f(x)$ | $\frac{199}{200}$ | $\frac{1}{200}$ | 1 |

**Example 2.**   A roulette wheel consists of 18 black slots, 18 red slots, and 2 green slots. If a gambler bets $10 on red, what is the gambler's expected gain or loss?

**Discussion.**   Define the r.v. $X$ by: $X = 10$ with probability 18/38 and $X = -10$ with probability 20/38, or in a tabular form

| $x$ | 10 | $-10$ | Total |
|------|----|-------|-------|
| $f(x)$ | $\frac{18}{38}$ | $\frac{20}{38}$ | 1 |

Then $EX = 10 \times \frac{18}{38} - 10 \times \frac{20}{38} = -\frac{10}{19} \simeq -0.526$. Thus, the gambler's expected loss is about 53 cents.

From the definition of the expectation and familiar properties of summations or integrals, it follows that:

$$E(cX) = cEX, \quad E(cX + d) = cEX + d, \quad \text{where } c \text{ and } d \text{ are constants.} \tag{4}$$

Also (see Exercise 1.18),

$$X \geq c \text{ constant, implies } EX \geq c, \text{ and, in particlar, } X \geq 0 \text{ implies } EX \geq 0. \tag{5}$$

Now if $Y$ is a r.v. which is a function of $X$, $Y = g(X)$, then, in principle, one may be able to determine the p.d.f. of $Y$ and proceed to defining its expectation by the appropriate version of formulas (1), (2), (3). It can be shown, however, that this is not necessary. Instead, the expectation of $Y$ is defined by using the p.d.f. of $X$; namely,

$$EY = \sum_{i=1}^{n} g(x_i)f(x_i) \quad \text{or} \quad EY = \sum_{i=1}^{\infty} g(x_i)f(x_i) \quad \text{or} \quad EY = \int_{-\infty}^{\infty} g(x)f(x)\, dx, \tag{6}$$

under provisions similar to the ones mentioned in connection with (2) and (3). By taking $g(x) = x^k$, where $k$ is a positive integer, we obtain the $k$th *moment* of $X$:

$$EX^k = \sum_{i=1}^{n} x_i^k f(x_i) \quad \text{or} \quad EX^k = \sum_{i=1}^{\infty} x_i^k f(x_i) \quad \text{or} \quad EX^k = \int_{-\infty}^{\infty} x^k f(x)\, dx. \tag{7}$$

For $k = 1$, we revert to the expectation of $X$, and for $k = 2$, we get its *second moment*. Moments are important, among other things, in that, in certain circumstances, a number of them completely determine the distribution of $X$. This will be illustrated by concrete cases in Section 3.3 (see also Remark 4).

The following simple example illustrates that the expectation, as a measure of location of the center of the distribution, may reveal very little about the entire distribution. Indeed, let the r.v. $X$ take on the values $-1$, $1$, and $2$ with corresponding probabilities $\frac{5}{8}$, $\frac{1}{8}$, and $\frac{2}{8}$, so that $EX = 0$. Also, let the r.v. $Y$ take on the values $-10$, $10$, and $20$ with respective probabilities $\frac{5}{8}$, $\frac{1}{8}$, and $\frac{2}{8}$; then again $EY = 0$. The distribution of $X$ is over an interval of length 3, whereas the distribution of $Y$ is over an interval of length 10 times as large. Yet, they have the same location of their center. This simple example, clearly, indicates that the expectation by itself is not an adequate measure of description of a distribution, and an additional measure is needed to be associated with the spread of a distribution. Such a measure exists and is the variance of a r.v. or of its distribution.

**Definition 2.**    The *variance* of a r.v. $X$ is denoted by $\mathrm{Var}(X)$ and is defined by:

$$\mathrm{Var}(X) = E(X - EX)^2. \tag{8}$$

The explicit expression of the right-hand side in (8) is taken from (6) for $g(x) = (x - EX)^2$. The alternative notations $\sigma^2(X)$ and $\sigma_X^2$ are also often used for the $\mathrm{Var}(X)$.

For the r.v.'s $X$ and $Y$ mentioned before, we have $\mathrm{Var}(X) = 1.75$ and $\mathrm{Var}(Y) = 175$. Thus, the variance does convey adequately the difference in size of the range of the distributions of the r.v.'s $X$ and $Y$. More generally, for a r.v. $X$ taking on finitely many values $x_1, \ldots, x_n$ with respective probabilities $f(x_1), \ldots, f(x_n)$, the variance is $\mathrm{Var}(X) = \sum_{i=1}^{n}(x_i - EX)^2 f(x_i)$ and represents the sum of the weighted squared distances of the points $x_i, i = 1, \ldots, n$, from the center of the distribution, $EX$. Thus, the further from $EX$ the $x_i$'s are located, the larger the variance, and vice versa. The same interpretation holds for the case that $X$ takes on (countably) infinite many values or is of the continuous type. Because of this characteristic property of the variance, the variance is referred to as a measure of *dispersion* of the underlying distribution. In mechanics, the variance is referred to as the moment of *inertia*.

The positive square root of the $\mathrm{Var}(X)$ is called the *standard deviation* (s.d.) of $X$. Unlike the variance, the s.d. is measured in the same units as $X$ (and $EX$) and serves as a yardstick of measuring deviations of $X$ from $EX$.

From (8), (6), and familiar properties of summations and integrals, one obtains

$$\mathrm{Var}(X) = EX^2 - (EX)^2. \tag{9}$$

This formula often facilitates the actual calculation of the variance. From (8), it also follows immediately that

$$\mathrm{Var}(cX) = c^2\,\mathrm{Var}(X), \quad \mathrm{Var}(cX + d) = c^2\,\mathrm{Var}(X),$$

$$\text{where } c \text{ and } d \text{ are constants.} \tag{10}$$

For a r.v. $Y$ which is a function of $X$, $Y = g(X)$, the calculation of the Var$[g(X)]$ reduces to calculating expectations as in (6) because, by means of (8) and (9):

$$\text{Var}[g(X)] = \text{Var}(Y) = E(Y - EY)^2 = EY^2 - (EY)^2 = Eg^2(X) - [Eg(X)]^2. \quad (11)$$

Formulas (8) and (9) are special cases of (11).

In reference to Examples 1 and 2, the variances and the s.d.'s of the r.v.'s involved are $\sigma^2(X) = 4975, \sigma(X) \simeq 70.534$, and $\sigma^2(X) = \frac{36,000}{361} \simeq 99.723$, $\sigma(X) \simeq 9.986$, respectively.

**Example 3.** Let $X$ be a r.v. with p.d.f. $f(x) = 3x^2$, $0 < x < 1$. Then:

**(i)** Calculate the quantities: $EX, EX^2$, and Var$(X)$.
**(ii)** If the r.v. $Y$ is defined by: $Y = 3X - 2$, calculate the $EY$ and the Var$(Y)$.

**Discussion.**

**(i)** By (3), $EX = \int_0^1 x \times 3x^2 \, dx = \frac{3}{4}x^4 \big|_0^1 = \frac{3}{4} = 0.75$, whereas by (7), applied with $k = 2, EX^2 = \int_0^1 x^2 \times 3x^2 \, dx = \frac{3}{5} = 0.60$, so that, by (9), Var$(X) = 0.60 - (0.75)^2 = 0.0375$.
**(ii)** By (4) and (6), $EY = E(3X - 2) = 3EX - 2 = 3 \times 0.75 - 2 = 0.25$, whereas by (10), Var$(Y) = \text{Var}(3X - 2) = 9 \text{ Var}(X) = 9 \times 0.0375 = 0.3375$.

In (6), the $EY$ was defined for $Y = g(X)$, some function of $X$. In particular, we may take $Y = e^{tX}$ for an arbitrary but fixed $t \in \Re$. Assuming that there exist $t$'s in $\Re$ for which $Ee^{tX}$ is finite, then this expectation defines a function in $t$. This function is denoted by $M(t)$ and is called the moment generating function of $X$. That is,

**Definition 3.** The function $M(t) = Ee^{tX}$, defined for all those $t$ in $\Re$ for which $Ee^{tX}$ is finite, is called the *moment generating function* (m.g.f.) of $X$.

Sometimes the notation $M_X(t)$ is also used to emphasize the fact that the m.g.f. under discussion is that of the r.v. $X$. The m.g.f. of any r.v. always exists for $t = 0$, since $Ee^{0X} = E1 = 1$; it may exist only for $t = 0$, or for $t$ in a proper subset (interval) in $\Re$, or for every $t$ in $\Re$. All these points will be demonstrated by concrete examples later on (see, e.g., relations (20) through (46)). The following properties of $M(t)$ follow immediately from its definition:

$$M_{cX}(t) = M_X(ct), \quad M_{cX+d}(t) = e^{dt} M_X(ct), \quad \text{where } c \text{ and } d \text{ are constants.}$$
$$(12)$$

Indeed,

$$M_{cX}(t) = Ee^{t(cX)} = Ee^{(ct)X} = M_X(ct),$$

and

$$M_{cX+d}(t) = Ee^{t(cX+d)} = E\big[e^{dt} \times e^{(ct)X}\big] = e^{dt} Ee^{(ct)X} = e^{dt} M_X(ct).$$

Under certain conditions, it is also true that:

$$\frac{\mathrm{d}}{\mathrm{d}t}M_X(t)\bigg|_{t=0} = EX \quad \text{and} \quad \frac{\mathrm{d}^n}{\mathrm{d}t^n}M_X(t)\bigg|_{t=0} = EX^n, \quad n = 2, 3, \ldots. \tag{13}$$

For example, for the first property, we have

$$\frac{\mathrm{d}}{\mathrm{d}t}M_X(t)\bigg|_{t=0} = \left(\frac{\mathrm{d}}{\mathrm{d}t}Ee^{tX}\right)\bigg|_{t=0} = E\left(\frac{\mathrm{d}}{\mathrm{d}t}e^{tX}\bigg|_{t=0}\right)$$
$$= E(Xe^{tX}|_{t=0}) = EX.$$

What is required for this derivation to be legitimate is that the order in which the operators $\frac{\mathrm{d}}{\mathrm{d}t}$ and $E$ operate on $e^{tX}$ can be interchanged. The justification of the property in (13) for $n \geq 2$ is quite similar. On account of property (13), $M_X(t)$ *generates* the moments of $X$ through differentiation and evaluation of the derivatives at $t = 0$. It is from this property that the m.g.f. derives its name.

The m.g.f. is also a valuable mathematical tool in many other cases, some of which be dealt with in subsequent chapters. Presently, it suffices only to state one fundamental property of the m.g.f. in the form of a proposition.

**Proposition 1.** *Under certain conditions, the m.g.f. $M_X$ of a r.v. X uniquely determines the distribution of X.*

This proposition is, actually, a rather deep probability result and it is referred to as the *inversion formula*.

Some forms of such a formula for characteristic functions, which are a version of a m.g.f. may be found, e.g., in pages 141-145 in *A Course in Mathematical Statistics*, 2nd edition (1997), Academic Press, by G. G. Roussas.

Still another important result associated with m.g.f.'s is stated (but not proved) in the following proposition.

**Proposition 2.** *If for the r.v. X all moments $EX^n, n = 1, 2, \ldots$, are finite, then, under certain conditions, these moments uniquely determine the m.g.f. $M_X$ of X, and hence (by Proposition 1) the distribution of X.*

Exercise 3.49 provides an example of an application of the proposition just stated.

For Examples 1 and 2, the m.g.f.'s of the r.v.'s involved are: $M_X(t) = 0.995 + 0.005e^{-1000t}, t \in \mathfrak{R}$, and $M_X(t) = \frac{1}{19}(9e^{10t} + 10e^{-10t}), t \in \mathfrak{R}$. Then, by differentiation, we get: $\frac{\mathrm{d}}{\mathrm{d}t}M_X(t)|_{t=0} = -5 = EX$, $\frac{\mathrm{d}^2}{\mathrm{d}t^2}M_X(t)|_{t=0} = 5000 = EX^2$, so that $\sigma^2(X) = 4975$; and $\frac{\mathrm{d}}{\mathrm{d}t}M_X(t)|_{t=0} = -\frac{10}{19} = EX$, $\frac{\mathrm{d}^2}{\mathrm{d}t^2}M_X(t)|_{t=0} = 100 = EX^2$, so that $\sigma^2(X) = \frac{36,000}{361} \simeq 99.723$.

**Example 4.** Let $X$ be a r.v. with p.d.f. $f(x) = e^{-x}, x > 0$. Then:

**(i)** Find the m.g.f. $M_X(t)$ for the $t$'s for which it is finite.

(ii) Using $M_X$, obtain the quantities: $EX, EX^2$, and Var$(X)$.

(iii) If the r.v. $Y$ is defined by: $Y = 2 - 3X$, determine $M_Y(t)$ for the $t$'s for which it is finite.

**Discussion.**

(i) By Definition 3,

$$M_X(t) = Ee^{tX} = \int_0^\infty e^{tx} \times e^{-x}\, dx = \int_0^\infty e^{-(1-t)x}\, dx$$

$$= -\frac{1}{1-t}e^{-(1-t)x}\Big|_0^\infty \quad \text{(provided } t \neq 1\text{)}$$

$$= -\frac{1}{1-t}(0 - 1) = \frac{1}{1-t} \quad \text{(provided } 1 - t > 0 \text{ or } t < 1\text{)}.$$

Thus, $M_X(t) = \frac{1}{1-t}$, $t < 1$.

(ii) By (13), $\frac{d}{dt}M_X(t)|_{t=0} = \frac{d}{dt}(\frac{1}{1-t})|_{t=0} = \frac{1}{(1-t)^2}|_{t=0} = 1 = EX$, $\frac{d^2}{dt^2}M_X(t)|_{t=0} = \frac{d}{dt}(\frac{1}{(1-t)^2})|_{t=0} = \frac{2}{(1-t)^3}|_{t=0} = 2 = EX^2$, so that, by (9), Var$(X) = 2 - 1^2 = 1$.

(iii) By (12), $M_Y(t) = M_{2-3X}(t) = M_{-3X+2}(t) = e^{2t}M_X(-3t) = e^{2t} \times \frac{1}{1-(-3t)} = \frac{e^{2t}}{1+3t}$, provided $t > -\frac{1}{3}$.

---

## EXERCISES

**Remark.** In several calculations required in solving some exercises in this section, the formulas #4 and #5 in Table 8 in the Appendix may prove useful.

**1.1**    Refer to Exercise 2.1 in Chapter 2 and calculate the quantities: $EX$, Var$(X)$, and the s.d. of $X$.

**1.2**    For the r.v. $X$ for which $P(X = -c) = P(X = c) = 1/2$ (for some $c > 0$):

(i) Calculate the $EX$ and the Var$(X)$.

(ii) Show that $P(|X - EX| \leq c) = \text{Var}(X)/c^2$.

**1.3**    A chemical company currently has in stock 100 lb of a certain chemical, which it sells to customers in 5 lb packages. Let $X$ be the r.v. denoting the number of packages ordered by a randomly chosen customer, and suppose that the p.d.f. of $X$ is given by: $f(1) = 0.2, f(2) = 0.4, f(3) = 0.3, f(4) = 0.1$.

| x | 1 | 2 | 3 | 4 |
|---|---|---|---|---|
| f(x) | 0.2 | 0.4 | 0.3 | 0.1 |

(i) Compute the following quantities: $EX, EX^2$, and Var$(X)$.

(ii) Compute the expected number of pounds left after the order of the customer in question has been shipped, as well as the s.d. of the number of pounds around the expected value.

**Hint.** For part (ii), observe that the leftover number of pounds is the r.v. $Y = 100 - 5X$.

**1.4** Let $X$ be a r.v. denoting the damage incurred (in \$) in a certain type of accident during a given year, and suppose that the distribution of $X$ is given by the following table:

| $x$ | 0 | 1000 | 5000 | 10,000 |
|-----|-----|------|------|--------|
| $f(x)$ | 0.8 | 0.1 | 0.08 | 0.02 |

A particular company offers a \$500 deductible policy. If the company's expected profit on a given policy is \$100, what premium amount should it charge?

**Hint.** If $Y = X - 500$ is the net loss to the insurance company, then: (Premium) $-EY = 100$.

**1.5** Let $X$ be the r.v. denoting the number in the uppermost side of a fair die when rolled once.
(i) Determine the m.g.f. of $X$.
(ii) Use the m.g.f. to calculate: $EX, EX^2, \text{Var}(X)$, and the s.d. of $X$.

**1.6** For any r.v. $X$, for which the $EX$ and the $EX^2$ are finite, show that:

$$\text{Var}(X) = EX^2 - (EX)^2 = E[X(X-1)] + EX - (EX)^2.$$

**1.7** Suppose that for a r.v. $X$ it is given that: $EX = 5$ and $E[X(X-1)] = 27.5$. Calculate:
(i) $EX^2$.
(ii) $\text{Var}(X)$ and s.d. of $X$.
**Hint.** Refer to Exercise 1.6.

**1.8** For the r.v. $X$ with p.d.f. $f(x) = (1/2)^x, x = 1, 2, \ldots$:
(i) Calculate the $EX$ and the $E[X(X-1)]$.
(ii) Use part (i) and Exercise 1.6 to compute the $\text{Var}(X)$.
**Hint.** See #5 in Table 8 in the Appendix.

**1.9** The p.d.f. $f$ of a r.v. $X$ is given by: $f(x) = c(1/3)^x$, for $x = 0, 1, \ldots$ ($c$ some positive constant).
(i) Calculate the $EX$.
(ii) Determine the m.g.f. $M_X$ of $X$ and specify the range of its argument.
(iii) Employ the m.g.f. in order to derive the $EX$.
**Hint.** For part (i), see #5 in Table 8 in the Appendix, and for part(ii), use #4 in Table 8 in the Appendix. Also, Exercise 2.10 in Chapter 2.

**1.10** For the r.v. $X$ with p.d.f. $f(x) = 0.5x$, for $0 \le x \le 2$, calculate: $EX$, $\text{Var}(X)$, and the s.d. of $X$.

**1.11** If the r.v. $X$ has p.d.f. $f(x) = 3x^2 - 2x + 1$, for $0 < x < 1$, compute the expectation and variance of $X$.

**1.12** If the r.v. $X$ has p.d.f. $f$ given by:

$$f(x) = \begin{cases} c_1 x, & -2 < x < 0, \\ c_2 x, & 0 \le x < 1, \\ 0, & \text{otherwise,} \end{cases}$$

and if we suppose that $EX = \frac{1}{3}$, determine the constants $c_1$ and $c_2$.
**Hint.** Two relations are needed for the determination of $c_1$ and $c_2$.

**1.13** The lifetime in hours of electric tubes is a r.v. $X$ with p.d.f. $f$ given by:
$f(x) = \lambda^2 x e^{-\lambda x}$, for $x > 0$ $(\lambda > 0)$. Calculate the expected life of such tubes.

**1.14** Let $X$ be a r.v. whose $EX = \mu \in \Re$. Then:
  **(i)** For any constant $c$, show that:

$$E(X - c)^2 = E(X - \mu)^2 + (\mu - c)^2 = \text{Var}(X) + (\mu - c)^2.$$

  **(ii)** Use part (i) to show that $E(X - c)^2$, as a function of $c$, is minimized for $c = \mu$.
  **Hint.** In part (i), add and subtract $\mu$.

**1.15** Let $X$ be a r.v. with p.d.f. $f(x) = \frac{|x|}{c^2}$, for $-c < x < c, c > 0$. Then:
  **(i)** Show that $f(x)$ is, indeed, a p.d.f.
  **(ii)** For any $n = 1, 2, \ldots$, calculate the $EX^n$, and as a special case, derive the $EX$ and the $\text{Var}(X)$.
  **Hint.** Split the integration from $-c$ to 0 and from 0 to $c$.

**1.16** Let $X$ be a r.v. with p.d.f. given by: $f(x) = \frac{1}{\pi} \times \frac{1}{1+x^2}, x \in \Re$. Show that:
  **(i)** $f$ is, indeed, a p.d.f. (called the *Cauchy* p.d.f.).
  **(ii)** $\int_{-\infty}^{\infty} x f(x)\, dx = \infty - \infty$, so that the $EX$ does *not* exist.
  **Hint.** In part (i), use the transformation $x = \tan u$.

**1.17** If $X$ is a r.v. for which all moments $EX^n, n = 0, 1, \ldots$ are finite, show that

$$M_X(t) = \sum_{n=0}^{\infty} (EX^n) \frac{t^n}{n!}.$$

**Hint.** Use the expansion $e^x = \sum_{n=0}^{\infty} \frac{x^n}{n!}$.
**Remark.** The result in this exercise says that the moments of a r.v. determine (under certain conditions) the m.g.f. of the r.v., and hence its distribution.

**1.18** Establish the inequalities stated in relation (5) for both the discrete and the continuous case.

**1.19** Establish relations (9), (10), and (11).

**1.20** The claim sizes of an auto insurance company are the values $x$ of a r.v. $X$ with respective probabilities given below:

| $x$ | $f(x)$ | $x$ | $f(x)$ |
|-----|--------|-----|--------|
| 10 | 0.13 | 60 | 0.09 |
| 20 | 0.12 | 70 | 0.11 |
| 30 | 0.10 | 80 | 0.08 |
| 40 | 0.13 | 90 | 0.08 |
| 50 | 0.10 | 100 | 0.06 |

   (i) What is the probability $P(10 \leq X \leq 60)$?
   (ii) Compute the $EX = \mu$ and the s.d. of $X$, $\sigma = \sqrt{\text{Var}(X)}$.
   (iii) What is the proportion of claims lying within $\sigma$, $1.5\sigma$ and $2\sigma$ from $\mu$?

**1.21** The claims submitted to an insurance company over a specified period of time $t$ is a r.v. $X$ with p.d.f. $f(x) = \frac{c}{(1+x)^4}, x > 0, (c > 0)$.
   (i) Determine the constant $c$.
   (ii) Compute the probability $P(1 \leq X \leq 4)$.
   (iii) Compute the expected number of claims over the period of time $t$.

**1.22** Starting today, someone buys two bananas every day until the store sells him at least one rotten banana when he stops. If the probability of getting at least one rotten banana each day is 0.1, and the events in successive days are independent, what is the expected number of bananas bought?

**1.23** From medical records, it follows that a rare disease will cause partial disability, complete disability, and no disability with respective probabilities 0.6, 0.3, and 0.1. It is known that only 1 in 10,000 will become afflicted with the disease in a given year.
   (i) For a person chosen at random from the target population, compute the probabilities that the person will become:
     (a) Partially disabled (event $D_p$)
     (b) Completely disabled (event $D_c$)
     (c) Not disabled (event $D_n$)
   **Hint.** It is assumed that a nondiseased person is not disabled.
   (ii) If a life insurance company pays $20,000 for partial disability, and $50,000 for total disability, what is the premium to be charged, so that the company breaks even?
   **Hint.** Introduce a suitable r.v. and compute its expectation.

**1.24** Let $X$ be a r.v. with p.d.f. $f$ and $EX$ given by: $f(x) = ax + b, 0 < x < 1$ (and 0 elsewhere); $EX = 0.6$. Determine $a$ and $b$.

**1.25** For the r.v.'s $X$ and its p.d.f. $f$ given below:

| $x$ | 0 | 3 | 6 |
|---|---|---|---|
| $f(x)$ | 1/2 | 1/3 | 1/6 |

   (i) Compute the $EX$, $Var(X)$, and s.d. of $X$.

  (ii) Calculate the probability $P(|X - EX| < 3)$.

 (iii) By using the Tchebichev inequality, compute a lower bound for the probability in part (ii) and compare it with the exact value found in (ii).

**1.26** Let $\sigma^2 = Var(X)$ and $Y = aX + b$, where $a$ and $b$ are constants.

   (i) Determine the standard deviation (s.d.) $\sigma_Y$ of the r.v.'s $Y$ in terms of $a$, $b$, and $\sigma$.

  (ii) What is the value of $\sigma_Y$ for $\sigma = 3$ and $a = -4$?

**1.27** A r.v. $X$ takes the values $-1$ and $1$ with probability $1/2$ each.

   (i) Compute $EX$, $Var(X)$, $\sigma(X)$.

  (ii) Find the m.g.f. of $X$.

 (iii) Use part (ii) to re-derive the quantities in part (i).

**1.28** For any two r.v.'s $X$ and $Y$ with finite expectations, it is always true that $X \geq Y$ implies that $EX \geq EY$. By a simple example, show that $X \geq Y$ (with $EX$, $EY$, $\sigma^2(X)$, and $\sigma^2(Y)$) need not imply that $\sigma^2(X) \geq \sigma^2(Y)$.

**1.29** Show that if $E|X|^r < \infty$ $(r > 0)$, then $E|X|^{r'} < \infty, 0 < r' < r$.

**1.30** For an event $A$, set $X = I_A$ and show that $EX = P(A)$, $\sigma^2(X) = P(A)P(A^c)$.

**1.31** Let the r.v. $X$ have p.d.f. $f(x) = \left(\frac{x}{c}\right)\left(\frac{c}{x}\right)^{\alpha+1}, x > c \, (> 0), \alpha > 0$. Then show that $EX = \frac{\alpha c}{\alpha - 1}$ for $\alpha > 1$, and $EX = \infty$ for $\alpha \leq 1$.

**1.32** If the r.v. $X$ takes the values $1, 2, \ldots, n$ with equal probabilities, show that

$$EX = \frac{n + 1}{2}, \quad \sigma^2(X) = \frac{(n + 1)(n - 1)}{12}.$$

**1.33** The r.v. $X$, denoting the lifetime of a TV tube, has the p.d.f. $f(x) = \lambda e^{-\lambda x}$, $x > 0$ $(\lambda > 0)$. The TV tube is sold with the following guaranty: If $X < c$ (some positive constant), then the buyer is reimbursed \$7, whereas if $X \geq c$, the company gains \$4 from the sale. Let $Y$ be the loss to the company manufacturing the TV tubes. Then:

   (i) What is the expected loss incurred?

  (ii) For what value of $c$ (expressed in terms of $\lambda$) the expected loss is 0?

**1.34** Let $h : \mathfrak{R} \to \mathfrak{R}$ with a $\frac{d^2}{dx^2}h(x) \geq 0$ for all $x \in \mathfrak{R}$. Then:

   (i) If $X$ is a r.v. with finite $EX$ and $Eh(X)$, show that $h(EX) \leq Eh(X)$.

**(ii)** Apply the result in part (i) for $h(x) = x^n$ (for a positive integer $n > 2$) to obtain $(EX)^n \leq EX^n$,

**(iii)** Apply the result in part (i) for $h(x) = \left(\frac{1}{x}\right)^n$ $(x \neq 0)$ to obtain

$$\left(\frac{1}{EX}\right)^n \leq E\left(\frac{1}{X}\right)^n.$$

**1.35** If for a r.v. $X$ the $E|X|^k < \infty$ $(k > 0)$, then show that $n^{k-\delta}P(|X| \geq n) \xrightarrow[n \to \infty]{} 0$ $(0 < \delta < k)$.

**1.36** For the r.v. $X$, assume that $|EX^3| < \infty$, and set $\mu = EX$, $\sigma^2 = \sigma^2(X)$, $\gamma_1 = E\left(\frac{X-\mu}{\sigma}\right)^3$. Then, show that: If the p.d.f. $f$ of $X$ is symmetric about $\mu$, then $\gamma_1 = 0$.

**Remark:** The number $\gamma_1$ is called *skewness* of the distribution. So, for symmetric distributions, $\gamma_1 = 0$. Also, $\gamma_1 \neq 0$ implies that the distribution is *not* symmetric. For $\gamma_1 > 0$, the distribution is skewed to the right, and for $\gamma_1 < 0$, the distribution is skewed to the left.

**1.37** Let $X$ be a r.v. taking on the value $j$ with probability $p_j = P(X = j)$, $j = 0, 1, \ldots$ Set

$$G(t) = \sum_{j=0}^{\infty} p_j t^j, \quad -1 \leq t \leq 1.$$

The function $G$ is called the *generating function of the sequence* $\{p_j\}, j \geq 0$.

**(i)** Show that if $|EX| < \infty$, then $EX = \frac{d}{dt}G(t)|_{t=1}$.

**(ii)** Also show that if $|E[X(X-1)\ldots(X-k+1)]| < \infty$, then

$$E[X(X-1)\ldots(X-k+1)] = \frac{d^k}{dt^k}G(t)\bigg|_{t=1}.$$

**1.38** Let $X \sim U(0, 8)$, so that $f(x) = 1/8$, $0 < x < 8$.

**(i)** Compute $EX$, $EX^2$, and $\sigma^2(X)$.

**(ii)** Compute the probability $P(|X - 4| > 3)$.

**(iii)** Use the Tchebichev inequality to find an upper bound for the probability in part (ii).

**1.39** In a game of cards, you win \$10 if you draw an ace, \$5 if you draw a face card, and \$0 if you draw any one of the remaining 36 cards. If the price of entering the game is \$2, compute your expected gain or loss.

**1.40** If the r.v. $X$ has p.d.f. $f(x) = 3x^2$, $0 < x < 1$, let $Y = 1 - X^2$, and do the following:

**(i)** Find the p.d.f. of $Y$, by using the d.f. approach.

**(ii)** For $0 < a < b < 1$, express the probability $P(a < Y < b)$ in terms of $a$ and $b$, and evaluate it for $a = 1/3$ and $b = 2/3$.

**(iii)** Compute the $EY$ and the $\sigma^2(Y)$.

**1.41**  Let $X$ be a r.v. (of the continuous type) with d.f. $F$ given by:

$$F(x) = \begin{cases} 0, & x \geq 0 \\ x^3 - x^2 + x, & 0 < x < 1 \\ 1, & x \geq 1. \end{cases}$$

    **(i)**  Show that the p.d.f. of $X$ is given by: $f(x) = 3x^2 - 2x + 1$ for $0 < x < 1$, and 0 otherwise.
    **(ii)**  Calculate the expectation and the variance of $X$, $EX$, and $\sigma^2(X)$.
    **(iii)**  Calculate the probability $P(X > 1/2)$.

**1.42**  If the r.v. $X$ has a p.d.f. $f(x) = \left(\frac{1}{2}\right)^x, x = 1, 2, \ldots$, show that:
    **(i)**  $f(x)$ is, indeed, a p.d.f.
    **(ii)**  $EX = 2, E[X(X - 1)] = 4$.
    **(iii)**  From part (ii), conclude that $\sigma^2(X) = 2$.

**1.43**  If for the r.v. $X$ it holds $P(X = c) = P(X = -c) = \frac{1}{2}$ for some positive constant $c$, then show that: $EX = 0, \sigma^2(X) = c^2, P(|X - EX| \leq c) = 1$, so that Tchebichev's inequality upper bound is attained.

**1.44**  Refer to Exercise 2.24 (later in this chapter) and show that:
    **(i)**  $EX = \alpha e^{\beta^2/2}$.
    **(ii)**  $EX^2 = \alpha^2 e^{2\beta^2}$.
    **(iii)**  $\sigma^2(X) = \alpha^2 e^{\beta^2}\left(e^{\beta^2} - 1\right)$.

**1.45**  For any r.v. $X$ with $EX^4 < \infty$ (and hence $EX, EX^2, EX^3$ all finite), show that:
    **(i)**  $EX^2 = E[X(X - 1)] + EX$.
    **(ii)**  $EX^3 = E[X(X - 1)(X - 2)] + 3EX^2 - 2EX$.
    **(iii)**  $EX^4 = E[X(X - 1)(X - 2)(X - 3)] + 6EX^3 - 11EX^2 + 6EX$.

**1.46**  Let $X$ be a r.v. having the Double Exponential distribution with parameter $\mu \in \Re$; i.e.,

$$f(x) = \frac{1}{2}e^{-|x-\mu|}, \quad x \in \Re.$$

Show that:
    **(i)**  $f(x)$ is, indeed, a p.d.f.
    **(ii)**  $EX = \mu$.
    **(iii)**  $EX^2 = \mu^2 + 2$, so that $\sigma^2(X) = 2$.

**1.47**  In reference to Exercise 1.46, show that:
    **(i)**  $EX^3 = 6\mu + \mu^3$.
    **(ii)**  $EX^4 = 24$.
    **(iii)**  $\gamma_2 = 9$ (so that the distribution is leptokurtic).

**1.48** If the r.v. $X$ takes on the values $1, 2, \ldots$ (with positive probability) and $EX < \infty$, show that:

  (i) $EX = \sum_{n=1}^{\infty} P(X \geq n)$.

  (ii) $EX \geq 2 - P(X = 1)$.

  (iii) $EX \geq 3 - 2P(X = 1) - P(X = 2)$.

**1.49** If a r.v. $X$ with d.f. $F$ has $E|X| < \infty$, then:

  (i) Show that $EX = \int_0^\infty [1 - F(x)]dx - \int_{-\infty}^0 F(x)dx$.

  (ii) Give a Geometric explanation as to what the above integrals represents.

**1.50** If the r.v. $X$ has a p.d.f. symmetric about a constant $c$ and finite expectation, then:

  (i) Show that $EX = c$.

  (ii) If $c = 0$ and the moments $EX^{2n+1}$, $n \geq 0$ integer, exist, then show that $EX^{2n+1} = 0$, $n \geq 0$, integer.

**1.51** Let the r.v.'s $X$, $Y$ have the joint p.d.f.:

$$f(x, y) = \frac{2}{n(n+1)}, \quad y = 1, \ldots, x, \ x = 1, \ldots, n.$$

It can be seen that:

$$f_X(x) = \frac{2x}{n(n+1)}, \quad x = 1, \ldots, n; \quad f_Y(y) = \frac{2(n - y + 1)}{n(n+1)}, \quad y = 1, \ldots, n.$$

Then show that:

  (i) $EX = \frac{2n+1}{3}$, $EY = \frac{n+2}{3}$; $EX^2 = \frac{n(n+1)}{2}$, $EY^2 = \frac{n^2+3n+2}{6}$.

  (ii) $\sigma^2(X) = \frac{n^2+n-2}{18}$, $\sigma^2(Y) = \frac{n^2+n-2}{18}$, by using part (i).

**1.52** If the r.v. $X$ has a p.d.f. given by: $f(x) = \frac{\sigma}{\pi} \times \frac{1}{\sigma^2+(x-\mu)^2}$, $x \in \Re$, $\mu \in \Re$, $\sigma > 0$ (i.e., Cauchy with parameters $\mu$ and $\sigma^2$), then show that $E|X| = \infty$.

**1.53** Let $X$ be a r.v. distributed as $P(\lambda)$, and set $E = \{0, 2, \ldots\}$, $O = \{1, 3, \ldots\}$.

  (i) Express the probabilities $P(X \in E)$, $P(X \in O)$ in terms of the parameter $\lambda$.

  (ii) Compute the probabilities in part (i) for $\lambda = 5$.

  **Hint.** Recall that $e^x = \sum_{k=0}^{\infty} \frac{x^k}{k!}$, $x \in \Re$.

## 3.2 SOME PROBABILITY INEQUALITIES

If the r.v. $X$ has a known p.d.f. $f$, then, in principle, we can calculate probabilities $P(X \in B)$ for $B \subseteq \Re$. This, however, is easier said than done in practice. What one might be willing to settle for would be some suitable and computable bounds for such probabilities. This line of thought leads us to the inequalities discussed here.

**Theorem 1.**

(i) *For any nonnegative r.v. X and for any constant c > 0, it holds*

$$P(X \geq c) \leq EX/c.$$

(ii) *More generally, for any nonnegative function of any r.v. X, g(X), and for any constant c > 0, it holds*

$$P[g(X) \geq c] \leq Eg(X)/c. \tag{14}$$

(iii) *By taking g(X) = |X − EX| in part (ii), the inequality reduces to the Markov inequality; namely,*

$$P(|X - EX| \geq c) = P(|X - EX|^r \geq c^r) \leq E|X - EX|^r/c^r, \quad r > 0. \tag{15}$$

(iv) *In particular, for r = 2 in (15), we get the Tchebichev inequality; namely,*

$$P(|X - EX| \geq c) \leq \frac{E(X - EX)^2}{c^2} = \frac{\sigma^2}{c^2} \quad \text{or}$$

$$P(|X - EX| < c) \geq 1 - \frac{\sigma^2}{c^2}, \tag{16}$$

*where $\sigma^2$ stands for the Var(X). Furthermore, if c = kσ, where σ is the s.d. of X, then:*

$$P(|X - EX| \geq k\sigma) \leq \frac{1}{k^2} \quad \text{or} \quad P(|X - EX| < k\sigma) \geq 1 - \frac{1}{k^2}. \tag{17}$$

**Remark 2.** From the last expression, it follows that $X$ lies within $k$ s.d.'s from its mean with probability at least $1 - \frac{1}{k^2}$, regardless of the distribution of $X$. It is in this sense that the s.d. is used as a yardstick of deviations of $X$ from its mean, as already mentioned elsewhere.

Thus, for example, for $k = 2, 3$, we obtain, respectively,

$$P(|X - EX| < 2\sigma) \geq 0.75, \qquad P(|X - EX| < 3\sigma) \geq \frac{8}{9} \simeq 0.889. \tag{18}$$

*Proof of Theorem 1.* Clearly, all one has to do is to justify (14) and this only for the case that $X$ is continuous with p.d.f. $f$, because the discrete case is entirely analogous.

Indeed, let $A = \{x \in \Re; g(x) \geq c\}$, so that $A^c = \{x \in \Re; g(x) < c\}$. Then, clearly:

$$Eg(X) = \int_{-\infty}^{\infty} g(x)f(x)\,dx = \int_A g(x)f(x)\,dx + \int_{A^c} g(x)f(x)\,dx$$

$$\geq \int_A g(x)f(x)\,dx \quad (\text{since } g(x) \geq 0)$$

$$\geq c\int_A f(x)\,dx \quad (\text{since } g(x) \geq c \text{ on } A)$$

$$= cP(A) = cP[g(X) \geq c].$$

Solving for $P[g(X) \geq c]$, we obtain the desired result. ∎

**Example 5.** Let the r.v. $X$ take on the values $-2$, $-1/2$, $1/2$, and $2$ with respective probabilities $0.05, 0.45, 0.45$, and $0.05$. Then $EX = 0$ and $\sigma^2 = \text{Var}(X) = 0.625$, so that $2\sigma \simeq 1.582$. Then: $P(|X| < 2\sigma) = P(-1.582 < X < 1.582) = P(X = -\frac{1}{2}) + P(X = \frac{1}{2}) = 0.90$, compared to the lower bound of $0.75$.

**Example 6.** Let the r.v. $X$ take on the value $x$ with probability $f(x) = e^{-\lambda}\frac{\lambda^x}{x!}$, $x = 0, 1, \ldots$, some $\lambda > 0$. As will be seen later on (see relation (23)), this is a Poisson-distributed r.v. with parameter $\lambda$, and $EX = \text{Var}(X) = \lambda$. For selected values of $\lambda$, probabilities $P(X \leq k)$ are given by the Poisson tables. For illustrative purposes, let $\lambda = 9$. Then $\sigma = 3$ and therefore: $P(|X - 9| < 2 \times 3) = P(3 < X < 15) = 0.9373$, compared to $0.75$, and $P(|X - 9| < 3 \times 3) = P(0 < X < 18) = 0.9946$, compared to $0.889$.

## EXERCISES

**2.1** Suppose the distribution of the r.v. $X$ is given by the following table:

| $x$ | $-1$ | $0$ | $1$ |
|-----|------|-----|-----|
| $f(x)$ | 1/18 | 8/9 | 1/18 |

(i) Calculate the $EX$ (call it $\mu$), the $\text{Var}(X)$, and the s.d. of $X$ (call it $\sigma$).

(ii) Compute the probability: $P(|X - \mu| \geq k\sigma)$ for $k = 2, 3$.

(iii) By the Tchebichev inequality: $P(|X - \mu| \geq k\sigma) \leq 1/k^2$. Compare the exact probabilities computed in part (ii) with the respective upper bounds.

**2.2** If $X$ is a r.v. with expectation $\mu$ and s.d. $\sigma$, use the Tchebichev inequality:

(i) To determine $c$ in terms of $\sigma$ and $\alpha$, so that:

$$P(|X - \mu| < c) \geq \alpha \qquad (0 < \alpha < 1).$$

(ii) Give the numerical value of $c$ for $\sigma = 1$ and $\alpha = 0.95$.

**2.3** Let $X$ be a r.v. with p.d.f. $f(x) = c(1 - x^2)$, for $-1 \leq x \leq 1$. Refer to Exercise 2.11(i) in Chapter 2 for the determination of the constant $c$ (actually, $c = 3/4$), and then:

(i) Calculate the $EX$ and $\text{Var}(X)$.

(ii) Use the Tchebichev inequality to find a lower bound for the probability $P(-0.9 < X < 0.9)$, and compare it with the exact probability calculated in Exercise 2.11(ii) in Chapter 2.

**2.4** Let $X$ be a r.v. with (finite) mean $\mu$ and variance $0$. Then:

(i) Use the Tchebichev inequality to show that $P(|X - \mu| \geq c) = 0$ for all $c > 0$.

(ii) Use part (i) and Theorem 2 in Chapter 2 in order to conclude that $P(X = \mu) = 1$.

## 3.3 SOME SPECIAL RANDOM VARIABLES

### 3.3.1 THE DISCRETE CASE

In this section, we discuss seven distributions — four of the discrete and three of the continuous type, which occur often. These are the Binomial, the Geometric, the Poisson, the Hypergeometric, the Gamma (which includes the Negative Exponential and the Chi-square), the Normal, and the Uniform distributions. At this point, it should be mentioned that a p.d.f. is 0 for all the values of its argument not figuring in its definition.

### *Binomial distribution*

We first introduced the concept of *a Binomial experiment*, which is meant to be an experiment resulting in two possible outcomes, one termed as a *success*, denoted by $S$ and occurring with probability $p$, and the other termed a *failure*, denoted by $F$ and occurring with probability $q = 1 - p$. A Binomial experiment is performed $n$ independent times (with $p$ remaining the same), and let $X$ be the r.v. denoting the number of successes. Then, clearly, $X$ takes on the values $0, 1, \ldots, n$, with the respective probabilities:

$$P(X = x) = f(x) = \binom{n}{x} p^x q^{n-x}, \qquad x = 0, 1, \ldots, n, \ 0 < p < 1, \ q = 1 - p. \quad (19)$$

The r.v. $X$ is said to be *Binomially* distributed, its distribution is called *Binomial* with *parameters* $n$ and $p$, and the fact that $X$ is so distributed is denoted by $X \sim B(n, p)$. The graph of $f$ depends on $n$ and $p$; two typical cases, for $n = 12, p = \frac{1}{4}$, and $n = 10, p = \frac{1}{2}$ are given in Figures 3.1 and 3.2.

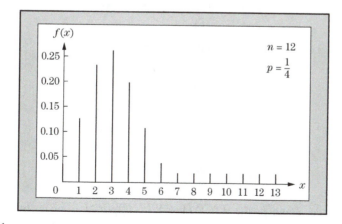

**FIGURE 3.1**

Graph of the p.d.f. of the Binomial distribution for $n = 12$, $p = \frac{1}{4}$.

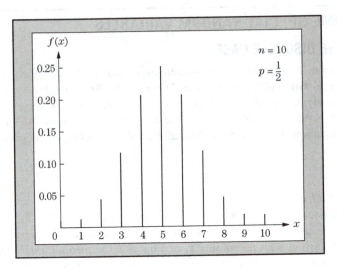

**FIGURE 3.2**

Graph of the p.d.f. of the Binomial distribution for $n = 10$, $p = \frac{1}{2}$.

Values of the p.d.f. $f$ of the $B(12, \frac{1}{4})$ distribution

$$f(0) = 0.0317, \qquad f(7) = 0.0115,$$
$$f(1) = 0.1267, \qquad f(8) = 0.0024,$$
$$f(2) = 0.2323, \qquad f(9) = 0.0004,$$
$$f(3) = 0.2581, \qquad f(10) = 0.0000,$$
$$f(4) = 0.1936, \qquad f(11) = 0.0000,$$
$$f(5) = 0.1032, \qquad f(12) = 0.0000,$$
$$f(6) = 0.0401.$$

Values of the p.d.f. $f$ of the $B(10, \frac{1}{2})$ distribution

$$f(0) = 0.0010, \qquad f(6) = 0.2051,$$
$$f(1) = 0.0097, \qquad f(7) = 0.1172,$$
$$f(2) = 0.0440, \qquad f(8) = 0.0440,$$
$$f(3) = 0.1172, \qquad f(9) = 0.0097,$$
$$f(4) = 0.2051, \qquad f(10) = 0.0010,$$
$$f(5) = 0.2460.$$

For selected $n$ and $p$, the d.f. $F(k) = \sum_{j=0}^{k} \binom{n}{j} p^j q^{n-j}$ is given by tables, the Binomial tables (see, however, Exercise 3.1). The individual probabilities $\binom{n}{j} p^j q^{n-j}$

may be found by subtraction. Alternatively, such probabilities can be calculated recursively (see Exercise 3.9).

For $n = 1$, the corresponding r.v. is known as the *Bernoulli*-distributed r.v. It is then clear that a $B(n, p)$ r.v. $X$ is the sum of $n$ $B(1, p)$ r.v.'s. More precisely, in $n$ independent Binomial experiments, associate with the $i$th performance of the experiment the r.v. $X_i$ defined by: $X_i = 1$ if the outcome is $S$ (a success) and $X_i = 0$ otherwise, $i = 1, \ldots, n$. Then, clearly, $\sum_{i=1}^{n} X_i$ is the number of 1's in the $n$ trials, or, equivalently, the number of $S$'s, which is exactly what the r.v. $X$ stands for. Thus, $X = \sum_{i=1}^{n} X_i$. Finally, it is mentioned here that, if $X \sim B(n, p)$, then:

$$EX = np, \quad \text{Var}(X) = npq, \quad \text{and} \quad M_X(t) = (pe^t + q)^n, \quad t \in \mathfrak{R}. \tag{20}$$

The relevant derivations are left as exercises (see Exercises 3.10 and 3.11). A brief justification of formula (19) is as follows: Think of the $n$ outcomes of the $n$ experiments as $n$ points on a straight line segment, where an $S$ or an $F$, is to be placed. By independence, the probability that there will be exactly $x$ $S$'s in $x$ *specified* positions (and therefore $n - x$ $F$'s in the remaining positions) is $p^x q^{n-x}$, and this probability is independent of the locations where the $x$ $S$'s occur. Because there are $\binom{n}{x}$ ways of selected $x$ points for the $S$'s, the conclusion follows.

Finally, for illustrative purposes, refer to Example 7 in Chapter 1. In that example, clearly $X \sim B(n, 0.75)$, and for the sake of specificity take $n = 25$, so that $X$ takes on the values $0, 1, \ldots, 25$. Next (see Exercise 3.1), $\binom{25}{x}(0.75)^x(0.25)^{25-x} = \binom{25}{y}(0.25)^y(0.75)^{25-y}$, where $y = 25 - x$. Therefore, for $a = 15$ and $b = 20$, for example, $P(15 \le X \le 25) = \sum_{y=5}^{10} \binom{25}{y}(0.25)^y(0.75)^{25-y} = 0.9703 - 0.2134 = 0.7569$.    Finally,    $EX = 25 \times 0.75 = 18.75, \text{Var}(X) = 25 \times 0.75 \times 0.25 = 4.6875$, so that $\sigma(X) \simeq 2.165$. Examples 8–10 in Chapter 1 fit into the same framework.

### Geometric distribution

This distribution arises in a Binomial experiment situation when trials are carried out independently (with constant probability $p$ of an $S$) until the *first* $S$ occurs. The r.v. $X$ denoting the number of required trials is a *Geometrically* distributed r.v. with *parameter $p$* and its distribution is the *Geometric* distribution with *parameter $p$*. It is clear that $X$ takes on the values $1, 2, \ldots$ with the respective probabilities:

$$P(X = x) = f(x) = pq^{x-1}, \quad x = 1, 2, \ldots, \quad 0 < p < 1, \quad q = 1 - p. \tag{21}$$

The justification of this formula is immediate because, if the first $S$ is to appear in the $x$th position, the overall outcome is $\underbrace{FF \ldots FS}_{x-1}$ whose probability (by independence) is $q^{x-1}p$.

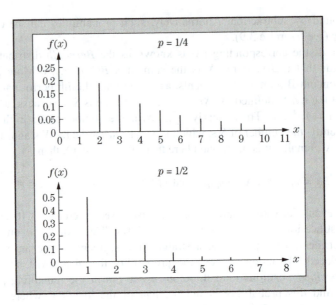

**FIGURE 3.3**

Graphs of the p.d.f.'s of the Geometric distribution with $p = \frac{1}{4}$ and $p = \frac{1}{2}$.

The graph of $f$ depends on $p$; two typical cases for $p = \frac{1}{4}$ and $p = \frac{1}{2}$ are given in Figure 3.3.

Values of $f(x) = (0.25)(0.75)^{x-1}$,    Values of $f(x) = (0.5)^{x}$,
$x = 1, 2, \ldots$    $x = 1, 2, \ldots$

$f(1) = 0.2500,$        $f(1) = 0.5000,$
$f(2) = 0.1875,$        $f(2) = 0.2500,$
$f(3) = 0.1406,$        $f(3) = 0.1250,$
$f(4) = 0.1055,$        $f(4) = 0.0625,$
$f(5) = 0.0791,$        $f(5) = 0.0313,$
$f(6) = 0.0593,$        $f(6) = 0.0156,$
$f(7) = 0.0445,$        $f(7) = 0.0078,$
$f(8) = 0.0334,$
$f(9) = 0.0250,$
$f(10) = 0.0188.$

If the r.v. $X$ is geometrically distributed with parameter $p$, then:

$$EX = \frac{1}{p}, \quad \text{Var}(X) = \frac{q}{p^2}, \quad M_X(t) = \frac{pe^t}{1 - qe^t}, \quad t < -\log q. \tag{22}$$

**Remark 3.**    Sometimes the p.d.f. of $X$ is given in the form: $f(x) = pq^x, x = 0, 1, \ldots$;
then $EX = \frac{q}{p}$, $\text{Var}(X) = \frac{q}{p^2}$ and $M_X(t) = \frac{p}{1 - qe^t}, t < -\log q$.

In reference to Example 11 in Chapter 1, assume for mathematical convenience that the number of cars passing by may be infinite. Then the r.v. $X$ described there has the Geometric distribution with some $p$. Here, probabilities are easily calculated. For example, $P(X \geq 20) = \sum_{x=20}^{\infty} pq^{x-1} = pq^{19} \sum_{x=0}^{\infty} q^x = pq^{19} \frac{1}{1-q} = q^{19}$; i.e., $p(X \geq 20) = q^{19}$. For instance, if $p = 0.01$, then $q = 0.99$ and $P(X \geq 20) \simeq 0.826$.

### Poisson Distribution

A r.v. $X$ taking on the values $0, 1, \ldots$ with respective probabilities given in (23) is said to have the *Poisson distribution* with *parameter* $\lambda$; its distribution is called the *Poisson distribution* with *parameter* $\lambda$. That $X$ is Poisson distributed with parameter $\lambda$ is denoted by $X \sim P(\lambda)$.

$$P(X = x) = f(x) = e^{-\lambda} \frac{\lambda^x}{x!}, \qquad x = 0, 1, \ldots, \lambda > 0. \tag{23}$$

The graph of $f$ depends on $\lambda$; for example, for $\lambda = 5$, it looks like that in Figure 3.4. That $f$ is a p.d.f. follows from the formula $\sum_{x=0}^{\infty} \frac{\lambda^x}{x!} = e^{\lambda}$.

Values of the p.d.f. $f$ of the $P(5)$ distribution

| | |
|---|---|
| $f(0) = 0.0067,$ | $f(9) = 0.0363,$ |
| $f(1) = 0.0337,$ | $f(10) = 0.0181,$ |
| $f(2) = 0.0843,$ | $f(11) = 0.0082,$ |
| $f(3) = 0.1403,$ | $f(12) = 0.0035,$ |
| $f(4) = 0.1755,$ | $f(13) = 0.0013,$ |
| $f(5) = 0.1755,$ | $f(14) = 0.0005,$ |
| $f(6) = 0.1462,$ | $f(15) = 0.0001,$ |
| $f(7) = 0.1044,$ | |
| $f(8) = 0.0653,$ | $f(n)$ is negligible for $n \geq 16.$ |

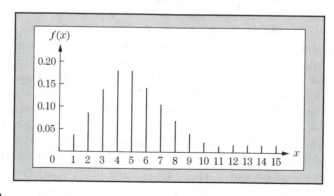

**FIGURE 3.4**

Graph of the p.d.f. of the Poisson distribution with $\lambda = 5$.

For selected values of $\lambda$, the d.f. $F(k) = \sum_{j=0}^{k} e^{-\lambda} \frac{\lambda^j}{j!}$ is given by tables, the Poisson tables. The individual values $e^{-\lambda} \frac{\lambda^j}{j!}$ are found by subtraction. Alternatively, such probabilities can be calculated recursively (see Exercise 3.20). It is not hard to see (see Exercises 3.21 and 3.22) that, if $X \sim P(\lambda)$, then:

$$EX = \lambda, \quad \text{Var}(X) = \lambda, \quad \text{and} \quad M_X(t) = e^{\lambda e^t - \lambda}, \qquad t \in \mathfrak{R}. \tag{24}$$

From these expressions, the parameter $\lambda$ acquires a special meaning: It is both the mean and the variance of the r.v. $X$.

Example 12 in Chapter 1 may serve as an illustration of the usage of the Poisson distribution. Assuming, for mathematical convenience, that the number of bacteria may be infinite, then the Poisson distribution may be used to describe the actual distribution of bacteria (for a suitable value of $\lambda$) quite accurately. There is a host of similar cases for the description of which the Poisson distribution is appropriate. These include the number of telephone calls served by a certain telephone exchange center within a certain period of time, the number of particles emitted by a radioactive source within a certain period of time, the number of typographical errors in a book, etc.

There is an intimate relationship between the Poisson and the Binomial distributions: The former may be obtained as the limit of the latter, as explained in the following. Namely, it is seen (see Exercise 3.23) that in the Binomial, $B(n, p)$, situation, if $n$ is large and $p$ is small, then the Binomial probabilities $\binom{n}{x} p^x (1-p)^{n-x}$ are close to the Poisson probabilities $e^{-np} \frac{(np)^x}{x!}$. More precisely, $\binom{n}{x} p_n^x (1 - p_n)^{n-x} \to e^{-\lambda} \frac{\lambda^x}{x!}$, provided $n \to \infty$ and $p_n \to 0$ so that $np_n \to \lambda \in (0, \infty)$. Here, $p_n$ is the probability of a success in the $n$th trial. Thus, for large values of $n$, $\binom{n}{x} p_n^x (1 - p_n)^{n-x} \simeq e^{-\lambda} \frac{\lambda^x}{x!}$; or, upon replacing $\lambda$ by $np_n$, we obtain the approximation mentioned before. A rough explanation as to why Binomial probabilities are approximated by Poisson probabilities is given next. To this end, suppose an event $A$ occurs once in a small time interval $h$ with approximate probability proportional to $h$ and coefficient of proportionally $\lambda$; i.e., $A$ occurs once in $h$ with approximate probability $\lambda h$. It occurs two or more times with probability approximately 0, so that it occurs zero times with probability approximately $1 - \lambda h$. Finally, occurrences in nonoverlapping intervals of length $h$ are independent. Next, consider the unit interval $(0, 1]$ and divide it into a large number $n$ of nonoverlapping subintervals of equal length $h$: $(t_{i-1}, t_i], i = 1, \ldots, n, t_0 = 0, t_n = 1, h = \frac{1}{n}$. With the $i$th interval $(t_{i-1}, t_i]$, associate the r.v. $X_i$ defined by: $X_i = 1$ with approximate probability $\lambda h$ and 0 with approximate probability $1 - \lambda h$. Then the r.v. $X = \sum_{i=1}^{n} X_i$ denotes the number of occurrences of $A$ over the unit $(0, 1]$ interval with approximate probabilities $\binom{n}{x} (\lambda h)^x (1 - \lambda h)^{n-x}$. The exact probabilities are found by letting $n \to \infty$ (which implies $h \to 0$). Because here $p_n = \lambda h$ and $np_n = n\lambda h = n\lambda \frac{1}{n} = \lambda$, we have that $\binom{n}{x} (\lambda h)^x (1 - \lambda h)^{n-x} \to e^{-\lambda} \frac{\lambda^x}{x!}$, as $n \to \infty$ (by Exercise 3.23), so that the exact probabilities are $e^{-\lambda} \frac{\lambda^x}{x!}$. So, the exact probability that $A$ occurs $x$ times in $(0, 1]$ is the Poisson probability $e^{-\lambda} \frac{\lambda^x}{x!}$, and the approximate probability that $A$ occurs the same number of times is the Binomial probability $\binom{n}{x}(\lambda h)^x(1 - \lambda h)^{n-x}$; these two probabilities are close to each other for large $n$.

The following example sheds some light on the approximation just discussed.

**Example 7.** If $X$ is a r.v. distributed as $B(25, \frac{1}{16})$, we find from the Binomial tables that $P(X = 2) = 0.2836$. Next, considering a r.v. $Y$ distributed as $P(\lambda)$ with $\lambda = \frac{25}{16} = 1.5625$, we have that $P(Y = 2) = e^{-1.5625}\frac{(1.5625)^2}{2!} \simeq 0.2556$. Thus, the exact probability is underestimated by the amount 0.028. The error committed is of the order of 9.87%. Given the small value of $n = 25$, the approximate value is not bad at all.

### Hypergeometric distribution

This distribution occurs quite often and is suitable in describing situations of the following type: $m$ identical objects (e.g., balls) are thoroughly mixed with $n$ identical objects (which again can be thought of as being balls) but distinct from the $m$ objects. From these $m + n$ objects, $r$ are drawn *without replacement*, and let $X$ be the number among the $r$ which come from the $m$ objects. Then the r.v. $X$ takes on the values $0, 1, \ldots, \min(r, m)$ with respective probabilities given below. Actually, by defining $\binom{m}{x} = 0$ for $x > m$, we have:

$$P(X = x) = f(x) = \frac{\binom{m}{x}\binom{n}{r-x}}{\binom{m+n}{r}}, \qquad x = 0, \ldots, r; \qquad (25)$$

$m$ and $n$ may be referred to as the *parameters* of the distribution. By assuming that the selections of $r$ objects out of the $m + n$ are all equally likely, there are $\binom{m+n}{r}$ ways of selecting these $r$ objects, whereas there are $\binom{m}{x}$ ways of selecting $x$ out of the $m$ objects, and $\binom{n}{r-x}$ ways of selecting the remaining $r - x$ objects out of $n$ objects. Thus, the probability that $X = x$ is as given in the preceding formula. The simple justification that these probabilities actually sum to 1 follows from Exercise 5.12 in Chapter 2. For large values of any one of $m, n$, and $r$, actual calculation of the probabilities in (25) may be quite involved. A recursion formula (see Exercise 3.26) facilitates significantly these calculations. The calculation of the expectation and of the variance of $X$ is based on the same ideas as those used in Exercise 3.10 in calculating the $EX$ and $\text{Var}(X)$ when $X \sim B(n, p)$. We omit the details and give the relevant formulas; namely,

$$EX = \frac{mr}{m+n}, \qquad \text{Var}(X) = \frac{mnr(m+n-r)}{(m+n)^2(m+n-1)}.$$

Finally, by utilizing ideas and arguments similar to those employed in Exercise 3.23, it is shown that as $m$ and $n \to \infty$ so that $\frac{m}{m+n} \to p \in (0, 1)$, then $\binom{m}{x}\binom{n}{r-x}/\binom{m+n}{r}$ tends to $\binom{r}{x}p^x(1 - p)^{r-x}$. Thus, for large values of $m$ and $n$, the Hypergeometric probabilities $\binom{m}{x}\binom{n}{r-x}/\binom{m+n}{r}$ may be approximated by the simpler Binomial probabilities $\binom{r}{x}p_{m,n}^x(1 - p_{m,n})^{r-x}$, where $p_{m,n} = \frac{m}{m+n}$.

**Example 8.** As an application of formula (25) and the approximation discussed, take $m = 70, n = 90, r = 25$ and $x = 10$. Then:

$$f(10) = \frac{\binom{70}{10}\binom{90}{25-10}}{\binom{70+90}{25}} = \frac{\binom{70}{10}\binom{90}{15}}{\binom{160}{25}} \simeq 0.166,$$

after quite a few calculations. On the other hand, since $\frac{m}{m+n} = \frac{70}{160} = \frac{7}{16}$, the Binomial tables give for the $B(25, \frac{7}{16})$ distribution: $\binom{25}{10}(\frac{7}{16})^{10}(\frac{9}{16})^{15} = 0.15$. Therefore, the approximation overestimates the exact probability by the amount 0.016. The error committed is of the order of 10.7%.

### 3.3.2 THE CONTINUOUS CASE

#### *Gamma distribution*

For its introduction, a certain function, the so-called Gamma function, is to be defined first. It is shown that the integral $\int_0^\infty y^{\alpha-1}e^{-y}\,dy$ is finite for $\alpha > 0$ and is thus defining a function (in $\alpha$); namely,

$$\Gamma(\alpha) = \int_0^\infty y^{\alpha-1}e^{-y}\,dy, \qquad \alpha > 0. \tag{26}$$

This is the *Gamma function*. By means of the Gamma function, the *Gamma distribution* is defined as follows through its p.d.f.:

$$f(x) = \frac{1}{\Gamma(\alpha)\beta^\alpha}x^{\alpha-1}e^{-x/\beta}, \qquad x > 0, \ \alpha > 0, \ \beta > 0; \tag{27}$$

$\alpha$ and $\beta$ are the *parameters* of the distribution. That the function $f$ integrates to 1 is an immediate consequence of the definition of $\Gamma(\alpha)$. A r.v. $X$ taking on values in $\Re$ and having p.d.f. $f$, given in (27), is said to be *Gamma distributed* with *parameters* $\alpha$ and $\beta$; one may choose the notation $X \sim \Gamma(\alpha, \beta)$ to express this fact. The graph of $f$ depends on $\alpha$ and $\beta$ but is, of course, always concentrated on $(0, \infty)$. Typical cases for several values of the pair $(\alpha, \beta)$ are given in Figures 3.5 and 3.6.

The Gamma distribution is suitable for describing waiting times between successive occurrences of a random event, occurring according to a certain distribution, and is also used for describing survival times. In both cases, it provides great flexibility through its two parameters $\alpha$ and $\beta$. For specific values of the pair $(\alpha, \beta)$, we obtain the Negative Exponential and Chi-square distributions to be studied subsequently. By integration by parts, one may derive the following useful recursive relation for the Gamma function (see Exercise 3.27):

$$\Gamma(\alpha) = (\alpha - 1)\Gamma(\alpha - 1)(\alpha > 1). \tag{28}$$

In particular, if $\alpha$ is an integer, repeated applications of recursive relation (28) produce

$$\Gamma(\alpha) = (\alpha - 1)(\alpha - 2)\ldots\Gamma(1).$$

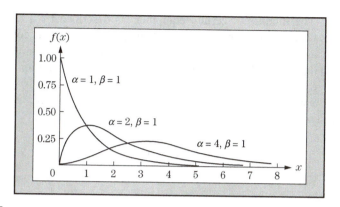

**FIGURE 3.5**

Graphs of the p.d.f. of the Gamma Distribution for several values of $\alpha, \beta$.

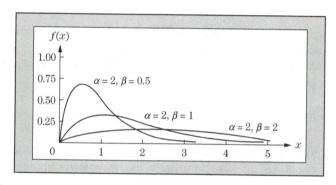

**FIGURE 3.6**

Graphs of the p.d.f. of the Gamma Distribution for several values of $\alpha, \beta$.

But $\Gamma(1) = \int_0^\infty e^{-y}\,dy = 1$, so that

$$\Gamma(\alpha) = (\alpha - 1)(\alpha - 2)\dots 1 = (\alpha - 1)! \tag{29}$$

For later reference, we mention here (see also Exercise 3.45) that, by integration, we obtain:

$$\Gamma\left(\frac{1}{2}\right) = \sqrt{\pi}, \tag{30}$$

and then, by means of this and the recursive formula (28), we can calculate $\Gamma(\frac{3}{2}), \Gamma(\frac{5}{2})$, etc. Finally, by integration (see Exercises 3.28 and 3.29), it is seen that:

$$EX = \alpha\beta, \quad \mathrm{Var}(X) = \alpha\beta^2, \quad \text{and} \quad M_X(t) = \frac{1}{(1 - \beta t)^\alpha}, \quad t < \frac{1}{\beta}. \tag{31}$$

**Example 9.** The lifetime of certain equipment is described by a r.v. $X$ whose distribution is Gamma with parameters $\alpha = 2$ and $\beta = \frac{1}{3}$, so that the corresponding p.d.f. is: $f(x) = 9xe^{-3x}$, for $x > 0$. Determine the expected lifetime, the variation around it, and the probability that the lifetime is at least 1 unit of time.

**Discussion.** Since $EX = \alpha\beta$ and $\text{Var}(X) = \alpha\beta^2$, we have here: $EX = \frac{2}{3}$ and $\text{Var}(X) = \frac{2}{9}$. Also,

$$P(X > 1) = \int_1^\infty 9xe^{-3x}\,dx = \frac{4}{e^3} \simeq 0.199.$$

### Negative Exponential distribution

In (27), set $\alpha = 1$ and $\beta = \frac{1}{\lambda}$ ($\lambda > 0$) to obtain:

$$f(x) = \lambda e^{-\lambda x}, \qquad x > 0, \ \lambda > 0. \tag{32}$$

This is the so-called *Negative Exponential* distribution with *parameter* $\lambda$. The graph of $f(x)$ depends on $\lambda$ but, typically, looks as in Figure 3.7.

For a r.v. $X$ having the *Negative Exponential* distribution with parameter $\lambda$, formulas (31) give

$$EX = \frac{1}{\lambda}, \quad \text{Var}(X) = \frac{1}{\lambda^2}, \quad \text{and} \quad M_X(t) = \frac{\lambda}{\lambda - t}, \qquad t < \lambda. \tag{33}$$

The expression $EX = \frac{1}{\lambda}$ provides special significance for the parameter $\lambda$: Its inverse value is the mean of $X$. This fact also suggests the *reparameterization* of $f$; namely, set $\frac{1}{\lambda} = \mu$, in which case:

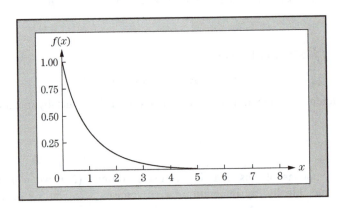

**FIGURE 3.7**

Graph of the Negative Exponential p.d.f. with $\lambda = 1$.

$$f(x) = \frac{1}{\mu}e^{-x/\mu}, \qquad x > 0, \quad EX = \mu, \quad \text{Var}(X) = \mu^2, \quad \text{and}$$

$$M_X(t) = \frac{1}{1 - \mu t}, \qquad t < \frac{1}{\mu}. \tag{34}$$

From (32), one finds by a simple integration:

$$F(x) = 1 - e^{-\lambda x}, \qquad x > 0, \quad \text{so that } P(X > x) = e^{-\lambda x}, \qquad x > 0. \tag{35}$$

The Negative Exponential distribution is used routinely as a survival distribution; namely, as describing the lifetime of an equipment, etc., put in service at what may be termed as time zero. As such, it exhibits a *lack of memory* property, which may not be desirable in this context. Namely, if one poses the following question: What is the probability that an equipment will last for $t$ additional units of time, given that it has already survived $s$ units of time, the answer (by means of the Negative Exponential distribution) is, by (35):

$$P(X > s + t \mid X > s) = \frac{P(X > s + t, X > s)}{P(X > s)} = \frac{P(X > s + t)}{P(X > s)} = \frac{e^{-\lambda(s+t)}}{e^{-\lambda s}}$$

$$= e^{-\lambda t} = P(X > t);$$

i.e., $P(X > s + t \mid X > s) = P(X > t)$ independent of $s$! Well, in real life, used pieces of equipment do not exactly behave as brand-new ones! Finally, it is to be mentioned that the Negative Exponential distribution is the waiting time distribution between the occurrence of any two successive events following the Poisson distribution (see also Exercise 3.20(ii) in Chapter 2).

**Example 10.**    The lifetime of an automobile battery is described by a r.v. $X$ having the Negative Exponential distribution with parameter $\lambda = \frac{1}{3}$. Then:

(i) Determine the expected lifetime of the battery and the variation around this mean.

(ii) Calculate the probability that the lifetime will be between 2 and 4 time units.

(iii) If the battery has lasted for 3 time units, what is the (conditional) probability that it will last for at least an additional time unit?

**Discussion.**

(i) Since $EX = \frac{1}{\lambda}$ and $\text{Var}(X) = \frac{1}{\lambda^2}$, we have here: $EX = 3$, $\text{Var}(X) = 9$, and s.d.$(X) = 3$.

(ii) Since, by (35), $F(x) = 1 - e^{-x/3}$ for $x > 0$, we have $P(2 < X < 4) = P(2 < X \leq 4) = P(X \leq 4) - P(X \leq 2) = F(4) - F(2) = (1 - e^{-4/3}) - (1 - e^{-\frac{2}{3}}) = e^{-2/3} - e^{-4/3} \simeq 0.252$.

(iii) The required probability is: $P(X > 4 \mid X > 3) = P(X > 1)$, by the memoryless property of this distribution, and $P(X > 1) = 1 - P(X \leq 1) = 1 - F(1) = e^{-1/3} \simeq 0.716$.

### Chi-square distribution

In formula (27), set $\alpha = \frac{r}{2}$ for a positive integer $r$ and $\beta = 2$ to obtain:

$$f(x) = \frac{1}{\Gamma\left(\frac{r}{2}\right)2^{r/2}}x^{(r/2)-1}e^{-x/2}, \qquad x > 0, \ r > 0 \text{ integer}. \tag{36}$$

The resulting distribution is known as the *Chi-square* distribution with $r$ *degrees of freedom* (d.f.). This distribution is used in certain statistical inference problems involving confidence intervals for variances and testing hypotheses about variances. The notation used for a r.v. $X$ having the Chi-square distribution with $r$ d.f. is $X \sim \chi_r^2$. For such a r.v., formulas (31) then become:

$$EX = r, \quad \text{Var}(X) = 2r \quad \text{(both easy to remember)}, \quad \text{and}$$

$$M_X(t) = \frac{1}{(1 - 2t)^{r/2}}, \qquad t < \frac{1}{2}. \tag{37}$$

The shape of the graph of $f$ depends on $r$, and, typically, looks like that in Figure 3.8.

Later on (see Corollary to Theorem 5 in Chapter 5), it will be seen why $r$ is referred to as the number of d.f. of the distribution.

### Normal distribution

This is by far the most important distribution, in both probability and statistics. The reason for this is twofold: First, many observations do follow to a very satisfactory degree a Normal distribution (see, for instance, Examples 13–17 in Chapter 1); and second, no matter what the underlying distribution of observations is, the sum of sufficiently many observations behaves pretty much as if it were Normally distributed, under very mild conditions. This second property is referred to as *Normal*

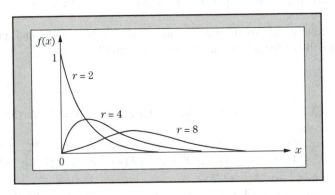

**FIGURE 3.8**

Graph of the p.d.f. of the Chi-Square distribution for several values of $r$.

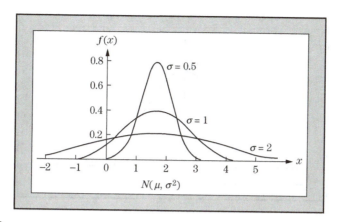

**FIGURE 3.9**

Graph of the p.d.f. of the Normal distribution with $\mu = 1.5$ and several values of $\sigma$.

*approximation* or as the *Central Limit Theorem* and will be revisited later on (see Section 7.2). The p.d.f. of a Normal distribution is given by:

$$f(x) = \frac{1}{\sqrt{2\pi}\sigma}e^{-(x-\mu)^2/2\sigma^2}, \qquad x \in \Re, \ \mu \in \Re, \ \sigma > 0; \tag{38}$$

$\mu$ and $\sigma^2$ (or $\sigma$) are referred to as the *parameters* of the distribution. The graph of $f$ depends on $\mu$ and $\sigma$; typical cases for $\mu = 1.5$ and various values of $\sigma$ are given in Figure 3.9.

No matter what $\mu$ and $\sigma$ are, the curve representing $f$ attains its maximum at $x = \mu$ and this maximum is equal to $1/\sqrt{2\pi}\sigma$, is symmetric around $\mu$ (i.e., $f(\mu - x) = f(\mu + x)$), and $f(x)$ tends to 0 as $x \to \infty$ or $x \to -\infty$. All these observations follow immediately from formula (38). That the function $f(x)$ integrates to 1 is seen through a technique involving a double integral and polar coordinates (see Exercise 3.44). For $\mu = 0$ and $\sigma = 1$, formula (38) is reduced to:

$$f(x) = \frac{1}{\sqrt{2\pi}}e^{-x^2/2}, \qquad x \in \Re, \tag{39}$$

and this is referred to as the *standard Normal* distribution (see Figure 3.10 for its graph).

The fact that a r.v. $X$ is *Normally distributed* with *parameters* $\mu$ and $\sigma^2$ (or $\sigma$) is conveniently denoted by: $X \sim N(\mu, \sigma^2)$. In particular, $X \sim N(0, 1)$ for $\mu = 0$, $\sigma = 1$. We often use the notation $Z$ for a $N(0, 1)$ distributed r.v.

The d.f. of the $N(0, 1)$-distribution is usually denoted by $\Phi$; i.e., if $Z \sim N(0, 1)$, then:

$$P(Z \le x) = \Phi(x) = \int_{-\infty}^{x} \frac{1}{\sqrt{2\pi}}e^{-t^2/2}\,dt, \qquad x \in \Re. \tag{40}$$

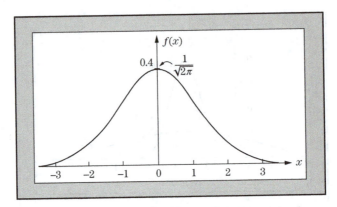

**FIGURE 3.10**

Graph of the p.d.f. of the standard Normal distribution.

Calculations of probabilities of the form $P(a < X < b)$ for $-\infty < a \le b < \infty$ are done through two steps: First, turn the r.v. $X \sim N(\mu, \sigma^2)$ into a $N(0, 1)$-distributed r.v., or, as we say, *standardize* it, and then use available tables, the Normal tables (see Proposition 3). The standardization is based on the following simple result.

**Proposition 3.**   *If* $X \sim N(\mu, \sigma^2)$, *then* $Z = \frac{X-\mu}{\sigma}$ *is* $\sim N(0, 1)$.

*Proof.* Indeed, for $y \in \Re$,

$$F_Z(y) = P(Z \le y) = P\left(\frac{X - \mu}{\sigma} \le y\right)$$

$$= P(X \le \mu + \sigma y)$$

$$= \int_{-\infty}^{\mu+\sigma y} \frac{1}{\sqrt{2\pi}\,\sigma} e^{-(t-\mu)^2/2\sigma^2}\, dt.$$

Set $\frac{t-\mu}{\sigma} = z$, so that $t = \mu + \sigma z$ with range from $-\infty$ to $y$, and $dt = \sigma\, dz$, to obtain:

$$F_Z(y) = \int_{-\infty}^{y} \frac{1}{\sqrt{2\pi}\,\sigma} e^{-z^2/2}\sigma\, dz$$

$$= \int_{-\infty}^{y} \frac{1}{\sqrt{2\pi}} e^{-z^2/2}\, dz, \quad \text{so that}$$

$$f_Z(y) = \frac{d}{dy} F_Z(y) = \frac{1}{\sqrt{2\pi}} e^{-y^2/2},$$

which is the p.d.f. of the $N(0, 1)$ distribution.  ∎

Thus, if $X \sim N(\mu, \sigma^2)$ and $a$ and $b$ are as above, then:

$$P(a < X < b) = P\left(\frac{a - \mu}{\sigma} < \frac{X - \mu}{\sigma} < \frac{b - \mu}{\sigma}\right) = P\left(\frac{a - \mu}{\sigma} < Z < \frac{b - \mu}{\sigma}\right)$$

$$= \Phi\left(\frac{b - \mu}{\sigma}\right) - \Phi\left(\frac{a - \mu}{\sigma}\right).$$

That is,

$$P(a < X < b) = \Phi\left(\frac{b - \mu}{\sigma}\right) - \Phi\left(\frac{a - \mu}{\sigma}\right). \tag{41}$$

Any other probabilities (involving intervals) can be found by way of probability (40) by exploiting the symmetry (around 0) of the $N(0, 1)$ curve.

Now, if $Z \sim N(0, 1)$, it is clear that $EZ^{2n+1} = 0$ for $n = 0, 1, \ldots$; by integration by parts, the following recursive relation is also easily established:

$$m_{2n} = (2n - 1)m_{2n-2}, \quad \text{where } m_k = \int_{-\infty}^{\infty} x^k \times \frac{1}{\sqrt{2\pi}} e^{-x^2/2} \, dx, \tag{42}$$

from which it follows that $EZ = 0$ and $EZ^2 = 1$, so that $\text{Var}(Z) = 1$. (For details, see Exercise 3.48.)

If $X \sim N(\mu, \sigma^2)$, then $Z = \frac{X - \mu}{\sigma} \sim N(0, 1)$, so that (by properties (9) and (10)):

$$0 = EZ = \frac{EX}{\sigma} - \frac{\mu}{\sigma}, \quad 1 = \text{Var}(Z) = \frac{1}{\sigma^2}\text{Var}(X), \quad \text{or} \quad EX = \mu \quad \text{and}$$

$$\text{Var}(X) = \sigma^2.$$

In other words:

$$\text{If } X \sim N(\mu, \sigma^2), \quad \text{then } EX = \mu \quad \text{and} \quad \text{Var}(X) = \sigma^2. \tag{43}$$

Thus, the parameters $\mu$ and $\sigma^2$ have specific interpretations: $\mu$ is the mean of $X$ and $\sigma^2$ is its variance (so that $\sigma$ is its s.d.).

If $Z \sim N(0, 1)$, it is seen from the Normal tables that:

$$P(-1 < Z < 1) = 0.68269, \quad P(-2 < Z < 2) = 0.95450,$$
$$P(-3 < Z < 3) = 0.99730,$$

so that almost all of the probability mass lies within three standard deviations from the mean. The same is true, by means of formula (41), applied with $a = \mu - k\sigma$ and $b = \mu + k\sigma$ with $k = 1, 2, 3$ in case $X \sim N(\mu, \sigma^2)$ (Figure 3.11). That is:

$$P(\mu - \sigma < X < \mu + \sigma) = 0.68269, \quad P(\mu - 2\sigma < X < \mu + 2\sigma) = 0.95450,$$
$$P(\mu - 3\sigma < X < \mu + 3\sigma) = 0.99730.$$

Finally, simple integration produces the m.g.f. of $X$ (see also Exercise 3.46); namely,

$$M_X(t) = e^{\mu t + \sigma^2 t^2/2}, \quad t \in \Re, \text{ for } X \sim N(\mu, \sigma^2), \tag{44}$$

$$M_Z(t) = e^{t^2/2}, \quad t \in \Re, \text{ for } Z \sim N(0, 1).$$

As will be seen in subsequent chapters, the Normal distribution is widely used in problems of statistical inference, involving point estimation, interval estimation, and testing hypotheses. Some instances where the Normal distribution is assumed as an

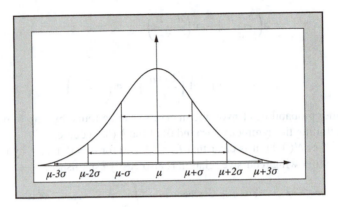

**FIGURE 3.11**

Probabilities within one, two, three s.d.'s $\sigma$ from the mean $\mu$ (approximately 0.68269, 0.95450, 0.99730, respectively).

appropriate (approximate) underlying distribution are described in Examples 13–17 in Chapter 1, as mentioned already.

**Example 11.** Suppose that numerical grades in a statistics class are values of a r.v. $X$ which is (approximately) Normally distributed with mean $\mu = 65$ and s.d. $\sigma = 15$. Furthermore, suppose that letter grades are assigned according to the following rule: The student receives an $A$ if $X \geq 85$; $B$ if $70 \leq X < 85$; $C$ if $55 \leq X < 70$; $D$ if $45 \leq X < 55$; and $F$ if $X \leq 45$.

   **(i)** If a student is chosen at random from that class, calculate the probability that the student will earn a given letter grade.
   **(ii)** Identify the expected proportions of letter grades to be assigned.

**Discussion.**

   **(i)** The student earns an $A$ with probability $P(X \geq 85) = 1 - P(X < 85) = 1 - P(\frac{X-\mu}{\sigma} < \frac{85-65}{15}) \simeq 1 - P(Z < 1.34) \simeq 1 - \Phi(1.34) = 1 - 0.909877 = 0.090123 \simeq 0.09$. Likewise, the student earns a $B$ with probability $P(70 \leq X < 85) = P(\frac{70-65}{15} \leq \frac{X-\mu}{\sigma} < \frac{85-65}{15}) \simeq P(0.34 \leq Z < 1.34) \simeq \Phi(1.34) - \Phi(0.34) = 0.909877 - 0.633072 = 0.276805 \simeq 0.277$. Similarly, the student earns a $C$ with probability $P(55 \leq X < 70) \simeq \Phi(0.34) + \Phi(0.67) - 1 = 0.381643 \simeq 0.382$. The student earns a $D$ with probability $P(45 \leq X < 55) \simeq \Phi(1.34) - \Phi(0.67) = 0.161306 \simeq 0.161$, and the student is assigned an F with probability $P(X < 45) \simeq \Phi(-1.34) = 1 - \Phi(1.34) = 0.09123 \simeq 0.091$.
   **(ii)** The respective expected proportions for $A$, $B$, $C$, $D$, and $F$ are: 9%, 28%, 38%, 16%, and 9%.

Indeed, suppose there are $n$ students, and let $X_A$ be the number of those whose numerical grades are $\geq 85$. By assuming that the $n$ events that the numerical grade of each one of the $n$ students is $\geq 85$ are independent, we have that $X_A \sim B(n, 0.09)$. Then, $\frac{X_A}{n}$ is the proportion of $A$ grades, and $E(\frac{X_A}{n}) = \frac{1}{n} \times n \times 0.09 = 0.09 = 9\%$ is the expected proportion of $A$'s. Likewise for the other grades.

This section is concluded with a simple distribution, the Uniform (or Rectangular) distribution.

### Uniform (or Rectangular) distribution

Such a distribution is restricted to finite intervals between the *parameters* $\alpha$ and $\beta$ with $-\infty < \alpha < \beta < \infty$, and its p.d.f. is given by:

$$f(x) = \frac{1}{\beta - \alpha}, \quad \alpha \leq x \leq \beta \quad (-\infty < \alpha < \beta < \infty). \tag{45}$$

Its graph is given in Figure 3.12, and it also justifies its name as rectangular.

The term "Uniform" is justified by the fact that intervals of equal length in $(\alpha, \beta)$ are assigned the same probability regardless of their location. The notation used for such a distribution is $U(\alpha, \beta)$ (or $R(\alpha, \beta)$), and the fact that the r.v. $X$ is distributed as such is denoted by $X \sim U(\alpha, \beta)$ (or $X \sim R(\alpha, \beta)$). Simple integrations give (see also Exercise 3.51):

$$EX = \frac{\alpha + \beta}{2}, \quad \text{Var}(X) = \frac{(\alpha - \beta)^2}{12}, \quad \text{and} \quad M_X(t) = \frac{e^{\beta t} - e^{\alpha t}}{(\beta - \alpha)t}, \quad t \in \Re. \tag{46}$$

**Example 12.**  A bus is supposed to arrive at a given bus stop at 10:00 a.m., but the actual time of arrival is a r.v. $X$ which is Uniformly distributed over the 16-min interval from 9:52 to 10:08. If a passenger arrives at the bus stop at exactly 9:50, what is the probability that the passenger will board the bus no later than 10 min from the time of his/her arrival?

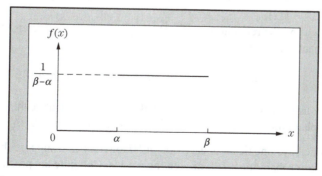

**FIGURE 3.12**

Graph of the p.d.f. of the $U(\alpha, \beta)$ distribution.

**Discussion.**    The p.d.f. of $X$ is $f(x) = 1/16$ for $x$ ranging between 9:52 and 10:08, and 0 otherwise. The passenger will board the bus no later than 10 min from the time of his/her arrival at the bus stop if the bus arrives at the bus stop between 9:52 and 10:00 (as the passenger will necessarily have to wait for 2 min, between 9:50 and 9:52). The probability for the bus to arrive between 9:52 and 10:00 is $8/16 = 0.5$. This is the required probability.

**Remark 4.**    It has been stated (see comments after relation (7)) that sometimes a number of moments of a r.v. $X$ completely determine the distribution of $X$. Actually, this has been the case in all seven distributions examined in this section. In the Binomial distribution, knowledge of the mean amounts to knowledge of $p$, and hence of $f$. The same is true in the Geometric distribution, as well as in the Poisson distribution. In the Hypergeometric distribution, knowledge of the first two moments, or equivalently, of the mean and variance of $X$ (see expressions for the expectation and variance), determine $m$ and $n$ and hence the distribution itself. The same is true of the Gamma distribution, as well as the Normal and Uniform distributions.

# EXERCISES

**3.1**    If $X \sim B(n, p)$ with $p > 0.5$, the Binomial tables (in this book) cannot be used, even if $n$ is suitable. This problem is resolved by the following result.
   **(i)**  If $X \sim B(n, p)$, show that $P(X = x) = P(Y = n - x)$, where $Y \sim$ $B(n, q)$ $(q = 1 - p)$.
   **(ii)**  Apply part (i) for $n = 20$, $p = 0.625$, and $x = 8$.

**3.2**    Let $X$ be a r.v. distributed as $B(n, p)$, and recall that
$P(X = x) = f(x) = \binom{n}{x} p^x q^{n-x}, x = 0, 1, \ldots, n$ $(q = 1 - p)$. Set
$B(n, p; x) = f(x)$.
   **(i)**  By using the relationship: $\binom{m+1}{y} = \binom{m}{y} + \binom{m}{y-1}$ (see Exercise 5.11 in Chapter 2), show that:

$$B(n + 1, p; x) = pB(n, p; x - 1) + qB(n, p; x).$$

   **(ii)**  By using this recursive relation of $B(n + 1, p; .)$, calculate the probabilities $B(n, p; x)$ for $n = 26$, $p = 0.25$, and $x = 10$.

**3.3**    Someone buys one ticket in each one of 50 lotteries, and suppose that each ticket has probability $1/100$ of winning a prize. Compute the probability that the person in question will win a prize:
   **(i)**  Exactly once.
   **(ii)**  At least once.

**3.4**  Suppose that 15 people, chosen at random from a target population, are asked
if they favor a certain proposal, and let $X$ be the r.v. denoting the number of
those who do favor the proposal. If 43.75% of the target population favor the
proposal, calculate the probability that:
  **(i)**  At least 5 of the 15 polled favor the proposal.
  **(ii)**  A majority of those polled favor the proposal.
  **(iii)**  Also, compute the $EX$, the $\mathrm{Var}(X)$, and the s.d. of $X$.

**3.5**  A fair die is tossed independently 18 times, and the appearance of a 6 is called
a success. Find the probability that:
  **(i)**  The number of successes is greater than the number of failures.
  **(ii)**  The number of successes is twice as large as the number of failures.
  **(iii)**  The number of failures is 3 times the number of successes.

**3.6**  Suppose you are throwing darts at a target and that you hit the bull's eye with
probability $p$. It is assumed that the trials are independent and that $p$ remains
constant throughout.
  **(i)**  If you throw darts 100 times, what is the probability that you hit the
bull's eye at least 40 times?
  **(ii)**  How does this expression become for $p = 0.25$?
  **(iii)**  What is the expected number of hits, and what is the s.d. around this
expected number?

**3.7**  If $X \sim B(100, 1/4)$, use the Tchebichev inequality to determine a lower bound
for the probability: $P(|X - 25| < 10)$.

**3.8**  A manufacturing process produces defective items at the constant (but
unknown to us) proportion $p$. Suppose that $n$ items are sampled independently
and let $X$ be the r.v. denoting the number of defective items among the $n$, so
that $X \sim B(n, p)$. Use the Tchebichev inequality in order to determine the
smallest value of the sample size $n$, so that:
$P(|\frac{X}{n} - p| < 0.05\sqrt{pq}) \geq 0.95$ ($q = 1 - p$).

**3.9**  If $X \sim B(n, p)$ show that $f(x + 1) = \frac{p}{q} \times \frac{n-x}{x+1}f(x)$, $x = 0, 1, \ldots, n - 1$,
($q = 1 - p$).

**3.10**  If $X \sim B(n, p)$:
  **(i)**  Calculate the $EX$ and the $E[X(X - 1)]$.
  **(ii)**  Use part (i) and Exercise 1.6 to calculate the $\mathrm{Var}(X)$.
  **Hint.** For part (i), observe that:

$$EX = \sum_{x=1}^{n} x \times \frac{n(n - 1)!}{x(x - 1)!(n - x)!}p^x q^{n-x} = np \sum_{x=1}^{n} \binom{n - 1}{x - 1}p^{x-1}q^{(n-1)-x}$$

$$= np \sum_{y=0}^{n-1} \binom{n - 1}{y}p^y q^{(n-1)-y} = np,$$

and $E[X(X-1)] = \sum_{x=2}^{n} x(x-1) \times \dfrac{n(n-1)(n-2)!}{x(x-1)(x-2)!(n-x)!} p^x q^{n-x}$

$$= n(n-1)p^2 \sum_{x=2}^{n} \binom{n-2}{x-2} p^{x-2} q^{(n-2)-(x-2)}$$

$$= n(n-1)p^2 \sum_{y=0}^{n-2} \binom{n-2}{y} p^y q^{(n-2)-y} = n(n-1)p^2.$$

**3.11** If $X \sim B(n,p)$:

   **(i)** Show that $M_X(t) = (pe^t + q)^n$, $t \in \Re$ $(q = 1 - p)$.

   **(ii)** Use part (i) to rederive the $EX$ and the $\text{Var}(X)$.

**3.12** Let the r.v. $X$ have the Geometric p.d.f.

   $f(x) = pq^{x-1}$, $x = 1, 2, \ldots (q = 1 - p)$.

   **(i)** What is the probability that the first success will occur by the 10th trial?

   **(ii)** What is the numerical value of this probability for $p = 0.2$?

**3.13** A manufacturing process produces defective items at the rate of 1%. Let $X$ be the r.v. denoting the number of trials required until the first defective item is produced. Then calculate the probability that $X$ is not larger than 10.

   **Hint.** See # 4 in Table 8 in the Appendix.

**3.14** A fair die is tossed repeatedly until a six appears for the first time. Calculate the probability that:

   **(i)** This happens on the third toss.

   **(ii)** At least five tosses will be needed.

   **Hint.** As in Exercise 3.13.

**3.15** A coin with probability $p$ of falling heads is tossed repeatedly and independently until the first head appears.

   **(i)** Determine the smallest number of tosses, $n$, required to have the first head appearing by the $n$th time with prescribed probability $\alpha$.

   **(ii)** Determine the value of $n$ for $\alpha = 0.95$, and $p = 0.25$ $(q = 0.75)$ and $p = 0.50(=q)$.

   **Hint.** As in Exercise 3.13.

**3.16** If $X$ has the Geometric distribution; i.e., $f(x) = pq^{x-1}$, for $x = 1, 2, \ldots (q = 1 - p)$:

   **(i)** Calculate the $EX$ and the $E[X(X-1)]$.

   **(ii)** Use part (i) and Exercise 1.6 to calculate the $\text{Var}(X)$.

   **Hint.** For part (i), refer to #5 in Table 8 in the Appendix.

**3.17** If $X$ has the Geometric distribution, then:

   **(i)** Derive the m.g.f. of $X$ and specify the range of its argument.

   **(ii)** Employ the m.g.f. in order to derive: $EX, EX^2$, and $\text{Var}(X)$.

**3.18** Suppose that the r.v. $X$ is distributed as $P(\lambda)$; i.e., $f(x) = e^{-\lambda}\frac{\lambda^x}{x!}$, for $x = 0, 1, \ldots$, and that $f(2) = 2f(0)$. Determine the value of the parameter $\lambda$.

**3.19** Let $X$ be a Poisson distributed r.v. with parameter $\lambda$, and suppose that $P(X = 0) = 0.1$. Calculate the probability $P(X = 5)$.

**3.20** If $X \sim P(\lambda)$, show that: $f(x + 1) = \frac{\lambda}{x+1}f(x)$, $x = 0, 1, \ldots$.

**3.21** If $X \sim P(\lambda)$:
   **(i)** Calculate the $EX$ and the $E[X(X - 1)]$.
   **(ii)** Use part (i) and Exercise 1.6 to calculate the Var$(X)$.
   **Hint.** For part (i), observe that:

$$EX = e^{-\lambda} \sum_{x=1}^{\infty} x \times \frac{\lambda \times \lambda^{x-1}}{x(x - 1)!} = \lambda e^{-\lambda} \sum_{y=0}^{\infty} \frac{\lambda^y}{y!} = \lambda e^{-\lambda} e^{\lambda} = \lambda, \quad \text{and}$$

$$E[X(X - 1)] = \lambda^2 e^{-\lambda} \sum_{x=2}^{\infty} x(x - 1) \times \frac{\lambda^{x-2}}{x(x - 1)(x - 2)!} = \lambda^2 e^{-\lambda} \sum_{y=0}^{\infty} \frac{\lambda^y}{y!} = \lambda^2.$$

**3.22** If $X \sim P(\lambda)$:
   **(i)** Show that $M_X(t) = e^{\lambda(e^t - 1)}$, $t \in \mathfrak{R}$.
   **(ii)** Use the m.g.f. to rederive the $EX$ and the Var$(X)$.

**3.23** For $n = 1, 2, \ldots$, let the r.v. $X_n \sim B(n, p_n)$ where, as $n \to \infty, 0 < p_n \to 0$, and $np_n \to \lambda \in (0, \infty)$. Then show that:

$$\binom{n}{x}p_n^x q_n^{n-x} \xrightarrow[n\to\infty]{} e^{-\lambda}\frac{\lambda^x}{x!} \quad (q_n = 1 - p_n).$$

**Hint.** Write $\binom{n}{x}$ as $n(n - 1) \ldots (n - x + 1)/x!$, set $np_n = \lambda_n$, so that $p_n = \frac{\lambda_n}{n} \xrightarrow[n\to\infty]{} 0$ and $q_n = 1 - p_n = 1 - \frac{\lambda_n}{n} \xrightarrow[n\to\infty]{} 1$. Group terms suitably, take the limit as $n \to \infty$, and use the calculus fact that $(1 + \frac{x_n}{n})^n \to e^x$ when $x_n \to x$ as $n \to \infty$.

**3.24** In an undergraduate statistics class of 80, 10 of the students are, actually, graduate students. If five students are chosen at random from the class, what is the probability that:
   **(i)** No graduate students are included?
   **(ii)** At least three undergraduate students are included?

**3.25** Suppose a geologist has collected 15 specimens of a certain rock, call it $R_1$, and 10 specimens of another rock, call it $R_2$. A laboratory assistant selects randomly 15 specimens for analysis, and let $X$ be the r.v. denoting the number of specimens of rock $R_1$ selected for analysis.
   **(i)** Specify the p.d.f. of the r.v. $X$.
   **(ii)** What is the probability that at least 10 specimens of the rock $R_1$ are included in the analysis?

(iii) What is the probability that all specimens come from the rock $R_2$?
**Hint.** For part (ii), just write down the right formula.

**3.26** If the r.v. $X$ has the Hypergeometric distribution; i.e.,

$$P(X = x) = f(x) = \frac{\binom{m}{x}\binom{n}{r-x}}{\binom{m+n}{r}}, \quad x = 0, 1, \ldots, r, \text{ then show that:}$$

$$f(x + 1) = \frac{(m - x)(r - x)}{(n - r + x + 1)(x + 1)} f(x).$$

**Hint.** Start with $f(x + 1)$ and write the numerator in terms of factorials. Then modify suitably some terms and regroup them to arrive at the expression on the right-hand side.

**3.27** By using the definition of $\Gamma(\alpha)$ by (26) and integrating by parts, show that:
$\Gamma(\alpha) = (\alpha - 1)\Gamma(\alpha - 1)$, $\alpha > 1$.
**Hint.** Use #13 in Table 8 in the Appendix.

**3.28** Let the r.v. $X$ have the Gamma distribution with parameters $\alpha$ and $\beta$. Then:
  (i) Show that: $EX = \alpha\beta$, $\text{Var}(X) = \alpha\beta^2$.
  (ii) As a special case of part (i), show that: If $X$ has the Negative Exponential distribution with parameter $\lambda$, then $EX = \frac{1}{\lambda}$, $\text{Var}(X) = \frac{1}{\lambda^2}$.
  (iii) If $X \sim \chi_r^2$, then $EX = r$, $\text{Var}(X) = 2r$.
**Hint.** For part (i), use formula (27) and a suitable transformation.

**3.29** If the r.v. $X$ is distributed as Gamma with parameters $\alpha$ and $\beta$, then:
  (i) Show that $M_X(t) = 1/(1 - \beta t)^\alpha$, provided $t < 1/\beta$.
  (ii) Use the m.g.f. to rederive the $EX$ and the $\text{Var}(X)$.
**Hint.** For part (i), use formula (27).

**3.30** Let $X$ be a r.v. denoting the lifetime of a certain component of a system, and suppose that $X$ has the Negative Exponential distribution with parameter $\lambda$. Also, let $g(x)$ be the cost of operating this equipment to time $X = x$.
  (i) Compute the expected cost of operation over the lifetime of the component under consideration, when:
    (a) $g(x) = cx$, where $c$ is a positive constant,
    (b) $g(x) = c(1 - 0.5 e^{-\alpha x})$, where $\alpha$ is a positive constant.
  (ii) Specify the numerical values in part (i) when $\lambda = 1/5, c = 2$, and $\alpha = 0.2$.
**Hint.** For part (i), use Definition 1, and compute the required expression in terms of $c$ and $\lambda$.

**3.31** If the r.v. $X$ has the Negative Exponential p.d.f. with parameter $\lambda$:
  (i) Calculate the *failure rate* $r(x)$ defined by: $r(x) = \frac{f(x)}{1-F(x)}$, for $x > 0$, where $F$ is the d.f. of $X$.
  (ii) Compute the (conditional) probability $P(X > s + t \mid X > s)$, for $s$ and $t$ positive, and comment on the result.

**3.32** Suppose that certain events occur in a time interval $t$ according to the Poisson distribution with parameter $\lambda t$. Then show that the waiting time between any two such successive events is a r.v. $T$ which has the Negative Exponential distribution with parameter $\lambda$, by showing that $P(T > t) = e^{-\lambda t}, t > 0$.

**3.33** Let $X$ be the r.v. denoting the number of particles arriving independently at a detector at the average rate of 3 per second, and let $Y$ be the r.v. denoting the waiting time between two successive arrivals. Refer to Exercise 3.32 in order to calculate:

(i) The probability that the first particle will arrive within 1 second.

(ii) Given that we have waited for 1 second since the arrival of the last particle without a new arrival, what is the probability that we have to wait for at least another second?

**3.34** Let $X$ be a r.v. with p.d.f. $f(x) = \alpha \beta x^{\beta-1} e^{-\alpha x^{\beta}}$, for $x > 0$ (where the parameters $\alpha$ and $\beta$ are $> 0$). This is the so-called *Weibull* distribution employed in describing the lifetime of living organisms or of mechanical systems.

(i) Show that $f$ is, indeed, a p.d.f.

(ii) For what values of the parameters does $f$ become a Negative Exponential p.d.f.?

(iii) Calculate the quantities: $EX, EX^2$, and $\text{Var}(X)$.

**Hint.** For part (i), observe that: $\int_0^\infty \alpha \beta x^{\beta-1} e^{-\alpha x^{\beta}} dx =$
$\int_0^\infty e^{-\alpha x^{\beta}} \times (\alpha \beta x^{\beta-1}) dx = -\int_0^\infty de^{-\alpha x^{\beta}} = -e^{-\alpha x^{\beta}} \big|_0^\infty = 1$.

For part (iii), set $\alpha x^{\beta} = t$, so that $x = t^{1/\beta}/\alpha^{1/\beta}$, $dx = (t^{\frac{1}{\beta}-1}/\beta \alpha^{1/\beta}) dt$ and $0 < t < \infty$. Then:

$$EX^n = \frac{1}{\alpha^{n/\beta}} \int_0^\infty t^{((n/\beta)+1)-1} e^{-t} dt.$$

Next, multiply and divide by the constant $\Gamma(\frac{n}{\beta} + 1)$ and observe that $\frac{1}{\Gamma(\frac{n}{\beta}+1)} t^{((n/\beta)+1)-1} e^{-t}$ $(t > 0)$ is a Gamma p.d.f. with parameters $\frac{n}{\beta} + 1$ and 1.

**3.35** In reference to Exercise 3.34, calculate:

(i) The *failure rate* $r(x) = \frac{f(x)}{1-F(x)}$, $x > 0$, where $F$ is the d.f. of the r.v. $X$.

(ii) The conditional probability $P(X > s+t \mid X > s)$, $s > 0$, $t > 0$.

(iii) Compare the results in parts (i) and (ii) with the respective results in Exercise 3.31 here, and Exercise 3.20(ii) in Chapter 2.

**3.36** If $\Phi$ is the d.f. of the r.v. $Z \sim N(0, 1)$, show that:

(i) For $0 \le a < b$, $P(a < Z < b) = \Phi(b) - \Phi(a)$.

(ii) For $a \le 0 < b$, $P(a < Z < b) = \Phi(-a) + \Phi(b) - 1$.

(iii) For $a \le b < 0$, $P(a < Z < b) = \Phi(-a) - \Phi(-b)$.

(iv) For $c > 0$, $P(-c < Z < c) = 2\Phi(c) - 1$.

**3.37** If the r.v. $Z \sim N(0, 1)$, use the Normal tables in the Appendix to verify that:
  (i) $P(-1 < Z < 1) = 0.68269$.
  (ii) $P(-2 < Z < 2) = 0.9545$.
  (iii) $P(-3 < Z < 3) = 0.9973$.

**3.38** (i) If the r.v. $X$ is distributed as $N(\mu, \sigma^2)$, identify the constant $c$, in terms of $\mu$ and $\sigma$, for which:

$$P(X < c) = 2 - 9P(X > c).$$

  (ii) What is the numerical value of $c$ for $\mu = 5$ and $\sigma = 2$?

**3.39** For any r.v. $X$ with expectation $\mu$ and variance $\sigma^2$ (both finite), use the Tchebichev inequality to determine a lower bound for the probabilities: $P(|X - \mu| < k\sigma)$, for $k = 1, 2, 3$. Compare these bounds with the respective probabilities when $X \sim N(\mu, \sigma^2)$ (see Exercise 3.37).

**3.40** The distribution of I.Q.'s of the people in a given group is approximated well described by the Normal distribution with $\mu = 105$ and $\sigma = 20$. What proportion of the individuals in the group in question has an I.Q. :
  (i) At least 50?
  (ii) At most 80?
  (iii) Between 95 and 125?

**3.41** A certain manufacturing process produces light bulbs whose life length (in hours) is a r.v. $X$ distributed as Normal with $\mu = 2000$ and $\sigma = 200$. A light bulb is supposed to be defective if its lifetime is less than 1800. If 25 light bulbs are tested, what is the probability that at most 15 of them are defective? **Hint.** Use the required independence and the Binomial distribution suitably. Just write down the correct formula.

**3.42** A manufacturing process produces 1/2-in. ball bearings, which are assumed to be satisfactory if their diameter lies in the interval $0.5 \pm 0.0006$ and defective otherwise. A day's production is examined, and it is found that the distribution of the actual diameters of the ball bearings is approximately Normal with $\mu = 0.5007$ in. and $\sigma = 0.0005$ in. What would you expect the proportion of defective ball bearings to be equal to? **Hint.** Use the required independence and the Binomial distribution suitably. Also, refer to the concluding part of the discussion in Example 11.

**3.43** Let $f$ be the p.d.f. of the $N(\mu, \sigma^2)$ distribution. Then show that:
  (i) $f$ is symmetric about $\mu$.
  (ii) $\max_{x \in \Re} f(x) = 1/\sqrt{2\pi}\sigma$.

**3.44**  (i)  Show that $f(x) = \frac{1}{\sqrt{2\pi}}e^{-x^2/2}, x \in \Re$, is a p.d.f.

   (ii)  Use part (i) in order to show that
$$f(x) = \frac{1}{\sqrt{2\pi}\sigma}e^{-(x-\mu)^2/2\sigma^2}, x \in \Re \ (\mu \in \Re, \sigma > 0) \text{ is also a p.d.f.}$$

   **Hint.** Set $I = \frac{1}{\sqrt{2\pi}}\int_{-\infty}^{\infty}e^{-x^2/2}dx$ and show that $I^2 = 1$, by writing $I^2$ as a product of two integrals and then as a double integral; at this point, use polar coordinates: $x = r\cos\theta, y = r\sin\theta, 0 < r < \infty, 0 \le \theta < 2\pi$. Part (ii) is reduced to part (i) by letting $\frac{x-\mu}{\sigma} = y$.

**3.45**  Refer to the definition of $\Gamma(\alpha)$ by (26) and show that $\Gamma(\frac{1}{2}) = \sqrt{\pi}$.

   **Hint.** Use the transformation $y^{1/2} = t/\sqrt{2}$, and observe that the outcome is a multiple of the $N(0, 1)$ p.d.f.

**3.46**  (i)  If $X \sim N(0, 1)$, show that $M_X(t) = e^{t^2/2}, t \in \Re$.

   (ii)  If $X \sim N(\mu, \sigma^2)$, use part (i) to show that $M_X(t) = e^{\mu t + \sigma^2 t^2/2}, t \in \Re$.

   (iii)  Employ the m.g.f. in part (ii) in order to show that $EX = \mu$ and $\text{Var}(X) = \sigma^2$.

   **Hint.** For part (i), complete the square in the exponent, and for part (ii), set $Z = (X - \mu)/\sigma$.

**3.47**  If the r.v. $X$ has m.g.f. $M_X(t) = e^{\alpha t + \beta t^2}$, where $\alpha \in \Re$ and $\beta > 0$, identify the distribution of $X$.

   **Hint.** Just go through the list of m.g.f.'s given in Table 7 in the Appendix, and then use Proposition 1.

**3.48**  If $X \sim N(0, 1)$, show that:

   (i)  $EX^{2n+1} = 0$ and $EX^{2n} = \frac{(2n)!}{2^n(n!)}, n = 0, 1, \ldots$

   (ii)  From part (i), derive that $EX = 0$ and $\text{Var}(X) = 1$.

   (iii)  Employ part (ii) in order to show that, if $X \sim N(\mu, \sigma^2)$, then $EX = \mu$ and $\text{Var}(X) = \sigma^2$.

   **Hint.** For part (i), that $EX^{2n+1} = 0$ follows by the fact that the integrand is an odd function. For $EX^{2n}$, establish a recursive relation, integrating by parts, and then multiply out the resulting recursive relations to find an expression for $EX^{2n}$. The final form follows by simple manipulations. For part (iii), recall that $X \sim N(\mu, \sigma^2)$ implies $\frac{X-\mu}{\sigma} \sim N(0, 1)$.

**3.49**  Let $X$ be a r.v. with moments given by:
$$EX^{2n+1} = 0, \quad EX^{2n} = \frac{(2n)!}{2^n(n!)}, \quad n = 0, 1, \ldots$$

   (i)  Use Exercise 1.17 in order to express the m.g.f. of $X$ in terms of the moments given.

   (ii)  From part (i) and Exercise 3.46(i), conclude that $X \sim N(0, 1)$.

**3.50** If the r.v. $X$ is distributed as $U(-\alpha, \alpha)$ ($\alpha > 0$), determine the parameter $\alpha$, so that each of the following equalities holds:

(i) $P(-1 < X < 2) = 0.75$.

(ii) $P(|X| < 1) = P(|X| > 2)$.

**3.51** If $X \sim U(\alpha, \beta)$, show that $EX = \frac{\alpha+\beta}{2}$, $\mathrm{Var}(X) = \frac{(\alpha-\beta)^2}{12}$.

**3.52** If the r.v. $X$ is distributed as $U(0, 1)$, compute the expectations:

(i) $E(3X^2 - 7X + 2)$.

(ii) $E(2e^X)$.

**Hint.** Use Definition 1 and properties (4).

**3.53** If $X \sim B(n, 0.15)$ determine the smallest value of $n$ for which
$P(X > 1) > P(X > 0)$.

**3.54** Suppose that in a preelection campaign, propositions #1 and #2 are favored by $100p_1\%$ and $100p_2\%$ of voters, respectively, whereas the remaining $100(1 - p_1 - p_2)\%$ of the voters are either undecided or refuse to respond $(0 < p_1 < 1, 0 < p_2 < 1, p_1 + p_2 \leq 1)$. A random sample of size $2n$ is taken. Then:

(i) What are the expected numbers of voters favoring each of the two propositions?

(ii) What is the probability that the number of voters favoring proposition #1 is at most $n$?

(iii) What is the probability that the number of voters favoring proposition #2 is at most $n$?

(iv) Give the numerical values in parts (i)-(iii) for $p_1 = 31.25\%$, $p_2 = 43.75\%$, and $2n = 24$.

**Hint.** For parts (i)-(iii), just write down the right formulas.

**3.55** A quality control engineer suggests that for each shipment of 1000 integrated circuit components, a sample of 100 be selected from the 1000 for testing. If either 0 or 1 defective is found in the sample, the entire shipment is declared acceptable.

(i) Compute the probability of accepting the entire shipment is there are 3 defective among the 1000 items.

(ii) What is the expected number of defective items (among the 100 to be tested)?

(iii) What is the variance and the s.d. of the defective items (among the 100 to be tested)?

**3.56** If the number of claims filed by policyholders over a period of time is a r.v. $X$ which has Poisson distribution; i.e., $X \sim P(\lambda)$, then, if $P(X = 2) = 3P(X = 4)$, determine:

(i) The $EX$ and $\mathrm{Var}(X)$.

(ii) The probabilities $P(2 \leq X \leq 4)$ and $P(X \geq 5)$.

**3.57** Refer to Exercise 3.55 and use the Poisson approximation to Binomial distribution (see Exercise 3.23 here) to find an approximate value for the probability computed in part (i) of Exercise 1.32.

The following exercise, Exercise 3.58, is recorded here mostly for reference purposes; its solution goes along the same lines as that of Exercise 3.23, but it is somewhat more involved. The interested reader is referred for details to Theorem 2, Chapter 3, in the book *A Course in Mathematical Statistics*, 2nd edition (1997), Academic Press, by G. G. Roussas.

**3.58** Let $X$ be a r.v. having Hypergeometric distribution with parameter $m$ and $n$, so that its p.d.f. is given by $f(x) = \binom{m}{x}\binom{n}{r-x}/\binom{m+n}{r}, x = 0, 1, \ldots, r$. Suppose that $m$ and $n \to \infty$, so that $\frac{m}{m+n} \to p \in (0, \infty)$. Then:

$$\frac{\binom{m}{x}\binom{n}{r-x}}{\binom{m+n}{r}} \to \binom{r}{x}p^x(1-p)^{r-x}, \quad x = 0, 1, \ldots, r.$$

Thus, for large $m$ and $n$, the Hypergeometric probabilities may be approximated by the (simpler) Binomial probabilities, as following

$$\frac{\binom{m}{x}\binom{n}{r-x}}{\binom{m+n}{r}} \simeq \binom{r}{x}\left(\frac{m}{m+n}\right)^x\left(1-\frac{m}{m+n}\right)^{r-x}, \quad x = 0, 1, \ldots, r.$$

**3.59** Refer to Exercise 3.55, and use the Binomial approximation to Hypergeometric distribution (see Exercise 3.58 here) to find an approximate value for the probability computed in part (i) of Exercise 3.55.

**3.60** Suppose that 125 out of 1000 California homeowners have earthquake insurance. If 40 such homeowners are chosen at random:
   **(i)** What is the expected number of earthquake insurance holders, and what is the s.d. around this expected number?
   **(ii)** What is the probability that the number of earthquake insurance holders is within one s.d. of the mean, inclusive?
   **(iii)** Use both the Binomial and the Poisson approximation in part (ii).
   **Hint.** For part (ii), just write down the correct formula.

**3.61** Refer to the Geometric distribution with parameter $p$ (see relation (21)), but now suppose that the predetermined number of successes $r \geq 1$. Then the number of (independent) trials required in order to obtain exactly $r$ successes is a r.v. $X$ which takes on the values $r, r+1, \ldots$ with respective probabilities

$$P(X = x) = f(x) = \binom{x-1}{r-1}q^{x-r}p^r = \binom{x-1}{x-r}q^{x-r}p^r \quad (q = 1 - p).$$

This distribution is called *Negative Binomial* (for obvious reasons) with parameter $p$.
   **(i)** Show that the p.d.f. of $X$ is, indeed, given by the above formula.

**(ii)** Use the identity

$$\frac{1}{(1-x)^n} = 1 + \binom{n}{1}x + \binom{n+1}{2}x^2 + \cdots \qquad (|x| < 1,\ n \geq 1 \text{ integer})$$

in order to show that the m.g.f. of $X$ is given by

$$M_X(t) = \left(\frac{pe^t}{1 - qe^t}\right)^r, \qquad t < -\log q.$$

**(iii)** For $t < -\log q$, show that

$$\frac{d}{dt}M_X(t) = \frac{r(pe^t)^r}{(1 - qe^t)^{r+1}}, \qquad \frac{d^2}{dt^2}M_X(t) = \frac{r(pe^t)^r(r + qe^t)}{(1 - qe^t)^{r+2}}.$$

**(iv)** Use part (iii) in order to find that $EX = \frac{r}{p}$, $EX^2 = \frac{r(r+q)}{p^2}$, and

$$\sigma^2(X) = \frac{rq}{p^2}.$$

**(v)** For $r = 5$ and $p = 0.25, 0.50, 0.75$, compute the $EX$, $\sigma^2(X)$, and $\sigma(X)$.

**Hint.**

**(i)** For part (i), derive the expression $f(x) = \binom{x-1}{r-1}q^{x-r}p^r$; the second expression follows from it and the identity $\binom{n}{m} = \binom{n}{n-m}$.

**(ii)** For part (ii), we have

$$M_X(t) = \sum_{x=r}^{\infty} e^{tx}\binom{x-1}{x-r}q^{x-r}p^r$$

$$= (pe^t)^r\left[1 + \binom{r}{1}(qe^t) + \binom{r+1}{2}(qe^t)^2 + \cdots\right].$$

**3.62** If the r.v. $X \sim U(\alpha, \beta)$, then $EX^n = \frac{\beta^{n+1}-\alpha^{n+1}}{(n+1)(\beta-\alpha)}$.

**3.63** The lifetime of an electrical component is a r.v. $X$ having the Negative Exponential distribution with parameter $\lambda$. If we know that $P(X \leq c) = p$:

**(i)** Express the parameter $\lambda$ in terms of $c$ and $p$.

**(ii)** Express $EX$, $\text{Var}(X)$, $\sigma(X)$, and $P(X > 2000)$ in terms of $c$ and $p$.

**(iii)** Give the numerical values in parts (i) and (ii) for $c = 1000$ and $p = 0.75$.

**3.64** There are $r$ openings in a certain line of a job, and $m$ men and $w$ women applicants are judged to be fully qualified for the job. The hiring officer decides to do the hiring by using a random lottery. Let $X$ be the r.v. denoting the number of women hired.

**(i)** Write down the p.d.f. of $X$, and its expectation.

**(ii)** Find the numerical values in part (i) for: $m = 6$, $w = 4$, and $r = 3$.

**3.65** A coin with probability $p$ of falling heads is tossed independently $n$ times.

**(i)** Find an expression for the probability that the number of heads is bigger than $k$ times the number of tails, where $k$ is a positive integer.

**(ii)** Compute the numerical answers in part (i) for a fair coin ($p = 0.5$), $n = 10$ and $k = 1$; also for $p = 0.25$, $n = 25$, $k = \frac{1}{4}$.

**3.66** **(i)** In reference to the variance of the Hypergeometric distribution, determine the size $r$ of the sample which maximizes the variance.
**(ii)** Determine the $EX$ and $\text{Var}(X)$ for the maximizer sample size.

**3.67** If the r.v. $X \sim B(n, p)$, then $\sigma^2(X) = np(1 - p)$, $0 < p < 1$. Show that $\sigma^2(X) \leq \frac{n}{4}$ and $\sigma^2(X) = \frac{n}{4}$ if and only if $p = \frac{1}{2}$.

**3.68** A fair coin is tossed repeatedly and independently until 10 heads appear.
**(i)** What is the expected number of tosses, and what is the standard deviation?
**(ii)** What is the probability that more than 15 tosses will be required?

**3.69** Recall that both roots $x_1$ and $x_2$ of the quadratic equation $ax^2 + bx + c = 0$ are real if and only if $b^2 - 4ac \geq 0$. Also, recall that $ax^2 + bx + c = a(x - x_1)(x - x_2)$. Thus, if $a > 0$, then $ax^2 + bx + c \geq 0$ if and only if either $x \leq \min\{x_1, x_2\}$ or $x \geq \max\{x_1, x_2\}$. Finally, also recall that $x_1 = (-b - \sqrt{b^2 - 4ac})/2a$, $x_2 = (-b + \sqrt{b^2 - 4ac})/2a$. With this in mind, consider the quadratic (in $x$) equation $4x^2 + 5Yx + (Y + 0.1) = 0$, where $Y \sim U(0, 1)$, and determine the probability that both its roots are real.

**3.70** The number of typos per page in a textbook is a r.v. $X$ distributed as Poisson with parameter $\lambda$; i.e., $X \sim P(\lambda)$. Also, the number of typos in $n$ pages is a r.v. $Y$ distributed as Poisson with parameter $n\lambda$; i.e., $Y \sim P(n\lambda)$. Thus, the expected number of typos per page is $\frac{Y}{n}$.
**(i)** Express $E\left(\frac{Y}{n}\right)$ and $\sigma^2\left(\frac{Y}{n}\right)$ in terms of $\lambda$.
**(ii)** Use Tchebichev's inequality to find the smallest number of pages $n$, in terms of $c$, $\lambda$, and $p$, for which

$$P\left(\left|\frac{Y}{n} - \lambda\right| \leq c\right) \geq p.$$

**(iii)** Find the numerical value of $n$, if $c = 0.25$, $\lambda = 2$, and $p = 0.90$.

**3.71** Let $X \sim U(0, 8)$, so that $f(x) = 1/8$, $0 < x < 8$.
**(i)** Compute $EX$, $EX^2$, and $\sigma^2(X)$.
**(ii)** Compute the probability $P(|X - 4| > 3)$.
**(iii)** Use the Tchebichev inequality to find an upper bound for the probability in part (ii).

**3.72** If a r.v. $X$ has a p.d.f. $f(x)$ which is 0 for $x < a$ and $x > b$ $(-\infty < a < b < \infty)$, then show that:
**(i)** $a \leq EX \leq b$.
**(ii)** $\sigma^2(X) \leq \frac{(b-a)^2}{4}$.

**3.73** If $X \sim N(\mu, \sigma^2)$:
**(i)** Express the $EX^3$ as a function of $\mu$ and $\sigma^2$.
**(ii)** Compute the numerical value in part (i) for $\mu = 1$ and $\sigma^2 = 4$.

**3.74** Your traveling time between your home and the university is a r.v. $X$ distributed as Normal with mean $\mu = 20$ and s.d. $\sigma = 4$.

(i) Denoting by 0 the starting time, for what value of $x$ (in minutes) is your traveling time $X$ within the interval $(0, x)$ with probability 0.95?

(ii) If you have a class at 10:00 am, at what time should you leave home in order to be in your class on time with probability 0.95?

**3.75** Certain events occur during a time interval of length $h$ according to a Poisson distribution with parameter $\lambda h$. Thus, if $X$ is the r.v. denoting the number of events which occurred in a time interval of length $h$, then $X \sim P(\lambda h)$ (so that $P(X = x) = f(x) = e^{-\lambda h} \frac{(\lambda h)^x}{x!}$, $x = 0, 1, \ldots$).
An event just occurred and call the time of its occurrence time 0. Let $Y$ be the r.v. denoting the time we have to wait until the next event occurs. Then $Y$ is a r.v. of the continuous type taking on all positive values.

(i) Show that $P(Y > y) = e^{-\lambda y}$, $y > 0$.

(ii) From the relation: $P(Y > y) = 1 - P(Y \leq y) = 1 - F_Y(y)$, find the p.d.f. $f_Y$.

(iii) Give the name of the distribution that corresponds to the p.d.f. found in part (ii).

**Hint.** Note that the event $\{Y > y\}$ means that no events happen in the interval $(0, y]$.

**3.76** Let the r.v. $X$ denote the number of traffic tickets a 25-year-old driver gets in 1 year, and suppose that $EX = 1$ and $\sigma(X) = 1/3$.

(i) Is it possible that $X \sim B(1, p)$ (for some $0 < p < 1$)?

(ii) Is it possible that $X \sim P(\lambda)$ (for some $\lambda > 0$)?

(iii) Is it possible that the p.d.f. of $X$ is given by: $f(x) = pq^x$, $x = 0, 1, \ldots$ (for some $0 < p < 1$, $q = 1 - p$)?

(iv) If $X$ takes on the values $0, 1, 2$ with respective probability $p_0, p_1$, and $p_2$, determine these probabilities.

**3.77** The temperature at midnight on January 1 in a certain location is r.v. $X$ distributed as Normal with $\mu = 10$ and $\sigma = 5$. Compute:

(i) $P(X < 0)$.

(ii) $P(X > 5)$.

(iii) $P(-5 < X < 5)$.

(iv) $P(X > 15)$.

**3.78** If the r.v. $X \sim P(\lambda)$, show that $E[X(X - 1) \ldots (X - k + 1)] = \lambda^k$.

**3.79** The r.v. $X \sim \chi^2_{40}$. Then:

(i) Compute $EX$, $\sigma^2(X)$, $E(X/n)$, $\sigma^2(X/n)$ for $n = 40$.

(ii) Use the Tchebichev equality to find a lower bound for the $P(|\frac{X}{n} - 1| \leq 0.5)$.

(iii) Compute the probability $P(|\frac{X}{n} - 1| \leq 0.5)$ by using the Chi-square tables.

**3.80** If the r.v. $X \sim P(\lambda)$:
   **(i)** Show that $E[X(X-1)(X-2)] = \lambda^3$, $EX^3 = \lambda^3 + 3\lambda^2 + \lambda$, and
   $E(X-\lambda)^3 = \lambda$.
   **(ii)** Use part (i), in order to show that $\gamma_1 = \frac{1}{\sqrt{\lambda}}$ (so that the Poisson
   distribution is always skewed to the right).

**3.81** In reference to Exercise 1.37 (in this chapter):
   **(i)** Find the generating function of the sequences

$$\left\{ \binom{n}{j} p^j q^{n-j} \right\}, \quad j \ge 0, \ 0 < p < 1, \ q = 1 - p,$$

   and

$$\left\{ e^{-\lambda} \frac{\lambda^j}{j!} \right\}, \quad j \ge 0, \quad \lambda > 0.$$

   **(ii)** Calculate the $k$th factorial moments of $X$ being $B(n,p)$ and $X$ being $P(\lambda)$.

**3.82** For the purpose of testing the effectiveness of a new medication for headache
relief, one group of $n$ patients is given the medication and another group of
patients, of the same size, is given a placebo. Let $p_1$ be the probability that a
patient in the first group finding relief, and let $p_2$ be the probability for a
patient in the second group finding relief. Also, let $X$ and $Y$ be independent
r.v.'s denoting the numbers of patients who found relief in the first and second
group, respectively. Then $X \sim B(n, p_1)$, $Y \sim B(n, p_2)$. The medication will be
superior to the placebo if $X > Y$.
   **(i)** Show that

$$P(X > Y) = 1 - \sum_{y=0}^{n} P(X \le y)P(Y = y).$$

   **(ii)** Compute the numerical value of the probability in part (i) for $n = 5$,
   $p_1 = 0.5(= \frac{8}{16})$, and $p_2 = 0.375(= \frac{6}{16})$.

**3.83** In reference to the Geometric distribution (see formula (21)), show that it has
the "memoryless" property. Namely,

$$P(X > x + y | X > x) = P(X > y).$$

**3.84** Consider the function $f$ defined as follows:

$$f(x) = \begin{cases} \frac{\Gamma(\alpha+\beta)}{\Gamma(\alpha)\Gamma(\beta)} x^{\alpha-1}(1-x)^{\beta-1}, & 0 < x < 1 \\ 0, & \text{otherwise,} \end{cases}$$

where $\alpha, \beta > 0$.
By a series of transformations, it can be shown that $\int_0^1 f(x)dx = 1$, so that $f$ is,
indeed, a p.d.f. It is referred as the *Beta* p.d.f. with parameters $\alpha$ and $\beta$. (For
more information about it, see, e.g., Section 3.3.6 in the book *A Course in*

*Mathematical Statistics*, 2nd editions (1997), Academic Press, by G.G. Roussas.)

Show that:

(i) For $\alpha = \beta = 1$, it becomes the $U(0, 1)$ distribution.

(ii) For $\alpha = \beta = 2$, the resulting p.d.f.

$$f(x) = 6x(1 - x), \quad 0 < x < 1$$

has a unique mode.

(iii) For $\alpha, \beta > 1$, it has the unique mode $(\alpha - 1)/(\alpha + \beta - 2)$.

**3.85** Let $X$ be a r.v. having the Geometric distribution with parameter $p$, and for $M$ integer with $M \geq 1$, define the r.v. $Y$ as follows:

$$Y = \begin{cases} X, & \text{if } X < M, \\ M, & \text{if } X \geq M. \end{cases}$$

(i) Then show that: $f_Y(y) = pq^{y-1}, y = 1, \ldots, M - 1$, and $f_Y(M) = q^{M-1}$.

(ii) Determine the $EY$ in terms of $M$ and $q$.

**3.86** If the r.v. $X \sim B(n, p)$, show that:

$E[X(X - 1) \ldots (X - k + 1)] = n(n - 1) \ldots (n - k + 1)p^k$ for $k \geq 1$ integer.

**3.87** If the r.v. $X$ is distributed as Gamma with $\beta = 1$, then show that:

(i) $EX^3 = \alpha(\alpha + 1)(\alpha + 2)$,

(ii) $E(X - \alpha)^3 = 2\alpha$.

**3.88** If the r.v. $X \sim \Gamma(n, 1)$ ($n \geq 1$ integer); i.e., $f(x) = \frac{1}{\Gamma(n)}x^{n-1}e^{-x}, x > 0$, then show that:

$$P(X > \lambda) = \sum_{x=0}^{n-1} e^{-\lambda} \frac{\lambda^x}{x!} \quad (\lambda > 0).$$

**3.89** Refer to Exercise 1.36 (in this chapter), and suppose that $X \sim B(n, p)$. Then:

(i) Show that $\gamma_1 = \frac{1-2p}{\sqrt{npq}}$ $(q = 1 - p)$.

(ii) From part (i), conclude that , if $p = 1/2$, then $\gamma_1 = 0$ (as it should be).

**3.90** Let the r.v. $X$ have the Negative Exponential distribution with parameter $\lambda$; i.e., $f_X(x) = \lambda e^{-\lambda x}, x > 0$ ($\lambda > 0$). Then:

(i) By using the change of variables $\lambda x = y$ and looking at the resulting integral as a Gamma function, show that: $EX^n = \Gamma(n + 1)/\lambda^n = \frac{n!}{\lambda^n}$, $n \geq 1$ integer.

(ii) From part (i), conclude that: $EX = 1/\lambda$, $EX^2 = 2/\lambda^2$, $EX^3 = 6/\lambda^3$.

(iii) Use part (ii) to show that $\gamma_1 = 2$ (so that the distribution is always skewed to the right).

**3.91** Let $X$ be a r.v. with $EX^4 < \infty$, and define (the pure number) $\gamma_2$ by:

$$\gamma_2 = E\left(\frac{X - \mu}{\sigma}\right)^4 - 3, \quad \text{where } \mu = EX, \sigma^2 = \sigma^2(X).$$

$\gamma_2$ is called the *kurtosis* of the distribution of the r.v. $X$, and is a measure of "peakedness" of the distribution, where the $N(0, 1)$ p.d.f. is a measure of reference. If $\gamma_2 > 0$, the distribution is called *leptokurtic*, and if $\gamma_2 < 0$, the distribution is called *platykurtic*. Then show that:

(i)   $\gamma_2 < 0$ if $X$ is distributed as $U(\alpha, \beta)$.

(ii)  $\gamma_2 > 0$ if $X$ has a Double Exponential distribution (see Exercise 1.46 in this chapter).

**3.92** If the r.v. $X$ has the Hypergeometric distribution; i.e., $f(x) = \dfrac{\binom{m}{x}\binom{n}{r-x}}{\binom{m+n}{r}}$,

$x = 0, 1, \ldots, r$, then show that:

(i)   $EX = \dfrac{mr}{m+n}$.

(ii)  $E[X(X - 1)] = \dfrac{mr(m-1)(r-1)}{(m+n)(m+n-1)}$, so that

$EX^2 = \dfrac{mr(mr+n-r)}{(m+n)(m+n-1)}$.

(iii) $\sigma^2(X) = \dfrac{mnr(m+n+r)}{(m+n)^2(m+n-1)}$.

**3.93** Let the r.v. $X$ have the Geometric distribution (i.e., $P(X = x) = f(x) = pq^{x-1}$, $x = 1, 2, \ldots$).

(i)   Then show that: $P(X > n + k | X > n) = P(X > k), n, k = 1, 2, \ldots$

(ii)  Compute the probability in part (i) for: $k = 5$ and $p = 1/2$.

## 3.4 MEDIAN AND MODE OF A RANDOM VARIABLE

Although the mean of a r.v. $X$ does specify the location of the center of the distribution of $X$ (in the sense of the center of gravity), sometimes this is not what we actually wish to know. A case in point is the distribution of yearly income in a community (e.g., in a state or in a country). For the sake of illustration, consider the following (rather) extreme example. A community consisting of 10 households comprises 1 household with yearly income $500,000 and 9 households with respective yearly incomes $x_i = \$20,000 + \$1000(i - 2)$, $i = 2, \ldots, 10$. Defining the r.v. $X$ to take the values $x = \$500,000$ and $x_i$, $i = 2, \ldots, 10$ with respective probabilities 0.10, we obtain: $EX = \$71,600$. Thus, the average yearly income in this community would be $71,600, significantly above the national average yearly income, which would indicate a rather prosperous community. The reality, however, is that this community is highly stratified, and the expectation does not reveal this characteristic. What is more appropriate for cases like this are numerical characteristics of a distribution known as median or, more generally, percentiles or quantiles.

The median of the distribution of a r.v. $X$ is usually defined as a point, denoted by $x_{0.50}$, for which

$$P(X \leq x_{0.50}) \geq 0.50 \quad \text{and} \quad P(X \geq x_{0.50}) \geq 0.50, \tag{47}$$

or, equivalently,

$$P(X < x_{0.50}) \leq 0.50 \quad \text{and} \quad P(X \leq x_{0.50}) \geq 0.50. \tag{48}$$

If the underlying distribution is continuous, the median is (essentially) unique and may be simply defined by:

$$P(X \leq x_{0.50}) = P(X \geq x_{0.50}) = 0.50 \tag{49}$$

However, in the discrete case, relation (47) (or 48) may not define the median in a unique manner, as the following example shows.

**Example 13.**    Examine the median of the r.v. $X$ distributed as follows:

| x | 1 | 2 | 3 | 4 | 5 | 6 | 7 | 8 | 9 | 10 |
|---|---|---|---|---|---|---|---|---|---|---|
| f(x) | 2/32 | 1/32 | 5/32 | 3/32 | 4/32 | 1/32 | 2/32 | 6/32 | 2/32 | 6/32 |

**Discussion.**    We have $P(X \leq 6) = 16/32 = 0.50 \geq 0.50$ and $P(X \geq 6) = 17/32 > 0.05 \geq 0.50$, so that (47) is satisfied. Also,

$$P(X \leq 7) = 18/32 > 0.50 \geq 0.50 \quad \text{and} \quad P(X \geq 7) = 16/32 = 0.50 \geq 0.50,$$

so that (47) is satisfied again. However, if we define the median as the point $(6+7)/2 = 6.5$, then $P(X \leq 6.5) = P(X \geq 6.5) = 0.50$, as (47) requires, and the median is uniquely defined.

Relations (47)–(49) and Example 13 suggest the following definition of the median.

**Definition 4.**    The *median* of the distribution of a continuous r.v. $X$ is the (essentially) unique point $x_{0.50}$ defined by (49). For the discrete case, consider two cases: Let $x_k$ be the value for which $P(X \leq x_k) = 0.50$, if such a value exists. Then the unique *median* is defined to be the midpoint between $x_k$ and $x_{k+1}$; i.e., $x_{0.50} = (x_k + x_{k+1})/2$. If there is no such value, the unique *median* is defined by the relations: $P(X < x_{0.50}) < 0.50$ and $P(X \leq x_{0.50}) > 0.50$ (or $P(X \leq x_{0.50}) \geq 0.50$ and $P(X \geq x_{0.50}) \geq 0.50$).

Thus, in Example 14, $x_{0.50} = 6$, because $P(X < 6) = P(X \leq 5) = 15/32 < 0.50$ and $P(X \leq 6) = 17/32 > 0.50$.

**Example 14.**    Determine the median of the r.v. $X$ distributed as follows.

| x | 1 | 2 | 3 | 4 | 5 | 6 | 7 | 8 | 9 | 10 |
|---|---|---|---|---|---|---|---|---|---|---|
| f(x) | 2/32 | 1/32 | 2/32 | 6/32 | 4/32 | 2/32 | 1/32 | 7/32 | 1/32 | 6/32 |

More generally, the $p$th quantile is defined as follows.

**Definition 5.**   For any $p$ with $0 < p < 1$, the $p$th *quantile* of the distribution of a r.v. $X$, denoted by $x_p$, is defined as follows: If $X$ is continuous, then the (essentially) unique $x_p$ is defined by:

$$P(X \leq x_p) = p \quad \text{and} \quad P(X \geq x_p) = 1 - p.$$

For the discrete case, consider two cases: Let $x_k$ be the value for which $P(X \leq x_k) = p$, if such a value exists. Then the unique $p$th *quantile* is defined to be the midpoint between $x_k$ and $x_{k+1}$; i.e., $x_p = (x_k + x_{k+1})/2$. If there is no such value, the unique $p$th *quantile* is defined by the relation: $P(X < x_p) < p$ and $P(X \leq x_p) > p$ (or $P(X \leq x_p) \geq p$ and $P(X \geq x_p) \geq 1 - p$).

Thus, the $p$th quantile is a point $x_p$, which divides the distribution of $X$ into two parts, and $(-\infty, x_p]$ contains exactly $100p\%$ (or at least $100p\%$) of the distribution, and $[x_p, \infty)$ contains exactly $100(1 - p)\%$ (or at least $100(1 - p)\%$) of the distribution of $X$. For $p = 0.50$, we obtain the median. For $p = 0.25$, we have the first quartile, and for $p = 0.75$, we have the third quartile. These concepts are illustrated further by the following examples.

**Example 15.**   Refer to Figure 3.1 ($B(12, 1/4)$) and determine $x_{0.25}, x_{0.50}$, and $x_{0.75}$.

**Discussion.**   Here, $x_{0.25} = 2$ since $P(X < 2) = P(X = 0) + P(X = 1) = 0.1584 \leq 0.25$ and $P(X \leq 2) = 0.1584 + P(X = 2) = 0.3907 \geq 0.25$. Likewise, $x_{0.50} = 3$ since $P(X < 3) = 0.3907 \leq 0.50$ and $P(X \leq 3) = 0.6488 \geq 0.50$. Finally, $x_{0.75} = 4$, since $P(X < 4) = 0.6488 \leq 0.75$ and $P(X \leq 4) = 0.8424 > 0.75$.

**Example 16.**   Refer to Figure 3.4 ($P(5)$) and determine $x_{0.25}, x_{0.50}$, and $x_{0.75}$.

As in the previous example, $x_{0.25} = 2, x_{0.50} = 4$, and $x_{0.75} = 6$.

**Example 17.**   If $X \sim U(0, 1)$, take $p = 0.10, 0.20, 0.30, 0.40, 0.50, 0.60, 0.70, 0.80$, and $0.90$ and determine the corresponding $x_p$.

Here, $F(x) = \int_0^x dt = x$, $0 \leq x \leq 1$. Therefore, $F(x_p) = p$ gives $x_p = p$.

**Example 18.**   If $X \sim N(0, 1)$, take $p$ as in the previous example and determine the corresponding $x_p$.

From the Normal tables, we obtain: $x_{0.10} = -x_{0.90} = -1.282, x_{0.20} = -x_{0.80} = -0.842, x_{0.30} = -x_{0.70} = -0.524, x_{0.40} = -x_{0.60} = -0.253$, and $x_{0.50} = 0$.

Another numerical characteristic which helps shed some light on the distribution of a r.v. $X$ is the so-called mode.

**Definition 6.**   A *mode* of the distribution of a r.v. $X$ is any point, if such points exist, which maximizes the p.d.f. of $X, f$.

A mode, being defined as a maximizing point, is subject to all shortcomings of maximization: It may not exist at all; it may exist but is not obtainable in closed form;

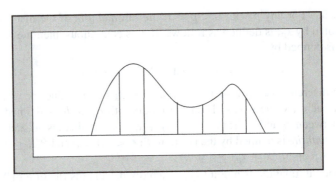

**FIGURE 3.13**

Graph of a unimodal (continuous) p.d.f. Probabilities corresponding to an interval *I* centered at the mode and another two locations.

there may be more than one mode (the distribution is a *multimodal* one). It may also happen that there is a unique mode (*unimodal* distribution). Clearly, if a mode exists, it will be of particular importance for discrete distributions, as the modes provide the values of the r.v. *X* which occur with the largest probability. For the case of continuous r.v.'s, see Figure 3.13. With this in mind, we restrict ourselves to two of the most popular discrete distributions: The Binomial and the Poisson distribution.

**Theorem 2.** *Let X be B(n, p); that is,*

$$f(x) = \binom{n}{x} p^x q^{n-x}, \qquad 0 < p < 1, \ q = 1 - p, \ x = 0, 1, \ldots, n.$$

*Consider the number $(n + 1)p$ and set $m = [(n + 1)p]$, where $[y]$ denotes the largest integer which is $\leq y$. Then, if $(n + 1)p$ is not an integer, $f(x)$ has a unique mode at $x = m$. If $(n + 1)p$ is an integer, then $f(x)$ has two modes obtained for $x = m$ and $x = m - 1$.*

*Proof.* For $x \geq 1$, we have

$$\frac{f(x)}{f(x-1)} = \frac{\binom{n}{x} p^x q^{n-x}}{\binom{n}{x-1} p^{x-1} q^{n-x+1}}$$

$$= \frac{\frac{n!}{x!(n-x)!} p^x q^{n-x}}{\frac{n!}{(x-1)!(n-x+1)!} p^{x-1} q^{n-x+1}} = \frac{n-x+1}{x} \times \frac{p}{q}.$$

That is,

$$\frac{f(x)}{f(x-1)} = \frac{n-x+1}{x} \times \frac{p}{q}.$$

Hence,

$$f(x) > f(x-1) \text{ if and only if } \frac{f(x)}{f(x-1)} > 1 \text{ if and only if}$$

$$\frac{n-x+1}{x} \times \frac{p}{q} > 1 \text{ if and only if } x < (n+1)p. \tag{50}$$

Also,

$$f(x) < f(x-1) \text{ if and only if } \frac{f(x)}{f(x-1)} < 1 \text{ if and only if}$$

$$\frac{n-x+1}{x} \times \frac{p}{q} < 1 \text{ if and only if } x > (n+1)p. \tag{51}$$

Finally,

$$f(x) = f(x-1) \text{ if and only if } \frac{f(x)}{f(x-1)} = 1 \text{ if and only if}$$

$$\frac{n-x+1}{x} \times \frac{p}{q} = 1 \text{ if and only if } x = (n+1)p. \tag{52}$$

First, consider the case that $(n+1)p$ is *not* an integer. Then we have the following diagram:

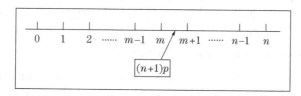

From relations (50) and (51), it follows that:

$$f(x) < f(m), \quad x = 0, 1, \ldots, m-1; \quad f(x) < f(m), \quad x = m+1, \ldots, n,$$

so that $f(x)$ attains its unique maximum at $x = m$.

Now, consider the case that $(n+1)p$ *is* an integer and look at the following diagram:

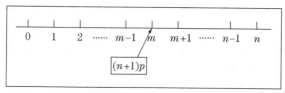

From the relations (52) we have $f(m) = f(m-1)$, whereas the relations (50) and (51), we have that:

$$f(x) < f(m-1), \quad x = 0, 1, \ldots, m-2; \quad f(x) < f(m), \quad x = m+1, \ldots, n,$$

so that $f(x)$ has two maxima at $x = m$ and $x = m-1$. ∎

**Theorem 3.** *Let X be P(λ); that is,*

$$f(x) = e^{-\lambda}\frac{\lambda^x}{x!}, \quad x = 0, 1, 2, \ldots, \quad \lambda > 0.$$

*Then, if λ is not an integer, f(x) has a unique mode at x = [λ]. If λ is an integer, then f(x) has two modes obtained for x = λ and x = λ − 1.*

*Proof.* For $x \geq 1$, we have

$$\frac{f(x)}{f(x-1)} = \frac{e^{-\lambda}\lambda^x/x!}{e^{-\lambda}\lambda^{x-1}/(x-1)!} = \frac{\lambda}{x}.$$

Hence, $f(x) > f(x-1)$ if and only if $\lambda > x$, $f(x) < f(x-1)$ if and only if $x > \lambda$, and $f(x) = f(x-1)$ if and only if $x = \lambda$ in case λ is an integer. Thus, if λ is *not* an integer, $f(x)$ keeps increasing for $x \leq [\lambda]$ and then decreases for $x > [\lambda]$. Thus the maximum of $f(x)$ occurs at $x = [\lambda]$. If λ *is* an integer, then a maximum occurs at $x = \lambda$. But in this case $f(x) = f(x-1)$, which implies that $x = \lambda - 1$ is a second point which gives the maximum value to the p.d.f. ∎

**Example 19.** Let $X \sim B(n, p)$ with $n = 20$ and $p = \frac{1}{4}$. Then $(n+1)p = \frac{21}{4}$ is not an integer and therefore there is a unique mode. Since $\frac{21}{4} = 5.25$, the mode is $[5.25] = 5$. The maximum probability is $\binom{20}{5}(0.25)^5(0.75)^{15} = 0.2024$. If $n = 15$ and $p = \frac{1}{4}$, then $(n+1)p = \frac{16}{4} = 4$ and therefore there are two modes; they are 4 and 3. The respective maximum probability is $\binom{15}{4}(0.25)^4(0.75)^{11} = 0.2252$.

**Example 20.** Let $X \sim P(\lambda)$ and let $\lambda = 4.5$. Then there is a unique mode which is $[4.5] = 4$. The respective maximum probability is 0.1898. If, on the other hand, $\lambda = 7$, then there are two modes 7 and 6. The respective maximum probability is 0.149.

---

## EXERCISES

**4.1** Let X be a r.v. with p.d.f. $f(x) = 3x^2$, for $0 \leq x \leq 1$.
  **(i)** Calculate the $EX$ and the median of X and compare them.
  **(ii)** Determine the 0.125-quantile of X.

**4.2** Let X be a r.v. with p.d.f. $f(x) = x^n$, for $0 \leq x \leq c$ (n positive integer), and let $0 < p < 1$. Determine:
  **(i)** The pth quantile $x_p$ of X in terms of n and p.
  **(ii)** The median $x_{0.50}$ for $n = 3$.

**4.3** **(i)** If the r.v. X has p.d.f. $f(x) = \lambda e^{-\lambda x}$, for $x > 0$ ($\lambda > 0$), determine the pth quantile $x_p$ in terms of λ and p.
  **(ii)** What is the numerical value of $x_p$ for $\lambda = \frac{1}{10}$ and $p = 0.25$?

**4.4**  Let $X$ be a r.v. with p.d.f. $f$ given by:

$$f(x) = \begin{cases} c_1 x^2, & -1 \le x \le 0, \\ c_2(1 - x^2), & 0 < x \le 1, \\ 0, & \text{otherwise.} \end{cases}$$

  **(i)**  If it is also given that $EX = 0$, determine the constants $c_1$ and $c_2$.
  **(ii)**  Determine the $\frac{1}{3}$-quantile of the distribution.
  **Hint.** In part(i), two relations are needed for the determination of $c_1$ and $c_2$.

**4.5**  Let $X$ be a r.v. with d.f. $F$ given in Exercise 2.2 of Chapter 2:
  **(i)**  Determine the mode of the respective p.d.f. $f$.
  **(ii)**  Show that $\frac{1}{2}$ is the $\frac{5}{32} = 0.15625$-quantile of the distribution.

**4.6**  Two fair and distinct dice are rolled once, and let $X$ be the r.v. denoting the sum of the numbers shown, so that the possible values of $X$ are: $2, 3, \ldots, 12$.
  **(i)**  Derive the p.d.f. $f$ of the r.v. $X$.
  **(ii)**  Compute the $EX$.
  **(iii)**  Find the median of $f$, as well as its mode.

**4.7**  Determine the modes of the following p.d.f.'s:
  **(i)**  $f(x) = (\frac{1}{2})^x, x = 1, 2, \ldots$.
  **(ii)**  $f(x) = (1 - \alpha)^x, x = 1, 2, \ldots (0 < \alpha < 1)$. Also, what is the value of $\alpha$?
  **(iii)**  $f(x) = \frac{2}{3^{x+1}}, x = 0, 1, \ldots$.

**4.8**  Let $X \sim B(100, 1/4)$ and suppose you were to bet on the observed value of $X$. On which value would you bet?

**4.9**  In reference to Exercise 3.33, which number(s) of particles arrives within 1 second with the maximum probability?

**4.10**  Let $X$ be a r.v. (of the continuous type) with p.d.f. $f$ symmetric about a constant $c$ (i.e., $f(c - x) = f(c + x)$ for all $x$; in particular, if $c = 0$, then $f(-x) = f(x)$ for all $x$). Then show that $c$ is the median of $X$. (As a by-product of it, we have, e.g., that the mean $\mu$ in the $N(\mu, \sigma^2)$ is also the median.)
  **Hint.** Start with $P(X \le c) = \int_{-\infty}^{c} f(x)dx$ and, by making a change of the variable $x$, show that this last integral equals $\int_0^{\infty} f(c - y)dy$. Likewise, $P(X \ge c) = \int_c^{\infty} f(x)dx$ and a change of the variable $x$ leads to the integral $\int_0^{\infty} f(c + y)dy$. Then the use of symmetry completes the proof.

**4.11**  Let $X$ be a r.v. of the continuous type with p.d.f. $f$, with finite expectation, and median $m$, and let $c$ be any constant. Then:
  **(i)**  Show that:

$$E|X - c| = E|X - m| + 2\int_m^c (c - x)f(x)\,dx.$$

(ii)  Use part (i) to conclude that the constant $c$ which minimizes the $E|X - c|$
is $c = m$.

**Hint.** For $m < c$, show that:

$$|x - c| - |x - m| = \begin{cases} c - m, & x < m, \\ c + m - 2x, & m \leq x \leq c, \\ m - c, & x > c. \end{cases}$$

Then

$$E|X - c| - E|X - m| = \int_{-\infty}^{m} (c - m)f(x)\,dx + \int_{m}^{c} (c + m - 2x)f(x)\,dx$$

$$+ \int_{c}^{\infty} (m - c)f(x)\,dx$$

$$= \frac{c - m}{2} + (c + m)\int_{m}^{c} f(x)\,dx - 2\int_{m}^{c} xf(x)\,dx$$

$$+ (m - c)\int_{m}^{\infty} f(x)\,dx - (m - c)\int_{m}^{c} f(x)\,dx$$

$$= \frac{c - m}{2} + \frac{m - c}{2} + 2c\int_{m}^{c} f(x)\,dx - 2\int_{m}^{c} xf(x)\,dx$$

$$= 2\int_{m}^{c} (c - x)f(x)\,dx.$$

For $m \geq c$, show that:

$$|x - c| - |x - m| = \begin{cases} c - m, & x < c, \\ -c - m + 2x, & c \leq x \leq m, \\ m - c, & x > m. \end{cases}$$

Then

$$E|X - c| - E|X - m| = \int_{-\infty}^{c} (c - m)f(x)\,dx + \int_{c}^{m} (-c - m + 2x)f(x)\,dx$$

$$+ \int_{m}^{\infty} (m - c)f(x)\,dx$$

$$= (c - m)\int_{-\infty}^{m} f(x)\,dx - (c - m)\int_{c}^{m} f(x)\,dx$$

$$- (c + m)\int_{c}^{m} f(x)\,dx + 2\int_{c}^{m} xf(x)\,dx$$

$$+ (m - c)\int_{m}^{\infty} f(x)\,dx$$

$$= \frac{c - m}{2} + \frac{m - c}{2} - 2c\int_{c}^{m} f(x)\,dx + 2\int_{c}^{m} xf(x)\,dx$$

$$= -2\int_{c}^{m} (c - x)f(x)\,dx = 2\int_{m}^{c} (c - x)f(x)\,dx.$$

Combining the two results, we get

$$E|X - c| = E|X - m| + 2 \int_m^c (c - x)f(x)\mathrm{d}x.$$

**4.12** Let $X$ be a continuous r.v. with $p$th quantile $x_p$, and let $Y = g(X)$, where $g$ is a strictly increasing function, so that the inverse $g^{-1}$ exists (and is also strictly increasing). Let $y_p$ be the $p$th quantile of the r.v. $Y$.

(i) Show that $y_p = g(x_p)$.
(ii) If $X$ has the Negative Exponential distribution with $\lambda = 1$, calculate $x_p$ in terms of $p$.
(iii) Use parts (i) and (ii) to determine $y_p$ without calculations, where $Y = e^X$.
(iv) How do parts (ii) and (iii) become for $p = 0.5$?

**4.13** Two Stores, $A$ and $B$, have to compete for a customer's attention. The customers will spend \$50 in the store he chooses (and \$0 in the other store). On Wednesday, store $A$ mails out a number of coupons to the customer. Each coupon is worth of \$1, and the customer will use all of them if he selects store $A$. The number of coupons is a r.v. $X$ with d.f. $F$ shown below. To compete, store $B$ also mails out coupons on Wednesday. Store $B$ knows $F$, but does not know how many coupons store $A$ mailed out that same week. The store picked by the customer is the store with the most coupons. In case of a tie, the customer will choose store $B$, because it is closer to his house.

| $x$ | 1 | 2 | 3 | 4 | 5 | 6 | 7 | 8 | 9 | 10 | 11 | 12 | 13 | 14 | 15 |
|---|---|---|---|---|---|---|---|---|---|---|---|---|---|---|---|
| $F(x)$ | 0.01 | 0.05 | 0.08 | 0.09 | 0.1 | 0.11 | 0.12 | 0.18 | 0.36 | 0.52 | 0.72 | 0.85 | 0.94 | 0.97 | 1 |

(i) Compute the p.d.f. $f$ of the r.v. $X$.
(ii) Find the median and the mode of $X$.
(iii) Compute the $EX$, $\sigma^2(X)$, and $\sigma(X)$.

**4.14** Let $X$ be a r.v. (of the continuous type) with p.d.f. given by:

$$f(x) = x^n, \quad 0 \le x \le c, \text{ and } 0 \text{ otherwise.}$$

(i) Determine the constant $c$, in terms of $n$, so that $f$ is, indeed, a p.d.f.
(ii) For $0 < p < 1$, determine the $p$th quantile of $X$, $x_p$, in terms of $n$ and $p$.
(iii) What is the numerical value of the constant $c$ and of the median $x_{0.5}$ for $n = 3$?

**4.15** Let $X$ be a r.v. (of the continuous type) with p.d.f. $f(x) = c(1 - x^2)$, $0 < x < 1$.

(i) Determine the constant $c$.
(ii) Compute the probability $P(X > 1/2)$.
(iii) What is the $11/16 = 0.6875$th quantile of $X$?

# Joint and conditional p.d.f.'s, conditional expectation and variance, moment generating function, covariance, and correlation coefficient

A brief description of the material discussed in this chapter is as follows. In the first section, two r.v.'s are considered and the concepts of their joint probability distribution function, joint d.f., and joint p.d.f. are defined. The basic properties of the joint d.f. are given, and a number of illustrative examples are provided. On the basis of a joint d.f., marginal d.f.'s are defined. Also, through a joint p.d.f., marginal and conditional p.d.f.'s are defined, and illustrative examples are supplied. By means of conditional p.d.f.'s, conditional expectations and conditional variances are defined and are applied to some examples. These things are done in the second section of the chapter.

In the following section, the expectation is defined for a function of two r.v.'s and some basic properties are listed. As a special case, one obtains the joint m.g.f. of the r.v.'s involved, and from this, marginal m.g.f.'s are derived. Also, as a special case, one obtains the covariance and the correlation coefficient of two r.v.'s. Their significance is explained, and a basic inequality is established, regarding the range of their values. Finally, a formula is provided for the calculation of the variance of the sum of two r.v.'s.

In the fourth section of the chapter, many of the concepts, defined for two r.v.'s in the previous sections, are generalized to $k$ r.v.'s. In the final section, three specific multidimensional distributions are introduced, the Multinomial, the Bivariate (or two-dimensional) Normal, and the Multivariate Normal. The derivation of marginal and conditional p.d.f.'s of the Multinomial and Bivariate Normal distributions is also presented. This section is concluded with a brief discussion of the Multivariate Normal distribution.

## 4.1 JOINT D.F. AND JOINT p.d.f. OF TWO RANDOM VARIABLES

In carrying out a random experiment, we are often interested simultaneously in two outcomes rather than one. Then with each one of these outcomes a r.v. is associated, and thus we are furnished with two r.v.'s or a *two-dimensional random vector.* Let us denote by $(X, Y)$ the two relevant r.v.'s or the two-dimensional random vector. Here are some examples where two r.v.'s arise in a natural way. The pair of r.v.'s $(X, Y)$ denotes, respectively: The SAT and GPA scores of a student from a specified student population; the number of customers waiting for service in two lines in your local favorite bank; the days of a given year that the Dow Jones Industrial Average index closed with a gain and the corresponding gains; the number of hours a student spends daily for studying and for other activities; the weight and the height of an individual from a targeted population; the amount of fertilizer used and the yield of a certain agricultural commodity; the lifetimes of two components used in an electronic system; the dosage of a drug used for treating a certain allergy and the number of days a patient enjoys relief.

We are going to restrict ourselves to the case where both $X$ and $Y$ are either discrete or of the continuous type. The concepts of probability distribution function, distribution function, and probability density function are defined by a straightforward generalization of the definition of these concepts in Section 2.2 of Chapter 2. Thus, the *joint probability distribution function* of $(X, Y)$, to be denoted by $P_{X,Y}$, is defined by: $P_{X,Y}(B) = P[(X, Y) \in B], B \subseteq \Re^2 = \Re \times \Re$, the two-dimensional Euclidean space, the plane. In particular, by taking $B = (-\infty, x] \times (-\infty, y]$, we obtain the *joint d.f.* of $X, Y$, to be denoted by $F_{X,Y}$; namely, $F_{X,Y}(x, y) = P(X \leq x, Y \leq y), x, y \in \Re$. The d.f. $F_{X,Y}$ has properties similar to the ones mentioned in the case of a single r.v., namely:

1. $0 \leq F_{X,Y}(x, y) \leq 1$ for all $x, y \in \Re$.
   Whereas it is, clearly, still true that $x_1 \leq x_2$ and $y_1 \leq y_2$ imply $F_{X,Y}(x_1, y_1) \leq F_{X,Y}(x_2, y_2)$, property #2 in the case of a single r.v. may be restated as follows: $x_1 < x_2$ implies $F_X(x_2) - F_X(x_1) \geq 0$. This property is replaced here by:
2. The *variation* of $F_{X,Y}$ over rectangles with sides parallel to the axes, given in Figure 4.1, is $\geq 0$.
3. $F_{X,Y}$ is continuous from the right (right-continuous); i.e., if $x_n \downarrow x$ and $y_n \downarrow y$, then $F_{X,Y}(x_n, y_n) \rightarrow F_{X,Y}(x, y)$ as $n \rightarrow \infty$.
4. $F_{X,Y}(+\infty, +\infty) = 1$ and $F_{X,Y}(-\infty, -\infty) = F_{X,Y}(-\infty, y) = F_{X,Y}(x, -\infty) = 0$ for any $x, y \in \Re$, where, of course, $F_{X,Y}(+\infty, +\infty)$ is defined to be the $\lim_{n \rightarrow \infty} F_{X,Y}(x_n, y_n)$ as $x_n \uparrow \infty$ and $y_n \uparrow \infty$, and similarly for the remaining cases.

Property #1 is immediate, and property #2 follows by the fact that the variation of $F_{X,Y}$ as described is simply the probability that the pair $(X, Y)$ lies in the rectangle of Figure 4.1, or, more precisely, the probability $P(x_1 < X \leq x_2, y_1 < Y \leq y_2)$, which, of course, is $\geq 0$; the justification of properties #3 and #4 is based on Theorem 2 in Chapter 2.

Now, suppose that the r.v.'s $X$ and $Y$ are discrete and take on the values $x_i$ and $y_j, i, j \geq 1$, respectively. Then the *joint p.d.f.* of $X$ and $Y$, to be denoted by $f_{X,Y}$, is defined by: $f_{X,Y}(x_i, y_j) = P(X = x_i, Y = y_j)$ and $f_{X,Y}(x, y) = 0$ when $(x, y) \neq$

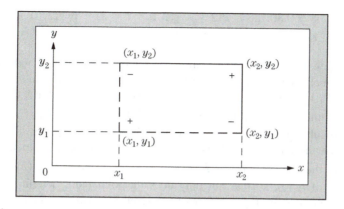

**FIGURE 4.1**

The Variation $V$ of $F_{X,Y}$ over the rectangle is $F_{X,Y}(x_1, y_1) + F_{X,Y}(x_2, y_2) - F_{X,Y}(x_1, y_2) - F_{X,Y}(x_2, y_1)$.

$(x_i, y_j)$ (i.e., at least one of $x$ or $y$ is not equal to $x_i$ or $y_j$, respectively). It is then immediate that for $B \subseteq \mathfrak{R}^2$, $P[(X, Y) \in B] = \sum_{(x_i, y_j) \in B} f_{X,Y}(x_i, y_j)$, and, in particular, $\sum_{(x_i, y_j) \in \mathfrak{R}^2} f_{X,Y}(x_i, y_j) = 1$, and $F_{X,Y}(x, y) = \sum_{x_i \le x, y_j \le y} f_{X,Y}(x_i, y_j)$. In the last relation, $F_{X,Y}$ is expressed in terms of $f_{X,Y}$. The converse is also possible (as it was in the case of a single r.v.), but we do not intend to indulge in it. A simple illustrative example, however, may be in order.

**Example 1.** Each one of the r.v.'s $X$ and $Y$ takes on four values only, 0, 1, 2, 3, with joint probabilities expressed best in a matrix form as in Table 4.1.

**Discussion.** The r.v.'s $X$ and $Y$ may represent, for instance, the number of customers waiting for service in two lines in a bank. Then, for example, for $(x, y)$ with $x = 2$ and $y = 1$, we have $F_{X,Y}(x, y) = F_{X,Y}(2, 1) = \sum_{u \le 2, v \le 1} f_{X,Y}(u, v) = f_{X,Y}(0, 0) + f_{X,Y}(0, 1) + f_{X,Y}(1, 0) + f_{X,Y}(1, 1) + f_{X,Y}(2, 0) + f_{X,Y}(2, 1) = 0.05 + 0.20 + 0.21 +$

**Table 4.1** The joint p.d.f of the r.v.'s in Example 1.

| y\x | 0 | 1 | 2 | 3 | Totals |
|-----|------|------|------|------|--------|
| 0 | 0.05 | 0.21 | 0 | 0 | 0.26 |
| 1 | 0.20 | 0.26 | 0.08 | 0 | 0.54 |
| 2 | 0 | 0.06 | 0.07 | 0.02 | 0.15 |
| 3 | 0 | 0 | 0.03 | 0.02 | 0.05 |
| Totals | 0.25 | 0.53 | 0.18 | 0.04 | 1 |

$0.26 + 0 + 0.08 = 0.80$; also, $P(2 \leq X \leq 3, 0 \leq Y \leq 2) = f_{X,Y}(2,0) + f_{X,Y}(2,1) + f_{X,Y}(2,2) + f_{X,Y}(3,0) + f_{X,Y}(3,1) + f_{X,Y}(3,2) = 0 + 0.08 + 0.07 + 0 + 0 + 0.02 = 0.17$.

Now, suppose that both $X$ and $Y$ are of the continuous type, and, indeed, a little bit more; namely, there exists a nonNegative function $f_{X,Y}$ defined on $\Re^2$ such that, for all $x$ and $y$ in $\Re$: $F_{X,Y}(x,y) = \int_{-\infty}^{y} \int_{-\infty}^{x} f_{X,Y}(s,t)\, ds\, dt$. Then for $B \subseteq \Re^2$ (interpret $B$ as a familiar Geometric figure in $\Re^2$): $P[(X,Y) \in B] = \int_B \int f_{X,Y}(x,y)\, dx\, dy$, and, in particular, $\int_{-\infty}^{\infty} \int_{-\infty}^{\infty} f_{X,Y}(x,y)\, dx\, dy \,(= P[(X,Y) \in \Re^2]) = 1$. The function $f_{X,Y}$ is called the *joint p.d.f.* of the r.v.'s $X$ and $Y$. Analogously to the case of a single r.v., the relationship $\frac{\partial^2}{\partial x\, \partial y} F_{X,Y}(x,y) = f_{X,Y}(x,y)$ holds true (for continuity points $(x,y)$ of $f_{X,Y}$), so that not only does the joint p.d.f. determine the joint d.f. through an integration process, but the converse is also true; i.e., the joint d.f. determines the joint p.d.f. through differentiation. Again, as in the case of a single r.v., $P(X = x, Y = y) = 0$ for all $x, y \in \Re$; also, if a nonNegative function $f$, defined on $\Re^2$, integrates to 1, then there exist two r.v.'s $X$ and $Y$ for which $f$ is their joint p.d.f.

This section is concluded with a reference to Example 37 in Chapter 1, where two continuous r.v.'s $X$ and $Y$ arise in a natural manner. Later on (see Subsection 4.5.2), it may be stipulated that the joint distribution of $X$ and $Y$ is the Bivariate Normal. For the sake of a simpler illustration, consider the following example.

**Example 2.** Let the r.v.'s $X$ and $Y$ have the joint p.d.f. $f_{X,Y}(x,y) = \lambda_1 \lambda_2 e^{-\lambda_1 x - \lambda_2 y}$, $x, y > 0$, $\lambda_1, \lambda_2 > 0$. For example, $X$ and $Y$ may represent the lifetimes of two components in an electronic system. Derive the joint d.f. $F_{X,Y}$.

**Discussion.**  The corresponding joint d.f. is

$$F_{X,Y}(x,y) = \int_0^y \int_0^x \lambda_1 \lambda_2 \times e^{-\lambda_1 s - \lambda_2 t}\, ds\, dt = \int_0^y \lambda_2 e^{-\lambda_2 t} \left( \int_0^x \lambda_1 e^{-\lambda_1 s}\, ds \right) dt$$

$$= \int_0^y \lambda_2 e^{-\lambda_2 t} (1 - e^{-\lambda_1 x})\, dt = (1 - e^{-\lambda_1 x})(1 - e^{-\lambda_2 y})$$

for $x > 0, y > 0$, and 0 otherwise. That is,

$$F_{X,Y}(x,y) = (1 - e^{-\lambda_1 x})(1 - e^{-\lambda_2 y}), \qquad x > 0, y > 0,$$

and   $F_{X,Y}(x,y) = 0$ otherwise. $\hspace{2cm}$ (1)

By letting $x$ and $y \to \infty$, we obtain $F_{X,Y}(\infty, \infty) = 1$, which also shows that $f_{X,Y}$, as given above, is, indeed, a p.d.f., since $F_{X,Y}(\infty, \infty) = \int_0^\infty \int_0^\infty \lambda_1 \lambda_2 \times e^{-\lambda_1 s - \lambda_2 t}\, ds\, dt$.

**Example 3.**  It is claimed that the function $F_{X,Y}$ given by: $F_{X,Y}(x,y) = \frac{1}{16} xy \times (x + y), 0 \leq x \leq 2, 0 \leq y \leq 2$, is the joint d.f. of the r.v.'s $X$ and $Y$. Then:

(i)  Verify that $F_{X,Y}$ is, indeed, a d.f.
(ii)  Determine the corresponding joint p.d.f. $f_{X,Y}$.

(iii) Verify that $f_{X,Y}$ found in part (ii) is, indeed, a p.d.f.
(iv) Calculate the probability: $P(0 \leq X \leq 1, 1 \leq Y \leq 2)$.

**Discussion.**

(i) We have to verify the validity of the defining relations (1)–(4). Clearly, $F_{X,Y}(x,y)$ attains its maximum for $x = y = 2$, which is 1. Since also $F_{X,Y}(x,y) \geq 0$, the first property holds. Next, for any rectangle as in Figure 4.1, we have:

$$16[F_{X,Y}(x_1,y_1) + F_{X,Y}(x_2,y_2) - F_{X,Y}(x_1,y_2) - F_{X,Y}(x_2,y_1)]$$
$$= x_1y_1(x_1 + y_1) + x_2y_2(x_2 + y_2) - x_1y_2(x_1 + y_2) - x_2y_1(x_2 + y_1)$$
$$= x_1^2y_1 + x_1y_1^2 + x_2^2y_2 + x_2y_2^2 - x_1^2y_2 - x_1y_2^2 - x_2^2y_1 - x_2y_1^2$$
$$= -x_1^2(y_2 - y_1) + x_2^2(y_2 - y_1) - y_1^2(x_2 - x_1) + y_2^2(x_2 - x_1)$$
$$= (x_2^2 - x_1^2)(y_2 - y_1) + (x_2 - x_1)(y_2^2 - y_1^2)$$
$$= (x_1 + x_2)(x_2 - x_1)(y_2 - y_1) + (x_2 - x_1)(y_2 + y_1)(y_2 - y_1)$$
$$\geq 0,$$

because $x_1 \leq x_2$ and $y_1 \leq y_2$, so that the second property also holds. The third property holds because $F_{X,Y}$ is continuous, and hence right continuous. Finally, the fourth property holds because as either $x \to -\infty$ or $y \to -\infty$ (in fact, if either one of them is 0), then $F_{X,Y}$ is 0, and if $x \to \infty$ and $y \to \infty$ (in fact, if $x = y = 1$), then $F_{X,Y}$ is 1.

(ii) For $0 \leq x \leq 2$ and $0 \leq y \leq 2$, $f_{X,Y}(x,y) = \frac{\partial^2}{\partial x \partial y}(\frac{1}{16}xy(x+y)) = \frac{1}{16}\frac{\partial^2}{\partial x \partial y}(x^2y + xy^2) = \frac{1}{16}\frac{\partial}{\partial y}\frac{\partial}{\partial x}(x^2y + xy^2) = \frac{1}{16}\frac{\partial}{\partial y}(2xy + y^2) = \frac{1}{16}(2x + 2y) = \frac{1}{8}(x + y)$; i.e., $f_{X,Y}(x,y) = \frac{1}{8}(x+y), 0 \leq x \leq 2, 0 \leq y \leq 2$. For $(x,y)$ outside the rectangle $[0,2] \times [0,2]$, $f_{X,Y}$ is 0, since $F_{X,Y}$ is constantly either 0 or 1.

(iii) Since $f_{X,Y}$ is nonNegative, all we have to show is that it integrates to 1. In fact,

$$\int_{-\infty}^{\infty}\int_{-\infty}^{\infty} f_{X,Y}(x,y)\,dx\,dy = \int_0^2\int_0^2 \frac{1}{8}(x+y)\,dx\,dy$$

$$= \frac{1}{8}\left(\int_0^2\int_0^2 x\,dx\,dy + \int_0^2\int_0^2 y\,dx\,dy\right)$$

$$= \frac{1}{8}(2 \times 2 + 2 \times 2)$$

$$= 1.$$

(iv) Here, $P(0 \leq X \leq 1, 1 \leq Y \leq 2) = \int_1^2\int_0^1 \frac{1}{8}(x+y)dx\,dy = \frac{1}{8}[\int_1^2(\int_0^1 x\,dx)dy + \int_1^2(y\int_0^1 dx)dy] = \frac{1}{8}(\frac{1}{2} \times 1 + 1 \times \frac{3}{2}) = \frac{1}{4}$.

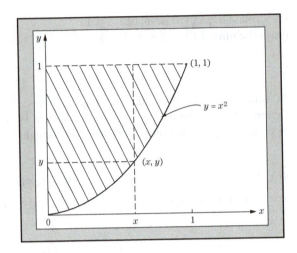

**FIGURE 4.2**

Range of the pair $(x, y)$.

**Example 4.**   If the function $f_{X,Y}$ is given by: $f_{X,Y}(x, y) = cx^2y$ for $0 < x^2 \leq y < 1$, $x > 0$ (and $0$ otherwise):

(i) Determine the constant $c$, so that $f_{X,Y}$ is a p.d.f.
(ii) Calculate the probability: $P(0 < X < \frac{3}{4}, \frac{1}{4} \leq Y < 1)$.

**Discussion.**

(i) Clearly, for the function to be nonNegative, $c$ must be $> 0$. The actual value of $c$ will be determined through the relationship below for $x > 0$:

$$\iint_{\{(x,y);0<x^2\leq y<1\}} cx^2y\, dx\, dy = 1.$$

The region over which the p.d.f. is positive is the shaded region in Figure 4.2, determined by a branch of the parabola $y = x^2$, the $y$-axis, and the line segment connecting the points $(0, 1)$ and $(1, 1)$. Since for each fixed $x$ with $0 < x < 1$, $y$ ranges from $x^2$ to $1$, we have: $\int \int_{\{x^2 \leq y < 1\}} cx^2y\, dx\, dy = c \int_0^1 (x^2 \int_{x^2}^1 y\, dy)\, dx = \frac{c}{2} \int_0^1 x^2(1 - x^4)\, dx = \frac{c}{2}(\frac{1}{3} - \frac{1}{7}) = \frac{2c}{21} = 1$ and $c = \frac{21}{2}$.

(ii) Since $y = x^2 = \frac{1}{4}$ for $x = \frac{1}{2}$, it follows that, for each $x$ with $0 < x \leq \frac{1}{2}$, the range of $y$ is from $\frac{1}{4}$ to $1$; on the other hand, for each $x$ with $\frac{1}{2} < x \leq \frac{3}{4}$, the range of $y$ is from $x^2$ to $1$ (see Figure 4.3).

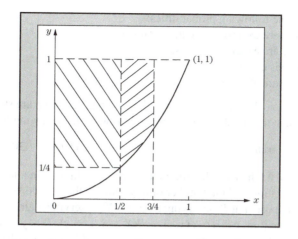

**FIGURE 4.3**

Diagram facilitating integration.

Thus,

$$P\left(0 < X \le \frac{3}{4}, \frac{1}{4} \le Y < 1\right) = c \int_0^{\frac{1}{2}} \int_{\frac{1}{4}}^1 x^2 y \, dy \, dx + c \int_{\frac{1}{2}}^{\frac{3}{4}} \int_{x^2}^1 x^2 y \, dy \, dx$$

$$= c \int_0^{\frac{1}{2}} \left(x^2 \int_{\frac{1}{4}}^1 y \, dy\right) dx + c \int_{\frac{1}{2}}^{\frac{3}{4}} \left(x^2 \int_{x^2}^1 y \, dy\right) dx$$

$$= \frac{c}{2} \int_0^{\frac{1}{2}} x^2 \left(1 - \frac{1}{16}\right) dx + \frac{c}{2} \int_{\frac{1}{2}}^{\frac{3}{4}} x^2 (1 - x^4) \, dx$$

$$= \frac{15c}{3 \times 2^8} + \frac{38c}{3 \times 2^8} - \frac{2059c}{7 \times 2^{15}} = c \times \frac{41{,}311}{21 \times 2^{15}}$$

$$= \frac{21}{2} \times \frac{41{,}311}{21 \times 2^{15}} = \frac{41{,}311}{2^{16}} = \frac{41{,}311}{65{,}536} \simeq 0.63.$$

## EXERCISES

**1.1** Let $X$ and $Y$ be r.v.'s denoting, respectively, the number of cars and buses lined up at a stoplight at a given point in time, and suppose their joint p.d.f. is given by the following table:

| y \ x | 0 | 1 | 2 | 3 | 4 | 5 |
|-------|-------|-------|-------|-------|-------|-------|
| 0 | 0.025 | 0.050 | 0.125 | 0.150 | 0.100 | 0.050 |
| 1 | 0.015 | 0.030 | 0.075 | 0.090 | 0.060 | 0.030 |
| 2 | 0.010 | 0.020 | 0.050 | 0.060 | 0.040 | 0.020 |

Calculate the following probabilities:
  (i) There are exactly four cars and no buses.
  (ii) There are exactly five cars.
  (iii) There is exactly one bus.
  (iv) There are at most three cars and at least one bus.

**1.2** In a sociological project, families with 0, 1, and 2 children are studied. Suppose that the numbers of children occur with the following frequencies:

$$0 \text{ children: } 30\%; \quad 1 \text{ child: } 40\%; \quad 2 \text{ children: } 30\%.$$

Let $X$ and $Y$ be the r.v.'s denoting the number of children in a family from the target population and the number of boys among those children, respectively. Finally, suppose that $P(\text{observing a boy}) = P(\text{observing a girl}) = 0.5$. Calculate the joint p.d.f. $f_{X,Y}(x, y) = P(X = x, Y = y)$, $0 \le y \le x$, $x = 0, 1, 2$. **Hint.** Tabulate the joint probabilities as indicated below by utilizing the formula:

$$P(X = x, Y = y) = P(Y = y \mid X = x)P(X = x).$$

| y \ x | 0 | 1 | 2 |
|-------|---|---|---|
| 0 | | | |
| 1 | | | |
| 2 | | | |

**1.3** If the r.v.'s $X$ and $Y$ have the joint p.d.f. given by:

$$f_{X,Y}(x, y) = x + y, \quad 0 < x < 1, \quad 0 < y < 1,$$

calculate the probability $P(X < Y)$.
**Hint.** Can you guess the answer without doing any calculations?

**1.4** The r.v.'s $X$ and $Y$ have the joint p.d.f. $f_{X,Y}$ given by:

$$f_{X,Y}(x, y) = \frac{6}{7}\left(x^2 + \frac{xy}{2}\right), \quad 0 < x \le 1, \ 0 < y \le 2.$$

  (i) Show that $f_{X,Y}$ is, indeed, a p.d.f.
  (ii) Calculate the probability $P(X > Y)$.

**1.5** The r.v.'s $X$ and $Y$ have the joint p.d.f. $f_{X,Y}(x, y) = e^{-x-y}$, $x > 0, y > 0$.
  (i) Calculate the probability $P(X \le Y \le c)$ for some $c > 0$.
  (ii) Find the numerical value in part (i) for $c = \log 2$, where log is the natural logarithm.
  **Hint.** The integration may be facilitated in part (i) by drawing the picture of the set for which $x \le y$.

**1.6**  If the r.v.'s $X$ and $Y$ have the joint p.d.f. $f_{X,Y}(x, y) = e^{-x-y}$, for $x > 0$ and $y > 0$, compute the following probabilities:

$$(i)\ P(X \leq x);\quad (ii)\ P(Y \leq y);\quad (iii)\ P(X < Y);\quad (iv)\ P(X + Y \leq 3).$$

**Hint.** For part (iii), draw the picture of the set for which $0 < x < y$, and for part(iv), draw the picture of the set for which $0 < x + y \leq 3$

**1.7**  Let $X$ and $Y$ be r.v.'s jointly distributed with p.d.f. $f_{X,Y}(x, y) = 2/c^2$, for $0 < x \leq y < c$.
Determine the constant $c$.
**Hint.** Draw the picture of the set for which $0 < x \leq y < c$.

**1.8**  The r.v.'s $X$ and $Y$ have the joint p.d.f. $f_{X,Y}$ given by:

$$f_{X,Y}(x, y) = cye^{-xy/2},\quad 0 < y < x.$$

Determine the constant $c$.
**Hint.** Draw the picture of the set for which $x > y > 0$.

**1.9**  The joint p.d.f. of the r.v.'s $X$ and $Y$ is given by:

$$f_{X,Y}(x, y) = xy^2,\quad 0 < x \leq c_1,\ 0 < y \leq c_2.$$

Determine the condition that $c_1$ and $c_2$ must satisfy so that $f_{X,Y}$ is, indeed, a p.d.f.
**Hint.** All that can be done here is to find a relation that $c_1$ and $c_2$ satisfy; $c_1$ and $c_2$ cannot be determined separately.

**1.10**  The joint p.d.f. of the r.v.'s $X$ and $Y$ is given by:

$$f_{X,Y}(x, y) = cx,\quad x > 0,\ y > 0,\ 1 \leq x + y < 2\quad (c > 0).$$

Determine the constant $c$.
**Hint.** The above diagram (Figure 4.4) should facilitate the calculations. The range of the pair $(x, y)$ is the shadowed area.

**1.11**  The r.v.'s $X$ and $Y$ have joint p.d.f. $f_{X,Y}$ given by:

$$f_{X,Y}(x, y) = c(y^2 - x^2)e^{-y},\quad -y < x < y,\ 0 < y < \infty.$$

Determine the constant $c$.
**Hint.** We have that $\int_0^\infty ye^{-y}\,dy = 1$, and the remaining integrals are computed recursively.

**1.12**  Consider the Uniform distribution over a square whose side length is one unit, and with center one of the apexes of the square, draw a circle with diameter one unit. What is the probability assigned to the part of the square covered by the circle?

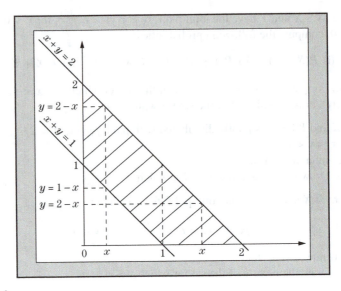

**FIGURE 4.4**

Diagram facilitating integration.

## 4.2 MARGINAL AND CONDITIONAL p.d.f.'s, CONDITIONAL EXPECTATION AND VARIANCE

In the case of two r.v.'s with joint d.f. $F_{X,Y}$ and joint p.d.f. $f_{X,Y}$, we may define quantities which were not available in the case of a single r.v. These quantities are marginal d.f.'s and p.d.f.'s, conditional p.d.f.'s, and conditional expectations and variances. To this end, consider the joint d.f. $F_{X,Y}(x, y) = P(X \leq x, Y \leq y)$, and let $y \to \infty$. Then we obtain $F_{X,Y}(x, \infty) = P(X \leq x, Y < \infty) = P(X \leq x) = F_X(x)$; thus, $F_X(x) = F_{X,Y}(x, \infty)$, and likewise, $F_Y(y) = F_{X,Y}(\infty, y)$. That is, the d.f.'s of the r.v.'s $X$ and $Y$ are obtained from their joint d.f. by eliminating one of the variables $x$ or $y$ through a limiting process. The d.f.'s $F_X$ and $F_Y$ are referred to as *marginal* d.f.'s. If the r.v.'s $X$ and $Y$ are discrete with joint p.d.f. $f_{X,Y}$, then $P(X = x_i) = P(X = x_i, -\infty < Y < \infty) = \sum_{y_j \in \Re} f_{X,Y}(x_i, y_j)$; i.e., $f_X(x_i) = \sum_{y_j \in \Re} f_{X,Y}(x_i, y_j)$, and likewise, $f_Y(y_j) = \sum_{x_i \in \Re} f_{X,Y}(x_i, y_j)$. Because of this marginalization process, the p.d.f.'s of the r.v.'s. $X$ and $Y$, $f_X$ and $f_Y$, are referred to as *marginal* p.d.f.'s. In the continuous case, $f_X$ and $f_Y$ are obtained by integrating out the "superfluous" variables; i.e., $f_X(x) = \int_{-\infty}^{\infty} f_{X,Y}(x, y) dy$ and $f_Y(y) = \int_{-\infty}^{\infty} f_{X,Y}(x, y) dx$. The marginal $f_X$ is, indeed, the p.d.f. of $X$ because $P(X \leq x) = P(X \leq x, -\infty < Y < \infty) = \int_{-\infty}^{x} \int_{-\infty}^{\infty} f_{X,Y}(s, t) dt\, ds = \int_{-\infty}^{x} [\int_{-\infty}^{\infty} f_{X,Y}(s, t) dt]\, ds = \int_{-\infty}^{x} f_X(s) ds$; i.e., $F_X(x) = P(X \leq x) = \int_{-\infty}^{x} f_X(s) ds$, so that $\frac{d}{dx} F_X(x) = f_X(x)$, and likewise, $\frac{d}{dy} F_Y(y) = f_Y(y)$ (for continuity points $x$ and $y$ of $f_X$ and $f_Y$, respectively).

In terms of the joint and the marginal p.d.f.'s, one may define formally the functions:

$$f_{X|Y}(x|y) = f_{X,Y}(x, y)/f_Y(y) \quad \text{for fixed } y \text{ with } f_Y(y) > 0,$$

and

$$f_{Y|X}(y|x) = f_{X,Y}(x, y)/f_X(x) \quad \text{for fixed } x \text{ with } f_X(x) > 0.$$

These nonNegative functions are, actually, p.d.f.'s. For example, for the continuous case:

$$\int_{-\infty}^{\infty} f_{X|Y}(x|y)\, dx = \frac{1}{f_Y(y)} \int_{-\infty}^{\infty} f_{X,Y}(x, y)\, dx = \frac{f_Y(y)}{f_Y(y)} = 1,$$

and similarly for $f_{Y|X}(y|x)$; in the discrete case, integrals are replaced by summation signs. The p.d.f. $f_{X|Y}(\cdot|y)$ is called the *conditional* p.d.f. of $X$, given $Y = y$, and $f_{Y|X}(\cdot \mid x)$ is the *conditional* p.d.f. of $Y$, given $X = x$. The motivation for this terminology is as follows: For the discrete case, $f_{X|Y}(x \mid y) = \frac{f_{X,Y}(x,y)}{f_Y(y)} = \frac{P(X=x,Y=y)}{P(Y=y)} = P(X = x \mid Y = y)$; i.e., $f_{X|Y}(x \mid y)$ does, indeed, stand for the conditional probability that $X = x$, given that $Y = y$. Likewise for $f_{Y|X}(\cdot \mid x)$. In the continuous case, the points $x$ and $y$ are to be replaced by "small" intervals around them.

The concepts introduced so far are now illustrated by means of examples.

**Example 5.**   Refer to Example 1 and derive the marginal and conditional p.d.f.'s involved.

**Discussion.**   From Table 4.1, we have: $f_X(0) = 0.25, f_X(1) = 0.53, f_X(2) = 0.18$, and $f_X(3) = 0.04$; also, $f_Y(0) = 0.26, f_Y(1) = 0.54, f_Y(2) = 0.15$, and $f_Y(3) = 0.05$. Thus, the probability that there are two people in line one, for instance, regardless of how many people are in the other line, is $P(X = 2) = f_X(2) = 0.18$. For $f_{X|Y}(x|y)$:

| $y\backslash x$ | 0 | 1 | 2 | 3 |
|---|---|---|---|---|
| 0 | $\frac{0.05}{0.26} = \frac{5}{26} \simeq 0.192$ | $\frac{0.21}{0.26} = \frac{21}{26} \simeq 0.808$ | 0 | 0 |
| 1 | $\frac{0.20}{0.54} = \frac{20}{54} \simeq 0.37$ | $\frac{0.26}{0.54} = \frac{26}{54} \simeq 0.482$ | $\frac{0.08}{0.54} = \frac{8}{54} \simeq 0.148$ | 0 |
| 2 | 0 | $\frac{0.06}{0.15} = \frac{6}{15} = 0.40$ | $\frac{0.07}{0.15} = \frac{7}{15} \simeq 0.467$ | $\frac{0.02}{0.15} = \frac{2}{15} \simeq 0.133$ |
| 3 | 0 | 0 | $\frac{0.03}{0.05} = \frac{3}{5} = 0.60$ | $\frac{0.02}{0.05} = \frac{2}{5} = 0.40$ |

Likewise, for $f_{Y|X}(y|x)$:

| $y\backslash x$ | 0 | 1 | 2 | 3 |
|---|---|---|---|---|
| 0 | 0.2 | $\frac{21}{53} \simeq 0.396$ | 0 | 0 |
| 1 | 0.8 | $\frac{26}{53} \simeq 0.491$ | $\frac{8}{18} \simeq 0.444$ | 0 |
| 2 | 0 | $\frac{6}{53} \simeq 0.113$ | $\frac{7}{18} \simeq 0.389$ | 0.5 |
| 3 | 0 | 0 | $\frac{3}{18} \simeq 0.167$ | 0.5 |

**Example 6.**   Refer to Example 2 and derive the marginal d.f.'s and p.d.f.'s, as well as the conditional p.d.f.'s, involved.

**Discussion.**   In (1), let $y \to \infty$ to obtain $F_X(x) = 1 - e^{-\lambda_1 x}, x > 0$, and likewise $F_Y(y) = 1 - e^{-\lambda_2 y}, y > 0$, by letting $x \to \infty$. Next, by differentiation, $f_X(x) = \lambda_1 e^{-\lambda_1 x}, x > 0$, and $f_Y(y) = \lambda_2 e^{-\lambda_2 y}, y > 0$, so that the r.v.'s $X$ and $Y$ have the Negative Exponential distribution with parameters $\lambda_1$ and $\lambda_2$, respectively. Finally, for $x > 0$ and $y > 0$:

$$f_{X|Y}(x|y) = \frac{\lambda_1 \lambda_2 e^{-\lambda_1 x - \lambda_2 y}}{\lambda_2 e^{-\lambda_2 y}} = \lambda_1 e^{-\lambda_1 x} = f_X(x), \quad \text{and likewise}$$

$$f_{Y|X}(y|x) = f_Y(y).$$

**Example 7.**   Refer to Example 4 and determine the marginal and conditional p.d.f.'s $f_X, f_Y, f_{X|Y}$, and $f_{Y|X}$.

**Discussion.**   We have:

$$f_X(x) = \int_{x^2}^1 c x^2 y \, dy = c x^2 \int_{x^2}^1 y \, dy = \frac{21}{4} x^2 (1 - x^4), \quad 0 < x < 1,$$

$$f_Y(y) = \int_0^{\sqrt{y}} c x^2 y \, dx = cy \int_0^{\sqrt{y}} x^2 \, dx = \frac{7}{2} y^2 \sqrt{y}, \quad 0 < y < 1,$$

and therefore

$$f_{X|Y}(x|y) = \frac{\frac{21}{2} x^2 y}{\frac{21}{6} y^2 \sqrt{y}} = \frac{3x^2}{y\sqrt{y}}, \quad 0 < x \le \sqrt{y}, \ 0 < y < 1,$$

$$f_{Y|X}(y|x) = \frac{\frac{21}{2} x^2 y}{\frac{21}{4} x^2 (1 - x^4)} = \frac{2y}{1 - x^4}, \quad x^2 \le y < 1, \ 0 < x < 1.$$

**Example 8.**   Consider the function $f_{X,Y}$ defined by:

$$f_{X,Y}(x, y) = 8xy, \quad 0 < x \le y < 1.$$

(i)   Verify that $f_{X,Y}$ is, indeed, a p.d.f.
(ii)  Determine the marginal and conditional p.d.f.'s.
(iii) Calculate the quantities: $EX, EX^2, \text{Var}(X), EY, EY^2, \text{Var}(Y)$, and $E(XY)$.

**Discussion.**

(i)   Since $f_{X,Y}$ is nonNegative, all we have to check is that it integrates to 1. In fact,

$$\int_0^1 \int_0^y 8xy \, dx \, dy = 8 \int_0^1 \left( y \int_0^y x \, dx \right) dy = 4 \int_0^1 y^3 \, dy = 1.$$

**(ii)**

$$f_X(x) = \int_x^1 8xy\,dy = 8x\int_x^1 y\,dy = 4x(1-x^2), \qquad 0 < x < 1,$$

$$f_Y(y) = \int_0^y 8xy\,dx = 8y\int_0^y x\,dx = 4y^3, \qquad 0 < y < 1,$$

and therefore

$$f_{X|Y}(x|y) = \frac{8xy}{4y^3} = \frac{2x}{y^2}, \qquad 0 < x \le y < 1,$$

$$f_{Y|X}(y|x) = \frac{8xy}{4x(1-x^2)} = \frac{2y}{1-x^2}, \qquad 0 < x \le y < 1.$$

**(iii)**

$$EX = \int_0^1 x \times 4x(1-x^2)\,dx = 4\int_0^1 x^2(1-x^2)\,dx = 4\left(\int_0^1 x^2 dx - \int_0^1 x^4 dx\right) = \frac{8}{15},$$

$$EX^2 = \int_0^1 x^2 \times 4x(1-x^2)\,dx = 4\left(\int_0^1 x^3 dx - \int_0^1 x^5 dx\right) = \frac{1}{3}, \quad \text{so that}$$

$$Var(X) = EX^2 - (EX)^2 = \frac{1}{3} - \frac{64}{225} = \frac{11}{225};$$

$$EY = \int_0^1 y \times 4y^3 dy = 4\int_0^1 y^4 dy = \frac{4}{5},$$

$$EY^2 = \int_0^1 y^2 \times 4y^3 dy = 4\int_0^1 y^5 dy = \frac{2}{3}, \quad \text{so that}$$

$$Var(Y) = EY^2 - (EY)^2 = \frac{2}{3} - \frac{16}{25} = \frac{2}{75} = \frac{6}{225}.$$

Finally,

$$E(XY) = \int_0^1 \int_0^y xy \times 8xy\,dx\,dy = 8\int_0^1 y^2\left(\int_0^y x^2 dx\right) dy$$

$$= \frac{8}{3}\int_0^1 y^5 dy = \frac{4}{9}.$$

Once a conditional p.d.f. is at hand, an expectation can be defined as done in relations (1), (2), and (3) of Chapter 3. However, a modified notation will be needed to reveal the fact that the expectation is calculated with respect to a *conditional* p.d.f. The resulting expectation is the *conditional* expectation of one r.v., given the other r.v., as specified below.

$$E(X \mid Y = y_j) = \sum_{x_i \in \Re} x_i f_{X|Y}(x_i \mid y_j) \quad \text{or} \quad E(X \mid Y = y) = \int_{-\infty}^{\infty} x f_{X|Y}(x \mid y)\,dx, \quad (2)$$

for the discrete and continuous case, respectively; similarly:

$$E(Y \mid X = x_i) = \sum_{y_j \in \Re} y_j f_{Y|X}(y_j \mid x_i) \quad \text{or} \quad E(Y \mid X = x) = \int_{-\infty}^{\infty} y f_{Y|X}(y \mid x) dy.$$

(3)

Of course, it is understood that the preceding expectations exist as explained right after relations (2) and (3) in Chapter 3 were defined. However, unlike the results in (1)–(3) in Chapter 3 which are numbers, in relations (2) and (3) above the outcomes depend on $y_j$ or $y$, and $x_i$ or $x$, respectively, which reflect the values that the "conditioning" r.v.'s assume. For illustrative purposes, let us calculate some conditional expectations.

**Example 9.** In reference to Example 1, calculate: $E(X \mid Y = 0)$ and $E(Y \mid X = 2)$.

**Discussion.** In Example 5, we have calculated the conditional p.d.f.'s $f_{X|Y}(\cdot \mid 0)$ and $f_{Y|X}(\cdot \mid 2)$. Therefore:

$$E(X \mid Y = 0) = 0 \times \frac{5}{26} + 1 \times \frac{21}{26} + 2 \times 0 + 3 \times 0 = \frac{21}{26} \simeq 0.808, \quad \text{and}$$

$$E(Y \mid X = 2) = 0 \times 0 + 1 \times \frac{8}{18} + 2 \times \frac{7}{18} + 3 \times \frac{3}{18} = \frac{31}{18} \simeq 1.722.$$

So, if in the $y$-line there are no customers waiting, the expected number of those waiting in the $x$-line will be about 0.81; likewise, if there are 2 customers waiting in the $x$-line, the expected number of those waiting in the $y$-line will be about 1.72.

**Example 10.** In reference to Example 2, calculate: $E(X \mid Y = y)$ and $E(Y \mid X = x)$.

**Discussion.** In Example 6, we have found that $f_{X|Y}(x \mid y) = f_X(x) = \lambda_1 e^{-\lambda_1 x}$ $(x > 0)$, and $f_{Y|X}(y \mid x) = f_Y(y) = \lambda_2 e^{-\lambda_2 y}$ $(y > 0)$, so that: $E(X \mid Y = y) = \int_0^\infty x \lambda_1 e^{-\lambda_1 x} dx = 1/\lambda_1$, and $E(Y \mid X = x) = \int_0^\infty y \lambda_2 e^{-\lambda_2 y} dy = 1/\lambda_2$, by integration by parts, or simply by utilizing known results.

**Example 11.** In reference to Example 4, calculate: $E(X \mid Y = y)$ and $E(Y \mid X = x)$.

**Discussion.** In Example 7, we have found that $f_{X|Y}(x \mid y) = \frac{3x^2}{y\sqrt{y}}$, $0 < x \leq \sqrt{y}$, so that

$$E(X \mid Y = y) = \int_0^{\sqrt{y}} x \times \frac{3x^2}{y\sqrt{y}} dx = \frac{3}{y\sqrt{y}} \int_0^{\sqrt{y}} x^3 dx = \frac{3\sqrt{y}}{4}, \quad 0 < y < 1.$$

Also, $f_{Y|X}(y \mid x) = \frac{2y}{1-x^4}$, $x^2 \leq y < 1$, so that

$$E(Y \mid X = x) = \int_{x^2}^1 y \times \frac{2y}{1-x^4} dy = \frac{2}{1-x^4} \int_{x^2}^1 y^2 dy = \frac{2(1-x^6)}{3(1-x^4)}, \quad 0 < x < 1.$$

**Example 12.**    In reference to Example 8, calculate: $E(X \mid Y = y)$ and $E(Y \mid X = x)$.

**Discussion.**    In Example 8(ii), we have found that $f_{X\mid Y}(x \mid y) = \frac{2x}{y^2}$, $0 < x \leq y < 1$, and $f_{Y\mid X}(y \mid x) = \frac{2y}{1-x^2}$, $0 < x \leq y < 1$, so that

$$E(X \mid Y = y) = \int_0^y x \times \frac{2x}{y^2}dx = \frac{2}{y^2}\int_0^y x^2\,dx = \frac{2y}{3}, \qquad 0 < y < 1,$$

and

$$E(Y \mid X = x) = \int_x^1 y \times \frac{2y}{1-x^2}dy = \frac{2}{1-x^2}\int_x^1 y^2\,dy = \frac{2(1-x^3)}{3(1-x^2)}, \quad 0 < x < 1.$$

Now, for the discrete case, set $g(y_j) = E(X \mid Y = y_j)$ and proceed to replace $y_j$ by the r.v. $Y$. We obtain the r.v. $g(Y) = E(X \mid Y)$, and then it makes sense to talk about its expectation $Eg(Y) = E[E(X \mid Y)]$. Although the $E(X \mid Y = y_j)$ depends on the particular values of $Y$, it turns out that its average does not, and, indeed, is the same as the $EX$. More precisely, it holds:

$$E[E(X \mid Y)] = EX \quad \text{and} \quad E[E(Y \mid X)] = EY. \tag{4}$$

That is, the expectation of the conditional expectation of $X$ is equal to its expectation, and likewise for $Y$. Relation (4) is true both for the discrete and the continuous case. Its justification for the continuous case, for instance, is as follows:

We have $g(Y) = E(X \mid Y)$ and therefore

$$Eg(Y) = \int_{-\infty}^{\infty} g(y)f_Y(y)dy = \int_{-\infty}^{\infty} E(X \mid y)f_Y(y)dy$$

$$= \int_{-\infty}^{\infty} \left[\int_{-\infty}^{\infty} xf_{X\mid Y}(x \mid y)dx\right]f_Y(y)dy$$

$$= \int_{-\infty}^{\infty}\int_{-\infty}^{\infty} [xf_{X\mid Y}(x \mid y)f_Y(y)dx]dy = \int_{-\infty}^{\infty}\int_{-\infty}^{\infty} xf_{X,Y}(x,y)dx\,dy$$

$$= \int_{-\infty}^{\infty} x\left[\int_{-\infty}^{\infty} f_{X,Y}(x,y)dy\right]dx = \int_{-\infty}^{\infty} xf_X(x)dx = EX; \text{ i.e.,}$$

$$Eg(Y) = E[E(X \mid Y)] = EX.$$

**Remark 1.**    However, $\mathrm{Var}[E(X \mid Y)] \leq \mathrm{Var}(X)$ with equality holding, if and only if $Y$ is a function of $X$ (with probability 1). A proof of this fact may be found in Section 5.3.1 in the book *A Course in Mathematical Statistics*, 2nd edition (1997), Academic Press, by G. G. Roussas.

**Example 13.**    Verify the first relation $E[E(X \mid Y)] = EX$, in (4) for Examples 4 and 8.

**Discussion.**    By Example 7, $f_X(x) = \frac{21}{4}x^2(1-x^4)$, $0 < x < 1$, so that

$$EX = \int_0^1 x \times \frac{21}{4}x^2(1-x^4)dx = \frac{21}{4}\left(\int_0^1 x^3dx - \int_0^1 x^7dx\right) = \frac{21}{32}.$$

From Example 11, $E(X \mid Y) = \frac{3\sqrt[3]{Y}}{4}$, $0 < Y < 1$, whereas, from Example 7, $f_Y(y) = \frac{21}{6}y^2\sqrt{y}$, $0 < y < 1$, so that

$$E[E(X \mid Y)] = \int_0^1 \frac{3\sqrt[3]{y}}{4} \times \frac{21}{6}y^2\sqrt{y}\,dy = \frac{21}{8}\int_0^1 y^3\,dy = \frac{21}{32} = EX.$$

(However,    $\text{Var}[E(X \mid Y)] = \text{Var}(\frac{3\sqrt[3]{Y}}{4}) = \frac{9}{16}\text{Var}(\sqrt{Y}) = \frac{9}{16}[EY - (E\sqrt{Y})^2] = \frac{9}{16}(\frac{4}{5} - \frac{49}{64}) = \frac{99}{5120} < \frac{2}{75} = \text{Var}(Y)$.)

Also, from Examples 12 and 8(ii), $E(X \mid Y) = \frac{2Y}{3}$, $0 < Y < 1$, and $f_Y(y) = 4y^3$, $0 < y < 1$, and $EX = \frac{8}{15}$ by Example 8(iii), so that

$$E[E(X \mid Y)] = \int_0^1 \frac{2y}{3} \times 4y^3\,dy = \frac{8}{3}\int_0^1 y^4\,dy = \frac{8}{15} = EX.$$

(However, $\text{Var}[E(X \mid Y)] = \text{Var}(\frac{2Y}{3}) = \frac{4}{9}\text{Var}(Y) < \text{Var}(Y)$.)

In addition to the conditional expectation of $X$, given $Y$, one may define the *conditional variance* of $X$, given $Y$, by utilizing the conditional p.d.f. and formula (8) in Chapter 3; the notation to be used is $\text{Var}(X \mid Y = y_j)$ or $\text{Var}(X \mid Y = y)$ for the discrete and continuous case, respectively. Thus:

$$\text{Var}(X \mid Y = y_j) = \sum_{x_i \in \Re} [x_i - E(X \mid Y = y_j)]^2 f_{X|Y}(x_i \mid y_j), \tag{5}$$

and

$$\text{Var}(X \mid Y = y) = \int_{-\infty}^{\infty} [x - E(X \mid Y = y)]^2 f_{X|Y}(x \mid y)dx, \tag{6}$$

for the discrete and the continuous case, respectively. The conditional variances depend on the values of the conditioning r.v., as was the case for the conditional expectations. From formulas (5) and (6), it is not hard to see (see also Exercise 2.20) that:

$$\text{Var}(X \mid Y = y_j) = E(X^2 \mid Y = y_j) - [E(X \mid Y = y_j)]^2 \quad \text{or}$$

$$\text{Var}(X \mid Y = y) = E(X^2 \mid Y = y) - [E(X \mid Y = y)]^2, \tag{7}$$

for the discrete and the continuous case, respectively.

**Example 14.**   In reference to Example 8, determine $\text{Var}(X \mid Y = y)$ by using the second formula in (7).

**Discussion.** By (7),

$$\text{Var}(X \mid Y = y) = E(X^2 \mid Y = y) - [E(X \mid Y = y)]^2$$

$$= \int_0^y x^2 \times \frac{2x}{y^2}dx - \left(\frac{2y}{3}\right)^2 \quad \text{(by Examples 8(ii) and 12)}$$

$$= \frac{2}{y^2}\int_0^y x^3\,dx - \frac{4y^2}{9} = \frac{y^2}{2} - \frac{4y^2}{9} = \frac{y^2}{18}, \quad 0 < y < 1.$$

## EXERCISES

**2.1**  Refer to Exercise 1.1 and calculate the marginal p.d.f.'s $f_X$ and $f_Y$.

**2.2**  Refer to Exercise 1.2 and calculate the marginal p.d.f.'s $f_X$ and $f_Y$.

**2.3**  If the joint p.d.f. of the r.v.'s $X$ and $Y$ is given by the following table, determine the marginal p.d.f.'s $f_X$ and $f_Y$.

| $y \setminus x$ | −4 | −2 | 2 | 4 |
|---|---|---|---|---|
| −2 | 0 | 0.25 | 0 | 0 |
| −1 | 0 | 0 | 0 | 0.25 |
| 1 | 0.25 | 0 | 0 | 0 |
| 2 | 0 | 0 | 0.25 | 0 |

**2.4**  The r.v.'s $X$ and $Y$ take on the values 1, 2, and 3, as indicated in the following table:

| $y \setminus x$ | 1 | 2 | 3 |
|---|---|---|---|
| 1 | 2/36 | 2/36 | 3/36 |
| 2 | 1/36 | 10/36 | 3/36 |
| 3 | 4/36 | 5/36 | 6/36 |

  **(i)**  Determine the marginal p.d.f.'s $f_X$ and $f_Y$.
  **(ii)**  Determine the conditional p.d.f.'s $f_{X|Y}(\cdot \mid y)$ and $f_{Y|X}(\cdot \mid x)$.

**2.5**  The r.v.'s $X$ and $Y$ have joint p.d.f. $f_{X,Y}$ given by the entries of the following table:

| $y \setminus x$ | 0 | 1 | 2 | 3 |
|---|---|---|---|---|
| 1 | 1/8 | 1/16 | 3/16 | 1/8 |
| 2 | 1/16 | 1/16 | 1/8 | 1/4 |

  **(i)**  Determine the marginal p.d.f.'s $f_X$ and $f_Y$, and the conditional p.d.f. $f_{X|Y}(\cdot \mid y)$, $y = 1, 2$.
  **(ii)**  Calculate: $EX, EY, E(X \mid Y = y)$, $y = 1, 2$, and $E[E(X \mid Y)]$.
  **(iii)**  Compare $EX$ and $E[E(X \mid Y)]$.
  **(iv)**  Calculate: $\text{Var}(X)$ and $\text{Var}(Y)$.

**2.6**  Let the r.v.'s $X$ and $Y$ have the joint p.d.f.:

$$f_{X,Y}(x, y) = \frac{2}{n(n+1)}, \qquad y = 1, \ldots, x; \ x = 1, \ldots, n.$$

Then compute:
  **(i)**  The marginal p.d.f.'s $f_X$ and $f_Y$.
  **(ii)**  The conditional p.d.f.'s $f_{X|Y}(\cdot \mid y)$ and $f_{Y|X}(\cdot \mid x)$.
  **(iii)**  The conditional expectations $E(X \mid Y = y)$ and $E(Y \mid X = x)$.
  **Hint.** For part (iii), use the appropriate part of #1 in Table 8 in the Appendix.

**2.7** In reference to Exercise 1.3, calculate the marginal p.d.f.'s $f_X$ and $f_Y$.

**2.8** Show that the marginal p.d.f.'s of the r.v.'s $X$ and $Y$ whose joint p.d.f. is given by:

$$f_{X,Y}(x, y) = \frac{6}{5}(x + y^2), \qquad 0 \le x \le 1, \quad 0 \le y \le 1,$$

are as follows:

$$f_X(x) = \frac{2}{5}(3x + 1), \quad 0 \le x \le 1, \ f_Y(y) = \frac{3}{5}(2y^2 + 1), \ 0 \le y \le 1.$$

**2.9** Let $X$ and $Y$ be two r.v.'s with joint p.d.f. given by:

$$f_{X,Y}(x, y) = ye^{-x}, \qquad 0 < y \le x < \infty.$$

- **(i)** Determine the marginal p.d.f.'s $f_X$ and $f_Y$, and specify the range of the arguments involved.
- **(ii)** Determine the conditional p.d.f.'s $f_{X|Y}(\cdot \mid y)$ and $f_{Y|X}(\cdot \mid x)$, and specify the range of the arguments involved.
- **(iii)** Calculate the (conditional) probability $P(X > 2\log 2 \mid Y = \log 2)$, where always log stands for the natural logarithm.

**2.10** The joint p.d.f. of the r.v.'s $X$ and $Y$ is given by:

$$f_{X,Y}(x, y) = xe^{-(x+y)}, \qquad x > 0, \quad y > 0.$$

- **(i)** Determine the marginal p.d.f.'s $f_X$ and $f_Y$.
- **(ii)** Determine the conditional p.d.f. $f_{Y|X}(\cdot \mid x)$.
- **(iii)** Calculate the probability $P(X > \log 4)$, whereas as always log stands for the natural logarithm.

**2.11** The joint p.d.f. of the r.v.'s $X$ and $Y$ is given by:

$$f_{X,Y}(x, y) = \frac{1}{2}ye^{-xy}, \qquad 0 < x < \infty, \quad 0 < y < 2.$$

- **(i)** Determine the marginal p.d.f. $f_Y$.
- **(ii)** Find the conditional p.d.f. $f_{X|Y}(\cdot \mid y)$, and evaluate it at $y = 1/2$.
- **(iii)** Compute the conditional expectation $E(X \mid Y = y)$, and evaluate it at $y = 1/2$.

**2.12** In reference to Exercise 1.4, calculate:
- **(i)** The marginal p.d.f.'s $f_X, f_Y$, and the conditional p.d.f. $f_{Y|X}(\cdot \mid x)$; in all cases, specify the range of the variables involved.
- **(ii)** $EY$ and $E(Y \mid X = x)$.
- **(iii)** $E[E(Y \mid X)]$ and observe that it is equal to $EY$.
- **(iv)** The probability $P(Y > \frac{1}{2} \mid X < \frac{1}{2})$.

**2.13** In reference to Exercise 1.7, calculate:
  **(i)** The marginal p.d.f.'s $f_X$ and $f_Y$.
  **(ii)** The conditional p.d.f.'s $f_{X|Y}(\cdot \mid y)$ and $f_{Y|X}(\cdot \mid x)$.
  **(iii)** The probability $P(X \leq 1)$.

**2.14** In reference to Exercise 1.8, determine the marginal p.d.f. $f_Y$ and the conditional p.d.f. $f_{X|Y}(\cdot \mid y)$.

**2.15** In reference to Exercise 1.9:
  **(i)** Determine the marginal p.d.f.'s $f_X$ and $f_Y$.
  **(ii)** Determine the conditional p.d.f. $f_{X|Y}(\cdot \mid y)$.
  **(iii)** Calculate the $EX$ and $E(X \mid Y = y)$.
  **(iv)** Show that $E[E(X \mid Y)] = EX$.

**2.16** In reference to Exercise 1.10, determine:
  **(i)** The marginal p.d.f. $f_X$.
  **(ii)** The conditional p.d.f. $f_{Y|X}(\cdot \mid x)$.
  **Hint.** Consider separately the cases $0 \leq x \leq 1$ and $1 < x \leq 2$, $x$ whatever else.

**2.17** In reference to Exercise 1.11, determine:
  **(i)** The marginal p.d.f. $f_Y$.
  **(ii)** The conditional p.d.f. $f_{X|Y}(\cdot \mid y)$.
  **(iii)** The marginal p.d.f. $f_X$.
  **Hint.** For part (iii), consider separately the cases that $x < 0$ (so that $-x < y$) and $x \geq 0$ (so that $x > y$).

**2.18** **(i)** For a fixed $y > 0$, consider the function $f(x, y) = e^{-y}\frac{y^x}{x!}$, $x = 0, 1, \ldots$ and show that it is the conditional p.d.f. of a r.v. $X$, given that another r.v. $Y = y$, $f_{X|Y}(\cdot \mid y)$.
  **(ii)** Now, suppose that the marginal p.d.f. of $Y$ is Negative Exponential with parameter $\lambda = 1$. Determine the joint p.d.f. of the r.v.'s $X$ and $Y$.
  **(iii)** Show that the marginal p.d.f. $f_X$ is given by:

$$f_X(x) = \left(\frac{1}{2}\right)^{x+1}, \qquad x = 0, 1, \ldots.$$

  **Hint.** For part (iii), observe that the integrand is essentially the p.d.f. of a Gamma distribution (except for constants). Also, use the fact that $\Gamma(x + 1) = x!$ for $x \geq 0$ integer.

**2.19** Suppose the r.v. $Y$ is distributed as $P(\lambda)$ and that the conditional p.d.f. of a r.v. $X$, given $Y = y$, is $B(y, p)$. Then:
  **(i)** Find the joint p.d.f. of the r.v.'s $X$ and $Y$, $f_{X,Y}$.
    Show that:
  **(ii)** The marginal p.d.f. $f_X$ is Poisson with parameter $\lambda p$.
  **(iii)** The conditional p.d.f. $f_{Y|X}(\cdot \mid x)$ is Poisson with parameter $\lambda q$ (with $q = 1 - p$) over the set: $x, x + 1, \ldots$.

**Hint.** For part (i), form first the joint p.d.f. of $X$ and $Y$. Also, use the appropriate part of #6 of Table 8 in the Appendix.

**2.20** **(i)** Let $X$ and $Y$ be two discrete r.v.'s with joint p.d.f. $f_{X,Y}$. Then show that the conditional variance of $X$, given $Y$, satisfies the following relation:

$$\text{Var}(X \mid Y = y_j) = E(X^2 \mid Y = y_j) - [E(X \mid Y = y_j)]^2.$$

**(ii)** Establish the same relation, if the r.v.'s $X$ and $Y$ are of the continuous type.

**2.21** In reference to Exercise 3.76(iii), (iv) (in this chapter), suppose that $P(Y = y|X = x) = pq^y$, $y = 0, 1, \ldots$, where $q = 1 - p$, so that $E(Y|X = x) = q/p$. Further, assume that $E(Y|X = x) = x^2$ as in Exercise 3.76. Then determine the values of $p = p(x)$ for $x = 0, 1, 2$.

**2.22** If for the r.v.'s $X$, $Y$, the functions $g$, $h$: $\Re \to \Re$ are such that $E[g(X)h(Y)]$, $Eh(Y)$ are finite, then show that: $E[g(X)h(Y)|X = x] = g(x)E[h(Y)|X = x]$.

**2.23** A couple has $X$ children, where $X \sim P(\lambda_1)$, and each one of the children produces $Y$ children with $Y \sim P(\lambda_2)$.
**(i)** What is the expected number of the grandchildren of the couple?
**(ii)** What is the numerical value in part (i) for $\lambda_1 = 4$ and $\lambda_2 = 3$?

**2.24** The r.v.'s $X$ and $Y$ (of the continuous type) have joint p.d.f. given by:
$f_{X,Y}(x, y) = 8xy$, $0 < x < y < 1$.
**(i)** Determine the marginal p.d.f. $f_X$.
**(ii)** Show that the conditional p.d.f. $f_{Y|X}(\cdot|x)$ is given by: $f_{Y|X}(y|x) = \dfrac{2y}{1 - x^2}$, $0 < x < y < 1$.
**(iii)** Compute the conditional expectation: $E(Y|X = x)$.
**(iv)** What is the value of $E(Y|X = x)$ for $x = 1/2$?

**2.25** The (discrete) r.v.'s $X$ and $Y$ have the joint p.d.f. $f_{X,Y}$ where values are given in the table below:

| Y\X | 1 | 2 |
|-----|-----|-----|
| 1 | 0.2 | 0.4 |
| 2 | 0.3 | 0.1 |

Then determine:
**(i)** The marginal p.d.f.'s $f_X, f_Y$.
**(ii)** The conditional p.d.f. $f_{X|Y}(\cdot|y)$, and the specific values $f_{X|Y}(1|1)$. $f_{X|Y}(2|1)$; $f_{X|Y}(1|2)$, $f_{X|Y}(2|2)$.
**(iii)** The $EX$, the $\sigma^2(X)$, and the $\sigma(X)$; likewise, $EY$, $\sigma^2(Y)$, and $\sigma(Y)$.
**(iv)** The $E(XY)$.
**(v)** The $\text{Cov}(X, Y)$ and the $\rho(X, Y)$.
**(vi)** The conditional expectation $E(X|Y = 1)$ and $E(X|Y = 2)$.
**(vii)** Verify that $E[E(X|Y)] = EX$.

**2.26** If the r.v.'s $X$ and $Y$ have the joint p.d.f.:

$$f_{X,Y}(x, y) = x + y, \ 0 < x < 1, 0 < y < 1,$$

then show that:

(i) $f_X(x) = x + 1/2, 0 < x < 1; f_Y(y) = y + 1/2, 0 < y < 1;$
$f_{X|Y}(x|y) = \frac{x+y}{y+1/2}, 0 < x < 1, 0 < y < 1.$

(ii) $EX = EY = \frac{7}{12}; EX^2 = EY^2 = \frac{5}{12}; \sigma^2(X) = \sigma^2(Y) = \frac{11}{144}.$

(iii) $E(X|Y = y) = \frac{3y+2}{6y+2}, 0 < y < 1; \sigma^2(X|Y = y) = \frac{6y^2+6y+1}{18(2y+1)^2}, 0 < y < 1.$

**2.27** The r.v. $X$ has p.d.f. $f(x) = 20x^3(1 - x), 0 < x < 1$, and the conditional p.d.f. of $Y$, given $X = x$, is $U(0, 1 - x)$. Derive:

(i) The joint p.d.f. of $X$ and $Y$.

(ii) The marginal p.d.f. of $Y$.

(iii) The conditional p.d.f. of $X$, given $Y = y$.

(iv) The conditional expectations and variances $E(Y|X = x), E(X|Y = y)$;
$\sigma^2(Y|X = x), \sigma^2(X|Y = y).$

**2.28** The joint p.d.f. of the r.v.'s $X$ and $Y$ is given by: $f_{X,Y}(x, y) = cxy^2$,
$0 < x < y < 1 \ (c > 0)$.

(i) Determine the constant $c$.

(ii) Find the marginal p.d.f.'s $f_X$ and $f_Y$.

(iii) Find the conditional p.d.f.'s $f_{X|Y}(\cdot|y)$ and $f_{Y|X}(\cdot|x)$.

(iv) Determine $f_{X|Y}(x|y = 1/2)$ and $f_{Y|X}(y|x = 1/3)$.

**2.29** The r.v.'s $X$ and $Y$ have joint p.d.f. given by:

$$f_{X,Y}(x, y) = c^2 e^{-cy}, \quad 0 < x < y \ (c > 0).$$

(i) Determine the constant $c$.

(ii) Find the marginal p.d.f.'s $f_X$ and $f_Y$, and identify them.

(iii) Write down the $EX, \sigma^2(X)$, and $EY, \sigma^2(Y)$ without any integrations.

(iv) Find the conditional p.d.f. $f_{X|Y}(\cdot|y)$, and write down the $E(X|Y = y)$ and $\sigma^2(X|Y = y)$ without any integrations.

(v) Find the conditional p.d.f. $f_{Y|X}(\cdot|x)$.

(vi) Integrating over the region specified by the straight lines $y = x$ and $y = 2x$, show that $P(Y < 2X) = \frac{1}{2}$.

**2.30** If the joint p.d.f. of the r.v.'s $X$ and $Y$ is given by: $f_{X,Y}(x, y) = \frac{cx^2}{y}$,
$0 < x < y < 1 \ (c > 0)$:

(i) Determine the constant $c$.

(ii) Find the marginal p.d.f.'s $f_X$ and $f_Y$.

(iii) Find the conditional p.d.f.'s $f_{X|Y}(\cdot|y)$ and $f_{Y|X}(\cdot|x)$.

**2.31** The joint p.d.f. of the r.v.'s $X$ and $Y$ is given by:

$$f_{X,Y}(x, y) = cxe^{-x(y+c)}, \quad x > 0, y > 0 \ (c > 0).$$

   (i) Determine the marginal p.d.f. $f_X$, and identify it. From this conclude that $c$ can be any positive constant.

   (ii) Determine the conditional p.d.f. $f_{Y|X}(\cdot|x)$, and identify it.

   (iii) Use part (ii) in order to write down the $E(Y|X = x)$ and the $\sigma^2(Y|X = x)$ without any integrations.

   (iv) Find the p.d.f. $f_Y$ without integration.

**2.32** The r.v.'s $X$ and $Y$ have joint p.d.f. given by:

$$f_{X,Y}(x, y) = cx(1 - x)y^2, \; 0 < x < 1, 0 < y < 1 \; (c > 0).$$

   (i) Determine the constant $c$.

   (ii) Derive the marginal p.d.f.'s $f_X$ and $f_Y$.

   (iii) Compute the $EX$ and $EY$.

   (iv) Compute the probability $P(X < Y)$.

**2.33** The r.v.'s $X$ and $Y$ have joint p.d.f. given by:

$$f_{X,Y}(x, y) = cy^2, \; 0 < y < x < 1 \; (c > 0).$$

   (i) Determine the constant $c$.

   (ii) Derive the marginal p.d.f.'s $f_X$ and $f_Y$.

   (iii) Derive the conditional p.d.f.'s $f_{Y|X}(\cdot|x)$ and $f_{X|Y}(\cdot|y)$.

   (iv) Find the conditional expectations $E(Y|X = x)$, and evaluate it at $x = 0.7$, $0.8, 0.9$.

**2.34** Let the r.v. $Y \sim P(\lambda)$, and suppose that the conditional distribution of the r.v. $X$, given $Y = y$, is $B(y, p)$; i.e., $X|Y = y \sim B(y, p)$. Then show that the (unconditional) p.d.f. of $X$ is $P(\lambda p)$; i.e., $X \sim P(\lambda p)$.

   **Hint.** For $x = 0, 1, \ldots$, write $P(X = x) = \sum_{y=0}^{\infty} P(X = x, Y = y) = \sum_{y=0}^{\infty} P(X = x|Y = y)P(Y = y)$, use what is given, recall that $\binom{y}{x} = 0$ for $y < x$, and that $\sum_{k=0}^{\infty} \frac{u^k}{k!} = e^u$.

**2.35** Consider the r.v.'s $X$ and $N$, where $X \sim B(n, p)$ and $N \sim P(\lambda)$. Next, let $X(k)$ be the r.v. defined as follows: $X(k)|N = k \sim B(k + 1, p)$. So, for $N = k$, $X(k) = Y_0 + Y_1 + \cdots + Y_k$, where the r.v.'s $Y_i$ are independent $\sim B(1, p)$, and $X(N) = Y_0 + Y_1 + \cdots + Y_N$.

   (i) Use this expression of $X(N)$ in order to compute the $EX(N)$, and the $P[X(N) = 0]$.

   (ii) State a condition in terms of $n$, $p$, and $\lambda$ for which $EX > EX(N)$.

   (iii) Do the same as in part (ii) for the inequality $P(X = 0) > P[X(N) = 0]$.

**2.36** In reference to Exercise 2.28:

   (i) Compute the $EX$ and $\sigma^2(X)$.

   (ii) Also, compute $EY$ and $\sigma^2(Y)$.

(iii) Furthermore, find $E(X|Y = y)$ and $E(Y|X = x)$, and evaluate them at $y = \frac{1}{3}$ and $x = \frac{1}{2}$.

(iv) Finally, verify that $E[E(X|Y)] = EX$, and $E[E(Y|X)] = EY$.

**2.37** The r.v.'s $X$ and $Y$ have a joint p.d.f. given by:

$$f_{X,Y}(x, y) = x + y, \quad 0 < x < 1, \, 0 < y < 1.$$

(i) Verify that this function is, indeed, a p.d.f.

(ii) Derive the marginal p.d.f.'s $f_X$ and $f_Y$.

(iii) Compute the $EX$, $EY$; and $\sigma^2(X)$, $\sigma^2(Y)$.

(iv) For $0 \le \alpha < \beta$, compute the probabilities:

$$P(X^\beta < Y < X^\alpha), \quad P(Y^\beta < X < Y^\alpha).$$

(v) Evaluate the probabilities in part (iv) for: $(\alpha = 1, \, \beta = 2)$ and $(\alpha = 0, \beta = 1/2)$.

**2.38** The joint p.d.f. of the r.v.'s $X$ and $Y$ is given by:

$$f_{X,Y}(x, y) = cx(1 - x)y^2, \; x > 0, \; y > 0, \; \text{and} \; x + y < 1 \; (c > 0).$$

(i) Determine the constant $c$.

(ii) Derive the marginal p.d.f.'s $f_X$ and $f_Y$.

(iii) Form the conditional p.d..f's $f_{X|Y}(\cdot|y)$ and $f_{Y|X}(\cdot|x)$.

(iv) Compute the $EX$ and $EY$.

(v) Find the $E(X|Y = y)$, and evaluate it at $y = 1/2$.

**2.39** Let $X$ and $Y$ be r.v.'s with joint p.d.f. given by:

$$f_{X,Y}(x, y) = \frac{2}{c^2}, \quad 0 < x < y < c \text{ (and 0 elsewhere)}.$$

It can be seen that $c$ can be any (positive) constant, and the p.d.f.'s $f_X, f_Y,$ $f_{X|Y}(\cdot|y), f_{Y|X}(\cdot|x)$ are given by:

$$f_X(x) = \frac{2(c - x)}{c^2}, \quad 0 < x < c; \quad f_Y(y) = \frac{2y}{c^2}, \quad 0 < y < c;$$

$$f_{X|Y}(x|y) = 1/y, \quad 0 < x < y < c; \quad f_{Y|X}(y|x) = \frac{1}{c - x}, \quad 0 < x < y < c$$

Now express in terms of $c$:

(i) $EX, EX^2, \sigma^2(X)$.

(ii) $EY, EY^2, \sigma^2(Y)$.

(iii) $E(X|Y = y), E(Y|X = x)$.

(iv) Verify that $E[E(X|Y)] = EX, E[E(Y|X)] = EY$.

(v) Derive the $\text{Var}[E(X|Y)], \text{Var}[E(Y|X)]$ and compare them with $\sigma^2(X)$ and $\sigma^2(Y)$, respectively.

## 4.3 EXPECTATION OF A FUNCTION OF TWO r.v.'s, JOINT AND MARGINAL m.g.f.'s, COVARIANCE, AND CORRELATION COEFFICIENT

In this section, a function of the r.v.'s $X$ and $Y$ is considered and its expectation and variance are defined. As a special case, one obtains the joint m.g.f. of $X$ and $Y$, the covariance of $X$ and $Y$, and their correlation coefficient. To this end, let $g$ be a real-valued function defined on $\Re^2$, so that $g(X, Y)$ is a r.v. Then the expectation of $g(X, Y)$ is defined as in (6) in Chapter 3 except that the joint p.d.f. of $X$ and $Y$ is to be used. Thus:

$$Eg(X, Y) = \sum_{x_i \in \Re, y_j \in \Re} g(x_i, y_j) f_{X,Y}(x_i, y_j) \quad \text{or} \quad \int_{-\infty}^{\infty} \int_{-\infty}^{\infty} g(x, y) f_{X,Y}(x, y) dx \, dy, \tag{8}$$

for the discrete and the continuous case, respectively, provided, of course, the quantities defined exist. Properties analogous to those in (4) in Chapter 3 apply here, too. Namely, for $c$ and $d$ constants:

$$E[cg(X, Y)] = cEg(X, Y), \qquad E[cg(X, Y) + d] = cEg(X, Y) + d. \tag{9}$$

Also, if $h$ is another real-valued function, then (see also Exercise 3.17):

$$g(X, Y) \le h(X, Y) \quad \text{implies } Eg(X, Y) \le Eh(X, Y), \tag{10}$$

and, in particular,

$$g(X) \le h(X) \quad \text{implies } Eg(X) \le Eh(X). \tag{11}$$

For the special choice of the function $g(x, y) = e^{t_1 x + t_2 y}$, $t_1, t_2$ reals, the expectation $E \exp(t_1 X + t_2 Y)$ defines a function in $t_1, t_2$ for those $t_1, t_2$ for which this expectation is finite. That is:

$$M_{X,Y}(t_1, t_2) = Ee^{t_1 X + t_2 Y}, \qquad (t_1, t_2) \in C \subseteq \Re^2. \tag{12}$$

Thus, for the discrete and the continuous case, we have, respectively,

$$M_{X,Y}(t_1, t_2) = \sum_{x_i \in \Re, y_j \in \Re} e^{t_1 x_i + t_2 y_j} f_{X,Y}(x_i, y_j), \tag{13}$$

and

$$M_{X,Y}(t_1, t_2) = \int_{-\infty}^{\infty} \int_{-\infty}^{\infty} e^{t_1 x + t_2 y} f_{X,Y}(x, y) dx \, dy. \tag{14}$$

The function $M_{X,Y}(\cdot, \cdot)$ so defined is called the *joint m.g.f.* of the r.v.'s $X$ and $Y$. Clearly, $M_{X,Y}(0, 0) = 1$ for any $X$ and $Y$, and it may happen that $C = \{(0, 0)\}$ or $C \subset \Re^2$ or $C = \Re^2$. Here are two examples of joint m.g.f.'s.

**Example 15.**   Refer to Example 1 and calculate the joint m.g.f. of the r.v.'s involved.

**Discussion.**   For any $t_1, t_2 \in \Re$, we have, by means of (13):

$$M_{X,Y}(t_1, t_2) = \sum_{x=0}^{3} \sum_{y=0}^{3} e^{t_1 x + t_2 y} f_{X,Y}(x, y)$$

$$= 0.05 + 0.20e^{t_2} + 0.21e^{t_1} + 0.26e^{t_1+t_2} + 0.06e^{t_1+2t_2} + 0.08e^{2t_1+t_2}$$
$$+ 0.07e^{2t_1+2t_2} + 0.03e^{2t_1+3t_2} + 0.02e^{3t_1+2t_2} + 0.02e^{3t_1+3t_2}. \tag{15}$$

**Example 16.**   Refer to Example 2 and calculate the joint m.g.f. of the r.v.'s involved.

**Discussion.**   By means of (14), we have here:

$$M_{X,Y}(t_1, t_2) = \int_0^\infty \int_0^\infty e^{t_1 x + t_2 y} \lambda_1 \lambda_2 e^{-\lambda_1 x - \lambda_2 y} dx\, dy$$

$$= \int_0^\infty \lambda_1 e^{-(\lambda_1 - t_1)x} dx \times \int_0^\infty \lambda_2 e^{-(\lambda_2 - t_2)y} dy.$$

But $\int_0^\infty \lambda_1 e^{-(\lambda_1 - t_1)x} dx = -\frac{\lambda_1}{\lambda_1 - t_1} e^{-(\lambda_1 - t_1)x}|_0^\infty = \frac{\lambda_1}{\lambda_1 - t_1}$, provided $t_1 < \lambda_1$, and likewise
$\int_0^\infty \lambda_2 e^{-(\lambda_2 - t_2)y} dy = \frac{\lambda_2}{\lambda_2 - t_2}$ for $t_2 < \lambda_2$. (We arrive at the same results without integration by recalling (Example 6) that the r.v.'s $X$ and $Y$ have the Negative Exponential distributions with parameters $\lambda_1$ and $\lambda_2$, respectively.) Thus,

$$M_{X,Y}(t_1, t_2) = \frac{\lambda_1}{\lambda_1 - t_1} \times \frac{\lambda_2}{\lambda_2 - t_2}, \qquad t_1 < \lambda_1, \quad t_2 < \lambda_2. \tag{16}$$

In (12), by setting successively $t_2 = 0$ and $t_1 = 0$, we obtain:

$$M_{X,Y}(t_1, 0) = Ee^{t_1 X} = M_X(t_1), \qquad M_{X,Y}(0, t_2) = Ee^{t_2 Y} = M_Y(t_2). \tag{17}$$

Thus, the m.g.f.'s of the individual r.v.'s $X$ and $Y$ are taken as marginals from the joint m.g.f. of $X$ and $Y$, and they are referred to as *marginal m.g.f.'s*. For example, in reference to (15) and (16), we obtain:

$$M_X(t_1) = 0.25 + 0.53\, e^{t_1} + 0.18\, e^{2t_1} + 0.04 e^{3t_1}, \qquad t_1 \in \mathfrak{R}, \tag{18}$$

$$M_Y(t_2) = 0.26 + 0.54\, e^{t_2} + 0.15\, e^{2t_2} + 0.05 e^{3t_2}, \qquad t_2 \in \mathfrak{R}, \tag{19}$$

and

$$M_X(t_1) = \frac{\lambda_1}{\lambda_1 - t_1}, \quad t_1 < \lambda_1, \qquad M_Y(t_2) = \frac{\lambda_2}{\lambda_2 - t_2}, \quad t_2 < \lambda_2. \tag{20}$$

The joint m.g.f., as defined in (12), has properties analogous to the ones stated in (12) of Chapter 3. Namely, for $c_1, c_2$ and $d_1, d_2$ constants:

$$M_{c_1 X + d_1, c_2 Y + d_2}(t_1, t_2) = e^{d_1 t_1 + d_2 t_2} M_{X,Y}(c_1 t_1, c_2 t_2). \tag{21}$$

Its simple justification is left as an exercise (see Exercise 3.2).

In the present context, a version of the properties stated in (13) of Chapter 3, is the following:

$$\frac{\partial}{\partial t_1} M_{X,Y}(t_1, t_2)|_{t_1 = t_2 = 0} = EX, \qquad \frac{\partial}{\partial t_2} M_{X,Y}(t_1, t_2)|_{t_1 = t_2 = 0} = EY, \tag{22}$$

and

$$\frac{\partial^2}{\partial t_1 \partial t_2} M_{X,Y}(t_1, t_2)|_{t_1 = t_2 = 0} = E(XY), \tag{23}$$

provided one may interchange the order of differentiating and taking expectations. For example, for (23), we have:

$$\frac{\partial^2}{\partial t_1 \partial t_2} M_{X,Y}(t_1, t_2)|_{t_1=t_2=0} = \frac{\partial^2}{\partial t_1 \partial t_2} E\, e^{t_1 X + t_2 Y}\Big|_{t_1=t_2=0}$$

$$= E\left(\frac{\partial^2}{\partial t_1 \partial t_2} e^{t_1 X + t_2 Y}\Big|_{t_1=t_2=0}\right)$$

$$= E\left(XY\, e^{t_1 X + t_2 Y}\Big|_{t_1=t_2=0}\right) = E(XY).$$

**Remark 2.** Although properties (21) and (22) allow us to obtain moments by means of the m.g.f.'s of the r.v.'s $X$ and $Y$, the most significant property of the m.g.f. is that it allows (under certain conditions) to retrieve the distribution of the r.v.'s $X$ and $Y$. This is done through the so-called *inversion formula*.

Now, select the function $g$ as follows: $g(x, y) = cx + dy$, where $c$ and $d$ are constants. Then, for the continuous case:

$$Eg(X, Y) = E(cX + dY) = \int_{-\infty}^{\infty} \int_{-\infty}^{\infty} (cx + dy) f_{X,Y}(x, y) dx\, dy$$

$$= c \int_{-\infty}^{\infty} \int_{-\infty}^{\infty} x f_{X,Y}(x, y) dx\, dy + d \int_{-\infty}^{\infty} \int_{-\infty}^{\infty} y f_{X,Y}(x, y) dx\, dy$$

$$= c \int_{-\infty}^{\infty} \left[ x \int_{-\infty}^{\infty} f_{X,Y}(x, y) dy \right] dx + d \int_{-\infty}^{\infty} \left[ y \int_{-\infty}^{\infty} f_{X,Y}(x, y) dx \right] dy$$

$$= c \int_{-\infty}^{\infty} x f_X(x) dx + d \int_{-\infty}^{\infty} y f_Y(x) dy = cEX + dEY;\ \text{i.e.,}$$

assuming the expectations involved exist:

$$E(cX + dY) = cEX + dEY, \quad \text{where } c \text{ and } d \text{ are constants.} \tag{24}$$

In the discrete case, integrals are replaced by summation signs. On account of the usual properties of integrals and summations, property (24) applies to a more general situation. Thus, for two functions $g_1$ and $g_2$, we have:

$$E[g_1(X, Y) + g_2(X, Y)] = Eg_1(X, Y) + Eg_2(X, Y), \tag{25}$$

provided the expectations involved exist.

Next, suppose the r.v.'s $X$ and $Y$ have finite expectations and take $g(x, y) = (x - EX)(y - EY)$. Then the $Eg(X, Y) = E[(X - EX)(Y - EY)]$ is called the *covariance* of the r.v.'s $X$ and $Y$ and is denoted by $\text{Cov}(X, Y)$. Thus:

$$\text{Cov}(X, Y) = E[(X - EX)(Y - EY)] = E(XY) - (EX)(EY). \tag{26}$$

The second equality in (26) follows by multiplying out $(X - EX)(Y - EY)$ and applying property (25).

The variance of a single r.v. has been looked upon as a measure of dispersion of the distribution of the r.v. Some motivation will be given subsequently to the effect that the $Cov(X, Y)$ may be thought of as a measure of the degree to which $X$ and $Y$ tend to increase or decrease simultaneously when $Cov(X, Y) > 0$ or to move toward opposite directions when $Cov(X, Y) < 0$. This point is sufficiently made by the following simple example.

**Example 17.** Consider the events $A$ and $B$ with $P(A)P(B) > 0$ and set $X = I_A$ and $Y = I_B$ for the *indicator* functions, where $I_A(s) = 1$ if $s \in A$ and $I_A(s) = 0$ if $s \in A^c$. Then, clearly, $EX = P(A)$, $EY = P(B)$, and $XY = I_{A \cap B}$, so that $E(XY) = P(A \cap B)$. It follows that $Cov(X, Y) = P(A \cap B) - P(A)P(B)$. Next,

$$P(A)[P(Y = 1 \mid X = 1) - P(Y = 1)] = P(A \cap B) - P(A)P(B)$$

$$= Cov(X, Y), \tag{27}$$

$$P(A^c)[P(Y = 0 \mid X = 0) - P(Y = 0)] = P(A^c \cap B^c) - P(A^c)P(B^c)$$

$$= P(A \cap B) - P(A)P(B) = Cov(X, Y), \tag{28}$$

$$P(A^c)[P(Y = 1 \mid X = 0) - P(Y = 1)] = P(A^c \cap B) - P(A^c)P(B)$$

$$= -[P(A \cap B) - P(A)P(B)] = -Cov(X, Y), \tag{29}$$

$$P(A)[P(Y = 0 \mid X = 1) - P(Y = 0)] = P(A \cap B^c) - P(A)P(B^c)$$

$$= -[P(A \cap B) - P(A)P(B)] = -Cov(X, Y), \tag{30}$$

(see also Exercise 3.3).

From (27) and (28), it follows that $Cov(X, Y) > 0$ if and only if $P(Y = 1 \mid X = 1) > P(Y = 1)$, or $P(Y = 0 \mid X = 0) > P(Y = 0)$. That is, $Cov(X, Y) > 0$ if and only if, given that $X$ has taken a "large" value (namely, 1), it is more likely that $Y$ does so as well than it otherwise would; also, given that $X$ has taken a "small" value (namely, 0), it is more likely that $Y$ does so too than it otherwise would. On the other hand, from relations (29) and (30), we see that $Cov(X, Y) < 0$ if and only if $P(Y = 1 \mid X = 0) > P(Y = 1)$, or $P(Y = 0 \mid X = 1) > P(Y = 0)$. That is, $Cov(X, Y) < 0$ if and only if, given that $X$ has taken a "small" value, it is more likely for $Y$ to take a "large" value than it otherwise would, and given that $X$ has taken a "large" value, it is more likely for $Y$ to take a "small" value than it otherwise would.

As a further illustration of the significance of the covariance we proceed to calculate the $Cov(X, Y)$ for the r.v.'s of Example 1.

**Example 18.** Refer to Example 1 and calculate the $Cov(X, Y)$.

**Discussion.** In Example 5, the (marginal) p.d.f.'s $f_X$ and $f_Y$ were calculated. Then: $EX = 1.01$ and $EY = 0.99$. Next, the r.v. $XY$ is distributed as follows, on the basis of Table 4.1:

| xy | 0 | 1 | 2 | 3 | 4 | 6 | 9 |
|---|---|---|---|---|---|---|---|
| $f_{XY}$ | 0.46 | 0.26 | 0.14 | 0 | 0.07 | 0.05 | 0.02 |

Therefore, $E(XY) = 1.3$ and then, by formula (26), $\text{Cov}(X, Y) = 1.3 - 1.01 \times 0.99 = 0.3001$.

Here, the covariance is positive, and by comparing the values of the conditional probabilities in Example 5 with the appropriate unconditional probabilities, we see that this is consonant with the observation just made that $X$ and $Y$ tend to take simultaneously either "large" values or "small" values. (See also Example 19 later.)

The result obtained next provides the range of values of the covariance of two r.v.'s; it is also referred to as a version of the *Cauchy-Schwarz inequality*.

**Theorem 1.**

(i) *Consider the r.v.'s $X$ and $Y$ with $EX = EY = 0$ and $\text{Var}(X) = \text{Var}(Y) = 1$. Then always $-1 \leq E(XY) \leq 1$, and $E(XY) = 1$ if and only if $P(X = Y) = 1$, and $E(XY) = -1$ if and only if $P(X = -Y) = 1$.*

(ii) *For any r.v.'s $X$ and $Y$ with finite expectations and positive variances $\sigma_X^2$ and $\sigma_Y^2$, it always holds:*

$$-\sigma_X \sigma_Y \leq \text{Cov}(X, Y) \leq \sigma_X \sigma_Y, \tag{31}$$

*and $\text{Cov}(X, Y) = \sigma_X \sigma_Y$ if and only if $P[Y = EY + \frac{\sigma_Y}{\sigma_X}(X - EX)] = 1, \text{Cov}(X, Y) = -\sigma_X \sigma_Y$ if and only if $P[Y = EY - \frac{\sigma_Y}{\sigma_X}(X - EX)] = 1$.*

*Proof.*

(i) Clearly, $0 \leq E(X - Y)^2 = EX^2 + EY^2 - 2E(XY) = 2 - 2E(XY)$, so that $E(XY) \leq 1$; also, $0 \leq E(X + Y)^2 = EX^2 + EY^2 + 2E(XY) = 2 + 2E(XY)$, so that $-1 \leq E(XY)$. Combining these results, we obtain $-1 \leq E(XY) \leq 1$. As for equalities, observe that, if $P(X = Y) = 1$, then $E(XY) = EX^2 = 1$, and if $P(X = -Y) = 1$, then $E(XY) = -EX^2 = -1$. Next, $E(XY) = 1$ implies $E(X - Y)^2 = 0$ or $\text{Var}(X - Y) = 0$. But then $P(X - Y = 0) = 1$ or $P(X = Y) = 1$ (see Exercise 2.4 in Chapter 3). Also, $E(XY) = -1$ implies $E(X + Y)^2 = 0$ or $\text{Var}(X + Y) = 0$, so that $P(X = -Y) = 1$ (by the exercise just cited).

(ii) Replace the r.v.'s $X$ and $Y$ by the r.v.'s $X^* = \frac{X - EX}{\sigma_X}$ and $Y^* = \frac{Y - EY}{\sigma_Y}$, for which $EX^* = EY^* = 0$ and $\text{Var}(X^*) = \text{Var}(Y^*) = 1$. Then the inequalities $-1 \leq E(X^* Y^*) \leq 1$ become

$$-1 \leq E\left[\left(\frac{X - EX}{\sigma_X}\right)\left(\frac{Y - EY}{\sigma_Y}\right)\right] \leq 1 \tag{32}$$

from which (31) follows. Also, $E(X^* Y^*) = 1$ if and only if $P(X^* = Y^*) = 1$ becomes $E[(X - EX)(Y - EY)] = \sigma_X \sigma_Y$ if and only if $P[Y = EY + \frac{\sigma_Y}{\sigma_X} \times (X - EX)] = 1$, and $E(X^* Y^*) = -1$ if and only if $P(X^* = -Y^*) = 1$ becomes $E[(X - EX)(Y - EY)] = -\sigma_X \sigma_Y$ if and only if $P[Y = EY - \frac{\sigma_Y}{\sigma_X} \times (X - EX)] = 1$. A restatement of the last two conclusions is: $\text{Cov}(X, Y) = \sigma_X \sigma_Y$ if and only if $P[Y = EY + \frac{\sigma_Y}{\sigma_X}(X - EX)] = 1$, and $\mathrm{Cov}(X, Y) = -\sigma_X\sigma_Y$ if and only if $P[Y = EY - \frac{\sigma_Y}{\sigma_X}(X - EX)] = 1$. ∎

From the definition of the $\mathrm{Cov}(X, Y)$ in (26), it follows that if $X$ is measured in units, call them $a$, and $Y$ is measured in units, call them $b$, then $\mathrm{Cov}(X, Y)$ is measured in units $ab$. Furthermore, because the variance of a r.v. ranges from 0 to $\infty$, it follows from (31) that $\mathrm{Cov}(X, Y)$ may vary from $-\infty$ to $\infty$. These two characteristics of a covariance are rather undesirable and are both eliminated through the standardization process of replacing $X$ and $Y$ by $\frac{X-EX}{\sigma_X}$ and $\frac{Y-EY}{\sigma_Y}$. By (32), the range of the covariance of these standardized r.v.'s is the interval $[-1, 1]$, and is a pure number. This covariance is called the *correlation coefficient* of the r.v.'s $X$ and $Y$ and is denoted by $\rho(X, Y)$. Thus:

$$\rho(X, Y) = E\left[\left(\frac{X - EX}{\sigma_X}\right)\left(\frac{Y - EY}{\sigma_Y}\right)\right]$$

$$= \frac{\mathrm{Cov}(X, Y)}{\sigma_X\sigma_Y} = \frac{E(XY) - (EX)(EY)}{\sigma_X\sigma_Y}. \tag{33}$$

Furthermore, by (32):

$$-1 \le \rho(X, Y) \le 1, \tag{34}$$

and, by part (ii) of Theorem 1:

$$\rho(X, Y) = 1 \text{ if and only if } P\left[Y = EY + \frac{\sigma_Y}{\sigma_X}(X - EX)\right] = 1, \tag{35}$$

$$\rho(X, Y) = -1 \text{ if and only if } P\left[Y = EY - \frac{\sigma_Y}{\sigma_X}(X - EX)\right] = 1. \tag{36}$$

The straight lines represented by $y = EY + \frac{\sigma_Y}{\sigma_X}(x - EX)$ and $y = EY - \frac{\sigma_Y}{\sigma_X}(x - EX)$ are depicted in Figure 4.5.

From relation (35), we have that $\rho(X, Y) = 1$ if and only if $(X, Y)$ are linearly related (with probability 1). On the other hand, from Example 17, we have that $\mathrm{Cov}(X, Y) > 0$ if and only if $X$ and $Y$ tend to take simultaneously either "large"

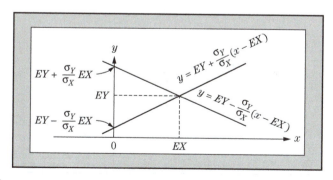

**FIGURE 4.5**

Lines of perfect linear relation of x and y.

values or "small" values. Since $\text{Cov}(X, Y)$ and $\rho(X, Y)$ have the same sign, the same statement can be made about $\rho(X, Y)$, being positive if and only if $X$ and $Y$ tend to take simultaneously either "large" values or "small" values. The same arguments apply for the case that $\text{Cov}(X, Y) < 0$ (equivalently, $\rho(X, Y) < 0$). This reasoning indicates that $\rho(X, Y)$ may be looked upon as a measure of *linear* dependence between $X$ and $Y$. The pair $(X, Y)$ lies on the line $y = EY + \frac{\sigma_Y}{\sigma_X}(x - EX)$ if $\rho(X, Y) = 1$; pairs identical to $(X, Y)$ tend to be arranged along this line, if $(0 <)\rho(X, Y) < 1$, and they tend to move further and further away from this line as $\rho(X, Y)$ gets closer to 0; the pairs bear no sign of linear tendency whatever, if $\rho(X, Y) = 0$. Rough arguments also hold for the reverse assertions. For $0 < \rho(X, Y) \leq 1$, the r.v.'s $X$ and $Y$ are said to be *positively correlated*, and *uncorrelated* if $\rho(X, Y) = 0$. Likewise, the pair $(X, Y)$ lies on the line $y = EY - \frac{\sigma_Y}{\sigma_X}(x - EX)$ if $\rho(X, Y) = -1$; pairs identical to $(X, Y)$ tend to be arranged along this line if $-1 < \rho(X, Y) < 0$. Again, rough arguments can also be made for the reverse assertions. For $-1 \leq \rho(X, Y) < 0$, the r.v.'s $X$ and $Y$ are said to be *Negatively correlated*. (See Figure 4.6.)

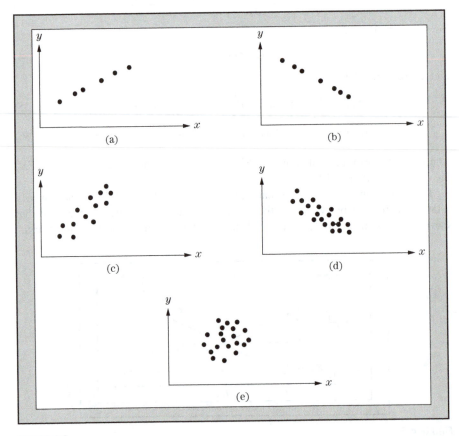

**FIGURE 4.6**

Depiction of the correlation coefficient of two r.v.'s.

Actually, a more precise argument to this effect can be made by considering the distance $D$ of the (random) point $(X, Y)$ from the lines $y = EY \pm \frac{\sigma_Y}{\sigma_X}(x - EX)$. It can be seen that:

$$ED^2 = \frac{2\sigma_X^2\sigma_Y^2}{\sigma_X^2 + \sigma_Y^2}(1 - |\rho(X, Y)|). \tag{37}$$

Then one may use the interpretation of the expectation as an average and exploit (37) in order to arrive at the same reasoning as above, but in a more rigorous way.

As an illustration, let us calculate the $\rho(X, Y)$ for Examples 1 and 8.

**Example 19.**   In reference to Example 1, calculate the Cov$(X, Y)$ and the $\rho(X, Y)$.

**Discussion.**   From Table 4.1, we find $EX^2 = 1.61, EY^2 = 1.59$. By Example 18, $EX = 1.01, EY = 0.99$, so that Var$(X) = EX^2 - (EX)^2 = 0.5899$, Var$(Y) = EY^2 - (EY)^2 = 0.6099$. Since Cov$(X, Y) = 0.3001$ (by Example 18), we have: $\rho(X, Y) = \frac{\text{Cov}(X,Y)}{\sqrt{\text{Var}(X)\text{Var}(Y)}} = \frac{0.3001}{\sqrt{0.5899 \times 0.6099}} \simeq 0.5$.

**Example 20.**   In reference to Example 8, calculate the Cov$(X, Y)$ and the $\rho(X, Y)$.

**Discussion.**   By Example 8(iii), Cov$(X, Y) = E(XY) - (EX)(EY) = \frac{4}{9} - \frac{8}{15} \times \frac{4}{5} = \frac{4}{225}$, $\sqrt{\text{Var}(X)} = \frac{\sqrt{11}}{15}$, $\sqrt{\text{Var}(Y)} = \frac{\sqrt{6}}{15}$, so that

$$\rho(X, Y) = \frac{\text{Cov}(X, Y)}{\sqrt{\text{Var}(X)}\sqrt{\text{Var}(Y)}} = \frac{\frac{4}{225}}{\frac{\sqrt{11}}{15} \times \frac{\sqrt{6}}{15}} = \frac{4}{\sqrt{66}} \simeq 0.492.$$

**Example 21.**   Let $X$ and $Y$ be two r.v.'s with finite expectations and equal (finite) variances, and set $U = X + Y$ and $V = X - Y$. Then the r.v.'s $U$ and $V$ are uncorrelated.

**Discussion.**   Indeed,

$$E(UV) = E[(X + Y)(X - Y)] = E(X^2 - Y^2) = EX^2 - EY^2,$$
$$(EU)(EV) = [E(X + Y)][E(X - Y)]$$
$$= (EX + EY)(EX - EY) = (EX)^2 - (EY)^2,$$

so that

$$\text{Cov}(U, V) = E(UV) - (EU)(EV) = [EX^2 - (EX)^2] - [EY^2 - (EY)^2]$$
$$= \text{Var}(X) - \text{Var}(Y) = 0.$$

Figure 4.6 illustrates the behavior of the correlation coefficient $\rho(X, Y)$ of the r.v.'s $X$ and $Y$. In (a), $\rho(X, Y) = 1$, the r.v.'s $X$ and $Y$ are perfectly positively linearly related. In (b), $\rho(X, Y) = -1$, the r.v.'s $X$ and $Y$ are perfectly Negatively linearly related. In (c), $0 < \rho(X, Y) < 1$, the r.v.'s $X$ and $Y$ are positively correlated. In (d), $-1 < \rho(X, Y) < 0$, the r.v.'s $X$ and $Y$ are Negatively correlated. In (e), $\rho(X, Y) = 0$, the r.v.'s $X$ and $Y$ are uncorrelated.

The following result presents an interesting property of the correlation coefficient.

**Theorem 2.**  *Let $X$ and $Y$ be r.v.'s with finite first and second moments and positive variances, and let $c_1, c_2, d_1, d_2$ be constants with $c_1 c_2 \neq 0$. Then:*

$$\rho(c_1 X + d_1, c_2 Y + d_2) = \pm \rho(X, Y), \quad \text{with} + \text{if } c_1 c_2 > 0 \text{ and} - \text{if } c_1 c_2 < 0.$$

$$(38)$$

*Proof.*  Indeed, $\text{Var}(c_1 X + d_1) = c_1^2 \text{Var}(X)$, $\text{Var}(c_2 Y + d_2) = c_2^2 \text{Var}(Y)$, and $\text{Cov}(c_1 X + d_1, c_2 Y + d_2) = E\{[(c_1 X + d_1) - E(c_1 X + d_1)][(c_2 Y + d_2) - E(c_2 Y + d_2)]\} = E[c_1(X - EX) \times c_2(Y - EY)] = c_1 c_2 E[(X - EX)(Y - EY)] = c_1 c_2 \text{Cov}(X, Y)$. Therefore $\rho(c_1 X + d_1, c_2 Y + d_2) = \frac{c_1 c_2 \text{Cov}(X,Y)}{|c_1 c_2| \sqrt{\text{Var}(X)\text{Var}(Y)}}$, and the conclusion follows. ∎

**Example 22.**  Let $X$ and $Y$ be temperatures in two localities measured in the Celsius scale, and let $U$ and $V$ be the same temperatures measured in the Fahrenheit scale. Then $\rho(X, Y) = \rho(U, V)$, as it should be. This is so because $U = \frac{9}{5} X + 32$ and $V = \frac{9}{5} Y + 32$, so that (38) applies with the $+$ sign.

This section is concluded with the following result and an example.

**Theorem 3.**  *For two r.v.'s $X$ and $Y$ with finite first and second moments, and (positive) standard deviations $\sigma_X$ and $\sigma_Y$, it holds:*

$$\text{Var}(X + Y) = \sigma_X^2 + \sigma_Y^2 + 2\text{Cov}(X, Y) = \sigma_X^2 + \sigma_Y^2 + 2\sigma_X \sigma_Y \rho(X, Y), \quad (39)$$

*and*

$$\text{Var}(X + Y) = \sigma_X^2 + \sigma_Y^2 \quad \text{if } X \text{ and } Y \text{ are uncorrelated.} \quad (40)$$

*Proof.*  Since (40) follows immediately from (39), and $\text{Cov}(X, Y) = \sigma_X \sigma_Y \times \rho(X, Y)$, it suffices to establish only the first equality in (39). Indeed,

$$\text{Var}(X + Y) = E[(X + Y) - E(X + Y)]^2 = E[(X - EX) + (Y - EY)]^2$$

$$= E(X - EX)^2 + E(Y - EY)^2 + 2E[(X - EX)(Y - EY)]$$

$$= \sigma_X^2 + \sigma_Y^2 + 2\text{Cov}(X, Y). \qquad \blacksquare$$

**Example 23.**  In reference to Examples 1 and 8 and by means of results obtained in Examples 19, 8(iii), and 20, respectively, calculate $\text{Var}(X + Y)$.

**Discussion.** By (39),

$$\text{Var}(X + Y) = \text{Var}(X) + \text{Var}(Y) + 2\text{Cov}(X, Y)$$

$$= 0.5899 + 0.6099 + 2 \times 0.3001 = 1.8 \quad \text{for Example 1, and}$$

$$= \frac{11}{225} + \frac{2}{75} + 2 \times \frac{4}{225} = \frac{1}{9} \quad \text{for Example 8.}$$

## EXERCISES

**3.1**  Let $X$ and $Y$ be the r.v.'s denoting the number of sixes when two fair dice are rolled independently 15 times each. Determine the $E(X + Y)$.

**3.2**  Show that the joint m.g.f. of two r.v.'s $X$ and $Y$ satisfies the following property, where $c_1, c_2, d_1$, and $d_2$ are constants.

$$M_{c_1 X + d_1, c_2 Y + d_2}(t_1, t_2) = e^{d_1 t_1 + d_2 t_2} M_{X,Y}(c_1 t_1, c_2 t_2).$$

**3.3**  Provide a justification of relations (28)–(30). That is:
  (i)   $P(A^c \cap B^c) - P(A^c)P(B^c) = P(A \cap B) - P(A)P(B)$.
  (ii)  $P(A^c \cap B) - P(A^c)P(B) = -P(A \cap B) + P(A)P(B)$.
  (iii) $P(A \cap B^c) - P(A)P(B^c) = -P(A \cap B) + P(A)P(B)$.

**3.4**  Let $X$ and $Y$ be two r.v.'s with $EX = EY = 0$. Then, if $\text{Var}(X - Y) = 0$, it follows that $P(X = Y) = 1$, and if $\text{Var}(X + Y) = 0$, then $P(X = -Y) = 1$. **Hint.** Use Exercise 2.4 in Chapter 3.

**3.5**  In reference to Exercise 2.1 (see also Exercise 1.1), calculate:
  (i)   $EX, EY, \text{Var}(X)$, and $\text{Var}(Y)$.
  (ii)  $\text{Cov}(X, Y)$ and $\rho(X, Y)$.
  (iii) Decide on the kind of correlation of the r.v.'s $X$ and $Y$.

**3.6**  Refer to Exercises 1.2 and 2.2 and calculate:
  (i)   $EX, EY, \text{Var}(X), \text{Var}(Y)$.
  (ii)  $E(XY), \text{Cov}(X, Y)$.
  (iii) $\rho(X, Y)$.
  (iv)  What kind of correlation, if any, do the r.v.'s $X$ and $Y$ exhibit?

**3.7**  In reference to Exercise 2.3:
  (i)   Calculate $EX, EY, \text{Var}(X)$, and $\text{Var}(Y)$.
  (ii)  Calculate $\text{Cov}(X, Y)$ and $\rho(X, Y)$.
  (iii) Plot the points $(-4, 1), (-2, -2), (2, 2)$, and $(4, -1)$, and reconcile this graph with the value of $\rho(X, Y)$ found in part (ii).

**3.8**  In reference to Exercise 2.4, calculate the following quantities:
  (i)   $EX, EY, \text{Var}(X)$, and $\text{Var}(Y)$.
  (ii)  $\text{Cov}(X, Y)$ and $\rho(X, Y)$.

**3.9** Refer to Exercise 2.5, and calculate the $\text{Cov}(X, Y)$ and the $\rho(X, Y)$.

**3.10** Let $X$ be a r.v. taking on the values $-2, -1, 1, 2$, each with probability $1/4$, and define the r.v. $Y$ by: $Y = X^2$. Then calculate the quantities: $EX$, $\text{Var}(X)$, $EY$, $\text{Var}(Y)$, $E(XY)$, $\text{Cov}(X, Y)$, and $\rho(X, Y)$. Are you surprised by the value of $\rho(X, Y)$? Explain.

**3.11** Refer to Example 8 and compute the covariance $\text{Cov}(X, Y)$ and the correlation coefficient $\rho(X, Y)$. Decide on the kind of correlation of the r.v.'s $X$ and $Y$.

**3.12** In reference to Exercise 2.7 (see also Exercise 1.3), calculate:
   **(i)** The expectations $EX$ and $EY$.
   **(ii)** The variances $\text{Var}(X)$ and $\text{Var}(Y)$.
   **(iii)** The covariance $\text{Cov}(X, Y)$ and the correlation coefficient $\rho(X, Y)$.
   **(iv)** On the basis of part (iii), decide on the kind of correlation of the r.v.'s $X$ and $Y$.

**3.13** In reference to Exercise 2.8, calculate:
   **(i)** The expectations $EX$ and $EY$.
   **(ii)** The variances $\text{Var}(X)$ and $\text{Var}(Y)$.
   **(iii)** The covariance $\text{Cov}(X, Y)$ and the correlation coefficient $\rho(X, Y)$.
   **(iv)** On the basis of part (iii), decide on the kind of correlation of the r.v.'s $X$ and $Y$.

**3.14** Let $X$ be a r.v. with finite expectation and finite and positive variance, and set $Y = aX + b$, where $a$ and $b$ are constants and $a \neq 0$. Then show that $|\rho(X, Y)| = 1$ and, indeed $\rho(X, Y) = 1$ if and only if $a > 0$, and $\rho(X, Y) = -1$ if and only if $a < 0$.

**3.15** For any two r.v.'s $X$ and $Y$, set $U = X + Y$ and $V = X - Y$. Then show that:
   **(i)** $P(UV < 0) = P(|X| < |Y|)$.
   **(ii)** If $EX^2 = EY^2 < \infty$, then $E(UV) = 0$.
   **(iii)** If $EX^2 < \infty, EY^2 < \infty$ and $\text{Var}(X) = \text{Var}(Y)$, then the r.v.'s $U$ and $V$ are uncorrelated.

**3.16** Let $X$ and $Y$ be r.v.'s with finite second moments $EX^2, EY^2$, and $\text{Var}(X) > 0$. Suppose we know $X$ and we wish to *predict* $Y$ in terms of $X$ through the *linear* relationship $\alpha X + \beta$, where $\alpha$ and $\beta$ are (unknown) constants. Further, suppose that there exist values $\hat{\alpha}$ and $\hat{\beta}$ of $\alpha$ and $\beta$, respectively, for which the expectation of the square difference $[Y - (\hat{\alpha}X + \hat{\beta})]^2$ is minimum. Then $\hat{Y} = \hat{\alpha}X + \hat{\beta}$ is called the *best linear predictor* of $Y$ in terms of $X$ (when the criterion of optimality is that of minimizing $E[Y - (\alpha X + \beta)]^2$ over all $\alpha$ and $\beta$). Then show that $\hat{\alpha}$ and $\hat{\beta}$ are given as follows:

$$\hat{\alpha} = \frac{\sigma_Y}{\sigma_X} \rho(X, Y), \qquad \hat{\beta} = EY - \hat{\alpha}EX,$$

where $\sigma_X$ and $\sigma_Y$ are the s.d.'s of the r.v.'s $X$ and $Y$, respectively.

**3.17** Justify the statement made in relation (10), for both the discrete and the continuous case.

**Hint.** Set $g(\alpha, \beta) = E[Y - (\alpha X - \beta)^2]^2$, carry out the operations on the right-hand side in order to get: $g(\alpha, \beta) = EY^2 + \alpha^2 EX^2 + \beta^2 + 2\alpha\beta EX - 2\alpha E(XY) - 2\beta EY$, minimize $g(\alpha, \beta)$ by equating to 0 the two partial derivatives in order to find the values $\hat{\alpha}$ and $\hat{\alpha}$ given above. Finally, show that these values $\hat{\alpha}$ and $\hat{\beta}$ do, indeed, minimize $g(\alpha, \beta)$ by showing that the $2 \times 2$ matrix of the second-order derivatives has its $1 \times 1$ and $2 \times 2$ determinants positive. Or that the matrix of the second-order derivatives is positive definite.

**3.18** Consider the (discrete) r.v. $X$ and its distribution given below:

| $x$ | -2 | -1 | 1 | 2 |
|-----|-----|-----|-----|-----|
| $f(x)$ | 1/4 | 1/4 | 1/4 | 1/4 |

Set $Y = X^2$ (so that $XY = X^3$), and:
  **(i)** Compute the $EX$.
  **(ii)** What are the values of the r.v. $XY (= X^3)$, and the respective probabilities?
  **(iii)** Compute the covariance $\text{Cov}(X, Y)$.
  **(iv)** How do you interpret the result in part (iii)? Are you surprised by it?

**3.19** In reference to Exercise 3.18, compute:
  **(i)** $EX^2$ and $\text{Var}(X)$.
  **(ii)** $EY$, $EY^2$, and $\text{Var}(Y)$.
  **(iii)** $\text{Var}(2X + 6Y)$.

**3.20** Let the (discrete) r.v.'s $X$ and $Y$ denote the amount of an automobile insurance policy and the amount of a house owner insurance policy, respectively, and suppose that their p.d.f. is given by the following table; the values of $X$ and $Y$ are expressed in thousands of dollars. That is:

| $Y\backslash X$ | 100 | 250 | |
|-----|-----|-----|-----|
| 0 | 0.20 | 0.05 | 0.25 |
| 100 | 0.10 | 0.15 | 0.25 |
| 200 | 0.20 | 0.30 | 0.50 |
| | 0.50 | 0.50 | 1 |

  **(i)** Determine the marginal p.d.f.'s $f_X$ and $f_Y$.
  **(ii)** Compute the $EX$, $EX^2$, and $\sigma^2(X)$.
  **(iii)** Compute the $EY$, $EY^2$, and $\sigma^2(Y)$.
  **(iv)** Compute the $E(XY)$, $\text{Cov}(X, Y)$, and $\rho(X, Y)$.

**3.21** If $X$ and $Y$ are two r.v.'s with (finite) variances $\sigma_1^2$, $\sigma_2^2$, respectively, and correlation coefficients $\rho$, show that $\mathrm{Var}(X + Y) \leq (\sigma_1 + \sigma_2)^2$.

**3.22** In reference to Exercise 1.51 (in Chapter 3), show that

    (i) $E(XY) = \frac{3n^2+7n+8}{12}$.

    (ii) $\mathrm{Cov}(X, Y) = \frac{n^2+n-2}{36}$, and $\rho(X, Y) = \frac{1}{2}$.

**3.23** If the r.v.'s $X$ and $Y$ have (positive and finite) variances $\sigma^2(X)$ and $\sigma^2(Y)$, show that the r.v.'s $X$ and $aX + Y$ are uncorrelated if and only if $a = -\mathrm{Cov}(X, Y)/\sigma^2(X)$. Likewise, the r.v.'s $Y$ and $X + bY$ are uncorrelated if and only if $b = -\mathrm{Cov}(X, Y)/\sigma^2(Y)$.

**3.24** In reference to Exercise 2.28 (in this chapter), compute:

    (i) $E(XY)$.

    (ii) $\mathrm{Cov}(X, Y)$.

    (iii) $\rho(X, Y)$.

    (iv) $E\left(\frac{X}{Y}\right)$, $E\left(\frac{X}{Y^2}\right)$, and $E\left(\frac{Y}{X}\right)$.

**3.25** In reference to Exercise 2.37 (in this chapter):

    (i) Show that $E(XY) = \frac{1}{3}$; $\mathrm{Cov}(X, Y) = -1/144$, and $\rho(X, Y) = -1/11$.

    (ii) Compute the variance of the r.v. $Z = aX + bY$ as a function of $a$ and $b$, and evaluated it for: $a = b = 1$ and $a = -b = 1$.

**3.26** Let $X$ and $Y$ be two r.v.'s with joint p.d.f. given by:

$$f_{X,Y}(x, y) = 8xy, \quad 0 < x < y < 1.$$

    (i) Show that the marginal p.d.f.'s $f_X$ and $f_Y$ are given by:

$$f_X(x) = 4x(1 - x^2), \quad 0 < x < 1; \quad f_Y(y) = 4y^3, \quad 0 < y < 1.$$

    (ii) Show that $EX = 8/15$, $EX^2 = 1/2$; $EY = 4/5$, $EY^2 = 2/3$.

    (iii) Compute the $\mathrm{Var}(X) = \sigma^2(X)$ and $\mathrm{Var}(Y) = \sigma^2(Y)$.

    (iv) Show that $E(XY) = 4/9$.

    (v) Compute the $\mathrm{Cov}(X, Y) = E(XY) - (EX)(EY)$.

    (vi) Compute the correlation coefficient $\rho(X, Y) = \dfrac{\mathrm{Cov}(X, Y)}{\sigma(X)\sigma(Y)}$.

**3.27** The pressures (in psi) of the front right and left tires of certain automobiles are r.v.'s $X$ and $Y$, respectively, with joint p.d.f. given by:

$$f_{X,Y}(x, y) = c(x^2 + y^2), \quad 20 \leq x \leq 30, \ 20 \leq y \leq 30 \ (c > 0).$$

    (i) Determine the constant $c$.

    (ii) Find the p.d.f.'s $f_X$ and $f_{Y|X}(\cdot|x)$.

    (iii) Compute the $EX$ and the $E(Y|X = x)$.

    (iv) Evaluate the $E(Y|X = x)$ for $x = 20, 25, 30$ (psi).

**3.28** In reference to Exercise 3.27:

    **(i)** Suppose that the Normal pressure is 26 psi. Then what is the probability that both tires are underfilled?

    **(ii)** Compute the $P(|Y - X| \leq 2)$.

## 4.4 SOME GENERALIZATIONS TO *k* RANDOM VARIABLES

If instead of two r.v.'s $X$ and $Y$ we have $k$ r.v.'s $X_1, \ldots, X_k$, most of the concepts defined and results obtained in the previous sections are carried over to the $k$-dimensional case in a straightforward way. Thus, the *joint probability distribution function* of $(X_1, \ldots, X_k)$, to be denoted by $P_{X_1,\ldots,X_k}$, is defined by: $P_{X_1,\ldots,X_k}(B) = P[(X_1, \ldots, X_k) \in B], B \subseteq \Re^k = \Re \times \cdots \times \Re$ ($k$ factors), and their *joint d.f.* is: $F_{X_1,\ldots,X_k}(x_1, \ldots, x_k) = P(X_1 \leq x_1, \ldots, X_k \leq x_k), x_1, \ldots, x_k \in \Re$. The obvious versions of properties #1 and #3 stated in Section 4.1 hold here too; also, a suitable version of property #2 holds, but we shall not insist on it. The *joint p.d.f.* of $X_1, \ldots, X_k$ is denoted by $f_{X_1,\ldots,X_k}$ and is defined in an obvious manner. Thus, for the case the r.v.'s $X_1, \ldots, X_k$ are discrete taking on respective values $x_{1i}, \ldots, x_{ki}$, with $x_{1i}$: $i = 1, \ldots, r_1(\leq \infty), \ldots, x_{ki}$: $i = 1, \ldots, r_k(\leq \infty)$, we have $f_{X_1,\ldots,X_k}(x_{1i}, \ldots, x_{ki}) = P(X_1 = x_{1i}, \ldots, X_k = x_{ki})$ and 0 otherwise. Then, for $B \subseteq \Re^k$, $P[(X_1, \ldots, X_k) \in B] = \sum f_{X_1,\ldots,X_k}(x_{1i}, \ldots, x_{ki})$, where the summation extends over all $(x_{1i}, \ldots, x_{ki}) \in B$. For the continuous case, the joint p.d.f. is a nonNegative function such that $F_{X_1,\ldots,X_k}(x_1, \ldots, x_k) = \int_{-\infty}^{x_k} \cdots \int_{-\infty}^{x_1} f_{X_1,\ldots,X_k}(t_1, \ldots, t_k) dt_1 \ldots dt_k, x_1, \ldots, x_k$ in $\Re$. It follows that for $B \subseteq \Re^k$ (where you may interpret $B$ as a familiar Geometric figure in $\Re^k$): $P[(X_1, \ldots, X_k) \in B] = \underbrace{\int \cdots \int}_{B} f_{X_1,\ldots,X_k}(x_1, \ldots, x_k) dx_1 \ldots dx_k$, and in particular,

$$\int_{-\infty}^{\infty} \cdots \int_{-\infty}^{\infty} f_{X_1,\ldots,X_k}(x_1, \ldots, x_k) dx_1 \ldots dx_k (= P[(X_1, \ldots, X_k) \in \Re^k]) = 1.$$ As in the two-dimensional case, $\frac{\partial^k}{\partial x_1 \ldots \partial x_k} F_{X_1,\ldots,X_k}(x_1, \ldots, x_k) = f_{X_1,\ldots,X_k}(x_1, \ldots, x_k)$ (for continuity points $(x_1, \ldots, x_k)$ of $f_{X_1,\ldots,X_k}$). In the next subsection, three concrete examples will be presented, one for the discrete case, and two for the continuous case. In the present $k$-dimensional case, there are many marginal d.f.'s and p.d.f.'s. Thus, if in $F_{X_1,\ldots,X_k}(x_1, \ldots, x_k)$, $t$ of the $x$'s, $x_{j_1}, \ldots, x_{j_t}$, are replaced by $+\infty$ (in the sense they are let to tend to $+\infty$), then what is left is the *marginal joint d.f.* of the r.v.'s $X_{i_1}, \ldots, X_{i_s}, F_{X_{i_1},\ldots,X_{i_s}}$, where $s + t = k$. Likewise, if in $f_{X_1,\ldots,X_k}(x_1, \ldots, x_k)$, $x_{j_1}, \ldots, x_{j_t}$ are eliminated through summation (for the discrete case) or integration (for the continuous case), what is left is the *marginal joint p.d.f.* of the r.v.'s $X_{i_1}, \ldots, X_{i_s}, f_{X_{i_1}}, \ldots, X_{i_s}$. Combining joint and marginal joint p.d.f.'s, as in the two-dimensional case, we obtain a variety of *conditional p.d.f.'s*. Thus, for example,

$$f_{X_{j_1},\ldots,X_{j_t}|X_{i_1},\ldots,X_{i_s}}(x_{j_1}, \ldots, x_{j_t} \mid x_{i_1}, \ldots, x_{i_s}) = \frac{f_{X_1,\ldots,X_k}(x_1, \ldots, x_k)}{f_{X_{i_1},\ldots,X_{i_s}}(x_{i_1}, \ldots, x_{i_s})}.$$

Utilizing conditional p.d.f.'s, we can define *conditional expectations* and *conditional variances*, as in the two-dimensional case (see relations (2), (3) and (5), (6)). For a (real-valued) function $g$ defined on $\Re^k$, the expectation of the r.v. $g(X_1, \ldots, X_k)$ is defined in a way analogous to that in (8) for the two-dimensional case, and the validity of properties (9) and (10) is immediate. In particular, provided the expectations involved exist:

$$E(c_1X_1 + \cdots + c_kX_k + d) = c_1EX_1 + \cdots + c_kEX_k + d,$$
$$c_1, \ldots, c_k, d \text{ constants.} \tag{41}$$

By choosing $g(x_1, \ldots, x_k) = \exp(t_1x_1 + \cdots + t_kx_k)$, $t_1, \ldots, t_k \in \Re$, the resulting expectation (assuming it is finite) is the *joint m.g.f.* of $X_1, \ldots, X_k$; i.e.,

$$M_{X_1,\ldots,X_k}(t_1, \ldots, t_k) = Ee^{t_1X_1 + \cdots + t_kX_k}, \qquad (t_1, \ldots, t_k) \in C \subseteq \Re^k. \tag{42}$$

The appropriate versions of properties (21) and (23) become here:

$$M_{c_1X_1+d_1,\ldots,c_kX_k+d_k}(t_1, \ldots, t_k) = e^{d_1t_1 + \cdots + d_kt_k}M_{X_1,\ldots,X_k}(c_1t_1, \ldots, c_kt_k), \tag{43}$$

where $c_1, \ldots, c_k$ and $d_1, \ldots, d_k$ are constants, and:

$$\frac{\partial^{n_1 + \cdots + n_k}}{\partial^{n_1}t_1 \ldots \partial^{n_k}t_k} M_{X_1,\ldots,X_k}(t_1, \ldots, t_k)|_{t_1 = \cdots = t_k = 0} = E\left(X_1^{n_1} \ldots X_k^{n_k}\right), \tag{44}$$

for $\geq 0$ integers $n_1, \ldots, n_k$.

**Remark 3.** Relation (44) demonstrates the joint moment generating property of the joint m.g.f. The joint m.g.f. can also be used for recovering the joint distribution of the r.v.'s $X_1, \ldots, X_k$ as indicated in Remark 2.

Finally, the appropriate versions of relations (39) and (40) become here, by setting $\sigma_{X_i}^2 = \text{Var}(X_i)$, $i = 1, \ldots, k$:

$$\text{Var}(X_1 + \cdots + X_k) = \sum_{i=1}^{k} \sigma_{X_i}^2 + 2 \sum_{1 \leq i < j \leq k} \text{Cov}(X_i, X_j)$$

$$= \sum_{i=1}^{k} \sigma_{X_i}^2 + 2 \sum_{1 \leq i < j \leq k} \sigma_{X_i}\sigma_{X_j}\rho(X_i, X_j), \tag{45}$$

and

$$\text{Var}(X_1 + \cdots + X_k) = \sum_{i=1}^{k} \sigma_{X_i}^2 \quad \text{if the } X_i\text{'s are pairwise uncorrelated;} \tag{46}$$

i.e., $\rho(X_i, X_j) = 0$ for $i \neq j$.

---

# EXERCISES

**4.1** If the r.v.'s $X_1, X_2, X_3$ have the joint p.d.f. $f_{X_1,X_2,X_3}(x_1, x_2, x_3) = c^3e^{-c(x_1+x_2+x_3)}$, $x_1 > 0, x_2 > 0, x_3 > 0$ $(c > 0)$, determine:

    **(i)** The constant $c$.
    **(ii)** The marginal p.d.f.'s $f_{X_1}, f_{X_2}$, and $f_{X_3}$.
    **(iii)** The conditional joint p.d.f. of $X_1$ and $X_2$, given $X_3$.
    **(iv)** The conditional p.d.f. of $X_1$, given $X_2$ and $X_3$.

**4.2**    Determine the joint m.g.f. of the r.v.'s $X_1, X_2, X_3$ with p.d.f.

$$f_{X_1,X_2,X_3}(x_1, x_2, x_3) = c^3 e^{-c(x_1+x_2+x_3)}, \quad x_1 > 0, \ x_2 > 0, \ x_3 > 0$$

where $c$ is a positive constant (see also Exercise 4.1).

**4.3**    (*Cramér-Wold devise*) Show that if we know the joint distribution of the r.v.'s $X_1, \ldots, X_n$, then we can determine the distribution of any linear combination $c_1 X_1 + \cdots + c_n X_n$ of $X_1, \ldots, X_n$, where $c_1, \ldots, c_n$ are constants. Conversely, if we know the distribution of all linear combinations just described, then we can determine the joint distribution of $X_1, \ldots, X_n$.
    **Hint.** Use the m.g.f. approach.

**4.4**    If the r.v.'s $X_1, \ldots, X_m$ and $Y_1, \ldots, Y_n$ have finite second moments, then show that:

$$\text{Cov}\left(\sum_{i=1}^{m} X_i, \sum_{j=1}^{n} Y_j\right) = \sum_{i=1}^{m}\sum_{j=1}^{n} \text{Cov}(X_i, Y_j).$$

    **Hint.** Use the definition of covariance and the linearity property of the expectation.

## 4.5 THE MULTINOMIAL, THE BIVARIATE NORMAL, AND THE MULTIVARIATE NORMAL DISTRIBUTIONS

In this section, we introduce and study to some extent three multidimensional distributions; they are the *Multinomial* distribution, the *two-dimensional Normal* or *Bivariate Normal* distribution, and the *k-dimensional Normal* distribution.

### 4.5.1 MULTINOMIAL DISTRIBUTION

A *Multinomial* experiment is a straightforward generalization of a Binomial experiment, where, instead of 2, there are $k$ (mutually exclusive) possible outcomes, $O_1, \ldots, O_k$, say, occurring with respective probabilities $p_1, \ldots, p_k$. Simple examples of Multinomial experiments are those of rolling a die (with 6 possible outcomes); selecting (with replacement) $r$ balls from a collection of $n_1 + \cdots + n_k$ balls, so that $n_i$ balls have the number $i$ written on them, $i = 1, \ldots, k$; selecting (with replacement) $r$ objects out of a collection of objects of which $n_1$ are in good condition, $n_2$ have minor defects, and $n_3$ have serious defects, etc. Suppose a Multinomial experiment is carried out independently $n$ times and the probabilities $p_1, \ldots, p_k$ remain the same throughout. Denote by $X_i$ the r.v. of the number of times outcome $O_i$ occurs, $i = 1, \ldots, k$. Then the joint p.d.f. of $X_1, \ldots, X_k$ is given by:

$$f_{X_1, \ldots, X_k}(x_1, \ldots, x_k) = \frac{n!}{x_1! \ldots x_k!} p_1^{x_1} \cdots p_k^{x_k}, \tag{47}$$

where $x_1, \ldots, x_k$ are $\geq 0$ integers with $x_1 + \cdots + x_k = n$, and, of course, $0 < p_i < 1$, $i = 1, \ldots, k$, $p_1 + \cdots + p_k = 1$. The distribution given by (47) is the *Multinomial* distribution with *parameters* $n$ and $p_1, \ldots, p_k$, and the r.v.'s $X_1, \ldots, X_k$ are said to have the *Multinomial* distribution with these parameters. That the right-hand side of (47) is the right formula for the joint probabilities $P(X_1 = x_1, \ldots, X_k = x_k)$ ensues as follows: By independence, the probability that $O_i$ occurs $n_i$ times, $i = 1, \ldots, k$, in specified positions, is given by: $p_1^{x_1} \cdots p_k^{x_k}$, regardless of the location of the specific positions of occurrence of $O_i$'s. The different ways of choosing the $n_i$ positions for the occurrence of $O_i$, $i = 1, \ldots, k$, is equal to: $\binom{n}{n_1}\binom{n-n_1}{n_2} \cdots \binom{n-n_1-\cdots-n_{k-1}}{n_k}$. Writing out each term in factorial form and making the obvious cancellations, we arrive at: $n!/(x_1! \ldots x_k!)$ (see also Exercise 5.1). For illustrative purposes, let us consider the following example.

**Example 24.** A fair die is rolled independently 10 times. Find the probability that faces #1 through #6 occur the following respective number of times: 2, 1, 3, 1, 2, and 1.

**Discussion.** By letting $X_i$ be the r.v. denoting the number of occurrences of face $i, i = 1, \ldots, 6$, we have

$$f_{X_1, \ldots, X_6}(2, 1, 3, 1, 2, 1) = \frac{10!}{2!1!3!1!2!1!}(1/6)^{10} = \frac{4725}{1,889,568} \simeq 0.003.$$

In a Multinomial distribution, all marginal p.d.f.'s and all conditional p.d.f.'s are also Multinomial. More precisely, we have the following result.

**Theorem 4.** *Let $X_1, \ldots, X_k$ be Multinomially distributed r.v.'s with parameters $n$ and $p_1, \ldots, p_k$, and for $1 \leq s < k$, let $1 \leq i_1 < i_2 < \cdots < i_s \leq k$, $Y = n - (X_{i_1} + \cdots + X_{i_s})$ and $q = 1 - (p_{i_1} + \cdots + p_{i_s})$. Then:*

**(i)** *The r.v.'s $X_{i_1}, \ldots, X_{i_s}, Y$ are distributed Multinomially with parameters $n$ and $p_{i_1}, \ldots, p_{i_s}, q$.*

**(ii)** *The conditional joint distribution of $X_{j_1}, \ldots, X_{j_t}$, given $X_{i_1} = x_{i_1}, \ldots, X_{i_s} = x_{i_s}$, is Multinomial with parameters $n - r$ and $p_{j_1}/q, \ldots, p_{j_t}/q$, where $r = x_{i_1} + \cdots + x_{i_s}$ and $t = k - s$.*

*Proof.*

**(i)** For $\geq 0$ integers $x_{i_1}, \ldots, x_{i_s}$ with $x_{i_1} + \cdots + x_{i_s} = r \leq n$, we have:

$$f_{X_{i_1}, \ldots, X_{i_s}}(x_{i_1}, \ldots, x_{i_s}) = P(X_{i_1} = x_{i_1}, \ldots, X_{i_s} = x_{i_s})$$
$$= P(X_{i_1} = x_{i_1}, \ldots, X_{i_s} = x_{i_s}, Y = n - r)$$

$$= \frac{n!}{x_{i_1}! \ldots x_{i_s}!(n-r)!} p_{i_1}^{x_{i_1}} \ldots p_{i_s}^{x_{i_s}} q^{n-r}.$$

(ii)  For $\geq 0$ integers $x_{j_1}, \ldots, x_{j_t}$ with $x_{j_1} + \cdots + x_{j_t} = n - r$, we have:

$$f_{X_{j_1}, \ldots, X_{j_t} | X_{i_1}, \ldots, X_{i_s}}(x_{j_1}, \ldots, x_{j_t} | x_{i_1}, \ldots, x_{i_s})$$
$$= P(X_{j_1} = x_{j_1}, \ldots, X_{j_t} = x_{j_t} | X_{i_1} = x_{i_1}, \ldots, X_{i_s} = x_{i_s})$$
$$= \frac{P(X_{j_1} = x_{j_1}, \ldots, X_{j_t} = x_{j_t}, X_{i_1} = x_{i_1}, \ldots, X_{i_s} = x_{i_s})}{P(X_{i_1} = x_{i_1}, \ldots, X_{i_s} = x_{i_s})}$$
$$= \frac{\frac{n!}{x_{j_1}! \cdots x_{j_t}! x_{i_1}! \cdots x_{i_s}!} p_{j_1}^{x_{j_1}} \cdots p_{j_t}^{x_{j_t}} \times p_{i_1}^{x_{i_1}} \cdots p_{i_s}^{x_{i_s}}}{\frac{n!}{x_{i_1}! \cdots x_{i_s}!(n-r)!} p_{i_1}^{x_{i_1}} \cdots p_{i_s}^{x_{i_s}} q^{n-r}}$$
$$= \frac{(n-r)!}{x_{j_1}! \ldots x_{j_t}!} (p_{j_1}/q)^{x_{j_1}} \ldots (p_{j_t}/q)^{x_{j_t}}.$$  ∎

**Example 25.**  In reference to Example 24, calculate: $P(X_2 = X_4 = X_6 = 2)$ and $P(X_1 = X_3 = 1, X_5 = 2 | X_2 = X_4 = X_6 = 2)$.

**Discussion.**  Here, $n = 10$, $r = 6$, $p_2 = p_4 = p_6 = \frac{1}{6}$ and $q = 1 - \frac{3}{6} = \frac{1}{2}$. Thus:

$$P(X_2 = X_4 = X_6 = 2) = \frac{10!}{2!2!2!4!} \left(\frac{1}{6}\right)^6 \left(\frac{1}{2}\right)^4 = \frac{4725}{186,624} \simeq 0.025,$$

and:

$$P(X_1 = X_3 = 1, X_5 = 2 | X_2 = X_4 = X_6 = 2) = \frac{4!}{1!1!2!} \left(\frac{1/6}{1/2}\right)^4 = \frac{4}{27} \simeq 0.148.$$

**Example 26.**  In a genetic experiment, two different varieties of a certain species are crossed and a specific characteristic of the offspring can occur only at three levels, $A$, $B$, and $C$, say. According to a proposed model, the probabilities for $A$, $B$, and $C$ are $\frac{1}{12}$, $\frac{3}{12}$, and $\frac{8}{12}$, respectively. Out of 60 offspring, calculate:

(i)  The probability that 6, 18, and 36 fall into levels A, B, and C, respectively.
(ii)  The (conditional) probability that 6 and 18 fall into levels A and B, respectively, given that 36 have fallen into level C.

**Discussion.**

(i)  Formula (47) applies with $n = 60, k = 3, p_1 = \frac{1}{12}, p_2 = \frac{3}{12}, p_3 = \frac{8}{12}, x_1 = 6, x_2 = 18, x_3 = 36$ and yields:

$$P(X_1 = 6, X_2 = 18, X_3 = 36) = \frac{60!}{6!18!36!} \left(\frac{1}{12}\right)^6 \left(\frac{3}{12}\right)^{18} \left(\frac{8}{12}\right)^{36} \simeq 0.011.$$

(ii)  Here, Theorem 4(ii) applies with $s = 1$, $t = 2$, $x_{i_1} = x_3 = 36$, $x_{j_1} = x_1 = 6$, $x_{j_2} = x_2 = 18$, $r = 36$, so that $n - r = 60 - 36 = 24$, $q = 1 - p_3 =$

$1 - \frac{8}{12} = \frac{4}{12}$, and yields:

$$P(X_1 = 6, X_2 = 18 \mid X_3 = 36) = \frac{(n-r)!}{x_1! x_2!} \left(\frac{p_1}{q}\right)^{x_1} \left(\frac{p_2}{q}\right)^{x_2}$$

$$= \frac{(24)!}{6! 18!} \left(\frac{\frac{1}{12}}{\frac{4}{12}}\right)^6 \left(\frac{\frac{3}{12}}{\frac{4}{12}}\right)^{18}$$

$$= \binom{24}{6} \left(\frac{1}{4}\right)^6 \left(\frac{3}{4}\right)^{18}$$

$$= 0.1852 \quad \text{(from the Binomial tables)}.$$

An application of formula (42) gives the joint m.g.f. of $X_1, \ldots, X_k$ as follows, where the summation is over all $\geq 0$ integers $x_1, \ldots, x_k$ with $x_1 + \cdots + x_k = n$:

$$M_{X_1,\ldots,X_k}(t_1, \ldots, t_k) = \sum e^{t_1 x_1 + \cdots + t_k x_k} \frac{n!}{x_1! \cdots x_k!} p_1^{x_1} \cdots p_k^{x_k}$$

$$= \sum \frac{n!}{x_1! \cdots x_k!} (p_1 e^{t_1})^{x_1} \cdots (p_k e^{t_k})^{x_k}$$

$$= (p_1 e^{t_1} + \cdots + p_k e^{t_k})^n; \quad \text{i.e.,}$$

$$M_{X_1,\ldots,X_k}(t_1, \ldots, t_k) = (p_1 e^{t_1} + \cdots + p_k e^{t_k})^n, \qquad t_1, \ldots, t_k \in \Re. \tag{48}$$

By means of (44) and (48), we can find the $\text{Cov}(X_i, X_j)$ and the $\rho(X_i, X_j)$ for any $1 \leq i < j \leq k$. Indeed, $EX_i = np_i$, $EX_j = np_j$, $\text{Var}(X_i) = np_i(1 - p_i)$, $\text{Var}(X_j) = np_j(1 - p_j)$ and $E(X_i X_j) = n(n-1)p_i p_j$. Therefore:

$$\text{Cov}(X_i, X_j) = -np_i p_j \quad \text{and} \quad \rho(X_i, X_j) = -[p_i p_j / ((1 - p_i)(1 - p_j))]^{1/2} \tag{49}$$

(see Exercise 5.4 for details).

## 4.5.2 BIVARIATE NORMAL DISTRIBUTION

The joint distribution of the r.v.'s $X$ and $Y$ is said to be the *Bivariate Normal* distribution with *parameters* $\mu_1, \mu_2$ in $\Re$, $\sigma_1, \sigma_2$ positive, and $\rho \in [-1, 1]$, if the joint p.d.f. is given by the formula:

$$f_{X,Y}(x, y) = \frac{1}{2\pi \sigma_1 \sigma_2 \sqrt{1 - \rho^2}} e^{-q/2}, \qquad x, y \in \Re, \tag{50}$$

where

$$q = \frac{1}{1 - \rho^2} \left[ \left(\frac{x - \mu_1}{\sigma_1}\right)^2 - 2\rho \left(\frac{x - \mu_1}{\sigma_1}\right)\left(\frac{y - \mu_2}{\sigma_2}\right) + \left(\frac{y - \mu_2}{\sigma_2}\right)^2 \right]. \tag{51}$$

This distribution is also referred to as *two-dimensional Normal*. The shape of $f_{X,Y}$ looks like a bell sitting on the $xy$-plane and whose highest point is located at the point $(\mu_1, \mu_2, 1/(2\pi\sigma_1\sigma_2\sqrt{1-\rho^2}))$ (see Figure 4.7).

That $f_{X,Y}$ integrates to 1 and therefore is a p.d.f. is seen by rewriting it in a convenient way. Specifically,

$$\left(\frac{x-\mu_1}{\sigma_1}\right)^2 - 2\rho\left(\frac{x-\mu_1}{\sigma_1}\right)\left(\frac{y-\mu_2}{\sigma_2}\right) + \left(\frac{y-\mu_2}{\sigma_2}\right)^2$$

$$= \left(\frac{y-\mu_2}{\sigma_2}\right)^2 - 2\left(\rho\frac{x-\mu_1}{\sigma_1}\right)\left(\frac{y-\mu_2}{\sigma_2}\right) + \left(\rho\frac{x-\mu_1}{\sigma_1}\right)^2 + (1-\rho^2)\left(\frac{x-\mu_1}{\sigma_1}\right)^2$$

$$= \left[\left(\frac{y-\mu_2}{\sigma_2}\right) - \left(\rho\frac{x-\mu_1}{\sigma_1}\right)\right]^2 + (1-\rho^2)\left(\frac{x-\mu_1}{\sigma_1}\right)^2. \tag{52}$$

Furthermore,

$$\frac{y-\mu_2}{\sigma_2} - \rho\frac{x-\mu_1}{\sigma_1} = \frac{y-\mu_2}{\sigma_2} - \frac{1}{\sigma_2} \times \rho\sigma_2\frac{x-\mu_1}{\sigma_1}$$

$$= \frac{1}{\sigma_2}\left\{y - \left[\mu_2 + \frac{\rho\sigma_2}{\sigma_1}(x-\mu_1)\right]\right\}$$

$$= \frac{y - b_x}{\sigma_2}, \quad \text{where } b_x = \mu_2 + \frac{\rho\sigma_2}{\sigma_1}(x-\mu_1)$$

(see also Exercise 5.6).

Therefore, the right-hand side of (52) is equal to:

$$\left(\frac{y-b_x}{\sigma_2}\right)^2 + (1-\rho^2)\left(\frac{x-\mu_1}{\sigma_1}\right)^2,$$

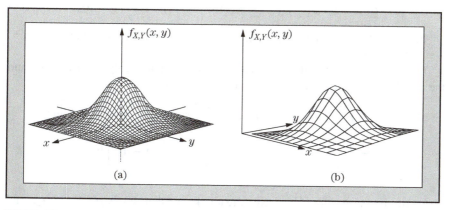

$f_{X,Y}(x,y)$

$f_{X,Y}(x,y)$

(a)

(b)

**FIGURE 4.7**

Graphs of the p.d.f. of the Bivariate Normal distribution: (a) Centered at the origin; (b) Centered elsewhere in the $(x, y)$-plane.

and hence the exponent becomes

$$-\frac{(x-\mu_1)^2}{2\sigma_1^2} - \frac{(y-b_x)^2}{2(\sigma_2\sqrt{1-\rho^2})^2}.$$

Then the joint p.d.f. may be rewritten as follows:

$$f_{X,Y}(x,y) = \frac{1}{\sqrt{2\pi}\sigma_1}e^{-(x-\mu_1)^2/2\sigma_1^2} \times \frac{1}{\sqrt{2\pi}(\sigma_2\sqrt{1-\rho^2})}e^{-(y-b_x)^2/2(\sigma_2\sqrt{1-\rho^2})^2}. \qquad (53)$$

The first factor on the right-hand side of (53) is the p.d.f. of $N(\mu_1, \sigma_1^2)$ and the second factor is the p.d.f. of $N(b_x, (\sigma_2\sqrt{1-\rho^2})^2)$. Therefore, integration with respect to $y$ produces the marginal $N(\mu_1, \sigma_1^2)$ distribution, which, of course, integrates to 1. So, we have established the following two facts: $\int_{-\infty}^{\infty}\int_{-\infty}^{\infty} f_{X,Y}(x,y)\, dx\, dy = 1$, and

$$X \sim N(\mu_1, \sigma_1^2), \quad \text{and, by symmetry, } Y \sim N(\mu_2, \sigma_2^2). \qquad (54)$$

The results recorded in (54) also reveal the special significance of the parameters $\mu_1, \sigma_1^2$ and $\mu_2, \sigma_2^2$. Namely, they are the means and the variances of the (Normally distributed) r.v.'s $X$ and $Y$, respectively. Relations (53) and (54) also provide immediately the conditional p.d.f. $f_{Y|X}$; namely,

$$f_{Y|X}(y|x) = \frac{1}{\sqrt{2\pi}(\sigma_2\sqrt{1-\rho^2})^2} \exp\left[-\frac{(y-b_x)^2}{2(\sigma_2\sqrt{1-\rho^2})^2}\right].$$

Thus, in obvious notation:

$$Y \mid X = x \sim N(b_x, (\sigma_2\sqrt{1-\rho^2})^2), \quad b_x = \mu_2 + \frac{\rho\sigma_2}{\sigma_1}(x-\mu_1), \qquad (55)$$

and, by symmetry:

$$X \mid Y = y \sim N(b_y, (\sigma_1\sqrt{1-\rho^2})^2), \quad b_y = \mu_1 + \frac{\rho\sigma_1}{\sigma_2}(y-\mu_2). \qquad (56)$$

In Figure 4.8, the conditional p.d.f. $f_{Y|X}(\cdot \mid x)$ is depicted for three values of $x$: $x = 5, 10$, and 15.

Formulas (53), (54), and (56) also allow us to calculate easily the covariance and the correlation coefficient of $X$ and $Y$. Indeed, by (53):

$$E(XY) = \int_{-\infty}^{\infty}\int_{-\infty}^{\infty} xy f_{X,Y}(x,y)dx\, dy = \int_{-\infty}^{\infty} x f_X(x)\left[\int_{-\infty}^{\infty} y f_{Y|X}(y \mid x)dy\right]dx$$

$$= \int_{-\infty}^{\infty} x f_X(x) b_x\, dx = \int_{-\infty}^{\infty} x f_X(x)\left[\mu_2 + \frac{\rho\sigma_2}{\sigma_1}(x-\mu_1)\right]dx$$

$$= \mu_1\mu_2 + \rho\sigma_1\sigma_2$$

(see also Exercise 5.7). Since we already know that $EX = \mu_1, EY = \mu_2$, and $\mathrm{Var}(X) = \sigma_1^2, \mathrm{Var}(Y) = \sigma_2^2$, we obtain:

$$\mathrm{Cov}(X,Y) = E(XY) - (EX)(EY) = \mu_1\mu_2 + \rho\sigma_1\sigma_2 - \mu_1\mu_2 = \rho\sigma_1\sigma_2,$$

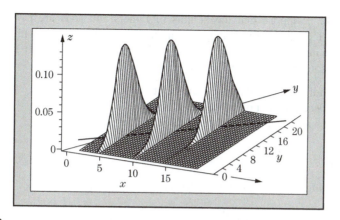

**FIGURE 4.8**

Conditional probability density functions of the bivariate Normal distribution.

and therefore $\rho(X, Y) = \frac{\rho\sigma_1\sigma_2}{\sigma_1\sigma_2} = \rho$. Thus, we have:

$$\text{Cov}(X, Y) = \rho\sigma_1\sigma_2 \quad \text{and} \quad \rho(X, Y) = \rho. \tag{57}$$

Relation (57) reveals that the parameter $\rho$ in (50) is, actually, the correlation coefficient of the r.v.'s $X$ and $Y$.

**Example 27.** If the r.v.'s $X_1$ and $X_2$ have the Bivariate Normal distribution with parameters $\mu_1, \mu_2, \sigma_1^2, \sigma_2^2$, and $\rho$:

**(i)** Calculate the quantities: $E(c_1X_1 + c_2X_2), \text{Var}(c_1X_1 + c_2X_2)$, where $c_1, c_2$ are constants.

**(ii)** How the expression in part (i) becomes for: $\mu_1 = -1, \mu_2 = 3, \sigma_1^2 = 4, \sigma_2^2 = 9$, and $\rho = \frac{1}{2}$?

**Discussion.**

**(i)** $E(c_1X_1 + c_2X_2) = c_1EX_1 + c_2EX_2 = c_1\mu_1 + c_2\mu_2$, since $X_i \sim N(\mu_i, \sigma_i^2)$, so that $EX_i = \mu_i$, $i = 1, 2$. Also,

$$\text{Var}(c_1X_1 + c_2X_2) = c_1^2\sigma_{X_1}^2 + c_2^2\sigma_{X_2}^2 + 2c_1c_2\sigma_{X_1}\sigma_{X_2}\rho(X_1, X_2) \quad \text{(by 39)}$$

$$= c_1^2\sigma_1^2 + c_2^2\sigma_2^2 + 2c_1c_2\sigma_1\sigma_2\rho,$$

since $X_i \sim N(\mu_i, \sigma_i^2)$, so that $\text{Var}(X_i) = \sigma_i^2, i = 1, 2$, and $\rho(X_1, X_2) = \rho$, by (57).

**(ii)** Here, $E(c_1X_1 + c_2X_2) = -c_1 + 3c_2$, and $\text{Var}(c_1X_1 + c_2X_2) = 4c_1 + 9c_2 + 2c_1c_2 \times 2 \times 3 \times \frac{1}{2} = 4c_1 + 9c_2 + 6c_1c_2$.

**Example 28.** Suppose that the heights of fathers and sons are r.v.'s $X$ and $Y$, respectively, having (approximately) Bivariate Normal distribution with parameters (expressed in inches) $\mu_1 = 70, \sigma_1 = 2, \mu_2 = 71, \sigma_2 = 2$ and $\rho = 0.90$. If for a given pair (father, son) it is observed that $X = x = 69$, determine:

(i) The conditional distribution of the height of the son.
(ii) The expected height of the son.
(iii) The probability of the height of the son to be more than 72 in.

**Discussion.**

(i) According to (55), $Y|X = x \sim N(b_x, (\sigma_2\sqrt{1 - \rho^2})^2)$, where

$$b_x = \mu_2 + \frac{\rho\sigma_2}{\sigma_1}(x - \mu_1) = 71 + 0.90 \times (69 - 70) = 70.1, \text{ and}$$

$$\sigma_2\sqrt{1 - \rho^2} = 2 \times \sqrt{1 - 0.90^2} \simeq 0.87.$$

That is, $Y|X = 69$ is distributed as $N(70.1, (0.87)^2)$.

(ii) The (conditional) expectation of $Y$, given $X = 69$, is equal to $b_{69} = 70.1$.
(iii) The required (conditional) probability is

$$P(Y > 72|X = 69) = P\left(\frac{Y - b_{69}}{\sigma_2\sqrt{1 - \rho^2}} > \frac{72 - 70.1}{0.87}\right) \simeq P(Z > 2.18)$$

$$= 1 - \Phi(2.18) = 1 - 0.985371 = 0.014629.$$

Finally, it can be seen by integration that the joint m.g.f. of $X$ and $Y$ is given by the formula:

$$M_{X,Y}(t_1, t_2) = \exp\left[\mu_1 t_1 + \mu_2 t_2 + \frac{1}{2}(\sigma_1^2 t_1^2 + 2\rho\sigma_1\sigma_2 t_1 t_2 + \sigma_2^2 t_2^2)\right], \quad t_1, t_2 \in \Re; \quad (58)$$

we choose not to pursue its justification (which can be found, e.g., in pages 158–159, in the book *A Course in Mathematical Statistics*, 2nd edition (1997), Academic Press, by G.G. Roussas). We see, however, easily that:

$$\frac{\partial}{\partial t_1}M_{X,Y}(t_1, t_2) = (\mu_1 + \sigma_1^2 t_1 + \rho\sigma_1\sigma_2 t_2)M_{X,Y}(t_1, t_2),$$

and hence:

$$\frac{\partial^2}{\partial t_1 \partial t_2}M_{X,Y}(t_1, t_2) = \rho\sigma_1\sigma_2 M_{X,Y}(t_1, t_2) + (\mu_1 + \sigma_1^2 t_1 + \rho\sigma_1\sigma_2 t_2)$$

$$\times (\mu_2 + \sigma_2^2 t_2 + \rho\sigma_1\sigma_2 t_1)M_{X,Y}(t_1, t_2),$$

which, evaluated at $t_1 = t_2 = 0$, yields $\rho\sigma_1\sigma_2 + \mu_1\mu_2 = E(XY)$, as we have already seen.

### 4.5.3 MULTIVARIATE NORMAL DISTRIBUTION

The Multivariate Normal distribution is a generalization of the Bivariate Normal distribution and can be defined in a number of ways; we choose the one given here. To this end, for $k \geq 2$, let $\boldsymbol{\mu} = (\mu_1, \ldots, \mu_k)$ be a vector of constants, and let $\boldsymbol{\Sigma}$ be a $k \times k$ nonsingular matrix of constants, so that the inverse $\boldsymbol{\Sigma}^{-1}$ exists and the determinant $|\boldsymbol{\Sigma}| \neq 0$. Finally, set $X$ for the vector of r.v.'s $X_1, \ldots, X_k$; i.e., $X = (X_1, \ldots, X_k)$ and $x = (x_1, \ldots, x_k)$ for any point in $\mathfrak{R}^k$. Then, the joint p.d.f. of the $X_i$'s, or the p.d.f. of the random vector $X$, is said to be *Multivariate Normal*, or *k-Variate Normal*, if it is given by the formula:

$$f_X(x) = \frac{1}{(2\pi)^{k/2}|\boldsymbol{\Sigma}|^{1/2}} \exp\left[-\frac{1}{2}(x - \boldsymbol{\mu})\boldsymbol{\Sigma}^{-1}(x - \boldsymbol{\mu})'\right], \qquad x \in \mathfrak{R}^k,$$

where, it is to be recalled that "'" stands for transpose.

It can be seen that: $EX_i = \mu_i$, $\text{Var}(X_i) = \sigma_i^2$ is the $(i, i)$th element of $\boldsymbol{\Sigma}$, and $\text{Cov}(X_i, X_j)$ is the $(i, j)$th element of $\boldsymbol{\Sigma}$, so that $\boldsymbol{\mu} = (EX_1, \ldots, EX_k)$ and $\boldsymbol{\Sigma} = (\text{Cov}(X_i, X_j))$, $i, j = 1, \ldots, k$. The quantities $\boldsymbol{\mu}$ and $\boldsymbol{\Sigma}$ are called the *parameters* of the distribution. It can also be seen that the joint m.g.f. of the $X_i$'s, or the m.g.f. of the random vector $X$, is given by:

$$M_X(t) = \exp\left(\boldsymbol{\mu}t' + \frac{1}{2}t\boldsymbol{\Sigma}t'\right), \qquad t \in \mathfrak{R}^k.$$

The $k$-Variate Normal distribution has properties similar to those of the two-dimensional Normal distribution, and the latter is obtained from the former by taking $\boldsymbol{\mu} = (\mu_1, \mu_2)$ and $\boldsymbol{\Sigma} = \begin{pmatrix} \sigma_1^2 & \rho\sigma_1\sigma_2 \\ \rho\sigma_1\sigma_2 & \sigma_2^2 \end{pmatrix}$, where $\rho = \rho(X_1, X_2)$.

More relevant information can be found, e.g., in Chapter 18 of the reference cited in the discussion of Example 27.

---

### EXERCISES

**5.1**  Show that for $n_1, \ldots, n_k \geq 0$ integers with $n_1 + \cdots + n_k = n$:

$$\binom{n}{n_1}\binom{n - n_1}{n_2} \cdots \binom{n - n_1 - \cdots - n_{k-1}}{n_k} = \frac{n!}{n_1!n_2!\ldots n_k!}.$$

**Hint.** Write out the terms on the left-hand side as factorials, and recall that $0! = 1$.

**5.2**  In a store selling TV sets, it is known that 25% of the customers will purchase a TV set of brand A, 40% will purchase a TV set of brand B, and 35% will just be browsing around. For a lot of 10 customers:

    **(i)**  What is the probability that two will purchase a TV set of brand A, three will purchase a TV set of brand B, and five will purchase neither?

**(ii)** If it is known that six customers did not purchase a TV set, what is the (conditional) probability that one of the rest will purchase a TV set of brand A and three will purchase a TV set of brand B?

**Hint.** Part (i) is an application of formula (47), and part (ii) is an application of Theorem 4(i).

**5.3** Human blood

occurs in four types termed A, B, AB, and O with respective frequencies $p_A = 0.40$, $p_B = 0.10$, $p_{AB} = 0.05$, and $p_O = 0.45$. If $n$ donors participate in a blood drive, denote by $X_A, X_B, X_{AB}$, and $X_O$ the numbers of donors with respective blood types A, B, AB, and O. Then $X_A, X_B, X_{AB}$, and $X_O$ are r.v.'s having the Multinomial distribution with parameters $n$ and $p_A, p_B, p_{AB}, p_O$. Write out the appropriate formulas for the following probabilities:

**(i)** $P(X_A = x_A, X_B = x_B, X_{AB} = x_{AB}, X_O = x_O)$ for
$x_A, x_B, x_{AB}$, and $x_O$ nonNegative integers with $x_A + x_B + x_{AB} + x_O = n$.
**(ii)** $P(X_A = x_A, X_B = x_B, X_{AB} = x_{AB})$.
**(iii)** $P(X_A = x_A, X_B = x_B)$.
**(iv)** $P(X_A = x_A)$.
**(v)** $P(X_A = x_A, X_B = x_B, X_{AB} = x_{AB} \mid X_O = x_O)$.
**(vi)** $P(X_A = x_A, X_B = x_B \mid X_{AB} = x_{AB}, X_O = x_O)$.
**(vii)** $P(X_A = x_A \mid X_B = x_B, X_{AB} = x_{AB}, X_O = x_O)$.
**(viii)** Give numerical answers
to parts (i)-(vii), if $n = 20$, and $x_A = 8, x_B = 2, x_{AB} = 1, x_O = 9$.

**Hint.** Part (i) is an
application of formula (47) and parts (ii)-(vii) are applications of Theorem 4.

**5.4** In conjunction with the Multinomial distribution, show that:

$$EX_i = np_i, EX_j = np_j, \quad \text{Var}(X_i) = np_i(1 - p_i), \quad \text{Var}(X_j) = np_j(1 - p_j),$$

$$\text{Cov}(X_i, X_j) = -np_i p_j \quad \text{and} \quad \rho(X_i, X_j) = -\frac{p_i p_j}{[p_i(1 - p_i)p_j(1 - p_j)]^{1/2}}.$$

**Hint.** Use the m.g.f. given in formula (48).

**5.5** Refer to Exercises 5.3 and 5.4, and for $n = 20$, calculate the quantities:

$$EX_A, \ EX_B, \ EX_{AB}, \ EX_O; \ \text{Var}(X_A), \ \text{Var}(X_B), \ \text{Var}(X_{AB}),$$

$$\text{Var}(X_O); \ \text{Cov}(X_A, X_B), \ \text{Cov}(X_A, X_{AB}), \ \text{Cov}(X_A, X_O);$$

$$\rho(X_A, X_B), \ \rho(X_A, X_{AB}), \ \rho(X_A, X_O).$$

**5.6** Elaborate on the expressions in (52), as well as the expressions following (52).

**5.7** If the r.v.'s $X$ and $Y$ have the Bivariate Normal distribution with parameters $\mu_1, \mu_2, \sigma_1^2, \sigma_2^2$, and $\rho$, show that $E(XY) = \mu_1 \mu_2 + \rho \sigma_1 \sigma_2$.
**Hint.** Write the joint p.d.f. $f_{X,Y}$ as $f_{Y|X}(y \mid x)f_X(x)$ and use the fact (see relation (55)) that $E(Y \mid X = x) = b_x = \mu_2 + \frac{\rho \sigma_2}{\sigma_1}(x - \mu_1)$.

**5.8**  If the r.v.'s $X$ and $Y$ have the Bivariate
Normal distribution, then, by using Exercise 5.7, show that the parameter
$\rho$ is, indeed, the correlation coefficient of the r.v.'s $X$ and $Y$, $\rho = \rho(X, Y)$.

**5.9**  If the r.v.'s $X$
and $Y$ have the Bivariate Normal distribution, and $c_1, c_2$ are constants, express
the expectation $E(c_1 X + c_2 Y)$ and the variance $\mathrm{Var}(c_1 X + c_2 Y)$ in terms
of $c_1, c_2, \mu_1 = EX, \mu_2 = EY, \sigma_1^2 = \mathrm{Var}(X), \sigma_2^2 = \mathrm{Var}(Y)$, and $\rho = \rho(X, Y)$.

**5.10**  If the r.v.'s $X$ and $Y$ have the Bivariate
Normal distribution, then it is known (see, e.g., relation (11), Chapter
6, in the book *A Course in Mathematical Statistics*, 2nd edition (1997),
Academic Press, by G.G. Roussas) that the joint m.g.f. of $X$ and $Y$ is given by:

$$M_{X,Y}(t_1, t_2) = \exp\left[\mu_1 t_1 + \mu_2 t_2 + \frac{1}{2}\left(\sigma_1^2 t_1^2 + 2\rho\sigma_1\sigma_2 t_1 t_2 + \sigma_2^2 t_2^2\right)\right], \quad t_1, t_2 \in \Re.$$

Use this m.g.f. in order to show that:

$$EX = \mu_1, \quad EY = \mu_2, \quad \mathrm{Var}(X) = \sigma_1^2, \quad \mathrm{Var}(Y) = \sigma_2^2,$$
$$\mathrm{Cov}(X, Y) = \rho\sigma_1\sigma_2, \quad \text{and} \quad \rho(X, Y) = \rho.$$

**5.11**  Use the joint m.g.f. of the r.v.'s $X$ and $Y$ having
a Bivariate Normal distribution (see Exercise 5.10) in order to show that:
  **(i)**  If $X$ and $Y$ have the Bivariate
  Normal distribution with parameters $\mu_1, \mu_2, \sigma_1^2, \sigma_2^2$, and $\rho$, then, for
  any constants $c_1$ and $c_2$, the r.v. $c_1 X + c_2 Y$ has the Normal distribution
  with parameters $c_1\mu_1 + c_2\mu_2$, and $c_1^2\sigma_1^2 + 2c_1 c_2\rho\sigma_1\sigma_2 + c_2^2\sigma_2^2$.
  **(ii)**  If the r.v. $c_1 X + c_2 Y$ is Normally distributed, then
  the r.v.'s. $X$ and $Y$ have the Bivariate Normal distribution with parameters
  $\mu_1 = EX, \mu_2 = EY, \sigma_1^2 = \mathrm{Var}(X), \sigma_2^2 = \mathrm{Var}(Y)$, and $\rho = \rho(X, Y)$.
  **Hint.** For part (i), use the m.g.f. given in Exercise 5.10 and
  regroup the terms appropriately. For part (ii), evaluate the m.g.f. of $t_1 X + t_2 Y$,
  for any $t_1$ and $t_2$ real, at 1, and plug in the $E(t_1 X + t_2 Y)$ and $\mathrm{Var}(t_1 X + t_2 Y)$.

**5.12**  Consider the function $f$ defined by:

$$f(x, y) = \begin{cases} \frac{1}{2\pi}e^{-\frac{x^2+y^2}{2}}, & \text{for } (x, y) \text{ outside the square } [-1, 1] \times [-1, 1] \\ \frac{1}{2\pi}e^{-\frac{x^2+y^2}{2}} + \frac{1}{2\pi e}x^3 y^3, & \text{for } (x, y) \text{ in the square } [-1, 1] \times [-1, 1]. \end{cases}$$

  **(i)**  Show that $f$ is a non-Bivariate Normal p.d.f.
  **(ii)**  Also, show that both marginals, call them $f_1$ and $f_2$, are $N(0, 1)$ p.d.f.'s.
  **Remark.**   We know that if $X, Y$ have
  the Bivariate Normal distribution, then the distributions of the r.v.'s $X$ and $Y$
  themselves are Normal. This exercise shows that the inverse need not be true.

**5.13**  Let the r.v.'s $X$ and $Y$ have the Bivariate Normal distribution with parameters
$\mu_1, \mu_2, \sigma_1^2, \sigma_2^2$, and $\rho$, and set $U = X + Y, V = X - Y$. Then show that:

(i) The r.v.'s $U$ and $V$ also have the Bivariate Normal distribution with parameters: $\mu_1 + \mu_2, \mu_1 - \mu_2, \tau_1^2 = \sigma_1^2 + 2\rho\sigma_1\sigma_2 + \sigma_2^2, \tau_2^2 = \sigma_1^2 - 2\rho\sigma_1\sigma_2 + \sigma_2^2$, and $\rho_0 = (\sigma_1^2 - \sigma_2^2)/\tau_1\tau_2$.

(ii) $U \sim N(\mu_1 + \mu_2, \tau_1^2)$, $V \sim N(\mu_1 - \mu_2, \tau_2^2)$.

(iii) The r.v.'s $U$ and $V$ are uncorrelated if and only if $\sigma_1^2 = \sigma_2^2$.

**Hint.** For part (i), start out with the joint m.g.f. of $U$ and $V$, and express it in terms of the m.g.f. of $X$ and $Y$. Then use the formula given in Exercise 5.10, and regroup the terms in the exponent suitably to arrive at the desired conclusion. Parts (ii) and (iii) follow from part (i) and known facts about Bivariate Normal distribution.

**5.14** Let the r.v.'s $X$ and $Y$ denote the scores in two tests $T_1$ and $T_2$, and suppose that they have Bivariate Normal distribution with the following parameter: $\mu_1 = 82$, $\sigma_1 = 10$, $\mu_2 = 90$, $\sigma_2 = 9$, $\rho = 0.75$. Compute the following quantities: $P(Y > 92|X = 84)$, $P(X > Y)$, $P(X + Y > 190)$.
**Hint.** The first probability if calculated by the fact that we know the conditional distribution of $Y$, given $X = x$. The last two probabilities are calculated by the fact that the distributions of $X - Y$ and $X + Y$ are given in Exercise 5.13.

**5.15** If $X$ and $Y$ have Bivariate Normal distribution with parameters $\mu_1 = 3.2$, $\sigma_1 = 12$, $\mu_2 = 1.44$, $\sigma_2 = 16$, and $\rho = 0.7$, determine the following quantities:
(i) $EX, EY, \sigma^2(X), \sigma^2(Y), \rho(X, Y)$, and $\text{Cov}(X, Y)$.
(ii) $E(X|Y = 10), E(Y|X = 3.8), \sigma^2(X|Y = 10), \sigma^2(Y|X = 3.8)$.
(iii) The distribution of $X$ and the distribution of $Y$.
(iv) The probabilities $P(0.8 < X < 4.2), P(Y > 14)$.
(v) The conditional distribution of $Y$, given $X = 3.8$.
(vi) The probabilities $P(X > 3.2|Y = 10), P(Y < 12|X = 3.8)$.

**5.16** Let the r.v.'s $X$ and $Y$ denote the heights of pairs (father, son) in a certain age bracket, and suppose that they have Bivariate Normal distribution with the following parameters (measured in inches as appropriate): $\mu_1 = 67, \sigma_1 = 2$, $\mu_2 = 68, \sigma_2 = 1, \rho = 0.8$. If it is observed that $X = 69$, compute the following quantities: $E(Y|X = 69)$, conditional s.d. of $Y|X = 69$, $P(Y > 70|X = 69)$ and $P(Y > X)$.

**5.17** Suppose the r.v.'s $X$ and $Y$ have Bivariate Normal distribution with parameters $\mu_1, \mu_2, \sigma_1^2, \sigma_2^2, \rho$, and set $Y = X + cY$.
(a) Compute the $\sigma^2(U)$ in terms of the parameters involved and $c$.
(b) Determine the value of $c$ which minimizes the variance in part (i).
(c) What is the minimum variance in part (i)?
(d) How do the variance and the minimum variance in parts (ii) and (iii) become when $X$ and $Y$ are independent?
**Hint.** For part (i), use Theorem 3 in Chapter 8.

**5.18** Show that the Bivariate Normal p.d.f. $f(x, y)$ given by relation (50) is maximized for $x = \mu_1$, $y = \mu_2$, and the maximum is equal to $1/(2\pi\sigma_1\sigma_2\sqrt{1-\rho^2})$.
**Hint.** Set $\frac{x-\mu_1}{\sigma_1} = u$, $\frac{y-\mu_2}{\sigma_2} = v$, so that the
exponent in the p.d.f. becomes, in terms of $u$ and $v$: $g(v, v) = u^2 - 2\rho uv + v^2$
(apart from the factor $(1 - \rho^2)^{-1}$). Then show that $g(u, v)$ is minimized for
$u = v = 0$, which would imply that $f(x, y)$ is maximized for $x = \mu_1$, $y = \mu_2$.

**5.19** Let the r.v.'s $X$ and $Y$ have
Bivariate Normal distribution with parameters $\mu_1$, $\mu_2$, $\sigma_1^2$, $\sigma_2^2$, $\rho$, and suppose
that all parameters are known and the r.v. $X$ is observable. Then we wish to
predict $Y$ in terms of a linear expression in $X$; namely, $\hat{Y} = a + b(X - \mu_1)$, by
determining $a$ and $b$, so that the (average) $E(Y - \hat{Y})^2$ becomes minimum.
   **(i)** Show that the values of $a$ and $b$ that
      minimize the expression $E(Y - \hat{Y})^2$ are given by: $\hat{a} = \mu_2$ and $\hat{b} = \rho\frac{\sigma_2}{\sigma_1}$.
   **(ii)** The predictor $\hat{Y}$ of $Y$ is given by: $\hat{Y} = \mu_2 + \rho\frac{\sigma_2}{\sigma_1}(x - \mu_1)$, which
      is the $E(Y|X = x)$. Hence, the $E(Y|X = x)$ is the best predictor of $Y$
      (among all predictors of the form $a + b(x - \mu_1)$) in the sense that when
      $x$ is replaced by $X$, $E(Y|X)$ minimizes the mean square error $E(Y - \hat{Y})^2$.
**Hint.** The required determination of $a$ and $b$ can be made
through the usual procedure of differentiations. Alternatively, we can write

$$E(Y - \hat{Y})^2 = E[Y - a - b(X - \mu_1)]^2 = E[(Y - \mu_2) + (\mu_2 - a) - b(X - \mu_1)]^2$$

and proceed in order to find:

$$E(Y - \hat{Y})^2 = \sigma_2^2 + (\mu_2 - a)^2 + b^2\sigma_1^2 - 2b\text{Cov}(X, Y)$$

$$= \sigma_2^2 + (\mu_2 - a^2) + \sigma_1^2\left[b - \frac{1}{\sigma_1^2}\text{Cov}(X, Y)\right]^2 - \frac{\text{Cov}^2(X, Y)}{\sigma_1^2}$$

$$= (\mu_2 - a)^2 + \sigma_1^2\left[b - \frac{1}{\sigma_1^2}\text{Cov}(X, Y)\right]^2 + \frac{\sigma_1^2\sigma_2^2 - \text{Cov}^2(X, Y)}{\sigma_1^2\sigma_2^2}.$$

Since $\text{Cov}^2(X, Y) \leq \sigma_1^2\sigma_2^2$, it follows
that $E(Y - \hat{Y})^2$ is minimized for $a = \mu_1$ and $b = \frac{1}{\sigma_1^2}\text{Cov}(X, Y) = \rho\frac{\sigma_2}{\sigma_1}$.

**5.20** Show that the intersection of the surface represented by a Bivariate Normal
p.d.f. by any plane perpendicular to the $z$-axis is an ellipse, and it is a circle
if and only if $\rho = 0$.

**5.21** For any two r.v.'s $X_1$, $X_2$ with finite expectations, it is known that:

$$E[E(X_2|X_1)] = EX_2, \quad E[E(X_1|X_2)] = EX_1.$$

Suppose now that the r.v.'s have jointly the Bivariate Normal distribution
with parameters $\mu_1$, $\mu_2$, $\sigma_1^2$, $\sigma_2^2$, $\rho$. Then show that:

$$\text{Var}[E(X_2|X_1)] \leq \sigma_2^2 = \sigma^2(X_2), \quad \text{Var}[E(X_1|X_2)] \leq \sigma_1^2 = \sigma^2(X_1).$$

**5.22** Let $L$ and $S$ be r.v.'s representing the temperatures on October 1 in Los Angeles and San Francisco, respectively, and suppose that these r.v.'s have the Bivariate Normal distribution with $EL = 70.0$, $\sigma_1 = \sigma(L) = 2$, and $ES = 66.8$, $\sigma_2 = \sigma(S) = 2$, and that $\rho = \rho(L, S) = 0.68$. Compute: $E(L - S)$, $\text{Var}(L - S)$, $\sigma(L - S)$, and the $P(S > L)$.

**Hint.** Use Exercise 5.11(i).

# Independence of random variables and some applications

This chapter consists of two sections. In the first section, we introduce the concept of independence of r.v.'s and establish criteria for proving or disproving independence. Also, its relationship to uncorrelatedness is discussed. In the second section, the sample mean and the sample variance are defined, and some of their moments are also produced. The main thrust of this section, however, is the discussion of the reproductive property of certain distributions. As a by-product, we also obtain the distribution of the sample mean and of a certain multiple of the sample variance for independent and Normally distributed r.v.'s.

## 5.1 INDEPENDENCE OF RANDOM VARIABLES AND CRITERIA OF INDEPENDENCE

In Section 2.4, the concept of independence of two events was introduced and it was suitably motivated and illustrated by means of examples. This concept was then generalized to more than two events. What is done in this section is, essentially, to carry over the concept of independence from events to r.v.'s. To this end, consider first two r.v.'s $X_1$ and $X_2$ and the events induced in the sample space $\mathcal{S}$ by each one of them separately as well as by both of them jointly. That is, for subsets $B_1$ and $B_2$ of $\Re$, let:

$$A_1 = (X_1 \in B_1) = X_1^{-1}(B_1) = \{s \in \mathcal{S}; X_1(s) \in B_1\}, \tag{1}$$

$$A_2 = (X_2 \in B_2) = X_2^{-1}(B_2) = \{s \in \mathcal{S}; X_2(s) \in B_2\}, \tag{2}$$

$$A_{12} = ((X_1, X_2) \in B_1 \times B_2) = (X_1 \in B_1 \,\&\, X_2 \in B_2) = (X_1, X_2)^{-1}(B_1 \times B_2)$$

$$= \{s \in \mathcal{S}; X_1(s) \in B_1 \,\&\, X_2(s) \in B_2\} = A_1 \cap A_2. \tag{3}$$

Then the r.v.'s $X_1$ and $X_2$ are said to be independent if, for any $B_1$ and $B_2$ as before, the corresponding events $A_1$ and $A_2$ are independent; that is, $P(A_1 \cap A_2) = P(A_1)P(A_2)$. By (1)–(3), clearly, this relation is equivalent to:

$$P(X_1 \in B_1, X_2 \in B_2) = P(X_1 \in B_1)P(X_2 \in B_2). \tag{4}$$

**An Introduction to Probability and Statistical Inference**
Copyright © 2015 Elsevier Inc. All rights reserved.

This relation states, in effect, that information regarding one r.v. has no effect on the probability distribution of the other r.v. For example,

$$P(X_1 \in B_1 | X_2 \in B_2) = \frac{P(X_1 \in B_1, X_2 \in B_2)}{P(X_2 \in B_2)}$$

$$= \frac{P(X_1 \in B_1)P(X_2 \in B_2)}{P(X_2 \in B_2)} = P(X_1 \in B_1).$$

Relation (4) is taken as the definition of independence of these two r.v.'s, which is then generalized in a straightforward way to $k$ r.v.'s.

**Definition 1.** Two r.v.'s $X_1$ and $X_2$ are said to be *independent* (*statistically* or *stochastically* or in the *probability sense*) if, for any subsets $B_1$ and $B_2$ of $\Re$,

$$P(X_1 \in B_1, X_2 \in B_2) = P(X_1 \in B_1)P(X_2 \in B_2).$$

The r.v.'s $X_1, \ldots, X_k$ are said to be *independent* (in the same sense as above) if, for any subsets $B_1, \ldots, B_k$ of $\Re$,

$$P(X_i \in B_i, i = 1, \ldots, k) = \prod_{i=1}^{k} P(X_i \in B_i). \tag{5}$$

Nonindependent r.v.'s are said to be *dependent*.

The practical question which now arises is how one checks independence of $k$ given r.v.'s, or lack thereof. This is done by means of the following criterion referred to as the factorization theorem because of the form of the expressions involved.

**Theorem 1** (Criterion of Independence, Factorization Theorem). *For $k \geq 2$, the r.v.'s $X_1, \ldots, X_k$ are independent if and only if any one of the following three relations holds:*

$$\text{(i)} \ F_{X_1,\ldots,X_k}(x_1, \ldots, x_k) = F_{X_1}(x_1) \cdots F_{X_k}(x_k) \tag{6}$$

*for all $x_1, \ldots, x_k$ in $\Re$.*

$$\text{(ii)} \ f_{X_1,\ldots,X_k}(x_1, \ldots, x_k) = f_{X_1}(x_1) \cdots f_{X_k}(x_k) \tag{7}$$

*for all $x_1, \ldots, x_k$ in $\Re$.*

$$\text{(iii)} \ M_{X_1,\ldots,X_k}(t_1, \ldots, t_k) = M_{X_1}(t_1) \cdots M_{X_k}(t_k) \tag{8}$$

*for all $t_1, \ldots, t_k$ in a nondegenerate interval containing 0.*

Before we proceed with the justification of this theorem, let us refer to Example 1 in Chapter 4 and notice that: $f_X(3) = 0.04, f_Y(2) = 0.15$, and $f_{X,Y}(3, 2) = 0.02$, so that $f_{X,Y}(3, 2) = 0.02 \neq 0.04 \times 0.15 = 0.006 = f_X(3)f_Y(2)$.

Accordingly, the r.v.'s $X$ and $Y$ are *not* independent. On the other hand, in reference to Example 2 (see also Example 6 in Chapter 4), we have, for all $x, y > 0$:

$$f_{X,Y}(x, y) = \lambda_1 \lambda_2 e^{-\lambda_1 x - \lambda_2 y} = (\lambda_1 e^{-\lambda_1 x})(\lambda_2 e^{-\lambda_2 y}) = f_X(x)f_Y(y),$$

so that $f_{X,Y}(x,y) = f_X(x)f_Y(y)$ for all $x$ and $y$, and consequently, the r.v.'s $X$ and $Y$ are independent. Finally, refer to the Bivariate Normal distribution whose p.d.f. is given by (49) of Chapter 4 and set $\rho = 0$. Then, from (49), (50), and (53), we have $f_{X,Y}(x,y) = f_X(x)f_Y(y)$ for all $x$ and $y$. Therefore, $\rho = 0$ implies that the r.v.'s $X$ and $Y$ are independent.

**Example 1.** Examine the r.v.'s $X$ and $Y$ from an independence viewpoint, if their joint p.d.f. is given by: $f_{X,Y}(x,y) = 4xy, 0 < x < 1, 0 < y < 1$ (and 0 otherwise).

**Discussion.** We will use part (ii) of Theorem 1 for which the marginal p.d.f.'s are needed. To this end, we have:

$$f_X(x) = 4x \int_0^1 y \, dy = 2x, \quad 0 < x < 1;$$

$$f_Y(y) = 4y \int_0^1 x \, dx = 2y, \quad 0 < y < 1.$$

Hence, for all $0 < x < 1$ and $0 < y < 1$, it holds that: $2x \times 2y = 4xy$, or $f_X(x)f_Y(y) = f_{X,Y}(x,y)$. This relation is also, trivially, true (both sides are equal to 0) for $x$ and $y$ not satisfying the inequalities $0 < x < 1$ and $0 < y < 1$. It follows that $X$ and $Y$ are independent.

Here are two examples where the r.v.'s involved are not independent.

**Example 2.** If the r.v.'s $X$ and $Y$ have joint p.d.f. given by: $f_{X,Y}(x,y) = 2, 0 < x < y < 1$ (and 0 otherwise), check whether these r.v.'s are independent or not.

**Discussion.** Reasoning as in the previous example, we find:

$$f_X(x) = 2 \int_x^1 dy = 2(1-x), \quad 0 < x < 1;$$

$$f_Y(y) = 2 \int_0^y dx = 2y, \quad 0 < y < 1.$$

Then independence of $X$ and $Y$ would require that: $4(1-x)y = 2$ for all $0 < x < y < 1$, which, clearly, need not hold. For example, for $x = \frac{1}{4}, y = \frac{1}{2}$, $4(1-x)y = 4 \times \frac{3}{4} \times \frac{1}{2} = \frac{3}{2} \neq 2$. Thus, the $X$ and $Y$ are not independent.

**Example 3.** In reference to Example 8 in Chapter 4, the r.v.'s $X$ and $Y$ have joint p.d.f. $f_{X,Y}(x,y) = 8xy, 0 < x \leq y < 1$ (and 0 otherwise), and

$$f_X(x) = 4x(1-x^2), \quad 0 < x < 1; \quad f_Y(y) = 4y^3, \quad 0 < y < 1.$$

Independence of $X$ and $Y$ would require that: $4x(1-x^2) \times 4y^3 = 8xy$ or $(1-x^2)y^2 = \frac{1}{2}, 0 < x \leq y < 1$. However, this relation need not be true, because, for example, for $x = \frac{1}{4}$ and $y = \frac{1}{2}$, we have: Left-hand side $= \frac{15}{64} \neq \frac{1}{2} =$ right-hand side. So, the r.v.'s $X$ and $Y$ are dependent.

**Remark 1.** On the basis of Examples 2 and 3, one may surmise the following rule of thumb: If the arguments $x$ and $y$ (for the case of two r.v.'s) do not vary independently of each other, the r.v.'s involved are likely to be dependent.

A special case of the following result will be needed for the proof of Theorem 1.

**Proposition 1.** *Consider the r.v.'s* $X_1, \ldots, X_k$, *the functions* $g_i: \Re \to \Re$, $i = 1, \ldots, k$, *and suppose all expectations appearing below are finite. Then independence of the r.v.'s* $X_1, \ldots, X_k$ *implies:*

$$E\left[\prod_{i=1}^{k} g_i(X_i)\right] = \prod_{i=1}^{k} Eg_i(X_i). \tag{9}$$

*Proof.* Suppose the r.v.'s are of the continuous type (so that we use integrals; replace them by summations, if the r.v.'s are discrete). Then:

$$E\left[\prod_{i=1}^{k} g_i(X_i)\right] = \int_{-\infty}^{\infty} \cdots \int_{-\infty}^{\infty} g_1(x_1) \cdots g_k(x_k) f_{X_1,\ldots,X_k}(x_1,\ldots,x_k)\, dx_1 \cdots dx_k$$

$$= \int_{-\infty}^{\infty} \cdots \int_{-\infty}^{\infty} g_1(x_1) \cdots g_k(x_k) f_{X_1}(x_1) \cdots f_{X_k}(x_k)\, dx_1 \cdots dx_k$$

(by independence)

$$= \left[\int_{-\infty}^{\infty} g_1(x_1) f_{X_1}(x_1)\, dx_1\right] \cdots \left[\int_{-\infty}^{\infty} g_k(x_k) f_{X_k}(x_k)\, dx_k\right]$$

$$= Eg_1(X_1) \cdots Eg_k(X_k) = \prod_{i=1}^{k} Eg_i(X_i). \qquad \blacksquare$$

**Corollary 1.** *By taking* $g_i(X_i) = e^{t_i X_i}$, $t_i \in \Re$, $i = 1, \ldots, k$, *relation (9) becomes:*

$$E \exp(t_1 X_1 + \cdots + t_k X_k) = \prod_{i=1}^{k} E \exp(t_i X_i), \quad \text{or}$$

$$M_{X_1,\ldots,X_k}(t_1,\ldots,t_k) = \prod_{i=1}^{k} M_{X_i}(t_i). \tag{10}$$

**Corollary 2.** *If the r.v.'s X and Y are independent, then they are uncorrelated. The converse is also true, if the r.v.'s have the Bivariate Normal distribution.*

*Proof.* In (9), take $k = 2$, identify $X_1$ and $X_2$ with $X$ and $Y$, respectively, and let $g_1(x) = g_2(x) = x$, $x \in \Re$. Then $E(XY) = (EX)(EY)$, which implies $\text{Cov}(X, Y) = 0$ and $\rho(X, Y) = 0$. The converse for the Bivariate Normal distribution follows by means of (50) and (53) in Chapter 4. $\blacksquare$

**Remark 2.**    That uncorrelated r.v.'s are not, in general, independent may be illustrated by means of examples (see, e.g., Exercise 1.20).

**Remark 3.**    If $X_1, \ldots, X_k$ are independent r.v.'s, then it is intuitively clear that independence should be preserved for suitable functions of the $X_i$'s. For example, if $Y_i = g_i(X_i)$, $i = 1, \ldots, k$, then the r.v.'s $Y_1, \ldots, Y_k$ are also independent. Independence is also preserved if we take several functions of $X_i$'s, provided these functions do not include the same $X_i$'s. For instance, if $Y = g(X_{i_1}, \ldots, X_{i_m})$ and $Z = h(X_{j_1}, \ldots, X_{j_n})$, where $1 \le i_1 < \cdots < i_m \le k, 1 \le j_1 < \cdots < j_n \le k$ and all $i_1, \ldots, i_m$ are distinct from all $j_1, \ldots, j_n$, then the r.v.'s $Y$ and $Z$ are independent (see Exercise 1.26).

*Proof of Theorem 1.*    The proof can be only partial but sufficient for the purposes of this book.

(i)    Independence of the r.v.'s $X_1, \ldots, X_k$ means that relation (5) is satisfied. In particular, this is true if $B_i = (-\infty, x_i]$, $i = 1, \ldots, k$ which is (6). That (6) implies (5) is a deep probabilistic result dealt with at a much higher level.

(ii)    Suppose the r.v.'s are independent and first assume they are discrete. Then, by taking $B_i = \{x_i\}, i = 1, \ldots, k$ in (5), we obtain (7). If the r.v.'s are continuous, then consider (6) and differentiate both sides with respect to $x_1, \ldots, x_k$, which, once again, leads to (7) (for continuity points $x_1, \ldots, x_k$). For the converse, suppose that (7) is true; that is, for all $t_1, \ldots, t_k$ in $\Re$,

$$f_{X_1, \ldots, X_k}(t_1, \ldots, t_k) = f_{X_1}(t_1) \cdots f_{X_k}(t_k).$$

Then, if the r.v.'s are discrete, sum over the $t_i$'s from $-\infty$ to $x_i$, $i = 1, \ldots, k$ to obtain (6); if the r.v.'s are continuous, replace the summation operations by integrations in order to obtain (6) again. In either case, independence follows.

(iii)    Independence of $X_1, \ldots, X_k$ implies (8) by means of Corollary 1 to Proposition 1 above.

The converse of part (iii) is also true but its proof will not be pursued here (it requires the use of the so-called inversion formula as indicated in Proposition 1 of Chapter 3 and Remarks 2 and 3 of Chapter 4). ■

Part (ii) of Theorem 1 has the following corollary, which provides still another useful criterion for independence of $k$ r.v.'s.

**Corollary.**    *The r.v.'s $X_1, \ldots, X_k$ are independent if and only if $f_{X_1, \ldots, X_k}(x_1, \ldots, x_k) = h_1(x_1) \cdots h_k(x_k)$ for all $x_1, \ldots, x_k$ in $\Re$, where $h_i$ is a nonNegative function of $x_i$ alone, $i = 1, \ldots, k$.*

*Proof.*    Suppose the r.v.'s $X_1, \ldots, X_k$ are independent. Then, by (7), $f_{X_1, \ldots, X_k}(x_1, \ldots, x_k) = f_{X_1}(x_1) \cdots f_{X_k}(x_1)$ for all $x_1, \ldots, x_k$ in $\Re$, so that the above factorization holds with $h_i = f_{X_i}$, $i = 1, \ldots, k$. Next, assume that the factorization holds, and suppose that the r.v.'s are continuous. For each fixed $i = 1, \ldots, k$, set

$$c_i = \int_{-\infty}^{\infty} h_i(x_i)\, dx_i,$$

so that
$$c_1 \ldots c_k = \int_{-\infty}^{\infty} h_1(x_1)\, dx_1 \cdots \int_{-\infty}^{\infty} h_k(x_k)\, dx_k$$

$$= \int_{-\infty}^{\infty} \cdots \int_{-\infty}^{\infty} h_1(x_1) \cdots h_k(x_k)\, dx_1 \cdots dx_k$$

$$= \int_{-\infty}^{\infty} \cdots \int_{-\infty}^{\infty} f_{X_1,\ldots,X_k}(x_1,\ldots,x_k)\, dx_1 \cdots dx_k$$

$$= 1.$$

Then, integrating $f_{X_1,\ldots,X_k}(x_1,\ldots,x_k)$ with respect to all $x_j$'s with $j \neq i$, we get

$$f_{X_i}(x_i) = c_1 \cdots c_{i-1}\, c_{i+1} \cdots c_k h_i(x_i)$$

$$= \frac{1}{c_i} h_i(x_i).$$

Hence

$$f_{X_1}(x_1)\ldots f_{X_k}(x_k) = \frac{1}{c_1 \cdots c_k} h_1(x_1) \cdots h_k(x_k)$$

$$= h_1(x_1) \cdots h_k(x_k) = f_{X_1,\ldots,X_k}(x_1,\ldots,x_k),$$

or $f_{X_1,\ldots,X_k}(x_1,\ldots,x_k) = f_{X_1}(x_1) \cdots f_{X_k}(x_k)$, for all $x_1,\ldots,x_k$ in $\Re$, so that the r.v.'s $X_1,\ldots,X_k$ are independent. The same conclusion holds in case the r.v.'s are discrete by using summations rather than integrations. ■

The significance of the corollary to Theorem 1 is that, in order to check for independence of the r.v.'s $X_1,\ldots,X_k$ all one has to do is to establish a factorization of $f_{X_1,\ldots,X_k}$ as stated in the corollary. One does not have to ascertain as to what the factors are. (From the proof above, it follows that they are multiples of the marginal p.d.f.'s.)

This section is concluded with the definition of what is known as a random sample. Namely, $n$ independent and identically distributed (i.i.d.) r.v.'s are referred to as forming a *random sample* of size $n$. Some of their properties are discussed in the next section.

---

## EXERCISES

**1.1**    In reference to Exercise 2.5 in Chapter 4, determine whether or not the r.v.'s $X$ and $Y$ are independent. Justify your answer.

**1.2**    In reference to Exercises 1.1 and 2.1 in Chapter 4, determine whether or not the r.v.'s $X$ and $Y$ are independent.

**1.3** The r.v.'s $X$, $Y$, and $Z$ have the joint p.d.f. given by: $f_{X,Y,Z}(x,y,z) = \frac{1}{4}$ if $x = 1$, $y = z = 0$; $x = 0$, $y = 1$, $z = 0$; $x = y = 0$, $z = 1$; $x = y = z = 1$.
  (i) Derive the marginal joint p.d.f.'s $f_{X,Y}$, $f_{X,Z}$, and $f_{Y,Z}$.
  (ii) Derive the marginal p.d.f.'s $f_X$, $f_Y$, and $f_Z$.
  (iii) Show that any two of the r.v.'s $X$, $Y$, and $Z$ are independent.
  (iv) Show that the r.v.'s $X$, $Y$, and $Z$ are dependent.

**1.4** In reference to Exercise 2.8 in Chapter 4, decide whether or not the r.v.'s $X$ and $Y$ are independent. Justify your answer.

**1.5** In reference to Examples 4 and 7 in Chapter 4, investigate whether or not the r.v.'s $X$ and $Y$ are independent and justify your answer.

**1.6** Let $X$ and $Y$ be r.v.'s with joint p.d.f. given by:

$$f_{X,Y}(x,y) = \frac{6}{5}(x^2 + y), \quad 0 \le x \le 1, \quad 0 \le y \le 1.$$

  (i) Determine the marginal p.d.f.'s $f_X$ and $f_Y$.
  (ii) Investigate whether or not the r.v.'s $X$ and $Y$ are independent. Justify your answer.

**1.7** The r.v.'s $X$ and $Y$ have joint p.d.f. given by:

$$f_{X,Y}(x,y) = 1, \quad 0 < x < 1, \quad 0 < y < 1.$$

Then:
  (i) Derive the marginal p.d.f.'s $f_X$ and $f_Y$.
  (ii) Show that $X$ and $Y$ are independent.
  (iii) Calculate the probability $P(X + Y < c)$.
  (iv) Give the numerical value of the probability in part (iii) for $c = 1/4$.
  **Hint.** For part (iii), you may wish to draw the picture of the set for which $x + y < c$, and compute the probability in terms of $c$.

**1.8** The r.v.'s $X$, $Y$, and $Z$ have joint p.d.f. given by:

$$f_{X,Y,Z}(x,y,z) = 8xyz, \quad 0 < x < 1, \quad 0 < y < 1, \quad 0 < z < 1.$$

  (i) Derive the marginal p.d.f.'s $f_X$, $f_Y$, and $f_Z$.
  (ii) Show that the r.v.'s $X$, $Y$, and $Z$ are independent.
  (iii) Calculate the probability $P(X < Y < Z)$.
  **Hint.** In part (iii), can you guess the answer without doing any calculations?

**1.9** The r.v.'s $X$ and $Y$ have joint p.d.f. given by:

$$f_{X,Y}(x,y) = c, \quad \text{for } x^2 + y^2 \le 9.$$

  (i) Determine the constant $c$.
  (ii) Derive the marginal p.d.f.'s $f_X$ and $f_Y$.
  (iii) Show that the r.v.'s $X$ and $Y$ are dependent.

**1.10** The r.v.'s $X$, $Y$, and $Z$ have joint p.d.f. given by:

$$f_{X,Y,Z}(x, y, z) = c^3 e^{-c(x+y+z)}, \quad x > 0, \quad y > 0, \quad z > 0.$$

- **(i)** Determine the constant $c$.
- **(ii)** Derive the marginal joint p.d.f.'s $f_{X,Y}$, $f_{X,Z}$, and $f_{Y,Z}$.
- **(iii)** Derive the marginal p.d.f.'s $f_X$, $f_Y$, and $f_Z$.
- **(iv)** Show that any two of the r.v.'s $X$, $Y$, and $Z$, as well as all three r.v.'s, are independent.

**1.11** The r.v.'s $X$ and $Y$ have joint p.d.f. given by the following product: $f_{X,Y}(x, y) = g(x)h(y)$, where $g$ and $h$ are nonNegative functions.
- **(i)** Derive the marginal p.d.f.'s $f_X$ and $f_Y$ as functions of $g$ and $h$, respectively.
- **(ii)** Show that the r.v.'s $X$ and $Y$ are independent.
- **(iii)** If $h = g$, then the r.v.'s are identically distributed.
- **(iv)** From part (iii), conclude that $P(X > Y) = 1/2$, provided the distribution is of the continuous type.

**Hint.** For part (i), we have $f_X(x) = cg(x)$, $f_Y(y) = \frac{1}{c}h(y)$, where $c = \int_{-\infty}^{\infty} h(y)dy$. Parts (ii) and (iii) follow from part (i). Part (iv) follows either by symetry or by calculations.

**1.12** The life of a certain part in a new automobile is a r.v. $X$ whose p.d.f. is Negative Exponential with parameter $\lambda = 0.005$ days.
- **(i)** What is the expected life of the part in question?
- **(ii)** If the automobile comes with a spare part whose life is a r.v. $Y$ distributed as $X$ and independent of it, find the p.d.f. of the combined life of the part and its spare.
- **(iii)** What is the probability that $X + Y \geq 500$ days?

**1.13** Let the r.v. $X$ be distributed as $U(0, 1)$ and set $Y = -\log X$.
- **(i)** Determine the d.f. of $Y$ and then its p.d.f.
- **(ii)** If the r.v.'s $Y_1, \ldots, Y_n$ are independently distributed as $Y$, and $Z = Y_1 + \cdots + Y_n$, determine the distribution of the r.v. $Z$.

**Hint.** For part (ii), use the m.g.f. approach.

**1.14** Let the independent r.v.'s $X$ and $Y$ be distributed as $N(\mu_1, \sigma_1^2)$ and $N(\mu_2, \sigma_2^2)$, respectively, and define the r.v.'s $U$ and $V$ by: $U = aX + b$, $V = cY + d$, where $a, b, c$, and $d$ are constants.
- **(i)** Use the m.g.f. approach in order to show that:

$$U \sim N(a\mu_1 + b, (a\sigma_1)^2), \quad V \sim N(c\mu_2 + d, (c\sigma_2)^2).$$

- **(ii)** Determine the joint m.g.f. of $U$ and $V$.
- **(iii)** From parts (i) and (ii), conclude that $U$ and $V$ are independent.

**1.15** Let $X$ and $Y$ be independent r.v.'s denoting the lifetimes of two batteries and having the Negative Exponential distribution with parameter $\lambda$. Set $T = X + Y$ and:

    **(i)** Determine the d.f. of $T$ by integration, and then the corresponding p.d.f.

    **(ii)** Determine the p.d.f. of $T$ by using the m.g.f. approach.

    **(iii)** For $\lambda = 1/3$, calculate the probability $P(T \leq 6)$.

    **Hint.** For part (i), you may wish to draw the picture of the set for which $x + y < t$ ($t > 0$).

**1.16** Let $X_1, \ldots, X_n$ be i.i.d. r.v.'s with m.g.f. $M$, and let $\bar{X} = \frac{1}{n}(X_1 + \cdots + X_n)$. Express the m.g.f. $M_{\bar{X}}$ in terms of $M$.

**1.17** In reference to Exercise 3.1 in Chapter 4:

    **(i)** Calculate the Var $(X + Y)$ and the s.d. of $X + Y$.

    **(ii)** Use the Tchebichev inequality to determine a lower bound for the probability: $P(X + Y \leq 10)$.

    **Hint.** For part (ii), use part (i) to bring it into the form required for the application of the Tchebichev inequality.

**1.18** Let $p$ be the proportion of defective computer chips in a very large lot of chips produced over a period of time by a certain manufacturing process. For $i = 1, \ldots, n$, associated with the $i$th chip the r.v. $X_i$, where $X_i = 1$ if the $i$th chip is defective, and $X_i = 0$ otherwise. Then $X_1, \ldots, X_n$ are independent r.v.'s distributed as $B(1, p)$, and let $\bar{X} = \frac{1}{n}(X_1 + \cdots + X_n)$.

    **(i)** Calculate the $E\bar{X}$ and the Var $(\bar{X})$ in terms of $p$ and $q = 1 - p$.

    **(ii)** Use the Tchebichev inequality to determine the smallest value of $n$ for which $P(|\bar{X} - p| < 0.1\sqrt{pq}) \geq 0.99$.

**1.19** Let the independent r.v.'s $X_1, \ldots, X_n$ be distributed as $P(\lambda)$, and set $\bar{X} = \frac{1}{n}(X_1 + \cdots + X_n)$.

    **(i)** Calculate the $E\bar{X}$ and the Var $(\bar{X})$ in terms of $\lambda$ and $n$.

    **(ii)** Use the Tchebichev inequality to determine the smallest $n$, in terms of $\lambda$ and $c$, for which $P(|\bar{X} - \lambda| < c) \geq 0.95$, for some $c > 0$.

    **(iii)** Give the numerical value of $n$ for $c = \sqrt{\lambda}$ and $c = 0.1\sqrt{\lambda}$.

**1.20** The joint distribution of the r.v.'s $X$ and $Y$ is given by:

| $y \setminus x$ | $-1$ | $0$ | $1$ |
|:---:|:---:|:---:|:---:|
| $-1$ | $\alpha$ | $\beta$ | $\alpha$ |
| $0$ | $\beta$ | $0$ | $\beta$ |
| $1$ | $\alpha$ | $\beta$ | $\alpha$ |

where $\alpha, \beta > 0$ with $\alpha + \beta = 1/4$.

    **(i)** Derive the marginal p.d.f.'s $f_X$ and $f_Y$.

    **(ii)** Calculate the $EX, EY$, and $E(XY)$.

    **(iii)** Show that Cov $(X, Y) = 0$.

    **(iv)** Show that the r.v.'s $X$ and $Y$ are dependent.

    **Remark.** Whereas independent r.v.'s are always uncorrelated, this exercise shows that the converse need not be true.

**1.21** Refer to Exercise 1.10 and calculate the following quantities without any integration: $E(XY)$, $E(XYZ)$, Var $(X + Y)$, Var $(X + Y + Z)$.

**1.22** The i.i.d. r.v.'s $X_1, \ldots, X_n$ have expectation $\mu \in \mathfrak{R}$ and variance $\sigma^2 < \infty$, and set $\bar{X} = \frac{1}{n}(X_1 + \cdots + X_n)$.
  (i) Determine the $E\bar{X}$ and the Var $(\bar{X})$ in terms of $\mu$ and $\sigma$.
  (ii) Use the Tchebichev inequality to determine the smallest value of $n$ for which $P(|\bar{X} - \mu| < k\sigma)$ is at least 0.99; take $k = 1, 2, 3$.

**1.23** A piece of equipment works on a battery whose lifetime is a r.v. $X$ with expectation $\mu$ and s.d. $\sigma$. If $n$ such batteries are used successively and independently of each other, denote by $X_1, \ldots, X_n$ their respective lifetimes, so that $\bar{X} = \frac{1}{n}(X_1 + \cdots + X_n)$ is the average lifetime of the batteries. Use the Tchebichev inequality to determine the smallest value of $n$ for which $P(|\bar{X} - \mu| < 0.5\sigma) \geq 0.99$.

**1.24** Let $X_1, \ldots, X_n$ be i.i.d. r.v.'s with $EX_1 = \mu \in \mathfrak{R}$ and Var $(X_1) = \sigma^2 < \infty$, and set $\bar{X} = \frac{1}{n}(X_1 + \cdots + X_n)$.
  (i) Calculate the $E\bar{X}$ and the Var $(\bar{X})$ in terms of $\mu$ and $\sigma$.
  (ii) Use the Tchebichev inequality in order to determine the smallest value of $n$, in terms of the positive constant $c$ and $\alpha$, so that

$$P(|\bar{X} - \mu| < c\sigma) \geq \alpha \ (0 < \alpha < 1).$$

  (iii) What is the numerical value of $n$ in part (ii) if $c = 0.1$ and $\alpha = 0.90, \alpha = 0.95, \alpha = 0.99$?
  **Remark.** See also Excercise 1.22.

**1.25** In reference to Exercise 5.13(iii) in Chapter 4, show that the r.v.'s $U$ and $V$ are independent if and only if $\sigma_1^2 = \sigma_2^2$.
  **Hint.** Use Corrolary 2 to Proposition 1.

**1.26** Refer to Remark 3 and show that independence of the r.v.'s $X_1, \ldots, X_n$ implies that of the r.v.'s $Y_1, \ldots, Y_n$, by using Theorem 1 (ii), (iii).

**1.27** In reference to Exercise 3.26 (in Chapter 4), decide whether or not the r.v.'s $X$ and $Y$ are independent, and justify your answer.

**1.28** Let the r.v.'s $X$ and $Y$ have the joint p.d.f.:

$$f_{X,Y}(x, y) = \frac{21}{2}x^2y, \quad 0 < x^2 < y < 1 \ (x > 0).$$

Then it is seen that, $f_X(x) = \frac{21}{4}x^2(1 - x^4), 0 < x < 1; f_Y(y) = \frac{21}{6}y^2\sqrt{y},$ $0 < y < 1$. Determine whether the r.v.'s $X$ and $Y$ are independent.

**1.29** Let $X$ and $Y$ be independent r.v.'s distributed as $N(0, 1)$, and set $U = X + Y$, $V = X - Y$. Use the m.g.f. approach to:

(i) Decide on the distribution of the r.v.'s $U$ and $V$.

(ii) Show that $U$ and $V$ are independent.

**1.30** If the independent r.v.'s $X$ and $Y$ are distributed as $N(\mu_1, \sigma_1^2)$ and $N(\mu_2, \sigma_2^2)$, respectively, then show that the pair $(X, Y)$ has the Bivariate Normal distribution, and specify the parameters involved.

**1.31** Refer to Exercise 3.76(iv) (in Chapter 3), and let $Y$ be a r.v. denoting the number of accidents caused in 1 year by a person as described in Exercise 3.76. Further, suppose that $E(Y|X = x) = x^2$, $x = 0, 1, 2$. Then:

(i) Compute $EY$ and $E(XY)$.

(ii) Decide whether $X$ and $Y$ are dependent.

**1.32** The r.v.'s $X$ and $Y$ are jointly Uniformly distributed over the unit circle centered at the origin.

(i) Compute the marginal p.d.f.'s.

(ii) Are the r.v.'s $X$ and $Y$ independent or not?

**1.33** The r.v. $X$ is distributed as $U(0, 1)$, whereas the conditional distribution of $Y$, given $X = x$, is $N(x, x^2)$; i.e., $Y|X \sim N(x, x^2)$. Set $U = Y/X$ and $V = X$, and then:

(i) Derive the joint p.d.f. $f_{U,V}$.

(ii) Find the marginal p.d.f. of $U$.

(iii) Show that the r.v.'s $U$ and $V$ are independent.

**1.34** Let $X$ and $Y$ be r.v.'s with joint p.d.f. given by:

$$f_{X,Y}(x, y) = 24xy, \quad 0 \le x \le 1, 0 \le y \le 1, 0 \le x + y \le 1 \text{ (and 0 otherwise)}.$$

(i) Show that the marginal p.d.f.'s $f_X$ and $f_Y$ are given by

$$f_X(x) = 12x(1 - x)^2, \quad 0 \le x \le 1; \quad f_Y(y) = 12y(1 - y)^2, \quad 0 \le y \le 1.$$

(ii) Are the r.v.'s $X$ and $Y$ independent or not? Justify your answer.

(iii) Find the conditional p.d.f. $f_{X|Y}(\cdot|y)$.

(iv) Compute the $E(X|Y = y)$, and find its value for $y = 1/2$.

**1.35** The joint probability distribution of the number $X$ of cars, and the number $Y$ of buses per signal cycle at a proposed left turn lane is displayed in the following table.

| | | $y$ | |
|---|---|---|---|
| $f(x,y)$ | 0 | 1 | 2 |
| 0 | 0.025 | 0.015 | 0.010 |
| 1 | 0.050 | 0.030 | 0.020 |
| $x$ 2 | 0.125 | 0.075 | 0.050 |
| 3 | 0.150 | 0.090 | 0.060 |
| 4 | 0.100 | 0.060 | 0.040 |
| 5 | 0.050 | 0.030 | 0.020 |

(i) Determine the marginal p.d.f.'s of $X$ and $Y$, $f_X$ and $f_Y$, respectively.
(ii) Are the r.v.'s $X$ and $Y$ independent or not? Justify your answer.
(iii) Calculate the probabilities that: There is exactly one car during a cycle; there is exactly on bus during a cycle.

**1.36** The breakdown voltage of a randomly chosen diode of a certain type is known to be Normally distributed with mean value $40V$ and standard deviation $1.5V$.
(i) What is the probability that the voltage of a single diode is between 39 and 42?
(ii) If four diodes are independently selected, what is the probability that at least one has a voltage exceeding 42?

**1.37** Let $X_1, \ldots, X_n$ be i.i.d. r.v.'s with finite second moment and variance $\sigma^2$, and let $\bar{X}$ be their sample mean. Then, show that, for each $i = 1, \ldots, n$, the r.v.'s $X_i - \bar{X}$ and $\bar{X}$ are uncorrelated.

**1.38** The number of tables in a restaurant served during lunch time is a r.v. $X$ distributed as $P(\lambda_1)$, and during dinner time is a r.v. $Y \sim P(\lambda_2)$, and it is assumed that $X$ and $Y$ are independent.
(i) What is the distribution of the total number of tables served in 1 day?
(ii) What is the expected number of these tables and the s.d.?
(iii) Give the numerical values in part (ii) for $\lambda_1 = 8$ and $\lambda_2 = 10$.

**1.39** The grades $X_1, \ldots, X_n$ in a class of size $n$ are independent r.v.'s $\sim N(75, 144)$.
(i) If $n = 25$, compute the probability $P(\bar{X}_{25} > 80)$.
(ii) If $n = 64$, compute the probability $P(\bar{X}_{64} > 80)$.
(iii) If the grades in the two classes in parts (i) and (ii) are independent r.v.'s, compute the probability $P(\bar{X}_{25} - \bar{X}_{64} > 3)$.

## 5.2 THE REPRODUCTIVE PROPERTY OF CERTAIN DISTRIBUTIONS

Independence plays a decisive role in the reproductive property of certain r.v.'s. Specifically, if $X_1, \ldots, X_k$ are r.v.'s having certain distributions, then, if they are also *independent*, it follows that the r.v. $X_1 + \cdots + X_k$ is of the same kind. This is, basically, the content of this section. The tool used in order to establish this assertion is the m.g.f., and the basic result employed is relation (13) below following from relation (8), which characterizes independence of r.v.'s. The conditions of applicability of (8) hold in all cases considered here.

First, we derive some general results regarding the sample mean and the sample variance of $k$ r.v.'s, which will be used, in particular, in the Normal distribution case discussed below. To this end, for any $k$ r.v.'s $X_1, \ldots, X_k$, their *sample mean*, denoted by $\bar{X}_k$ or just $\bar{X}$, is defined by:

$$\bar{X} = \frac{1}{k} \sum_{i=1}^{k} X_i. \tag{11}$$

The *sample variance* of the $X_i$'s, denoted by $S_k^2$ or just $S^2$, is defined by:

$$S^2 = \frac{1}{k} \sum_{i=1}^{k} (X_i - EX_i)^2,$$

provided the $EX_i$'s are finite. In particular, if $EX_1 = \cdots = EX_k = \mu$, say, then $S^2$ becomes:

$$S^2 = \frac{1}{k} \sum_{i=1}^{k} (X_i - \mu)^2. \tag{12}$$

The r.v.'s defined by (11) and (12) are most useful when the underlying r.v.'s form a random sample; that is, they are i.i.d.

**Proposition 2.** *Let $X_1, \ldots, X_k$ be i.i.d. r.v.'s with (finite) mean $\mu$. Then $E\bar{X} = \mu$. Furthermore, if the $X_i$'s also have (finite) variance $\sigma^2$, then $Var(\bar{X}) = \frac{\sigma^2}{k}$ and $ES^2 = \sigma^2$.*

*Proof.* The first result follows from (40) in Chapter 4 by taking $c_1 = \cdots = c_k = 1/k$. The second result follows from (44) in the same chapter, by way of Corollary 2 to Proposition 1 here, because independence of $X_i$ and $X_j$, for $i \neq j$, implies $\rho(X_i, X_j) = 0$. In order to check the third result, observe that:

$$ES^2 = E\left[ \frac{1}{k} \sum_{i=1}^{k} (X_i - \mu)^2 \right] = \frac{1}{k} E \sum_{i=1}^{k} (X_i - \mu)^2$$

$$= \frac{1}{k} \sum_{i=1}^{k} E(X_i - \mu)^2 = \frac{1}{k} \times k\sigma^2 = \sigma^2. \qquad \blacksquare$$

The general thrust of the following four results is to the effect that, if $X_1, \ldots, X_k$ are independent and have certain distributions, then their sum $X_1 + \cdots + X_k$ has a distribution of the same respective kind. The proof of this statement relies on following relation:

$$M_{X_1 + \cdots + X_k}(t) = \prod_{j=1}^{k} M_{X_j}(t), \text{ and } M_{X_1 + \cdots + X_k}(t) \tag{13}$$

$$= [M_{X_1}(t)]^k \quad \text{if the } X_i\text{'s are i.i.d.}$$

**Theorem 2.** *Let the r.v.'s $X_1, \ldots, X_k$ be independent and let $X_i \sim B(n_i, p)$ (the same $p$), $i = 1, \ldots, k$. Then $\sum_{i=1}^{k} X_i \sim B(\sum_{i=1}^{k} n_i, p)$.*

*Proof.* By relation (13) above, relation (20) in Chapter 3, and $t \in \mathfrak{R}$:

$$M_{\sum_{i=1}^{k} X_i}(t) = \prod_{i=1}^{k} M_{X_i}(t) = \prod_{i=1}^{k} (pe^t + q)^{n_i} = (pe^t + q)^{\sum_{i=1}^{k} n_i},$$

which is the m.g.f. of $B(\sum_{i=1}^{k} n_i, p)$. Then $\sum_{i=1}^{k} X_i \sim B(\sum_{i=1}^{k} n_i, p)$. $\blacksquare$

**Theorem 3.**  *Let the r.v.'s $X_1, \ldots, X_k$ be independent and let $X_i \sim P(\lambda_i), i = 1, \ldots, k$. Then $\sum_{i=1}^{k} X_i \sim P(\sum_{i=1}^{k} \lambda_i)$.*

*Proof.* As above, employ relation (13) and relation (24) in Chapter 3 in order to obtain:

$$M_{\sum_{i=1}^{k} X_i}(t) = \prod_{i=1}^{k} M_{X_i}(t) = \prod_{i=1}^{k} \exp(\lambda_i e^t - \lambda_i) = \exp\left[\left(\sum_{i=1}^{k} \lambda_i\right)e^t - \left(\sum_{i=1}^{k} \lambda_i\right)\right],$$

which is the m.g.f. of $P(\sum_{i=1}^{k} \lambda_i)$, so that $\sum_{i=1}^{k} X_i \sim P(\sum_{i=1}^{k} \lambda_i)$. ∎

**Theorem 4.**  *Let the r.v.'s $X_1, \ldots, X_k$ be independent and let $X_i \sim N(\mu_i, \sigma_i^2), i = 1, \ldots, k$. Then $\sum_{i=1}^{k} X_i \sim N(\sum_{i=1}^{k} \mu_i, \sum_{i=1}^{k} \sigma_i^2)$. In particular, if $\mu_1 = \cdots = \mu_k = \mu$ and $\sigma_1 = \cdots = \sigma_k = \sigma$, then $\sum_{i=1}^{k} X_i \sim N(k\mu, k\sigma^2)$.*

*Proof.* Use relation (13), and formula (44) in Chapter 3, for $t \in \mathfrak{R}$, in order to obtain:

$$M_{\sum_{i=1}^{k} X_i}(t) = \prod_{i=1}^{k} M_{X_i}(t) = \prod_{i=1}^{k} \exp\left(\mu_i t + \frac{\sigma_i^2}{2} t^2\right)$$

$$= \exp\left[\left(\sum_{i=1}^{k} \mu_i\right)t + \frac{\sum_{i=1}^{k} \sigma_i^2}{2} t^2\right],$$

which is the m.g.f. of $N(\sum_{i=1}^{k} \mu_i, \sum_{i=1}^{k} \sigma_i^2)$, so that $\sum_{i=1}^{k} X_i \sim N(\sum_{i=1}^{k} \mu_i, \sum_{i=1}^{k} \sigma_i^2)$. The special case is immediate. ∎

To this theorem, there are the following two corollaries.

**Corollary 1.**  *If the r.v.'s $X_1, \ldots, X_k$ are independent and distributed as $N(\mu, \sigma^2)$, then their sample mean $\bar{X} \sim N(\mu, \frac{\sigma^2}{k})$, and $\frac{\sqrt{k}(\bar{X} - \mu)}{\sigma} \sim N(0, 1)$.*

*Proof.* Here $\bar{X} = Y_1 + \cdots + Y_k$, where $Y_i = \frac{X_i}{k}, i = 1, \ldots, k$ are independent and $Y_i \sim N(\frac{\mu}{k}, \frac{\sigma^2}{k^2})$ by Example 3 in Chapter 6 applied with $a = 1/k$ and $b = 0$. Then the conclusion follows by Theorem 4. The second conclusion is immediate by the part just established and Proposition 3 in Chapter 3, since $\frac{\sqrt{k}(\bar{X} - \mu)}{\sigma} = \frac{\bar{X} - \mu}{\sqrt{\sigma^2/k}}$. ∎

**Corollary 2.**  *Let the r.v.'s $X_1, \ldots, X_k$ be independent, let $X_i \sim N(\mu_i, \sigma_i^2), i = 1, \ldots, k$, and let $c_i, i = 1, \ldots, k$ be constants. Then $\sum_{i=1}^{k} c_i X_i \sim N(\sum_{i=1}^{k} c_i \mu_i, \sum_{i=1}^{k} c_i^2 \sigma_i^2)$.*

*Proof.* As in Corollary 1, $X_i \sim N(\mu_i, \sigma_i^2)$ implies $c_i X_i \sim N(c_i \mu_i, c_i^2 \sigma_i^2)$, and the r.v.'s $c_i X_i, i = 1, \ldots, k$ are independent. Then the conclusion follows from the theorem. ∎

**Theorem 5.**  *Let the r.v.'s $X_1, \ldots, X_k$ be independent and let $X_i \sim \chi^2_{r_i}$, $i = 1, \ldots, k$. Then $\sum_{i=1}^{k} X_i \sim \chi^2_{r_1 + \cdots + r_k}$.*

**Proof.** Use relation (13), and formula (37) in Chapter 3, for $t < \frac{1}{2}$, to obtain:

$$M_{\sum_{i=1}^{k} X_i}(t) = \prod_{i=1}^{k} M_{X_i}(t) = \prod_{i=1}^{k} \frac{1}{(1 - 2t)^{r_i/2}} = \frac{1}{(1 - 2t)^{(r_1 + \cdots + r_k)/2}},$$

which is the m.g.f. of $\chi^2_{r_1 + \cdots + r_k}$. ■

**Corollary.**  *Let the r.v.'s $X_1, \ldots, X_k$ be independent and let $X_i \sim N(\mu_i, \sigma_i^2)$, $i = 1, \ldots, k$. Then $\sum_{i=1}^{k} (\frac{X_i - \mu_i}{\sigma_i})^2 \sim \chi^2_k$, and, in particular, if $\mu_1 = \cdots = \mu_k = \mu$ and $\sigma_1^2 = \cdots = \sigma_k^2 = \sigma^2$, then $\frac{kS^2}{\sigma^2} \sim \chi^2_k$, where $S^2$ is given in (12).*

**Proof.** The assumption $X_i \sim N(\mu_i, \sigma_i^2)$ implies that $\frac{X_i - \mu_i}{\sigma_i} \sim N(0, 1)$ by Proposition 3 in Chapter 3, and $(\frac{X_i - \mu_i}{\sigma_i})^2 \sim \chi^2_1$ by Exercise 1.9 in Chapter 6. Since independence of $X_i, i = 1, \ldots, k$ implies that of $(\frac{X_i - \mu_i}{\sigma_i})^2, i = 1, \ldots, k$, the theorem applies and yields the first assertion. The second assertion follows from the first by taking $\mu_1 = \cdots = \mu_k = \mu$ and $\sigma_1 = \cdots = \sigma_k = \sigma$, and using (12) to obtain $\frac{kS^2}{\sigma^2} = \sum_{i=1}^{k} (\frac{X_i - \mu}{\sigma})^2$. ■

**Remark 4.** From the fact that $\frac{kS^2}{\sigma^2} \sim \chi^2_k$ and formula (37) in Chapter 3, we have $E(\frac{kS^2}{\sigma^2}) = k$, $\text{Var}(\frac{kS^2}{\sigma^2}) = 2k$, or $ES^2 = \sigma^2$ and $\text{Var}(S^2) = 2\sigma^4/k$.

**Remark 5.** Knowing the distribution of $\sum_{i=1}^{k} X_i$ is of considerable practical importance. For instance, if $X_i$ is the number of defective items among $n_i$ in the $i$th lot of certain items, $i = 1, \ldots, k$, then $\sum_{i=1}^{k} X_i$ is the total number of defective items in the $k$ lots (and Theorem 2 applies). Likewise, if $X_i$ is the number of particles emitted by the $i$th radioactive source, $i = 1, \ldots, k$, then $\sum_{i=1}^{k} X_i$ is the total number of particles emitted by all $k$ radioactive sources (and Theorem 3 applies). Also, if $X_i$ is the rain (in inches, for example) which fell in the $i$th location over a specified period of time, $i = 1, \ldots, k$, then $\sum_{i=1}^{k} X_i$ is the total rainfall in all of $k$ locations under consideration over the specified period of time (and Theorem 4 applies). Finally, if $Y_i$ denotes the lifetime of the $i$th battery in a lot of $k$ identical batteries, whose lifetime is assumed to be Normally distributed, then $X_i = [(Y_i - \mu)/\sigma]^2$ measures a deviation from the mean lifetime $\mu$, and $\sum_{i=1}^{k} X_i$ is the totality of such deviations for the $k$ batteries (and Theorem 5 applies).

Here are some numerical applications.

**Example 4.** The defective items in two lots of sizes $n_1 = 10$ and $n_2 = 15$ occur independently at the rate of 6.25%. Calculate the probabilities that the total number of defective items: (i) Does not exceed 2; (ii) Is more than 5.

**Discussion.** If $X_1$ and $X_2$ are the r.v.'s denoting the numbers of defective items in the two lots, then $X_1 \sim B(10, 0.0625), X_2 \sim B(15, 0.0625)$ and they are independent. Then $X = X_1 + X_2 \sim B(25, 0.0625)$ and therefore: (i) $P(X \leq 2) = 0.7968$ and (ii) $P(X > 5) = 1 - P(X \leq 5) = 0.0038$ (from the Binomial tables).

**Example 5.** Five radioactive sources independently emit particles at the rate of 0.08 per certain time unit. What is the probability that the total number of particles does not exceed 3 in the time unit considered?

**Discussion.** In obvious notation, we have here the independent r.v.'s $X_i$ distributed as $P(0.08), i = 1, \ldots, 5$. Then $X = \sum_{i=1}^{5} X_i \sim P(0.4)$, and the required probability is: $P(X \leq 3) = 0.999224$ (from the Poisson tables).

**Example 6.** The rainfall in two locations is measured (in inches over a certain time unit) by two independent and Normally distributed r.v.'s $X_1$ and $X_2$ as follows: $X_1 \sim N(10, 9)$ and $X_2 \sim N(15, 25)$. What is the probability that the total rainfall: (i) Will exceed 30 in. (which may result in flooding)? (ii) Will be less than 8 in. (which will mean a drought)?

**Discussion.** If $X = X_1 + X_2$, then $X \sim N(25, 34)$, so that: (i) $P(X > 30) = 1 - P(X \leq 30) = 1 - P(Z \leq \frac{30-25}{\sqrt{34}}) \simeq 1 - \Phi(0.86) = 1 - 0.805105 = 0.194895$, and (ii) $P(X < 8) = P(Z < \frac{8-25}{\sqrt{34}}) \simeq \Phi(-2.92) = 1 - \Phi(2.92) = 1 - 0.99825 = 0.00175$.

In the definition of $S^2$ by (12), we often replace $\mu$ by the sample mean $\bar{X}$; this is done habitually in statistics as $\mu$ is not really known. Let us denote by $\bar{S}^2$ the resulting quantity; that is,

$$\bar{S}^2 = \frac{1}{k} \sum_{i=1}^{k} (X_i - \bar{X})^2. \tag{14}$$

Then it is easy to establish the following identity (see also Exercise 2.1(ii)):

$$\sum_{i=1}^{k} (X_i - \mu)^2 = \sum_{i=1}^{k} (X_i - \bar{X})^2 + k(\bar{X} - \mu)^2, \tag{15}$$

or

$$kS^2 = k\bar{S}^2 + [\sqrt{k}(\bar{X} - \mu)]^2. \tag{16}$$

Indeed,

$$\sum_{i=1}^{k} (X_i - \mu)^2 = \sum_{i=1}^{k} [(X_i - \bar{X}) + (\bar{X} - \mu)]^2 = \sum_{i=1}^{k} (X_i - \bar{X})^2 + k(\bar{X} - \mu)^2,$$

since $\sum_{i=1}^{k} (X_i - \bar{X})(\bar{X} - \mu) = (\bar{X} - \mu)(k\bar{X} - k\bar{X}) = 0$.

From (16), we have, dividing through by $\sigma^2$:

$$\frac{kS^2}{\sigma^2} = \frac{k\bar{S}^2}{\sigma^2} + \left[\frac{\sqrt{k}(\bar{X} - \mu)}{\sigma}\right]^2. \tag{17}$$

Now $\frac{kS^2}{\sigma^2} \sim \chi_k^2$ and $[\frac{\sqrt{k}(\bar{X}-\mu)}{\sigma}]^2 \sim \chi_1^2$ (by Propositions 1 and 2 in Chapter 3) when the r.v.'s $X_1, \ldots, X_k$ are independently distributed as $N(\mu, \sigma^2)$. Therefore, from (17), it appears quite feasible that $\frac{k\bar{S}^2}{\sigma^2} \sim \chi_{k-1}^2$. This is, indeed, the case and is the content of the following theorem. This theorem is presently established under an assumption to be justified later on (see Theorem 9 in Chapter 6). The *assumption* is this: If the r.v.'s $X_1, \ldots, X_k$ are independent and distributed as $N(\mu, \sigma^2)$, then the r.v.'s $\bar{X}$ and $\bar{S}^2$ are independent. (The independence of $\bar{X}$ and $\bar{S}^2$ implies then that of $[\frac{\sqrt{k}(\bar{X}-\mu)}{\sigma}]^2$ and $\frac{k\bar{S}^2}{\sigma^2}$.)

**Theorem 6.** *Let the r.v.'s $X_1, \ldots, X_k$ be independent and distributed as $N(\mu, \sigma^2)$, and let $\bar{S}^2$ be defined by (14). Then $\frac{k\bar{S}^2}{\sigma^2} \sim \chi_{k-1}^2$. Consequently, $E\bar{S}^2 = \frac{k-1}{k}\sigma^2$ and $Var(\bar{S}^2) = \frac{2(k-1)\sigma^4}{k^2}$.*

*Proof.* Consider relation (17), take the m.g.f.'s of both sides, and use the corollary to Theorem 5 and the assumption of independence made previously in order to obtain:

$$M_{kS^2/\sigma^2}(t) = M_{k\bar{S}^2/\sigma^2}(t) M_{[\sqrt{k}(\bar{X}-\mu)/\sigma]^2}(t),$$

so that

$$M_{k\bar{S}^2/\sigma^2}(t) = M_{kS^2/\sigma^2}(t) / M_{[\sqrt{k}(\bar{X}-\mu)/\sigma]^2}(t),$$

or

$$M_{k\bar{S}^2/\sigma^2}(t) = \frac{1/(1-2t)^{k/2}}{1/(1-2t)^{1/2}} = \frac{1}{(1-2t)^{(k-1)/2}},$$

which is the m.g.f. of the $\chi_{k-1}^2$ distribution. The second assertion follows immediately from the first and formula (37) in Chapter 3. ∎

This chapter is concluded with the following comment. Theorems 2–5 may be misleading in the sense that the sum of independent r.v.'s always has a distribution of the same kind as the summands. That this is definitely not so is illustrated by examples. For instance, if the independent r.v.'s $X$ and $Y$ are $U(0, 1)$, then their sum $X + Y$ is *not* Uniform; rather, it is triangular (see Example 4 (continued) in Chapter 6).

## EXERCISES

**2.1**    For any r.v.'s $X_1, \ldots, X_n$, set

$$\bar{X} = \frac{1}{n}\sum_{i=1}^{n} X_i \quad \text{and} \quad S^2 = \frac{1}{n}\sum_{i=1}^{n}(X_i - \bar{X})^2,$$

and show that:

(i)

$$ nS^2 = \sum_{i=1}^{n}(X_i - \bar{X})^2 = \sum_{i=1}^{n} X_i^2 - n\bar{X}^2. $$

(ii) If the r.v.'s have common (finite) expectation $\mu$, then (as in relation (14))

$$ \sum_{i=1}^{n}(X_i - \mu)^2 = \sum_{i=1}^{n}(X_i - \bar{X})^2 + n(\bar{X} - \mu)^2 = nS^2 + n(\bar{X} - \mu)^2. $$

**2.2** In reference to Exercise 3.1 in Chapter 4, specify the distribution of the sum $X + Y$, and write out the expression for the exact probability $P(X + Y \leq 10)$.

**2.3** If the independent r.v.'s $X$ and $Y$ are distributed as $B(m, p)$ and $B(n, p)$, respectively:
   (i) What is the distribution of the r.v. $X + Y$?
   (ii) If $m = 8, n = 12$, and $p = 0.25$, what is the numerical value of the probability: $P(5 \leq X + Y \leq 15)$?

**2.4** The independent r.v.'s $X_1, \ldots, X_n$ are distributed as $B(1, p)$, and let $S_n = X_1 + \cdots + X_n$.
   (i) Determine the distribution of the r.v. $S_n$.
   (ii) What is the $EX_i$ and the $\text{Var}(X_i)$, $i = 1, \ldots, n$?
   (iii) From part (ii) and the definition of $S_n$, compute the $ES_n$ and $\text{Var}(S_n)$.

**2.5** Let $X_1, \ldots, X_n$ be i.i.d. r.v.'s with p.d.f. $f$, and let $I$ be an interval in $\Re$. Let $p = P(X_1 \in I)$.
   (i) Express $p$ in terms of the p.d.f. $f$.
   (ii) For $k$ with $1 \leq k \leq n$, express the probability that at least $k$ of $X_1, \ldots, X_n$ take values in the interval $I$ in terms of $p$.
   (iii) Simplify the expression in part (ii), if $f$ is the Negative Exponential p.d.f. with parameter $\lambda$ and $I = (\frac{1}{\lambda}, \infty)$.
   (iv) Find the numerical value of the probability in part (iii) for $n = 4$ and $k = 2$.

**2.6** The breakdown voltage of a randomly chosen diode of a certain type is known to be Normally distributed with mean value 40 V and s.d. 1.5 V.
   (i) What is the probability that the voltage of a single diode is between 39 and 42?
   (ii) If five diodes are independently chosen, what is the probability that at least one has a voltage exceeding 42?

**2.7** Refer to Exercise 1.18 and set $X = X_1 + \cdots + X_n$.
   (i) Justify the statement that $X \sim B(n, p)$.
   (ii) Suppose that $n$ is large and $p$ is small (both assumptions quite appropriate in the framework of Exercise 1.18), so that:

$$f(x) = \binom{n}{x}p^x q^{n-x} \simeq e^{-np}\frac{(np)^x}{x!}, \quad x = 0, 1, \ldots$$

If $np = 2$, calculate the approximate values of the probabilities $f(x)$ for $x = 0, 1, 2, 3$, and 4.

**Hint.** See also Exercise 3.23 in Chapter 3.

**2.8** The r.v.'s $X_1, \ldots, X_n$ are independent and $X_i \sim P(\lambda_i)$:
   **(i)** What is the distribution of the r.v. $X = X_1 + \cdots + X_n$?
   **(ii)** If $\bar{X} = \frac{1}{n}(X_1 + \cdots + X_n)$, calculate the $E\bar{X}$ and the $\text{Var}(\bar{X})$ in terms of $\lambda_1, \ldots, \lambda_n$, and $n$.
   **(iii)** How do the $E\bar{X}$ and the $\text{Var}(\bar{X})$ become when the $X_i$'s in part (i) are distributed as $P(\lambda)$?

**2.9** Suppose that the number of no-shows for a scheduled airplane flight is a r.v. $X$ distributed as $P(\lambda)$, and it is known from past experience that, on the average, there are two no-shows. If there are five flights scheduled, compute the following probabilities for the total number of no-shows $X = X_1 + \cdots + X_5$:

| | | |
|---|---|---|
| (i)  0. | (v)  At most 10. | (ix)  15. |
| (ii)  At most 5. | (vi)  10. | (x)  At least 15. |
| (iii)  5. | (vii)  At least 10. | |
| (iv)  At least 5. | (viii)  At most 15. | |

**2.10** The r.v.'s $X_1, \ldots, X_n$ are independent and $X_i \sim P(\lambda_i), i = 1, \ldots, n$. Set $T = \sum_{i=1}^{n} X_i$ and $\lambda = \sum_{i=1}^{n} \lambda_i$, and show that:
   **(i)** The conditional p.d.f. of $X_i$, given $T = t$, is $B(t, \lambda_i/\lambda), i = 1, \ldots, n$.
   **(ii)** How does the distribution in part (i) become for $\lambda_1 = \cdots = \lambda_n = c$, say?

**2.11** If the independent r.v.'s $X$ and $Y$ are distributed as $N(\mu_1, \sigma_1^2)$ and $N(\mu_2, \sigma_2^2)$, respectively:
   **(i)** Specify the distribution of $X - Y$.
   **(ii)** Calculate the probability $P(X > Y)$ in terms of $\mu_1, \mu_2, \sigma_1$, and $\sigma_2$.
   **(iii)** If $\mu_1 = \mu_2$, conclude that $P(X > Y) = 0.5$.

**2.12** The $m + n$ r.v.'s $X_1, \ldots, X_m$ and $Y_1, \ldots, Y_n$ are independent and $X_i \sim N(\mu_1, \sigma_1^2), i = 1, \ldots, m, Y_j \sim N(\mu_2, \sigma_2^2), j = 1, \ldots, n$. Set $\bar{X} = \frac{1}{m}\sum_{i=1}^{m} X_i, \bar{Y} = \frac{1}{n}\sum_{j=1}^{n} Y_j$ and
   **(i)** Calculate the probability $P(\bar{X} > \bar{Y})$ in terms of $m, n, \mu_1, \mu_2, \sigma_1$, and $\sigma_2$.
   **(ii)** Give the numerical value of the probability in part (i) when $\mu_1 = \mu_2$ unspecified.

**2.13** Let the independent r.v.'s $X_1, \ldots, X_n$ be distributed as $N(\mu, \sigma^2)$ and set $X = \sum_{i=1}^{n} \alpha_i X_i, Y = \sum_{j=1}^{n} \beta_j X_j$, where the $\alpha_i$'s and the $\beta_j$'s are constants. Then:
   **(i)** Determine the p.d.f.'s of the r.v.'s $X$ and $Y$.

(ii) Show that the joint m.g.f. of $X$ and $Y$ is given by:

$$M_{X,Y}(t_1, t_2) = \exp\left[\mu_1 t_1 + \mu_2 t_2 + \frac{1}{2}\left(\sigma_1^2 t_1^2 + 2\rho\sigma_1\sigma_2 t_1 t_2 + \sigma_2^2 t_2^2\right)\right],$$

where $\mu_1 = \mu \sum_{i=1}^{n} \alpha_i$, $\mu_2 = \mu \sum_{j=1}^{n} \beta_j$, $\sigma_1^2 = \sigma^2 \sum_{i=1}^{n} \alpha_i^2$, $\sigma_2^2 = \sigma^2 \sum_{j=1}^{n} \beta_j^2$, $\rho = (\sum_{i=1}^{n} \alpha_i \beta_i)/\sigma_1\sigma_2$.

(iii) From part (ii), conclude that $X$ and $Y$ have the Bivariate Normal distribution with correlation coefficient

$$\rho(X, Y) = \rho = \left(\sum_{i=1}^{n} \alpha_i \beta_i\right)/\sigma_1\sigma_2.$$

(iv) From part (iii), conclude that $X$ and $Y$ are independent if and only if $\sum_{i=1}^{n} \alpha_i \beta_i = 0$.

**Hint.** For part (iii), refer to relation (58) in Chapter 4.

**2.14** Let $X$ and $Y$ be independent r.v.'s distributed as $N(0, \sigma^2)$.
  (i) Set $R = \sqrt{X^2 + Y^2}$ and determine the probability: $P(R \leq r)$, for $r > 0$.
  (ii) What is the numerical value of $P(R \leq r)$ for $\sigma = 1$ and
    $r = 1.665, r = 2.146, r = 2.448, r = 2.716, r = 3.035$, and $r = 3.255$?

**Hint.** For part (ii), use the Chi-square tables.

**2.15** Computer chips are manufactured independently by three factories, and let $X_i, i = 1, 2, 3$ be the r.v.'s denoting the total numbers of defective items in a day's production by the three factories, respectively. Suppose that the m.g.f. of $X_i$ is given by $M_i(t) = 1/(1 - \beta t)^{\alpha_i}, t < \frac{1}{\beta}, i = 1, 2, 3$, and let $X$ be the combined number of defective chips in a day's production.
  (i) Express the $EX$ and the $\sigma^2(X)$ in terms of the $\alpha_i$'s and $\beta$.
  (ii) Compute the numerical values of $EX$ and $\sigma(X)$ for $\beta = 2$ and $\alpha_i = 10^i$, $i = 1, 2, 3$.

**2.16** The blood pressure of an individual taken by an instrument used at home is a r.v. $X$ distributed as $N(\mu, 2\sigma^2)$, whereas the blood pressure of the same individual taken in a doctor's office by a more sophisticated device is a r.v. $Y$ distributed as $N(\mu, \sigma^2)$. If the r.v.'s $X$ and $Y$ are independent, compute the probability that the average $\frac{X+Y}{2}$ lies within $1.5\sigma$ from the mean $\mu$.

# Transformation of random variables

This chapter is devoted to transforming a given set of r.v.'s to another set of r.v.'s. The practical need for such transformations will become apparent by means of concrete examples to be cited and/or discussed. This chapter consists of five sections. In the first section, a single r.v. is transformed into another single r.v. In the following section, the number of available r.v.'s is at least two, and they are to be transformed into another set of r.v.'s of the same or smaller number. Two specific applications produce two new distributions, the $t$-distribution and the $F$-distribution, which are of great applicability in statistics. A brief account of specific kinds of transformations is given in the subsequent two sections, and the chapter is concluded with a section on order statistics.

## 6.1 TRANSFORMING A SINGLE RANDOM VARIABLE

**Example 1.** Suppose that the r.v.'s $X$ and $Y$ represent the temperature in a certain locality measured in degrees Celsius and Fahrenheit, respectively. Then it is known that $X$ and $Y$ are related as follows: $Y = \frac{9}{5}X + 32$.

This simple example illustrates the need for transforming a r.v. $X$ into another r.v. $Y$, if Celsius degrees are to be transformed into Fahrenheit degrees.

**Example 2.** As another example, let the r.v. $X$ denotes the velocity of a molecule of mass $m$. Then it is known that the kinetic energy of the molecule is a r.v. $Y$ related to $X$ in the following manner: $Y = \frac{1}{2}mX^2$.

Thus, determining the distribution of the kinetic energy of the molecule involves transforming the r.v. $X$ as indicated above.

The formulation of the general problem is as follows: Let $X$ be a r.v. of the continuous type with p.d.f. $f_X$, and let $h$ be a real-valued function defined on $\Re$. Define the r.v. $Y$ by $Y = h(X)$ and determine its p.d.f. $f_Y$. Under suitable regularity conditions, this problem can be resolved in two ways. One is to determine first the d.f. $F_Y$ and then obtain $f_Y$ by differentiation, and the other is to obtain $f_Y$ directly.

**Theorem 1.** *Let $S \subseteq \Re$ be the set over which $f_X$ is strictly positive, let $h : S \to T$ (the image of $S$ under $h$) $\subseteq \Re$ be one-to-one (i.e., to distinct $x$'s in $S$ there correspond*

**207**

*distinct y's in T) and strictly monotone, and let* $Y = h(X)$. *For* $x \in S$, *set* $y = h(x) \in T$. *Then* $F_Y(y) = F_X[h^{-1}(y)]$, *if h is increasing, and* $F_Y(y) = 1 - F_X[h^{-1}(y)]$, *if h is decreasing.*

*Proof.* Inverting the function $y = h(x)$, we get $x = h^{-1}(y)$. Then for increasing $h$ (which implies increasing $h^{-1}$), we have:

$$F_Y(y) = P(Y \leq y) = P[h(X) \leq y] = P\{h^{-1}[h(X)] \leq h^{-1}(y)\}$$
$$= P[X \leq h^{-1}(y)] = F_X[h^{-1}(y)]. \tag{1}$$

If $h$ is decreasing, then so is $h^{-1}$ and therefore:

$$F_Y(y) = P[h(X) \leq y] = P\{h^{-1}[h(X)] \geq h^{-1}(y)\}$$
$$= P[X \geq h^{-1}(y)] = 1 - P[X < h^{-1}(y)]$$
$$= 1 - P[X \leq h^{-1}(y)] = 1 - F_X[h^{-1}(y)]. \tag{2}$$

$\blacksquare$

As an illustration, consider the case $Y = \frac{9}{5}X + 32$ in Example 1. Here, $y = h(x) = \frac{9}{5}x + 32$ is one-to-one and strictly increasing. Therefore,

$$F_Y(y) = F_X\left[\frac{5}{9}(y - 32)\right] \quad \text{and, at continuity points of } f_X, f_Y(y) = \frac{5}{9}f_X\left[\frac{5}{9}(y - 32)\right]. \tag{3}$$

The function $y = h(x)$ may not be one-to-one and strictly increasing on the entire $S$ but it is so on subsets of it. Then $F_Y$ can still be determined. Example 2 illustrates in part the point. Let $Y = \frac{1}{2}mX^2$ as mentioned above. Then proceed as follows: For $y > 0$:

$$F_Y(y) = P(Y \leq y) = P\left(\frac{1}{2}mX^2 \leq y\right) = P\left(X^2 \leq \frac{2y}{m}\right)$$

$$= P\left(-\sqrt{\frac{2y}{m}} \leq X \leq \sqrt{\frac{2y}{m}}\right) = P\left(X \leq \sqrt{\frac{2y}{m}}\right) - P\left(X < -\sqrt{\frac{2y}{m}}\right)$$

$$= P\left(X \leq \sqrt{\frac{2y}{m}}\right), \quad \text{since } X \geq 0. \tag{4}$$

Assuming that we can differentiate in (1) or (2) (and depending on whether $h$ is increasing or decreasing), we obtain the p.d.f. of $Y$; namely, at continuity points of $f_X$,

$$f_Y(y) = \frac{d}{dy}F_Y(y) = \frac{d}{dx}F_X(x)\Big|_{x=h^{-1}(y)} \times \left|\frac{dx}{dy}\right| = f_X[h^{-1}(y)] \times \left|\frac{d}{dy}h^{-1}(y)\right|, \quad y \in T. \tag{5}$$

In the case of formula (3), relation (5) gives: $f_Y(y) = \frac{5}{9}f_X\left[\frac{5}{9}(y - 32)\right]$ at continuity points of $f_X$, as has already been seen. In the case of formula (4), and at continuity points of $f_X$,

$$f_Y(y) = (1/\sqrt{2my})f_X\left(\sqrt{\frac{2y}{m}}\right). \tag{6}$$

A further complete illustration is provided by the case where $Y = X^2$ and $X \sim N(0, 1)$ (see also Exercise 1.9). Here, for $y > 0$,

$$F_Y(y) = P(Y \le y) = P(X^2 \le y) = P(-\sqrt{y} \le X \le \sqrt{y})$$
$$= P(X \le \sqrt{y}) - P(X \le -\sqrt{y}) = F_X(\sqrt{y}) - F_X(-\sqrt{y}).$$

Therefore,

$$f_Y(y) = \frac{d}{dy} F_Y(y) = f_X(\sqrt{y}) \left| \frac{d}{dy} \sqrt{y} \right| - f_X(-\sqrt{y}) \left| \frac{d}{dy}(-\sqrt{y}) \right|$$

$$= \frac{1}{\sqrt{2\pi}} e^{-y/2} \frac{1}{2\sqrt{y}} + \frac{1}{\sqrt{2\pi}} e^{-y/2} \frac{1}{2\sqrt{y}} = \frac{1}{\sqrt{2\pi}} e^{-y/2} \frac{1}{\sqrt{y}}$$

$$= \frac{1}{\Gamma(1/2)\, 2^{1/2}} y^{(1/2)-1} e^{-y/2}, \quad \text{which is the p.d.f. of } \chi_1^2.$$

Instead of going through the d.f. (which process requires monotonicity of the transformation $y = h(x)$), under certain conditions, $f_Y$ may be obtained directly from $f_X$. Such conditions are described in the following theorem.

**Theorem 2.** *Let $X$ be a r.v. with positive and continuous p.d.f. $f_X$ on the set $S \subseteq \Re$, and let $h : S \to T$ (the image of $S$ under $h$) be a one-to-one transformation, so that the inverse $x = h^{-1}(y)$, $y \in T$, exists. Suppose that, for $y \in T$, the derivative $\frac{d}{dy} h^{-1}(y)$ exists, is continuous, and $\ne 0$. Then the p.d.f. of the r.v. $Y = h(X)$ is given by:*

$$f_Y(y) = f_X[h^{-1}(y)] \left| \frac{d}{dy} h^{-1}(y) \right|, \quad y \in T \text{ (and } = 0 \text{ for } y \notin T). \tag{7}$$

*Proof (rough outline).* Let $B = [c, d]$ be an interval in $T$ and suppose $B$ is transformed into the interval $A = [a, b]$ by the inverse transformation $x = h^{-1}(y)$. Then:

$$P(Y \in B) = P[h(X) \in B] = P(X \in A) = \int_A f_X(x)\, dx.$$

When transforming $x$ into $y$ through the transformation $x = h^{-1}(y)$, $\int_A f_X(x)dx = \int_B f_X[h^{-1}(y)] | \frac{d}{dy} h^{-1}(y) | dy$, according to the theory of changing variables in integrals. Thus,

$$P(Y \in B) = \int_B f_X[h^{-1}(y)] \left| \frac{d}{dy} h^{-1}(y) \right| dy,$$

which implies that the integrand is the p.d.f. of $Y$. ∎

Relation (7) has already been illustrated by Example 1. A slightly more general case is the following one.

**Example 3.** Determine the p.d.f. of the r.v. $Y$ defined by: $Y = aX + b$ $(a \ne 0)$. In particular, determine $f_Y$, if $X \sim N(\mu, \sigma^2)$.

**Discussion.**   The transformation $y = ax + b$ gives $x = h^{-1}(y) = \frac{y-b}{a}$, so that $\frac{dx}{dy} = \frac{d}{dy}h^{-1}(y) = \frac{1}{a}$. Therefore, $f_Y(y) = f_X(\frac{y-b}{a})\frac{1}{|a|}$. For the special case:

$$f_Y(y) = \frac{1}{\sqrt{2\pi}|a|\sigma}\exp\left[-\frac{\left(\frac{y-b}{a}-\mu\right)^2}{2\sigma^2}\right] = \frac{1}{\sqrt{2\pi}|a|\sigma}\exp\left\{-\frac{[y-(a\mu+b)]^2}{2(a\sigma)^2}\right\}.$$

Thus, if $X \sim N(\mu,\sigma^2)$, then $Y = aX + b \sim N(a\mu + b, (a\sigma)^2)$.

A modification of Theorem 2 when the assumption that $h : S \to T$ is one-to-one is not satisfied, but a version of it is, is stated in the following result. This result has already been illustrated by (6) in connection with Example 2.

**Theorem 3.**   *Let $X$ be a r.v. with positive and continuous p.d.f. $f_X$ on the set $S \subseteq \Re$, and suppose that the transformation $h : S \to T$ is not one-to-one. Suppose further that when $S$ is partitioned into the pairwise disjoint subsets $S_1,\ldots,S_r$ and $h$ is restricted to $S_j$ and takes values in $T_j$ (the image of $S_j$ under $h$), then $h$ is one-to-one. Denoting by $h_j$ this restriction, we have then: $h_j : S_j \to T_j$ is one-to-one, so that the inverse $x = h_j^{-1}(y)$, $y \in T_j$, exists, $j = 1,\ldots,r$. Finally, we suppose that, for any $y \in T_j$, $j = 1,\ldots,r$, the derivatives $\frac{d}{dy}h_j^{-1}(y)$ exist, are continuous, and $\neq 0$. Then the p.d.f. of the r.v. $Y = h(X)$ is determined as follows: Set*

$$f_{Y_j} = f_X\big[h_j^{-1}(y)\big]\left|\frac{d}{dy}h_j^{-1}(y)\right|, \quad y \in T_j, \quad j = 1,\ldots,r,$$

*and for $y \in T$, suppose that $y$ belongs to $k$ of the $r$ $T_j$'s, $1 \leq k \leq r$. Then $f_Y(y)$ is the sum of the corresponding $k$ $f_{Y_j}(y)$'s. Alternatively,*

$$f_Y(y) = \sum_{j=1}^{r}\delta_j(y)f_{Y_j}(y), \quad y \in T \quad (\text{and} = 0 \text{ for } y \notin T), \tag{8}$$

*where $\delta_j(y) = 1$, if $y \in T_j$ and $\delta_j(y) = 0$, if $y \notin T_j$, $j = 1,\ldots,r$.*

**Remark 1.**   It is to be noticed that, whereas the subsets $S_1,\ldots,S_r$ are pairwise disjoint, their images $T_1,\ldots,T_r$ need not be so. For instance, in Example 2, $S_1 = (0,\infty)$, $S_2 = (-\infty,0)$ but $T_1 = T_2 = (0,\infty)$.

## EXERCISES

**1.1**   The r.v. $X$ has p.d.f. $f_X(x) = (1-\alpha)\alpha^x$, $x = 0, 1, \ldots$ $(0 < \alpha < 1)$, and set $Y = X^3$. Determine the p.d.f. $f_Y$.

**1.2**   Let the r.v.'s $X$ and $Y$ represent the temperature of a certain object in degrees Celsius and Fahrenheit, respectively. Then, it is known that $Y = \frac{9}{5}X + 32$, so that $X = \frac{5}{9}Y - \frac{160}{9}$.

    (i) If $Y \sim N(\mu, \sigma^2)$, determine the distribution of $X$.

    (ii) If $P(90 \le Y \le 95) = 0.95$, then also $P(a \le X \le b) = 0.95$, for some $a < b$. Determine the numbers $a$ and $b$.

    (iii) We know that: $P(\mu - \sigma \le Y \le \mu + \sigma) \simeq 0.6827 = p_1$, $P(\mu - 2\sigma \le Y \le \mu + 2\sigma) \simeq 0.9545 = p_2$, and $P(\mu - 3\sigma \le Y \le \mu + 3\sigma) \simeq 0.9973 = p_3$. Calculate the intervals $[a_k, b_k], k = 1, 2, 3$ for which $P(a_k \le X \le b_k)$ is, respectively, equal to $p_k, k = 1, 2, 3$.

**1.3**    Let the r.v. $X$ have p.d.f. $f_X$ positive on the set $S \subseteq \Re$, and set $U = aX + b$, where $a$ and $b$ are constants and $a > 0$.

    (i) Use Theorem 2 in order to derive the p.d.f. $f_U$.

    (ii) If $X$ has the Negative Exponential distribution over the interval $(b, \infty)$ with parameter $\lambda$, show that $U$ has the same kind of distribution over the interval $(b, \infty)$ with parameter $\lambda/a$.

    (iii) If $X \sim U(c, d)$, then show that $U \sim U(ac + b, ad + b)$.

**1.4**    If the r.v. $X$ has the Negative Exponential distribution with parameter $\lambda$, set $Y = e^X$ and $Z = \log X$ and determine the p.d.f.'s $f_Y$ and $f_Z$.

**1.5**    Let $X \sim U(\alpha, \beta)$ and set $Y = e^X$. Then determine the p.d.f. $f_Y$. If $\alpha > 0$, set $Z = \log X$ and determine the p.d.f. $f_Z$.

**1.6**    (i) If the r.v. $X$ is distributed as $U(0, 1)$ and $Y = -2 \log X$, show that $Y$ is distributed as $\chi_2^2$.

    (ii) If $X_1, \ldots, X_n$ is a random sample from the $U(0, 1)$ distribution and $Y_i = -2 \log X_i$, use part (i) and the m.g.f. approach in order to show that $\sum_{i=1}^{n} Y_i$ is distributed as $\chi_{2n}^2$.

**1.7**    If the r.v. $X$ has the p.d.f. $f_X(x) = \frac{1}{\sqrt{2\pi}} x^{-2} e^{-1/2x^2}, x \in \Re$, show that the r.v. $Y = \frac{1}{X} \sim N(0, 1)$.

**1.8**    Suppose that the velocity of a molecule of mass $m$ is a r.v. $X$ with p.d.f. $f_X(x) = \sqrt{\frac{2}{\pi}} x^2 e^{-x^2/2}, x > 0$ (the so-called Maxwell distribution). Derive the p.d.f. of the r.v. $Y = \frac{1}{2} mX^2$, which is the kinetic energy of the molecule. **Hint.** Use relations (28) and (30) in Chapter 3.

**1.9**    If the r.v. $X \sim N(0, 1)$, use Theorem 3 in order to show that the r.v. $Y = X^2 \sim \chi_1^2$.
**Hint.** Use relation (30) in Chapter 3.

**1.10**   If $X$ has the Negative Exponential distribution with parameter $\lambda = 1$, determine the p.d.f. of $Y = \sqrt{X}$.

**1.11**   The r.v. $X$ is said to have the two-parameter Pareto distribution with parameters $\alpha$ and $\beta$, if its p.d.f. is given by:

$$f_X(x) = \frac{\alpha \beta^\alpha}{x^{\alpha+1}}, \quad x > \beta \; (\alpha > 0, \beta > 0).$$

(i) Show that the function just given is, indeed, a p.d.f.

(ii) Set $Y = X/\beta$ and show that its p.d.f. is given by:

$$f_Y(y) = \frac{\alpha}{y^{\alpha+1}}, \quad x > 1 \ (\alpha > 0),$$

which is referred to as the one-parameter Pareto distribution.

(iii) Show that $EX = \frac{\alpha\beta}{\alpha-1}$ $(\alpha > 1, \beta > 0)$, and that $\sigma^2(X) = \frac{\alpha\beta^2}{(\alpha-1)^2(\alpha-2)}$ $(\alpha > 2, \beta > 0)$, and deduce that $EY = \frac{\alpha}{\alpha-1}$ $(\alpha > 1)$, $\sigma^2(Y) = \frac{\alpha}{(\alpha-1)^2(\alpha-2)}$ $(\alpha > 2)$.

**Note:** The Pareto distribution is used in Economics to describe, e.g., distribution of income, and in Engineering and Biostatistics as a lifetime distribution.

## 6.2 TRANSFORMING TWO OR MORE RANDOM VARIABLES

Often the need arises to transform two or more given r.v.'s to another set of r.v.'s. The following examples illustrate the point.

**Example 4.** The times of arrival of a bus at two successive bus stops are r.v.'s $X_1$ and $X_2$ distributed as $U(\alpha, \beta)$, for two time points $\alpha < \beta$. Calculate the probabilities $P(X_1 + X_2 > x)$ for $2\alpha < x < 2\beta$.

Clearly, this question calls for the determination of the distribution of the r.v. $X_1 + X_2$.

Or more generally (and more realistically), suppose that a bus makes $k$ stops between its depot and its terminal, and that the arrival time at the $i$th stop is a r.v. $X_i \sim U(\alpha_i, \beta_i)$, $\alpha_i < \beta_i$, $i = 1, \ldots, k+1$ (where $X_{k+1}$ is the time of arrival at the terminal). Determine the distribution of the duration of the trip $X_1 + \cdots + X_{k+1}$.

**Example 5.** Consider certain events occurring in every time interval $[t_1, t_2]$ $(0 \le t_1 < t_2)$ according to the Poisson distribution $P(\lambda(t_2 - t_1))$. Then the waiting times between successive occurrences are independent r.v.'s distributed according to the Negative Exponential distribution with parameter $\lambda$. Let $X_1$ and $X_2$ be two such times. What is the probability that one would have to wait at least twice as long for the second occurrence than the first? That is, what is the probability $P(X_2 > 2X_1)$?

Here, one would have to compute the distribution of the r.v. $X_2 - 2X_1$.

Below, a brief outline of the theory underpinning the questions posed in the examples is presented. First, consider the case of two r.v.'s $X_1$ and $X_2$ having the joint p.d.f. $f_{X_1,X_2}$. Often the question posed is that of determining the distribution of a function of $X_1$ and $X_2, h_1(X_1, X_2)$. The general approach is to set $Y_1 = h_1(X_1, X_2)$ and also consider another (convenient) transformation $Y_2 = h_2(X_1, X_2)$. Next, determine the joint p.d.f. of $Y_1$ and $Y_2, f_{Y_1,Y_2}$, and, finally, compute the (marginal) p.d.f. $f_{Y_1}$. Conditions under which $f_{Y_1,Y_2}$ is determined by way of $f_{X_1,X_2}$ are given below.

**Theorem 4.** *Consider the r.v.'s $X_1$ and $X_2$ with joint p.d.f. $f_{X_1,X_2}$ positive and continuous on the set $S \subseteq \Re^2$, and let $h_1, h_2$ be two real-valued transformations defined on $S$; that is, $h_1, h_2 : S \to \Re$, and let $T$ be the image of $S$ under the transformation $(h_1, h_2)$. Suppose that $h_1$ and $h_2$ are one-to-one from $S$ onto $T$. Thus, if we set $y_1 = h_1(x_1, x_2)$ and $y_2 = h_2(x_1, x_2)$, we can solve uniquely for $x_1, x_2 : x_1 = g_1(y_1, y_2), x_2 = g_2(y_1, y_2)$. Suppose further that the partial derivatives $g_{1i}(y_1, y_2) = \frac{\partial}{\partial y_i} g_1(y_1, y_2)$ and $g_{2i}(y_1, y_2) = \frac{\partial}{\partial y_i} g_2(y_1, y_2)$, $i = 1, 2$ exist and are continuous for $(y_1, y_2) \in T$. Finally, suppose that the Jacobian $J = \begin{vmatrix} g_{11}(y_1, y_2) & g_{12}(y_1, y_2) \\ g_{21}(y_1, y_2) & g_{22}(y_1, y_2) \end{vmatrix}$ is $\neq 0$ on $T$. Then the joint p.d.f. of the r.v.'s $Y_1 = h_1(X_1, X_2)$ and $Y_2 = h_2(X_1, X_2), f_{Y_1,Y_2}$, is given by:*

$$f_{Y_1,Y_2}(y_1, y_2) = f_{X_1,X_2}[g_1(y_1, y_2), g_2(y_1, y_2)]|J|, \quad (y_1, y_2) \in T \tag{9}$$

*(and $= 0$ for $(y_1, y_2) \notin T$).*

The justification of this theorem is entirely analogous to that of Theorem 2 and will be omitted.

In applying Theorem 4, one must be careful in checking that the underlying assumptions hold and in determining correctly the set $T$. As an illustration, let us discuss Examples 4 and 5.

**Example 4** (continued).

**Discussion.** We have $y_1 = x_1 + x_2$ and let $y_2 = x_2$. Then $x_1 = y_1 - y_2$ and $x_2 = y_2$, so that $\frac{\partial x_1}{\partial y_1} = 1, \frac{\partial x_1}{\partial y_2} = -1, \frac{\partial x_2}{\partial y_1} = 0, \frac{\partial x_2}{\partial y_2} = 1$, and $J = \begin{vmatrix} 1 & -1 \\ 0 & 1 \end{vmatrix} = 1$. For the determination of $S$ and $T$, see Figures 6.1 and 6.2.

Since $f_{X_1,X_2}(x_1, x_2) = \frac{1}{(\beta-\alpha)^2}$ for $(x_1, x_2) \in S$, we have $f_{Y_1,Y_2}(y_1, y_2) = \frac{1}{(\beta-\alpha)^2}$ for $(y_1, y_2) \in T$; that is, for $2\alpha < y_1 < 2\beta, \alpha < y_2 < \beta, \alpha < y_1 - y_2 < \beta$ (and $= 0$ for $(y_1, y_2) \notin T$).

Thus, we get:

$$f_{Y_1,Y_2}(y_1, y_2) = \begin{cases} \frac{1}{(\beta-\alpha)^2}, & 2\alpha < y_1 < 2\beta, \alpha < y_2 < \beta, \alpha < y_1 - y_2 < \beta \\ 0, & \text{otherwise.} \end{cases}$$

Therefore,

$$f_{Y_1}(y_1) = \begin{cases} \frac{1}{(\beta-\alpha)^2} \int_\alpha^{y_1-\alpha} dy_2 = \frac{y_1-2\alpha}{(\beta-\alpha)^2}, & \text{for } 2\alpha < y_1 \le \alpha + \beta \\ \frac{1}{(\beta-\alpha)^2} \int_{y_1-\beta}^\beta dy_2 = \frac{2\beta-y_1}{(\beta-\alpha)^2}, & \text{for } \alpha + \beta < y_1 \le 2\beta \\ 0, & \text{otherwise.} \end{cases}$$

The graph of $f_{Y_1}$ is given in Figure 6.3.

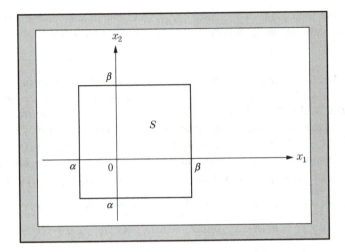

**FIGURE 6.1**

$S = \{(x_1, x_2) \in \Re^2; f_{X_1, X_2}(x_1, x_2) > 0\}.$

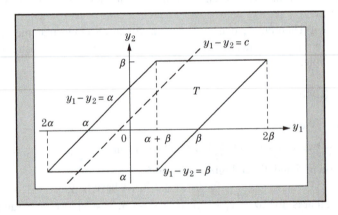

**FIGURE 6.2**

$T =$ Image of $S$ under the transformation used.

**Example 5 (continued).**

**Discussion.**   Here, $y_1 = x_2 - 2x_1 = -2x_1 + x_2$ and let $y_2 = x_2$. Then $x_1 = -\frac{1}{2}y_1 + \frac{1}{2}y_2$ and $x_2 = y_2$, so that $J = \begin{vmatrix} -\frac{1}{2} & \frac{1}{2} \\ 0 & 1 \end{vmatrix} = -\frac{1}{2}$ and $|J| = \frac{1}{2}$. Clearly, $S$ is the first quadrant. As for $T$, we have $y_2 = x_2$, so that $y_2 > 0$. Also, $-\frac{1}{2}y_1 + \frac{1}{2}y_2 = x_1$, so that $-\frac{1}{2}y_1 + \frac{1}{2}y_2 > 0$ or $-y_1 + y_2 > 0$ or $y_2 > y_1$. The conditions $y_2 > 0$ and $y_2 > y_1$ determine $T$ (see Figure 6.4).

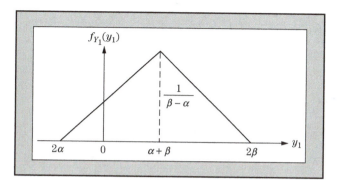

**FIGURE 6.3**

This density is known as the *triangular p.d.f.*

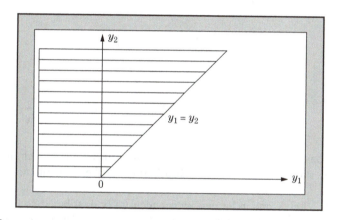

**FIGURE 6.4**

The shaded area is $T$, the range of $(y_1, y_2)$.

$T$ is the part of the plane above the $y_1$ axis and also to the left of the main diagonal $y_1 = y_2$.

Since $f_{X_1,X_2}(x_1, x_2) = \lambda^2 e^{-\lambda(x_1+x_2)}$ $(x_1, x_2 > 0)$, we have $f_{Y_1,Y_2}(y_1, y_2) = \frac{\lambda^2}{2}e^{(\lambda/2)y_1-(3\lambda/2)y_2}$, $(y_1, y_2) \in T$ (and $= 0$ otherwise). Therefore, $f_{Y_1}(y_1)$ is taken by integrating out $y_2$. More precisely, for $y_1 < 0$:

$$f_{Y_1}(y_1) = \frac{\lambda^2}{2}e^{(\lambda/2)y_1} \int_0^\infty e^{-(3\lambda/2)y_2}\,dy_2 = -\frac{\lambda^2}{2} \times \frac{2}{3\lambda}e^{(\lambda/2)y_1} \times e^{-(3\lambda/2)y_2}\Big|_0^\infty$$

$$= -\frac{\lambda}{3}e^{(\lambda/2)y_1}(0 - 1) = \frac{\lambda}{3}e^{(\lambda/2)y_1},$$

whereas for $y_1 > 0$:

$$f_{Y_1}(y_1) = \frac{\lambda^2}{2}e^{(\lambda/2)y_1}\int_{y_1}^{\infty} e^{-(3\lambda/2)y_2}\,dy_2 = -\frac{\lambda}{3}e^{(\lambda/2)y_1} \times e^{-(3\lambda/2)y_2}\Big|_{y_1}^{\infty}$$

$$= -\frac{\lambda}{3}e^{(\lambda/2)y_1}(0 - e^{-(3\lambda/2)y_1}) = \frac{\lambda}{3}e^{-\lambda y_1}.$$

To summarize:

$$f_{Y_1}(y_1) = \begin{cases} \frac{\lambda}{3}e^{(\lambda/2)y_1}, & y_1 < 0 \\ \frac{\lambda}{3}e^{-\lambda y_1}, & y_1 \geq 0. \end{cases}$$

Therefore, $P(X_2 > 2X_1) = P(X_2 - 2X_1 > 0) = P(Y_1 > 0) = \frac{\lambda}{3}\int_0^{\infty} e^{-\lambda y_1}\,dy_1 = \frac{1}{3}$.

**Remark 2.**   To be sure, the preceding probability is also calculated as follows:

$$P(X_2 > 2X_1) = \iint\limits_{(x_2 > 2x_1)} \lambda^2 e^{-\lambda x_1 - \lambda x_2}\,dx_1\,dx_2$$

$$= \int_0^{\infty} \lambda e^{-\lambda x_2}\left(\int_0^{x_2/2} \lambda e^{-\lambda x_1}\,dx_1\right)dx_2$$

$$= \int_0^{\infty} \lambda e^{-\lambda x_2}(1 - e^{-(\lambda/2)x_2})dx_2$$

$$= \int_0^{\infty} \lambda e^{-\lambda x_2}\,dx_2 - \frac{2}{3}\int_0^{\infty} \frac{3\lambda}{2}e^{-(3\lambda/2)x_2}\,dx_2 = 1 - \frac{2}{3} = \frac{1}{3}.$$

Applications of Theorem 4 lead to two new distributions, which are of great importance in statistics. They are the $t$-distribution and the $F$-distribution.

**Definition 1.**   Let $X$ and $Y$ be two independent r.v.'s distributed as follows: $X \sim N(0, 1)$ and $Y \sim \chi_r^2$, and define the r.v. $T$ by: $T = X/\sqrt{Y/r}$. The r.v. $T$ is said to have the (Student's) $t$-distribution with $r$ degrees of freedom (d.f.). The notation used is: $T \sim t_r$.

The p.d.f. of $T, f_T$, is given by the formula:

$$f_T(t) = \frac{\Gamma\left[\frac{1}{2}(r+1)\right]}{\sqrt{\pi r}\Gamma(r/2)} \times \frac{1}{[1 + (t^2/r)]^{(1/2)(r+1)}}, \quad t \in \mathfrak{R}, \tag{10}$$

and its graph (for $r = 5$) is presented in Figure 6.5.

From formula (10), it is immediate that $f_T$ is symmetric about 0 and tends to 0 as $t \to \pm\infty$. It can also be seen (see Exercise 2.9) that $f_T(t)$ tends to the p.d.f. of the $N(0, 1)$ distribution as the number $r$ of d.f. tends to $\infty$. This is depicted in Figure 6.5 by means of the curve denoted by $t_\infty$. Also, it is seen (see Exercise 2.9) that $ET = 0$ for $r \geq 2$, and $\text{Var}(T) = \frac{r}{r-2}$ for $r \geq 3$. Finally, the probabilities $P(T \leq t)$, for selected values of $t$ and $r$, are given by tables (the $t$-tables). For $r \geq 91$, one may use the tables for the standard Normal distribution.

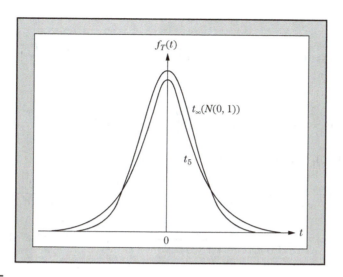

**FIGURE 6.5**

Two curves of the $t$ probability density function.

## DERIVATION OF THE P.D.F. OF $T, f_T$

Regarding the derivation of $f_T$, we have:

$$f_X(x) = \frac{1}{\sqrt{2\pi}} e^{-(1/2)x^2}, \quad x \in \Re,$$

$$f_Y(y) = \begin{cases} \frac{1}{\Gamma(\frac{1}{2}r)2^{(1/2)r}} y^{(r/2)-1} e^{-y/2}, & y > 0 \\ 0, & y \le 0. \end{cases}$$

Set $U = Y$ and consider the transformation

$$(h_1, h_2) : \begin{cases} t = \frac{x}{\sqrt{y/r}} \\ u = y \end{cases} ; \quad \text{then} \quad \begin{cases} x = \frac{1}{\sqrt{r}} t \sqrt{u} \\ y = u, \end{cases}$$

and

$$J = \begin{vmatrix} \frac{\sqrt{u}}{\sqrt{r}} & \frac{t}{2\sqrt{u}\sqrt{r}} \\ 0 & 1 \end{vmatrix} = \frac{\sqrt{u}}{\sqrt{r}}.$$

Therefore, for $t \in \Re, u > 0$, we get

$$f_{T,U}(t,u) = \frac{1}{\sqrt{2\pi}} e^{-t^2 u/2r} \times \frac{1}{\Gamma(r/2)2^{r/2}} u^{(r/2)-1} e^{-u/2} \times \frac{\sqrt{u}}{\sqrt{r}}$$

$$= \frac{1}{\sqrt{2\pi r}\Gamma(r/2)2^{r/2}} u^{(1/2)(r+1)-1} \exp\left[-\frac{u}{2}\left(1 + \frac{t^2}{r}\right)\right].$$

Hence

$$f_T(t) = \int_0^\infty \frac{1}{\sqrt{2\pi r}\Gamma(r/2)2^{r/2}} u^{(1/2)(r+1)-1} \exp\left[-\frac{u}{2}\left(1+\frac{t^2}{r}\right)\right] du.$$

Set

$$\frac{u}{2}\left(1+\frac{u^2}{r}\right) = z, \quad \text{so that} \quad u = 2z\left(1+\frac{t^2}{r}\right)^{-1}, \quad du = 2\left(1+\frac{t^2}{r}\right)^{-1} dz,$$

and $z \in [0, \infty)$. Therefore we continue as follows:

$$f_T(t) = \int_0^\infty \frac{1}{\sqrt{2\pi r}\Gamma(r/2)2^{r/2}} \left[\frac{2z}{1+(t^2/r)}\right]^{(1/2)(r+1)-1} \times e^{-z} \frac{2}{1+(t^2/r)} dz$$

$$= \frac{1}{\sqrt{2\pi r}\Gamma(r/2)2^{r/2}} \frac{2^{(1/2)(r+1)}}{[1+(t^2/r)]^{(1/2)(r+1)}} \times \int_0^\infty z^{(1/2)(r+1)-1}e^{-z} dz$$

$$= \frac{1}{\sqrt{\pi r}\Gamma(r/2)} \frac{1}{[1+(t^2/r)]^{(1/2)(r+1)}} \times \Gamma\left[\frac{1}{2}(r+1)\right],$$

since $\frac{1}{\Gamma[\frac{1}{2}(r+1)]}z^{(1/2)(r+1)-1}e^{-z}$ $(z > 0)$ is the p.d.f. of the Gamma distribution with parameters $\alpha = \frac{r+1}{2}$ and $\beta = 1$; that is,

$$f_T(t) = \frac{\Gamma[\frac{1}{2}(r+1)]}{\sqrt{\pi r}\Gamma(r/2)} \times \frac{1}{[1+(t^2/r)]^{(1/2)(r+1)}}, \quad t \in \Re.$$

Now, we proceed with the definition of the $F$-distribution.

**Definition 2.**　Let $X$ and $Y$ be two independent r.v.'s distributed as follows: $X \sim \chi^2_{r_1}$ and $Y \sim \chi^2_{r_2}$, and define the r.v. $F$ by: $F = \frac{X/r_1}{Y/r_2}$. The r.v. $F$ is said to have the $F$-distribution with $r_1$ and $r_2$ degrees of freedom (d.f.). The notation often used is: $F \sim F_{r_1,r_2}$.

The p.d.f. of $F$, $f_F$, is given by the formula:

$$f_F(f) = \begin{cases} \frac{\Gamma[\frac{1}{2}(r_1+r_2)](r_1/r_2)^{r_1/2}}{\Gamma(\frac{1}{2}r_1)\Gamma(\frac{1}{2}r_2)} \times \frac{f^{(r_1/2)-1}}{[1+(r_1/r_2)f]^{(1/2)(r_1+r_2)}}, & \text{for } f > 0 \\ 0, & \text{for } f \le 0, \end{cases} \quad (11)$$

and its graphs (for $r_1 = 10, r_2 = 4$, and $r_1 = r_2 = 10$) are given in Figure 6.6. The probabilities $P(F \le f)$, for selected values of $f$ and $r_1, r_2$, are given by tables (the $F$-tables).

### DERIVATION OF THE P.D.F. OF $F$, $f_F$

The derivation of $f_F$ is based on Theorem 4 and is as follows. For $x$ and $y > 0$, we have:

$$f_X(x) = \frac{1}{\Gamma(\frac{1}{2}r_1)2^{r_1/2}} x^{(r_1/2)-1}e^{-x/2}, \quad x > 0,$$

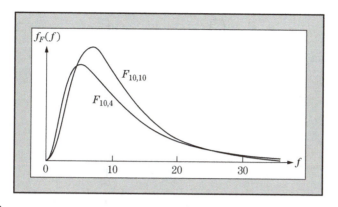

**FIGURE 6.6**

Two curves of the $F$ probability density function.

$$f_Y(y) = \frac{1}{\Gamma(\frac{1}{2}r_2)2^{r_2/2}} y^{(r_2/2)-1} e^{-y/2}, \quad y > 0.$$

We set $Z = Y$, and consider the transformation

$$(h_1, h_2) : \begin{cases} f = \frac{x/r_1}{y/r_2} \\ z = y \end{cases} ; \quad \text{then} \quad \begin{cases} x = \frac{r_1}{r_2} fz \\ y = z, \end{cases}$$

and

$$J = \begin{vmatrix} \frac{r_1}{r_2}z & \frac{r_1}{r_2}f \\ 0 & 1 \end{vmatrix} = \frac{r_1}{r_2}z, \quad \text{so that } |J| = \frac{r_1}{r_2}z.$$

For $f, z > 0$, we get:

$$f_{F,Z}(f,z) = \frac{1}{\Gamma(\frac{1}{2}r_1)\Gamma(\frac{1}{2}r_2)2^{(1/2)(r_1+r_2)}} \left(\frac{r_1}{r_2}\right)^{(r_1/2)-1} f^{(r_1/2)-1} z^{(r_1/2)-1} z^{(r_2/2)-1}$$

$$\times \exp\left(-\frac{r_1}{2r_2}fz\right) e^{-z/2} \frac{r_1}{r_2}z$$

$$= \frac{(r_1/r_2)^{r_1/2} f^{(r_1/2)-1}}{\Gamma(\frac{1}{2}r_1)\Gamma(\frac{1}{2}r_2)2^{(1/2)(r_1+r_2)}} z^{(1/2)(r_1+r_2)-1} \exp\left[-\frac{z}{2}\left(\frac{r_1}{r_2}f+1\right)\right].$$

Therefore

$$f_F(f) = \int_0^\infty f_{F,Z}(f,z)dz$$

$$= \frac{(r_1/r_2)^{r_1/2} f^{(r_1/2)-1}}{\Gamma(\frac{1}{2}r_1)\Gamma(\frac{1}{2}r_2)2^{(1/2)(r_1+r_2)}} \int_0^\infty z^{(1/2)(r_1+r_2)-1} \exp\left[-\frac{z}{2}\left(\frac{r_1}{r_2}f+1\right)\right]dz.$$

Set

$$\frac{z}{2}\left(\frac{r_1}{r_2}f+1\right) = t, \quad \text{so that } z = 2t\left(\frac{r_1}{r_2}f+1\right)^{-1},$$

$$dz = 2\left(\frac{r_1}{r_2}f + 1\right)^{-1} dt, \quad t \in [0, \infty).$$

Thus continuing, we have

$$f_F(f) = \frac{(r_1/r_2)^{r_1/2} f^{(r_1/2)-1}}{\Gamma(\frac{1}{2}r_1)\Gamma(\frac{1}{2}r_2)2^{(1/2)(r_1+r_2)}} 2^{(1/2)(r_1+r_2)-1}\left(\frac{r_1}{r_2}f + 1\right)^{-(1/2)(r_1+r_2)+1}$$

$$\times 2\left(\frac{r_1}{r_2}f + 1\right)^{-1} \int_0^\infty t^{(1/2)(r_1+r_2)-1} e^{-t} dt$$

$$= \frac{\Gamma[\frac{1}{2}(r_1+r_2)](r_1/r_2)^{r_1/2}}{\Gamma(\frac{1}{2}r_1)\Gamma(\frac{1}{2}r_2)} \times \frac{f^{(r_1/2)-1}}{[1+(r_1/r_2)f]^{(1/2)(r_1+r_2)}},$$

since $\frac{1}{\Gamma[\frac{1}{2}(r_1+r_2)]} t^{(1/2)(r_1+r_2)-1} e^{-t}$ $(t > 0)$ is the p.d.f. of the Gamma distribution with parameters $\alpha = \frac{r_1+r_2}{2}$ and $\beta = 1$. Therefore

$$f_F(f) = \begin{cases} \frac{\Gamma[\frac{1}{2}(r_1+r_2)](r_1/r_2)^{r_1/2}}{\Gamma(\frac{1}{2}r_1)\Gamma(\frac{1}{2}r_2)} \times \frac{f^{(r_1/2)-1}}{[1+(r_1/r_2)f]^{(1/2)(r_1+r_2)}}, & \text{for } f > 0 \\ 0, & \text{for } f \leq 0. \end{cases}$$

**Remark 3.**

(i) From the definition of the $F$-distribution, it follows that, if $F \sim F_{r_1,r_2}$, then $\frac{1}{F} \sim F_{r_2,r_1}$.

(ii) If $T \sim t_r$, then $T^2 \sim F_{1,r}$. Indeed, $T = X/\sqrt{Y/r}$, where $X$ and $Y$ are independent, and $X \sim N(0,1)$, $Y \sim \chi_r^2$. But then $T^2 = \frac{X^2}{Y/r} = \frac{X^2/1}{Y/r} \sim F_{1,r}$, since $X^2 \sim \chi_1^2$ and $X^2$ and $Y$ are independent.

(iii) If $F \sim F_{r_1,r_2}$, then it can be shown (see Exercise 2.11) that

$$EF = \frac{r_2}{r_2 - 2}, \text{ for } r_2 \geq 3, \quad \text{and} \quad \text{Var}(F) = \frac{2r_2^2(r_1 + r_2 - 2)}{r_1(r_2 - 2)^2(r_2 - 4)}, \text{ for } r_2 \geq 5.$$

One can formulate a version of Theorem 4 for $k$ (> 2) r.v.'s $X_1, \ldots, X_k$, as well as a version of Theorem 3. In the following, such versions are formulated for reference purposes.

**Theorem 5.** *Consider the r.v.'s $X_1, \ldots, X_k$ with joint p.d.f. $f_{X_1,\ldots,X_k}$ positive and continuous on the set $S \subseteq \mathfrak{R}^k$, and let $h_1, \ldots, h_k$ be real-valued transformations defined on $S$; that is, $h_1, \ldots, h_k : S \to \mathfrak{R}$, and let $T$ be the image of $S$ under the transformation $(h_1, \ldots, h_k)$. Suppose that $h_1, \ldots, h_k$ are one-to-one from $S$ onto $T$. Thus, if we set $y_i = h_i(x_1, \ldots, x_k)$, $i = 1, \ldots, k$, then we can solve uniquely for $x_i$, $i = 1, \ldots, k : x_i = g_i(y_1, \ldots, y_k)$, $i = 1, \ldots, k$. Suppose further that the partial derivatives $g_{ij}(y_1, \ldots, y_k) = \frac{\partial}{\partial y_j} g_i(y_1, \ldots, y_k), i,j = 1, \ldots, k$ exist and are continuous for $(y_1, \ldots, y_k) \in T$. Finally, suppose that the Jacobian*

$$J = \begin{vmatrix} g_{11}(y_1, \ldots, y_k) & \cdots & g_{1k}(y_1, \ldots, y_k) \\ \cdots & \cdots & \cdots \\ g_{k1}(y_1, \ldots, y_k) & \cdots & g_{kk}(y_1, \ldots, y_k) \end{vmatrix}$$

is $\neq 0$ on $T$. Then the joint p.d.f. of the r.v.'s $Y_i = h_i(X_1, \ldots, X_k)$, $i = 1, \ldots, k, f_{Y_1, \ldots, Y_k}$, is given by:

$$f_{Y_1, \ldots, Y_k}(y_1, \ldots, y_k) = f_{X_1, \ldots, X_k}[g_1(y_1, \ldots, y_k), \ldots, g_k(y_1, \ldots, y_k)]|J|,$$
$$(y_1, \ldots, y_k) \in T \quad (\text{and } = 0 \text{ for } (y_1, \ldots, y_k) \notin T). \tag{12}$$

A suitable version of the previous result when the transformations $h_1, \ldots, h_k$ are not one-to-one is stated below; it will be employed in Theorem 12.

**Theorem 6.** *Let $X_1, \ldots, X_k$ be r.v.'s with joint p.d.f. $f_{X_1, \ldots, X_k}$ positive and continuous on the set $S \subseteq \Re^k$, and let $h_1, \ldots, h_k$ be real-valued transformations defined on $S$; that is, $h_1, \ldots, h_k : S \to \Re$, and let $T$ be the image of $S$ under the transformation $(h_1, \ldots, h_k)$. Suppose that $h_1, \ldots, h_k$ are not one-to-one from $S$ onto $T$ but there is a partition of $S$ into (pairwise disjoint) subsets $S_1, \ldots, S_r$ such that when $(h_1, \ldots, h_k)$ is restricted to $S_j$ and takes values in $T_j$ (the image of $S_j$ under $(h_1, \ldots, h_k)$), $j = 1, \ldots, r$, then $(h_1, \ldots, h_k)$ is one-to-one. Denoting by $(h_{1j}, \ldots, h_{kj})$ this restriction, we have then: $(h_{1j}, \ldots, h_{kj}) : S_j \to T_j$ is one-to-one, so that we can solve uniquely for $x_i$, $i = 1, \ldots, k : x_i = g_{ji}(y_1, \ldots, y_k)$, $i = 1, \ldots, k$, for each $j = 1, \ldots, r$. Suppose further that the partial derivatives $g_{jil}(y_1, \ldots, y_k) = \frac{\partial}{\partial y_l} g_{ji}(y_1, \ldots, y_k)$, $i, l = 1, \ldots, k, j = 1, \ldots, r$ exist and are continuous for $(y_1, \ldots, y_k) \in T_j$, $j = 1, \ldots, r$, and the Jacobian*

$$J_j = \begin{vmatrix} g_{j11}(y_1, \ldots, y_k) & \cdots & g_{j1k}(y_1, \ldots, y_k) \\ \cdots & \cdots & \cdots \\ g_{jk1}(y_1, \ldots, y_k) & \cdots & g_{jkk}(y_1, \ldots, y_k) \end{vmatrix}$$

*is $\neq 0$ on $T_j$ for $j = 1, \ldots, r$.*
  Set

$$f_{Y_j}(y_1, \ldots, y_k) = f_{X_1, \ldots, X_k}[g_{j1}(y_1, \ldots, y_k), \ldots, g_{jk}(y_1, \ldots, y_k)]|J_j|,$$
$$(y_1, \ldots, y_k) \in T_j, \quad j = 1, \ldots, r.$$

*Then the joint p.d.f. of the r.v.'s $Y_i = h_i(X_1, \ldots, X_k)$, $i = 1, \ldots, k$, $f_{Y_1, \ldots, Y_k}$, is given by:*

$$f_{Y_1, \ldots, Y_k}(y_1, \ldots, y_k) = \sum_{j=1}^r \delta_j(y_1, \ldots, y_k) f_{Y_j}(y_1, \ldots, y_k), \quad (y_1, \ldots, y_k) \in T$$
$$(\text{and } = 0 \text{ for } (y_1, \ldots, y_k) \notin T), \tag{13}$$

*where $\delta_j(y_1, \ldots, y_k) = 1$, if $(y_1, \ldots, y_k) \in T_j$ and $\delta_j(y_1, \ldots, y_k) = 0$, if $(y_1, \ldots, y_k) \notin T_j, j = 1, \ldots, r$.*

---

## EXERCISES

**2.1**  The r.v.'s $X$ and $Y$ denote the outcomes of one independent throw of two fair dice, and let $Z = X + Y$. Determine the distribution of $Z$.

**2.2**  Let the independent r.v.'s $X$ and $Y$ have the Negative Exponential distribution with $\lambda = 1$, and set $U = X + Y$, $V = X/Y$.

(i) Derive the joint p.d.f. $f_{U,V}$.
(ii) Then derive the marginal p.d.f.'s $f_U$ and $f_V$.
(iii) Show that the r.v.'s $U$ and $V$ are independent.

**2.3** Let the independent r.v.'s $X$ and $Y$ have the Negative Exponential distribution with $\lambda = 1$, and set $U = \frac{1}{2}(X + Y), V = \frac{1}{2}(X - Y)$.
(i) Show that the joint p.d.f. of the r.v.'s $U$ and $V$ is given by:

$$f_{U,V}(u, v) = 2e^{-2u}, \quad -u < v < u, \ u > 0.$$

(ii) Also, show that the marginal p.d.f.'s $f_U$ and $f_V$ are given by:

$$f_U(u) = 4ue^{-2u}, \ u > 0; \quad f_V(v) = e^{-2v}, \ \text{for} \ v > 0,$$
$$f_V(v) = e^{2v}, \quad \text{for} \ v < 0.$$

**2.4** Let the independent r.v.'s $X$ and $Y$ have the joint p.d.f. $f_{X,Y}$ positive on a set $S$, subset of $\Re^2$, and set $U = aX + b, V = cY + d$, where $a, b, c$, and $d$ are constants with $ac \neq 0$.
(i) Use Theorem 4 in order to show that the joint p.d.f. of $U$ and $V$ is given by:

$$f_{U,V}(u, v) = \frac{1}{|ac|} f_{X,Y}\left(\frac{u - b}{a}, \frac{v - c}{d}\right)$$
$$= \frac{1}{|ac|} f_X\left(\frac{u - b}{c}\right) f_Y\left(\frac{v - c}{d}\right), \quad (u, v) \in T,$$

the image of $S$ under the transformations $u = ax + b, v = cy + d$.
(ii) If $X \sim N(\mu_1, \sigma_1^2)$ and $Y \sim N(\mu_2, \sigma_2^2)$, show that $U$ and $V$ are independently distributed as $N(a\mu_1 + b, (a\sigma_1^2))$ and $N(c\mu_2 + b, (c\sigma_2^2))$, respectively.

**2.5** If the independent r.v.'s $X$ and $Y$ are distributed as $N(0, 1)$, set $U = X + Y, V = X - Y$, and
(i) Determine the p.d.f.'s of $U$ and $V$.
(ii) Show that $U$ and $V$ are independent.
(iii) Compute the probability $P(U < 0, V > 0)$.

**2.6** Let $X$ and $Y$ be independent r.v.'s distributed as $N(0, 1)$, and set

$$U = \frac{1}{\sqrt{2}}(X + Y), \quad V = \frac{1}{\sqrt{2}}(X - Y).$$

(i) Determine the joint p.d.f. of $U$ and $V$.
(ii) From the joint p.d.f. $f_{U,V}$, infer $f_U$ and $f_V$ without integration.
(iii) Conclude that $U$ and $V$ are also independent.
(iv) How else could you arrive at the p.d.f.'s $f_U$ and $f_V$?

**2.7** Let $X$ and $Y$ be independent r.v.'s distributed as $N(0, \sigma^2)$. Then show that the r.v. $V = X^2 + Y^2$ has the Negative Exponential distribution with parameter $\lambda = 1/2\sigma^2$.
**Hint.** Use Exercise 1.9 in Chapter 6 and Theorem 5 in Chapter 5.

**2.8**  The independent r.v.'s $X$ and $Y$ have a joint p.d.f. given by: $f_{X,Y}(x, y) = \frac{1}{\pi}$, for
$x, y \in \Re$ with $x^2 + y^2 \leq 1$, and let $Z^2 = X^2 + Y^2$. Use polar coordinates to
determine the p.d.f. $f_{Z^2}$.
**Hint.** Let $Z = +\sqrt{Z^2}$ and set $X = Z \cos \Theta$, $Y = Z \sin \Theta$, where $Z \geq 0$ and
$0 < \Theta \leq 2\pi$. First, determine the joint p.d.f. $f_{Z,\Theta}$ and then the marginal p.d.f.
$f_Z$. Finally, by means of $f_Z$ and the transformation $U = Z^2$, determine the
p.d.f. $f_U = f_{Z^2}$.

**2.9**  Let $X_r$ be a r.v. distributed as $t$ with $r$ degrees of freedom: $X_r \sim t_r$
$(r = 1, 2, \ldots)$ whose p.d.f. is given in relation (10). Then show that:
  **(i)** $EX_r$ does not exist for $r = 1$.
  **(ii)** $EX_r = 0$ for $r \geq 2$.
  **(iii)** $Var(X_r) = \frac{r}{r-2}$ for $r \geq 3$.
**Hint.** That $EX_r$ does not exist for $r = 1$ is, actually, reduced to Exercise 1.16
in Chapter 3. That $EX_r = 0$ for $r \geq 2$ follows by a simple integration. So, all
that remains to calculate is $EX_r^2$. For this purpose, first reduce the original
integral to an integral over the interval $(0, \infty)$, by symmetry of the region of
integration and the fact that the integrand is an even function. Then, use the
transformation $\frac{t^2}{r} = x$, and next the transformation $\frac{1}{1+x} = y$. Except for
constants, the integral is then reduced to the form

$$\int_0^1 y^{\alpha-1}(1-y)^{\beta-1}dy \quad (\alpha > 0, \beta > 0).$$

At this point, use the following fact:

$$\int_0^1 y^{\alpha-1}(1-y)^{\beta-1}dy = \frac{\Gamma(\alpha)\Gamma(\beta)}{\Gamma(\alpha + \beta)}.$$

(A proof of this fact may be found, e.g., in Section 3.3.6, Chapter 3, of the
book *A Course in Mathematical Statistics*, 2nd edition (1997), Academic
Press, by G. G. Roussas.) The proof is concluded by using the recursive
relation of the Gamma function $(\Gamma(\gamma) = (\gamma - 1)\Gamma(\gamma - 1))$ and the fact that
$\Gamma(\frac{1}{2}) = \sqrt{\pi}$.

**2.10**  If the r.v. $X_r \sim t_r$, then the $t$-tables (at least the ones in this book) do not give
probabilities for $r > 90$. For such values, we can use instead the Normal
tables. The reason for this is that the p.d.f. of $X_r$ converges to the p.d.f. of the
$N(0, 1)$ distribution as $r \to \infty$. More precisely,

$$f_{X_r}(t) = \frac{\Gamma\left(\frac{r+1}{2}\right)}{\sqrt{\pi r}\Gamma\left(r/2\right)} \times \frac{1}{\left(1 + \frac{t^2}{r}\right)^{(r+1)/2}} \xrightarrow[r \to \infty]{} \frac{1}{\sqrt{2\pi}}e^{-t^2/2} \quad (t > 0).$$

**Hint.** In proving this convergence, first observe that

$$\left(1 + \frac{t^2}{r}\right)^{(r+1)/2} = \left[\left(1 + \frac{t^2}{r}\right)^r\right]^{1/2} \times \left(1 + \frac{t^2}{r}\right)^{1/2} \xrightarrow[r \to \infty]{} e^{t^2/2},$$

and then show that

$$\frac{\Gamma\left(\frac{r+1}{2}\right)}{\Gamma(r/2)} \xrightarrow[r \to \infty]{} \frac{1}{\sqrt{2}},$$

by utilizing the *Stirling formula*. This formula states that:

$$\frac{\Gamma(n)}{\sqrt{2\pi} n^{(2n-1)/2} e^{-n}} \to 1 \quad \text{as } n \to \infty.$$

**2.11** Let $X_{r_1,r_2}$ be a r.v. having the $F$-distribution with d.f. $r_1$ and $r_2$; i.e., $X_{r_1,r_2} \sim F_{r_1,r_2}$. Then show that:

$$EX_{r_1,r_2} = \frac{r_2}{r_2 - 2}, \quad r_2 \geq 3; \quad \text{Var}(X_{r_1,r_2}) = \frac{2r_2^2(r_1 + r_2 - 2)}{r_1(r_2 - 2)^2(r_2 - 4)}, \quad r_2 \geq 5.$$

**Hint.** Start out with the $k$th moment $EX_{r_1,r_2}^k$, use first the transformation $\frac{r_1}{r_2}f = x$, and next the transformation $\frac{1}{1+x} = y$. Then observe that the integral is expressed in terms of the Gamma function (see Hint in Exercise 2.9). Thus, the $EX_{r_1,r_2}^k$ is expressed in terms of the Gamma function without carrying out any integrations. Specifically, we find:

$$EX_{r_1,r_2}^k = \left(\frac{r_2}{r_1}\right)^k \frac{\Gamma\left(\frac{r_1+2k}{2}\right)\Gamma\left(\frac{r_2-2k}{2}\right)}{\Gamma(r_1/2)\Gamma(r_2/2)}, \quad r_2 > 2k.$$

Applying this formula for $k = 1$ (which requires that $r_2 \geq 3$), and $k = 2$ (which requires that $r_2 \geq 5$), and using the recursive property of the Gamma function, we determine the required expressions.

**2.12** Let $Z_i$, $i = 1, \ldots, 4$, be independent r.v.'s distributed as $N(0, 1)$, and set: $X = Z_1^2 + Z_2^2$, $Y = Z_3^2 + Z_4^2$, and $U = \frac{X}{X+Y}$. Show that $U \sim U(0, 1)$.
**Hint.** You may wish to write $U = \frac{1}{1+\frac{Y}{X}}$, where $\frac{Y}{X} = \frac{Y/2}{X/2} = V \sim F_{2,2}$.

**2.13** If the r.v.'s $X$ and $Y$ have the joint p.d.f. given by:

$$f_{X,Y}(x, y) = x + y, \quad 0 < x < 1, 0 < y < 1,$$

then:
  **(i)** Set $U = XY$ and $V = Y$ and show that their joint p.d.f. is given by

$$f_{U,V}(u, v) = \frac{u}{v^2} + 1, \quad 0 < u < v < 1.$$

  **(ii)** From part (i), derive the p.d.f. $f_U$.

**2.14** If the r.v.'s $X$, $Y$, and $Z$ have the joint p.d.f.

$$f_{X,Y,Z}(x, y, z) = 8xyz, \quad 0 < x, y, z < 1,$$

then:
  **(i)** Set $U = X$, $V = XY$, and $W = XYZ$, and show that the joint p.d.f. of $U$, $V$, $W$ is given by:

$$f_{U,V,W}(u, v, w) = \frac{8w}{uv}, \quad 0 < w < v < u < 1.$$

  **(ii)** Verify that $f_{U,V,W}$ is, indeed, a p.d.f.

**2.15** The r.v.'s $X$ and $Y$ are jointly Uniformly distributed over the circle of radius $R$; i.e., their joint p.d.f. is given by:

$$f_{X,Y}(x, y) = \frac{1}{\pi R^2}, \quad x^2 + y^2 \le R^2 \quad \text{(and 0 elsewhere)}.$$

  **(i)** If a point is selected at random within this circle, what is the probability that the point will be no further than $R/2$ from the point $(0, 0)$?
  **(ii)** Compute the probability: $P(|X| \le R/2, |Y| \le R/2)$.
  **(iii)** Derive the marginal p.d.f.'s $f_X$, $f_Y$.

**2.16** Let $X$, $Y$, $Z$ be r.v.'s with joint d.f. given by:

$$F(x, y, z) = \left(1 - \frac{1}{1+x}\right) \times \left(1 - \frac{1}{1+y}\right) \times \left(1 - \frac{1}{1+z}\right)$$

$$\times \left(1 - \frac{1}{1+x+y+z}\right),$$

for $x, y, z > 0$, and 0 otherwise.
  **(i)** Find the marginal d.f.'s $F_X$, $F_Y$, $F_Z$, and the marginal p.d.f.'s $f_X$, $f_Y$, $f_Z$.
  **(ii)** Find the marginal joint d.f.'s $F_{X,Y}$, $F_{X,Z}$, $F_{Y,Z}$, and the marginal joint p.d.f.'s $f_{X,Y}$, $f_{X,Z}$, $f_{Y,Z}$.
  **(iii)** Show that any two of the r.v.'s $X$, $Y$, and $Z$ are independent, but taken all three are not independent.
  **(iv)** Compute the following probabilities:
    **(a)** $P(X \le 1)$;
    **(b)** $P(X \le 1, Y \le 1)$;
    **(c)** $P(X \le 1, Y \le 1, Z \le 1)$;
    **(d)** $P(X > 1)$;
    **(e)** $P(X \le 1, Y > 1)$, $P(X > 1, Y \le 1)$;
    **(f)** $P(X > 1, Y > 1)$;
    **(g)** $P(X \le 1, Y \le 1, Z > 1)$, $P(X \le 1, Y > 1, Z \le 1)$, $P(X > 1, Y \le 1, Z \le 1)$;
    **(h)** $P(X \le 1, Y > 1, Z > 1)$, $P(X > 1, Y \le 1, Z > 1)$, $P(X > 1, Y > 1, Z \le 1)$;
    **(i)** $P(X > 1, Y > 1, Z > 1)$.

## 6.3 LINEAR TRANSFORMATIONS

In this section, a brief discussion is presented for a specific kind of transformation, linear transformations. The basic concepts and results used here can be found in any textbook on linear algebra.

**Definition 3.** Suppose the variables $x_1, \ldots, x_k$ are transformed into the variables $y_1, \ldots, y_k$ in the following manner:

$$y_i = \sum_{j=1}^{k} c_{ij} x_j, \quad i = 1, \ldots, k, \tag{14}$$

where the $c_{ij}$'s are real constants. Such a transformation is called a *linear transformation* (all the $x_i$'s enter into the transformation in a linear way, in the first power).

Some terminology and elementary facts from matrix algebra will be used here. Denote by $\mathbf{C}$ the $k \times k$ matrix of the $c_{ij}$, $i, j = 1, \ldots, k$ constants; that is, $\mathbf{C} = (c_{ij})$, and by $|\mathbf{C}|$ or $\Delta$ its determinant. Then it is well known that if $\Delta \neq 0$, one can uniquely solve for $x_i$ in (14):

$$x_i = \sum_{j=1}^{k} d_{ij} y_j, \quad i = 1, \ldots, k, \tag{15}$$

for suitable constants $d_{ij}$. Denote by $\mathbf{D}$ the $k \times k$ matrix of the $d_{ij}$'s and by $\Delta^*$ its determinant: $\mathbf{D} = (d_{ij})$, $\Delta^* = |\mathbf{D}|$. Then it is known that $\Delta^* = 1/\Delta$. Among the linear transformations, a specific class is of special importance; it is the class of orthogonal transformations.

A linear transformation is said to be *orthogonal*, if

$$\sum_{j=1}^{k} c_{ij}^2 = 1 \quad \text{and} \quad \sum_{j=1}^{k} c_{ij} c_{i'j} = 0, \quad i, i' = 1, \ldots, k, \ i \neq i'; \tag{16}$$

or, equivalently (see also Exercise 3.9),

$$\sum_{i=1}^{k} c_{ij}^2 = 1 \quad \text{and} \quad \sum_{i=1}^{k} c_{ij} c_{ij'} = 0, \quad j, j' = 1, \ldots, k, \ j \neq j'. \tag{16'}$$

Relations (16) and (16') simply state that the row (column) vectors of the matrix $\mathbf{C}$ have norm (length) 1, and any two of them are perpendicular. The matrix $\mathbf{C}$ itself is also called *orthogonal*. For an orthogonal matrix $\mathbf{C}$, it is known that $|\mathbf{C}| = \pm 1$. Also, in the case of an orthogonal matrix $\mathbf{C}$, it happens that $d_{ij} = c_{ji}$, $i, j = 1, \ldots, k$; or in matrix notation: $\mathbf{D} = \mathbf{C}'$, where $\mathbf{C}'$ is the *transpose* of $\mathbf{C}$ (the rows of $\mathbf{C}'$ are the same as the columns of $\mathbf{C}$). Thus, in this case:

$$x_i = \sum_{j=1}^{k} c_{ji} y_j, \quad i = 1, \ldots, k. \tag{17}$$

Also, under orthogonality, the vectors of the $x_i$'s and $y_j$'s have the same norm. To put it differently:

$$\sum_{i=1}^{k} x_i^2 = \sum_{j=1}^{k} y_j^2. \tag{18}$$

For example, the transformations $y_1 = x_1 + x_2$ and $y_2 = x_1 - x_2$ are linear but not orthogonal, whereas the transformations $y_1 = \frac{1}{\sqrt{2}}x_1 + \frac{1}{\sqrt{2}}x_2$ and $y_2 = \frac{1}{\sqrt{2}}x_1 - \frac{1}{\sqrt{2}}x_2$ are both linear and orthogonal. Also, observe that $\begin{vmatrix} 1/\sqrt{2} & 1/\sqrt{2} \\ 1/\sqrt{2} & -1/\sqrt{2} \end{vmatrix} = -1$, and $y_1^2 + y_2^2 = x_1^2 + x_2^2$.

Some of these concepts and results are now to be used in connection with r.v.'s.

**Theorem 7.** *Suppose the r.v.'s $X_1, \ldots, X_k$ are transformed into the r.v.'s $Y_1, \ldots, Y_k$ through a linear transformation with the matrix $C = (c_{ij})$ and $|C| = \Delta \neq 0$. Let $S \subseteq \Re^k$ be the set over which the joint p.d.f. of $X_1, \ldots, X_k, f_{X_1,\ldots,X_k}$, is positive, and let $T$ be the image of $S$ under the linear transformation. Then:*

**(i)** *The joint p.d.f. of $Y_1, \ldots, Y_k, f_{Y_1,\ldots,Y_k}$, is given by:*

$$f_{Y_1,\ldots,Y_k}(y_1,\ldots,y_k) = f_{X_1,\ldots,X_k}\left(\sum_{j=1}^{k} d_{1j}y_j, \ldots, \sum_{j=1}^{k} d_{kj}y_j\right)\frac{1}{|\Delta|}, \tag{19}$$

*for $(y_1, \ldots, y_k) \in T$ (and $= 0$ otherwise), where the $d_{ij}$'s are as in (15).*

**(ii)** *In particular, if $C$ is orthogonal, then:*

$$f_{Y_1,\ldots,Y_k}(y_1,\ldots,y_k) = f_{X_1,\ldots,X_k}\left(\sum_{j=1}^{k} c_{j1}y_j, \ldots, \sum_{j=1}^{k} c_{jk}y_j\right), \tag{20}$$

*for $(y_1, \ldots, y_k) \in T$ (and $= 0$ otherwise); also,*

$$\sum_{j=1}^{k} Y_j^2 = \sum_{i=1}^{k} X_i^2. \tag{21}$$

*Proof.*

**(i)** Relation (19) follows from Theorem 5.

**(ii)** Relation (20) follows from (19) and (17), and (21) is a restatement of (18). ∎

Next, we specialize this result to the case that the r.v.'s $X_1, \ldots, X_k$ are Normally distributed and independent.

**Theorem 8.** *Let the independent r.v.'s $X_1, \ldots, X_k$ be distributed as follows: $X_i \sim N(\mu_i, \sigma^2)$, $i = 1, \ldots, k$, and suppose they are transformed into the r.v.'s $Y_1, \ldots, Y_k$ by means of an orthogonal transformation $C$. Then the r.v.'s $Y_1, \ldots, Y_k$ are also independent and Normally distributed as follows:*

$$Y_i \sim N\left(\sum_{j=1}^{k} c_{ij}\mu_j, \sigma^2\right), \quad i = 1, \ldots, k. \tag{22}$$

*Furthermore,*

$$\sum_{j=1}^{k} Y_j^2 = \sum_{i=1}^{k} X_i^2.$$

*Proof.* From the transformations $Y_i = \sum_{j=1}^{k} c_{ij}X_j$, it is immediate that each $Y_i$ is Normally distributed with mean $EY_i = \sum_{j=1}^{k} c_{ij}\mu_j$ and variance $\mathrm{Var}(Y_i) = \sum_{j=1}^{k} c_{ij}^2\sigma^2 = \sigma^2 \sum_{j=1}^{k} c_{ij}^2 = \sigma^2$. So the only thing to be justified is the assertion of independence. From the normality assumption on the $X_i$'s, we have:

$$f_{X_1,\ldots,X_k}(x_1,\ldots,x_k) = \left(\frac{1}{\sqrt{2\pi}\sigma}\right)^k \exp\left[-\frac{1}{2\sigma^2}\sum_{i=1}^{k}(x_i - \mu_i)^2\right]. \tag{23}$$

Then, since $\mathbf{C}$ is orthogonal, (20) applies and gives, by means of (23):

$$f_{Y_1,\ldots,Y_k}(y_1,\ldots,y_k) = \left(\frac{1}{\sqrt{2\pi}\sigma}\right)^k \exp\left[-\frac{1}{2\sigma^2}\sum_{i=1}^{k}\left(\sum_{j=1}^{k}c_{ji}y_j - \mu_i\right)^2\right]. \tag{24}$$

Thus, the proof is completed by establishing the following algebraic relation:

$$\sum_{i=1}^{k}\left(\sum_{j=1}^{k}c_{ji}y_j - \mu_i\right)^2 = \sum_{i=1}^{k}\left(y_i - \sum_{j=1}^{k}c_{ij}\mu_j\right)^2 \tag{25}$$

(see Exercise 3.1). ∎

Finally, suppose the orthogonal matrix $\mathbf{C}$ in Theorem 8 is chosen to be as follows:

$$\mathbf{C} = \begin{pmatrix} 1/\sqrt{k} & 1/\sqrt{k} & \cdots & \cdots & \cdots & 1/\sqrt{k} \\ 1/\sqrt{2 \times 1} & -1/\sqrt{2 \times 1} & 0 & \cdots & \cdots & 0 \\ 1/\sqrt{3 \times 2} & 1/\sqrt{3 \times 2} & -2/\sqrt{3 \times 2} & 0 & \cdots & 0 \\ \cdots & \cdots & \cdots & \cdots & \cdots & \cdots \\ 1/\sqrt{k(k-1)} & 1/\sqrt{k(k-1)} & \cdots & \cdots & 1/\sqrt{k(k-1)} & -(k-1)/\sqrt{k(k-1)} \end{pmatrix}.$$

That is, the elements of $\mathbf{C}$ are given by the expressions:

$$c_{1j} = 1/\sqrt{k}, \quad j = 1, \ldots, k,$$

$$c_{ij} = 1/\sqrt{i(i-1)}, \quad \text{for } i = 2, \ldots, k \text{ and } j = 1, \ldots, i-1,$$

$$\text{and } 0 \text{ for } j = i+1, \ldots, k,$$

$$c_{ii} = -(i-1)/\sqrt{i(i-1)}, \quad i = 2, \ldots, k.$$

From these expressions, it readily follows that $\sum_{j=1}^{k} c_{ij}^2 = 1$ for all $i = 1, \ldots, k$, and $\sum_{j=1}^{k} c_{ij}c_{i'j} = 0$ for all $i, i' = 1, \ldots, k$, with $i \neq i'$, so that $\mathbf{C}$ is, indeed, orthogonal (see also Exercise 3.2). By means of $\mathbf{C}$ and Theorem 6, we may now establish the following result.

**Lemma 1.** *If $Z_1, \cdots, Z_k$ are independent r.v.'s distributed as $N(0, 1)$, then $\bar{Z}$ and $\sum_{i=1}^{k}(Z_i - \bar{Z})^2$ are independent, where $\bar{Z} = \frac{1}{k}\sum_{i=1}^{k} Z_i$.*

*Proof.* Transfrom $Z_1, \ldots, Z_k$ into the r.v.'s $Y_1, \ldots, Y_k$ by means of $\mathbf{C}$; that is,

$$Y_1 = \frac{1}{\sqrt{k}}Z_1 + \frac{1}{\sqrt{k}}Z_2 + \cdots + \frac{1}{\sqrt{k}}Z_k,$$

$$Y_2 = \frac{1}{\sqrt{2 \times 1}}Z_1 - \frac{1}{\sqrt{2 \times 1}}Z_2,$$

$$Y_3 = \frac{1}{\sqrt{3 \times 2}}Z_1 + \frac{1}{\sqrt{3 \times 2}}Z_2 - \frac{2}{\sqrt{3 \times 2}}Z_3,$$

$$\vdots$$

$$Y_k = \frac{1}{\sqrt{k(k-1)}}Z_1 + \frac{1}{\sqrt{k(k-1)}}Z_2 + \cdots + \frac{1}{\sqrt{k(k-1)}}Z_{k-1} - \frac{k-1}{\sqrt{k(k-1)}}Z_k.$$

Then, by Theorem 8, the r.v.'s $Y_1, \ldots, Y_k$ are independently distributed as $N(0, 1)$, whereas by (21),

$$\sum_{j=1}^{k} Y_j^2 = \sum_{i=1}^{k} Z_i^2.$$

However, $Y_1 = \sqrt{k}\bar{Z}$, so that

$$\sum_{j=2}^{k} Y_j^2 = \sum_{j=1}^{k} Y_j^2 - Y_1^2 = \sum_{i=1}^{k} Z_i^2 - (\sqrt{k}\bar{Z})^2 = \sum_{i=1}^{k} Z_i^2 - k\bar{Z}^2 = \sum_{i=1}^{k}(Z_i - \bar{Z})^2.$$

On the other hand, $\sum_{j=2}^{k} Y_j^2$ and $Y_1$ are independent; equivalently, $\sum_{i=1}^{k}(Z_i - \bar{Z})^2$ and $\sqrt{k}\bar{Z}$ are independent, or

$$\bar{Z} \quad \text{and} \quad \sum_{i=1}^{k}(Z_i - \bar{Z})^2 \quad \text{are independent.} \tag{26}$$

$\blacksquare$

This last conclusion is now applied as follows.

**Theorem 9.** *Let $X_1, \ldots, X_k$ be independent r.v.'s distributed as $N(\mu, \sigma^2)$. Then the sample mean $\bar{X} = \frac{1}{k}\sum_{i=1}^{k} X_i$ and the sample variance $S^2 = \frac{1}{k}\sum_{i=1}^{k}(X_i - \bar{X})^2$ are independent.*

*Proof.* The assumption that $X_i \sim N(\mu, \sigma^2)$ implies that $\frac{X_i - \mu}{\sigma} \sim N(0, 1)$. By setting $Z_i = (X_i - \mu)/\sigma$, $i = 1, \ldots, k$, the $Z_i$'s are as in Lemma 1 and therefore (26) applies. Since

$$\bar{Z} = \frac{1}{k} \sum_{i=1}^{k} \left( \frac{X_i - \mu}{\sigma} \right) = \frac{1}{\sigma}(\bar{X} - \mu), \quad \text{and}$$

$$\sum_{i=1}^{k} (Z_i - \bar{Z})^2 = \sum_{i=1}^{k} \left( \frac{X_i - \mu}{\sigma} - \frac{\bar{X} - \mu}{\sigma} \right)^2 = \frac{1}{\sigma^2} \sum_{i=1}^{k} (X_i - \bar{X})^2,$$

it follows that $\frac{1}{\sigma}(\bar{X} - \mu)$ and $\frac{1}{\sigma^2} \sum_{i=1}^{k} (X_i - \bar{X})^2$ are independent or that $\bar{X}$ and $\frac{1}{k} \sum_{i=1}^{k} (X_i - \bar{X})^2$ are independent. ∎

---

## EXERCISES

**3.1**  Establish relation (25) in the proof of Theorem 8.
**Hint.** Expand the left-hand side and the right-hand side in (25), use orthogonality, and show that the common value of both sides is:

$$\sum_{j=1}^{k} y_j^2 + \sum_{j=1}^{k} \mu_j^2 - 2 \sum_{j=1}^{k} \sum_{i=1}^{k} c_{ji} \mu_i y_j.$$

**3.2**  Show that the matrix with row elements given by:

$$c_{1j} = 1/\sqrt{k}, \quad j = 1, \ldots, k,$$
$$c_{ij} = 1/\sqrt{i(i-1)}, \quad i = 2, \ldots, k \text{ and } j = 1, \ldots, i-1,$$
$$\text{and } 0 \text{ for } j = i+1, \ldots, k,$$
$$c_{ii} = -(i-1)/\sqrt{i(i-1)}, \quad i = 2, \ldots, k \text{ is orthogonal.}$$

**3.3**  Let $X_1, X_2$, and $X_3$ be independent r.v.'s such that $X_i \sim N(\mu_i, \sigma^2)$, $i = 1, 2, 3$, and set

$$Y_1 = -\frac{1}{\sqrt{2}} X_1 + \frac{1}{\sqrt{2}} X_2,$$

$$Y_2 = -\frac{1}{\sqrt{3}} X_1 - \frac{1}{\sqrt{3}} X_2 + \frac{1}{\sqrt{3}} X_3,$$

$$Y_3 = \frac{1}{\sqrt{6}} X_1 + \frac{1}{\sqrt{6}} X_2 + \frac{2}{\sqrt{6}} X_3.$$

Then
**(i)**  Show that the r.v.'s $Y_1, Y_2$, and $Y_3$ are independent Normally distributed with variance $\sigma^2$ and respective means:

$$EY_1 = \frac{1}{\sqrt{2}}(-\mu_1 + \mu_2), \quad EY_2 = \frac{1}{\sqrt{3}}(-\mu_1 - \mu_2 + \mu_3),$$

$$EY_3 = \frac{1}{\sqrt{6}}(\mu_1 + \mu_2 + 2\mu_3).$$

(ii) If $\mu_1 = \mu_2 = \mu_3 = 0$, then show that $\frac{1}{\sigma^2}(Y_1^2 + Y_2^2 + Y_3^2) \sim \chi_3^2$,
**Hint.** For part (i), prove that the transformation employed is orthogonal and then use Theorem 8 to conclude independence of $Y_1, Y_2, Y_3$. That the means and the variance are as described follows either from Theorem 8 or directly. Part (ii) follows from part (i) and the assumption that $\mu_1 = \mu_2 = \mu_3 = 0$.

**3.4** If the r.v.'s $X$ and $Y$ have the Bivariate Normal distribution with parameters $\mu_1, \mu_2, \sigma_1^2, \sigma_2^2$, and $\rho$, then show that the r.v.'s $U = \frac{X-\mu_1}{\sigma_1}, V = \frac{Y-\mu_2}{\sigma_2}$ have the Bivariate Normal distribution with parameters $0, 0, 1, 1$, and $\rho$; and vice versa.
**Hint.** For the converse part, you just reverse the process.

**3.5** If the r.v.'s $X$ and $Y$ have the Bivariate Normal distribution with parameters $0, 0, 1, 1$, and $\rho$, then the r.v.'s $cX$ and $dY$ have the Bivariate Normal distribution with parameters $0, 0, c^2, d^2$, and $\rho_0$, where $\rho_0 = \rho$ if $cd > 0$, and $\rho_0 = -\rho$ if $cd < 0$; $c$ and $d$ are constants with $cd \neq 0$.

**3.6** Let the r.v.'s $X$ and $Y$ have the Bivariate Normal distribution with parameters $0, 0, 1, 1$, and $\rho$, and set: $U = X + Y, V = X - Y$. Then show that:
   (i) The r.v.'s $U$ and $V$ also have the Bivariate Normal distribution with parameters $0, 0, 2(1 + \rho), 2(1 - \rho)$, and $0$.
   (ii) From part (i), conclude that the r.v.'s $U$ and $V$ are independent.
   (iii) From part (i), also conclude that:
   $U \sim N(0, 2(1 + \rho)), V \sim N(0, 2(1 - \rho))$.

**3.7** Let the r.v.'s $X$ and $Y$ have the Bivariate Normal distribution with parameters $\mu_1, \mu_2, \sigma_1^2, \sigma_2^2$, and $\rho$, and set:
$$U = \frac{X - \mu_1}{\sigma_1}, \quad V = \frac{Y - \mu_2}{\sigma_2}.$$
Then
   (i) Determine the joint distribution of the r.v.'s $U$ and $V$.
   (ii) Show that $U + V$ and $U - V$ have the Bivariate Normal distribution with parameters $0, 0, 2(1 + \rho), 2(1 - \rho)$, and $0$ and are independent. Also, $U + V \sim N(0, 2(1 + \rho)), U - V \sim N(0, 2(1 - \rho))$.
   (iii) For $\sigma_1^2 = \sigma_2^2 = \sigma^2$, say, conclude that the r.v.'s $X + Y$ and $X - Y$ are independent.
**Remark.** Actually, the converse of part (iii) is also true; namely, if $X$ and $Y$ have the Bivariate Normal distribution $N(\mu_1, \mu_2, \sigma_1^2, \sigma_2^2, \rho)$, then independence of $X + Y$ and $X - Y$ implies $\sigma_1^2 = \sigma_2^2$. The justification of this statement is easier by means of m.g.f.'s, and it was, actually, discussed in Exercise 5.13 of Chapter 4.

**3.8** Let the independent r.v.'s $X_1, \ldots, X_n$ be distributed as $N(\mu, \sigma^2)$ and suppose that $\mu = k\sigma \ (k > 0)$. Set
$$\bar{X} = \frac{1}{n}\sum_{i=1}^{n} X_i, \quad S^2 = \frac{1}{n-1}\sum_{i=1}^{n}(X_i - \bar{X})^2.$$

Then

(i) Determine an expression for the probability:

$$P(a\mu < \bar{X} < b\mu, \quad 0 < S^2 < c\sigma^2),$$

where $a, b$, and $c$ are constants, $a < b$ and $c > 0$.

(ii) Give the numerical value of the probability in part (i) if
$a = \frac{1}{2}$, $b = \frac{3}{2}$, $c = 1.487$, $k = 1.5$, and $n = 16$.

**Hint.** Use independence of $\bar{X}$ and $S^2$ provided by Theorem 9. Also, use the fact that $\frac{(n-1)S^2}{\sigma^2} \sim \chi^2_{n-1}$ by Theorem 6 in Chapter 5 (where $S^2$ is denoted by $\bar{S}^2$).

**3.9**    Use matrix notation (and usual properties of matrices) in order to establish equivalence of relations (16) and (16′).

## 6.4 THE PROBABILITY INTEGRAL TRANSFORM

In this short section, a very special type of transformation is considered, the so-called *probability integral transform*. By means of this transformation, two results are derived. Roughly, these results state that, if $X \sim F$ and $Y = F(X)$, then, somewhat surprisingly, $Y$ is always distributed as $U(0, 1)$. Furthermore, for a given d.f. $F$, there is always a r.v. $X \sim F$; this r.v. is given by $X = F^{-1}(Y)$, where $Y \sim U(0, 1)$ and $F^{-1}$ is the inverse function of $F$. To facilitate the derivations, $F$ will be assumed to be (strictly) increasing.

**Theorem 10.**    *For a continuous and (strictly) increasing d.f. $F$, let $X \sim F$ and set $Y = F(X)$. Then $Y \sim U(0, 1)$.*

*Proof.*  Since $0 \le F(X) \le 1$, it suffices to consider $y \in [0, 1]$. Then

$$P(Y \le y) = P[F(X) \le y] = P\{F^{-1}[F(X)] \le F^{-1}(y)\}$$
$$= P[X \le F^{-1}(y)] = F[F^{-1}(y)] = y,$$

so that $Y \sim U(0, 1)$.  ∎

**Theorem 11.**    *Let $F$ be a given continuous and (strictly) increasing d.f., and let the r.v. $Y \sim U(0, 1)$. Define the r.v. $X$ by: $X = F^{-1}(Y)$. Then $X \sim F$.*

*Proof.*  For $x \in \Re$,

$$P(X \le x) = P[F^{-1}(Y) \le x] = P\{F[F^{-1}(Y)] \le F(x)\}$$
$$= P[Y \le F(x)] = F(x),$$

as was to be seen.  ∎

In the form of a verification of Theorems 10 and 11, consider the following simple examples.

**Example 6.** Let the r.v. $X$ have the Negative Exponential distribution with parameter $\lambda$. Then, for $x > 0$, $F(x) = 1 - e^{-\lambda x}$. Let $Y$ be defined by: $Y = 1 - e^{-\lambda X}$. Then $Y$ should be $\sim U(0, 1)$.

**Discussion.** Indeed, for $0 < y < 1$,

$$P(Y \leq y) = P(1 - e^{-\lambda X} \leq y) = P(e^{-\lambda X} \geq 1 - y) = P[-\lambda X \geq \log(1 - y)]$$

$$= P\left[X \leq -\frac{1}{\lambda}\log(1 - y)\right]$$

$$= 1 - \exp\left\{(-\lambda)\left[-\frac{1}{\lambda}\log(1 - y)\right]\right\}$$

$$= 1 - \exp[\log(1 - y)] = 1 - (1 - y) = y,$$

as was to be seen.

**Example 7.** Let $F$ be the d.f. of the Negative Exponential distribution with parameter $\lambda$, so that $F(x) = 1 - e^{-\lambda x}$, $x > 0$. Let $y = 1 - e^{-\lambda x}$ and solve for $x$ to obtain $x = -\frac{1}{\lambda}\log(1 - y)$, $0 < y < 1$. Let $Y \sim U(0, 1)$ and define the r.v. $X$ by: $X = -\frac{1}{\lambda}\log(1 - Y)$. Then $X$ should have the d.f. $F$.

**Discussion.** Indeed,

$$P(X \leq x) = P\left[-\frac{1}{\lambda}\log(1 - Y) \leq x\right] = P[\log(1 - Y) \geq -\lambda x]$$

$$= P(1 - Y \geq e^{-\lambda x}) = P(Y \leq 1 - e^{-\lambda x}) = 1 - e^{-\lambda x},$$

as was to be seen.

---

## EXERCISES

**4.1**   (i) Let $X$ be a r.v. with continuous and (strictly) increasing d.f. $F$, and define the r.v. $Y$ by $Y = F(X)$. Then use Theorem 2 in order to show that $Z = -2 \log(1 - Y) \sim \chi_2^2$.

(ii) If $X_1, \ldots, X_n$ is a random sample with d.f. $F$ as described in part (i) and if $Y_i = F(X_i)$, $i = 1, \ldots, n$, then show that the r.v. $U = \sum_{i=1}^{n} Z_i \sim \chi_{2n}^2$, where $Z_i = -2 \log(1 - Y_i)$, $i = 1, \ldots, n$.

**Hint.** For part (i), use Theorem 10, according to which $Y \sim U(0, 1)$. For part (ii), use part (i) and Theorem 5 in Chapter 5.

---

## 6.5 ORDER STATISTICS

In this section, an unconventional kind of transformation is considered, which, when applied to r.v.'s, leads to the so-called *order statistics*. For the definition of the transformation, consider $n$ distinct numbers $x_1, \ldots, x_n$ and order them in ascending order. Denote by $x_{(1)}$ the smallest number: $x_{(1)} = $ smallest of $x_1, \ldots, x_n$; by $x_{(2)}$ the second smallest, and so on until $x_{(n)}$ is the $n$th smallest or, equivalently, the largest of the $x_i$'s. In a summary form, we write: $x_{(j)} =$ the $j$th smallest of the numbers $x_1, \ldots, x_n$, where $j = 1, \ldots, n$. Then, clearly, $x_{(1)} < x_{(2)} < \cdots < x_{(n)}$. For simplicity, set $y_j = x_{(j)}$, $j = 1, \ldots, n$, so that again $y_1 < y_2 < \cdots < y_n$. The transformation under consideration is the one which transforms the $x_i$'s into the $y_j$'s in the way just described.

This transformation now applies to $n$ r.v.'s as follows.

Let $X_1, X_2, \ldots, X_n$ be i.i.d. r.v.'s with d.f. $F$. The *$j$th order statistic of* $X_1, X_2, \ldots, X_n$ is denoted by $X_{(j)}$, or $Y_j$ for easier writing, and is defined as follows:

$$Y_j = j\text{th smallest of the } X_1, X_2, \ldots, X_n, \quad j = 1, \ldots, n,$$

(i.e., for each $s \in \mathcal{S}$, look at $X_1(s), X_2(s), \ldots, X_n(s)$, and then $Y_j(s)$ is defined to be the $j$th smallest among the numbers $X_1(s), X_2(s), \ldots, X_n(s), j = 1, 2, \ldots, n$). It follows that $Y_1 \leq Y_2 \leq \cdots \leq Y_n$, and, in general, the $Y_j$'s are not independent.

We assume now that the $X_i$'s are of the continuous type with p.d.f. $f$ such that $f(x) > 0, (-\infty \leq)a < x < b(\leq \infty)$ and zero otherwise. One of the problems we are concerned with is that of finding the joint p.d.f. of the $Y_j$'s. By means of Theorem 6, it will be established that:

**Theorem 12.** *If $X_1, \ldots, X_n$ are i.i.d. r.v.'s with p.d.f. $f$ which is positive and continuous for $-\infty \leq a < x < b \leq \infty$ and 0 otherwise, then the joint p.d.f. of the order statistics $Y_1, \ldots, Y_n$ is given by:*

$$g(y_1, \ldots, y_n) = \begin{cases} n!f(y_1) \cdots f(y_n), & a < y_1 < y_2 < \cdots < y_n < b \\ 0, & \text{otherwise.} \end{cases} \tag{27}$$

*Proof.* The proof is carried out explicitly for $n = 2$ and $n = 3$, but it is easily seen, with the proper change in notation, to be valid in the general case as well. First, consider the case $n = 2$. Since

$$P(X_1 = X_2) = \int \int_{(x_1 = x_2)} f(x_1)f(x_2)dx_1\,dx_2 = \int_a^b \int_{x_2}^{x_2} f(x_1)f(x_2)dx_1\,dx_2 = 0,$$

we may assume that the joint p.d.f. $f(\cdot, \cdot)$ of $X_1$ and $X_2$ is 0 for $x_1 = x_2$. Thus, $f(\cdot, \cdot)$ is positive on the rectangle $ABCD$ except for its diagonal $DB$, call this set $S$.

Write $S = S_1 \cup S_2$ as in the Figure 6.7. Points $(x_1, x_2)$ in $S_1$ are mapped into the region $T$ consisting of the triangle $EFG$ (except for the side $GF$) depicted in Figure 6.7. This is so, because $x_1 < x_2$, so that $y_1 = x_1, y_2 = x_2$. For points $(x_1, x_2)$ in $S_2$, we have $x_1 > x_2$, so that $y_1 = x_2, y_2 = x_1$, and the point $(x_1, x_2)$ is also mapped into $T$ as indicated in the figures. On $S_1$: $y_1 = x_1$, $y_2 = x_2$, so that $x_1 = y_1$, $x_2 = y_2$, and $J = \begin{vmatrix} 1 & 0 \\ 0 & 1 \end{vmatrix} = 1$. Since $f(x_1, x_2) = f(x_1)f(x_2)$, it follows that $f_{Y_1, Y_2}(y_1, y_2) = f(y_1)f(y_2)$. On $S_2$: $y_1 = x_2, y_2 = x_1$, so that $x_1 = y_2, x_2 = y_1$, and $J = \begin{vmatrix} 0 & 1 \\ 1 & 0 \end{vmatrix} = -1$.

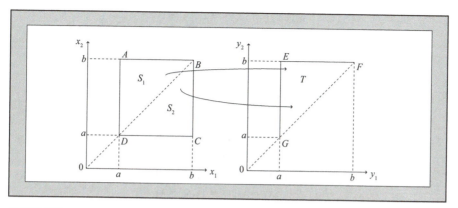

**FIGURE 6.7**

The set *ABCD* of positivity of the joint p.d.f. of $X_1$ and $X_2$ is partitioned into the disjoint set $S_1$ and $S_2$. Both sets $S_1$ and $S_2$ are mapped onto $T$ under the transformation of ordering $X_1$ and $X_2$.

It follows that $f_{Y_1,Y_2}(y_1, y_2) = f(y_2)f(y_1)$. Therefore, by Theorem 6, $f_{Y_1,Y_2}(y_1, y_2) = 2f(y_1)f(y_2) = 2!f(y_1)f(y_2)$, or $g(y_1, y_2) = 2!f(y_1)f(y_2)$, $a < y_1 < y_2 < b$.

For $n = 3$, again since for $i \neq j$,

$$P(X_i = X_j) = \int\int_{(x_i = x_j)} f(x_i)f(x_j) \, dx_i \, dx_j = \int_a^b \int_{x_j}^{x_j} f(x_i)f(x_j) \, dx_i \, dx_j = 0,$$

and therefore $P(X_i = X_j = X_k) = 0$ for $i \neq j \neq k$, we may assume that the joint p.d.f., $f(\cdot, \cdot, \cdot)$, of $X_1, X_2$, and $X_3$ is zero, if at least two of the arguments $x_1, x_2, x_3$ are equal. Thus, we have:

$$f(x_1, x_2, x_3) = \begin{cases} f(x_1)f(x_2)f(x_3), & a < x_1 \neq x_2 \neq x_3 < b \\ 0, & \text{otherwise.} \end{cases}$$

Therefore $f(x_1, x_2, x_3)$ is positive on the set $S$, where

$$S = \{(x_1, x_2, x_3) \in \Re^3; \quad a < x_i < b, \quad i = 1, 2, 3, \quad x_1, x_2, x_3 \text{ all different}\}.$$

Let $S_{ijk} \subset S$ be defined by:

$$S_{ijk} = \{(x_1, x_2, x_3); \quad a < x_i < x_j < x_k < b\}, \quad i, j, k = 1, 2, 3, \quad i \neq j \neq k.$$

Then we have that these six sets are pairwise disjoint and (essentially)

$$S = S_{123} \cup S_{132} \cup S_{213} \cup S_{231} \cup S_{312} \cup S_{321}.$$

Now on each one of the $S_{ijk}$'s there exists a one-to-one transformation from the $x_i$'s to the $y_i$'s defined as follows:

$$S_{123} : y_1 = x_1, \ y_2 = x_2, y_3 = x_3$$
$$S_{132} : y_1 = x_1, \ y_2 = x_3, y_3 = x_2$$

$$S_{213} : y_1 = x_2, y_2 = x_1, y_3 = x_3$$
$$S_{231} : y_1 = x_2, y_2 = x_3, y_3 = x_1$$
$$S_{312} : y_1 = x_3, y_2 = x_1, y_3 = x_2$$
$$S_{321} : y_1 = x_3, y_2 = x_2, y_3 = x_1.$$

Solving for the $x_i$'s, we have then:

$$S_{123} : x_1 = y_1, x_2 = y_2, x_3 = y_3$$
$$S_{132} : x_1 = y_1, x_2 = y_3, x_3 = y_2$$
$$S_{213} : x_1 = y_2, x_2 = y_1, x_3 = y_3$$
$$S_{231} : x_1 = y_3, x_2 = y_1, x_3 = y_2$$
$$S_{312} : x_1 = y_2, x_2 = y_3, x_3 = y_1$$
$$S_{321} : x_1 = y_3, x_2 = y_2, x_3 = y_1.$$

The Jacobians are thus given by:

$$S_{123} : J_{123} = \begin{vmatrix} 1 & 0 & 0 \\ 0 & 1 & 0 \\ 0 & 0 & 1 \end{vmatrix} = 1, \quad S_{231} : J_{231} = \begin{vmatrix} 0 & 0 & 1 \\ 1 & 0 & 0 \\ 0 & 1 & 0 \end{vmatrix} = 1,$$

$$S_{132} : J_{132} = \begin{vmatrix} 1 & 0 & 0 \\ 0 & 0 & 1 \\ 0 & 1 & 0 \end{vmatrix} = -1, \quad S_{312} : J_{312} = \begin{vmatrix} 0 & 1 & 0 \\ 0 & 0 & 1 \\ 1 & 0 & 0 \end{vmatrix} = 1,$$

$$S_{213} : J_{213} = \begin{vmatrix} 0 & 1 & 0 \\ 1 & 0 & 0 \\ 0 & 0 & 1 \end{vmatrix} = -1, \quad S_{321} : J_{321} = \begin{vmatrix} 0 & 0 & 1 \\ 0 & 1 & 0 \\ 1 & 0 & 0 \end{vmatrix} = -1.$$

Hence $|J_{123}| = \cdots = |J_{321}| = 1$, and Theorem 6 gives

$$g(y_1, y_2, y_3) = \begin{cases} f(y_1)f(y_2)f(y_3) + f(y_1)f(y_3)f(y_2) \\ + f(y_2)f(y_1)f(y_3) + f(y_3)f(y_1)f(y_2) \\ + f(y_2)f(y_3)f(y_1) + f(y_3)f(y_2)f(y_1), & a < y_1 < y_2 < y_3 < b \\ \\ 0, & \text{otherwise.} \end{cases}$$

That is,

$$g(y_1, y_2, y_3) = \begin{cases} 3! f(y_1)f(y_2)f(y_3), & a < y_1 < y_2 < y_3 < b \\ 0, & \text{otherwise.} \end{cases} \qquad ■$$

Notice that the proof in the general case is exactly the same. One has $n!$ regions forming $S$, one for each permutation of the integers 1 through $n$. From the definition of a determinant and the fact that each row and column contains exactly one 1 and the rest all 0, it follows that the $n!$ Jacobians are either 1 or $-1$ and the remaining part of the proof is identical to the one just given except that one adds up $n!$ like terms instead of 3!.

The theorem is illustrated by the following two examples.

**Example 8.** Let $X_1, \ldots, X_n$ be i.i.d. r.v.'s distributed as $N(\mu, \sigma^2)$. Then the joint p.d.f. of the order statistics $Y_1, \ldots, Y_n$ is given by

$$g(y_1, \ldots, y_n) = n! \left(\frac{1}{\sqrt{2\pi}\sigma}\right)^n \exp\left[-\frac{1}{2\sigma^2}\sum_{j=1}^{n}(y_j - \mu)^2\right],$$

if $-\infty < y_1 < \cdots < y_n < \infty$, and zero otherwise.

**Example 9.** Let $X_1, \ldots, X_n$ be i.i.d. r.v.'s distributed as $U(\alpha, \beta)$. Then the joint p.d.f. of the order statistics $Y_1, \ldots, Y_n$ is given by

$$g(y_1, \ldots, y_n) = \frac{n!}{(\beta - \alpha)^n},$$

if $\alpha < y_1 < \cdots < y_n < \beta$, and zero otherwise.

From the joint p.d.f. in (27), it is relatively easy to derive the p.d.f. of $Y_j$ for any $j$, as well as the joint p.d.f. of $Y_i$ and $Y_j$ for any $1 \leq i < j \leq n$. We restrict ourselves to the derivation of the distributions of $Y_1$ and $Y_n$ alone.

**Theorem 13.** *Let $X_1, \ldots, X_n$ be i.i.d. r.v.'s with d.f. $F$ and p.d.f. $f$ which is positive and continuous for $(-\infty \leq)a < x < b(\leq \infty)$ and zero otherwise, and let $Y_1, \ldots, Y_n$ be the order statistics. Then the p.d.f.'s $g_1$ and $g_n$ of $Y_1$ and $Y_n$, respectively, are given by:*

$$g_1(y_1) = \begin{cases} n[1 - F(y_1)]^{n-1}f(y_1), & a < y_1 < b \\ 0, & \text{otherwise,} \end{cases} \tag{28}$$

*and*

$$g_n(y_n) = \begin{cases} n[F(y_n)]^{n-1}f(y_n), & a < y_n < b \\ 0, & \text{otherwise.} \end{cases} \tag{29}$$

*Proof.* First, derive the d.f.'s involved and then differentiate them to obtain the respective p.d.f.'s. To this end,

$$\begin{aligned} G_n(y_n) &= P(Y_n \leq y_n) = P[\max(X_1, \ldots, X_n) \leq y_n] \\ &= P(\text{all } X_1, \ldots, X_n \leq y_n) = P(X_1 \leq y_n, \ldots, X_n \leq y_n) \\ &= P(X_1 \leq y_n) \cdots P(X_n \leq y_n) \quad \text{(by the independence of the } X_i\text{'s)} \\ &= [F(y_n)]^n. \end{aligned}$$

That is, $G_n(y_n) = [F(y_n)]^n$, so that

$$g_n(y_n) = \frac{d}{dy_n}G_n(y_n) = n[F(y_n)]^{n-1}\frac{d}{dy_n}F(y_n) = n[F(y_n)]^{n-1}f(y_n).$$

Likewise,

$$\begin{aligned} 1 - G_1(y_1) &= P(Y_1 > y_1) = P[\min(X_1, \ldots, X_n) > y_1] \\ &= P(\text{all } X_1, \ldots, X_n > y_1) = P(X_1 > y_1, \ldots, X_n > y_1) \end{aligned}$$

$$= P(X_1 > y_1) \cdots P(X_n > y_1) \quad \text{(by the independence of the } X_i\text{'s)}$$

$$= [1 - P(X_1 \le y_1)] \cdots [1 - P(X_1 \le y_1)] = [1 - F(y_1)]^n.$$

That is, $1 - G_1(y_1) = [1 - F(y_1)]^n$, so that

$$-g_1(y_1) = \frac{d}{dy_1}[1 - G_1(y_1)] = n[1 - F(y_1)]^{n-1}\frac{d}{dy_1}[1 - F(y_1)]$$

$$= n[1 - F(y_1)]^{n-1}[-f(y_1)] = -n[1 - F(y_1)]^{n-1}f(y_1),$$

and hence

$$g_1(y_1) = n[1 - F(y_1)]^{n-1}f(y_1). \qquad \blacksquare$$

As an illustration of the theorem, consider the following example.

**Example 10.** Let the independent r.v.'s $X_1, \ldots, X_n$ be distributed as $U(0, 1)$. Then, for $0 < y_1, y_n < 1$:

$$g_1(y_1) = n(1 - y_1)^{n-1} \quad \text{and} \quad g_n(y_n) = ny_n^{n-1}.$$

**Discussion.** Here, for $0 < x < 1$, $f(x) = 1$ and $F(x) = x$. Therefore relations (28) and (29) give, for $0 < y_1, y_n < 1$:

$$g_1(y_1) = n(1 - y_1)^{n-1} \times 1 = n(1 - y_1)^{n-1} \quad \text{and} \quad g_n(y_n) = ny_n^{n-1} \times 1 = ny_n^{n-1},$$

as asserted.

As a further illustration of the theorem, consider the following example, which is of interest in its own right.

**Example 11.** If $X_1, \ldots, X_n$ are independent r.v.'s having the Negative Exponential distribution with parameter $\lambda$, then $Y_1$ has also the Negative Exponential distribution with parameter $n\lambda$.

**Discussion.** Here $f(x) = \lambda e^{-\lambda x}$ and $F(x) = 1 - e^{-\lambda x}$ for $x > 0$. Then, for $y_1 > 0$, formula (28) yields:

$$g_1(y_1) = n(e^{-\lambda y_1})^{n-1} \times \lambda e^{-\lambda y_1} = (n\lambda)e^{-(n-1)y_1}e^{-\lambda y_1} = (n\lambda)e^{-(n\lambda)y_1},$$

as was to be seen.

**Example 12.**

(i) In a complex system, $n$ identical components are connected serially, so that the system works, if and only if all $n$ components function. If the lifetime of said components is described by a r.v. $X$ with d.f. $F$ and p.d.f. $f$ positive and continuous in $(\infty \le)a < x < b(\le \infty)$, write out the expression for the probability that the system functions for at least $t$ time units.

(ii)  Do the same as in part (i), if the components are connected in parallel, so that the system functions, if and only if at least one of the components works.

(iii)  Simplify the expressions in parts (i) and (ii), if $f$ is the Negative Exponential with parameter $\lambda$.

**Discussion.**

(i)  Clearly, $P$ (system works for at least $t$ time units)

$$= P(X_1 \geq t, \ldots, X_n \geq t) \qquad \text{(where } X_i \text{ is the lifetime of the } i\text{th component)}$$

$$= P(Y_1 \geq t) \qquad \text{(where } Y_1 \text{ is the smallest order statistic)}$$

$$= \int_t^\infty g_1(y)\, dy \qquad \text{(where } g_1 \text{ is the p.d.f. of } Y_1)$$

$$= \int_t^\infty n[1 - F(y)]^{n-1} f(y)\, dy \qquad \text{(by (28))}.$$

$$(30)$$

(ii)  Here

$P$ (system works for at least $t$ time units)

$$= P(\text{at least one of } X_1, \ldots, X_n \geq t)$$

$$= P(Y_n \geq t) \quad \text{(where } Y_n \text{ is the largest order statistic)}$$

$$= \int_t^\infty g_n(y)\, dy \quad \text{(where } g_n \text{ is the p.d.f. of } Y_n)$$

$$= \int_t^\infty n[F(y)]^{n-1} f(y)\, dy \quad \text{(by (29))}. \qquad (31)$$

(iii)  Here $F(y) = 1 - e^{-\lambda y}$ and $f(y) = \lambda e^{-\lambda y}$ $(y > 0)$ from Example 9. Also, from the same example, the p.d.f. of $Y_1$ is $g_1(y) = (n\lambda)e^{-(n\lambda)y}$, so that (30) gives:

$$P(Y_1 \geq t) = \int_t^\infty (n\lambda)e^{-(n\lambda)y}\, dy$$

$$= -\int_t^\infty de^{-(n\lambda)y}$$

$$= -e^{-(n\lambda)y}\Big|_t^\infty = e^{-n\lambda t},$$

and, by (31), the p.d.f. of $Y_n$ is $g_n(y) = n(1 - e^{-\lambda y})^{n-1}\lambda e^{-\lambda y}$, so that

$$P(Y_n \geq t) = \int_t^\infty n(1 - e^{-\lambda y})^{n-1}\lambda e^{-\lambda y}\, dy.$$

However,

$$\int_t^\infty n(1 - e^{-\lambda y})^{n-1}\lambda e^{-\lambda y}\, dy = -n\int_t^\infty (1 - e^{-\lambda y})^{n-1} de^{-\lambda y}$$

$$= n\int_t^\infty (1 - e^{-\lambda y})^{n-1} d(1 - e^{-\lambda y})$$

$$= \int_t^\infty d(1 - e^{-\lambda y})^n$$
$$= (1 - e^{-\lambda y})^n |_t^\infty = 1 - (1 - e^{-\lambda t})^n.$$

For example, for $n = 2$,

$$P(Y_2 \geq t) = 2e^{-\lambda t} - e^{-2\lambda t}.$$

## EXERCISES

**5.1**  Let $X_1, \ldots, X_n$ be independent r.v.'s with p.d.f. $f(x) = cx^{-(c+1)}, x > 1 \ (c > 0)$, and set $U = Y_1 = \min (X_1, \ldots, X_n), V = Y_n = \max (X_1, \ldots, X_n)$.

  **(i)**  Determine the d.f. $F$ corresponding to the p.d.f. $f$.

  **(ii)**  Use Theorem 13 to determine the p.d.f.'s $f_U$ and $f_V$.

**5.2**  Refer to Example 8 and calculate the expectations $EY_1$ and $EY_n$, and also determine the $\lim EY_1$ and the $\lim EY_n$ as $n \to \infty$.

**5.3**  Let $Y_1$ and $Y_n$ be the smallest and the largest order statistics based on a random sample $X_1, \ldots, X_n$ from the $U(\alpha, \beta) \ (\alpha < \beta)$ distribution.

  **(i)**  For $n = 3$ and $n = 4$, show that the joint p.d.f. of $Y_1$ and $Y_n$ is given, respectively, by:

$$g_{13}(y_1, y_3) = \frac{3 \times 2}{(\beta - \alpha)^3}(y_3 - y_1), \quad \alpha < y_1 < y_3 < \beta,$$

$$g_{14}(y_1, y_4) = \frac{4 \times 3}{(\beta - \alpha)^4}(y_4 - y_1)^2, \quad \alpha < y_1 < y_4 < \beta.$$

  **(ii)**  Generalize the preceding results and show that:

$$g_{1n}(y_1, y_n) = \frac{n(n-1)}{(\beta - \alpha)^n}(y_n - y_1)^{n-2}, \quad \alpha < y_1 < y_n < \beta.$$

**Hint.** For part (ii), all one has to do is to calculate the integrals:

$$\int_{y_1}^{y_n} \int_{y_1}^{y_{n-1}} \cdots \int_{y_1}^{y_4} \int_{y_1}^{y_3} dy_2 \, dy_3 \cdots dy_{n-2} \, dy_{n-1},$$

which is done one at a time; also, observe the pattern emerging.

**5.4**  Let $Y_1$ and $Y_n$ be the smallest and the largest order statistics based on a random sample $X_1, \ldots, X_n$ from the $U(0, 1)$ distribution. Then show that:

$$\text{Cov}(Y_1, Y_n) = \frac{1}{(n + 1)^2(n + 2)}.$$

**Hint.** Use the joint p.d.f. taken from Exercise 5.3 (ii) for $\alpha = 0$ and $\beta = 1$.

**5.5**  If $Y_1$ and $Y_n$ are the smallest and the largest order statistics based on a random sample $X_1, \ldots, X_n$ from the $U(0, 1)$ distribution:

**(i)** Show that the p.d.f. of the *sample range* $R = Y_n - Y_1$ is given by:

$$f_R(r) = n(n-1)r^{n-2}(1-r), \quad 0 < r < 1.$$

**(ii)** Also, calculate the expectation $ER$, and determine its limit as $n \to \infty$.
**Hint.** Use Execise 5.3 (ii) with $\alpha = 0$ and $\beta = 1$.

**5.6** Refer to Example 11 and set $Z = nY_1$. Then show that $Z$ is distributed as the $X_i$'s.

**5.7** The lifetimes of two batteries are independent r.v.'s $X$ and $Y$ with the Negative Exponential distribution with parameter $\lambda$. Suppose that the two batteries are connected serially, so that the system works if and only if both work.
  **(i)** Use Example 12 (with $n = 2$) to calculate the probability that the system works beyond time $t > 0$.
  **(ii)** What is the expected lifetime of the system?
  **(iii)** How do parts (i) and (ii) become for $\lambda = 1/3$?

**5.8** Let $Y_1$ and $Y_n$ be the smallest and the largest order statistics based on a random sample $X_1, \dots, X_n$ from the Negative Exponential distribution with parameter $\lambda$. Then, by Example 11, $g_1(y_1) = (n\lambda)e^{-(n\lambda)y_1}$, $y_1 > 0$.
  **(i)** Use relation (29) (with $a = 0$ and $b = \infty$) to determine the p.d.f. $g_n$ of the r.v. $Y_n$.
  **(ii)** Calculate the $EY_n$ for $n = 2$ and $n = 3$.

**5.9**  **(i)** Refer to Exercise 5.8 (i) and show that:

$$EY_n = \frac{n}{\lambda} \sum_{r=0}^{n-1} (-1)^{n-r-1} \frac{\binom{n-1}{r}}{(n-r)^2}.$$

  **(ii)** Apply part (i) for $n = 2$ and $n = 3$ to recover the values found in Exercise 5.8 (ii).
**Hint.** Consider the Binomial expansion: $(a+b)^k = \sum_{r=0}^k \binom{k}{r} a^r b^{k-r}$ and apply it to: $(1 - e^{-\lambda y})^{n-1}$ for $a = 1$, $b = -e^{-\lambda y}$, and $k = n-1$. Then carry out the multiplications indicated and integrate term by term.

**5.10** Let $X_1, \dots, X_n$ be a random sample of size $n$ of the continuous type with d.f. $F$ and p.d.f. $f$, positive and continuous in $(-\infty \le)a < x < b(\le \infty)$, and let $Y_1$ and $Y_n$ be the smallest and the largest order statistics of the $X_i$'s. Use relation (27) in order to show that the joint p.d.f. $g_{1n}$ of the r.v.'s $Y_1$ and $Y_n$ is given by the expression:

$$g_{1n}(y_1, y_n) = n(n-1)[F(y_n) - F(y_1)]^{n-2} f(y_1)f(y_n), \quad a < y_1 < y_n < b.$$

**Hint.** The p.d.f. $g_{1n}$ is obtained by integrating $g(y_1, \dots, y_n)$ in (27) with respect to $y_{n-1}, y_{n-2}, \dots, y_2$ as indicated below:

$$g_{1n}(y_1, y_n) = n!f(y_1)f(y_n) \int_{y_1}^{y_n} \cdots \int_{y_{n-3}}^{y_n} \int_{y_{n-2}}^{y_n} f(y_{n-1})f(y_{n-2})$$
$$\times \cdots f(y_2)dy_{n-1}dy_{n-2} \cdots dy_2.$$

However,

$$\int_{y_{n-2}}^{y_n} f(y_{n-1})dy_{n-1} = F(y_n) - F(y_{n-2}) = \frac{[F(y_n) - F(y_{n-2})]^1}{1!},$$

$$\int_{y_{n-3}}^{y_n} \frac{[F(y_n) - F(y_{n-2})]^1}{1!} f(y_{n-2})\, dy_{n-2}$$

$$= -\int_{y_{n-3}}^{y_n} \frac{[F(y_n) - F(y_{n-2})]^1}{1!} d[F(y_n)]$$

$$-F(y_{n-2})] = -\frac{[F(y_n) - F(y_{n-2})]^2}{2!}\Big|_{y_{n-3}}^{y_n} = \frac{[F(y_n) - F(y_{n-3})]^2}{2!},$$

and continuing on like this, we finally get:

$$\int_{y_1}^{y_n} \frac{[F(y_n) - F(y_2)]^{n-3}}{(n-3)!} f(y_2)\, dy_2$$

$$= -\int_{y_1}^{y_n} \frac{[F(y_n) - F(y_2)]^{n-3}}{(n-3)!} d[F(y_n) - F(y_2)]$$

$$= -\frac{[F(y_n) - F(y_2)]^{n-2}}{(n-2)!}\Big|_{y_1}^{y_n} = \frac{[F(y_n) - F(y_1)]^{n-2}}{(n-2)!}.$$

Since $\frac{n!}{(n-2)!} = n(n-1)$, the result follows.

**5.11** In reference of Exercise 5.10, determine the joint d.f. $g_{1n}$ if the r.v.'s $X_1, \ldots, X_k$ are distributed as $U(0,1)$.

**5.12** Let $X_1, \ldots, X_n$ be i.i.d. r.v.'s with d.f. $F$ and p.d.f. $f$ positive for $-\infty \le a < x < b \le \infty$, and let $Y_n$ be the largest order statistics for the $X_i$'s. Then, if $G_n$ and $g_n$ are the d.f. and the p.d.f. of $Y_n$, it is known that:

$$G_n(y) = [F(y)]^n, \quad a < y < b; \quad g_n(y) = n[F(y)]^{n-1}f(y), \quad a < y < b,$$

and if the $X_i$'s are distributed as $U(0,1)$, then:

$$g_n(y) = ny^{n-1}, \quad 0 < y < 1.$$

For this last case:

(i) Compute the $EY_n$ and the $Var(Y_n)$.

(ii) Show that, as $n \to \infty$, $EY_n \to 1$ and $Var(Y_n) \to 0$.

**5.13** Let $X_1, \ldots, X_5$ be a random sample from the $U(0,1)$ distribution, and let $Y_1, \ldots, Y_5$ be the order statistics. Then their joint distribution is given by (see Example 9 in Chapter 6):

$$g(y_1, \ldots, y_5) = 120, \quad 0 < y_1 < \cdots < y_5 < 1.$$

(i) Show that the p.d.f. of $Y_3$ is given by:

$$f_{Y_3}(y) = 30y^2(1 - 2y + y^2), \quad 0 < y < 1.$$

(ii) Also, compute the probability $P(1/3 < Y_3 < 2/3)$.

**5.14** Recall that the p.d.f. of the Weibull distribution with parameter $\alpha$ and $\beta$ is given by (Exercise 3.34, Chapter 3):

$$f(x) = \alpha\beta x^{\beta-1}e^{-\alpha x^{\beta}}, \quad x > 0 \ (\alpha, \beta > 0).$$

Now, let $X_1, \ldots, X_n$ be a random sample from this distribution, and set $Y = X_{(1)}$.

(i) Obtain the p.d.f. $f_Y$. Do you recognize it?

(ii) Use Exercise 3.34 in Chapter 3 in order to find the $EY$ and the $\sigma^2(Y)$.

**5.15** Let $X_1, \ldots, X_n$ be i.i.d. r.v.'s with p.d.f. given by:

$$f_X(x) = \frac{1}{1 - e^{-c}}e^{-x}, \quad 0 < x < c\,(> 0) \text{ (and 0 otherwise)},$$

and let $Y_1, \ldots, Y_n$ be the order statistics.

(i) Derive the p.d.f.'s $g_1$ and $g_n$ of $Y_1$ and $Y_n$, respectively.

(ii) Also, derive the joint p.d.f. of $Y_1$ and $Y_n$.

(iii) Use part (ii) in order to derive the p.d.f. $f_R$, say, of the r.v. $Y_n - Y_1 = R$ for $n = 3$.

**5.16** Let $X_1, \ldots, X_5$ be a random sample of size 5 from the Negative Exponential distribution with parameter $\lambda = 1$, and let $Y_1, \ldots, Y_5$ be the order statistics. Then the joint p.d.f. of the $Y_i$'s is given by:

$$g(y_1, \ldots, y_5) = 120e^{-(y_1 + \cdots + y_5)}, \quad 0 < y_1 < \cdots < y_5.$$

(i) Show that the joint p.d.f. of $Y_1$ and $Y_5$ is given by

$$g(y_1, y_5) = 20(e^{-y_1} - e^{-y_5})^3 e^{-(y_1 + y_5)}, \quad 0 < y_1 < y_5.$$

(ii) Set $R = Y_5 - Y_1$ for the *sample range* of the distribution (see also Exercise 5.5 in this chapter), and also let $S = Y_1$. Then show that the joint p.d.f. of $R$ and $S$ is given by:

$$f_{R,S}(r, s) = 20(e^{-r} - 3e^{-2r} + 3e^{-3r} - e^{-4r})e^{-5s}, \quad r, s > 0.$$

(iii) From part (ii), derive the marginal p.d.f. $f_R$. Specifically, show that

$$f_R(r) = 4(e^{-r} - 3e^{-2r} + 3e^{-3r} - e^{-4r}), \quad r > 0.$$

# Some modes of convergence of random variables, applications

The first thing which is done in this chapter is to introduce two modes of convergence for sequences of r.v.'s, convergence in distribution and convergence in probability, and then to investigate their relationship.

A suitable application of these convergences leads to the most important results in this chapter, which are the weak law of large numbers (WLLN) and the central limit theorem (CLT). These results are illustrated by concrete examples, including numerical examples in the case of the CLT.

In the final section of the chapter, it is shown that convergence in probability is preserved under continuity. This is also the case for convergence in distribution, but there will be no elaboration here. These statements are illustrated by two general results and a specific application.

The proofs of some of the theorems stated are given in considerable detail; in some cases, only a rough outline is presented, whereas in other cases, we restrict ourselves to the statements of the theorems alone.

## 7.1 CONVERGENCE IN DISTRIBUTION OR IN PROBABILITY AND THEIR RELATIONSHIP

In all that follows, $X_1, \ldots, X_n$ are i.i.d. r.v.'s, which may be either discrete or continuous. In applications, these r.v.'s represent $n$ independent observations on a r.v. $X$, associated with an underlying phenomenon which is of importance to us. In a probabilistic/statistical environment, our interest lies in knowing the distribution of $X$, whether it is represented by the probabilities $P(X \in B), B \subseteq \Re$, or the d.f. $F$ of $X$, or its p.d.f. $f$. In practice, this distribution is unknown to us. Something then that would be desirable would be to approximate the unknown distribution, in some sense, by a known distribution. In this section, the foundation is set for such an approximation (Figure 7.1).

**Definition 1.** Let $Y_1, \ldots, Y_n$ be r.v.'s with respective d.f.'s. $F_1, \ldots, F_n$. The r.v.'s may be either discrete or continuous and need be neither independent nor identically distributed. Also, let $Y$ be a r.v. with d.f. $G$. We say that the sequence of r.v.'s

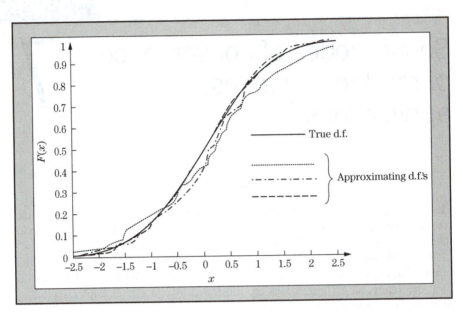

**FIGURE 7.1**

The d.f. represented by the solid curve is approximated by the d.f.'s represented by the
........, .—.—.—., and — — — — — curves.

$\{Y_n\}, n \geq 1$, *converges in distribution* to the r.v. $Y$ as $n \to \infty$ and write $Y_n \xrightarrow[n \to \infty]{d} Y$, if $F_n(x) \xrightarrow[n \to \infty]{} G(x)$ for all continuity points $x$ of $G$.

The following example illustrates the definition.

**Example 1.** For $n \geq 1$, let the d.f.'s $F_n$ and the d.f. $G$ be given by:

$$F_n(x) = \begin{cases} 0, & \text{if } x < 1 - \frac{1}{n} \\ \frac{1}{2}, & \text{if } 1 - \frac{1}{n} \leq x < 1 + \frac{1}{n} \\ 1, & \text{if } x \geq 1 + \frac{1}{n}, \end{cases} \qquad G(x) = \begin{cases} 0, & \text{if } x < 1 \\ 1, & \text{if } x \geq 1, \end{cases}$$

and discuss whether or not $F_n(x)$ converges to $G(x)$ as $n \to \infty$ (Figure 7.2).

**Discussion.** The d.f. $G$ is continuous everywhere except for the point $x = 1$. For $x < 1$, let $n_0 > 1/(1 - x)$. Then $x < 1 - \frac{1}{n_0}$ and also $x < 1 - \frac{1}{n}$ for all $n \geq n_0$. Thus, $F_n(x) = 0, n \geq n_0$. For $x > 1$, let $n_0 \geq 1/(x - 1)$. Then $x \geq 1 + \frac{1}{n_0}$ and also $x \geq 1 + \frac{1}{n}$ for all $n \geq n_0$, so that $F_n(x) = 1, n \geq n_0$. Thus, for $x \neq 1, F_n(x) \to G(x)$, so, if $Y_n$ and $Y$ are r.v.'s such that $Y_n \sim F_n$ and $Y \sim G$, then $Y_n \xrightarrow[n \to \infty]{d} Y$.

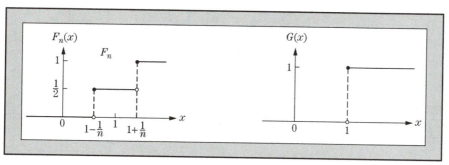

**FIGURE 7.2**

The d.f. $G$ is approximated by the d.f.'s $F_n$ at all points $x \neq 1$.

**Remark 1.** The example also illustrates the point that, if $x$ is a discontinuity point of $G$, then $F_n(x)$ need not converge to $G(x)$. In Example 1, $F_n(1) = \frac{1}{2}$ for all $n$, and $G(1) = 1$.

The idea, of course, behind Definition 1 is the approximation of the (presumably unknown) probability $P(Y \leq x) = G(x)$ by the (presumably known) probabilities $P(Y_n \leq x) = F_n(x)$, for large enough $n$. Convergence in distribution also allows the approximation of probabilities of the form $P(x < Y \leq y)$ by the probabilities $P(x < Y_n \leq y)$, for $x$ and $y$ continuity points of $G$. This is so because

$$P(x < Y_n \leq y) = P(Y_n \leq y) - P(Y_n \leq x) = F_n(y) - F_n(x) \xrightarrow[n \to \infty]{} G(y) - G(x) = P(x < Y \leq y).$$

Whereas convergence in distribution allows the comparison of certain probabilities, calculated in terms of the *individual* r.v.'s $Y_n$ and $Y$, it does not provide evaluation of probabilities calculated on the *joint* behavior of $Y_n$ and $Y$. This is taken care of to a satisfactory extent by the following mode of convergence.

**Definition 2.** The sequence of r.v.'s $\{Y_n\}, n \geq 1$, *converges in probability* to the r.v. $Y$ as $n \to \infty$, if, for every $\varepsilon > 0, P(|Y_n - Y| > \varepsilon) \xrightarrow[n \to \infty]{} 0$; equivalently, $P(|Y_n - Y| \leq \varepsilon) \xrightarrow[n \to \infty]{} 1$. The notation used is: $Y_n \xrightarrow[n \to \infty]{P} Y$.

Thus, if the event $A_n(\varepsilon)$ is defined by: $A_n(\varepsilon) = \{s \in S; Y(s) - \varepsilon \leq Y_n(s) \leq Y(s) + \varepsilon\}$ (i.e., the event for which the r.v. $Y_n$ is within $\varepsilon$ from the r.v. $Y$), then $P(A_n(\varepsilon)) \xrightarrow[n \to \infty]{} 1$ for every $\varepsilon > 0$. The probability that $Y_n$ lies within a small neighborhood around $Y$, such as $[Y_n - \varepsilon, Y_n + \varepsilon]$, is as close to 1 as one pleases, provided $n$ is sufficiently large (Figure 7.3). Equivalently, $P(A_n^c(\varepsilon)) = P(\{s \in S; Y_n(s) < Y(s) - \varepsilon \text{ or } Y_n(s) > Y(s) + \varepsilon\}) \xrightarrow[n \to \infty]{} 0$.

It is rather clear that convergence in probability is stronger than convergence in distribution. That this is, indeed, the case is illustrated by the following example, where we have convergence in distribution but not in probability.

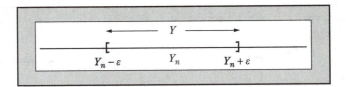

**FIGURE 7.3**

$Y$ lies in $[Y_n - \varepsilon, Y_n + \varepsilon]$ with high probability for large $n$.

**Example 2.**   Let $\mathcal{S} = \{1, 2, 3, 4\}$, and on the subsets of $\mathcal{S}$, let $P$ be the discrete Uniform probability function. Define the following r.v.'s:

$$X_n(1) = X_n(2) = 1, \quad X_n(3) = X_n(4) = 0, \quad n = 1, 2, \ldots,$$

and

$$X(1) = X(2) = 0, \quad X(3) = X(4) = 1.$$

**Discussion.**   Then

$$|X_n(s) - X(s)| = 1 \quad \text{for all } s \in \mathcal{S}.$$

Hence, $X_n$ does *not* converge in probability to $X$, as $n \to \infty$. Now,

$$F_n(x) = \begin{cases} 0, & x < 0 \\ \frac{1}{2}, & 0 \le x < 1 \\ 1, & x \ge 1, \end{cases} \qquad G(x) = \begin{cases} 0, & x < 0 \\ \frac{1}{2}, & 0 \le x < 1 \\ 1, & x \ge 1, \end{cases}$$

so that $F_n(x) = G(x)$ for all $x \in \mathfrak{R}$. Thus, trivially, $F_n(x) \underset{n \to \infty}{\longrightarrow} G(x)$ for all continuity points of $G$; that is, $X_n \underset{n \to \infty}{\overset{d}{\longrightarrow}} X$, but $X_n$ does not converge in probability to $X$.

The precise relationship between convergence in distribution and convergence in probability is stated in the following theorem.

**Theorem 1.**   *Let $\{Y_n\}, n \ge 1$, be a sequence of r.v.'s and let $Y$ be a r.v. Then $Y_n \underset{n \to \infty}{\overset{P}{\longrightarrow}} Y$ always implies $Y_n \underset{n \to \infty}{\overset{d}{\longrightarrow}} Y$. The converse is not true in general (as illustrated by Example 2). However, it is true if $P(Y = c) = 1$, where $c$ is a constant. That is, $Y_n \underset{n \to \infty}{\overset{d}{\longrightarrow}} c$ implies $Y_n \underset{n \to \infty}{\overset{P}{\longrightarrow}} c$, so that $Y_n \underset{n \to \infty}{\overset{P}{\longrightarrow}} c$ if and only if $Y_n \underset{n \to \infty}{\overset{d}{\longrightarrow}} c$.*

*Proof (outline).* That $Y_n \underset{n \to \infty}{\overset{P}{\longrightarrow}} Y$ implies $Y_n \underset{n \to \infty}{\overset{d}{\longrightarrow}} Y$ is established by employing the concepts of lim inf (limit inferior) and lim sup (limit superior) of a sequence of numbers, and we choose not to pursue it. For the proof of the fact that

$Y_n \xrightarrow[n\to\infty]{d} c$ implies $Y_n \xrightarrow[n\to\infty]{P} c$, observe that $F(x) = 0$, for $x < c$ and $F(x) = 1$ for $x \geq c$, where $F$ is the d.f. of $c$, so that $c - \varepsilon$ and $c + \varepsilon$ are continuity points of $F$ for all $\varepsilon > 0$. But $P(|Y_n - c| \leq \varepsilon) = P(c - \varepsilon \leq Y_n \leq c + \varepsilon) = P(Y_n \leq c + \varepsilon) - P(Y_n < c - \varepsilon) = F_n(c + \varepsilon) - P(Y_n < c - \varepsilon)$. However, $F_n(c + \varepsilon) \xrightarrow[n\to\infty]{} 1$ and $P(Y_n < c - \varepsilon) \leq P(Y_n \leq c - \varepsilon) = F_n(c - \varepsilon) \xrightarrow[n\to\infty]{} 0$, so that $P(Y_n < c - \varepsilon) \xrightarrow[n\to\infty]{} 0$. Thus, $P(|Y_n - c| \leq \varepsilon) \xrightarrow[n\to\infty]{} 1$ or $Y_n \xrightarrow[n\to\infty]{P} c$. ∎

According to Definition 1, in order to establish that $Y_n \xrightarrow[n\to\infty]{d} Y$, all one has to do is to prove the (pointwise) convergence $F_n(x) \xrightarrow[n\to\infty]{} F(x)$ for every continuity point $x$ of $F$. As is often the case, however, definitions do not lend themselves to checking the concepts defined. This also holds here. Accordingly, convergence in distribution is delegated to convergence of m.g.f.'s, which, in general, is a much easier task to perform. That this can be done is based on the following deep probabilistic result. Its justification is omitted entirely.

**Theorem 2** (Continuity Theorem).  *For $n = 1, 2, \ldots$, let $Y_n$ and $Y$ be r.v.'s with respective d.f.'s $F_n$ and $F$, and respective m.g.f.'s $M_n$ and $M$ (which are assumed to be finite at least in an interval $(-c, c)$, some $c > 0$). Then:*
*(i) If $F_n(x) \xrightarrow[n\to\infty]{} F(x)$ for all continuity points $x$ of $F$, it follows that $M_n(t) \xrightarrow[n\to\infty]{} M(t)$ for all $t \in (-c, c)$.*
*(ii) Let $M_n(t) \xrightarrow[n\to\infty]{} M(t)$, $t \in (-c, c)$. It follows that $F_n(x) \xrightarrow[n\to\infty]{} F(x)$ for all continuity points $x$ of $F$.*

Thus, according to this result, $Y_n \xrightarrow[n\to\infty]{d} Y$ or, equivalently, $F_n(x) \xrightarrow[n\to\infty]{} F(x)$ for all continuity points $x$ of $F$, if and only if $M_n(t) \xrightarrow[n\to\infty]{} M(t), t \in (-c, c)$, some $c > 0$. The fact that convergence of m.g.f.'s implies convergence of the respective d.f.'s is the most useful part from a practical viewpoint.

## EXERCISES

**1.1**  For $n = 1, 2, \ldots$, let $X_n$ be a r.v. with d.f. $F_n$ defined by: $F_n(x) = 0$ for $x < n$, and $F_n(x) = 1$ for $x \geq n$. Then show that $F_n(x) \xrightarrow[n\to\infty]{} F(x)$, which is identically 0 in $\Re$ and hence it is *not* a d.f. of a r.v.

**1.2**  Let $\{X_n\}, n \geq 1$, be r.v.'s with $X_n$ taking the values 1 and 0 with respective probabilities $p_n$ and $1 - p_n$; i.e., $P(X_n = 1) = p_n$ and $P(X_n = 0) = 1 - p_n$. Then show that $X_n \xrightarrow[n\to\infty]{P} 0$, if and only if $p_n \xrightarrow[n\to\infty]{} 0$.
**Hint.** Just elaborate on Definition 2.

**1.3**   For $n = 1, 2, \ldots$, let $X_n$ be a r.v. distributed as $B(n, p_n)$ and suppose that $np_n \xrightarrow[n \to \infty]{} \lambda \in (0, \infty)$. Then show that $X_n \xrightarrow[n \to \infty]{d} X$, where $X$ is a r.v. distributed as $P(\lambda)$, by showing that $M_{X_n}(t) \xrightarrow[n \to \infty]{} M_X(t)$, $t \in \mathfrak{R}$.

**Remark.** This is an application of Theorem 2(ii).

**1.4**   Let $Y_{1,n}$ and $Y_{n,n}$ be the smallest and the largest order statistics based on the random sample $X_1, \ldots, X_n$ from the $U(0, 1)$ distribution. Then show that:

(i) $Y_{1,n} \xrightarrow[n \to \infty]{P} 0$;   (ii) $Y_{n,n} \xrightarrow[n \to \infty]{P} 1$.

**Hint.** For $\varepsilon > 0$, calculate the probabilities: $P(|Y_{1,n}| > \varepsilon)$ and $P(|Y_{n,n} - 1| > \varepsilon)$ and show that they tend to 0 as $n \to \infty$. Use the p.d.f.'s of $Y_{1,n}$ and $Y_{n,n}$ determined in Example 10 of Chapter 6.

**1.5**   Refer to Exercise 1.4 and set: $U_n = nY_{1,n}$, $V_n = n(1 - Y_{n,n})$, and let $U$ and $V$ be r.v.'s having the Negative Exponential distribution with parameter $\lambda = 1$. Then:

(i)   Derive the p.d.f.'s of the r.v.'s $U_n$ and $V_n$.

(ii)   Derive the d.f.'s of the r.v.'s $U_n$ and $V_n$, and show that $U_n \xrightarrow[n \to \infty]{d} U$ by showing that

$$F_{U_n}(u) \xrightarrow[n \to \infty]{} F_U(u), \quad u \in \mathfrak{R}.$$

Likewise for $V_n$.

**Hint.** For part (ii), refer to #6 in Table 8 in the Appendix.

**1.6**   We say that a sequence $\{X_n\}, n \geq 1$, of r.v.'s converges to a r.v. $X$ in *quadratic mean* and write

$$X_n \xrightarrow[n \to \infty]{q.m.} X \quad \text{or} \quad X_n \xrightarrow[n \to \infty]{(2)} X, \quad \text{if } E(X_n - X)^2 \xrightarrow[n \to \infty]{} 0.$$

Now, if $X_1, \ldots, X_n$ are i.i.d. r.v.'s with (finite) expectation $\mu$ and (finite) variance $\sigma^2$, show that the sample mean $\bar{X}_n \xrightarrow[n \to \infty]{q.m.} \mu$.

**1.7**   In Theorem 1 of Chapter 4, the following, version of the *Cauchy-Schwarz* inequality was established: For any two r.v.'s $X$ and $Y$ with $EX = EY = 0$ and $Var(X) = Var(Y) = 1$, it holds: $|E(XY)| \leq 1$. (This is, actually, part only of said inequality.) Another more general version of this inequality is the following: For any two r.v.'s $X$ and $Y$ with finite expectations and variances, it holds: $|E(XY)| \leq E|XY| \leq E^{1/2}|X|^2 \times E^{1/2}|Y|^2$.

(i)   Prove the inequality in this setting.

(ii)   For any r.v. $X$, show that $|EX| \leq E|X| \leq E^{1/2}|X|^2$.

**Hint.** For part (i), use the obvious result $(x \pm y)^2 = x^2 + y^2 \pm 2xy \geq 0$ in order to conclude that $\pm xy \leq \frac{1}{2}(x^2 + y^2)$ and hence $|xy| \leq \frac{1}{2}(x^2 + y^2)$. Next, replace $x$ by $X/E^{1/2}|X|^2$, and $y$ by $Y/E^{1/2}|Y|^2$ (assuming, of course, that $E|X|^2 > 0, E|Y|^2 > 0$, because otherwise the inequality is, trivially, true), and take the expectations of both sides to arrive at the desirable result.

**1.8**  Let $\{X_n\}$ and $\{Y_n\}, n \geq 1$, be two sequences of r.v.'s such that: $X_n \xrightarrow[n\to\infty]{q.m.} X$, some r.v., and $X_n - Y_n \xrightarrow[n\to\infty]{q.m.} 0$. Then show that $Y_n \xrightarrow[n\to\infty]{q.m.} X$.

**Hint.** Use appropriately the Cauchy-Schwarz inequality discussed in Exercise 1.7.

**1.9**  For $n \geq 1$, consider the sequences of r.v.'s $\{X_n\}$, $\{Y_n\}$ such that, as $n \to \infty$,

$$X_n \xrightarrow[n\to\infty]{d} X \text{ a r.v.,} \quad Y_n \xrightarrow[n\to\infty]{P} c \text{ constant.}$$

Consider the random vector $(X, c)$ and observe that its joint d.f. is given by:

$$F_{X,c}(x, y) = P(X \leq x, c \leq y) = \begin{cases} 0, & x \in \Re, \ y < c, \\ P(X \leq x) = F_X(x), & x \in \Re, \ y \geq c. \end{cases}$$

Then show that the sequence of random vectors $\{(X_n, Y_n)\}$ converges in distribution to $(X, c)$; i.e.,

$$F_{X_n, Y_n}(x, y) \xrightarrow[n\to\infty]{} F_{X,c}(x, c) \text{ for continuity points } x \text{ of } F_X.$$

**1.10**  Let $X_1, \ldots, X_n$ be i.i.d. r.v.'s with expectation 0 and variance 1, and define the r.v.'s $Y_n$ and $Z_n$ by:

$$Y_n = n\sqrt{n}\frac{\sum_{i=1}^{n} X_i}{\sum_{i=1}^{n} X_i^2}, \quad Z_n = n\frac{\sum_{i=1}^{n} X_i}{\left(\sum_{i=1}^{n} X_i^2\right)^{1/2}}.$$

Then show that, as $n \to \infty$, $Y_n \xrightarrow{d} Z \sim N(0, 1)$ and $Z_n \xrightarrow{d} Z \sim N(0, 1)$.

## 7.2 SOME APPLICATIONS OF CONVERGENCE IN DISTRIBUTION: WLLN AND CLT

As a first application of the concept of convergence in distribution, we have the so-called *weak law of large numbers*. This result is stated and proved, an interpretation is provided, and then a number of specific applications are presented.

**Theorem 3** (Weak Law of Large Numbers).  *Let $X_1, X_2, \ldots$ be i.i.d. r.v.'s with (common) finite expectation $\mu$, and let $\bar{X}_n$ be the sample mean of $X_1, \ldots, X_n$. Then $\bar{X}_n \xrightarrow[n\to\infty]{d} \mu$, or (on account of Theorem 1) $\bar{X}_n \xrightarrow[n\to\infty]{P} \mu$.*

Thus, the probability that $\mu$ lies within a small neighborhood around $\bar{X}_n$, such as $[\bar{X}_n - \varepsilon, \bar{X}_n + \varepsilon]$, is as close to 1 as one pleases, provided $n$ is sufficiently large (Figure 7.4).

*Proof.* The proof is a one-line proof, if it happens that the $X_i$'s also have a (common) finite variance $\sigma^2$ (which they are not required to have for the validity of the theorem).

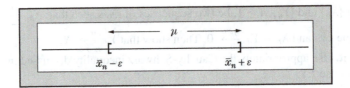

**FIGURE 7.4**

$\bar{X}_n - \varepsilon \le \mu \le \bar{X}_n + \varepsilon$. Parameter $\mu$ lies in the interval $[\bar{X}_n - \varepsilon, \bar{X}_n + \varepsilon]$ with high probability for all sufficiently large $n$.

Since $E\bar{X}_n = \mu$ and $\mathrm{Var}(\bar{X}_n) = \frac{\sigma^2}{n}$, the Tchebichev inequality gives, for every $\varepsilon > 0$,
$P(|\bar{X}_n - \mu| > \varepsilon) \le \frac{1}{\varepsilon^2} \times \frac{\sigma^2}{n} \xrightarrow[n\to\infty]{} 0$, so that $\bar{X}_n \xrightarrow[n\to\infty]{P} \mu$.

Without reference to the variance, one would have to show that $M_{\bar{X}_n}(t) \xrightarrow[n\to\infty]{} M_\mu(t)$
(for $t \in (-c, c)$, some $c > 0$). Let $M$ stand for the (common) m.g.f. of the $X_i$'s. Then use familiar properties of the m.g.f. and independence of the $X_i$'s in order to obtain:

$$M_{\bar{X}_n}(t) = M_{\sum_{i=1}^n X_i}\left(\frac{t}{n}\right) = \prod_{i=1}^n M_{X_i}\left(\frac{t}{n}\right) = \left[M\left(\frac{t}{n}\right)\right]^n.$$

Consider the function $M(z)$, and expand it around $z = 0$ according to Taylor's formula up to terms of first order to get:

$$M(z) = M(0) + \frac{z}{1!} \times \frac{\mathrm{d}}{\mathrm{d}z} M(z)|_{z=0} + R(z) \quad \left(\frac{1}{z}R(z) \to 0 \text{ as } z \to 0\right)$$

$$= 1 + z\mu + R(z),$$

since $M(0) = 1$ and $\frac{\mathrm{d}}{\mathrm{d}z}M(z)|_{z=0} = EX_1 = \mu$. Replacing $z$ by $t/n$, for fixed $t$, the last formula becomes:

$$M\left(\frac{t}{n}\right) = 1 + \frac{t}{n}\mu + R\left(\frac{t}{n}\right), \quad \text{where } nR\left(\frac{t}{n}\right) \to 0 \text{ as } n \to \infty.$$

Therefore

$$M_{\bar{X}_n}(t) = \left[1 + \frac{\mu t + nR(t/n)}{n}\right]^n,$$

and this converges to $e^{\mu t}$, as $n \to \infty$, by Remark 2 below. Since $e^{\mu t}$ is the m.g.f. of (the degenerate r.v.) $\mu$, we have shown that $M_{\bar{X}_n}(t) \xrightarrow[n\to\infty]{} M_\mu(t)$, as was to be seen. ∎

**Remark 2.** For every $z \in \mathfrak{R}$, one way of defining the exponential function $e^z$ is: $e^z = \lim_{n\to\infty}(1 + \frac{z}{n})^n$. It is a consequence of this result that, as $n \to \infty$, also $(1 + \frac{z_n}{n})^n \to e^z$ whenever $z_n \to z$.

The interpretation and most common use of the WLLN is that, if $\mu$ is an unknown entity, which is typically the case in statistics, then $\mu$ may be approximated (in the sense of distribution or probability) by the known entity $\bar{X}_n$, for sufficiently large $n$.

## 7.2.1 APPLICATIONS OF THE WLLN

**1.** If the independent $X_i$'s are distributed as $B(1, p)$, then $EX_i = p$ and therefore
$$\bar{X}_n \xrightarrow[n \to \infty]{P} p.$$

**2.** If the independent $X_i$'s are distributed as $P(\lambda)$, then $EX_i = \lambda$ and therefore
$$\bar{X}_n \xrightarrow[n \to \infty]{P} \lambda.$$

**3.** If the independent $X_i$'s are distributed as $N(\mu, \sigma^2)$, then $EX_i = \mu$ and therefore
$$\bar{X}_n \xrightarrow[n \to \infty]{P} \mu.$$

**4.** If the independent $X_i$'s are distributed as Negative Exponential with parameter $\lambda$,
$f(x) = \lambda e^{-\lambda x}, x > 0$, then $EX_i = 1/\lambda$ and therefore $\bar{X}_n \xrightarrow[n \to \infty]{P} 1/\lambda$.

A somewhat more involved application is that of the approximation of an entire d.f. by the so-called empirical d.f. To this effect:

**5.** Let $X_1, X_2, \ldots, X_n$ be i.i.d. r.v.'s with d.f. $F$, and define the *empirical* d.f. $F_n$ as follows. For each $x \in \Re$ and each $s \in S$,

$$F_n(x, s) = \frac{1}{n}[\text{number of } X_1(s), \ldots, X_n(s) \leq x].$$

From this definition, it is immediate that, for each fixed $x \in \Re$, $F_n(x, s)$ is a r.v. as a function of $s$, and for each fixed $s \in S$, $F_n(x, s)$ is a d.f. as a function of $x$. Actually, if we set $Y_i(x, s) = 1$ when $X_i(s) \leq x$, and $Y_i(x, s) = 0$ when $X_i(s) > x$, then $Y_i(x, \cdot), \ldots, Y_i(x, \cdot)$ are r.v.'s which are independent and distributed as $B(1, F(x))$, since $P[Y_i(x, \cdot) = 1] = P(X_i \leq x) = F(x)$. Also, $EY_i(x, \cdot) = F(x)$. Then $F_n(x, s)$ may be rewritten as:

$$F_n(x, s) = \frac{1}{n} \sum_{i=1}^{n} Y_i(x, s), \quad \text{the sample mean of } Y_1(x, s), \ldots, Y_n(x, s).$$

By omitting the sample point $s$, as is usually the case, we write $F_n(x)$ and $Y_i(x), i = 1, \ldots, n$ rather than $F_n(x, s)$ and $Y_i(x, s), i = 1, \ldots, n$, respectively. Then $F_n(x) \xrightarrow[n \to \infty]{P} F(x)$ for each $x \in \Re$. Thus, for every $x \in \Re$, the value of $F(x)$ of the (potentially unknown) d.f. $F$ is approximated by the (known) values $F_n(x)$ of the r.v.'s $F_n(x)$.

**Remark 3.** Actually, it can be shown that the convergence $F_n(x) \xrightarrow[n \to \infty]{P} F(x)$ is *Uniform* in $x \in \Re$. This implies that, for every $\varepsilon > 0$, there is a positive integer $N(\varepsilon)$ (*independent* of $x \in \Re$) such that $F_n(x) - \varepsilon < F(x) < F_n(x) + \varepsilon$ with probability as close to 1 as one pleases *simultaneously* for all $x \in \Re$, provided $n > N(\varepsilon)$.

As another application of the concept of convergence in distribution, we obtain, perhaps, the most celebrated theorem of probability theory; it is the so-called *central limit theorem*, which is stated and proved below. Comments on the significance of the CLT follow, and the section is concluded by applications and numerical examples.

**Theorem 4** (Central Limit Theorem). *Let $X_1, X_2, \ldots$ be i.i.d. r.v.'s with finite expectation $\mu$ and finite and positive variance $\sigma^2$, and let $\bar{X}_n$ be the sample mean of $X_1, \ldots, X_n$. Then:*

$$\frac{\bar{X}_n - E\bar{X}_n}{\sqrt{\mathrm{Var}(\bar{X}_n)}} = \frac{\bar{X}_n - \mu}{\frac{\sigma}{\sqrt{n}}} = \frac{\sqrt{n}(\bar{X}_n - \mu)}{\sigma} \xrightarrow[n\to\infty]{d} Z \sim N(0,1),$$

*or*

$$P\left[\frac{\sqrt{n}(\bar{X}_n - \mu)}{\sigma} \le z\right] \xrightarrow[n\to\infty]{} \Phi(z) = \int_{-\infty}^{z} \frac{1}{\sqrt{2\pi}} e^{-\frac{x^2}{2}} dx, \quad z \in \Re. \tag{1}$$

*(Also, see Remark 4 (iii).)*

**Remark 4.**

(i) Denote by $S_n$ the partial sum $\sum_{i=1}^{n} X_i$, $S_n = \sum_{i=1}^{n} X_i$, so that $ES_n = n\mu$ and $\mathrm{Var}(S_n) = n\sigma^2$. Then:

$$\frac{S_n - ES_n}{\sqrt{\mathrm{Var}(S_n)}} = \frac{S_n - n\mu}{\sigma\sqrt{n}} = \frac{\bar{X}_n - \mu}{\sigma/\sqrt{n}} = \frac{\sqrt{n}(\bar{X}_n - \mu)}{\sigma}.$$

Therefore, by (1):

$$P\left(\frac{S_n - n\mu}{\sigma\sqrt{n}} \le z\right) \xrightarrow[n\to\infty]{} \Phi(z), \quad z \in \Re. \tag{2}$$

(Although the notation $S_n$ has been used before (relation (12) in Chapter 5) to denote the sample standard deviation of $X_1, \ldots, X_n$, there should be no confusion; from the context, it should be clear what $S_n$ stands for.)

(ii) An interpretation of (1) and (2) is that, for sufficiently large $n$:

$$P\left[\frac{\sqrt{n}(\bar{X}_n - \mu)}{\sigma} \le z\right] = P\left(\frac{S_n - n\mu}{\sigma\sqrt{n}} \le z\right) \simeq \Phi(z), \quad z \in \Re. \tag{3}$$

Often this approximation is also denoted (rather loosely) as follows:

$$\frac{\sqrt{n}(\bar{X}_n - \mu)}{\sigma} \simeq N(0,1) \quad \text{or} \quad \bar{X}_n \simeq N\left(\mu, \frac{\sigma^2}{n}\right) \quad \text{or} \quad S_n \simeq N(n\mu, n\sigma^2). \tag{4}$$

(iii) Actually, it can be shown that the convergence in (1) or (2) is *Uniform* in $z \in \Re$. That is to say, if we set

$$F_n(z) = P\left[\frac{\sqrt{n}(\bar{X}_n - \mu)}{\sigma} \le z\right] = P\left(\frac{S_n - n\mu}{\sigma\sqrt{n}} \le z\right), \tag{5}$$

then

$$F_n(z) \xrightarrow[n\to\infty]{} \Phi(z) \quad \text{Uniformly in } z \in \Re. \tag{6}$$

To be more precise, for every $\varepsilon > 0$, there exists a positive integer $N(\varepsilon)$ *independent* of $z \in \Re$, such that

$$|F_n(z) - \Phi(z)| < \varepsilon \quad \text{for } n \geq N(\varepsilon) \text{ and all } z \in \Re. \tag{7}$$

**(iv)** The approximation of the probability $F_n(z)$ by $\Phi(z)$, provided by the CLT, is also referred to as *Normal approximation* for obvious reasons.

**(v)** On account of (3), the CLT also allows for the approximation of probabilities of the form $P(a < S_n \leq b)$ for any $a < b$. Indeed,

$$P(a < S_n \leq b) = P(S_n \leq b) - P(S_n \leq a)$$

$$= P\left(\frac{S_n - n\mu}{\sigma\sqrt{n}} \leq \frac{b - n\mu}{\sigma\sqrt{n}}\right) - P\left(\frac{S_n - n\mu}{\sigma\sqrt{n}} \leq \frac{a - n\mu}{\sigma\sqrt{n}}\right)$$

$$= P\left(\frac{S_n - n\mu}{\sigma\sqrt{n}} \leq b_n^*\right) - P\left(\frac{S_n - n\mu}{\sigma\sqrt{n}} \leq a_n^*\right),$$

where

$$a_n^* = \frac{a - n\mu}{\sigma\sqrt{n}} \quad \text{and} \quad b_n^* = \frac{b - n\mu}{\sigma\sqrt{n}}. \tag{8}$$

By (3),

$$P\left(\frac{S_n - n\mu}{\sigma\sqrt{n}} \leq b_n^*\right) \simeq \Phi(b_n^*) \quad \text{and} \quad P\left(\frac{S_n - n\mu}{\sigma\sqrt{n}} \leq a_n^*\right) \simeq \Phi(a_n^*),$$

so that

$$P(a < S_n \leq b) \simeq \Phi(b_n^*) - \Phi(a_n^*). \tag{9}$$

The uniformity referred to in Remark 4(iii) is what, actually, validates many of the applications of the CLT. This is the case, for instance, in Remark 4(v). Its justification is provided by Lemma 1, Section 8.6*, in the book "A Course in Mathematical Statistics," 2nd edition (1997), Academic Press, by G. G. Roussas.

**(vi)** So, the convergence in (1) is a special case of the convergence depicted in Figure 7.1, where the limiting d.f. is $\Phi$ and $F_n$ is the d.f. of $\frac{\sqrt{n}(\bar{X}_n - \mu)}{\sigma}$. This convergence holds for all $x \in \Re$ since $\Phi$ is a continuous function in $\Re$.

**Example 3.**    From a large collection of bolts which is known to contain 3% defective bolts, 1000 are chosen at random. If $X$ is the number of the defective bolts among those chosen, what is the (approximate) probability that $X$ does not exceed 5% of 1000?

**Discussion.**    With the selection of the $i$th bolt, associate the r.v. $X_i$ to take the value 1, if the bolt is defective, and 0 otherwise. Then it may be assumed that the r.v.'s $X_i, i = 1, \ldots, 1000$ are independently distributed as $B(1, 0.03)$. Furthermore, it is clear that $X = \sum_{i=1}^{1000} X_i$. Since 5% of 1000 is 50, the required probability is: $P(X \leq 50)$. Since $EX_i = 0.03$, $\text{Var}(X_i) = 0.03 \times 0.97 = 0.0291$, the CLT gives:

$$P(X \leq 50) = P(0 \leq X \leq 50) = P(-0.5 < X \leq 50)$$
$$= P(X \leq 50) - P(X \leq -0.5) \simeq \Phi(b_n^*) - \Phi(a_n^*),$$

where 
$$a_n^* = \frac{-0.5 - 1000 \times 0.03}{\sqrt{1000 \times 0.03 \times 0.97}} = -\frac{30.5}{\sqrt{29.1}} \simeq -\frac{30.5}{5.394} \simeq -5.65,$$

$$b_n^* = \frac{50 - 1000 \times 0.03}{\sqrt{1000 \times 0.03 \times 0.97}} = \frac{20}{\sqrt{29.1}} \simeq \frac{20}{5.394} \simeq 3.71,$$

so that
$$P(X \leq 50) \simeq \Phi(3.71) - \Phi(-5.65) = \Phi(3.71) = 0.999896.$$

**Example 4.**   A certain manufacturing process produces vacuum tubes whose life-times in hours are independent r.v.'s with Negative Exponential distribution with mean 1500 h. What is the probability that the total life of 50 tubes will exceed 80,000 h?

**Discussion.**   If $X_i$ is the r.v. denoting the lifetime of the $i$th vacuum tube, then $X_i, i = 1, \ldots, 50$ are independent Negative Exponentially distributed with $EX_i = \frac{1}{\lambda} = 1500$ and $\mathrm{Var}(X_i) = \frac{1}{\lambda^2} = 1500^2$. Since $nEX_i = 50 \times 1500 = 75,000, \sigma\sqrt{n} = 1500\sqrt{50}$, if we set $S_{50} = \sum_{i=1}^{50} X_i$, then the required probability is:

$$P(S_{50} > 80,000) = 1 - P(S_{50} \leq 80,000) \simeq 1 - \Phi\left(\frac{80,000 - 75,000}{1500\sqrt{50}}\right)$$

$$= 1 - \Phi\left(\frac{\sqrt{50}}{15}\right) \simeq 1 - \Phi(0.47)$$

$$= 1 - 0.680822 = 0.319178 \simeq 0.319.$$

The proof of the theorem is based on the same ideas as those used in the proof of the WLLN and goes as follows.

*Proof of Theorem 4.* Set $Z_i = \frac{X_i - \mu}{\sigma}$, so that $Z_1, \ldots, Z_n$ are i.i.d. r.v.'s with $EZ_i = 0$ and $\mathrm{Var}(Z_i) = 1$. Also,

$$\frac{1}{\sqrt{n}} \sum_{i=1}^{n} Z_i = \frac{1}{\sigma\sqrt{n}}(S_n - n\mu) = \frac{\sqrt{n}(\bar{X}_n - \mu)}{\sigma}. \tag{10}$$

With $F_n$ defined by (5), we wish to show that (6) holds (except for the uniformity assertion, with which we will not concern ourselves). By Theorem 2, it suffices to show that, for all $t$,

$$M_{\sqrt{n}(\bar{X}_n - \mu)/\sigma}(t) \xrightarrow[n \to \infty]{} M_Z(t) = e^{t^2/2}. \tag{11}$$

By means of (10), and with $M$ standing for the (common) m.g.f. of the $Z_i$'s, we have:

$$M_{\sqrt{n}(\bar{X}_n - \mu)/\sigma}(t) = M_{(1/\sqrt{n})\sum_{i=1}^{n} Z_i}(t) = M_{\sum_{i=1}^{n} Z_i}\left(\frac{t}{\sqrt{n}}\right)$$

$$= \prod_{i=1}^{n} M_{Z_i}\left(\frac{t}{\sqrt{n}}\right) = \left[M\left(\frac{t}{\sqrt{n}}\right)\right]^n. \tag{12}$$

Expand the function $M(z)$ around $z = 0$ according to Taylor's formula up to terms of second order to get:

$$M(z) = M(0) + \frac{z}{1!} \times \frac{d}{dz}M(z)|_{z=0} + \frac{z^2}{2!} \times \frac{d^2}{dz^2}M(z)|_{z=0} + R(z)$$

$$= 1 + zEZ_1 + \frac{z^2}{2}EZ_1^2 + R(z)$$

$$= 1 + \frac{z^2}{2} + R(z), \quad \text{where } \frac{1}{z^2}R(z) \to 0 \quad \text{as } z \to 0.$$

In this last formula, replace $z$ by $t/\sqrt{n}$, for fixed $t$, in order to obtain:

$$M\left(\frac{t}{\sqrt{n}}\right) = 1 + \frac{t^2}{2n} + R\left(\frac{t}{\sqrt{n}}\right), \quad nR\left(\frac{t}{\sqrt{n}}\right) \to 0 \quad \text{as } n \to \infty.$$

Therefore (12) becomes:

$$M_{\sqrt{n}(\bar{X}_n - \mu)/\sigma}(t) = \left[1 + \frac{t^2}{2n} + R\left(\frac{t}{\sqrt{n}}\right)\right]^n = \left\{1 + \frac{\frac{t^2}{2}\left[1 + \frac{2n}{t^2}R\left(\frac{t}{\sqrt{n}}\right)\right]}{n}\right\}^n,$$

and this converges to $e^{t^2/2}$, as $n \to \infty$, by Remark 2. This completes the proof of the theorem. ∎

## 7.2.2 APPLICATIONS OF THE CLT

In all of the following applications, it will be assumed that $n$ is sufficiently large, so that the CLT will apply.

**1.** Let the independent $X_i$'s be distributed as $B(1, p)$, set $S_n = \sum_{i=1}^{n} X_i$, and let $a, b$ be integers such that $0 \le a < b \le n$. By an application of the CLT, we wish to find an approximate value to the probability $P(a < S_n \le b)$.

If $p$ denotes the proportion of defective items in a large lot of certain items, then $S_n$ is the number of the actually defective items among the $n$ sampled. Then approximation of the probability $P(a < S \le b)$ is meaningful when the Binomial tables are not usable (either because of $p$ or because of $n$ or, perhaps, because of both). Here $EX_i = p$, $\text{Var}(X_i) = pq$ $(q = 1 - p)$, and therefore by (9):

$$P(a < S_n \le b) \simeq \Phi(b_n^*) - \Phi(a_n^*), \quad \text{where } a_n^* = \frac{a - np}{\sqrt{npq}}, \quad b_n^* = \frac{b - np}{\sqrt{npq}}. \tag{13}$$

**Remark 5.** If the required probability is of any one of the forms: $P(a \le S_n \le b)$ or $P(a \le S_n < b)$ or $P(a < S_n < b)$, then formula (9) applies again, provided the necessary adjustments are first made; namely, $P(a \le S_n \le b) = P(a - 1 < S_n \le b)$, $P(a \le S_n < b) = P(a - 1 < S_n \le b - 1)$, $P(a < S_n < b) = P(a < S_n \le b - 1)$. However, if the underlying distribution is continuous, then $P(a < S_n \le b) = P(a \le S_n \le b) = P(a \le S_n < b) = P(a < S_n < b)$, and no adjustments are required for the approximation in (9) to hold.

**Example 5** (Numerical).   For $n = 100$ and $p = \frac{1}{2}$ or $p = \frac{5}{16}$, find the probability $P(45 \le S_n \le 55)$.

**Discussion.**

(i) For $p = \frac{1}{2}$, it is seen (from tables) that the exact value is equal to 0.7288. For the Normal approximation, we have: $P(45 \le S_n \le 55) = P(44 < S_n \le 55)$ and, by (13):

$$a^* = \frac{44 - 100 \times \frac{1}{2}}{\sqrt{100 \times \frac{1}{2} \times \frac{1}{2}}} = -\frac{6}{5} = -1.2, \quad b^* = \frac{55 - 100 \times \frac{1}{2}}{\sqrt{100 \times \frac{1}{2} \times \frac{1}{2}}} = \frac{5}{5} = 1.$$

Therefore,    $\Phi(b^*) - \Phi(a^*) = \Phi(1) - \Phi(-1.2) = \Phi(1) + \Phi(1.2) - 1 =$ $0.841345 + 0.884930 - 1 = 0.7263$. So:

*Exact value:* 0.7288,    *Approximate value:* 0.7263,

and the exact probability is underestimated by 0.0025, or the approximating probability is about 99.66% of the exact probability.

(ii) For $p = \frac{5}{16}$, the exact probability is almost 0; 0.0000. For the approximate probability, we find $a^* = 2.75$ and $b^* = 4.15$, so that $\Phi(b^*) - \Phi(a^*) = 0.0030$. Thus:

*Exact value:* 0.0000,    *Approximate value:* 0.0030,

and the exact probability is overestimated by 0.0030.

**2.** If the underlying distribution is $P(\lambda)$, then $ES_n = \mathrm{Var}(S_n) = n\lambda$ and formulas (8) and (9) become:

$$P(a < S_n \le b) \simeq \Phi(b_n^*) - \Phi(a_n^*), \quad a_n^* = \frac{a - n\lambda}{\sqrt{n\lambda}}, \quad b_n^* = \frac{b - n\lambda}{\sqrt{n\lambda}}.$$

The comments made in Remark 4 apply here also.

**Example 6** (Numerical).   In the Poisson distribution $P(\lambda)$, let $n$ and $\lambda$ be so that $n\lambda = 16$ and find the probability $P(12 \le S_n \le 21)(= P(11 < S_n \le 21))$.

**Discussion.**   The exact value (found from tables) is 0.7838. For the Normal approximation, we have:

$$a^* = \frac{11 - 16}{\sqrt{16}} = -\frac{5}{4} = -1.25, \quad b^* = \frac{21 - 16}{\sqrt{16}} = \frac{5}{4} = 1.25,$$

so that $\Phi(b^*) - \Phi(a^*) = \Phi(1.25) - \Phi(-1.25) = 2\Phi(1.25) - 1 = 2 \times 0.894350 - 1 = 0.7887$. So:

*Exact value:* 0.7838,    *Approximate value:* 0.7887,

and the exact probability is overestimated by 0.0049, or the approximating probability is about 100.63% of the exact probability.

### 7.2.3 THE CONTINUITY CORRECTION

When a discrete distribution is approximated by the Normal distribution, the error committed is easy to see in a Geometric picture. This is done, for instance in Figure 7.5, where the p.d.f. of the $B(10, 0.2)$ distribution is approximated by the p.d.f. of the $N(10 \times 0.2, 10 \times 0.2 \times 0.8) = N(2, 1.6)$ distribution (see relation (4)). From the same figure, it is also clear how the approximation may be improved.

Now

$$P(1 < S_n \leq 3) = P(2 \leq S_n \leq 3) = f_n(2) + f_n(3)$$
$$= \text{shaded area,}$$

while the approximation without correction is the area bounded by the Normal curve, the horizontal axis, and the abscissas 1 and 3. Clearly, the correction given by the area bounded by the Normal curve, the horizontal axis, and the abscissas 1.5 and 3.5, is closer to the exact area.

To summarize, under the conditions of the CLT, and for discrete r.v.'s, $P(a < S_n \leq b) \simeq \Phi(b^*) - \Phi(a^*)$, where $a^* = \frac{a-n\mu}{\sigma\sqrt{n}}$ and $b^* = \frac{b-n\mu}{\sigma\sqrt{n}}$ *without* continuity correction, and $P(a < S_n \leq b) \simeq \Phi(b') - \Phi(a')$, where $a' = \frac{a+0.5-n\mu}{\sigma\sqrt{n}}$ and $b' = \frac{b+0.5-n\mu}{\sigma\sqrt{n}}$ *with* continuity correction.

For integer-valued r.v.'s and probabilities of the form $P(a \leq S_n \leq b)$, we first rewrite the expression as follows:

$$P(a \leq S_n \leq b) = P(a - 1 < S_n \leq b),$$

and then apply the preceding approximations in order to obtain:

$$P(a \leq S_n \leq b) \simeq \Phi(b^*) - \Phi(a^*),$$

where $a^* = \frac{a-1-n\mu}{\sigma\sqrt{n}}$ and $b^* = \frac{b-n\mu}{\sigma\sqrt{n}}$ *without* continuity correction, and $P(a \leq S_n \leq b) \simeq \Phi(b') - \Phi(a')$, where $a' = \frac{a-0.5-n\mu}{\sigma\sqrt{n}}$ and $b' = \frac{b+0.5-n\mu}{\sigma\sqrt{n}}$ *with* continuity correction. Similarly for the intervals $[a, b)$ and $(a, b)$.

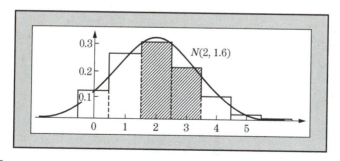

**FIGURE 7.5**

Exact and approximate values for the probability $P(a \leq S_n \leq b) = P(a - 1 < S_n \leq b) = P(1 < S_n \leq 3)$.

The improvement brought about by the continuity correction is demonstrated by the following numerical examples.

**Example 5** (continued).

**Discussion.**

(i) For $p = \frac{1}{2}$, we get:

$$a' = \frac{44 + 0.5 - 100 \times \frac{1}{2}}{\sqrt{100 \times \frac{1}{2} \times \frac{1}{2}}} = -\frac{5.5}{5} = -1.1,$$

$$b' = \frac{55 + 0.5 - 100 \times \frac{1}{2}}{\sqrt{100 \times \frac{1}{2} \times \frac{1}{2}}} = \frac{5.5}{5} = 1.1,$$

so that:

$$\Phi(b') - \Phi(a') = \Phi(1.1) - \Phi(-1.1) = 2\Phi(1.1) - 1$$
$$= 2 \times 0.864334 - 1 = 0.7286.$$

Thus, we have:

*Exact value: 0.7288,*

*Approximate value with continuity correction: 0.7286,*

and the approximation underestimates the probability by only 0.0002, or the approximating probability (with continuity correction) is about 99.97% of the exact probability.

(ii) For $p = \frac{5}{16}$, we have $a' = 2.86, b' = 5.23$ and $\Phi(b') - \Phi(a') = 0.0021$. Then:

*Exact value: 0.0000,*

*Approximate value with continuity correction: 0.0021,*

and the probability is overestimated by only 0.0021.

**Example 6** (continued).

**Discussion.** Here:

$$a' = \frac{11 + 0.5 - 16}{\sqrt{16}} = -\frac{4.5}{4} = -1.125, \quad b' = \frac{21 + 0.5 - 16}{\sqrt{16}} = \frac{5.5}{4} = 1.375,$$

so that:

$$\Phi(b') - \Phi(a') = \Phi(1.375) - \Phi(-1.125) = \Phi(1.375) + \Phi(1.125) - 1 = 0.7851.$$

Thus:

*Exact value: 0.7838,*

*Approximate value with continuity correction: 0.7851,*

and the approximation overestimates the probability by only 0.0013, or the approximating probability (with continuity correction) is about 100.17% of the exact probability.

## EXERCISES

**2.1**   Let $X_1, \ldots, X_n$ be i.i.d. r.v.'s, and for a positive integer $k$, suppose that $EX_1^k$ is finite. Form the $k$th *sample mean* $\bar{X}_n^{(k)}$ defined by

$$\bar{X}_n^{(k)} = \frac{1}{n} \sum_{i=1}^{n} X_i^k.$$

Then show that:

$$\bar{X}_n^{(k)} \xrightarrow[n \to \infty]{P} EX_1^k.$$

**2.2**   Let $X$ be a r.v. with p.d.f. $f_X(x) = c\alpha^x, x = 0, 1, \ldots (0 < \alpha < 1)$. Then $c = 1 - \alpha$ by Exercise 2.8 in Chapter 2.
  **(i)**   Show that the m.g.f. of $X$ is: $M_X(t) = \frac{1-\alpha}{1-\alpha e^t}$,    $t < -\log \alpha$.
  **(ii)**  Use the m.g.f. to show that $EX = \frac{\alpha}{1-\alpha}$.
  **(iii)** If $X_1, \ldots, X_n$ is a random sample from $f_X$, show that the WLLN holds by showing that

$$M_{\bar{X}_n}(t) \xrightarrow[n \to \infty]{} e^{\alpha t/(1-\alpha)} = M_{EX}(t), \quad t < -\log \alpha.$$

**Hint.** Expand $e^t$ around 0 up to second term, according to Taylor's formula, $e^t = 1 + t + R(t)$, where $\frac{1}{t} R(t) \xrightarrow[t \to 0]{} 0$, replace $t$ by $\frac{t}{n}$, and use the fact that $(1 + \frac{x_n}{n})^n \to e^x$, if $x_n \to x$ as $n \to \infty$.

**2.3**   Let the r.v. $X$ be distributed as $B(150, 0.6)$. Then:
  **(i)**   Write down the formula for the exact probability $P(X \le 80)$.
  **(ii)**  Use the CLT in order to find an approximate value for the above probability. (Do not employ the continuity correction.)
  **Hint.** Write $P(X \le 80) = P(-0.5 < X \le 80)$ and then apply the CLT.

**2.4**   A Binomial experiment with probability $p$ of a success is repeated independently 1000 times, and let $X$ be the r.v. denoting the number of successes. For $p = \frac{1}{2}$ and $p = \frac{1}{4}$, find:
  **(i)**   The exact probability $P(1000p - 50 \le X \le 1000p + 50)$.
  **(ii)**  Use the CLT to find an approximate value for this probability.
  **Hint.** For part (i), just write down the right formula. For part (ii), first bring it under the form $P(a < X \le b)$, and then compute the approximate probability without continuity correction.

**2.5**   Let $X_1, \ldots, X_{100}$ be independent r.v.'s distributed as $B(1, p)$. Then:
  **(i)**   Write out the expression for the exact probability $P(\sum_{i=1}^{100} X_i = 50)$.
  **(ii)**  Use of CLT in order to find an approximate value for this probability.
  **(iii)** What is the numerical value of the probability in part (ii) for $p = 0.5$?
  **Hint.** For part (ii), first observe that $P(X = 50) = P(49.5 < X \le 50)$, and then apply the CLT, without continuity correction.

**2.6** Fifty balanced dice are tossed once, and let $X$ be the r.v. denoting the sum of the upturned spots. Use the CLT to find an approximate value of the probability $P(150 \leq X \leq 200)$.

**Hint.** With the $i$th die, associate the r.v. $X_i$ which takes on the values 1 to 6, each with probability $1/6$. These r.v.'s may be assumed to be independent and $X = \sum_{i=1}^{50} X_i$.

**2.7** One thousand cards are drawn (with replacement) from a standard deck of 52 playing cards, and let $X$ be the r.v. denoting the total number of aces drawn. Use the CLT to find an approximate value of the probability $P(65 \leq X \leq 90)$.

**Hint.** Write $P(65 \leq X \leq 90) = P(64 < X \leq 90)$ and then apply the CLT, without continuity correction.

**2.8** From a large collection of bolts which is known to contain 3% defective bolts, 1000 are chosen at random, and let $X$ be the r.v. denoting the number of defective bolts among those chosen. Use the CLT to find an approximate value of the probability that $X$ does not exceed 5% of 1000.

**Hint.** With the $i$th bolt drawn, associate the r.v. $X_i$ which takes on the value 1, if the bolt drawn is defective, and 0 otherwise. Since the collection of bolts is large, we may assume that after each drawing, the proportion of the remaining defective bolts remains (approximately) the same. This implies that the independent r.v.'s $X_1, \ldots, X_{1000}$ are distributed as $B(1, 0.03)$ and that $X = \sum_{i=1}^{1000} X_i \sim B(1000, 0.3)$. Next, write $P(X \leq 50) = P(-0.5 < X \leq 50)$ and use the CLT, without continuity correction.

**2.9** A manufacturing process produces defective items at the constant (but unknown to us) proportion $p$. Suppose that $n$ items are sampled independently, and let $X$ be the r.v. denoting the number of defective items among the $n$, so that $X \sim B(n, p)$. Determine the smallest value of the sample size $n$, so that

$$P\left(\left|\frac{X}{n} - p\right| \leq 0.05\sqrt{pq}\right) \geq 0.95 \quad (q = 1 - p):$$

**(i)** By utilizing the CLT.

**(ii)** By using the Tchebichev inequality.

**(iii)** Compare the answers in parts (i) and (ii).

**2.10** Suppose that 53% of the voters favor a certain legislative proposal. How many voters must be sampled so that the observed relative frequency of those favoring the proposal will not differ from the assumed frequency by more than 2% with probability 0.99?

**Hint.** With the $i$th voter sampled, associate the r.v. $X_i$ which takes on the value 1, if the voter favors the proposal, and 0 otherwise. Then it may be assumed that the r.v.'s $X_1, \ldots, X_n$ are independent and their common distribution is $B(1, 0.53)$. Furthermore, the number of the voters favoring the proposal is $X = \sum_{i=1}^{n} X_i$. Use the CLT in order to find the required probability.

**2.11** In playing a game, you win or lose $1 with probability 0.5, and you play the game independently 1000 times. Use the CLT to find an approximate value of the probability that your fortune (i.e., the total amount you won or lost) is at least $10.

**Hint.** With the $i$th game, associate the r.v. $X_i$ which takes on the value 1 if $1 is won, and $-1$ if $1 is lost. Then the r.v.'s $X_1, \ldots, X_{1000}$ are independent, and the fortune $X$ is given by $\sum_{i=1}^{1000} X_i$. Then write

$P(X \geq 10) = P(10 \leq X \leq 1000) = P(9 < X \leq 1000)$ and use the CLT, without continunity correction.

**2.12** It is known that the number of misprints in a page of a certain publication is a r.v. $X$ having the Poisson distribution with parameter $\lambda$. If $X_1, \ldots, X_n$ are the misprints counted in $n$ pages, use the CLT to determine the (approximate) probability that the total number of misprints is:

   **(i)** Not more than $\lambda n$.
   **(ii)** Not less than $\lambda n$.
   **(iii)** Not less than $\lambda n/2$, but not more than $3\lambda n/4$.
   **(iv)** Give the numerical values in parts (i)–(iii) for $\lambda n = 100$ (which may be interpreted, e.g., as one misprint per 4 pages ($\lambda = 0.25$) in a book of 400 pages).

**Hint.** In all cases, first bring it under the form $P(a < S_n \leq b)$, and then use the CLT, without continuity correction. For part (ii), consider two cases: $\lambda n$ is an integer and $\lambda n$ is not an integer. For part (iii), do the same for $\lambda n/2$.

**2.13** Let the r.v. $X$ be distributed as $P(100)$. Then:

   **(i)** Write down the formula for the exact probability $P(X \leq 116)$.
   **(ii)** Use the CLT appropriately in order to find an approximate value for the above probability. (Do not use the continuity correction.)

**Hint.** Select $n$ large and $\lambda$ small, so that $n\lambda = 100$ and look at $X$ as the sum $\sum_{i=1}^{n} X_i$ of $n$ independent r.v.'s $X_1, \ldots, X_n$ distributed as $P(\lambda)$.

**2.14** A certain manufacturing process produces vacuum tubes whose lifetimes in hours are independently distributed r.v.'s with Negative Exponential distribution with mean 1500 hours. Use the CLT in order to find an approximate value for the probability that the total lifetime of 50 tubes will exceed 80,000 hours.

**2.15** The lifespan of an electronic component in a (complicated) system is a r.v. $X$ having the Negative Exponential distribution with parameter $\lambda$.

   **(i)** What is the probability that said lifespan will be at least $t$ time units?
   **(ii)** If the independent r.v.'s $X_1, \ldots, X_n$ represent the lifespans of $n$ spare items such as the one described above, then $Y = \sum_{i=1}^{n} X_i$ is the combined lifespan of these $n$ items. Use the CLT in order to find an approximate value of the probability $P(t_1 \leq Y \leq t_2)$, where $0 < t_1 < t_2$ are given time units.

(iii) Compute the numerical answer in part (i), if $t = -\log(0.9)/\lambda$, (where log is the natural logarithm).

(iv) Do the same for part (ii), if $\lambda = 1/10, n = 36, t_1 = 300$, and $t_2 = 420$.

**2.16** Let the independent r.v.'s $X_1, \ldots, X_n$ be distributed as $U(0, 1)$.

(i) Use the CLT to find an approximate value for the probability $P(a \leq \bar{X} \leq b)$ $(a < b)$.

(ii) What is the numerical value of this probability for $n = 12, a = 7/16$, and $b = 9/16$?

**2.17** If the independent r.v.'s $X_1, \ldots, X_{12}$ are distributed as $U(0, \theta)$ $(\theta > 0)$, use the CLT to show that the probability $P(\frac{\theta}{4} < \bar{X} < \frac{3\theta}{4})$ is approximately equal to 0.9973.

**2.18** Refer to Exercise 3.42 in Chapter 3 and let $X_i, i = 1, \ldots, n$ be the diameters of $n$ ball bearings. If $EX_i = \mu = 0.5$ inch and s.d. $(X_i) = \sigma = 0.0005$ inch, use the CLT to determine the smallest value of $n$ for which $P(|\bar{X} - \mu| \leq 0.0001)$ $= 0.99$, where $\bar{X}$ is the sample mean of the $X_i$'s.

**2.19** The i.i.d. r.v.'s $X_1, \ldots, X_{100}$ have (finite) mean $\mu$ and variance $\sigma^2 = 4$. Use the CLT to determine the value of the constant $c$ for which $P(|\bar{X} - \mu| \leq c)$ $= 0.90$, where $\bar{X}$ is the sample means of the $X_i$'s.

**2.20** Let $X_1, \ldots, X_n$ be i.i.d. r.v.'s with (finite) expectation $\mu$ and (finite and positive) variance $\sigma^2$, and let $\bar{X}_n$ be the sample mean of the $X_i$'s. Determine the smallest value of the sample size $n$, in terms of $k$ and $p$, for which $P(|\bar{X}_n - \mu| \leq k\sigma) \geq p$, where $p \in (0, 1), k > 0$. Do so by using:

(i) The CLT.

(ii) The Tchebichev inequality.

(iii) Find the numerical values of $n$ in parts (i) and (ii) for $p = 0.90, 0.95, 0.99$ and $k = 0.50, 0.25, 0.10$ for each value of $p$.

(iv) Use the CLT approach in order to compute the (approximate) probability $P(|\bar{X}_n - \mu| \leq k\sigma)$ for the value of $n$ found in part (ii).

**2.21** Refer to Exercise 3.41 in Chapter 3, and suppose that the r.v. $X$ considered there has $EX = 2000$ and s.d.$(X) = 200$, but is not necessarily Normally distributed. Also, consider another manufacturing process producing light bulbs whose mean lifespan is claimed to be 10% higher than the mean lifespan of the bulbs produced by the existing process; it is assumed that the s.d. remains the same for the new process. How many bulbs manufactured by the new process must be examined to establish the claim of their superiority (should that be the case) with probability 0.95?

**Hint.** Let $Y$ be the r.v. denoting the lifespan of a light bulb manufactured by the new process. We do not necessarily assume that $Y$ is Normally distributed. If the claim made is correct, then $EY = 2000 + 10\% \times 2000 = 2200$, whereas s.d.$(Y) = 200$. A random sample from $Y$ produces the sample mean $\bar{Y}_n$ for which $E\bar{Y}_n = 2200$ (under the claim) and $\text{Var}(\bar{Y}_n) = 200^2/n$, and we must determine $n$, so that $P(\bar{Y}_n > 2000) = 0.95$. If the new process was the same as the old one, then, for all sufficiently large $n$, $P(\bar{Y}_n > 2000) \simeq 0.50$. So, if $P(\bar{Y}_n > 2000) = 0.95$, the claim made would draw support.

**2.22** (i) Consider the i.i.d. r.v.'s $X_1, \ldots, X_n$ and $Y_1, \ldots, Y_n$ with expectation $\mu$ and variance $\sigma^2$, both finite, and let $\bar{X}_n$ and $\bar{Y}_n$ be the respective sample means. Use the CLT in order to determine the sample size $n$, so that
$$P(|\bar{X}_n - \bar{Y}_n| \leq 0.25\sigma) = 0.95.$$
(ii) Let the random samples $X_1, \ldots, X_n$ and $Y_1, \ldots, Y_n$ be as in part (i), but we do not assume that they are coming from the same distribution. We do assume, however, that they have the same mean $\mu$ and the same variance $\sigma^2$, both finite. Then determine $n$ as required above by using the Tchebichev inequality.

**Hint.** Set $Z_i = X_i - Y_i$ and then work as in Exercise 2.20(ii) with the i.i.d. r.v.'s $Z_1, \ldots, Z_n$. Finally, revert to the $X_i$'s and the $Y_i$'s.

**2.23** Let $X_i$, $i = 1, \ldots, n$, $Y_i$, $i = 1, \ldots, n$ be independent r.v.'s such that the $X_i$'s are identically distributed with $EX_i = \mu_1$, $\text{Var}(X_i) = \sigma^2$, both finite, and the $Y_i$'s are identically distributed with $EY_i = \mu_2$ and $\text{Var}(Y_i) = \sigma^2$, both finite. If $\bar{X}_n$ and $\bar{Y}_n$ are the respective sample means of the $X_i$'s and the $Y_i$'s, then:
(i) Show that $E(\bar{X}_n - \bar{Y}_n) = \mu_1 - \mu_2, \text{Var}(\bar{X}_n - \bar{Y}_n) = \frac{2\sigma^2}{n}$.
(ii) Use the CLT in order to show that $\frac{\sqrt{n}[(\bar{X}_n - \bar{Y}_n) - (\mu_1 - \mu_2)]}{\sigma\sqrt{2}}$ is asymptotically distributed as $N(0, 1)$.

**Hint.** Set $Z_i = X_i - Y_i$ and work with the i.i.d. r.v.'s $Z_1, \ldots, Z_n$; then revert to the $X_i$'s and the $Y_i$'s.

**2.24** Within a certain period of time, let $n$ be the number of health claims submitted to an insurance company, and suppose that the sizes of the claims are independent r.v.'s $X_1, \ldots, X_n$ whose p.d.f. is the Negative Exponential with parameter $\lambda$; that is, $f(x) = \lambda e^{-\lambda x}, x > 0$. Let $P$ be the premium charged for each policy, and set $S_n = X_1 +, \ldots, +X_n$. If the total amount of claims is not to exceed the total premium of the $n$ policies sold with probability $p$:
(i) Express the premium $P$ in terms of $n$, $\lambda$, and $p$.
(ii) What is the value of $P$ is part (i) for $n = 10,000$, $\lambda = 1/1000$, and $p = 0.99$?

**Hint.** Employ the CLT.

**2.25** The lifetime of a light bulb is a r.v. $X$ having Negative Exponential distribution with parameter $\lambda = 0.2$ hour (i.e., the p.d.f. of $X$ is given by $f(x) = \lambda e^{-\lambda x}, x > 0$ ($\lambda = 0.2$)). If $X_1, \ldots, X_n$ are the independent lifetimes of $n$ such light bulbs:
(i) Determine the smallest value of $n$ (in terms of the constant $c > 0$ and $p$), so that
$$P(|\bar{X}_n - EX_1| \leq c) \geq p.$$
(ii) What is the numerical value of $n$ for $c = 1$ and $p = 0.950$?

**Hint.** Use the CLT.

**2.26** Certain measurements are rounded up to the nearest integer, and let $X$ be the r.v. denoting the difference between an actual measurement and its rounded-up value. It is assumption that $X \sim U(-0.5, 0.5)$. For a random sample of size $n = 100$, compute the probability that the sample mean and the true mean do not differ in absolute value by more than 0.1.

**Hint.** Use the CLT.

**2.27** A given police department collects \$100 per traffic ticket. If there are $n$ drivers in a city, and the $i$th driver gets $X_i$ traffic tickets in 1 year, then the city collects $T_n = 100(X_1 + \cdots + X_n)$ dollars in 1 year. If $X_1, \ldots, X_n$ are independent and identically distributed as follows: $P(X_1 = 0) = 1/18$, $P(X_1 = 1) = 16/18$, $P(X_1 = 2) = 1/18$:

   **(i)** Determine the limit $P(T_n > 97n)$, as $n \to \infty$.
   **(ii)** Find an approximate value for the probability
          $P(995,000 < T_n \leq 1,005,000)$ if $n = 10,000$.

**Hint.** Use the CLT.

**2.28** The independent r.v.'s $X_1, \ldots, X_{100}$ are distributed as $B(1, 0.1)$, and let $S_{100} = X_1 + \cdots + X_{100}$.

   **(i)** What is the distribution of $S_{100}$ and why? (Just give the statement and justification; do not prove anything.)
   **(ii)** Justify the statements that $ES_{100} = 10$ and $\mathrm{Var}(S_{100}) = 9$.
   **(iii)** Give the expression for the exact probability: $P(S_{100} = 16)$.
   **(iv)** Apply the Normal approximation (CLT) to the expression
          $P(15 < S_{100} < 16)$ in order to compute an approximate value to the probability $P(S_{100} = 16)$.

**2.29** A fair coin is tossed independently a large number of $n$ times, and each time you win or lose \$1, depending on whether heads or tails appear, respectively, so that your total gain or loss is $S_n = \sum_{i=1}^{n} X_i$, where $X_i = 1$ when heads appear, and $X_i = -1$ when tails appear, $i = 1, \ldots, n$.

   **(i)** For a positive integer $k$, find an approximate value of the $P(S_n = k)$, expressed in terms of $n$, $k$, and $p$.
   **(ii)** Let $p = \frac{1}{2}$. Then for $n = 50$, compute the (approximate) probabilities: $P(S_{50} = k)$ when $k = 10, 15, 20, 25$. Do the same when $n = 100$, and $k = 25, 40, 50, 60$.

**2.30** The lifetime of a light bulb is a r.v. $X$ distributed as Negative Exponential with parameter $\lambda$. Consider $n$ such light bulbs which are used successively, a burned out one being replaced immediately by another one, so that the total lifetime of these light bulbs is $S_n = \sum_{i=1}^{n} X_i$. Assuming that $n$ is sufficiently large (so that the CLT will apply):

   **(i)** Find an approximate expression for the probability $P(S_n > k \times \frac{n}{\lambda})$, where $k$ is a positive constant.
   **(ii)** Compute the (approximate) probability in part (i) for $k = 1.2, 1.4, 1.5$, and $n = 40$.

**2.31** When playing roulette, there are three possible outcomes (in terms of colors) red ($R$), black ($B$), and green ($G$) with respective probabilities $9/19, 9/19$, and $1/19$. Suppose you play independently $n$ times by betting \$1 each time on $R$. Then your total return (gain or loss) is $S_n = \sum_{i=1}^{n} X_i$, where $X_i = 1$ with probability $9/19$ (when $R$ occurs), and $X_i = -1$ with probability $10/19$ (when either $B$ or $G$ occurs).

    **(i)** For $c > 0$ integer, express the approximate value of the probability $P(S_n > c)$, assuming that $n$ is sufficiently large, so that the CLT will apply.

    **(ii)** How does this probability become for $n = 2k, k > 0$ integer, and $c = k - 1$?

    **(iii)** What is the numerical value of the probability in part (ii) for $k = 20$?

**2.32** Suppose the r.v. $X$ denotes the yearly claim submitted to an insurance company by each automobile insured by it. Suppose $EX = \mu \in \Re$ and $\sigma^2(X) = \sigma^2 \in (0, \infty)$. Let $X_i, i = 1, \ldots, n$ be $n$ independent claims distributed as $X$, so that the total amount of claims over a year's period is $S_n = \sum_{i=1}^{n} X_i$, and the total amount of expected claims is $n\mu$.

    **(i)** Derive an expression for the approximate probability that $S_n$ exceeds $n\mu$ by the proportion $cn\mu, 0 < c < 1$, i.e.,

$$P[S_n > (c+1)n\mu].$$

    **(ii)** How does the probability in part (i) become for $\mu = \$500, \sigma = \$100$, and $c = 0.01$?

    **(iii)** Compute the numerical value of the probability in part (ii) for $n = 1000, 5000, 15,000$.

**2.33** The ideal size of the freshman class in a college is 200, and from past experience, it is known that (on the average) 25% of those admitted actually enroll. If admission is offered to 750 students, denote by $X$ the number of those actually enrolled. Then:

    **(i)** Write down the formula for computing the exact probability $P(X > 200)$.

    **(ii)** Use the CLT to find an approximate value to the probability in part (i).

**Hint.** Set $X_i = \begin{cases} 1, & \text{if the } i\text{th admitted student enrolls} \\ 0, & \text{otherwise} \end{cases}, i = 1, \ldots, 750,$

make suitable assumptions on the $X_i$'s, and observe that $X = \sum_{i=1}^{750} X_i$.

## 7.3 FURTHER LIMIT THEOREMS

Convergence in probability enjoys some of the familiar properties of the usual pointwise convergence. One such property is stated below in the form of a theorem whose proof is omitted.

**Theorem 5.**

(i) *For $n \geq 1$, let $X_n$ and $X$ be r.v.'s such that $X_n \xrightarrow[n \to \infty]{P} X$, and let $g$ be a continuous real-valued function; that is, $g : \Re \to \Re$ continuous. Then the r.v.'s $g(X_n), n \geq 1$, also converge in probability to $g(X)$; that is, $g(X_n) \xrightarrow[n \to \infty]{P} g(X)$. More generally:*

(ii) *For $n \geq 1$, let $X_n, Y_n, X,$ and $Y$ be r.v.'s such that $X_n \xrightarrow[n \to \infty]{P} X$, $Y_n \xrightarrow[n \to \infty]{P} Y$, and let $g$ be a continuous real-valued function; that is, $g : \Re^2 \to \Re$ continuous. Then the r.v.'s $g(X_n, Y_n), n \geq 1$, also converge in probability to $g(X, Y)$; that is, $g(X_n, Y_n) \xrightarrow[n \to \infty]{P} g(X, Y)$. (This part also generalizes in an obvious manner to $k$ r.v.'s $\{X_n^{(i)}\}, n \geq 1, X_i, i = 1, \ldots, k.$)*

To this theorem, there is the following important corollary.

**Corollary.**  *If $X_n \xrightarrow[n \to \infty]{P} X$ and $Y_n \xrightarrow[n \to \infty]{P} Y$, then:*

(i) *$aX_n + bY_n \xrightarrow[n \to \infty]{P} aX + bY$, where $a$ and $b$ are constants; and, in particular, $X_n + Y_n \xrightarrow[n \to \infty]{P} X + Y$.*

(ii) *$X_n Y_n \xrightarrow[n \to \infty]{P} XY$.*

(iii) *$\frac{X_n}{Y_n} \xrightarrow[n \to \infty]{P} \frac{X}{Y}$, provided $P(Y_n \neq 0) = P(Y \neq 0) = 1$.*

*Proof.* Although the proof of the theorem was omitted, the corollary can be proved. Indeed, all one has to do is to take: $g : \Re^2 \to \Re$ as follows, respectively, for parts (i)–(iii), and observe that it is continuous: $g(x, y) = ax + by$ (and, in particular, $g(x, y) = x + y$); $g(x, y) = xy$; $g(x, y) = x/y, \, y \neq 0$.  ∎

Actually, a special case of the preceding corollary also holds for convergence in distribution. Specifically, we have

**Theorem 6** (Slutsky Theorem).  *Let $X_n \xrightarrow[n \to \infty]{d} X$ and let $Y_n \xrightarrow[n \to \infty]{d} c$, a constant $c$ rather than a (proper) r.v. $Y$. Then:*

(i) $X_n + Y_n \xrightarrow[n \to \infty]{d} X + c$; (ii) $X_n Y_n \xrightarrow[n \to \infty]{d} cX$; (iii) $\frac{X_n}{Y_n} \xrightarrow[n \to \infty]{d} \frac{X}{c}$, *provided $P(Y_n \neq 0) = 1$ and $c \neq 0$.*

*In terms of d.f.'s, these convergences are written as follows, always as $n \to \infty$ and for all $z \in \Re$ for which: $z - c$ is a continuity point of $F_X$ for part (i); $z/c$ is a continuity point of $F_X$ for part (ii); $cz$ is a continuity point of $F_X$ for part (iii):*

$$P(X_n + Y_n \leq z) \to P(X + c \leq z) = P(X \leq z - c), \quad \text{or}$$
$$F_{X_n+Y_n}(z) \to F_X(z - c);$$

$$P(X_n Y_n \leq z) \to P(cX \leq z) = \begin{cases} P(X \leq \frac{z}{c}), & c > 0 \\ P(X \geq \frac{z}{c}), & c < 0 \end{cases}, \quad \text{or}$$

$$F_{X_n Y_n}(z) \to \begin{cases} F_X(\frac{z}{c}), & c > 0 \\ 1 - P(X < \frac{z}{c}) = 1 - F_X(\frac{z}{c}), & c < 0; \end{cases}$$

$$P\left(\frac{X_n}{Y_n} \leq z\right) \to P\left(\frac{X}{c} \leq z\right) = \begin{cases} P(X \leq cz), & c > 0 \\ P(X \geq cz), & c < 0 \end{cases}, \quad \text{or}$$

$$F_{\frac{X_n}{Y_n}}(z) \to \begin{cases} F_X(cz), & c > 0 \\ 1 - P(X < cz) = 1 - F_X(cz), & c < 0. \end{cases}$$

The proof of this theorem, although conceptually not complicated, is, nevertheless, long and is omitted. Recall, however, that $Y_n \xrightarrow[n\to\infty]{d} c$ if and only if $Y_n \xrightarrow[n\to\infty]{P} c$, and this is the way the convergence of $Y_n$ is often stated in the theorem.

As a simple concrete application of Theorem 5, consider the following example.

**Example 7.** Suppose $X_n \xrightarrow[n\to\infty]{d} X \sim N(\mu, \sigma^2)$, and let $c_n, c, d_n$, and $d$ be constants such that $c_n \to c$ and $d_n \to d$. Then $c_n X_n + d_n \xrightarrow[n\to\infty]{d} Y \sim N(c\mu + d, c^2\sigma^2)$.

**Discussion.** Trivially, $c_n \xrightarrow[n\to\infty]{d} c$ and $d_n \xrightarrow[n\to\infty]{d} d$, so that, by Theorem 6(ii), $c_n X_n \xrightarrow[n\to\infty]{d} cX$, and by Theorem 6(i), $c_n X_n + d_n \xrightarrow[n\to\infty]{d} cX + d$. However, $X \sim N(\mu, \sigma^2)$ implies that $cX + d \sim N(c\mu + d, c^2\sigma^2)$. Thus, $c_n X_n + d_n \xrightarrow[n\to\infty]{d} cX + d = Y \sim N(c\mu + d, c^2\sigma^2)$.

The following result is an application of Theorems 5 and 6 and is of much use in statistical inference. For its formulation, let $X_1, \ldots, X_n$ be i.i.d. r.v.'s with finite mean $\mu$ and finite and positive variance $\sigma^2$, and let $\bar{X}_n$ and $S_n^2$ be the sample mean and the "adjusted" (in the sense that $\mu$ is replaced by $\bar{X}_n$) sample variance (which we have denoted by $\bar{S}_n^2$ in relation (13) of Chapter 5); that is, $\bar{X}_n = \frac{1}{n}\sum_{i=1}^n X_i$, $S_n^2 = \frac{1}{n}\sum_{i=1}^n (X_i - \bar{X}_n)^2$.

**Theorem 7.** *Under the assumptions just made and the notation introduced, it holds:*
*(i) $S_n^2 \xrightarrow[n\to\infty]{P} \sigma^2$; (ii) $\frac{\sqrt{n}(\bar{X}_n - \mu)}{S_n} \xrightarrow[n\to\infty]{d} Z \sim N(0, 1).$*

*Proof.* (i) Recall that $\sum_{i=1}^n (X_i - \bar{X}_n)^2 = \sum_{i=1}^n X_i^2 - n\bar{X}_n^2$, so that $S_n^2 = \frac{1}{n}\sum_{i=1}^n X_i^2 - \bar{X}_n^2$. Since $EX_i^2 = \text{Var}(X_i) + (EX_i)^2 = \sigma^2 + \mu^2$, the WLLN applies to the i.i.d. r.v.'s

$X_1^2, \ldots, X_n^2$ and gives: $\frac{1}{n} \sum_{i=1}^{n} X_i^2 \xrightarrow[n\to\infty]{P} \sigma^2 + \mu^2$. Also, $\bar{X}_n \xrightarrow[n\to\infty]{P} \mu$, by the WLLN again,

and then $\bar{X}_n^2 \xrightarrow[n\to\infty]{P} \mu^2$ by Theorem 5(i). Hence, by Theorem 5(ii),

$$\frac{1}{n} \sum_{i=1}^{n} X_i^2 - \bar{X}_n^2 \xrightarrow[n\to\infty]{P} (\sigma^2 + \mu^2) - \mu^2 = \sigma^2,$$

which is what part (i) asserts.

(ii) Part (i) and Theorem 5(i) imply that $S_n \xrightarrow[n\to\infty]{P} \sigma$, or $\frac{S_n}{\sigma} \xrightarrow[n\to\infty]{P} 1$. By Theorem 4,

$\frac{\sqrt{n}(\bar{X}_n - \mu)}{\sigma} \xrightarrow[n\to\infty]{d} Z \sim N(0, 1)$. Then Theorem 6(iii) applies and gives:

$$\frac{\sqrt{n}(\bar{X}_n - \mu)/\sigma}{S_n/\sigma} = \frac{\sqrt{n}(\bar{X}_n - \mu)}{S_n} \xrightarrow[n\to\infty]{d} Z \sim N(0, 1). \qquad \blacksquare$$

**Remark 6.** Part (ii) of the theorem states, in effect, that for sufficiently large $n$, $\sigma$ may be replaced in the CLT by the adjusted sample standard deviation $S_n$, and the resulting expression still has a distribution which is close to the $N(0, 1)$ distribution.

The WLLN states that $\bar{X}_n \xrightarrow[n\to\infty]{d} \mu$, which, for a real-valued continuous function $g$, implies that

$$g(\bar{X}_n) \xrightarrow[n\to\infty]{P} g(\mu).$$

On the other hand, the CLT states that:

$$\frac{\sqrt{n}(\bar{X}_n - \mu)}{\sigma} \xrightarrow[n\to\infty]{d} N(0, 1) \quad \text{or} \quad \sqrt{n}(\bar{X}_n - \mu) \xrightarrow[n\to\infty]{d} N(0, \sigma^2). \qquad (14)$$

The question then arises what happens to the distribution of $g(\bar{X}_n)$. In other words, is there a result analogous to (14) when the distribution of $g(\bar{X}_n)$ is involved? The question is answered by the following result.

**Theorem 8.** Let $X_1, \ldots, X_n$ be i.i.d. r.v.'s with finite mean $\mu$ and variance $\sigma^2 \in (0, \infty)$, and let $g : \mathfrak{R} \to \mathfrak{R}$ be differentiable with derivative $g'$ continuous at $\mu$. Then:

$$\sqrt{n}[g(\bar{X}_n) - g(\mu)] \xrightarrow[n\to\infty]{d} N(0, [\sigma g'(\mu)]^2). \qquad (15)$$

The proof of this result involves the employment of some of the theorems established in this chapter, including the CLT, along with a Taylor expansion. The proof itself will not be presented, and this section will be concluded with an application to Theorem 8. The method of establishing asymptotic normality for $g(\bar{X}_n)$ is often referred to as the delta method, and it also applies in cases more general than the one described here.

**Application 1.** Let the independent r.v.'s $X_1, \ldots, X_n$ be distributed as $B(1, p)$. Then:

$$\sqrt{n}[\bar{X}_n(1 - \bar{X}_n) - pq] \xrightarrow[n\to\infty]{d} N(0, pq(1 - 2p)^2) \quad (q = 1 - p). \qquad (16)$$

*Proof.* Here $\mu = p$, $\sigma^2 = pq$, and $g(x) = x(1-x)$, so that $g'(x) = 1 - 2x$ continuous for all $x$. Since $g(\bar{X}_n) = \bar{X}_n(1 - \bar{X}_n)$, $g(\mu) = p(1-p) = pq$, and $g'(\mu) = 1 - 2p$, the convergence in (15) becomes as stated in (16). ∎

## EXERCISES

**3.1**   Let $X_1, \ldots, X_n$ be i.i.d. r.v.'s with finite $EX_i = \mu$, and $\mathrm{Var}(X_i) = \sigma^2 \in (0, \infty)$, so that the CLT holds; that is,

$$\frac{\sqrt{n}(\bar{X}_n - \mu)}{\sigma} \xrightarrow[n \to \infty]{d} Z \sim N(0, 1), \quad \text{where } \bar{X}_n = \frac{1}{n}\sum_{i=1}^n X_i.$$

Then use Theorem 6 in order to show that the WLLN also holds.

**3.2**   Let $X_1, \ldots, X_n$ be i.i.d. r.v.'s with finite $EX_i = \mu$ and finite $\mathrm{Var}(X_i) = \sigma^2$. Then use the identity (see Exercise 2.1(i) in Chapter 5)

$$\sum_{i=1}^n (X_i - \bar{X}_n)^2 = \sum_{i=1}^n X_i^2 - n\bar{X}_n^2,$$

the WLLN, and Theorems 5, 6 and 1 in order to show that

$$\frac{1}{n-1}\sum_{i=1}^n (X_i - \bar{X}_n)^2 \xrightarrow[n \to \infty]{P} \sigma^2.$$

# An overview of statistical inference

A review of the previous chapters reveals that the main objectives throughout have been those of calculating probabilities or certain summary characteristics of a distribution, such as mean, variance, median, and mode. However, for these calculations to result in numerical answers, it is a prerequisite that the underlying distributions be completely known. Typically, this is rarely, if ever, the case. The reason for this is that the parameters which appear, for example, in the functional form of the p.d.f. of a distribution are simply unknown to us. The only thing known about them is that they lie in specified sets of possible values for these parameters, the *parameter space*.

It is at this point where statistical inference enters the picture. Roughly speaking, the aim of statistical inference is to make certain determinations with regard to the unknown constants (*parameters*) figuring in the underlying distribution. This is to be done on the basis of data, represented by the observed values of a random sample drawn from said distribution. Actually, this is the so-called *parametric statistical inference* as opposed to the *nonparametric statistical inference*. The former is applicable to distributions, which are completely determined by the knowledge of a finite number of parameters. The latter applies to distributions not determined by any finite number of parameters.

The remaining part of this book is, essentially, concerned with statistical inference and mostly with parametric statistical inference. Within the framework of parametric statistical inference, there are three main objectives, depending on what kind of determinations we wish to make with regard to the parameters. If the objective is to arrive at a number, by means of the available data, as the value of an unknown parameter, then we are talking about *point estimation*. If, on the other hand, we are satisfied with the statement that an unknown parameter lies within a known random interval (i.e., an interval with r.v.'s as its end-points) with high prescribed probability, then we are dealing with *interval estimation* or *confidence intervals*. Finally, if the objective is to decide that an unknown parameter lies in a specified subset of the parameter space, then we are in the area of *testing hypotheses*.

These three subjects — point estimation, interval estimation, and testing hypotheses — are briefly discussed in the following three sections. In the subsequent three sections, it is pointed out what the statistical inference issues are in specific models — a *regression model* and two *analysis of variance models*. The final section touches upon some aspects of *nonparametric statistical inference*.

## 8.1 THE BASICS OF POINT ESTIMATION

The problem here, briefly stated, is as follows. Let $X$ be a r.v. with a p.d.f. $f$ which, however, involves a parameter. This is the case, for instance, in the Binomial distribution $B(1,p)$, the Poisson distribution $P(\lambda)$, the Negative Exponential $f(x) = \lambda e^{-\lambda x}, x > 0$, distribution, the Uniform distribution $U(0,\alpha)$, and the Normal distribution $N(\mu, \sigma^2)$ with one of the quantities $\mu$ and $\sigma^2$ known. The *parameter* is usually denoted by $\theta$, and the set of its possible values is denoted by $\Omega$ and is called the *parameter space*. In order to emphasize the fact that the p.d.f. depends on $\theta$, we write $f(\cdot; \theta)$. Thus, in the distributions mentioned above, we have for the respective p.d.f.'s and the parameter spaces:

$$f(x; \theta) = \theta^x (1 - \theta)^{1-x}, \quad x = 0, 1, \ \theta \in \Omega = (0, 1).$$

The situations described in Examples 5, 6, 8, 9, and 10 of Chapter 1 may be described by a Binomial distribution.

$$f(x; \theta) = \frac{e^{-\theta}\theta^x}{x!}, \quad x = 0, 1, \ldots, \ \theta \in \Omega = (0, \infty).$$

The Poisson distribution can be used appropriately in the case described in Example 12 of Chapter 1.

$$f(x; \theta) = \theta e^{-\theta x}, \quad x > 0, \ \theta \in \Omega = (0, \infty).$$

$$f(x; \theta) = \begin{cases} \frac{1}{\theta}, & 0 < x < \theta \\ 0, & \text{otherwise,} \end{cases} \quad \theta \in \Omega = (0, \infty).$$

$$f(x; \theta) = \frac{1}{\sqrt{2\pi}\sigma}e^{-(x-\theta)^2/2\sigma^2}, \quad x \in \Re, \ \theta \in \Omega = \Re, \ \sigma^2 \text{ known,}$$

and

$$f(x; \theta) = \frac{1}{\sqrt{2\pi\theta}}e^{-(x-\mu)^2/2\theta}, \quad x \in \Re, \ \theta \in \Omega = (0, \infty), \ \mu \text{ known.}$$

Normal distributions are suitable for modeling the situations described in Examples 16 and 17 of Chapter 1.

Our objective is to draw a random sample of size $n, X_1, \ldots, X_n$, from the underlying distribution, and on the basis of it to construct a *point estimate (or estimator)* for $\theta$; that is, a statistic $\hat{\theta} = \hat{\theta}(X_1, \ldots, X_n)$, which is used for estimating $\theta$, where a *statistic* is a known function of the random sample $X_1, \ldots, X_n$. If $x_1, \ldots, x_n$ are the actually observed values of the r.v.'s $X_1, \ldots, X_n$, respectively, then the observed value of our estimate has the numerical value $\hat{\theta}(x_1, \ldots, x_n)$. The observed values $x_1, \ldots, x_n$ are also referred to as *data*. Then, on the basis of the available data, it is declared that the value of $\theta$ is $\hat{\theta}(x_1, \ldots, x_n)$ from among all possible points in $\Omega$. A point estimate is often referred to just as an estimate, and the notation $\hat{\theta}$ is often used indiscriminately, both for the estimate $\hat{\theta}(X_1, \ldots, X_n)$ (which is a r.v.) and for its observed value $\hat{\theta}(x_1, \ldots, x_n)$ (which is just a number).

The only obvious restriction on $\hat{\theta}(x_1, \ldots, x_n)$ is that it lies in $\Omega$ for all possible values of $X_1, \ldots, X_n$. Apart from it, there is any number of estimates one may construct — thus, the need to assume certain principles and/or invent methods for constructing $\hat{\theta}$. Perhaps, the most widely accepted principle is the so-called principle of *maximum likelihood*. This principle dictates that we form the joint p.d.f. of the $x_i$'s, for the observed values of the $X_i$'s, look at this joint p.d.f. as a function of $\theta$ (and call it the *likelihood function*), and maximize the likelihood function with respect to $\theta$. The maximizing point (assuming it exists and is unique) is a function of $x_1, \ldots, x_n$, and is what we call the *maximum likelihood estimate* (MLE) of $\theta$. The notation used for the likelihood function is $L(\theta \mid x_1, \ldots, x_n)$. Then, we have that:

$$L(\theta \mid x_1, \ldots, x_n) = f(x_1; \theta) \cdots f(x_n; \theta), \quad \theta \in \Omega.$$

The MLE will be studied fairly extensively in Chapter 9.

Another principle often used in constructing an estimate for $\theta$ is the principle of *unbiasedness*. In this context, an estimate is usually denoted by $U = U(X_1, \ldots, X_n)$. Then the principle of unbiasedness dictates that $U$ should be constructed so as to be *unbiased*; that is, its expectation (mean value) should always be $\theta$, no matter what the value of $\theta$ in $\Omega$ is. More formally, $E_\theta U = \theta$ for all $\theta \in \Omega$. (In the expectation sign $E$, the parameter $\theta$ was inserted to indicate that this expectation does depend on $\theta$, since it is calculated by using the p.d.f. $f(\cdot; \theta)$.) Now, it is intuitively clear that, in comparing two unbiased estimates, one would pick the one with the smaller variance, since it would be more closely concentrated around its mean $\theta$. Envision the case that, within the class of all unbiased estimates, there exists one which has the smallest variance (and that is true for all $\theta \in \Omega$). Such an estimate is called a *uniformly minimum variance unbiased* (UMVU) estimate and is, clearly, a desirable estimate. In the next chapter, we will see how we go about constructing such estimates.

The principle (or rather the method) based on sample moments is another way of constructing estimates. The *method of moments*, in the simplest case, dictates to form the sample mean $\bar{X}$ and equate it with the (theoretical) mean $E_\theta X$. Then solve for $\theta$ (assuming it can be done, and, indeed, uniquely) in order to arrive at a *moment estimate* of $\theta$.

A much more sophisticated method of constructing estimates of $\theta$ is the so-called *decision-theoretic* method. This method calls for the introduction of a host of concepts, terminology, and notation, and it will be taken up in the next chapter.

Finally, another relatively popular method (in particular, in the context of certain models) is the method of *least squares* (LS). The method of LS leads to the construction of an estimate for $\theta$, the *least squares estimate* (LSE) of $\theta$, through a minimization (with respect to $\theta$) of the sum of certain squares. This sum of squares represents squared deviations between what we actually observe after experimentation is completed and what we would expect to have on the basis of an assumed model. Once again, details will be presented later on, more specifically, in Chapter 13.

In all of the preceding discussion, it was assumed that the underlying p.d.f. depended on a single parameter, which was denoted by $\theta$. It may very well be the case that there are two or more parameters involved. This may happen, for instance, in the Uniform distribution $U(\alpha, \beta)$, $-\infty < \alpha < \beta < \infty$, where both $\alpha$ and $\beta$ are unknown; the Normal distribution, $N(\mu, \sigma^2)$, where both $\mu$ and $\sigma^2$ are unknown;

and it does happen in the Multinomial distribution, where the number of parameters is $k, p_1, \ldots, p_k$ (or more precisely, $k - 1$, since the $k$th parameter, e.g., $p_k = 1 - p_1 - \cdots - p_{k-1}$). For instance, Examples 20 and 21 of Chapter 1 refer to situations where a Multinomial distribution is appropriate. In such multiparameter cases, one simply applies to each parameter separately what was said above for a single parameter. The alternative option, to use the vector notation for the parameters involved, does simplify things in a certain way but also introduces some complications in other ways.

## 8.2 THE BASICS OF INTERVAL ESTIMATION

Suppose we are interested in constructing a point estimate of the mean $\mu$ in the Normal distribution $N(\mu, \sigma^2)$ with known variance; this is to be done on the basis of a random sample of size $n, X_1, \ldots, X_n$, drawn from the underlying distribution. This amounts to constructing a suitable statistic of the $X_i$'s, call it $V = V(X_1, \ldots, X_n)$, which for the observed values $x_i$ of $X_i$, $i = 1, \ldots, n$ is a numerical entity, and declare it to be the (unknown) value of $\mu$. This looks somewhat presumptuous, since from the set of possible values for $\mu$, $-\infty < \mu < \infty$, just one is selected as its value. Thinking along these lines, it might be more reasonable to aim instead at a random interval which will contain the (unknown) value of $\mu$ with high (prescribed) probability. This is exactly what a confidence interval does.

To be more precise and in casting the problem in a general setting, let $X_1, \ldots, X_n$ be a random sample from the p.d.f. $f(\cdot; \theta)$, $\theta \in \Omega \subseteq \mathfrak{R}$, and let $L = L(X_1, \ldots, X_n)$ and $U = U(X_1, \ldots, X_n)$ be two statistics of the $X_i$'s such that $L < U$. Then the interval with end-points $L$ and $U$, $[L, U]$, is called a *random interval*. Let $\alpha$ be a small number in (0, 1), such as 0.005, 0.01, 0.05, and suppose that the random interval $[L, U]$ contains $\theta$ with probability equal to $1 - \alpha$ (such as 0.995, 0.99, 0.95) no matter what the true value of $\theta$ in $\Omega$ is. In other words, suppose that:

$$P_\theta(L \leq \theta \leq U) = 1 - \alpha \quad \text{for all } \theta \in \Omega. \tag{1}$$

If relation (1) holds, then we say that the random interval $[L, U]$ is a *confidence interval* for $\theta$ with *confidence coefficient* $1 - \alpha$.

The significance of a confidence interval is based on the relative frequency interpretation of the concept of probability, and it goes like this: Suppose $n$ independent r.v.'s are drawn from the p.d.f. $f(\cdot; \theta)$, and let $x_1, \ldots, x_n$ be their observed values. Also, let $[L_1, U_1]$ be the interval resulting from the observed values of $L = L(X_1, \ldots, X_n)$ and $U = U(X_1, \ldots, X_n)$; that is, $L_1 = L(x_1, \ldots, x_n)$ and $U_1 = U(x_1, \ldots, x_n)$. Proceed to draw independently a second set of $n$ r.v.'s as above, and let $[L_2, U_2]$ be the resulting interval. Repeat this process independently a large number of times, $N$, say, with the corresponding interval being $[L_N, U_N]$. Then the interpretation of (1) is that, on the average, about $100(1 - \alpha)\%$ of the above $N$ intervals will, actually, contain the true value of $\theta$. For example, for $\alpha = 0.05$ and $N = 1000$, the proportion of such intervals will be 95%; that is, one would expect 950 out of the 1000 intervals constructed as above to contain the true value of $\theta$. Empirical evidence shows that such an expectation is valid.

We may also define an *upper confidence limit* for $\theta, U = U(X_1, \ldots, X_n)$, and a *lower confidence limit* for $\theta, L = L(X_1, \ldots, X_n)$, both with *confidence* coefficient $1 - \alpha$, if, respectively, the intervals $(-\infty, U]$ and $[L, \infty)$ are confidence intervals for $\theta$ with confidence coefficient $1 - \alpha$. That is to say:

$$P_\theta(-\infty < \theta \leq U) = 1 - \alpha, \quad P_\theta(L \leq \theta < \infty) = 1 - \alpha \text{ for all } \theta \in \Omega. \qquad (2)$$

Confidence intervals and upper and/or lower confidence limits can be sought, for instance, in Examples 5, 6, 8, 9, and 10 (Binomial distribution), 12 (Poisson distribution), and 16 and 17 (Normal distribution) in Chapter 1.

There are some variations of (1) and (2). For example, when the underlying p.d.f. is discrete, then equalities in (1) and (2) rarely obtain for given $\alpha$ and have to be replaced by inequalities $\geq$. Also, except for special cases, equalities in (1) and (2) are valid only approximately for large values of the sample size $n$ (even in cases where the underlying r.v.'s are continuous). In such cases, we say that the respective confidence intervals (confidence limits) have *confidence coefficient approximately* $1 - \alpha$.

Finally, the parameters of interest may be two (or more) rather than one, as we have assumed so far. In such cases, the concept of a confidence interval is replaced by that of a *confidence region* (in the multidimensional parameter space $\Omega$). This concept will be illustrated by an example in Chapter 10. In the same chapter, we will also expand considerably on what was briefly discussed here.

## 8.3 THE BASICS OF TESTING HYPOTHESES

Often, we are not interested in a point estimate of a parameter $\theta$ or even a confidence interval for it, but rather whether said parameter lies or does not lie in a specified subset $\omega$ of the parameter space $\Omega$. To clarify this point, we refer to some of the examples described in Chapter 1. Thus, in Example 5, all we might be interested in is whether Jones has ESP at all or not and not to what degree he does. In statistical terms, this amounts to taking $n$ independent observations from a $B(1, \theta)$ distribution and, on the basis of these observations, deciding whether $\theta \in \omega = (0, 0.5]$ (as opposed to $\theta \in \omega^c = (0.5, 1)$); here $\theta$ is the probability that Jones correctly identifies certain objects. The situation in Example 6 is similar, and the objective might be to decide whether or not $\theta \in \omega = (\theta_0, 1)$; here $\theta$ is the true proportion of unemployed workers and $\theta_0$ is a certain desirable or guessed value of $\theta$. Examples 8–10 in Chapter 1 fall into the same category.

In Example 12, the stipulated model is a Poisson distribution $P(\theta)$ and, on the basis of $n$ independent observations, we might wish to decide whether or not $\theta \in (\theta_0, \infty)$, where $\theta_0$ is a known value of $\theta$.

In Example 16, the stipulated underlying models may be Normal distributions $N(\mu_1, \sigma^2)$ and $N(\mu_2, \sigma^2)$ for the survival times $X$ and $Y$, respectively, and then the question of interest may be to decide whether or not $\mu_2 \leq \mu_1$; $\sigma^2$ may be assumed to be either known or unknown. Of course, we are going to arrive at the desirable decision on the basis of two independent random samples drawn from the underlying distributions. Example 17 is of the same type.

In Example 20, the statistical problem is that of comparing two Multinomial populations, by making appropriate statements about the probabilities $p_{AE}, p_{AA}, p_{AP}$ and $p_{BE}, p_{BA}, p_{BP}$; here $p_{AE}$ is the probability that any one of the 80 infants, subjected to diet $A$, is of "excellent" health, and similarly for the remaining probabilities. Example 21 is of a similar type.

On the basis of the preceding discussion and examples, we may now proceed with the formulation of the general problem. To this effect, let $X_1, \ldots, X_n$ be i.i.d. r.v.'s with p.d.f. $f(\cdot; \boldsymbol{\theta}), \boldsymbol{\theta} \in \Omega \subseteq \mathfrak{R}^r, r \geq 1$, and by means of this random sample, suppose we are interested in checking whether $\boldsymbol{\theta} \in \omega$, a proper subset of $\Omega$, or $\boldsymbol{\theta} \in \omega^c$, the complement of $\omega$ with respect to $\Omega$. The statements that $\boldsymbol{\theta} \in \omega$ and $\boldsymbol{\theta} \in \omega^c$ are called (*statistical*) *hypotheses* (about $\boldsymbol{\theta}$), and are denoted thus: $H_0 : \boldsymbol{\theta} \in \omega, H_A : \boldsymbol{\theta} \in \omega^c$. The hypothesis $H_0$ is called a *null* hypothesis and the hypothesis $H_A$ is called *alternative* (to $H_0$) hypothesis. The hypotheses $H_0$ and $H_A$ are called *simple*, if $\omega$ and $\omega^c$, respectively, contain a single point, and *composite* otherwise. More generally, the hypotheses $H_0$ and $H_A$ are called *simple*, if they completely determine the underlying distribution of the r.v.'s $X_1, \ldots, X_n$, and *composite* otherwise. The procedure of checking whether $H_0$ is true or not, on the basis of the observed values $x_1, \ldots, x_n$ of $X_1, \ldots, X_n$, is called *testing* the hypothesis $H_0$ against the alternative $H_A$.

In the special case that $\Omega \subseteq \mathfrak{R}$, some null hypotheses and the respective alternatives are as follows:

$$H_0 : \theta = \theta_0 \text{ against } H_A : \theta > \theta_0; \quad H_0 : \theta = \theta_0 \text{ against } H_A : \theta < \theta_0;$$
$$H_0 : \theta \leq \theta_0 \text{ against } H_A : \theta > \theta_0; \quad H_0 : \theta \geq \theta_0 \text{ against } H_A : \theta < \theta_0;$$
$$H_0 : \theta = \theta_0 \text{ against } H_A : \theta \neq \theta_0.$$

The testing is carried out by means of a function $\varphi : \mathfrak{R}^n \to [0, 1]$ which is called a *test function* or just a *test*. The number $\varphi(x_1, \ldots, x_n)$ represents the probability of rejecting $H_0$, given that $X_i = x_i, i = 1, \ldots, n$. In its simplest form, $\varphi$ is the indicator of a set $B$ in $\mathfrak{R}^n$, which is called the *critical* or *rejection region*; its complement $B^c$ is called the *acceptance region*. Thus, $\varphi(x_1, \ldots, x_n) = 1$ if $(x_1, \ldots, x_n)$ is in $B$, and $\varphi(x_1, \ldots, x_n) = 0$, otherwise. Actually, such a test is called a *nonrandomized* test as opposed to tests which also take values strictly between 0 and 1 and are called *randomized* tests. In the case of continuous distributions, nonrandomized tests suffice, but in discrete distributions, a test will typically be required to take on one or two values strictly between 0 and 1.

By using a test $\varphi$, suppose that our data $x_1, \ldots, x_n$ lead us to the rejection of $H_0$. This will happen, for instance, if the test $\varphi$ is nonrandomized with rejection region $B$, and the $(x_1, \ldots, x_n)$ lies in $B$. By rejecting the hypothesis $H_0$, we may be doing the correct thing, because $H_0$ is false (i.e., $\boldsymbol{\theta} \notin \omega$). On the other hand, we may be taking the wrong action, because it may happen that $H_0$ is, indeed, true (i.e., $\boldsymbol{\theta} \in \omega$), only the test and the data do not reveal it. Clearly, in so doing, we commit an error, which is referred to as *type I error*. Of course, we would like to find ways of minimizing the frequency of committing this error. To put it more mathematically, this means searching for a rejection region $B$, which will minimize the above frequency. In our framework, frequencies are measured by probabilities, and this leads to a determination of $B$ so that

$$P(\text{of type I error}) = P(\text{of rejecting } H_0 \text{ whereas } H_0 \text{ is true})$$
$$= P_\theta[(X_1, \ldots, X_n) \text{ lies in } B \text{ whereas } \theta \in \omega]$$
$$= P_\theta[(X_1, \ldots, X_n) \text{ lies in } B \mid \theta \in \omega] \overset{\text{def}}{=} \alpha(\theta). \tag{3}$$

Clearly, the probabilities $\alpha(\theta)$ in (3) must be minimized for each $\theta \in \omega$, since we don't know which value in $\omega$ is the true $\theta$. This will happen if we minimize the $\max_{\theta \in \omega} \alpha(\theta) \overset{\text{def}}{=} \alpha$. This maximum probability of type I error is called the *level of significance* (or *size*) of the test employed. Thus, we are led to selecting the rejection region $B$ so that its level of significance $\alpha$ will be minimum. Since $\alpha \geq 0$, its minimum value would be 0, and this would happen if (essentially) $B = \emptyset$. But then (essentially) $(x_1, \ldots, x_n)$ would always be in $B^c = \Re^n$, and this would happen with probability

$$P_\theta[(X_1, \ldots, X_n) \text{ in } \Re^n] = 1 \quad \text{for all } \theta. \tag{4}$$

This, however, creates a problem for the following reason. If the rejection region $B$ is $\emptyset$, then the acceptance region is $\Re^n$; that is, we always accept $H_0$. As long as $H_0$ is true (i.e., $\theta \in \omega$), this is exactly what we wish to do, but what about the case that $H_0$ is false (i.e., $\theta \in \omega^c$)? When we accept a false hypothesis $H_0$, we commit an error, which is called the *type II error*. As in (3), this error is also measured in terms of probabilities; namely,

$$P(\text{of type II error}) = P(\text{of accepting } H_0 \text{ whereas } H_0 \text{ is false})$$
$$= P_\theta[(X_1, \ldots, X_n) \text{ lies in } B^c \text{ whereas } \theta \in \omega^c]$$
$$= P_\theta[(X_1, \ldots, X_n) \text{ lies in } B^c \mid \theta \in \omega^c]$$
$$\overset{\text{def}}{=} \beta(\theta). \tag{5}$$

According to (5), these probabilities would be 1 for all $\theta \in \omega^c$ (actually, for all $\theta \in \Omega$), if $B = \emptyset$. Clearly, this is undesirable. The preceding discussion then leads to the conclusion that the rejection region $B$ must be different from $\emptyset$ and then $\alpha$ will be $> 0$. The objective then becomes that of choosing $B$ so that $\alpha$ will have a preassigned acceptable value (such as 0.005, 0.01, 0.05) and, subject to this restriction, the probabilities of type II error are minimized. That is,

$$\beta(\theta) = P_\theta[(X_1, \ldots, X_n) \text{ lies in } B^c] \text{ is minimum for each } \theta \in \omega^c. \tag{6}$$

Since $P_\theta[(X_1, \ldots, X_n) \text{ lies in } B^c] = 1 - P_\theta[(X_1, \ldots, X_n) \text{ lies in } B]$, the minimization in (6) is equivalent to the maximization of

$$P_\theta[(X_1, \ldots, X_n) \text{ lies in } B] = 1 - P_\theta[(X_1, \ldots, X_n) \text{ lies in } B^c] \quad \text{for all } \theta \in \omega^c.$$

The function $\pi(\theta), \theta \in \omega^c$, defined by:

$$\pi(\theta) = P_\theta[(X_1, \ldots, X_n) \text{ lies in } B], \quad \theta \in \omega^c, \tag{7}$$

is called the *power* of the test employed. So, power of a test $= 1-$ probability of a type II error, and we may summarize our objective as follows: Choose a test with a preassigned level of significance $\alpha$, which has maximum power among all tests with level of significance $\leq \alpha$. In other words, if $\varphi$ is the desirable test, then it should satisfy the requirements.

The level of significance of $\varphi$ is $\alpha$, and its power, to be denoted by $\pi_\varphi(\theta), \theta \in \omega^c$, satisfies the inequality $\pi_\varphi(\theta) \geq \pi_{\varphi^*}(\theta)$ for all $\theta \in \omega^c$ and any test $\varphi^*$ with level of significance $\leq \alpha$.

Such a test $\varphi$, should it exist, is called *uniformly most powerful* (UMP) for obvious reasons. (The term "most powerful" is explained by the inequality $\pi_\varphi(\theta) \geq \pi_{\varphi^*}(\theta)$, and the term "uniformly" is due to the fact that this inequality must hold for *all* $\theta \in \omega^c$.) If $\omega^c$ consists of a single point, then the concept of uniformity is void, and we talk simply of a *most powerful* test.

The concepts introduced so far hold for a parameter of any (finite) dimensionality. However, UMP tests can be constructed only when $\theta$ is a real-valued parameter, and then only for certain forms of $H_0$ and $H_A$ and specific p.d.f.'s, $f(\cdot; \theta)$. If the parameter is multidimensional, desirable tests can still be constructed; they are not going to be, in general, UMP tests, but they are derived, nevertheless, on the basis of principles which are intuitively satisfactory. Preeminent among such tests are the so-called *likelihood ratio* (LR) tests. Another class of tests are the so-called *goodness-of-fit* tests, and still others are constructed on the basis of *decision-theoretic* concepts. Some of the tests mentioned above will be discussed more extensively in Chapters 11 and 12. Here, we conclude this section with the introduction of a LR test.

On the basis of the random sample $X_1, \ldots, X_n$ with p.d.f. $f(\cdot; \theta), \theta \in \Omega \subseteq \mathfrak{R}^r, r \geq 1$, suppose we wish to test the hypothesis $H_0 : \theta \in \omega$ (a proper) subset of $\Omega$. It is understood that the alternative is $H_A : \theta \in \omega^c$, but in the present framework it is not explicitly stated. Let $x_1, \ldots, x_n$ be the observed values of $X_1, \ldots, X_n$ and form the likelihood function $L(\theta) = L(\theta \mid x_1, \ldots, x_n) = \prod_{i=1}^n f(x_i; \theta)$. Maximize $L(\theta)$ and denote the resulting maximum by $L(\hat{\Omega})$. This maximization happens when $\theta$ is equal to the MLE $\hat{\theta} = \hat{\theta}(x_1, \ldots, x_n)$, so that $L(\hat{\Omega}) = L(\hat{\theta})$. Next, maximize the likelihood $L(\theta)$ under the restriction that $\theta \in \omega$, and denote the resulting maximum by $L(\hat{\omega})$. Denote by $\hat{\theta}_\omega$ the MLE of $\theta$ subject to the restriction that $\theta \in \omega$. Then $L(\hat{\omega}) = L(\hat{\theta}_\omega)$. Assume now that $L(\theta)$ is continuous (in $\theta$), and suppose that the true value of $\theta$, call it $\theta_0$, is in $\omega$. It is a property of an MLE that it gets closer and closer to the true parameter as the sample size $n$ increases. Under the assumption that $\theta_0 \in \omega$, it follows that both $\hat{\theta}$ and $\hat{\theta}_\omega$ will be close to $\theta_0$ and therefore close to each other. Then, by the assumed continuity of $L(\theta)$, the quantities $L(\hat{\theta})$ and $L(\hat{\theta}_\omega)$ are close together, so that the ratio

$$\lambda(x_1, \ldots, x_n) = \lambda = L(\hat{\theta}_\omega)/L(\hat{\theta}) \tag{8}$$

(which is always $\leq 1$) is close to 1. On the other hand, if $\theta_0 \in \omega^c$, then $\hat{\theta}$ and $\hat{\theta}_\omega$ are not close together, and therefore $L(\hat{\theta})$ and $L(\hat{\theta}_\omega)$ need not be close either. Thus, the ratio $L(\hat{\theta}_\omega)/L(\hat{\theta})$ need not be close to 1. These considerations lend to the following test:

Reject $H_0$ when $\lambda < \lambda_0$,    where $\lambda_0$ is a constant to be determined.    (9)

By the monotonicity of the function $y = \log x$, the inequality $\lambda < \lambda_0$ is equivalent to $-2 \log \lambda(x_1, \ldots, x_n) > C(= -2 \log \lambda_0)$. It is seen in Chapter 12 that an approximate determination of $C$ is made by the fact that, under certain conditions, the distribution of $-2 \log \lambda(X_1, \ldots, X_n)$ is $\chi_f^2$, where $f = $ dimension of $\Omega-$ dimension of $\omega$. Namely:

$$\text{Reject } H_0 \text{ when } -2 \log \lambda > C, \quad \text{where } C \simeq \chi_{f; \alpha}^2. \quad (10)$$

In closing this section, it is to be mentioned that the concept of $P$-value is another way of looking at a test in an effort to assess how strong (or weak) the rejection of a hypothesis is. The *P-value* (*probability value*) of a test is defined to be the smallest probability at which the hypothesis tested would be rejected for the data at hand. Roughly put, the $P$-value of a test is the probability, calculated under the null hypothesis, when the observed value of the test statistic is used as if it was the cutoff point of the test. The $P$-value (or *significance probability*) of a test often accompanies a null hypothesis which is rejected, as an indication of the strength or weakness of rejection. The smaller the $P$-value, the stronger the rejection of the null hypothesis, and vice versa. Strong rejection of the null hypothesis is also referred to as the result being *highly statistically significant*. More about it in Chapter 11.

## 8.4 THE BASICS OF REGRESSION ANALYSIS

In the last three sections, we discussed the general principles of point estimation, interval estimation, and testing hypotheses in a general setup. These principles apply, in particular, in specific models. Two such models are *regression models* and *analysis of variance models*.

A regression model arises in situations such as those described in Examples 22 and 23 in Chapter 1. Its simplest form is as follows: At fixed points $x_1, \ldots, x_n$, respective measurements $y_1, \ldots, y_n$ are taken, which may be subject to an assortment of random errors $e_1, \ldots, e_n$. Thus, the $y_i$'s are values of r.v.'s $Y_i$'s, which may often be assumed to have the structure: $Y_i = \beta_1 + \beta_2 x_i + e_i, i = 1, \ldots, n$; here $\beta_1$ and $\beta_2$ are parameters (unknown constants) of the model. For the random errors $e_i$, it is not unreasonable to assume that $Ee_i = 0$; we also assume that they have the same variance, $\text{Var}(e_i) = \sigma^2 \in (0, \infty)$. Furthermore, it is reasonable to assume that the $e_i$'s are i.i.d. r.v.'s, which implies independence of the r.v.'s $Y_1, \ldots, Y_n$. It should be noted, however, that the $Y_i$'s are *not* identically distributed, since, for instance, they have different expectations: $EY_i = \beta_1 + \beta_2 x_i, i = 1, \ldots, n$. Putting these assumptions together, we arrive at the following simple *linear regression model*.

$Y_i = \beta_1 + \beta_2 x_i + e_i$, the $e_i$'s are i.i.d. with $Ee_i = 0$ and
$$\text{Var}(e_i) = \sigma^2, \quad i = 1, \ldots, n. \quad (11)$$

The quantities $\beta_1, \beta_2$, and $\sigma^2$ are the *parameters* of the model; the $Y_i$'s are independent but not identically distributed; also, $EY_i = \beta_1 + \beta_2 x_i$ and $\text{Var}(Y_i) = \sigma^2$, $i = 1, \ldots, n$.

The term "regression" derives from the way the $Y_i$'s are produced from the $x_i$'s, and the term "linear" indicates that the parameters $\beta_1$ and $\beta_2$ enter into the model raised to the first power.

The main problems in connection with model (11) are to estimate the parameters $\beta_1, \beta_2$, and $\sigma^2$; construct confidence intervals for $\beta_1$ and $\beta_2$; test hypotheses about $\beta_1$ and $\beta_2$; and predict the expected value $EY_{i_0}$ (or the value itself $Y_{i_0}$) corresponding to an $x_{i_0}$, distinct, in general, from $x_1, \ldots, x_n$. Estimates of $\beta_1$ and $\beta_2$, the LSE's, can be constructed without any further assumptions; the same hold for an estimate of $\sigma^2$. For the remaining parts, however, there is a need to stipulate a distribution for the $e_i$'s. Since the $e_i$'s are random errors, it is reasonable to assume that they are Normally distributed; this then implies Normal distribution for the $Y_i$'s. Thus, model (11) now becomes:

$$Y_i = \beta_1 + \beta_2 x_i + e_i, \quad \text{the } e_i\text{'s are independently}$$
$$\text{distributed as } N(0, \sigma^2), \quad i = 1, \ldots, n. \tag{12}$$

Under model (12), the MLE's of $\beta_1, \beta_2$, and $\sigma^2$ are derived, and their distributions are determined. This allows us to pursue the resolution of the parts of constructing confidence intervals, testing hypotheses, and of prediction. The relevant discussion is presented in Chapter 13.

## 8.5 THE BASICS OF ANALYSIS OF VARIANCE

*Analysis of variance* (ANOVA) is a powerful technique, which provides the means of assessing and/or comparing several entities. ANOVA can be used effectively in many situations; in particular, it can be used in assessing and/or comparing crop yields corresponding to different soil treatments; crop yields corresponding to different soils and fertilizers; for the comparison of a certain brand of gasoline with or without an additive by using it in several cars; the comparison of different brands of gasoline by using them in several cars; the comparison of the wearing of different materials; the comparison of the effect of different types of oil on the wear of several piston rings, etc.; the comparison of the yields of a chemical substance by using different catalytic methods; the comparison of the strengths of certain objects made from different batches of some material; the comparison of test scores from different schools and different teachers, etc.; and identification of the melting point of a metal by using different thermometers. Example 24 in Chapter 1 provides another case where ANOVA techniques are appropriate.

Assessment and comparisons are done by way of point estimation, interval estimation, and testing hypotheses, as these techniques apply to the specific ANOVA models to be considered. The more factors involved in producing an outcome, the more complicated the model becomes. However, the basic ideas remain the same throughout.

For the sake of illustrating the issues involved, consider the so-called *one-way layout* or *one-way classification* model. Consider one kind of gasoline, for example, unleaded regular gasoline, and suppose we supply ourselves with amounts of this gasoline, purchased from $I$ different companies. The objective is to compare these $I$ brands of gasoline from yield viewpoint. To this end, a car (or several but pretty similar cars) operates under each one of the $I$ brands of gasoline for $J$ runs in each case. Let $Y_{ij}$ be the number of miles per hour for the $j$th run when the $i$th brand of gasoline is used. Then the $Y_{ij}$'s are r.v.'s for which the following structure is assumed: For a given $i$, the actual number of miles per hour for the $j$th run varies around a *mean* value $\mu_i$, and these variations are due to an assortment of random errors $e_{ij}$. In other words, it makes sense to assume that $Y_{ij} = \mu_i + e_{ij}$. It is also reasonable to assume that the random errors $e_{ij}$ are independent r.v.'s distributed as $N(0, \sigma^2)$, some unknown variance $\sigma^2$. Thus, we have stipulated the following model:

$$Y_{ij} = \mu_i + e_{ij}, \quad \text{where the } e_{ij}\text{'s are independently}$$

$$\sim N(0, \sigma^2), \quad i = 1, \ldots, I(\geq 2), \ j = 1, \ldots, J(\geq 2). \tag{13}$$

The quantities $\mu_i, i = 1, \ldots, I$, and $\sigma^2$ are the *parameters* of the model.

It follows that the r.v.'s $Y_{ij}$ are independent and $Y_{ij} \sim N(\mu_i, \sigma^2), j = 1, \ldots, J,$ $i = 1, \ldots, I.$

The issues of interest here are those of estimating the $\mu_i$'s (mean number of miles per hour for the $i$th brand of gasoline) and $\sigma^2$. Also, we wish to test the hypothesis that there is really no difference between these $I$ different brands of gasoline; in other words, test $H_0 : \mu_1 = \cdots = \mu_I(= \mu,$ say, unknown). Should this hypothesis be rejected, we would wish to identify the brands of gasoline which cause the rejection. This can be done by constructing a confidence interval for certain linear combinations of the $\mu_i$'s called *contrasts*. That is, $\sum_{i=1}^{I} c_i \mu_i$, where $c_1, \ldots, c_I$ are constants with $\sum_{i=1}^{I} c_i = 0.$

Instead of having one factor (gasoline brand) affecting the outcome (number of miles per hour), there may be two (or more) such factors. For example, there might be some chemical additives meant to enhance the mileage. In this framework, suppose there are $J$ such chemical additives, and let us combine each one of the $I$ brands of gasoline with each one of the $J$ chemical additives. For simplicity, suppose we take just one observation, $Y_{ij}$, on each one of the $IJ$ pairs. Then it makes sense to assume that the r.v. $Y_{ij}$ is the result of the following additive components: A basic quantity (*grand mean*) $\mu$, the same for all $i$ and $j$; an *effect* $\alpha_i$ due to the $i$th brand of gasoline (the $i$th *row effect*); an *effect* $\beta_j$ due to the $j$th chemical additive (the $j$th *column effect*); and, of course, the random error $e_{ij}$ due to a host of causes. So, the assumed model is then: $Y_{ij} = \mu + \alpha_i + \beta_j + e_{ij}$. As usually, we assume that the $e_{ij}$'s are independent $\sim N(0, \sigma^2)$ with some (unknown) variance $\sigma^2$, which implies that the $Y_{ij}$'s are independent r.v.'s and $Y_{ij} \sim N(\mu + \alpha_i + \beta_j, \sigma^2)$. We further assume that some of $\alpha_i$ effects are $\geq 0$, some are $< 0$, and on the whole $\sum_{i=1}^{I} \alpha_i = 0$; and likewise for the $\beta_j$ effects: $\sum_{j=1}^{J} \beta_j = 0$. Summarizing these assumptions, we have then:

$$Y_{ij} = \mu + \alpha_i + \beta_j + e_{ij}, \quad \text{where the } e_{ij}\text{'s are independently}$$
$$\sim N(0, \sigma^2), i = 1, \ldots, I(\geq 2), \ j = 1, \ldots, J(\geq 2),$$

$$\sum_{i=1}^{I} \alpha_i = 0, \quad \sum_{j=1}^{J} \beta_j = 0. \tag{14}$$

The quantities $\mu, \alpha_i, i = 1, \ldots, I, \beta_j, j = 1, \ldots, J$ and $\sigma^2$ are the *parameters* of the model.

As already mentioned, the implication is that the r.v.'s $Y_{ij}$ are independent and $Y_{ij} \sim N(\mu + \alpha_i + \beta_j, \sigma^2), i = 1, \ldots, I, j = 1, \ldots, J$.

The model described by (14) is called *two-way layout* or *two-way classification*, as the observations are affected by two factors.

The main statistical issues are those of estimating the parameters involved and testing irrelevance of either one of the factors involved — that is, testing $H_{0A} : \alpha_1 = \cdots = \alpha_I = 0, H_{0B} : \beta_1 = \cdots = \beta_J = 0$. Details will be presented in Chapter 14. There, an explanation of the term "ANOVA" will also be given.

## 8.6 THE BASICS OF NONPARAMETRIC INFERENCE

All of the problems discussed in the previous sections may be summarized as follows: On the basis of a random sample of size $n, X_1, \ldots, X_n$, drawn from the p.d.f. $f(\cdot; \theta), \theta \in \Omega \subseteq \Re$, construct a point estimate for $\theta$, a confidence interval for $\theta$, and test hypotheses about $\theta$. In other words, the problems discussed were those of making (*statistical*) *inference* about $\theta$. These problems are suitably modified for a multidimensional parameter. The fundamental assumption in this framework is that the functional form of the p.d.f. $f(\cdot; \theta)$ is known; the only thing which does not render $f(\cdot; \theta)$ completely known is the presence of the (unknown constant) parameter $\theta$.

In many situations, stipulating a functional form for $f(\cdot; \theta)$ either is dictated by circumstances or is the product of accumulated experience. In the absence of these, we must still proceed with the problems of estimating important quantities, either by points or by intervals, and testing hypotheses about them. However, the framework now is *nonparametric*, and the relevant inference is referred to as *nonparametric inference*.

Actually, there have been at least three cases so far where *nonparametric estimation* was made without referring to it as such. Indeed, if $X_1, \ldots, X_n$ are i.i.d. r.v.'s with unknown mean $\mu$, then the sample mean $\bar{X}_n$ may be taken as an estimate of $\mu$, regardless of what the underlying distribution of the $X_i$'s is. This estimate is recommended on the basis of at least three considerations. First, it is unbiased, $E\bar{X}_n = \mu$ no matter what the underlying distribution is; second, $\bar{X}_n$ is the moment estimate of $\mu$; and third, by the WLLN, $\bar{X}_n \xrightarrow[n \to \infty]{P} \mu$, so that $\bar{X}_n$ is close to $\mu$, in the sense of probability, for all sufficiently large $n$. Now suppose that the $X_i$'s also have (an unknown) variance $\sigma^2 \in (0, \infty)$. Then (suitably modified) the sample variance $S_n^2$ can be used as an estimate of $\sigma^2$, because it is unbiased (Section 8.1) and also $S_n^2 \xrightarrow[n \to \infty]{P} \sigma^2$. Furthermore, by combining $\bar{X}_n$ and $S_n^2$ and using Theorem 7(ii) in Chapter 7, we have that $\sqrt{n}(\bar{X}_n - \mu)/S_n \simeq N(0, 1)$ for large $n$. Then, for such $n$, $[\bar{X}_n - z_{\alpha/2}\frac{S_n}{\sqrt{n}}, \bar{X}_n + z_{\alpha/2}\frac{S_n}{\sqrt{n}}]$ is a confidence interval for $\mu$ with confidence coefficient approximately $1 - \alpha$.

Also, the (unknown) d.f. $F$ of the $X_i$'s has been estimated at every point $x \in \Re$ by the empirical d.f. $F_n$ (see Application 7.2.1(5) in Chapter 7). The estimate $F_n$ has at least two desirable properties. For all $x \in \Re$ and regardless of the form of the d.f. $F$:

$$EF_n(x) = F(x) \text{ and } F_n(x) \xrightarrow[n \to \infty]{P} F(x).$$

What has not been done so far is to estimate the p.d.f. $f(x)$ at each $x \in \Re$, under certain regularity conditions, which do not include postulation of a functional form for $f$. There are several ways of doing this; in Chapter 15, we are going to adopt the so-called *kernel method* of estimating $f$. Some desirable results of the proposed estimate will be stated without proofs.

Regarding testing hypotheses, the problems to be addressed in Chapter 15 will be to test the hypothesis that the (unknown) d.f. $F$ is, actually, equal to a known one $F_0$; that is $H_0 : F = F_0$, the alternative $H_A$ being that $F(x) \neq F_0(x)$ for at least one $x \in \Re$. Actually, from a practical viewpoint, it is more important to compare two (unknown) d.f.'s $F$ and $G$, by stipulating $H_0 : F = G$. The alternative can be any one of the following: $H_A : F \neq G, H'_A : F > G, H''_A : F < G$, in the sense that $F(x) \geq G(x)$ or $F(x) \leq G(x)$, respectively, for all $x \in \Re$, and strict inequality for at least one $x$. In carrying out the appropriate tests, one has to use some pretty sophisticated asymptotic results regarding empirical d.f.'s. An alternative approach to using empirical d.f.'s is to employ the concept of a *rank* test or the concept of a *sign* test. These things will be discussed to some extent in Chapter 15. That chapter is concluded with the basics of regression estimation but in a nonparametric framework. In such a situation, what is estimated is an entire function rather than a few parameters. Some basic results are stated in Chapter 15.

# Point estimation

In the previous chapter, the basic terminology and concepts of parametric point estimation were introduced briefly. In the present chapter, we are going to elaborate extensively on this matter. For brevity, we will use the term estimation rather than parametric point estimation. The methods of estimation to be discussed here are those listed in the first section of the previous chapter; namely, maximum likelihood estimation, estimation through the concepts of unbiasedness and minimum variance (which lead to Uniformly minimum variance unbiased estimates), estimation based on decision-theoretic concepts, and estimation by the method of moments. The method of estimation by way of the principle of least squares is commonly used in the so-called linear models. Accordingly, it is deferred to Chapter 13.

Before we embark on the mathematical derivations, it is imperative to keep in mind the big picture; namely, why do we do what we do? A brief description is as follows. Let $X$ be a r.v. with p.d.f. $f(\cdot; \theta)$, where $\theta$ is a parameter lying in a parameter space $\Omega$. It is assumed that the functional form of the p.d.f. is completely known. So, if $\theta$ was known, then the p.d.f. would be known, and consequently we could calculate, in principle, all probabilities related to $X$, the expectation of $X$, its variance, etc. The problem, however, is that most often in practice (and in the present context) $\theta$ is not known. Then the objective is to estimate $\theta$ on the basis of a random sample of size $n$ from $f(\cdot; \theta), X_1, \ldots, X_n$. Then, replacing $\theta$ in $f(\cdot; \theta)$ by a "good" estimate of it, one would expect to be able to use the resulting p.d.f. for the purposes described above to a satisfactory degree.

## 9.1 MAXIMUM LIKELIHOOD ESTIMATION: MOTIVATION AND EXAMPLES

The following simple example is meant to shed light to the intuitive, yet quite logical, principle of maximum likelihood estimation.

**Example 1.** Let $X_1, \ldots, X_{10}$ be i.i.d. r.v.'s from the $B(1, \theta)$ distribution, $0 < \theta < 1$, and let $x_1, \ldots, x_{10}$ be the respective observed values. For convenience, set $t = x_1 + \cdots + x_{10}$. Further, suppose that in the 10 trials, six resulted in successes, so that $t = 6$. Then the likelihood function involved is: $L(\theta \mid x) = \theta^6 (1 - \theta)^4, 0 < \theta < 1, x = (x_1, \ldots, x_{10})$. Thus, $L(\theta \mid x)$ is the probability of observing exactly six successes in 10 independent Binomial trials, the successes occurring on those trials for which $x_i = 1, i = 1, \ldots, 10$; this probability is a function of the (unknown) parameter $\theta$. Let us calculate the values of this probability, for $\theta$ ranging from 0.1 to 0.9. We find:

| Values of $\theta$ | Values of $L(\theta \mid x)$ |
|---|---|
| 0.1 | 0.000006656 |
| 0.2 | 0.000026200 |
| 0.3 | 0.000175000 |
| 0.4 | 0.000531000 |
| 0.5 | 0.000976000 |
| 0.6 | 0.003320000 |
| 0.7 | 0.003010000 |
| 0.8 | 0.000419000 |
| 0.9 | 0.000053000 |

We observe that the values of $L(\theta \mid x)$ keep increasing, it attains its maximum value at $\theta = 0.6$, and then the values keep decreasing. Thus, if these nine values were the only possible values for $\theta$ (which they are not!), one would reasonably enough choose the value of 0.6 as the value of $\theta$. The value $\theta = 0.6$ has the distinction of maximizing (among the nine values listed) the probability of attaining the six already observed successes.

We observe that $0.6 = \frac{6}{10} = \frac{t}{n}$, where $n$ is the number of trials and $t$ is the number of successes. It will be seen in Example 2 that the value $\frac{t}{n}$, actually, maximizes the likelihood function among *all* values of $\theta$ with $0 < \theta < 1$. Then $\frac{t}{n}$ will be the maximum likehood estimate (MLE) of $\theta$ to be denoted by $\hat{\theta}$; i.e., $\hat{\theta} = \frac{t}{n}$.

In a general setting, let $X_1, \ldots, X_n$ be i.i.d. r.v.'s with p.d.f. $f(\cdot; \theta)$ with $\theta \in \Omega$, and let $x_1, \ldots, x_n$ be the respective observed values and $x = (x_1, \ldots, x_n)$. The likelihood function, $L(\theta \mid x)$, is given by $L(\theta \mid x) = \prod_{i=1}^{n} f(x_i; \theta)$, and a value of $\theta$ which maximizes $L(\theta \mid x)$ is called a *maximum likelihood estimate* of $\theta$. Clearly, the MLE depends on $x$, and we usually write $\hat{\theta} = \hat{\theta}(x)$. Thus,

$$L(\hat{\theta} \mid x) = \max\{L(\theta \mid x); \theta \in \Omega\}. \tag{1}$$

The justification for choosing an estimate as the value of the parameter which maximizes the likelihood function is the same as that given in Example 1, when the r.v.'s are discrete. The same interpretation holds true for r.v.'s of the continuous type, by considering small intervals around the observed values.

Once we decide to adopt the *maximum likelihood principle* (i.e., the principle of choosing an estimate of the parameter through the process of maximizing the likelihood function), the actual identification of an MLE is a purely mathematical problem; namely, that of maximizing a function. This maximization, if possible at all, often (but not always) is done through differentiation. Examples to be discussed below will illustrate various points. Before embarking on specific examples, it must be stressed that, whenever a maximum is sought by differentiation, the second-order derivative(s) must also be examined in search of a maximum. Also, it should be mentioned that maximization of the likelihood function, which is the product of $n$ factors, is equivalent to maximization of its logarithm (always with base $e$), which is the sum of $n$ summands, thus much easier to work with.

**Remark 1.**   Let us recall that a function $y = g(x)$ attains a maximum at a point $x = x_0$, if $\frac{d}{dx}g(x)|_{x=x_0} = 0$ and $\frac{d^2}{dx^2}g(x)|_{x=x_0} < 0$.

**Example 2.**   In terms of a random sample of size $n$, $X_1, \ldots, X_n$ from the $B(1, \theta)$ distribution with observed values $x_1, \ldots, x_n$, determine the MLE $\hat{\theta} = \hat{\theta}(x)$ of $\theta \in (0, 1)$, $x = (x_1, \ldots, x_n)$.

**Discussion.**   Since $f(x_i; \theta) = \theta^{x_i}(1 - \theta)^{1-x_i}$, $x_i = 0$ or $1$, $i = 1, \ldots, n$, the likelihood function is

$$L(\theta \mid x) = \prod_{i=1}^{n} f(x_i; \theta) = \theta^t (1 - \theta)^{n-t}, \quad t = x_1 + \cdots + x_n,$$

so that $t = 0, 1, \ldots, n$. Hence, $\log L(\theta \mid x) = t \log \theta + (n - t) \log(1 - \theta)$. From the *likelihood equation* $\frac{\partial}{\partial \theta} \log L(\theta \mid x) = \frac{t}{\theta} - \frac{n-t}{1-\theta} = 0$, we obtain $\theta = \frac{t}{n}$. Next, $\frac{\partial^2}{\partial \theta^2} \log L(\theta \mid x) = -\frac{t}{\theta^2} - \frac{n-t}{(1-\theta)^2}$, which is negative for all $\theta$ and hence for $\theta = t/n$. Therefore, the MLE of $\theta$ is $\hat{\theta} = \frac{t}{n} = \bar{x}$.

**Example 3.**   Determine the MLE $\hat{\theta} = \hat{\theta}(x)$ of $\theta \in (0, \infty)$ in the $P(\theta)$ distribution in terms of the random sample $X_1, \ldots, X_n$ with observed values $x_1, \ldots, x_n$.

**Discussion.**   Here $f(x_i; \theta) = \frac{e^{-\theta}\theta^{x_i}}{x_i!}$, $x_i = 0, 1, \ldots$, $i = 1, \ldots, n$, so that

$$\log L(\theta \mid x) = \log \left( \prod_{i=1}^{n} \frac{e^{-\theta}\theta^{x_i}}{x_i!} \right) = \log \left( e^{-n\theta} \prod_{i=1}^{n} \frac{\theta^{x_i}}{x_i!} \right)$$

$$= -n\theta + (\log \theta) \sum_{i=1}^{n} x_i - \log \left( \prod_{i=1}^{n} x_i! \right)$$

$$= -n\theta + (n \log \theta)\bar{x} - \log \left( \prod_{i=1}^{n} x_i! \right).$$

Hence $\frac{\partial}{\partial \theta} \log L(\theta \mid x) = -n + \frac{n\bar{x}}{\theta} = 0$, which gives $\theta = \bar{x}$, and $\frac{\partial^2}{\partial \theta^2} \log L(\theta \mid x) = -\frac{n\bar{x}}{\theta^2} < 0$ for all $\theta$ and hence for $\theta = \bar{x}$. Therefore the MLE of $\theta$ is $\hat{\theta} = \bar{x}$.

**Example 4.**   Determine the MLE $\hat{\theta} = \hat{\theta}(x)$ of $\theta \in (0, \infty)$ in the Negative Exponential distribution $f(x; \theta) = \theta e^{-\theta x}$, $x > 0$, on the basis of the random sample $X_1, \ldots, X_n$ with observed values $x_1, \ldots, x_n$.

**Discussion.**   Since $f(x_i; \theta) = \theta e^{-\theta x_i}$, $x_i > 0$, $i = 1, \ldots, n$, we have

$$\log L(\theta \mid x) = \log(\theta^n e^{-n\theta\bar{x}}) = n \log \theta - n\bar{x}\theta, \text{ so that}$$

$\frac{\partial}{\partial \theta} \log L(\theta \mid x) = \frac{n}{\theta} - n\bar{x} = 0$, and hence $\theta = 1/\bar{x}$. Furthermore, $\frac{\partial^2}{\partial \theta^2} \log L(\theta \mid x) = -\frac{n}{\theta^2} < 0$ for all $\theta$ and hence for $\theta = 1/\bar{x}$. It follows that $\hat{\theta} = 1/\bar{x}$.

**Example 5.** Let $X_1, \ldots, X_n$ be a random sample from the $N(\mu, \sigma^2)$ distribution, where only one of the parameters is known. Determine the MLE of the other (unknown) parameter.

**Discussion.** With $x_1, \ldots, x_n$ being the observed values of $X_1, \ldots, X_n$, we have:

**(i)** *Let $\mu$ be unknown.* Then

$$\log L(\mu \mid \boldsymbol{x}) = \log \left\{ \prod_{i=1}^{n} \frac{1}{\sqrt{2\pi}\sigma} \exp \left[ -\frac{1}{2\sigma^2} \sum_{i=1}^{n} (x_i - \mu)^2 \right] \right\}$$

$$= -n \log \left( \sqrt{2\pi}\sigma \right) - \frac{1}{2\sigma^2} \sum_{i=1}^{n} (x_i - \mu)^2, \text{ so that}$$

$\frac{\partial}{\partial \mu} \log L(\mu \mid \boldsymbol{x}) = \frac{n(\bar{x}-\mu)}{\sigma^2} = 0$, and hence $\mu = \bar{x}$. Furthermore,
$\frac{\partial^2}{\partial \mu^2} \log L(\mu \mid \boldsymbol{x}) = -\frac{n}{\sigma^2} < 0$ for all $\mu$ and hence for $\mu = \bar{x}$. It follows that the MLE of $\mu$ is $\hat{\mu} = \bar{x}$.

**(ii)** *Let $\sigma^2$ be unknown.* Then

$$\log L(\sigma^2 \mid \boldsymbol{x}) = \log \left\{ \prod_{i=1}^{n} \frac{1}{\sqrt{2\pi\sigma^2}} \exp \left[ -\frac{1}{2\sigma^2} \sum_{i=1}^{n} (x_i - \mu)^2 \right] \right\}$$

$$= -\frac{n}{2} \log(2\pi) - \frac{n}{2} \log \sigma^2 - \frac{1}{2\sigma^2} \sum_{i=1}^{n} (x_i - \mu)^2, \text{ so that}$$

$\frac{\partial}{\partial \sigma^2} \log L(\sigma^2 \mid \boldsymbol{x}) = -\frac{n}{2\sigma^2} + \frac{1}{2\sigma^4} \sum_{i=1}^{n} (x_i - \mu)^2 = 0$, and hence $\sigma^2 = \frac{1}{n} \sum_{i=1}^{n} (x_i - \mu)^2$; set $\frac{1}{n} \sum_{i=1}^{n} (x_i - \mu)^2 = s^2$. Then

$$\frac{\partial^2}{\partial (\sigma^2)^2} \log L(\sigma^2 \mid \boldsymbol{x}) = \frac{n}{2\sigma^4} - \frac{2n}{2\sigma^6} \times \frac{1}{n} \sum_{i=1}^{n} (x_i - \mu)^2$$

$$= \frac{n}{2(\sigma^2)^2} - \frac{2n}{2(\sigma^2)^3} s^2, \text{ so that}$$

$\frac{\partial^2}{\partial(\sigma^2)^2} \log L(\sigma^2 \mid \boldsymbol{x})|_{\sigma^2=s^2} = \frac{n}{2(s^2)^2} - \frac{2ns^2}{2(s^2)^3} = -\frac{n}{2s^4} < 0$. It follows that the MLE of $\sigma^2$ is $\hat{\sigma}^2 = \frac{1}{n} \sum_{i=1}^{n} (x_i - \mu)^2$.

In all of the preceding examples, the MLE's were determined through differentiation. Below is a case where this method does not apply because, simply, the derivative does not exist. As an introduction to the problem, let $X \sim U(0, \theta)$ $(\theta > 0)$, so that the likelihood function is $L(\theta \mid x) = \frac{1}{\theta} I_{[0,\theta]}(x)$ where it is to be recalled that the *indicator* function $I_A$ is defined by $I_A(x) = 1$ if $x \in A$, and $I_A(x) = 0$ if $x \in A^c$. The picture of $L(\cdot \mid x)$ is shown in Figure 9.1.

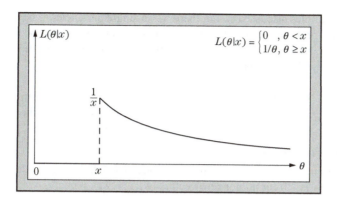

**FIGURE 9.1**

Picture of the likelihood function for the U(0, $\theta$) distribution.

**Example 6.** Let $X_1, \ldots, X_n$ be a random sample from the Uniform $U(\alpha, \beta)$ ($\alpha < \beta$) distribution, where only one of $\alpha$ and $\beta$ is unknown. Determine the MLE of the (unknown) parameter.

**Discussion.**

**(i)** *Let $\alpha$ be unknown.* Since

$$f(x_i; \alpha) = \frac{1}{\beta - \alpha} I_{[\alpha,\beta]}(x_i), \quad i = 1, \ldots, n, \text{ it follows that}$$

$$L(\alpha \mid x) = \frac{1}{(\beta - \alpha)^n} \prod_{i=1}^{n} I_{[\alpha,\beta]}(x_i) = \frac{1}{(\beta - \alpha)^n} I_{[\alpha,\beta]}(x_{(1)}) I_{[\alpha,\beta]}(x_{(n)}),$$

where $x_{(1)} = \min(x_1, \ldots, x_n), x_{(n)} = \max(x_1, \ldots, x_n)$; or

$$L(\alpha \mid x) = \frac{1}{(\beta - \alpha)^n} I_{[\alpha,\infty)}(x_{(1)}) I_{(-\infty,\beta]}(x_{(n)}). \tag{2}$$

Maximization of $L(\alpha \mid x)$ with respect to $\alpha$ means two things: Maximization of $I_{[\alpha,\infty)}(x_{(1)})$ and maximization of $1/(\beta - \alpha)^n$. The maximum value of the former quantity is 1 and occurs as long as $\alpha \leq x_{(1)}$. The latter quantity gets larger and larger as $\alpha$ gets closer and closer to $\beta$. But always $\alpha \leq x_{(1)} \leq \beta$, and $\alpha$ is subject to the restriction $\alpha \leq x_{(1)}$. Thus, $\alpha$ gets closest to $\beta$, if $\alpha = x_{(1)}$. In other words, the MLE of $\alpha$ is $\hat{\alpha} = x_{(1)}$.

**(ii)** *Let $\beta$ be unknown.* Relation (2) then becomes

$$L(\beta \mid x) = \frac{1}{(\beta - \alpha)^n} I_{[\alpha,\infty)}(x_{(1)}) I_{(-\infty,\beta]}(x_{(n)}),$$

whereas always $\alpha \leq x_{(n)} \leq \beta$. Then, arguing as in the first case, we have that the MLE of $\beta$ is $\hat{\beta} = x_{(n)}$.

In the examples discussed so far, there was a single parameter to be estimated. In the examples presented below, the parameters to be estimated will be two or more. If the maximization is to be done through differentiation, then the following remark reminds us how this method is implemented.

**Remark 2.** The function $y = g(x_1, \ldots, x_r)$ attains a maximum at a point $(x_{01}, \ldots, x_{0r})$, if the point $(x_{01}, \ldots, x_{0r})$ satisfies the system of the $r$ equations $\frac{\partial}{\partial x_i} g(x_1, \ldots, x_r) = 0$, $i = 1, \ldots, r$, and, in addition, the point $(x_{01}, \ldots, x_{0r})$ renders the $r \times r$ matrix of the second-order derivatives $(\frac{\partial^2}{\partial x_i \partial x_j} g(x_1, \ldots, x_r))$, $i, j = 1, \ldots, r$, negative definite. What is meant by the term "negative definite" is that the real-valued quantity below is $< 0$ for all nonzero vectors $(\lambda_1, \ldots, \lambda_r)$; namely,

$$(\lambda_1, \ldots, \lambda_r) \left( \frac{\partial^2}{\partial x_i \partial x_j} g(x_1, \ldots, x_r) \Big|_{(x_1, \ldots, x_r) = (x_{01}, \ldots, x_{0r})} \right) \begin{pmatrix} \lambda_1 \\ \vdots \\ \lambda_r \end{pmatrix} < 0.$$

**Example 7.**   Refer to Example 5 and suppose that both $\mu$ and $\sigma^2$ are unknown. Determine their MLE's.

**Discussion.**   Here

$$\log L(\mu, \sigma^2 \mid \boldsymbol{x}) = -\frac{n}{2} \log(2\pi) - \frac{n}{2} \log \sigma^2 - \frac{1}{2\sigma^2} \sum_{i=1}^{n} (x_i - \mu)^2,$$

and then the two likelihood equations produce the unique solution $\mu = \bar{x}$ and $\sigma^2 = \frac{1}{n} \sum_{i=1}^{n} (x_i - \bar{x})^2$, which we may denote by $s^2$. Next, the $2 \times 2$ matrix of the second-order derivatives, evaluated at $(\bar{x}, s^2)$, becomes $\begin{pmatrix} -\frac{n}{s^2} & 0 \\ 0 & -\frac{n}{2s^4} \end{pmatrix}$, which is negative definite (see Exercise 1.2). Thus, $\hat{\mu} = \bar{x}$ and $\hat{\sigma}^2 = \frac{1}{n} \sum_{i=1}^{n} (x_i - \bar{x})^2$ are the MLE's of $\mu$ and $\sigma^2$, respectively.

**Example 8.**   A Multinomial experiment is carried out independently $n$ times, so that the likelihood function is

$$L(p_1, \ldots, p_r \mid \boldsymbol{x}) = \frac{n!}{x_1! \cdots x_r!} p_1^{x_1} \cdots p_r^{x_r},$$

where $x_i \geq 0$, integers, $i = 1, \ldots, r$, with $x_1 + \cdots + x_r = n$, and $0 < p_i < 1, i = 1, \ldots, r$ with $p_1 + \cdots + p_r = 1$. Determine the MLE's of $p_i, i = 1, \ldots, r$.

**Discussion.**   The number of independent parameters is $r - 1$, since, for example, $p_r = 1 - p_1 - \cdots - p_{r-1}$. Looking at the log $L(p_1, \ldots, p_r \mid \boldsymbol{x})$ and taking partial derivatives with respect to $p_i, i = 1, \ldots, r - 1$ (and remembering that $p_r = 1 - p_1 - \cdots - p_{r-1}$) we obtain

$$x_i \frac{1}{p_i} - x_r \frac{1}{p_r} = 0, \quad i = 1, \ldots, r-1.$$

From these relations, the unique solution $p_i = \frac{x_i}{n}$, $i = 1, \ldots, r$ follows.

Next, the $(r-1) \times (r-1)$ matrix of the second-order derivatives, evaluated at $p_i = x_i/n$, $i = 1, \ldots, r$, is given by

$$\begin{pmatrix} -\frac{n^2}{x_1} - \frac{n^2}{x_r} & -\frac{n^2}{x_r} & \cdots & -\frac{n^2}{x_r} & -\frac{n^2}{x_r} \\ -\frac{n^2}{x_r} & -\frac{n^2}{x_2} - \frac{n^2}{x_r} & \cdots & -\frac{n^2}{x_r} & -\frac{n^2}{x_r} \\ \cdots & \cdots & \cdots & \cdots & \cdots \\ -\frac{n^2}{x_r} & -\frac{n^2}{x_r} & \cdots & -\frac{n^2}{x_r} & -\frac{n^2}{x_{r-1}} - \frac{n^2}{x_r} \end{pmatrix},$$

which is seen to be negative definite. Consequently, the MLE's of $p_i$, $i = 1, \ldots, r$ are $\hat{p}_i = \frac{x_i}{n}$, $i = 1, \ldots, r$ (also, see Exercise 1.3).

**Example 9.**   Refer to Example 6, assume that both $\alpha$ and $\beta$ are unknown, and determine their MLE's.

**Discussion.**   Expression (1) becomes here as follows:

$$L(\alpha, \beta \mid x) = \frac{1}{(\beta - \alpha)^n} I_{[\alpha,\infty)}(x_{(1)}) I_{(-\infty,\beta]}(x_{(n)}). \tag{3}$$

Since always $\alpha \le x_{(1)} \le x_{(n)} \le \beta$, the right-hand side of (3) is maximized if $I_{[\alpha,\infty)}(x_{(1)}) = 1$ and $I_{(-\infty,\beta]}(x_{(n)}) = 1$, which happen if $\alpha \le x_{(1)}, x_{(n)} \le \beta$ and also if $\alpha$ and $\beta$ are as close together as possible. Clearly, this happens for $\alpha = x_{(1)}$ and $\beta = x_{(n)}$. In other words, the MLE's of $\alpha$ and $\beta$ are $\hat{\alpha} = x_{(1)}$ and $\hat{\beta} = x_{(n)}$.

---

# EXERCISES

**1.1**   Refer to Example 6(ii) and justify the statement made there that $x_{(n)}$ is, indeed, the MLE of $\beta$.

**1.2**   Show that the matrix $\begin{pmatrix} -n/s^2 & 0 \\ 0 & -n/2s^4 \end{pmatrix}$ in Example 7 is, indeed, negative definite.

**1.3**   In reference to Example 8, show that:
   (i)   $p_i = \frac{x_i}{n}$, $i = 1, \ldots, r$, is, indeed, the unique solution of the system of equations considered there.
   (ii)   The $(r-1) \times (r-1)$ matrix exhibited there is the matrix of the second-order derivatives as stated.
   (iii)   The matrix in part (ii) is negative definite.

**1.4**   In reference to Example 18 below, show that $\mathrm{Var}_\theta(S^2) = \frac{2\sigma^4}{n-1}$ as stated there.

**1.5**   If $X_1, \ldots, X_n$ are independent r.v.'s distributed as $B(k, \theta)$, $\theta \in \Omega = (0, 1)$, with respective observed values $x_1, \ldots, x_n$, show that $\hat{\theta} = \frac{\bar{x}}{k}$ is the MLE of $\theta$, where $\bar{x}$ is the sample mean of the $x_i$'s.

**1.6**   If the independent r.v.'s. $X_1, \ldots, X_n$ have the Geometric p.d.f.
$f(x; \theta) = \theta(1 - \theta)^{x-1}$, $x = 1, 2, \ldots, \theta \in \Omega = (0, 1)$, and respective observed values $x_1, \ldots, x_n$, then show that $\hat{\theta} = 1/\bar{x}$ is the MLE of $\theta$.

**1.7**   On the basis of a random sample of size $n$ from the p.d.f.
$f(x; \theta) = (\theta + 1)x^{\theta}$, $0 < x < 1$, $\theta \in \Omega = (-1, \infty)$, derive the MLE of $\theta$.

**1.8**   On the basis of a random sample of size $n$ from the p.d.f.
$f(x; \theta) = \theta x^{\theta - 1}$, $0 < x < 1$, $\theta \in \Omega = (0, \infty)$, derive the MLE of $\theta$.

**1.9**   (i)  Show that the function $f(x; \theta) = \frac{1}{2\theta}e^{-|x|/\theta}$, $x \in \mathfrak{R}$, $\theta \in \Omega = (0, \infty)$ is a p.d.f. (the so-called *double Exponential* p.d.f.), and draw its picture.
   (ii)  On the basis of a random sample of size $n$ from this p.d.f., derive the MLE of $\theta$.

**1.10**   (i)  Verify that the function $f(x; \theta) = \theta^2 x e^{-\theta x}$, $x > 0$, $\theta \in \Omega = (0, \infty)$ is a p.d.f., by observing that it is the Gamma p.d.f. with parameters $\alpha = 2$, $\beta = 1/\theta$.
   (ii)  On the basis of a random sample of size $n$ from this p.d.f., derive the MLE of $\theta$.

**1.11**   (i)  Show that the function $f(x; \alpha, \beta) = \frac{1}{\beta}e^{-(x-\alpha)/\beta}$, $x \geq \alpha$, $\alpha \in \mathfrak{R}$, $\beta > 0$, is a p.d.f., and draw its picture.
   On the basis of a random sample of size $n$ from this p.d.f., determine the MLE of:
   (ii)  $\alpha$ when $\beta$ is known.
   (iii)  $\beta$ when $\alpha$ is known.
   (iv)  $\alpha$ and $\beta$ when both are unknown.

**1.12**   Refer to the Bivariate Normal distribution discussed in Section 4.5, whose p.d.f. is given by:

$$f_{X,Y}(x, y) = \frac{1}{2\pi\sigma_1\sigma_2\sqrt{1 - \rho^2}}e^{-q/2}, \quad x, y \in \mathfrak{R},$$

where

$$q = \frac{1}{1 - \rho^2}\left[\left(\frac{x - \mu_1}{\sigma_1}\right)^2 - 2\rho\left(\frac{x - \mu_1}{\sigma_1}\right)\left(\frac{y - \mu_2}{\sigma_2}\right) + \left(\frac{y - \mu_2}{\sigma_2}\right)^2\right],$$

$\mu_1, \mu_2 \in \mathfrak{R}, \sigma_1^2, \sigma_2^2 > 0$ and $-1 \leq \rho \leq 1$ are the parameters of the distribution. The objective here is to find the MLE's of these parameters. This is done in two stages, in the present exercise and the exercises following. For convenient writing, set $\theta = (\mu_1, \mu_2, \sigma_1^2, \sigma_2^2, \rho)$, and form the likelihood function for a sample of size $n$, $(X_i, Y_i)$, $i = 1, \ldots, n$, from the underlying distribution; i.e.,

$$L(\theta \mid x,y) = \left(\frac{1}{2\pi\sigma_1\sigma_2\sqrt{1-\rho^2}}\right)^n \exp\left(-\frac{1}{2}\sum_{i=1}^{n} q_i\right),$$

where

$$q_i = \frac{1}{1-\rho^2}\left[\left(\frac{x_i-\mu_1}{\sigma_1}\right)^2 - 2\rho\left(\frac{x_i-\mu_1}{\sigma_1}\right)\left(\frac{y_i-\mu_2}{\sigma_2}\right) + \left(\frac{y_i-\mu_2}{\sigma_2}\right)^2\right],$$

and $x = (x_1,\ldots,x_n), y = (y_1,\ldots,y_n)$, the observed values of the $X_i$'s and the $Y_i$'s. Also, set

$$\lambda(\theta) = \lambda(\theta \mid x,y) = \log L(\theta \mid x,y)$$

$$= -n\log(2\pi) - \frac{n}{2}\log\sigma_1^2 - \frac{n}{2}\log\sigma_2^2 - \frac{n}{2}\log(1-\rho^2) - \frac{1}{2}\sum_{i=1}^{n} q_i.$$

**(i)** Show that the first-order partial derivatives of $q$ given above are provided by the following expressions:

$$\frac{\partial q}{\partial \mu_1} = -\frac{2(x-\mu_1)}{\sigma_1^2(1-\rho^2)} + \frac{2\rho(y-\mu_2)}{\sigma_1\sigma_2(1-\rho^2)},$$

$$\frac{\partial q}{\partial \mu_2} = -\frac{2(y-\mu_2)}{\sigma_2^2(1-\rho^2)} + \frac{2\rho(x-\mu_1)}{\sigma_1\sigma_2(1-\rho^2)},$$

$$\frac{\partial q}{\partial \sigma_1^2} = -\frac{(x-\mu_1)^2}{\sigma_1^4(1-\rho^2)} + \frac{\rho(x-\mu_1)(y-\mu_2)}{\sigma_1^3\sigma_2(1-\rho^2)},$$

$$\frac{\partial q}{\partial \sigma_2^2} = -\frac{(y-\mu_2)^2}{\sigma_2^4(1-\rho^2)} + \frac{\rho(x-\mu_1)(y-\mu_2)}{\sigma_1\sigma_2^3(1-\rho^2)},$$

$$\frac{\partial q}{\partial \rho} = \frac{2}{(1-\rho^2)^2}\left\{\rho\left[\left(\frac{x-\mu_1}{\sigma_1}\right)^2 + \left(\frac{y-\mu_2}{\sigma_2}\right)^2\right]\right.$$
$$\left. - (1+\rho^2)\left(\frac{x-\mu_1}{\sigma_1}\right)\left(\frac{y-\mu_2}{\sigma_2}\right)\right\}.$$

**(ii)** Use the above obtained expressions and $\lambda(\theta)$ in order to show that:

$$\frac{\partial\lambda(\theta)}{\partial\mu_1} = \frac{n}{\sigma_1^2(1-\rho^2)}(\bar{x}-\mu_1) - \frac{n\rho}{\sigma_1\sigma_2(1-\rho^2)}(\bar{y}-\mu_2),$$

$$\frac{\partial\lambda(\theta)}{\partial\mu_2} = \frac{n}{\sigma_2^2(1-\rho^2)}(\bar{y}-\mu_2) - \frac{n\rho}{\sigma_1\sigma_2(1-\rho^2)}(\bar{x}-\mu_1),$$

$$\frac{\partial\lambda(\theta)}{\partial\sigma_1^2} = -\frac{n}{2\sigma_1^2} + \frac{\sum_{i=1}^{n}(x_i-\mu_1)^2}{2\sigma_1^4(1-\rho^2)} - \frac{\rho\sum_{i=1}^{n}(x_i-\mu_1)(y_i-\mu_2)}{2\sigma_1^3\sigma_2(1-\rho^2)},$$

$$\frac{\partial\lambda(\theta)}{\partial\sigma_2^2} = -\frac{n}{2\sigma_2^2} + \frac{\sum_{i=1}^{n}(y_i-\mu_2)^2}{2\sigma_2^4(1-\rho^2)} - \frac{\rho\sum_{i=1}^{n}(x_i-\mu_1)(y_i-\mu_2)}{2\sigma_1\sigma_2^3(1-\rho^2)},$$

$$\frac{\partial\lambda(\theta)}{\partial\sigma} = \frac{n\rho}{1-\rho^2} - \frac{\rho\sum_{i=1}^{n}(x_i-\mu_1)^2}{\sigma_1^2(1-\rho^2)} - \frac{\rho\sum_{i=1}^{n}(y_i-\mu_2)^2}{\sigma_2^2(1-\rho^2)}$$
$$+ \frac{(1+\rho^2)\sum_{i=1}^{n}(x_i-\mu_1)(y_i-\mu_2)}{\sigma_1\sigma_2(1-\rho^2)^2}.$$

**(iii)** Setting $\frac{\partial\lambda(\theta)}{\partial\mu_1} = \frac{\partial\lambda(\theta)}{\partial\mu_2} = 0$, and solving for $\mu_1$ and $\mu_2$, show that there is a unique solution given by: $\tilde{\mu}_1 = \bar{x}$ and $\tilde{\mu}_2 = \bar{y}$.

**(iv)** By setting $\frac{\partial\lambda(\theta)}{\partial\sigma_1^2} = \frac{\partial\lambda(\theta)}{\partial\sigma_2^2} = \frac{\partial\lambda(\theta)}{\partial\rho} = 0$ and replacing $\mu_1$ and $\mu_2$ by the respective expressions $\bar{x}$ and $\bar{y}$, show that we arrive at the equations:

$$\frac{S_x}{\sigma_1^2} - \frac{\rho S_{xy}}{\sigma_1\sigma_2} = 1 - \rho^2, \quad \frac{S_y}{\sigma_2^2} - \frac{\rho S_{xy}}{\sigma_1\sigma_2} = 1 - \rho^2,$$

$$\frac{S_x}{\sigma_1^2} + \frac{S_y}{\sigma_2^2} - \frac{(1+\rho^2)S_{xy}}{\rho\sigma_1\sigma_2} = 1 - \rho^2,$$

where $S_x = \frac{1}{n}\sum_{i=1}^n (x_i - \bar{x})^2, S_y = \frac{1}{n}\sum_{i=1}^n (y_i - \bar{y})^2$, and $S_{xy} = \frac{1}{n}\sum_{i=1}^n (x_i - \bar{x})(y_i - \bar{y})$.

**(v)** In the equations obtained in part (iv), solve for $\sigma_1^2, \sigma_2^2$, and $\rho$ in order to obtain the unique solution:

$$\tilde{\sigma}_1^2 = S_x, \quad \tilde{\sigma}_2^2 = S_y, \quad \tilde{\rho} = S_{xy}/S_x^{1/2}S_y^{1/2}.$$

**1.13** The purpose of this exercise is to show that the values $\tilde{\mu}_1, \tilde{\mu}_2, \tilde{\sigma}_1^2, \tilde{\sigma}_2^2$, and $\tilde{\rho}$ are actually the MLE's of the respective parameters. To this end:

**(i)** Take the second-order partial derivatives of $\lambda(\theta)$, as indicated below, and show that they are given by the following expressions:

$$\frac{\partial^2\lambda(\theta)}{\partial\mu_1^2} = -\frac{n}{\sigma_1^2(1-\rho^2)} \overset{\text{def}}{=} d_{11}, \quad \frac{\partial^2\lambda(\theta)}{\partial\mu_1\,\partial\mu_2} = \frac{n\rho}{\sigma_1\sigma_2(1-\rho^2)} \overset{\text{def}}{=} d_{12},$$

$$\frac{\partial^2\lambda(\theta)}{\partial\mu_1\,\partial\sigma_1^2} = -\frac{n(\bar{x}-\mu_1)}{\sigma_1^4(1-\rho^2)} + \frac{n\rho(\bar{y}-\mu_2)}{2\sigma_1^3\sigma_2(1-\rho^2)} \overset{\text{def}}{=} d_{13},$$

$$\frac{\partial^2\lambda(\theta)}{\partial\mu_1\,\partial\sigma_2^2} = \frac{n\rho(\bar{y}-\mu_2)}{2\sigma_1\sigma_2^3(1-\rho^2)} \overset{\text{def}}{=} d_{14},$$

$$\frac{\partial^2\lambda(\theta)}{\partial\mu_1\,\partial\rho} = \frac{2\rho n(\bar{x}-\mu_1)}{\sigma_1^2(1-\rho^2)^2} - \frac{n(1+\rho^2)(\bar{y}-\mu_2)}{\sigma_1\sigma_2(1-\rho^2)^2} \overset{\text{def}}{=} d_{15}.$$

**(ii)** In $d_{1i}, i = 1, \ldots, 5$, replace the parameters involved by their respective estimates, and denote the resulting expressions by $\tilde{d}_{1i}, i = 1, \ldots, 5$. Then show that:

$$\tilde{d}_{11} = -\frac{nS_y}{S_xS_y - S_{xy}^2}, \quad \tilde{d}_{12} = \frac{nS_{xy}}{S_xS_y - S_{xy}^2}, \quad \tilde{d}_{13} = \tilde{d}_{14} = \tilde{d}_{15} = 0,$$

where $S_x, S_y$, and $S_{xy}$ are given in Exercise 1.12(iv).

**(iii)** Work as in part (i) in order to show that:

$$d_{21} \overset{\text{def}}{=} \frac{\partial^2\lambda(\theta)}{\partial\mu_2\,\partial\mu_1} = \frac{\partial^2\lambda(\theta)}{\partial\mu_1\,\partial\mu_2} = d_{12}, \quad d_{22} \overset{\text{def}}{=} \frac{\partial^2\lambda(\theta)}{\partial\mu_2^2} = -\frac{n}{\sigma_2^2(1-\rho^2)},$$

$$d_{23} \overset{\text{def}}{=} \frac{\partial^2\lambda(\theta)}{\partial\mu_2\,\partial\sigma_1^2} = \frac{n\rho(\bar{x}-\mu_1)}{2\sigma_1^3\sigma(1-\rho^2)},$$

$$d_{24} \stackrel{\text{def}}{=} \frac{\partial^2 \lambda(\boldsymbol{\theta})}{\partial \mu_2 \partial \sigma_2^2} = -\frac{n(\bar{y} - \mu_2)}{\sigma_2^4(1 - \rho^2)} + \frac{n \rho(\bar{x} - \mu_1)}{2\sigma_1 \sigma_2^3(1 - \rho^2)},$$

$$d_{25} \stackrel{\text{def}}{=} \frac{\partial^2 \lambda(\boldsymbol{\theta})}{\partial \mu_2 \partial \rho} = \frac{2\rho n(\bar{y} - \mu_2)}{\sigma_2^2(1 - \rho^2)^2} - \frac{n(1 + \rho^2)(\bar{x} - \mu_1)}{\sigma_1 \sigma_2(1 - \rho^2)^2}.$$

**(iv)** In part (iii), replace the parameters by their respective estimates and denote by $\tilde{d}_{2i}$, $i = 1, \ldots, 5$ the resulting expressions. Then, show that:

$$\tilde{d}_{21} = \tilde{d}_{12}, \quad \tilde{d}_{22} = -\frac{nS_x}{S_x S_y - S_{xy}^2}, \quad \text{and} \quad \tilde{d}_{23} = \tilde{d}_{24} = \tilde{d}_{25} = 0.$$

**(v)** Work as in parts (i) and (iii), and use analogous notation in order to obtain:

$$d_{31} \stackrel{\text{def}}{=} \frac{\partial^2 \lambda(\boldsymbol{\theta})}{\partial \sigma_1^2 \partial \mu_1} = d_{13}, \quad d_{32} \stackrel{\text{def}}{=} \frac{\partial^2 \lambda(\boldsymbol{\theta})}{\partial \sigma_1^2 \partial \mu_2} = d_{23},$$

$$d_{33} \stackrel{\text{def}}{=} \frac{\partial^2 \lambda(\boldsymbol{\theta})}{\partial (\sigma_1^2)^2} = \frac{n}{2\sigma_1^4} - \frac{\sum_{i=1}^n (x_i - \mu_1)^2}{\sigma_1^6(1 - \rho^2)} + \frac{3\rho \sum_{i=1}^n (x_i - \mu_1)(y_i - \mu_2)}{4\sigma_1^5 \sigma_2(1 - \rho^2)},$$

$$d_{34} \stackrel{\text{def}}{=} \frac{\partial^2 \lambda(\boldsymbol{\theta})}{\partial \sigma_1^2 \partial \sigma_2^2} = \frac{\rho \sum_{i=1}^n (x_i - \mu_1)(y_i - \mu_2)}{4\sigma_1^3 \sigma_2^3(1 - \rho^2)},$$

$$d_{35} \stackrel{\text{def}}{=} \frac{\partial^2 \lambda(\boldsymbol{\theta})}{\partial \sigma_1^2 \partial \rho} = \frac{\rho \sum_{i=1}^n (x_i - \mu_1)^2}{\sigma_1^4(1 - \rho^2)^2} - \frac{(1 + \rho^2)\sum_{i=1}^n (x_i - \mu_1)(y_i - \mu_2)}{2\sigma_1^3 \sigma_2(1 - \rho^2)^2}.$$

**(vi)** Work as in parts (ii) and (iv), and use analogous notation in order to obtain: $\tilde{d}_{31} = \tilde{d}_{32} = 0$, and

$$\tilde{d}_{33} = -\frac{n(2S_x S_y - S_{xy}^2)}{4S_x^2(S_x S_y - S_{xy}^2)}, \quad \tilde{d}_{34} = \frac{nS_{xy}^2}{4S_x S_y(S_x S_y - S_{xy}^2)},$$

$$\tilde{d}_{35} = \frac{nS_y^{1/2}S_{xy}}{2S_x^{1/2}(S_x S_y - S_{xy}^2)}.$$

**(vii)** Work as in part (v) in order to obtain:

$$d_{41} \stackrel{\text{def}}{=} \frac{\partial^2 \lambda(\boldsymbol{\theta})}{\partial \sigma_2^2 \partial \mu_1} = d_{14}, \quad d_{42} \stackrel{\text{def}}{=} \frac{\partial^2 \lambda(\boldsymbol{\theta})}{\partial \sigma_2^2 \partial \mu_2} = d_{24}, \quad d_{43} \stackrel{\text{def}}{=} \frac{\partial^2 \lambda(\boldsymbol{\theta})}{\partial \sigma_2^2 \partial \sigma_1^2} = d_{34},$$

$$d_{44} \stackrel{\text{def}}{=} \frac{\partial^2 \lambda(\boldsymbol{\theta})}{\partial (\sigma_2^2)^2} = \frac{n}{2\sigma_2^4} - \frac{\sum_{i=1}^n (y_i - \mu_2)^2}{\sigma_2^6(1 - \rho^2)} + \frac{3\rho \sum_{i=1}^n (x_i - \mu_1)(y_i - \mu_2)}{4\sigma_1 \sigma_2^5(1 - \rho^2)},$$

$$d_{45} \stackrel{\text{def}}{=} \frac{\partial^2 \lambda(\boldsymbol{\theta})}{\partial \sigma_2^2 \partial \rho} = \frac{\rho \sum_{i=1}^n (y_i - \mu_2)^2}{\sigma_2^4(1 - \rho^2)^2} - \frac{(1 + \rho^2)\sum_{i=1}^n (x_i - \mu_1)(y_i - \mu_2)}{2\sigma_1 \sigma_2^3(1 - \rho^2)^2}.$$

**(viii)** Work as in part (vi) in order to get:
$$\tilde{d}_{41} = \tilde{d}_{42} = 0, \quad \tilde{d}_{43} = \tilde{d}_{34}, \text{ and}$$

$$\tilde{d}_{44} = -\frac{n\left(2S_x S_y - S_{xy}^2\right)}{4S_y^2\left(S_x S_y - S_{xy}^2\right)}, \quad \tilde{d}_{45} = \frac{nS_x^{1/2} S_{xy}}{2S_y^{1/2}\left(S_x S_y - S_{xy}^2\right)}.$$

**(ix)** Work as in part (v), and use analogous notation in order to get:
$$d_{51} = d_{15}, \quad d_{52} = d_{25}, \quad d_{53} = d_{35}, \quad d_{54} = d_{45}, \text{ and}$$

$$d_{55} = \frac{n(1 + \rho^2)}{(1 - \rho^2)^2} - \frac{1 + 3\rho^2}{(1 - \rho^2)^3} \left[ \frac{1}{\sigma_1^2} \sum_{i=1}^{n} (x_i - \mu_1)^2 + \frac{1}{\sigma_2^2} \sum_{i=1}^{n} (y_i - \mu_2)^2 \right.$$

$$\left. + \frac{2\rho(3 + \rho^2)}{\sigma_1 \sigma_2 (1 - \rho^2)^3} \sum_{i=1}^{n} (x_i - \mu_1)(y_i - \mu_2) \right].$$

**(x)** Work as in part (vi) in order to obtain:
$$\tilde{d}_{51} = \tilde{d}_{15}, \quad \tilde{d}_{52} = \tilde{d}_{25}, \quad \tilde{d}_{53} = \tilde{d}_{35}, \quad \tilde{d}_{54} = \tilde{d}_{45}, \text{ and}$$

$$\tilde{d}_{55} = -\frac{nS_x S_y\left(S_x S_y + S_{xy}^2\right)}{\left(S_x S_y - S_{xy}^2\right)^2}.$$

**1.14** In this exercise, it is shown that the solution values of $\tilde{\mu}_1 = \bar{x}, \tilde{\mu}_2 = \bar{y},$ $\tilde{\sigma}_1^2 = S_x, \tilde{\sigma}_2^2 = S_y,$ and $\tilde{\rho} = S_{xy}/S_x^{1/2} S_y^{1/2}$ are, indeed, the MLE's of the respective parameters. To this effect, set $\tilde{D}$ for the determinant

$$\tilde{D} = \begin{vmatrix} \tilde{d}_{11} & \tilde{d}_{12} & \tilde{d}_{13} & \tilde{d}_{14} & \tilde{d}_{15} \\ \tilde{d}_{21} & \tilde{d}_{22} & \tilde{d}_{23} & \tilde{d}_{24} & \tilde{d}_{25} \\ \tilde{d}_{31} & \tilde{d}_{32} & \tilde{d}_{33} & \tilde{d}_{34} & \tilde{d}_{35} \\ \tilde{d}_{41} & \tilde{d}_{42} & \tilde{d}_{43} & \tilde{d}_{44} & \tilde{d}_{45} \\ \tilde{d}_{51} & \tilde{d}_{52} & \tilde{d}_{53} & \tilde{d}_{54} & \tilde{d}_{55} \end{vmatrix},$$

and let $D_i$ be the determinants taken from $\tilde{D}$ by eliminating the last $5 - i, i = 1, \ldots, 5$ rows and columns; also, set $\tilde{D}_0 = 1$. Thus,

$$\tilde{D}_1 = \tilde{d}_{11}, \quad \tilde{D}_2 = \begin{vmatrix} \tilde{d}_{11} & \tilde{d}_{12} \\ \tilde{d}_{21} & \tilde{d}_{22} \end{vmatrix},$$

$$\tilde{D}_3 = \begin{vmatrix} \tilde{d}_{11} & \tilde{d}_{12} & \tilde{d}_{13} \\ \tilde{d}_{21} & \tilde{d}_{22} & \tilde{d}_{23} \\ \tilde{d}_{31} & \tilde{d}_{32} & \tilde{d}_{33} \end{vmatrix}, \quad \tilde{D}_4 = \begin{vmatrix} \tilde{d}_{11} & \tilde{d}_{12} & \tilde{d}_{13} & \tilde{d}_{14} \\ \tilde{d}_{21} & \tilde{d}_{22} & \tilde{d}_{23} & \tilde{d}_{24} \\ \tilde{d}_{31} & \tilde{d}_{32} & \tilde{d}_{33} & \tilde{d}_{34} \\ \tilde{d}_{41} & \tilde{d}_{42} & \tilde{d}_{43} & \tilde{d}_{44} \end{vmatrix},$$

and $\tilde{D}_5 = \tilde{D}.$

**(i)** Use parts (ii), (iv), (vi), (viii), and (x) in Exercise 1.13 in order to conclude that the determinants $\tilde{D}(= \tilde{D}_5)$ and $\tilde{D}_i, i = 1, \ldots, 4$ take the following forms:

$$\tilde{D} = \begin{vmatrix} \tilde{d}_{11} & \tilde{d}_{12} & 0 & 0 & 0 \\ \tilde{d}_{21} & \tilde{d}_{22} & 0 & 0 & 0 \\ 0 & 0 & \tilde{d}_{33} & \tilde{d}_{34} & \tilde{d}_{35} \\ 0 & 0 & \tilde{d}_{43} & \tilde{d}_{44} & \tilde{d}_{45} \\ 0 & 0 & \tilde{d}_{53} & \tilde{d}_{54} & \tilde{d}_{55} \end{vmatrix},$$

$$\tilde{D}_1 = \tilde{d}_{11}, \quad \tilde{D}_2 = \begin{vmatrix} \tilde{d}_{11} & \tilde{d}_{12} \\ \tilde{d}_{21} & \tilde{d}_{22} \end{vmatrix},$$

$$\tilde{D}_3 = \begin{vmatrix} \tilde{d}_{11} & \tilde{d}_{12} & 0 \\ \tilde{d}_{21} & \tilde{d}_{22} & 0 \\ 0 & 0 & \tilde{d}_{33} \end{vmatrix}, \quad \tilde{D}_4 = \begin{vmatrix} \tilde{d}_{11} & \tilde{d}_{12} & 0 & 0 \\ \tilde{d}_{21} & \tilde{d}_{22} & 0 & 0 \\ 0 & 0 & \tilde{d}_{33} & \tilde{d}_{34} \\ 0 & 0 & \tilde{d}_{43} & \tilde{d}_{44} \end{vmatrix}.$$

**(ii)** Expand the determinants $\tilde{D}_i$, $i = 1, \ldots, 4$, and also use parts (iv) and (viii) of Exercise 1.13 in order to obtain:

$$\tilde{D}_1 = \tilde{d}_{11}, \quad \tilde{D}_2 = \tilde{d}_{11}\tilde{d}_{22} - (\tilde{d}_{12})^2, \quad \tilde{D}_3 = \tilde{d}_{33}\tilde{D}_2,$$
$$\tilde{D}_4 = [\tilde{d}_{33}\tilde{d}_{44} - (\tilde{d}_{34})^2]\tilde{D}_2.$$

**(iii)** Expand the determinant $\tilde{D}_5 (= \tilde{D})$, and also use parts (viii) and (x) of Exercise 1.13 in order to get:

$$\tilde{D}_5 = \tilde{D}_2(\tilde{d}_{33}A - \tilde{d}_{34}B + \tilde{d}_{35}C),$$

where

$$A = \tilde{d}_{44}\tilde{d}_{55} - (\tilde{d}_{45})^2, \quad B = \tilde{d}_{34}\tilde{d}_{55} - \tilde{d}_{45}\tilde{d}_{35}, \quad C = \tilde{d}_{34}\tilde{d}_{45} - \tilde{d}_{44}\tilde{d}_{35}.$$

**(iv)** For convenience, set: $S_x = \alpha, S_y = \beta, S_{xy} = \gamma$, and $S_x S_y - S_{xy}^2 = \alpha\beta - \gamma^2 = \delta$, so that $\alpha, \beta > 0$ and also $\delta > 0$ by the Cauchy-Schwarz inequality (see Theorem 1(ii) in Chapter 4). Then use parts (ii), (vi), and (viii) in Exercise 1.13 in order to express the determinants $\tilde{D}_i$, $i = 1, \ldots, 4$ in part (ii) in terms of $\alpha, \beta, \gamma$, and $\delta$ and obtain:

$$\tilde{D}_1 = -\frac{n\beta}{\delta}, \quad \tilde{D}_2 = \frac{n^2}{\delta}, \quad \tilde{D}_3 = -\frac{n^3(2\alpha\beta - \gamma^2)}{4\alpha^2\delta^2} = -\frac{n^3(\alpha\beta + \delta)}{4\alpha^2\delta^2},$$
$$\tilde{D}_4 = \frac{n^4}{4\alpha\beta\delta^2}.$$

**(v)** Use the definition of $A, B$, and $C$ in part (iii), as well as the expressions of $\tilde{d}_{34}, \tilde{d}_{35}, \tilde{d}_{44}, \tilde{d}_{45}$, and $\tilde{d}_{55}$ given in parts (vi), (viii), and (x) of Exercise 1.13, in conjunction with the notation introduced in part (iv) of the present exercise, in order to show that:

$$A = \frac{\alpha^3\beta n^2}{2\delta^3}, \quad B = -\frac{\alpha\beta\gamma^2 n^2}{2\delta^3}, \quad C = \frac{\alpha^{1/2}\gamma n^2}{4\beta^{1/2}\delta^2}.$$

**(vi)** Use parts (iii) and (v) here, and parts (ii) and (iv) in Exercise 1.13 in order to obtain:

$$\tilde{D}_5 = \tilde{D}_2(\tilde{d}_{33}A - \tilde{d}_{34}B + \tilde{d}_{35}C) = \frac{n^2}{\delta}\left(-\frac{\alpha\beta n^3}{4\delta^3}\right) = -\frac{\alpha\beta n^5}{4\delta^4}.$$

**(vii)** From parts (iv) and (vi) and the fact that $\tilde{D}_0 = 1$, conclude that:

$$\tilde{D}_0 > 0, \quad \tilde{D}_1 < 0, \quad \tilde{D}_2 > 0, \quad \tilde{D}_3 < 0, \quad \tilde{D}_4 > 0, \quad \text{and} \quad \tilde{D}_5 < 0.$$

Then use a calculus result about the maximum of a function in more than one variable (see, e.g., Theorem 7.9, pages 151–152, in the book *Mathematical Analysis*, Addison-Wesley (1957), by T. M. Apostol) in order to conclude that $\tilde{\mu}_1, \tilde{\mu}_2, \tilde{\sigma}_1^2, \tilde{\sigma}_2^2$, and $\tilde{\rho}$ are, indeed, the MLE's of the respective parameters; i.e.,

$$\hat{\mu}_1 = \bar{x}, \quad \hat{\mu}_2 = \bar{y}, \quad \hat{\sigma}_1^2 = S_x, \quad \hat{\sigma}_2^2 = S_y, \quad \hat{\rho} = S_{xy}/S_x^{1/2}S_y^{1/2},$$

where

$$S_x = \frac{1}{n}\sum_{i=1}^{n}(x_i - \bar{x})^2, \quad S_y = \frac{1}{n}\sum_{i=1}^{n}(y_i - \bar{y})^2,$$

$$S_{xy} = \frac{1}{n}\sum_{i=1}^{n}(x_i - \bar{x})(y_i - \bar{y}).$$

## 9.2 SOME PROPERTIES OF MLE's

Refer to Example 3 and suppose that we are interested in estimating the probability that 0 events occur; that is, $P_\theta(X = 0) = \frac{e^{-\theta}\theta^0}{0!} = e^{-\theta}$, call it $g_1(\theta)$. Thus, the estimated quantity is a function of $\theta$ rather than $\theta$ itself. Next, refer to Example 4 and recall that, if $X \sim f(x; \theta) = \theta e^{-\theta x}$, $x > 0$, then $E_\theta X = 1/\theta$. Thus, in this case it would be, perhaps, more reasonable to estimate $1/\theta$ and call it $g_2(\theta)$, rather than $\theta$. Finally, refer to Examples 5(ii) and 7, and consider the problem of estimating the s.d. $\sigma = +\sqrt{\sigma^2}$ rather than the variance $\sigma^2$. This is quite meaningful, since, as we know, the s.d. is used as the yardstick for measuring distances from the mean. In this last case, set $g_3(\sigma) = +\sqrt{\sigma^2}$.

The functions $g_1, g_2$, and $g_3$ have the common characteristic that they are one-to-one functions of the parameter involved. The estimation problems described above are then formulated in a unified way as follows.

**Theorem 1.** *Let $\hat{\theta} = \hat{\theta}(x)$ be the MLE of $\theta$ on the basis of the observed values $x_1, \ldots, x_n$ of the random sample $X_1, \ldots, X_n$ from the p.d.f. $f(\cdot; \theta)$, $\theta \in \Omega \subseteq \mathfrak{R}$. Also, let $\theta^* = g(\theta)$ be a one-to-one function defined on $\Omega$ onto $\Omega^* \subseteq \mathfrak{R}$. Then the MLE of $\theta^*, \hat{\theta}^*(x)$, is given by $\hat{\theta}^*(x) = g[\hat{\theta}(x)]$.*

*Proof.* The equation $\theta^* = g(\theta)$ can be solved for $\theta$, on the basis of the assumption made, and let $\theta = g^{-1}(\theta^*)$. Then

$$L(\theta \mid x) = L[g^{-1}(\theta^*) \mid x] = L^*(\theta^* \mid x), \text{ say.}$$

Thus,

$$\max\{L(\theta \mid x); \quad \theta \in \Omega\} = \max\{L^*(\theta^* \mid x); \quad \theta^* \in \Omega^*\}. \tag{4}$$

Since the left-hand side in (4) is maximized for $\hat{\theta} = \hat{\theta}(x)$, clearly, the right-hand side is maximized for $\hat{\theta}^* = g(\hat{\theta})$. ∎

**Remark 3.** On the basis of Theorem 1, then, we have: The MLE of $\exp(-\theta)$ is $\exp(-\bar{x})$; the MLE of $1/\theta$ is $\bar{x}$; and the MLE of $\sigma$ is $\left[\frac{1}{n}\sum_{i=1}^{n}(x_i - \mu)^2\right]^{1/2}$ for Example 5(ii), and $\left[\frac{1}{n}\sum_{i=1}^{n}(x_i - \bar{x})^2\right]^{1/2}$ for Example 7.

However, in as simple a case as that of the $B(1, \theta)$ distribution, the function $g$ in Theorem 1 may not be one-to-one, and yet we can still construct the MLE of $\theta^* = g(\theta)$. This is the content of the next theorem.

**Theorem 2.** *Let $\hat{\theta} = \hat{\theta}(x)$ be the MLE of $\theta$ on the basis of the observed values $x_1, \ldots, x_n$ of the random sample $X_1, \ldots, X_n$ from the p.d.f. $f(\cdot; \theta)$, $\theta \in \Omega \subseteq \mathfrak{R}$. Also, let $\theta^* = g(\theta)$ be an arbitrary function defined on $\Omega$ into $\Omega^* \subseteq \mathfrak{R}$, where, without loss of generality, we may assume that $\Omega^*$ is the range of $g$, so that the function $g$ is defined on $\Omega$ onto $\Omega^*$. Then the MLE of $\theta^*, \hat{\theta}^*(x)$, is still given by $\hat{\theta}^*(x) = g[\hat{\theta}(x)]$. The same is true if $\theta$ is an $r$-dimensional parameter $\boldsymbol{\theta}$.*

*Proof.* For each $\theta^* \in \Omega^*$, there may be several $\theta$ in $\Omega$ mapped to the same $\theta^*$ under $g$. Let $\Omega_{\theta^*}$ be the set of all such $\theta$'s; i.e.,

$$\Omega_{\theta^*} = \{\theta \in \Omega; g(\theta) = \theta^*\}.$$

On $\Omega^*$, define the real-valued function $L^*$ by:

$$L^*(\theta^*) = \sup\{L(\theta); \theta \in \Omega_{\theta^*}\}.$$

The function $L^*$ may be called the *likelihood function induced in $\Omega^*$ by $g$*. Now, since $g$ is a function, it follows that $g(\hat{\theta}) = \hat{\theta}^*$ for a unique $\hat{\theta}^*$ in $\Omega^*$, and $L^*(\hat{\theta}^*) = L(\hat{\theta})$ from the definition of $L^*$. Finally, for every $\theta^* \in \Omega^*$,

$$L^*(\theta^*) = \sup\{L(\theta); \theta \in \Omega_{\theta^*}\} \leq \max\{L(\theta); \theta \in \Omega\} = L(\hat{\theta}) = L^*(\hat{\theta}^*).$$

This last inequality justifies calling $\hat{\theta}^* = g(\hat{\theta})$ the MLE of $\theta^*$. Observe that the arguments employed do not depend on the dimensionality of $\theta$. Thus, a suitable version of Theorem 2 holds for multidimensional parameters. ∎

(Theorem 2 was adapted from a result established by Peter W. Zehna in the *Annals of Mathematical Statistics*, Vol. 37 (1966), page 744.)

**Example 10.** Refer to Example 2 and determine the MLE $\hat{g}(\theta)$ of the function $g(\theta) = \theta(1 - \theta)$.

**Discussion.** Here, the function $g : (0, 1) \to (0, \frac{1}{4})$ is not one-to-one. However, by Theorem 2, the MLE $\hat{g}(\theta) = \bar{x}(1 - \bar{x})$, since $\hat{\theta} = \bar{x}$.

**Remark 4.** Theorem 1 is, of course, a special case of Theorem 2. This property of the MLE is referred to as the *invariance property* of the MLE for obvious reasons.

Reviewing the examples in the previous section, we see that the data $x_1, \ldots, x_n$ are entering into the MLE's in a compactified form, more, precisely, as a real-valued quantity. From this point on, this is all we have at our disposal; or, perhaps, that is all that was revealed to us by those who collected the data. In contemplating this situation, one cannot help but wonder what we are missing by knowing, for example, only $\bar{x}$ rather than the complete array of data $x_1, \ldots, x_n$. The almost shocking fact of the matter is that, in general, we are missing absolutely nothing, in terms of information carried by the data $x_1, \ldots, x_n$, provided the data are condensed in the right way. This is precisely the concept of sufficiency to be introduced below. For a motivation of the definition, consider the following example.

**Example 11.** In Example 1, each $x_i, i = 1, \ldots, 10$ takes on the value either 0 or 1, and we are given that $\bar{x} = 0.6$. There are $2^{10} = 1024$ arrangements of 10 0's or 1's with respective probabilities given by $\theta^{x_i}(1 - \theta)^{1-x_i}$ for each one of the 1024 arrangements of 0's and 1's. These probabilities, of course, depend on $\theta$. Now, restrict attention to the $\binom{10}{6} = 210$ arrangements of 0's and 1's only, which produce a sum of 6 or an average of 0.6; their probability is $210\theta^6(1 - \theta)^4$. Finally, calculate the conditional probability of each one of these arrangements, given that the average is 0.6 or that the sum is 6. In other words, calculate

$$P_\theta(X_i = x_i, i = 1, \ldots, 10 \mid T = 6), \quad T = \sum_{i=1}^{10} X_i. \tag{5}$$

Suppose that all these conditional probabilities have the same value, which, in addition, is independent of $\theta$. This would imply two things: First, given that the sum is 6, all possible arrangements, summing up to 6, have the same probability independent of the location of occurrences of 1's; and second, this probability has the same numerical value for all values of $\theta$ in $(0, 1)$. So, from a probabilistic viewpoint, given the information that the sum is 6, it does not really matter either what the arrangement is or what the value of $\theta$ is; we can reconstruct each one of all those arrangements giving sum 6 by choosing each one of the 210 possible arrangements, with probability $1/210$ each. It is in this sense that, restricting ourselves to the sum and ignoring or not knowing the individual values, we deprive ourselves of no information about $\theta$.

We proceed now with the calculation of the probabilities in (5). Although we can refer to existing results, let us derive the probabilities here.

$$P_\theta(X_i = x_i, \; i = 1, \ldots, 10 \mid T = 6) = P_\theta(X_i = x_i, i = 1, \ldots, 10, T = 6)/P_\theta(T = 6)$$

$$= P_\theta(X_i = x_i, \; i = 1, \ldots, 10)/P_\theta(T = 6)$$

$$(\text{since } X_i = x_i, \; i = 1, \ldots, 10 \text{ implies } T = 6)$$

$$= \theta^6(1 - \theta)^4 \Big/ \binom{10}{6} \theta^6(1 - \theta)^4 \quad (\text{since } T \sim B(10, \theta))$$

$$= 1 \Big/ \binom{10}{6} = 1/210 \; (\simeq 0.005).$$

Thus, what was assumed above is, actually, true.

This example and the elaboration associated with it lead to the following definition of sufficiency.

**Definition 1.** Let $X_1, \ldots, X_n$ be a random sample with p.d.f. $f(\cdot; \theta)$, $\theta \in \Omega \subseteq \mathfrak{R}$, and let $T = T(X_1, \ldots, X_n)$ be a statistic (i.e., a known function of the $X_i$'s). Then, if the conditional distribution of the $X_i$'s, given $T = t$, does not depend on $\theta$, we say that $T$ is a *sufficient* statistic for $\theta$.

**Remark 5.** If $g$ is a real-valued one-to-one function defined on the range of $T$, it is clear that knowing $T$ is equivalent to knowing $T^* = g(T)$, and vice versa. Thus, if $T$ is a sufficient statistic for $\theta$, so is $T^*$. In particular, if $T = \sum_{i=1}^n X_i$ or $\frac{1}{n}\sum_{i=1}^n(X_i - \mu)^2$ (or $\frac{1}{n}\sum_{i=1}^n(X_i - \bar{X})^2$) is a sufficient statistic for $\theta$ so is $\bar{X}$ or $\sum_{i=1}^n(X_i - \mu)^2$ (or $\sum_{i=1}^n(X_i - \bar{X})^2$).

**Remark 6.** The definition given for one parameter also applies for more than one parameter, but then we also need a multidimensional sufficient statistic, usually, with dimensionality equal to the number of the parameters. In all cases, we often use simply the term *sufficient* instead of *sufficient statistic(s) for* $\theta$, if no confusion is possible.

As is often the case, definitions do not lend themselves easily to identifying the quantity defined. This is also the case in Definition 1. A sufficient statistic is, actually, found by way of the theorem stated below.

**Theorem 3** (Fisher-Neyman Factorization Theorem). *Let* $X_1, \ldots, X_n$ *be a random sample with p.d.f.* $f(\cdot; \theta)$, $\theta \in \Omega \subseteq \mathfrak{R}$, *and let* $T = T(X_1, \ldots, X_n)$ *be a statistic. Then* $T$ *is a sufficient statistic for* $\theta$, *if and only if the joint p.d.f. of the* $X_i$'s *may be written as follows:*

$$f_{X_1, \ldots, X_n}(x_1, \ldots, x_n; \theta) = g[T(x_1, \ldots, x_n); \theta]h(x_1, \ldots, x_n). \tag{6}$$

The way this theorem applies is the following: One writes out the joint p.d.f. of the $X_i$'s and then one tries to rewrite it as the product of two factors, one factor, $g[T(x_1, \ldots, x_n); \theta]$, which contains the $x_i$'s only through the function $T(x_1, \ldots, x_n)$

and the parameter $\theta$, and another factor, $h(x_1, \ldots, x_n)$, which involves the $x_i$'s in whatever form but not $\theta$ in any form.

**Remark 7.** The theorem just stated also holds for multidimensional parameters $\theta$, but then the statistic $T$ is also multidimensional, usually of the same dimension as that of $\theta$. A rigorous proof of the theorem can be given, at least for the case of discrete $X_i$'s, but we choose to omit it.

In all of the Examples 2–10, the MLE's are, actually, sufficient statistics or functions thereof, as demonstrated below. This fact should certainly reinforce our appreciation for these MLE's.

**Application.** In Example 2, the p.d.f. is written as follows in a compact form $f(x_i; \theta) = \theta^{x_i}(1 - \theta)^{1-x_i} I_{\{0,1\}}(x_i)$, so that the joint p.d.f. becomes:

$$L(\theta \mid x) = \theta^t (1 - \theta)^{n-t} \times \prod_{i=1}^{n} I_{\{0,1\}}(x_i), \quad t = \sum_{i=1}^{n} x_i.$$

Then $g[T(x_1, \ldots, x_n); \theta] = \theta^t (1 - \theta)^{n-t}$, and $h(x_1, \ldots, x_n) = \prod_{i=1}^{n} I_{\{0,1\}}(x_i)$. It follows that $T = \sum_{i=1}^{n} X_i$ is sufficient and so is $\frac{T}{n} = \bar{X}$.

Examples 3 and 4 are treated similarly.

In Example 5(i),

$$L(\theta \mid x) = \exp\left[\frac{n\mu(2\bar{x} - \mu)}{2\sigma^2}\right] \times \left(\frac{1}{\sqrt{2\pi}\sigma}\right)^n \exp\left(-\frac{1}{2\sigma^2} \sum_{i=1}^{n} x_i^2\right),$$

so that $\bar{X}$ is sufficient for $\mu$. Likewise, in Example 5(ii),

$$L(\sigma^2 \mid x) = \left(\frac{1}{\sqrt{2\pi\sigma^2}}\right)^n \exp\left[-\frac{1}{2\sigma^2} \sum_{i=1}^{n} (x_i - \mu)^2\right] \times 1,$$

so that $\sum_{i=1}^{n} (X_i - \mu)^2$ is sufficient for $\sigma^2$ and so is $\frac{1}{n} \sum_{i=1}^{n} (X_i - \mu)^2$.

Example 6 is treated similarly.

In Example 7,

$$L(\mu, \sigma^2 \mid x) = \left(\frac{1}{\sqrt{2\pi\sigma^2}}\right)^n \exp\left[-\frac{1}{2\sigma^2} \sum_{i=1}^{n} (x_i - \bar{x})^2 - \frac{1}{2\sigma^2} n(\bar{x} - \mu)^2\right] \times 1$$

because $\sum_{i=1}^{n} (x_i - \mu)^2 = \sum_{i=1}^{n} [(x_i - \bar{x}) + (\bar{x} - \mu)]^2 = \sum_{i=1}^{n} (x_i - \bar{x})^2 + n(\bar{x} - \mu)^2$. It follows that the pair of statistics $(\bar{X}, \sum_{i=1}^{n} (X_i - \bar{X})^2)$ is sufficient for the pair of parameters $(\mu, \sigma^2)$.

Examples 8–10 are treated similarly.

**Remark 8.** Under certain regularity conditions, it is always the case that an MLE is only a function of a sufficient statistic.

Here is another example, in four parts, where a sufficient statistic is determined by way of Theorem 3.

**Example 12.**   On the basis of a random sample of size $n, X_1, \ldots, X_n$, from each one of the p.d.f.'s given below with observed values $x_1, \ldots, x_n$, determine a sufficient statistic for $\theta$.

(i) $f(x; \theta) = \frac{\theta}{x^{\theta+1}}$, $\quad x \geq 1$, $\quad \theta \in \Omega = (0, \infty)$.

(ii) $f(x; \theta) = \frac{x}{\theta} e^{-x^2/2\theta}$, $\quad x > 0$, $\quad \theta \in \Omega = (0, \infty)$.

(iii) $f(x; \theta) = (1 + \theta)x^\theta$, $\quad 0 < x < 1$, $\quad \theta \in \Omega = (-1, \infty)$.

(iv) $f(x; \theta) = \frac{\theta}{x^2}$, $\quad x \geq \theta$, $\quad \theta \in \Omega = (0, \infty)$.

**Discussion.**   In the first place, the functions given above are, indeed, p.d.f.'s (see Exercise 2.12). Next, rewriting each one of the p.d.f.'s by using the indicator function, we have:

(i) $f(x; \theta) = \frac{\theta}{x^{\theta+1}} I_{[1,\infty)}(x)$, so that:

$$\prod_{i=1}^{n} f(x_i; \theta) = \frac{\theta^n}{\left(\prod_{i=1}^{n} x_i\right)^{\theta+1}} \times \prod_{i=1}^{n} I_{[1,\infty)}(x_i) = \frac{\theta^n}{\left(\prod_{i=1}^{n} x_i\right)^{\theta+1}} \times I_{[1,\infty)}(x_{(1)}),$$

and therefore $\prod_{i=1}^{n} X_i$ is sufficient for $\theta$.

(ii) $f(x; \theta) = \frac{x}{\theta} e^{-x^2/2\theta} I_{(0,\infty)}(x)$, so that:

$$\prod_{i=1}^{n} f(x_i; \theta) = \frac{1}{\theta^n} \prod_{i=1}^{n} x_i e^{-(1/2\theta) \sum_{i=1}^{n} x_i^2} \prod_{i=1}^{n} I_{(0,\infty)}(x_i)$$

$$= \frac{1}{\theta^n} e^{-(1/2\theta) \sum_{i=1}^{n} x_i^2} \times \left(\prod_{i=1}^{n} x_i\right) I_{(0,\infty)}(x_{(1)}),$$

and therefore $\sum_{i=1}^{n} X_i^2$ is sufficient for $\theta$.

(iii) $f(x; \theta) = (1 + \theta)x^\theta I_{(0,1)}(x)$, so that

$$\prod_{i=1}^{n} f(x_i; \theta) = (1 + \theta)^n \left(\prod_{i=1}^{n} x_i\right)^\theta \prod_{i=1}^{n} I_{(0,1)}(x_i)$$

$$= (1 + \theta)^n \left(\prod_{i=1}^{n} x_i\right)^\theta \times I_{(0,1)}(x_{(1)}) I_{(0,1)}(x_{(n)}),$$

and therefore $\prod_{i=1}^{n} X_i$ is sufficient for $\theta$.

(iv) $f(x; \theta) = \frac{\theta}{x^2} I_{[\theta,\infty)}(x)$, so that:

$$\prod_{i=1}^{n} f(x_i; \theta) = \frac{\theta^n}{\left(\prod_{i=1}^{n} x_i^2\right)} \prod_{i=1}^{n} I_{[\theta,\infty)}(x_i) = \theta^n I_{[\theta,\infty)}(x_{(1)}) \times \frac{1}{\left(\prod_{i=1}^{n} x_i^2\right)},$$

and therefore $X_{(1)}$ is sufficient for $\theta$.

This section is concluded with two desirable asymptotic properties of a MLE. The first is consistency (in the probability sense), and the other is asymptotic Normality. However, we will not bother either to list the conditions needed or to justify the results stated.

**Theorem 4.** *Let $\hat{\theta}_n = \hat{\theta}_n(X_1,\ldots,X_n)$ be the MLE of $\theta \in \Omega \subseteq \Re$ based on the random sample $X_1,\ldots,X_n$ with p.d.f. $f(\cdot,\theta)$. Then, under certain regularity conditions, $\{\hat{\theta}_n\}$ is consistent in the probability sense; that is, $\hat{\theta}_n \to \theta$ in $P_\theta$-probability as $n \to \infty$.*

The usefulness of this result is, of course, that, for sufficiently large $n$, $\hat{\theta}_n$ is as close to (the unknown) $\theta$ as we please with probability as close to 1 as we desire. The tool usually employed in establishing Theorem 4 is either the Weak Law of Large Numbers (WLLN) or the Tchebichev inequality. In exercises at the end of this section, the validity of Theorem 4 is illustrated in some of the examples of the previous section.

**Theorem 5.** *In the notation of Theorem 4, and under suitable regularity conditions, the MLE $\hat{\theta}_n$ is asymptotically Normal. More precisely, under $P_\theta$-probability,*

$$\sqrt{n}(\hat{\theta}_n - \theta) \xrightarrow{d} N(0, \sigma_\theta^2), \quad \text{as } n \to \infty,$$

*where $\sigma_\theta^2 = 1/I(\theta)$ and $I(\theta) = E_\theta\left[\dfrac{\partial}{\partial\theta}\log f(X;\theta)\right]^2$, $X \sim f(\cdot;\theta)$.* (7)

To state it loosely, $\hat{\theta}_n \simeq N(\theta, \sigma_\theta^2/n)$ for sufficiently large $n$. That is, the MLE $\hat{\theta}_n$ is approximately Normally distributed around $\theta$, and therefore various probabilities related to it may be approximately calculated in principle. The justification of the theorem is done by using a Taylor expansion of the derivative $\frac{\partial}{\partial\theta}\log L(\theta \mid X)$ up to terms of third order, employing the fact that $\frac{\partial}{\partial\theta}\log L(\theta \mid X)\mid_{\theta=\hat{\theta}_n} = 0$, and suitably utilizing the WLLN and the Central Limit Theorem. For some applications of this theorem (see, e.g., Example 21, Section 12.10, of the book *A Course in Mathematical Statistics*, 2nd edition, Academic Press (1977), by G. G. Roussas).

**Remark 9.** The quantity $I(\theta)$ is referred to as the *Fisher information* carried by the random sample $X_1,\ldots,X_n$ about the parameter $\theta$. A justification for the "information" stems by the fact that $\sigma_\theta^2 = 1/I(\theta)$, so that the larger $I(\theta)$ is the smaller the variance $\sigma_\theta^2$ is, and therefore the more concentrated $\hat{\theta}$ is about $\theta$. The opposite happens for small values of $I(\theta)$.

## EXERCISES

**2.1** Let $X_1,\ldots,X_n$ be i.i.d. r.v.'s with the Negative Exponential p.d.f.,
$f(x; \theta) = \theta e^{-\theta x}$, $x > 0$, $\theta \in \Omega = (0, \infty)$. Then
(i) Show that $1/\bar{X}$ is the MLE of $\theta$.
(ii) Use Theorem 1 in order to conclude that the MLE of $\theta^*$ in the reparametrized form $f(x; \theta^*) = \frac{1}{\theta^*}e^{-x/\theta^*}$, $x > 0$, is $\bar{X}$.

**2.2**  Let $X$ be a r.v. denoting the life span of an equipment. Then the *reliability* of the equipment at time $x$, $R(x)$, is defined as the probability that $X > x$; i.e., $R(x) = P(X > x)$. Now, suppose that $X$ has the Negative Exponential p.d.f., $f(x; \theta) = \frac{1}{\theta}e^{-x/\theta}$, $x > 0$, $\theta \in \Omega = (0, \infty)$. Then
   **(i)**  Calculate the reliability $R(x; \theta)$ based on this r.v. $X$.
   **(ii)**  Use Theorem 1 in order to determine the MLE of $R(x; \theta)$, on the basis of a random sample $X_1, \ldots, X_n$ from the underlying p.d.f.

**2.3**  Let $X$ be a r.v. describing the lifetime of a certain equipment, and suppose that the p.d.f. of $X$ is $f(x; \theta) = \theta e^{-\theta x}$, $x > 0$, $\theta \in \Omega = (0, \infty)$.
   **(i)**  Show that the probability that $X$ is greater than or equal to $t$ time units is $g(\theta) = e^{-t\theta}$.
   **(ii)**  We know (see Exercise 2.1) that the MLE of $\theta$, based on a random sample of size $n$ from the above p.d.f., is $\hat{\theta} = 1/\bar{X}$. Then determine the MLE of $g(\theta)$.

**2.4**  Consider the independent r.v.'s $X_1, \ldots, X_n$ with the Weibull p.d.f.
   $f(x; \theta) = \frac{\gamma}{\theta}x^{\gamma-1}\exp(-x^{\gamma}/\theta)$, $x > 0$, $\theta \in \Omega = (0, \infty)$, $\gamma > 0$ known, and:
   **(i)**  Show that $\hat{\theta} = (\sum_{i=1}^{n} X_i^{\gamma})/n$ is the MLE of $\theta$.
   **(ii)**  Take $\gamma = 1$ and relate the result in part (i) to the result in Exercise 2.1(ii).

**2.5**  Let $X_1, \ldots, X_n$ be a random sample of size $n$ from the $N(\mu, \sigma^2)$ distribution, where both $\mu$ and $\sigma^2$ are unknown. Set $\theta = (\mu, \sigma^2)$ and let $p$ be a (known) number with $0 < p < 1$. Then
   **(i)**  Show that the point $c$ for which $P_{\theta}(\bar{X} \leq c) = p$ is given by:
   $c = \mu + \frac{\sigma}{\sqrt{n}}\Phi^{-1}(p)$.
   **(ii)**  Given that the MLE's of $\mu$ and $\sigma^2$ are, respectively, $\hat{\mu} = \bar{X}$ and $\hat{\sigma}^2 = S^2 = \frac{1}{n}\sum_{i=1}^{n}(X_i - \bar{X})^2$, determine the MLE of $c$, call it $\hat{c}$.
   **(iii)**  Express $\hat{c}$ in terms of the $X_i$'s, if $n = 25$ and $p = 0.95$.

**2.6**  **(i)**  Show that the function $f(x; \theta) = \theta x^{-(\theta+1)}$, $x \geq 1$, $\theta \in \Omega = (0, \infty)$ is a p.d.f.
   **(ii)**  On the basis of a random sample of size $n$ from this p.d.f., show that the statistic $X_1 \cdots X_n$ is sufficient for $\theta$, and so is the statistic $\sum_{i=1}^{n} \log X_i$.

**2.7**  Let $X$ be a r.v. having the Geometric p.d.f.,
   $f(x; \theta) = \theta(1 - \theta)^{x-1}$, $x = 1, 2, \ldots, \theta \in \Omega = (0, 1)$. Then show that $X$ is sufficient for $\theta$.

**2.8**  In reference to Exercise 1.9, show that $\sum_{i=1}^{n} |X_i|$ is a sufficient statistic for $\theta$.

**2.9**  **(i)**  In reference to Example 3, use Theorem 3 in order to find a sufficient statistic for $\theta$.
   **(ii)**  Do the same in reference to Example 4.
   **(iii)**  Do the same in reference to Example 6(i) and (ii).

**2.10**  Same as in Exercise 2.9 in reference to Examples 8 and 9.

**2.11** Refer to Exercise 1.11, and determine:
   (i) A sufficient statistic for $\alpha$ when $\beta$ is known.
   (ii) Also, a sufficient statistic for $\beta$ when $\alpha$ is known.
   (iii) A set of sufficient statistics for $\alpha$ and $\beta$ when they are both unknown.

**2.12** Show that the functions (i)–(iv) given in Example 12 are, indeed, p.d.f.'s.

## 9.3 UNIFORMLY MINIMUM VARIANCE UNBIASED ESTIMATES

Perhaps the second most popular method of estimating a parameter is that based on the concepts of unbiasedness and variance. This method will be discussed here to a certain extent and will also be illustrated by specific examples.

To start with, let $X_1, \ldots, X_n$ be a random sample with p.d.f. $f(\cdot; \theta)$, $\theta \in \Omega \subseteq \Re$, and let us introduce the notation $U = U(X_1, \ldots, X_n)$ for an estimate of $\theta$.

**Definition 2.** The estimate $U$ is said to be *unbiased* if $E_\theta U = \theta$ for all $\theta \in \Omega$.

Some examples of unbiased estimates follow.

**Example 13.** Let $X_1, \ldots, X_n$ be a random sample coming from any one of the following distributions:

(i) $B(1, \theta), \theta \in (0, 1)$. Then the sample mean $\bar{X}$ is an unbiased estimate of $\theta$.
(ii) $P(\theta), \theta > 0$. Then again $\bar{X}$ is an unbiased estimate of $\theta$. Here, since $E_\theta X_1 = \text{Var}_\theta(X_1) = \theta, \bar{X}$ is also an unbiased estimate of the variance.
(iii) $N(\theta, \sigma^2), \theta \in \Re, \sigma$ known. Then, once again, $\bar{X}$ is an unbiased estimate of $\theta$.
(iv) $N(\mu, \theta), \mu$ known, $\theta > 0$. Then the sample variance $\frac{1}{n} \sum_{i=1}^{n} (X_i - \mu)^2$ is an unbiased estimate of $\theta$. This is so because $\sum_{i=1}^{n} \left(\frac{X_i - \mu}{\sqrt{\theta}}\right)^2 \sim \chi_n^2$, so that

$$E_\theta \left[ \sum_{i=1}^{n} \left(\frac{X_i - \mu}{\sqrt{\theta}}\right)^2 \right] = n \quad \text{or} \quad E_\theta \left[ \frac{1}{n} \sum_{i=1}^{n} (X_i - \mu)^2 \right] = \theta.$$

(v) Gamma with $\alpha = \theta$ and $\beta = 1$. Then $\bar{X}$ is an unbiased estimate of $\theta$. This is so because, in the Gamma distribution, the expectation is $\alpha\beta$, so that for $\alpha = \theta$ and $\beta = 1$, $E_\theta X_1 = \theta$ and hence $E_\theta \bar{X} = \theta$.
(vi) Gamma with $\alpha = 1$ and $\beta = \theta, \theta > 0$ (which gives the reparametrized Negative Exponential distribution). Then $\bar{X}$ is an unbiased estimate of $\theta$ as explained in part (v).

**Example 14.** Let $X_1, \ldots, X_n$ be a random sample from the $U(0, \theta)$ ($\theta > 0$). Determine an unbiased estimate of $\theta$.

**Discussion.** Let $Y_n = \max(X_1, \ldots, X_n)$; i.e., the largest order statistic of the $X_i$'s. Then, by (29) in Chapter 6, the p.d.f. of $Y_n$ is given by:

$$g(y) = n[F(y)]^{n-1}f(y), \quad \text{for } 0 < y < \theta \text{ (and 0 otherwise)},$$

where

$$f(y) = \frac{1}{\theta}, \quad 0 < y < \theta, \quad F(y) = \begin{cases} 0, & y \le 0 \\ \frac{y}{\theta}, & 0 < y < \theta \\ 1, & y \ge \theta. \end{cases}$$

Then, for $0 < y < \theta$, $g(y) = n\left(\frac{y}{\theta}\right)^{n-1}\left(\frac{1}{\theta}\right) = \frac{n}{\theta^n}y^{n-1}$, so that

$$E_\theta Y_n = \int_0^\theta y \cdot \frac{n}{\theta^n}y^{n-1}\,dy = \frac{n}{\theta^n}\int_0^\theta y^n\,dy = \frac{n}{(n+1)\theta^n} \times y^{n+1}\Big|_0^\theta = \frac{n}{n+1}\theta.$$

It follows that $E_\theta\left(\frac{n+1}{n}Y_n\right) = \theta$, so that $\frac{n+1}{n}Y_n$ is an unbiased estimate of $\theta$.

The desirability of *unbiasedness* of an estimate stems from the interpretation of the expectation as an average value. Typically, one may construct many unbiased estimates for the same parameter $\theta$. This fact then raises the question of selecting one such estimate from the class of all unbiased estimates. Here is where the concept of the variance enters the picture. From two unbiased estimates $U_1 = U_1(X_1, \ldots, X_n)$ and $U_2 = U_2(X_1, \ldots, X_n)$ of $\theta$, one would select the one with the smaller variance. This estimate will be more concentrated around $\theta$ than the other. Pictorially, this is illustrated by Figure 9.2.

The next natural step is to look for an unbiased estimate which has the smallest variance in the class of all unbiased estimates, and this should happen for all $\theta \in \Omega$. Thus, we are led to the following concept.

**Definition 3.** The unbiased estimate $U = U(X_1, \ldots, X_n)$ of $\theta$ is said to be *Uniformly Minimum Variance Unbiased* (UMVU), if for any other unbiased estimate $V = V(X_1, \ldots, X_n)$, it holds that:

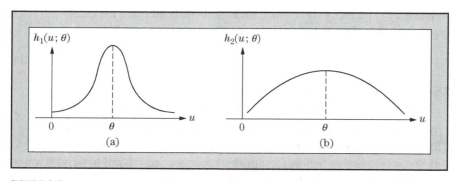

**FIGURE 9.2**

(a) p.d.f. of $U_1$ (for a fixed $\theta$) and (b) p.d.f. of $U_2$ (for a fixed $\theta$).

$$\text{Var}_\theta(U) \le \text{Var}_\theta(V) \quad \text{for all } \theta \in \Omega.$$

That a UMVU estimate is desirable is more or less indisputable (see, however, Exercise 3.18). The practical question which then arises is how one goes about finding such an estimate. The process of seeking a UMVU estimate is facilitated by the Cramér-Rao inequality stated next. First, this inequality is stated, and then we describe how it is used.

**Theorem 6** (Cramér-Rao Inequality). *Let $X_1, \ldots, X_n$ be a random sample with p.d.f. $f(\cdot; \theta)$, $\theta \in \Omega \subseteq \Re$, and suppose certain regularity conditions are met. Then, for any unbiased estimate $U = U(X_1, \ldots, X_n)$ of $\theta$, it holds that:*

$$\text{Var}_\theta(U) \ge 1/nI(\theta), \quad \text{for all } \theta \in \Omega, \tag{8}$$

*where*

$$I(\theta) = E_\theta \left[ \frac{\partial}{\partial \theta} \log f(X; \theta) \right]^2, \quad X \sim f(\cdot; \theta). \tag{9}$$

**Remark 10.** The unspecified conditions mentioned in the formulation of the theorem include the assumption that the domain of $x$ in the p.d.f. $f(x; \theta)$ does not depend on $\theta$; thus, the $U(0, \theta)$ distribution, for example, is left out. Also, the conditions include the validity of interchanging the operations of differentiation and integration in certain expressions. The proof of Theorem 6 is relatively long and involves an extensive list of regularity conditions. It may be found in considerable detail in Section 12.4.1 of Chapter 12 of the book *A Course in Mathematical Statistics*, 2nd edition, Academic Press (1997), by G. G. Roussas. Actually, in the reference just cited what is proved is a generalized version of Theorem 6, where the estimated function is a real-valued function of $\theta$, $g(\theta)$, rather than $\theta$ itself.

**Remark 11.** It can by shown that, under suitable conditions, the quantity $I(\theta)$ in (9) may also be calculated as follows:

$$I(\theta) = -E_\theta \left[ \frac{\partial^2}{\partial \theta^2} \log f(X; \theta) \right]. \tag{10}$$

This expression is often easier to calculate.

The Cramér-Rao (C-R) inequality is used in the following way.

  **(i)** Calculate the Fisher information either through (9) or by way of (10).
  **(ii)** Form the Cramér-Rao (C-R) lower bound figuring in inequality (8).
  **(iii)** Try to identify an unbiased estimate whose variance is equal to the C-R lower bound (for all $\theta \in \Theta$). If such an estimate is found,
  **(iv)** Declare the estimate described in (iii) as the UMVU estimate of $\theta$.

In connection with steps (i)–(iv), it should be noted that it is possible that a UMVU estimate exists and yet such an estimate is not located through this process. The reason for such a failure is that the C-R inequality provides, simply, a lower bound for the variances of unbiased estimates, which may be strictly smaller than the variance of a UMVU estimate. It is, nevertheless, a good try!

The use of the inequality will be illustrated by two examples; other cases are left as exercises.

**Example 15.** Refer to Example 2 and seek a UMVU estimate of $\theta$ through the C-R inequality.

**Discussion.** Here $f(x; \theta) = \theta^x(1-\theta)^{1-x}$,   $x = 0, 1$, so that:

(i) $\log f(X; \theta) = X \log \theta + (1 - X) \log(1 - \theta)$,

$$\frac{\partial}{\partial \theta} \log f(X; \theta) = \frac{X}{\theta} - \frac{1-X}{1-\theta} \quad \text{and} \quad \frac{\partial^2}{\partial \theta^2} \log f(X; \theta) = -\frac{X}{\theta^2} - \frac{1-X}{(1-\theta)^2},$$

$$I(\theta) = -E_\theta \left[ \frac{\partial^2}{\partial \theta^2} \log f(X; \theta) \right] = \frac{\theta}{\theta^2} + \frac{1-\theta}{(1-\theta)^2} = \frac{1}{\theta(1-\theta)}, \quad \text{since } E_\theta X = \theta.$$

(ii) The C-R lower bound $= \frac{1}{nI(\theta)} = \theta(1-\theta)/n$.

(iii) Consider $U = \bar{X}$. Then $E_\theta \bar{X} = \theta$, so that $\bar{X}$ is unbiased, and next $\sigma_\theta^2(\bar{X}) = \theta(1-\theta)/n = 1/nI(\theta)$, since $\sigma_\theta^2(X) = \theta(1-\theta)$.

(iv) The estimate $\bar{X}$ is UMVU.

**Example 16.** Refer to Example 4 and use the following parametrization:

$$X \sim f(x; \theta) = \frac{1}{\theta} e^{-(x/\theta)}, \quad x > 0, \text{ so that } E_\theta X = \theta, \quad \sigma_\theta^2(X) = \theta^2.$$

Then seek a UMVU estimate for $\theta$ through the C-R inequality.

**Discussion.** Here

(i) $\log f(X; \theta) = -\log \theta - \frac{X}{\theta}$,

$$\frac{\partial}{\partial \theta} \log f(X; \theta) = -\frac{1}{\theta} + \frac{X}{\theta^2} \quad \text{and} \quad \frac{\partial^2}{\partial \theta^2} \log f(X; \theta) = \frac{1}{\theta^2} - \frac{2X}{\theta^3},$$

$$I(\theta) = -E_\theta \left[ \frac{\partial^2}{\partial \theta^2} \log f(X; \theta) \right] = -\frac{1}{\theta^2} + \frac{2\theta}{\theta^3} = \frac{1}{\theta^2}.$$

(ii) The C-R lower bound $= \frac{1}{nI(\theta)} = \theta^2/n$.

(iii) Consider $U = \bar{X}$. Then $E_\theta U = \theta$, so that $\bar{X}$ is unbiased, and $\sigma_\theta^2(\bar{X}) = \frac{\theta^2}{n} = 1/nI(\theta)$.

(iv) The estimate $\bar{X}$ is UMVU.

There is an alternative way of looking for UMVU estimates, in particular, when the approach by means of the Cramér-Rao inequality fails to produce such an estimate. This approach hinges heavily on the concept of sufficiency already introduced and also an additional technical concept, the so-called *completeness*. The concept of completeness is a technical concept, and it says, in effect, that the only unbiased estimate of 0 is essentially the 0 statistics (i.e., statistics which are identically equal to 0). More technically, if $T$ is a r.v. with p.d.f. $f_T(\cdot; \theta), \theta \in \Omega \subseteq \Re$, the family $\{f_T(\cdot; \theta); \theta \in \Omega\}$ (or the r.v. $T$) is said to be *complete* if, for $h : \Re \to \Re, E_\theta h(T) = 0$ for all $\theta \in \Omega$ implies $h(t)$ is, essentially, equal to 0. For the precise definition and a number of illustrative examples, the reader is referred to Section 11.2 of Chapter 11 in the reference cited in Remark 10 here. The concepts of sufficiency and completeness combined lead constructively to a UMVU estimate by means of two theorems, the Rao-Blackwell and the Lehmann-Scheffé theorem. The procedure is summarized in the following result.

**Theorem 7** (Rao-Blackwell and Lehmann-Scheffé). *Let $X_1, \ldots, X_n$ be a random sample with p.d.f. $f(\cdot; \theta)$, $\theta \in \Omega \subseteq \Re$, and let $T = T(X_1, \ldots, X_n)$ be a sufficient statistic for $\theta$ (which is complete). Let $U = U(X_1, \ldots, X_n)$ be any unbiased estimate of $\theta$, and define the statistic $\varphi(T)$ by the relation:*

$$\varphi(T) = E_\theta(U \mid T). \tag{11}$$

*Then $\varphi(T)$ is unbiased, $Var_\theta[\varphi(T)] \leq Var_\theta(U)$ for all $\theta \in \Omega$, and, indeed, $\varphi(T)$ is a UMVU estimate of $\theta$. If $U$ is already a function of $T$ only, then the conditioning in (11) is superfluous.*

*Proof (rough outline).* That $\varphi(T)$ is independent of $\theta$ (and hence a statistic), despite the fact that we use quantities depending on $\theta$ in forming the conditional expectation, is due to the sufficiency of $T$. Recall that sufficiency of $T$ means that the conditional distribution of $U$, given $T$, is independent of $\theta$, and hence so is the expectation of $U$ formed by using this conditional distribution. That $\varphi(T)$ is unbiased is due to a property of the conditional expectation (namely, for two r.v.'s $X$ and $Y$: $E[E(X \mid Y)] = EX$), and the inequality involving the variances is also due to a property of the variance for conditional expectations (namely, $Var[E(X \mid Y)] \leq Var(X)$). The concept of completeness guarantees, through the Lehmann-Scheffé theorem, that no matter which unbiased estimate $U$ we start out with, we end up (essentially) with the same UMVU estimate $\varphi(T)$ through the procedure (11), which is known as *Rao-Blackwellization.* ∎

**Remark 12.** This theorem also applies suitably to multidimensional parameters $\theta$, although, it must be stated here that a version of the Cramér-Rao inequality also exists for such parameters.

The following examples illustrate how one goes about applying Theorem 7 in concrete cases.

**Example 17.**   Determine the UMVU estimate of $\theta$ on the basis of the random sample $X_1, \ldots, X_n$ from the distribution $P(\theta)$.

**Discussion.**   Perhaps, the simplest unbiased estimate of $\theta$ is $X_1$, and we already know that $T = X_1 + \cdots + X_n$ is sufficient for $\theta$. For the Rao-Blackwellization of $X_1$, we need the conditional distribution of $X_1$, given $T = t$. It is known, however (see Exercise 2.10(ii), in Chapter 5), that this conditional distribution is $B(t, \frac{1}{n})$; that is, $P_\theta(X_1 = x \mid T = t) = \binom{t}{x} (\frac{1}{n})^x (1 - \frac{1}{n})^{t-x}$. It follows that $E_\theta(X_1 \mid T = t) = \frac{t}{n}$, so that $\varphi(T) = E_\theta(X_1 \mid T) = \frac{T}{n} = \bar{X}$. It so happens that the conditions of Theorem 7 hold (see Exercise 3.17) and therefore $\bar{X}$ is the UMVU estimate of $\theta$.
   A well-known case where Theorem 7 works whereas Theorem 6 does not (or more properly, their suitable versions for two parameters do or do not) is illustrated by the following example.

**Example 18.**   Let $X_1, \ldots, X_n$ be a random sample from the $N(\mu, \sigma^2)$ distribution, where both $\mu$ and $\sigma$ are unknown.
By an application of either theorem, it is seen that $\bar{X}$ is the UMVU estimate of $\mu$. Working with the Cramér-Rao inequality regarding $\sigma^2$ as the estimated parameter, it is seen that the C-R lower bound is equal to $2\sigma^4/n$. Next, Theorem 7 leads to the UMVU estimate of $\sigma^2$, $S^2 = \frac{1}{n-1} \sum_{i=1}^n (X_i - \bar{X})^2$. Furthermore, it has been seen (Exercise 1.4) that $\text{Var}_{\sigma^2}(S^2) = \frac{2\sigma^4}{n-1}$, which is strictly larger than $\frac{2\sigma^4}{n}$. This is the reason the Cramér-Rao inequality approach fails. For a little more extensive discussion on this example, see, e.g., Example 9, Section 12,4, in the reference cited in Remark 10.

# EXERCISES

**3.1**   If $X$ is a r.v. distributed as $B(n, \theta), \theta \in \Omega = (0, 1)$, show that there is no unbiased estimate of $1/\theta$.
   **Hint.** If $h(X)$ were such an estimate, then $E_\theta h(X) = \frac{1}{\theta}$ for all $\theta (\in (0, 1))$. Write out the expectation, set $\frac{\theta}{1-\theta} = t$, and by expanding the right-hand side, conclude that $\binom{n+1}{n+1} (= 1) = 0$, which, of course, is a contradiction.

**3.2**   Let $X_1, \ldots, X_n$ be independent r.v.'s with p.d.f.
   $f(x; \theta) = \theta e^{-\theta x}, x > 0, \theta \in \Omega = (0, \infty)$, and let $Y_1$ be the smallest order statistic of the $X_i$'s. Then, by Example 11 in Chapter 6, the p.d.f. of $Y_1$ is $g_1(y) = (n\theta) e^{-(n\theta)y}, y > 0$.
   **(i)**  Show that both $\bar{X}$ and $nY_1$ are unbiased estimates of $1/\theta$.
   **(ii)** On the basis of variance considerations, which of these two estimates would you prefer?

**3.3** Let $X_1, \ldots, X_n$ be a random sample of size $n$ from the $U(0, \theta)$ distribution, $\theta \in \Omega = (0, \infty)$, and let $Y_n$ be the largest order statistic of the $X_i$'s. Then:

(i) Employ formula (29) in Chapter 6 in order to obtain the p.d.f. of $Y_n$.

(ii) Use part (i) in order to construct an unbiased estimate of $\theta$ depending only on $Y_n$.

(iii) By Example 6 here (with $\alpha = 0$ and $\beta = \theta$) in conjunction with Theorem 3, show that the unbiased estimate in part (ii) depends only on a sufficient statistic for $\theta$.

**3.4** Let $X_1, \ldots, X_n$ be a random sample of size $n$ from the $U(\theta_1, \theta_2)$ distribution, $\theta_1 < \theta_2$, and let $Y_1$ and $Y_n$ be the smallest and the largest order statistics of the $X_i$'s.

(i) Use formulas (28) and (29) in Chapter 6 to obtain the p.d.f.'s of $Y_1$ and $Y_n$, and then, by calculating the $E_\theta Y_1$ and $E_\theta Y_n$, construct unbiased estimates of the mean $(\theta_1 + \theta_2)/2$ and of the range $\theta_2 - \theta_1$ depending only on $Y_1$ and $Y_n$.

(ii) Employ Example 6 here (with $\alpha = \theta_1$ and $\beta = \theta_2$) in conjunction with Theorem 3 and Remark 7 in order to show that the unbiased estimates in part (i) depend only on a set of sufficient statistics for $(\theta_1, \theta_2)$.

**3.5** Let $X_1, \ldots, X_n$ be a random sample of size $n$ from the $U(\theta, 2\theta)$ distribution, $\theta \in \Omega = (0, \infty)$, and set:

$$U_1 = \frac{n+1}{2n+1} Y_n \quad \text{and} \quad U_2 = \frac{n+1}{5n+4}(2Y_n + Y_1),$$

where $Y_1$ and $Y_n$ are the smallest and the largest order statistics, respectively, of the $X_i$'s.

(i) Use relations (28) and (29) in Chapter 6 in order to obtain the p.d.f.'s $g_1$ and $g_n$ of $Y_1$ and $Y_n$, respectively.

(ii) By using part (i), show that:

$$E_\theta Y_1 = \frac{n+2}{n+1}\theta, \quad E_\theta Y_n = \frac{2n+1}{n+1}\theta.$$

(iii) By means of part (ii), conclude that both $U_1$ and $U_2$ are unbiased estimates of $\theta$.

**3.6** Refer to Exercise 3.5, and show that:

(i) $E_\theta Y_1^2 = \frac{n^2+5n+8}{(n+1)(n+2)}\theta^2$, $\quad \text{Var}_\theta(Y_1) = \frac{n}{(n+1)^2(n+2)}\theta^2$.

(ii) $E_\theta Y_n^2 = \frac{2(2n^2+4n+1)}{(n+1)(n+2)}\theta^2$, $\quad \text{Var}_\theta(Y_n) = \frac{n}{(n+1)^2(n+2)}\theta^2$.

**3.7** Refer to Exercise 3.5, and:

(i) Use Exercise 5.3 (ii) in Chapter 6 in order to show that the joint p.d.f., $g_{1n}$, of $Y_1$ and $Y_n$ is given by:

$$g_{1n}(y_1, y_n) = \frac{n(n-1)}{\theta^n}(y_n - y_1)^{n-2}, \quad \theta \le y_1 < y_n \le 2\theta.$$

(ii) Employ part (i) here and also part (ii) of Exercise 3.5 in order to show that:

$$E_\theta(Y_1 Y_n) = \frac{2n^2 + 7n + 5}{(n+1)(n+2)}\theta^2, \quad \text{Cov}_\theta(Y_1, Y_n) = \frac{\theta^2}{(n+1)^2(n+2)}.$$

**3.8** Refer to Exercise 3.5, and:
  (i) Use Exercises 3.6 and 3.7 (ii) in order to show that:

$$\text{Var}_\theta(U_1) = \frac{n}{(2n+1)^2(n+2)}\theta^2, \quad \text{Var}_\theta(U_2) = \frac{1}{(5n+4)(n+2)}\theta^2.$$

  (ii) From part (i), conclude that $\text{Var}_\theta(U_2) \le \text{Var}_\theta(U_1)$ for all $\theta$ (with equality holding only for $n = 1$), so that the (unbiased) estimate $U_2$ is Uniformly better (in terms of variance) than the (unbiased) estimate $U_1$.

**3.9** Let $X_1, \ldots, X_m$ and $Y_1, \ldots, Y_n$ be independent random samples with the same mean $\theta$ and known variances $\sigma_1^2$ and $\sigma_2^2$, respectively. For any fixed $c$ with $0 \le c \le 1$, set $U_c = c\bar{X} + (1-c)\bar{Y}$, where $\bar{X}$ and $\bar{Y}$ are the sample means of the $X_i$'s and of the $Y_i$'s, respectively. Then
  (i) Show that $U_c$ is an unbiased estimate of $\theta$ for every $c$ as specified above.
  (ii) Calculate the variance of $U_c$, and show that it is minimized for
$c = c_0 = m\sigma_2^2 / (n\sigma_1^2 + m\sigma_2^2)$.

**3.10** Let $X_1, \ldots, X_n$ be i.i.d. r.v.'s with mean $\mu$ and variance $\sigma^2$, both unknown. Then for any known constants $c_1, \ldots, c_n$, consider the *linear* estimate of $\mu$ defined by: $U_c = \sum_{i=1}^n c_i X_i$.
  (i) Identify the condition that the $c_i$'s must satisfy, so that $U_c$ is an unbiased estimate of $\mu$.
  (ii) Show that the sample mean $\bar{X}$ is the unbiased linear estimate of $\mu$ with the smallest variance (among all unbiased linear estimates of $\mu$).
  **Hint.** For part (ii), one has to minimize the expression $\sum_{i=1}^n c_i^2$ subject to the side restriction that $\sum_{i=1}^n c_i = 1$. For this minimization, use the Lagrange multipliers method, which calls for the minimization of the function
$\phi(c_1, \ldots, c_n) = \sum_{i=1}^n c_i^2 + \lambda(\sum_{i=1}^n c_i - 1)$ with respect to $c_1, \ldots, c_n$, where $\lambda$ is a constant (Lagrange multiplier). Alternatively, one may employ a Geometric argument to the same effect.
  In all of the following Exercises 3.11–3.15, employ steps (i)–(iv) listed after Remark 11 in an attempt to determine UMVU estimates. Use relation (10) whenever possible.

**3.11** If $X_1, \ldots, X_n$ is a random sample from the Poisson distribution $P(\theta)$, show that the sample mean $\bar{X}$ is the UMVU estimate of $\theta$.

**3.12** Let $X_1, \ldots, X_n$ be i.i.d. r.v.'s distributed as $N(\mu, \sigma^2)$.
  (i) If $\mu = \theta \in \Omega = \Re$ and $\sigma$ is known, show that $\bar{X}$ is the UMVU estimate of $\theta$.

    **(ii)** If $\sigma^2 = \theta \in \Omega = (0, \infty)$ and $\mu$ is known show that
$S^2 = \frac{1}{n} \sum_{i=1}^{n} (X_i - \mu)^2$ is the UMVU estimate of $\sigma^2$.

    **Hint.** For part (ii), recall that $Y = \sum_{i=1}^{n} \left( \frac{X_i - \mu}{\sqrt{\theta}} \right)^2 \sim \chi_n^2$, and hence
$E_\theta Y = n$, $\text{Var}_\theta(Y) = 2n$.

**3.13** Let $X_1, \ldots, X_n$ be i.i.d. r.v.'s from the Gamma distribution with parameters $\alpha$
known and $\beta = \theta \in \Omega = (0, \infty)$ unknown.
    **(i)** Determine the Fisher information $I(\theta)$.
    **(ii)** Show that the estimate $U = U(X_1, \ldots, X_n) = \frac{1}{n\alpha} \sum_{i=1}^{n} X_i$ is unbiased
    and calculate its variance.
    **(iii)** Show that $\text{Var}_\theta(U) = 1/nI(\theta)$, so that $U$ is the UMVU estimate of $\theta$.

**3.14** Let $X_1, \ldots, X_n$ be i.i.d. r.v.'s from the Negative Exponential p.d.f. in the
following parametric form: $f(x; \theta) = \frac{1}{\theta} e^{-x/\theta}$, $x > 0$, $\theta \in \Omega = (0, \infty)$, so that
$E_\theta X_1 = \theta$. Use Exercise 3.13 in order to show that $\bar{X}$ is the UMVU estimate
of $\theta$.

    **Hint.** Recall that the Negative Exponential distribution is a special case of the
Gamma distribution.

**3.15** Let $X$ be a r.v. with p.d.f. $f(x; \theta) = \frac{1}{2\theta} e^{-|x|/\theta}$, $x \in \mathfrak{R}$, $\theta \in \Omega = (0, \infty)$. Then
    **(i)** Show that $E_\theta |X| = \theta$, and $E_\theta X^2 = 2\theta^2$.
    **(ii)** Show that the statistic $U = U(X_1, \ldots, X_n) = \frac{1}{n} \sum_{i=1}^{n} |X_i|$ is an unbiased
    estimate of $\theta$, and calculate its variance.
    **(iii)** Show that the Fisher information number
$$I(\theta) = -E_\theta [\frac{\partial^2}{\partial \theta^2} \log f(X; \theta)] = 1/\theta^2.$$
    **(iv)** Conclude that $U$ is the UMVU estimate of $\theta$.
In Exercises 3.16–3.20, the purpose is to construct the UMVU estimates of
the parameters involved by invoking Theorem 7. Sufficiency can always be
established through Theorem 3; completeness sometimes will be established,
but it will always be assumed when appropriate.

**3.16** If $X_1, \ldots, X_n$ are independent r.v.'s distributed as $B(1, \theta), \theta \in \Omega = (0, 1)$, then
    **(i)** Show that $T = \sum_{i=1}^{n} X_i$ is sufficient for $\theta$.
    **(ii)** Also, show that $T$ is complete.
    **(iii)** From parts (i) and (ii), conclude that $\bar{X}$ is the UMVU estimate of $\theta$.

**3.17** Let $X_1, \ldots, X_n$ be a random sample of size $n$ from the $P(\theta)$ distribution,
$\theta \in \Omega = (0, \infty)$. With $T = \sum_{i=1}^{n} X_i$, it has been seen that the conditional
distribution $P_\theta(X_1 = x \mid T = t)$ is $B(t, \frac{1}{n})$ (see Exercise 2.10(ii) in Chapter 5)
and that $T$ is sufficient for $\theta$ (see Exercise 2.9 here). Show that $T$ is complete,
so that the conclusion reached in Example 17 will be fully
justified.

**3.18** Let the r.v. $X$ has the Geometric p.d.f.
$$f(x; \theta) = \theta(1 - \theta)^{x-1}, \; x = 1, 2, \ldots, \; \theta \in \Omega = [0, 1].$$

(i) Show that $X$ is both sufficient and complete.

(ii) Show that the estimate $U$ defined by: $U(X) = 1$ if $X = 1$, and $U(X) = 0$ if $X \geq 2$, is an unbiased estimate of $\theta$.

(iii) Conclude that $U$ is the UMVU estimate of $\theta$ and also an entirely unreasonable estimate.

(iv) Prove that the variance of $U$ is Uniformly bigger than the Cramér-Rao lower bound. (So, on account of this, the Cramér-Rao inequality could not produce this UMVU estimate.)

**Remark.** We have stipulated that an estimate always takes values in the appropriate parameter $\Omega$. In order to be consistent with this stipulation, we take $\Omega = [0, 1]$ in part (ii).

**3.19** Let $X_1, \ldots, X_n$ be independent r.v.'s with the Negative Exponential p.d.f. $f(x; \theta) = \frac{1}{\theta}e^{-(1/\theta)x}$, $x > 0$, $\theta \in \Omega = (0, \infty)$. Then

(i) $\bar{X}$ is sufficient (and complete, although completeness will not be established here).

(ii) $\bar{X}$ is the UMVU estimate of $\theta$.

**3.20** Let $X_1, \ldots, X_n$ be independent r.v.'s distributed as $U(0, \theta)$, $\theta \in \Omega = (0, \infty)$. Then:

(i) The largest order statistic $Y_n$ of the $X_i$'s is sufficient (and complete, although completeness will not be established here).

(ii) The unbiased estimate $U = \frac{n+1}{n}Y_n$ is the UMVU estimate of $\theta$.

(iii) Explain why the Cramér-Rao inequality approach is not applicable here; take $n = 1$.

## 9.4 DECISION-THEORETIC APPROACH TO ESTIMATION

In this section, a brief discussion is presented of still another approach to parameter estimation, which, unlike the previous two approaches, is kind of penalty driven. In order to introduce the relevant concepts and notation, let $X_1, \ldots, X_n$ be a random sample with p.d.f. $f(\cdot; \theta)$, $\theta \in \Omega \subseteq \mathfrak{R}$, and let $\delta$ be a function defined on $\mathfrak{R}^n$ into $\Omega$; i.e., $\delta : \mathfrak{R}^n \to \Omega$. If $x_1, \ldots, x_n$ are the observed values of $X_1, \ldots, X_n$, then the value $\delta(x_1, \ldots, x_n)$ is the proposed estimate of $\theta$. The quality of this estimate is usually measured by its squared distance from the estimated quantity $\theta$; that is, $[\theta - \delta(x_1, \ldots, x_n)]^2$. Denote it by $L[\theta; \delta(x_1, \ldots, x_n)]$ and call it a *loss function*. So $L[\theta; \delta(x_1, \ldots, x_n)] = [\theta - \delta(x_1, \ldots, x_n)]^2$. The closer the estimate $\delta(x_1, \ldots, x_n)$ is to $\theta$ (on either side of it) the smaller is the loss we suffer, and vice versa. The objective here is to select $\delta$ in some optimal way to be discussed below. The first step to this effect is that $\delta$ be selected so that it minimizes the average loss we suffer by using this estimate. For this purpose, consider the r.v. $L[\theta; \delta(X_1, \ldots, X_n)] = [\theta - \delta(X_1, \ldots, X_n)]^2$ and take its expectation to be denoted by $R(\theta; \delta)$; namely,

$$R(\theta; \delta) = E_\theta[\theta - \delta(X_1,\ldots,X_n)]^2$$

$$= \begin{cases} \int_{-\infty}^{\infty} \cdots \int_{-\infty}^{\infty} [\theta - \delta(x_1,\ldots,x_n)]^2 f(x_1; \theta) \cdots f(x_n; \theta) dx_1 \cdots dx_n, \\ \text{for the continuous case,} \\ \sum_{x_1} \cdots \sum_{x_n} [\theta - \delta(x_1,\ldots,x_n)]^2 f(x_1; \theta) \cdots f(x_n; \theta), \\ \text{for the discrete case.} \end{cases} \tag{12}$$

The average loss $R(\theta; \delta)$ is called the *risk function*, corresponding to $\delta$. The value $R(\theta; \delta)$ is the average loss suffered corresponding to the point $\theta$, when $\delta$ is used. At this point, there are two options available to us in pursuing the issue of selecting $\delta$. One is to choose $\delta$, so as to minimize the worst thing which can happen to us. More formally, choose $\delta$ so that, for any other estimate $\delta^*$, it holds that:

$$\sup[R(\theta; \delta); \quad \theta \in \Omega] \leq \sup[R(\theta; \delta^*); \quad \theta \in \Omega].$$

Such an estimate, if it exists, is called *minimax* (by the fact that it minimizes the maximum risk). The second option would be to average $R(\theta; \delta)$, with respect to $\theta$, and then choose $\delta$ to minimize this average. The implementation of this plan goes as follows. Let $\lambda(\theta)$ be a p.d.f. on $\Omega$ and average $R(\theta; \delta)$ by using this p.d.f.; let $r(\delta)$ be the resulting average. Thus,

$$r(\delta) = E_\lambda R(\theta; \delta) = \begin{cases} \int_\Omega R(\theta; \delta)\lambda(\theta)\, d\theta, & \text{for the continuous case,} \\ \sum_{\theta \in \Omega} R(\theta; \delta)\lambda(\theta), & \text{for the discrete case.} \end{cases} \tag{13}$$

Then select $\delta$ to minimize $r(\delta)$. Such an estimate is called a *Bayes estimate*, corresponding to the p.d.f. $\lambda$, and it may be denoted by $\delta_\lambda$. In this context, the parameter $\theta$ is interpreted as a r.v. taking values in $\Omega$ according to the p.d.f. $\lambda(\theta)$, which is called a *prior* or *a priori* p.d.f. It so happens that, under minimal assumptions, a Bayes estimate always exists and is given by an explicit formula.

**Theorem 8.** *Suppose $\theta$ is a r.v. of the continuous type with prior p.d.f. $\lambda(\theta)$, and that the three quantities $\int_\Omega f(x_1; \theta) \cdots f(x_n; \theta)\lambda(\theta)\, d\theta$, $\int_\Omega \theta f(x_1; \theta) \times \cdots \times f(x_n; \theta)\lambda(\theta)\, d\theta$, and $\int_\Omega \theta^2 f(x_1; \theta) \cdots f(x_n; \theta)\lambda(\theta)\, d\theta$ are finite. Then the Bayes estimate (for square loss function) corresponding to $\lambda(\theta), \delta_\lambda$, is given by the expression*

$$\delta_\lambda(x_1,\ldots,x_n) = \frac{\int_\Omega \theta f(x_1; \theta) \cdots f(x_n; \theta)\lambda(\theta)d\theta}{\int_\Omega f(x_1; \theta) \cdots f(x_n; \theta)\lambda(\theta)d\theta}. \tag{14}$$

*If $\theta$ is a discrete r.v., all integrals above are to be replaced by summation signs.*

This theorem has the following corollary.

**Corollary.** *The Bayes estimate $\delta_\lambda(x_1,\ldots,x_n)$ defined in relation (14) can also be calculated thus:*

$$\delta_\lambda(x_1,\ldots,x_n) = \int_\Omega \theta h(\theta \mid x_1,\ldots,x_n)\, d\theta, \tag{15}$$

*where $h(\theta \mid x_1,\ldots,x_n)$ is the conditional p.d.f. of $\theta$, given $X_i = x_i, i = 1,\ldots,n$, which is also called the posterior p.d.f. (of $\theta$). The integral is to be replaced by a summation sign in the discrete case.*

*Proof.* Observe that $f(x_1; \theta) \cdots f(x_n; \theta)$ is, actually, the joint conditional p.d.f. of $X_1, \ldots, X_n$, given $\theta$, so that the product $f(x_1; \theta) \cdots f(x_n; \theta)\lambda(\theta)$ is the joint p.d.f. of $X_1, \ldots, X_n$ and $\theta$. Then its integral over $\Omega$ is the marginal (joint) p.d.f. of $X_1, \ldots, X_n$, and therefore

$$f(x_1; \theta) \cdots f(x_n; \theta)\lambda(\theta) / \int_\Omega f(x_1; \theta) \cdots f(x_n; \theta)\lambda(\theta)\, d\theta$$

is $h(\theta \mid x_1, \ldots, x_n)$ as described above. Then expression (14) completes the proof. ■

So, in computing $\delta_\lambda(x_1, \ldots, x_n)$, one may use relation (14) or, alternatively, first calculate $h(\theta \mid x_1, \ldots, x_n)$ and then apply formula (15).

We now proceed with the justification of (14).

*Proof of Theorem 8.* All operations below are valid without any further explanation. The derivations are carried out for the continuous case; in the discrete case, the integrals are to be replaced by summation signs. By relation (13),

$$r(\delta) = \int_\Omega R(\theta; \delta)\lambda(\theta)d\theta = \int_\Omega \left\{ \int_{-\infty}^{\infty} \cdots \int_{-\infty}^{\infty} [\theta - \delta(x_1, \ldots, x_n)]^2 \times \right.$$

$$\left. f(x_1; \theta) \cdots f(x_n; \theta)dx_1 \cdots dx_n \right\} \lambda(\theta)d\theta$$

$$= \int_{-\infty}^{\infty} \cdots \int_{-\infty}^{\infty} \left\{ \int_\Omega [\theta - \delta(x_1, \ldots, x_n)]^2 \lambda(\theta)f(x_1; \theta) \cdots f(x_n; \theta)d\theta \right\}$$

$$\times dx_1 \ldots dx_n.$$

Then, in order to minimize $r(\delta)$, it suffices to minimize the inner integral for each $x_1, \ldots, x_n$. However,

$$\int_\Omega [\theta - \delta(x_1, \ldots, x_n)]^2 \lambda(\theta)f(x_1; \theta) \cdots f(x_n; \theta)d\theta$$

$$= \delta^2(x_1, \ldots, x_n) \left[ \int_\Omega f(x_n; \theta) \cdots f(x_n; \theta)\lambda(\theta)d\theta \right]$$

$$- 2\delta(x_1, \ldots, x_n) \left[ \int_\Omega \theta f(x_1; \theta) \cdots f(x_n; \theta)\lambda(\theta)d\theta \right]$$

$$+ \left[ \int_\Omega \theta^2 f(x_1; \theta) \cdots f(x_n; \theta)\lambda(\theta)d\theta \right],$$

and this is of the form: $g(t) = at^2 - 2bt + c$, where

$$a = \int_\Omega f(x_1; \theta) \cdots f(x_n; \theta)\lambda(\theta)d\theta,$$

$$b = \int_\Omega \theta f(x_1; \theta) \cdots f(x_n; \theta)\lambda(\theta)d\theta,$$

$$c = \int_\Omega \theta^2 f(x_1; \theta) \cdots f(x_n; \theta)\lambda(\theta)d\theta,$$

and

$$t = \delta(x_1, \dots, x_n).$$

The quadratic expression $g(t) = at^2 - 2bt + c$ is minimized for $t = \frac{b}{a}$, since $g'(t) = 2at - 2b = 0$ gives $t = \frac{b}{a}$ and $g''(t) = 2a > 0$. But $\frac{b}{a}$ is equal to the right-hand side in expression (14). The proof is complete. ■

**Remark 13.** In the context of the present section, the function $\delta$ and the estimate $\delta(x_1, \dots, x_n)$ are also referred to as a *decision function* and a *decision* (associated with the specific outcome $x_1, \dots, x_n$). Hence the title of the section.

**Remark 14.** The Bayes approach presents us with both advantages and disadvantages. An issue which often arises is how the prior p.d.f. $\lambda(\theta)$ is to be chosen. People have given various considerations in selecting $\lambda(\theta)$, including mathematical convenience. Perhaps the most significant advantage of this approach is that, in selecting the prior $\lambda(\theta)$, we have flexibility in incorporating whatever information we may have about the parameter $\theta$.

The theorem is illustrated with one example.

**Example 19.** Let $X_1, \dots, X_n$ be a random sample from the $B(1, \theta)$ distribution, $\theta \in \Omega = (0, 1)$, and choose $\lambda(\theta)$ to be the so-called *Beta* density with parameters $\alpha$ and $\beta$; that is,

$$\lambda(\theta) = \begin{cases} \frac{\Gamma(\alpha+\beta)}{\Gamma(\alpha)\Gamma(\beta)} \theta^{\alpha-1}(1-\theta)^{\beta-1}, & \text{if } \theta \in (0, 1) \\ 0, & \text{otherwise.} \end{cases} \quad (\alpha, \beta > 0) \qquad (16)$$

(For a proof that $\lambda$ is, indeed, p.d.f., see, e.g., Section 3.3.6 in the book *A Course in Mathematical Statistics*, 2nd edition (1997), Academic Press, by G. G. Roussas.) Then the Bayes estimate is given by relation (20) below.

**Discussion.** Now, from the definition of the p.d.f. of a beta distribution with parameters $\alpha$ and $\beta$, we have

$$\int_0^1 x^{\alpha-1}(1-x)^{\beta-1}\, dx = \frac{\Gamma(\alpha)\Gamma(\beta)}{\Gamma(\alpha+\beta)}, \qquad (17)$$

and, of course, $\Gamma(\gamma) = (\gamma - 1)\Gamma(\gamma - 1)$. Then, for simplicity, writing $\sum_j x_j$ rather than $\sum_{j=1}^n x_j$ when this last expression appears as an exponent, we have

$$I_1 = \int_\Omega f(x_1; \theta) \cdots f(x_n; \theta)\lambda(\theta)\, d\theta$$

$$= \frac{\Gamma(\alpha+\beta)}{\Gamma(\alpha)\Gamma(\beta)} \int_0^1 \theta^{\sum_j x_j}(1-\theta)^{n-\sum_j x_j}\theta^{\alpha-1}(1-\theta)^{\beta-1}\, d\theta$$

$$= \frac{\Gamma(\alpha+\beta)}{\Gamma(\alpha)\Gamma(\beta)} \int_0^1 \theta^{\left(\alpha+\sum_j x_j\right)-1}(1-\theta)^{\left(\beta+n-\sum_j x_j\right)-1}\, d\theta,$$

which, by means of (17), becomes as follows:

$$I_1 = \frac{\Gamma(\alpha + \beta)}{\Gamma(\alpha)\Gamma(\beta)} \times \frac{\Gamma\left(\alpha + \sum_{j=1}^{n} x_j\right)\Gamma\left(\beta + n - \sum_{j=1}^{n} x_j\right)}{\Gamma(\alpha + \beta + n)}. \tag{18}$$

Next,

$$
\begin{aligned}
I_2 &= \int_{\Omega} \theta f(x_1; \theta) \cdots f(x_n; \theta) \lambda(\theta) \, d\theta \\
&= \frac{\Gamma(\alpha + \beta)}{\Gamma(\alpha)\Gamma(\beta)} \int_0^1 \theta \theta^{\sum_j x_j}(1 - \theta)^{n - \sum_j x_j} \theta^{\alpha - 1}(1 - \theta)^{\beta - 1} \, d\theta \\
&= \frac{\Gamma(\alpha + \beta)}{\Gamma(\alpha)\Gamma(\beta)} \int_0^1 \theta^{\left(\alpha + \sum_j x_j + 1\right) - 1}(1 - \theta)^{\left(\beta + n - \sum_j x_j\right) - 1} \, d\theta.
\end{aligned}
$$

Once more, relation (17) gives

$$I_2 = \frac{\Gamma(\alpha + \beta)}{\Gamma(\alpha)\Gamma(\beta)} \times \frac{\Gamma\left(\alpha + \sum_{j=1}^{n} x_j + 1\right)\Gamma\left(\beta + n - \sum_{j=1}^{n} x_j\right)}{\Gamma(\alpha + \beta + n + 1)}. \tag{19}$$

Relations (18) and (19) imply, by virtue of (14),

$$\delta(x_1, \ldots, x_n) = \frac{\Gamma(\alpha + \beta + n)\Gamma\left(\alpha + \sum_{j=1}^{n} x_j + 1\right)}{\Gamma(\alpha + \beta + n + 1)\Gamma\left(\alpha + \sum_{j=1}^{n} x_j\right)} = \frac{\alpha + \sum_{j=1}^{n} x_j}{\alpha + \beta + n};$$

that is,

$$\delta(x_1, \ldots, x_n) = \frac{\sum_{j=1}^{n} x_j + \alpha}{n + \alpha + \beta}. \tag{20}$$

**Remark 15.**   When $\alpha = \beta = 1$, the beta distribution becomes $U(0, 1)$, as follows from (16), since $\Gamma(2) = 1 \times \Gamma(1) = 1$. In this case, the corresponding Bayes estimate is $\delta(x_1, \ldots, x_n) = \left(\sum_{i=1}^{n} x_i + 1\right)/(n + 2)$.

A minimax estimate is usually found indirectly, by showing that a Bayes estimate is also minimax. The following theorem tells the story.

**Theorem 9.**   *Let $\delta_\lambda(x_1, \ldots, x_n)$ be the Bayes estimate corresponding to the prior p.d.f. $\lambda(\theta)$, and suppose its risk $R(\theta; \delta_\lambda)$, as given in (12), is independent of $\theta \in \Omega$. Then $\delta_\lambda(x_1, \ldots, x_n)$ is minimax.*

*Proof.* The justification is straightforward and goes like this. Set $R(\theta; \delta_\lambda) = c$, and let $\delta^* = \delta^*(x_1, \ldots, x_n)$ be any other estimate. Then

$$
\begin{aligned}
\sup[R(\theta; \delta_\lambda); \ \theta \in \Omega] &= c = \int_\Omega c\lambda(\theta) \, d\theta \\
&= \int_\Omega R(\theta; \delta_\lambda)\lambda(\theta) \, d\theta \leq \int_\Omega R(\theta; \delta^*)\lambda(\theta) \, d\theta \quad \text{(since } \delta_\lambda \text{ is Bayes)} \\
&\leq \sup[R(\theta; \delta^*); \ \theta \in \Omega] \quad \text{(since } \lambda \text{ is a p.d.f. on } \Omega\text{)}.
\end{aligned}
$$

This completes the proof.   ∎

The following example illustrates this theorem.

**Example 20.**    Let $X_1, \ldots, X_n$ and $\lambda(\theta)$ be as in Example 19. Then the corresponding Bayes estimate $\delta$ is given by (20), and the estimate $\delta^*$ given in (21) is minimax.

**Discussion.**    By setting $X = \sum_{j=1}^{n} X_j$ and taking into consideration that $E_\theta X = n\theta$ and $E_\theta X^2 = n\theta(1 - \theta + n\theta)$, we obtain

$$R(\theta; \delta_\lambda) = E_\theta \left( \theta - \frac{X + \alpha}{n + \alpha + \beta} \right)^2$$

$$= \frac{1}{(n + \alpha + \beta)^2} \{ [(\alpha + \beta)^2 - n]\theta^2 - (2\alpha^2 + 2\alpha\beta - n)\theta + \alpha^2 \}.$$

By taking $\alpha = \beta = \frac{1}{2}\sqrt{n}$ and denoting by $\delta^*$ the resulting estimate, we have

$$(\alpha + \beta)^2 - n = 0, \quad 2\alpha^2 + 2\alpha\beta - n = 0,$$

so that

$$R(\theta; \delta^*) = \frac{\alpha^2}{(n + \alpha + \beta)^2} = \frac{n}{4(n + \sqrt{n})^2} = \frac{1}{4(1 + \sqrt{n})^2}.$$

Since $R(\theta; \delta^*)$ is independent of $\theta$, Theorem 9 implies that

$$\delta^*(x_1, \ldots, x_n) = \frac{\sum_{j=1}^{n} x_j + \frac{1}{2}\sqrt{n}}{n + \sqrt{n}} = \frac{2\sqrt{n}\bar{x} + 1}{2(1 + \sqrt{n})} \tag{21}$$

is minimax.

# EXERCISES

**4.1**    Consider one observation from the p.d.f.
$f(x; \theta) = (1 - \theta)\theta^{x-1}$, $x = 1, 2, \ldots, \theta \in \Omega = (0, 1)$, and let the prior p.d.f. $\lambda$ on $(0, 1)$ be the $U(0, 1)$ distribution. Then, determine:
  (i)  The posterior p.d.f. of $\theta$, given $X = x$.
  (ii)  The Bayes estimate of $\theta$, by using relation (15).

**4.2**    If the r.v. $X$ has the Beta distribution with parameters $\alpha$ and $\beta$; i.e., its p.d.f. is given by expression (16), then without integration and by using the recursive property of the Gamma function ($\Gamma(\gamma) = (\gamma - 1)\Gamma(\gamma - 1), \gamma > 1$), show that $EX = \alpha/(\alpha + \beta)$.

**4.3**    In reference to Example 19:
  (i)  Show that the marginal p.d.f., $h(x_1, \ldots, x_n)$, defined by
  $h(x_1, \ldots, x_n) = \int_0^1 f(x_1; \theta) \cdots f(x_n; \theta) \lambda(\theta) \, d\theta$ with
  $f(x; \theta) = \theta^x (1 - \theta)^{1-x}$, $x = 0, 1$, and $\lambda(\theta)$ as in relation (16), is given by:
  $$h(x_1, \ldots, x_n) = \frac{\Gamma(\alpha + \beta)\Gamma(\alpha + t)\Gamma(\beta + n - t)}{\Gamma(\alpha)\Gamma(\beta)\Gamma(\alpha + \beta + n)},$$

where $t = x_1 + \cdots + x_n$. Do it without, actually, carrying out any integrations, by taking notice of the form of a Beta p.d.f.

(ii) Show that the posterior p.d.f. of $\theta$, given $X_1 = x_1, \ldots, X_n = x_n, h(\theta \mid x_1, \ldots, x_n)$, is the Beta p.d.f. with parameters $\alpha + t$ and $\beta + n - t$.

(iii) Use the posterior p.d.f. obtained in part (ii) in order to rederive the Bayes estimate $\delta(x_1, \ldots, x_n)$ given in (20) by utilizing relation (15). Do it without carrying out any integrations, by using Exercise 4.2.

**4.4** Let $X_1, \ldots, X_n$ be independent r.v.'s from the $N(\theta, 1)$ distribution, $\theta \in \Omega = \mathfrak{R}$, and on $\mathfrak{R}$, consider the p.d.f. $\lambda$ to be that of $N(\mu, 1)$ with $\mu$ known. Then show that the Bayes estimate of $\theta$, $\delta_\lambda(x_1, \ldots, x_n)$, is given by: $\delta(x_1, \ldots, x_n) = \frac{n\bar{x}+\mu}{n+1}$.

**Hint.** By (14), we have to find suitable expressions for the integrals:

$$I_1 = \int_{-\infty}^{\infty} \frac{1}{(\sqrt{2\pi})^n} \exp\left[-\frac{1}{2}\sum_{i=1}^{n}(x_i - \theta)^2\right] \times \frac{1}{\sqrt{2\pi}} \exp\left[-\frac{(\theta - \mu)^2}{2}\right] d\theta,$$

$$I_2 = \int_{-\infty}^{\infty} \theta \times \frac{1}{(\sqrt{2\pi})^n} \exp\left[-\frac{1}{2}\sum_{i=1}^{n}(x_i - \theta)^2\right] \times \frac{1}{\sqrt{2\pi}} \exp\left[-\frac{(\theta - \mu)^2}{2}\right] d\theta.$$

The integrand of $I_1$ is equal to:

$$\frac{1}{(\sqrt{2\pi})^n} \exp\left[-\frac{1}{2}\left(\sum_{i=1}^{n}x_i^2 + \mu^2\right)\right]$$

$$\times \frac{1}{\sqrt{2\pi}} \exp\left\{-\frac{1}{2}[(n + 1)\theta^2 - 2(n\bar{x} + \mu)\theta]\right\}.$$

However,

$$(n + 1)\theta^2 - 2(n\bar{x} + \mu)\theta = (n + 1)\left(\theta^2 - 2\frac{n\bar{x} + \mu}{n + 1}\theta\right)$$

$$= (n + 1)\left[\left(\theta - \frac{n\bar{x} + \mu}{n + 1}\right)^2 - \left(\frac{n\bar{x} + \mu}{n + 1}\right)^2\right]$$

$$= \frac{\left(\theta - \frac{n\bar{x}+\mu}{n+1}\right)^2}{1/(\sqrt{n + 1})^2} - \frac{(n\bar{x} + \mu)^2}{n + 1},$$

so that

$$\frac{1}{\sqrt{2\pi}} \exp\left\{-\frac{1}{2}[(n + 1)\theta^2 - 2(n\bar{x} + \mu)\theta]\right\}$$

$$= \frac{1}{\sqrt{n + 1}} \exp\left[\frac{1}{2} \times \frac{(n\bar{x} + \mu)^2}{n + 1}\right] \times \frac{1}{\sqrt{2\pi}(1/\sqrt{n + 1})} \exp\left[-\frac{\left(\theta - \frac{n\bar{x}+\mu}{n+1}\right)^2}{2/(\sqrt{n + 1})^2}\right],$$

and the second factor is the p.d.f. of $N(\frac{n\bar{x}+\mu}{n+1}, \frac{1}{n+1})$. Therefore, the integration produces the constant:

$$\frac{1}{\sqrt{n+1}} \times \frac{1}{(\sqrt{2\pi})^n} \exp\left\{-\frac{1}{2}\left[\sum_{i=1}^{n} x_i^2 + \mu^2 - \frac{(n\bar{x} + \mu)^2}{n+1}\right]\right\}.$$

Likewise, the integrand in $I_2$ is rewritten thus:

$$\frac{1}{\sqrt{n+1}} \times \frac{1}{(\sqrt{2\pi})^n} \exp\left\{-\frac{1}{2}\left[\sum_{i=1}^{n} x_i^2 + \mu^2 - \frac{(n\bar{x} + \mu)^2}{n+1}\right]\right\}$$

$$\times \frac{1}{\sqrt{2\pi}(1/\sqrt{n+1})}\theta \exp\left[-\frac{\left(\theta - \frac{n\bar{x}+\mu}{n+1}\right)^2}{2/(\sqrt{n+1})^2}\right],$$

and the second factor, when integrated with respect to $\theta$, is the mean of $N\left(\frac{n\bar{x}+\mu}{n+1}, \frac{1}{n+1}\right)$ distribution, which is $\frac{n\bar{x}+\mu}{n+1}$. Dividing then $I_2$ by $I_1$, we obtain the desired result.

**4.5** Refer to Exercise 4.4, and by utilizing the derivations in the hint, derive the posterior p.d.f. $h(\theta \mid x_1, \ldots, x_n)$.

**4.6** Let $X_1, \ldots, X_n$ be independent r.v.'s distributed as $P(\theta)$, $\theta \in \Omega = (0, \infty)$, and consider the estimate $\delta(x_1, \ldots, x_n) = \bar{x}$ and the loss function
$L(\theta; \delta) = [\theta - \delta(x_1, \ldots, x_n)]^2/\theta$.
  **(i)** Calculate the risk $R(\theta; \delta) = \frac{1}{\theta}E_\theta[\theta - \delta(X_1, \ldots, X_n)]^2$, and show that it is independent of $\theta$.
  **(ii)** Can you conclude that the estimate is minimax by using Theorem 9?

## 9.5 OTHER METHODS OF ESTIMATION

In addition to the methods of estimation discussed so far, there are also other methods and approaches, such as the so-called *minimum Chi-square method*, the method of *least squares*, and the *method of moments*. The method of least squares is usually associated with the so-called linear models, and therefore we defer its discussion to a later chapter (see Chapter 13). Here, we are going to present only a brief outline of the method of moments, and illustrate it with three examples.

To this end, let $X_1, \ldots, X_n$ be a random sample with p.d.f. $f(\cdot; \theta)$, $\theta \in \Omega \subseteq \mathfrak{R}$, and suppose that $E_\theta X_1 = m_1(\theta)$ is finite. The objective is to estimate $\theta$ by means of the random sample at hand. By the WLLN,

$$\frac{1}{n}\sum_{i=1}^{n} X_i = \bar{X}_n \xrightarrow[n\to\infty]{P_\theta} m_1(\theta). \tag{22}$$

Therefore, for large $n$, it would make sense to set $\bar{X}_n = m_1(\theta)$ (since it will be approximately so with probability as close to 1 as one desires), and make an attempt to solve for $\theta$. Assuming that this can be done and that there is a unique solution, we declare that solution as the *moment estimate* of $\theta$.

This methodology applies in principle also in the case that there are $r$ parameters involved, $\theta_1, \ldots, \theta_r, r \geq 1$. In such a case, we have to assume that the $r$ first moments of the $X_i$'s are finite; that is,

$$E_\theta X_1^k = m_k(\theta_1, \ldots, \theta_r) \in \Re, \quad k = 1, \ldots, r, \quad \boldsymbol{\theta} = (\theta_1, \ldots, \theta_r).$$

Then form the first $r$ sample moments $\frac{1}{n}\sum_{i=1}^n X_i^k$, $k = 1, \ldots, r$, and equate them to the corresponding (population) moments; that is,

$$\frac{1}{n}\sum_{i=1}^n X_i^k = m_k(\theta_1, \ldots, \theta_r), \quad k = 1, \ldots, r. \tag{23}$$

The reasoning for doing this is the same as the one explained above in conjunction with (22). Assuming that we can solve for $\theta_1, \ldots, \theta_r$ in (23), and that the solutions are unique, we arrive at what we call the *moment estimates* of the parameters $\theta_1, \ldots, \theta_r$.

The following examples should help shed some light on the above exposition.

**Example 21.**  On the basis of the random sample $X_1, \ldots, X_n$ from the $B(1, \theta)$ distribution, find the moment estimate of $\theta$.

**Discussion.**  Here, $E_\theta X_1 = \theta$ and there is only one parameter. Thus, it suffices to set $\bar{X} = \theta$, so that the moment estimate of $\theta$ is the same as the MLE and the UMVU estimate, but (slightly) different from the Bayes (and the minimax estimate).

**Example 22.**  On the basis of the random sample $X_1, \ldots, X_n$ from the $N(\mu, \sigma^2)$ distribution with both $\mu$ and $\sigma^2$ unknown, determine the moment estimates of $\mu$ and $\sigma^2$.

**Discussion.**  The conditions referred to above are satisfied here, and, specifically, $E_\theta X_1 = \mu$, $E_\theta X_1^2 = \sigma^2 + \mu^2$, $\boldsymbol{\theta} = (\mu, \sigma^2)$. Here we need the first two sample moments, $\bar{X}$ and $\frac{1}{n}\sum_{i=1}^n X_i^2$. We have then: $\bar{X} = \mu$ and $\frac{1}{n}\sum_{i=1}^n X_i^2 = \sigma^2 + \mu^2$. Hence $\mu = \bar{X}$ and $\sigma^2 = \frac{1}{n}\sum_{i=1}^n X_i^2 - \bar{X}^2 = \frac{1}{n}(\sum_{i=1}^n X_i^2 - n\bar{X}^2) = \frac{1}{n}\sum_{i=1}^n (X_i - \bar{X})^2$. Thus, the moment estimates are $\tilde{\mu} = \bar{X}$ and $\tilde{\sigma}^2 = \frac{1}{n}\sum_{i=1}^n (X_i - \bar{X})^2$. The estimate $\tilde{\mu}$ is identical with the MLE and the UMVU estimate, whereas $\tilde{\sigma}^2$ is the same as the MLE, but (slightly) different from the UMVU estimate.

**Example 23.**  Let the random sample $X_1, \ldots, X_n$ be from the $U(\alpha, \beta)$ distribution, where both $\alpha$ and $\beta$ are unknown. Determine their moment estimates.

**Discussion.**  Recall that $E_\theta X_1 = \frac{\alpha+\beta}{2}$ and $\sigma_\theta^2(X_1) = \frac{(\alpha-\beta)^2}{12}$, $\boldsymbol{\theta} = (\alpha, \beta)$, so that:

$$\bar{X} = \frac{\alpha+\beta}{2}, \quad \frac{1}{n}\sum_{i=1}^n X_i^2 = \frac{(\alpha-\beta)^2}{12} + \left(\frac{\alpha+\beta}{2}\right)^2.$$

Hence $\frac{(\alpha-\beta)^2}{12} = \frac{1}{n}\sum_{i=1}^{n}X_i^2 - \bar{X}^2 = \frac{1}{n}\sum_{i=1}^{n}(X_i - \bar{X})^2$, call it $S^2$. Thus, $\beta + \alpha = 2\bar{X}$ and $(\alpha - \beta)^2 = 12S^2$, or $\beta - \alpha = 2S\sqrt{3}$, so that the moment estimates of $\alpha$ and $\beta$ are: $\tilde{\alpha} = \bar{X} - S\sqrt{3}$ and $\tilde{\beta} = \bar{X} + S\sqrt{3}$. These estimates are entirely different from the MLE's of these parameters.

## EXERCISES

**5.1** Refer to Exercise 1.6, and derive the moment estimate of $\theta$. Also, compare it with the MLE $\hat{\theta} = 1/\bar{X}$.

**5.2** (i) Refer to Exercise 1.7, and derive the moment estimate of $\theta$, $\tilde{\theta}$.
    (ii) Find the numerical values of $\tilde{\theta}$ and of the MLE $\hat{\theta}$ (see Exercise 1.7), if:
       $n = 10$ and

$$x_1 = 0.92, \quad x_2 = 0.79, \quad x_3 = 0.90, \quad x_4 = 0.65, \quad x_5 = 0.86,$$
$$x_6 = 0.47, \quad x_7 = 0.73, \quad x_8 = 0.97, \quad x_9 = 0.94, \text{ and } x_{10} = 0.77.$$

**5.3** Refer to Exercise 1.8, and derive the moment estimate of $\theta$.

**5.4** Refer to Exercise 1.9, and show that $E_\theta|X| = \theta$, and therefore the moment estimate of $\theta$ is $\tilde{\theta} = \frac{1}{n}\sum_{i=1}^{n}|X_i|$.

**5.5** Refer to Exercise 1.10, and find the expectation of the given p.d.f. by recalling that, if $X \sim$ Gamma with parameters $\alpha$ and $\beta$, then $EX = \alpha\beta$ (and $\mathrm{Var}(X) = \alpha\beta^2$). Then derive the moment estimate of $\theta$, and compare it with the MLE found in Exercise 1.10(ii).

**5.6** Refer to Exercise 1.11, and:
    (i) Show that $EX = \alpha + \beta$ and $EX^2 = \alpha^2 + 2\alpha\beta + 2\beta^2$, where $X$ is a r.v. with the p.d.f. given in the exercise cited. Also, calculate the $\mathrm{Var}(X)$.
    (ii) Derive the moment estimates of $\alpha$ and $\beta$.

**5.7** Let $X_1, \ldots, X_n$ be independent r.v.'s from the $U(\theta - a, \theta + b)$ distribution, where $a$ and $b$ are (known) positive constants and $\theta \in \Omega = \Re$. Determine the moment estimate $\tilde{\theta}$ of $\theta$, and compute its expectation and variance.

**5.8** If the independent r.v.'s $X_1, \ldots, X_n$ have the $U(-\theta, \theta)$ distribution, $\theta \in \Omega = (0, \infty)$, how can one construct a moment estimate of $\theta$?

**5.9** If the independent r.v.'s $X_1, \ldots, X_n$ have the Gamma distribution with parameters $\alpha$ and $\beta$, show that the moment estimates of $\alpha$ and $\beta$ are: $\tilde{\alpha} = \bar{X}^2/S^2$ and $\tilde{\beta} = S^2/\bar{X}$, where $S^2 = \frac{1}{n}\sum_{i=1}^{n}(X_i - \bar{X})^2$.
**Hint.** Recall that, if $X \sim$ Gamma with parameters $\alpha$ and $\beta$, then $EX = \alpha\beta$, $\mathrm{Var}(X) = \alpha\beta^2$.

**5.10** Let $X$ be a r.v. with p.d.f. $f(x; \theta) = \frac{2}{\theta^2}(\theta - x)$, $0 < x < \theta, \theta \in \Omega = (0, \infty)$. Then:

   (i) Show that $f(\cdot; \theta)$ is, indeed, a p.d.f.

   (ii) Show that $E_\theta X = \frac{\theta}{3}$ and $\text{Var}_\theta(X) = \frac{\theta^2}{18}$.

   (iii) On the basis of a random sample of size $n$ from $f(\cdot; \theta)$, find the moment estimate of $\theta, \tilde{\theta}$, and show that it is unbiased. Also, calculate the variance of $\tilde{\theta}$.

**5.11** Let $X$ be a r.v. having the Beta p.d.f. with parameters $\alpha$ and $\beta$; i.e.,
$f(x; \alpha, \beta) = \frac{\Gamma(\alpha+\beta)}{\Gamma(\alpha)\Gamma(\beta)} x^{\alpha-1}(1-x)^{\beta-1}$, $0 < x < 1$ $(\alpha, \beta > 0)$. Then, by Exercise 4.2, $EX = \alpha/(\alpha+\beta)$.

   (i) Follow the same approach used in discussing Exercise 4.2 in order to establish that $EX^2 = \alpha(\alpha+1)/(\alpha+\beta)(\alpha+\beta+1)$.

   (ii) On the basis of a random sample of size $n$ from the underlying p.d.f., determine the moment estimates of $\alpha$ and $\beta$.

**5.12** Let $X$ and $Y$ be *any* two r.v.'s with finite second moments, so that their correlation coefficient, $\rho(X, Y)$, is given by $\rho(X, Y) = \text{Cov}(X, Y)/\sigma(X)\sigma(Y)$. Let $X_i$ and $Y_i, i = 1, \ldots, n$ be i.i.d. r.v.'s distributed as the r.v.'s $X$ and $Y$, respectively. From the expression, $\text{Cov}(X, Y) = E[(X - EX) \times (Y - EY)]$, it makes sense to estimate $\rho(X, Y)$ by $\hat{\rho}_n(X, Y)$ given by:

$$\hat{\rho}(X, Y) = \frac{1}{n} \sum_{i=1}^{m} (X_i - \bar{X})(Y_i - \bar{Y})/\hat{\sigma}(X)\hat{\sigma}(Y),$$

where

$$\hat{\sigma}(x) = \sqrt{\frac{1}{n} \sum_{i=1}^{n} (X_i - \bar{X})^2} \quad \text{and} \quad \hat{\sigma}(Y) = \sqrt{\frac{1}{n} \sum_{i=1}^{n} (Y_i - \bar{Y})^2}.$$

Then set $EX = \mu_1, EY = \mu_2, \text{Var}(X) = \sigma_1^2, \text{Var}(Y) = \sigma_2^2, \rho(X, Y) = \rho$, and show that:

   (i) $\frac{1}{n} \sum_{i=1}^{n} (X_i - \bar{X})(Y_i - \bar{Y}) = \frac{1}{n} \sum_{i=1}^{n} (X_i Y_i) - \bar{X}\bar{Y}$.

   (ii) $E(XY) = \text{Cov}(X, Y) + \mu_1\mu_2 = \rho\sigma_1\sigma_2 + \mu_1\mu_2$.

   (iii) Use the WLLN (Theorem 3 in Chapter 7) in conjunction with the Corollary to Theorem 5 in Chapter 7 in order to show that
$\hat{\rho}_n(X, Y) \xrightarrow[n\to\infty]{P} \rho(X, Y) = \rho$, so that $\hat{\rho}_n(X, Y)$ is a consistent (in the probability sense) estimate of $\rho$.
(Notice that $\hat{\rho}_n(X, Y)$ is the same as the MLE of $\rho$ for the case that the pair $(X, Y)$ has the Bivariate Normal distribution; see Exercise 1.14 in this chapter.)

**5.13** (i) For any $n$ pairs of real numbers $(\alpha_i, \beta_i), i = 1, \ldots, n$, show that:
$(\sum_{i=1}^{n} \alpha_i\beta_i)^2 \leq (\sum_{i=1}^{n} \alpha_i^2)(\sum_{i=1}^{n} \beta_i^2)$.
**Hint.** One way of proving it is to consider the function in
$\lambda, g(\lambda) = \sum_{i=1}^{n} (\alpha_i - \lambda\beta_i)^2$, and observe that $g(\lambda) \geq 0$ for all real $\lambda$, and, in

particular, for $\lambda = (\sum_{i=1}^{n} \alpha_i \beta_i)/(\sum_{i=1}^{n} \beta_i)$, which is actually the minimizing value for $g(\lambda)$.

**(ii)** Use part (i) in order to show that $[\hat{\rho}_n(X, Y)]^2 \leq 1$.

**5.14** In reference to Example 25 in Chapter 1, denote by $x_i$ and $y_i$, $i = 1, \ldots, 15$, respectively, the observed measurements for the cross-fertilized and the self-fertilized pairs. Then calculate the (observed) sample means $\bar{x}, \bar{y}$; sample variances $s_x^2, s_y^2$; and the sample s.d.'s $s_x, s_y$, where $s_x^2 = \frac{1}{n} \sum_{i=1}^{n} (x_i - \bar{x})^2$ and similarly for $s_y^2$.

# Confidence intervals and confidence regions

10

In Section 8.2, the basic concepts about confidence intervals, etc. were introduced; the detailed discussion was deferred to the present chapter. The point estimation problem, in its simplest form, discussed extensively in the previous chapter, is as follows: On the basis of a random sample $X_1, \ldots, X_n$ with p.d.f. $f(\cdot; \theta)$, $\theta \in \Omega \subseteq \Re$, and its observed values $x_1, \ldots, x_n$, construct a point estimate of $\theta$, call it $\hat{\theta} = \hat{\theta}(x_1, \ldots, x_n)$. Thus, for example, in the $N(\theta, 1)$ case, we are invited to pinpoint a value of $\theta \in \Re$ as the (unknown to us but) true value of $\theta$. Such estimates were, actually, constructed by way of at least three methods. Also, certain desirable properties of estimates (fixed sample size properties, as well as asymptotic properties) were established or stated.

Now, declaring that (the unknown value of) $\theta$ is, actually, $\bar{x}$ may look quite unreasonable. How is it possible to single out one value out of $\Re$, $\bar{x}$, and identify it as the true value of $\theta$? The concept of a confidence interval with a given confidence coefficient mitigates this seemingly unreasonable situation. It makes much more sense to declare that $\theta$ lies within an interval in $\Re$ with high confidence. This is, in effect, what we are doing in this chapter by formulating the questions and problems in rigorous probabilistic/statistical terms.

The chapter consists of four sections. The first section concerns itself with confidence intervals of one real-valued parameter. The following section considers the same kind of a problem when *nuisance* (unknown but of no interest to us) parameters are present. In the third section, an example is discussed, where a confidence region of two parameters is constructed; no general theory is developed. (See, however, Theorem 4 in Chapter 12.) In the final section, some confidence intervals are constructed with given approximate confidence coefficient.

## 10.1 CONFIDENCE INTERVALS

We formalize in the form of a definition some concepts already introduced in the second section of Chapter 8.

**Definition 1.** Let $X_1, \ldots, X_n$ be a random sample with p.d.f. $f(\cdot; \theta)$, $\theta \in \Omega \subseteq \Re$. Then:

**(i)** A *random interval* is an interval whose end-points are r.v.'s.

**(ii)** A *confidence interval* for $\theta$ with *confidence coefficient* $1 - \alpha$ $(0 < \alpha < 1,$ $\alpha$ small) is a random interval whose end-points are statistics $L(X_1, \ldots, X_n)$ and $U(X_1, \ldots, X_n)$, say, such that $L(X_1, \ldots, X_n) < U(X_1, \ldots, X_n)$ and

$$P_\theta[L(X_1, \ldots, X_n) \leq \theta \leq U(X_1, \ldots, X_n)] \geq 1 - \alpha, \quad \text{for all } \theta \in \Omega. \quad (1)$$

**(iii)** The statistic $L(X_1, \ldots, X_n)$ is called a *lower confidence limit* for $\theta$ with *confidence coefficient* $1 - \alpha$, if the interval $[L(X_1, \ldots, X_n), \infty)$ is a confidence interval for $\theta$ with confidence coefficient $1 - \alpha$. Likewise, $U(X_1, \ldots, X_n)$ is said to be an *upper confidence limit* for $\theta$ with *confidence coefficient* $1 - \alpha$, if the interval $(-\infty, U(X_1, \ldots, X_n)]$ is a confidence interval for $\theta$ with confidence coefficient $1 - \alpha$.

**Remark 1.** The significance of a confidence interval stems from the relative frequency interpretation of probability. Thus, on the basis of the observed values $x_1, \ldots, x_n$ of $X_1, \ldots, X_n$, construct the interval with end-points $L(x_1, \ldots, x_n)$ and $U(x_1, \ldots, x_n)$, and denote it by $[L_1, U_1]$. Repeat the underlying random experiment independently another $n$ times and likewise form the interval $[L_2, U_2]$. Repeat this process a large number of times, $N$, independently each time, and let $[L_N, U_N]$ be the corresponding interval. Then the fact that $[L(X_1, \ldots, X_n), U(X_1, \ldots, X_n)]$ is a confidence interval for $\theta$ with confidence coefficient $1 - \alpha$ means that approximately $100(1 - \alpha)\%$ of the above $N$ intervals will cover $\theta$, no matter what its value is.

**Remark 2.** When the underlying r.v.'s are of the continuous type, the inequalities in the above definition, regarding the confidence coefficient $1 - \alpha$, become equalities.

**Remark 3.** If $L(X_1, \ldots, X_n)$ is a lower confidence limit for $\theta$ with confidence coefficient $1 - \frac{\alpha}{2}$, and $U(X_1, \ldots, X_n)$ is an upper confidence limit for $\theta$ with confidence coefficient $1 - \frac{\alpha}{2}$, then $[L(X_1, \ldots, X_n), U(X_1, \ldots, X_n)]$ is a confidence interval for $\theta$ with confidence coefficient $1 - \alpha$.

Indeed, writing $L$ and $U$ instead of $L(X_1, \ldots, X_n)$ and $U(X_1, \ldots, X_n)$, and keeping in mind that $L < U$, we have:

$$P_\theta(L \leq \theta) = P_\theta(L \leq \theta, U \geq \theta) + P_\theta(L \leq \theta, \ U < \theta)$$

$$= P_\theta(L \leq \theta \leq U) + P_\theta(U < \theta), \quad \text{since } (U < \theta) \subseteq (L \leq \theta),$$

and

$$P_\theta(\theta \leq U) = P_\theta(U \geq \theta, \ L \leq \theta) + P_\theta(U \geq \theta, \ L > \theta)$$

$$= P_\theta(L \leq \theta \leq U) + P_\theta(L > \theta), \quad \text{since } (L > \theta) \subseteq (U \geq \theta).$$

Summing them up, we have then

$$P_\theta(L \leq \theta) + P_\theta(U \geq \theta) = 2P_\theta(L \leq \theta \leq U) + P_\theta(U < \theta) + P_\theta(L > \theta),$$

or

$$2P_\theta(L \le \theta \le U) = P_\theta(L \le \theta) + P_\theta(U \ge \theta) - P_\theta(U < \theta) - P_\theta(L > \theta)$$
$$= P_\theta(L \le \theta) + P_\theta(U \ge \theta) - 1 + P_\theta(U \ge \theta) - 1 + P_\theta(L \le \theta)$$
$$= 2[P_\theta(L \le \theta) + P_\theta(U \ge \theta) - 1],$$

or

$$P_\theta(L \le \theta \le U) = P_\theta(L \le \theta) + P_\theta(U \ge \theta) - 1$$
$$\ge 1 - \frac{\alpha}{2} + 1 - \frac{\alpha}{2} - 1 = 1 - \alpha,$$

as was to be seen.

This section is concluded with the construction of confidence intervals in some concrete examples. In so doing, we draw heavily on distribution theory and point estimates. It would be, perhaps, helpful to outline the steps we usually follow in constructing a confidence interval.

**(a)** Think of a r.v. which contains the parameter $\theta$, the r.v.'s $X_1, \ldots, X_n$, preferably in the form of a sufficient statistic, and whose distribution is (exactly or at least approximately) known.
**(b)** Determine suitable points $a < b$ such that the r.v. in step (a) lies in $[a, b]$ with $P_\theta$-probability $\ge 1 - \alpha$.
**(c)** In the expression of step (b), rearrange the terms to arrive at an interval with the end-points being statistics and containing $\theta$.
**(d)** The interval in step (c) is the required confidence interval.

**Example 1.** Let $X_1, \ldots, X_n$ be a random interval from the $N(\mu, \sigma^2)$ distribution, where only one of $\mu$ or $\sigma^2$ is unknown. Construct a confidence interval for it with confidence coefficient $1 - \alpha$.

**Discussion.**

**(i)** *Let $\mu$ be unknown.* The natural r.v. to think of is $\sqrt{n}(\bar{X} - \mu)/\sigma$, which satisfies the requirements in step (a). Next, determine any two points $a < b$ from the Normal tables for which $P(a \le Z \le b) = 1 - \alpha$ where $Z \sim N(0, 1)$. (See, however, Exercise 1.1 for the best choice of $a$ and $b$.) Since $\frac{\sqrt{n}(\bar{X} - \mu)}{\sigma} \sim N(0, 1)$, it follows that

$$P_\mu\left[a \le \frac{\sqrt{n}(\bar{X} - \mu)}{\sigma} \le b\right] = 1 - \alpha, \quad \text{for all } \mu,$$

so that step (b) is satisfied. Rearranging the terms inside the square brackets, we obtain:

$$P_\mu\left(\bar{X} - \frac{b\sigma}{\sqrt{n}} \le \mu \le \bar{X} - \frac{a\sigma}{\sqrt{n}}\right) = 1 - \alpha, \quad \text{for all } \mu,$$

so that step (c) is fulfilled. In particular, for $b = z_{\alpha/2}$ (recall $P(Z \geq z_{\alpha/2}) = \frac{\alpha}{2}$) and $a = -z_{\alpha/2}$, we have

$$P_\mu\left(\bar{X} - z_{\alpha/2}\frac{\sigma}{\sqrt{n}} \leq \mu \leq \bar{X} + z_{\alpha/2}\frac{\sigma}{\sqrt{n}}\right) = 1 - \alpha, \quad \text{for all } \mu.$$

It follows that

$$\left[\bar{X} - z_{\alpha/2}\frac{\sigma}{\sqrt{n}}, \bar{X} + z_{\alpha/2}\frac{\sigma}{\sqrt{n}}\right] = \bar{X} \pm z_{\alpha/2}\frac{\sigma}{\sqrt{n}} \quad \text{(for brevity)} \tag{2}$$

is the required confidence interval.

**(ii)** *Let $\sigma^2$ be unknown.* Set $S^2 = \frac{1}{n}\sum_{i=1}^{n}(X_i - \mu)^2$ and recall that $\frac{nS^2}{\sigma^2} = \sum_{i=1}^{n}$ $(\frac{X_i-\mu}{\sigma})^2 \sim \chi_n^2$. The r.v. $\frac{nS^2}{\sigma^2}$ satisfies the requirements of step (a). From the Chi-square tables, determine any pair $0 < a < b$ for which $P(a \leq X \leq b) = 1 - \alpha$, where $X \sim \chi_n^2$. Then

$$P_{\sigma^2}\left(a \leq \frac{nS^2}{\sigma^2} \leq b\right) = 1 - \alpha, \quad \text{for all } \sigma^2, \quad \text{or } P_{\sigma^2}\left(\frac{nS^2}{b} \leq \sigma^2 \leq \frac{nS^2}{a}\right)$$

$$= 1 - \alpha, \quad \text{for all } \sigma^2 \text{ and steps (b) and (c) are satisfied.}$$

In particular,

$$P_{\sigma^2}\left(\frac{nS^2}{\chi_{n;\,\alpha/2}^2} \leq \sigma^2 \leq \frac{nS^2}{\chi_{n;\,1-\alpha/2}^2}\right) = 1 - \alpha, \quad \text{for all } \sigma^2,$$

where $P(X \leq \chi_{n;\,1-\alpha/2}^2) = P(X \geq \chi_{n;\,\alpha/2}^2) = \frac{\alpha}{2}$. It follows that

$$\left[\frac{nS^2}{\chi_{n;\,\alpha/2}^2}, \frac{nS^2}{\chi_{n;\,1-\alpha/2}^2}\right], \quad S^2 = \frac{1}{n}\sum_{i=1}^{n}(X_i - \mu)^2 \tag{3}$$

is the required confidence interval.

*Numerical Example.* Let $n = 25$ and $1 - \alpha = 0.95$. For part (i), we have $z_{\alpha/2} = z_{0.025} = 1.96$, so that $\bar{X} \pm z_{\alpha/2}\frac{\sigma}{\sqrt{n}} = \bar{X} \pm 1.96 \times \frac{\sigma}{5} = \bar{X} \pm 0.392\sigma$. For $\sigma = 1$, for example, the required interval is then: $\bar{X} \pm 0.392$. For the second part, we have $\chi_{n;\,\alpha/2}^2 = \chi_{25;0.025}^2 = 40.646$, and $\chi_{n;\,1-\alpha/2}^2 = \chi_{25;0.975}^2 = 13.120$. The required interval is then:

$$\left[\frac{25S^2}{40.646}, \frac{25S^2}{13.120}\right] \simeq [0.615S^2, 1.905S^2].$$

**Example 2.** On the basis of the random sample $X_1, \ldots, X_n$ from the $U(0, \theta)$ ($\theta > 0$) distribution, construct a confidence interval for $\theta$ with confidence coefficient $1 - \alpha$.

**Discussion.** It has been seen (just apply Example 6 (ii) in Chapter 9 with $\alpha = 0$ and $\beta = \theta$) that $X = X_{(n)}$ is a sufficient statistic for $\theta$. Also, the p.d.f. of $X$ is given by (see Example 14 in Chapter 9) $f_X(x; \theta) = \frac{n}{\theta^n}x^{n-1}$, $0 \leq x \leq \theta$. Setting $Y = X/\theta$, it is easily seen that the p.d.f. of $Y$ is: $f_Y(y) = ny^{n-1}$, $0 \le y \le 1$. The r.v. $Y$ satisfies the requirements of step (a). Next, determine any $0 \le a < b < 1$ such that $\int_a^b f_Y(y)dy = \int_a^b ny^{n-1}dy = \int_a^b dy^n = y^n|_a^b = b^n - a^n = 1 - \alpha$. Then

$$P_\theta(a \le Y \le b) = P_\theta\left(a \le \frac{X}{\theta} \le b\right) = P_\theta\left(\frac{X}{b} \le \theta \le \frac{X}{a}\right) = 1 - \alpha, \quad \text{for all } \theta,$$

so that steps (b) and (c) are satisfied. It follows that $[\frac{X}{b}, \frac{X}{a}] = [\frac{X_{(n)}}{b}, \frac{X_{(n)}}{a}]$ is the required confidence interval.

Looking at the length of this interval, $X_{(n)}(\frac{1}{a} - \frac{1}{b})$, setting $a = a(b)$ and minimizing with respect to $b$, we find that the shortest interval is taken for $b = 1$ and $a = \alpha^{1/n}$. That is, $[X_{(n)}, \frac{X_{(n)}}{\alpha^{1/n}}]$. (See also Exercise 1.5.)

*Numerical Example.* For $n = 32$ and $1 - \alpha = 0.95$, we get (approximately) $[X_{(32)}, 1.098X_{(32)}]$.

---

## EXERCISES

**1.1** Let $\Phi$ be the d.f. of the $N(0, 1)$ distribution, and let $a < b$ be such that $\Phi(b) - \Phi(a) = \gamma$, some fixed number with $0 < \gamma < 1$. Show that the length $b - a$ of the interval $(a, b)$ is minimum, if, for some $c > 0$, $b = c$ and $a = -c$. Also, identify $c$.

**1.2** If $X_1, \ldots, X_n$ are independent r.v.'s distributed as $N(\mu, \sigma^2)$ with $\mu$ unknown and $\sigma$ known, then a $100(1 - \alpha)\%$ confidence interval for $\mu$ is given by $\bar{X}_n \pm z_{\frac{\alpha}{2}} \frac{\sigma}{\sqrt{n}}$ (see Example 1(i)). Suppose that the length of this interval is 7.5 and we wish to halve it. What sample size $m = m(n)$ will be needed?
**Hint.** Set $m = cn$ and determine $c$.

**1.3** The stray-load loss (in watts) for a certain type of induction motor, when the line current is held at 10 amps for a speed of 1500 rpm, is a r.v. $X \sim N(\mu, 9)$. On the basis of the observed values $x_1, \ldots, x_n$ of a random sample of size $n$:
   (i) Compute a 99% observed confidence interval for $\mu$ when $n = 100$ and $\bar{x} = 58.3$.
   (ii) Determine the sample size $n$, if the length of the 99% confidence interval is required to be 1.

**1.4** If the independent r.v.'s $X_1, \ldots, X_n$ are distributed as $N(\theta, \sigma^2)$ with $\sigma$ known, the $100(1 - \alpha)\%$ confidence interval for $\theta$ is given by $\bar{X}_n \pm z_{\frac{\alpha}{2}} \frac{\sigma}{\sqrt{n}}$ (see Example 1(i)).
   (i) If the length of the confidence interval is to be equal to a preassigned number $l$, determine the sample size $n$ as a function of $l, \sigma$, and $\alpha$.
   (ii) Compute the numerical value of $n$, if $l = 0.1$, $\sigma = 1$, and $\alpha = 0.05$.

**1.5** Refer to Example 2 and show that the shortest length of the confidence interval is, indeed, $[X_{(n)}, X_{(n)}/\alpha^{1/n}]$ as asserted.

**Hint.** Set $a = a(b)$, differentiate $g(b) = \frac{1}{a} - \frac{1}{b}$, with respect to $b$, and use the derivative of $b^n - a^n = 1 - \alpha$ in order to show that $\frac{dg(b)}{db} < 0$, so that $g(b)$ is decreasing. Conclude that $g(b)$ is minimized for $b = 1$.

**1.6** Let $X_1, \ldots, X_n$ be independent r.v.'s with the Negative Exponential p.d.f. given in the form $f(x; \theta) = \frac{1}{\theta}e^{-x/\theta}$, $x > 0$, $\theta \in \Omega = (0, \infty)$. Then:

(i) By using the m.g.f. approach, show that the r.v. $U = \sum_{i=1}^{n} X_i$ has the Gamma distribution with parameters $\alpha = n$ and $\beta = \theta$.

(ii) Also, show that the r.v. $V = \frac{2U}{\theta}$ is distributed as $\chi^2_{2n}$.

(iii) By means of part (ii), construct a confidence interval for $\theta$ with confidence coefficient $1 - \alpha$.

**1.7** If $X$ is a r.v. with the Negative Exponential p.d.f. $f(x; \theta) = \frac{1}{\theta}e^{-x/\theta}$, $x > 0$, $\theta \in \Omega = (0, \infty)$, then, by Exercise 2.2 in Chapter 9, the reliability $R(x; \theta) = P_\theta(X > x) = e^{-x/\theta}$. If $X_1, \ldots, X_n$ is a random sample of size $n$ from this p.d.f., use Exercise 1.6(iii) in order to construct a confidence interval for $R(x; \theta)$ with confidence coefficient $1 - \alpha$.

**1.8** Let $X_1, \ldots, X_n$ be a random sample of size $n$ from the p.d.f. $f(x; \theta) = e^{-(x-\theta)}$, $x > \theta, \theta \in \Omega = \mathfrak{R}$, and let $Y_1$ be the smallest order statistic of the $X_i$'s.

(i) Use formula (28) of Chapter 6 in order to show that the p.d.f. of $Y_1$, call it $g$, is given by: $g(y) = ne^{-n(y-\theta)}$, $y > \theta$.

(ii) Set $T(\theta) = 2n(Y_1 - \theta)$ and show that $T \sim \chi^2_2$.

(iii) Use part (ii) in order to show that a $100(1 - \alpha)\%$ confidence interval for $\theta$, based on $T(\theta)$, is given by: $[Y_1 - \frac{b}{2n}, Y_1 - \frac{a}{2n}]$, for suitable $0 < a < b$; a special choice of $a$ and $b$ is: $a = \chi^2_{2;1-\frac{\alpha}{2}}$ and $b = \chi^2_{2;\frac{\alpha}{2}}$.

**1.9** Let the independent r.v.'s $X_1, \ldots, X_n$ have the Weibull distribution with parameters $\gamma$ and $\theta$ with $\theta \in \Omega = (0, \infty)$ and $\gamma > 0$ known; i.e., their p.d.f. $f(\cdot; \theta)$ is given by:

$$f(x; \theta) = \frac{\gamma}{\theta}x^{\gamma-1}e^{-x^\gamma/\theta}, \quad x > 0.$$

(i) For $i = 1, \ldots, n$, set $Y_i = X_i^\gamma$ and show that the p.d.f. of $Y_i$, $g_T(\cdot; \theta)$, is Negative Exponential parameterized as follows: $g(y; \theta) = \frac{1}{\theta}e^{-y/\theta}$, $y > 0$.

(ii) For $i = 1, \ldots, n$, set $T_i(\theta) = \frac{2Y_i}{\theta}$ and show that the p.d.f. of $T_i(\theta)$, $g_T(\cdot; \theta)$, is that of a $\chi^2_2$ distributed r.v., and conclude that the r.v. $T(\theta) = \sum_{i=1}^{n} T_i(\theta) \sim \chi^2_{2n}$.

(iii) Show that a $100(1 - \alpha)\%$ confidence interval for $\theta$, based on $T(\theta)$, is of the form $[\frac{2Y}{b}, \frac{2Y}{a}]$, for suitable $0 < a < b$, where $Y = \sum_{i=1}^{n} X_i^\gamma$. In particular, $a$ and $b$ may be chosen to be $\chi^2_{2n;1-\frac{\alpha}{2}}$ and $\chi^2_{2n;\frac{\alpha}{2}}$, respectively.

**1.10** If the independent r.v.'s $X_1, \ldots, X_n$ have p.d.f.
$f(x; \theta) = \frac{1}{2\theta}e^{-|x|/\theta}$, $x \in \mathfrak{R}$, $\theta \in \Omega = (0, \infty)$, then show that:

  **(i)** The independent r.v.'s $Y_i = |X_i|$, $i = 1, \ldots, n$ have the Negative Exponential p.d.f. $g(y; \theta) = \frac{1}{\theta}e^{-y/\theta}$, $y > 0$.

  **(ii)** The independent r.v.'s $T_i(\theta) = \frac{2Y_i}{\theta}$, $i = 1, \ldots, n$ are $\chi_2^2$-distributed, so that the r.v. $T(\theta) = \sum_{i=1}^{n} T_i(\theta) = \frac{2}{\theta}\sum_{i=1}^{n} Y_i = \frac{2Y}{\theta} \sim \chi_{2n}^2$, where $Y = \sum_{i=1}^{n} Y_i = \sum_{i=1}^{n} |X_i|$.

  **(iii)** A $100(1 - \alpha)\%$ confidence interval for $\theta$, based on $T(\theta)$, is given by $[\frac{2Y}{b}, \frac{2Y}{a}]$, for suitable $0 < a < b$. In particular, $a$ and $b$ may be chosen to be $a = \chi_{2n;1-\frac{\alpha}{2}}^2$, $b = \chi_{2n;\frac{\alpha}{2}}^2$.

**1.11** Consider the p.d.f. $f(x; \alpha, \beta) = \frac{1}{\beta}e^{-(x-\alpha)/\beta}$, $x \geq \alpha$, $\alpha \in \mathfrak{R}$, $\beta > 0$ (see Exercise 1.11 in Chapter 9), and suppose that $\beta$ is known and $\alpha$ is unknown, and denote it by $\theta$. Thus, we have here:

$$f(x; \theta) = \frac{1}{\beta}e^{-(x-\theta)/\beta}, \quad x \geq \theta, \quad \theta \in \Omega = \mathfrak{R}.$$

  **(i)** Show that the corresponding d.f., $F(\cdot; \theta)$, is given by:
  $F(x; \theta) = 1 - e^{-(x-\theta)/\beta}$, $x \geq \theta$, so that
  $1 - F(x; \theta) = e^{-(x-\theta)/\beta}$, $x \geq \theta$.

  **(ii)** Let $X_1, \ldots, X_n$ be independent r.v.'s drawn from the p.d.f. $f(\cdot; \theta)$, and let $Y_1$ be the smallest order statistic. Use relation (28) in Chapter 6 in order to show that the p.d.f. of $Y_1$ is given by:

$$f_{Y_1}(y; \theta) = \frac{n}{\beta}e^{-n(y-\theta)/\beta}, \quad y \geq \theta.$$

  **(iii)** Consider the r.v. $T = T_n(\theta)$ defined by: $T = n(Y_1 - \theta)/\beta$, and show that its p.d.f. is given by: $f_T(t) = e^{-t}$, $t \geq 0$.

**1.12** In reference to Exercise 1.11:

  **(i)** Determine $0 \leq a < b$, so that $P(a \leq T \leq b) = 1 - \alpha$, for some $0 < \alpha < 1$.

  **(ii)** By part (i), $P_\theta[a \leq \frac{n(Y_1-\theta)}{\beta} \leq b] = 1 - \alpha$, since $T$ has the p.d.f. $f_T(t) = e^{-t}$, $t \geq 0$. Use this relation to conclude that $[Y_1 - \frac{b\beta}{n}, Y_1 - \frac{a\beta}{n}]$ is a $100(1 - \alpha)\%$ confidence interval of $\theta$.

  **(iii)** The length $l$ of the confidence interval in part (ii) is $l = \frac{\beta}{n}(b - a)$. Set $b = b(a)$ and show that the shortest confidence interval is given by: $[Y_1 + \frac{\beta \log \alpha}{n}, Y_1]$.

  **Hint.** For part (iii), set $b = b(a)$, and from $e^{-a} - e^{-b} = 1 - \alpha$, obtain $\frac{db}{da} = e^{b-a}$ by differentiation. Then replace $\frac{db}{da}$ in $\frac{dl}{da}$ and observe that it is $> 0$. This implies that $l$ obtains its minimum at $a = 0$.

**1.13** Let $X_1, \ldots, X_n$ be independent r.v.'s with d.f. $F$ and p.d.f. $f$ with $f(x) > 0$ for $-\infty \leq a < x < b \leq \infty$, and let $Y_1$ and $Y_n$ be, respectively, the smallest and the largest order statistics of the $X_i$'s.

(i) By using the hint given below, show that the joint p.d.f., $f_{Y_1, Y_n}$, of $Y_1$ and $Y_n$ is given by:

$$f_{Y_1, Y_n}(y_1, y_n) = n(n-1)[F(y_n) - F(y_1)]^{n-2} f(y_1) f(y_n),$$
$$a < y_1 < y_n < b.$$

**Hint.** $P(Y_n \leq y_n) = P(Y_1 \leq y_1, Y_n \leq y_n) + P(Y_1 > y_1, Y_n \leq y_n) = F_{Y_1, Y_n}(y_1, y_n) + P(y_1 < Y_1 < Y_n \leq y_n)$. But $P(Y_n \leq y_n) = P(\text{all } X_i\text{'s} \leq y_n) = P(X_1 \leq y_n, \ldots, X_n \leq y_n) = P(X_1 \leq y_n) \ldots P(X_n \leq y_n) = [F(y_n)]^n$, and:
$P(y_1 < Y_1 < Y_n \leq y_n) = P(\text{all } X_i\text{'s are} > y_1 \text{ and also} \leq y_n)$
$= P(y_1 < X_1 \leq y_n, \ldots, y_1 < X_n \leq y_n) = P(y_1 < X_1 \leq y_n) \ldots P(y_1 < X_n \leq y_n) = [P(y_1 < X_1 \leq y_n)]^n = [F(y_n) - F(y_1)]^n$. Thus,

$$[F(y_n)]^n = F_{Y_1, Y_n}(y_1, y_n) + [F(y_n) - F(y_1)]^n, \quad a < y_1 < y_n < b.$$

Solving for $F_{Y_1, Y_n}(y_1, y_n)$ and taking the partial derivatives with respect to $y_1$ and $y_n$, we get the desired result.

(ii) Find the p.d.f. $f_{Y_1, Y_n}$ when the $X_i$'s are distributed as $U(0, \theta)$, $\theta \in \Omega = (0, \infty)$.

(iii) Do the same for the case the $X_i$'s have the Negative Exponential p.d.f. $f(x; \theta) = \frac{1}{\theta} e^{-x/\theta}$, $x > 0$, $\theta \in \Omega = (0, \infty)$.

**1.14** Refer to Exercise 1.13(ii), and show that the p.d.f. of the *range* $R = Y_n - Y_1$ is given by:

$$f_R(r; \theta) = \frac{n(n-1)}{\theta^n} r^{n-2}(\theta - r), \quad 0 < r < \theta.$$

**1.15** Refer to Exercise 1.13(iii), and show that the p.d.f. of the range $R = Y_n - Y_1$ is given by:

$$f_R(r; \theta) = \frac{n-1}{\theta} e^{-\frac{r}{\theta}} (1 - e^{-\frac{r}{\theta}})^{n-2}, \quad r > 0.$$

**1.16** In reference to Exercise 1.14:

(i) Set $T = \frac{R}{\theta}$ and show that $f_T(t) = n(n-1)t^{n-2}(1-t)$, $0 < t < 1$.

(ii) Take $0 < c < 1$ such that $P_\theta(c \leq T \leq 1) = 1 - \alpha$, and construct a confidence interval for $\theta$, based on the range $R$, with confidence coefficient $1 - \alpha$. Also, show that $c$ is a root of the equation $c^{n-1} \times [n - (n-1)c] = \alpha$.

**1.17** Consider the independent random samples $X_1, \ldots, X_m$ from the $N(\mu_1, \sigma_1^2)$ distribution and $Y_1, \ldots, Y_n$ from the $N(\mu_2, \sigma_2^2)$ distribution, where $\mu_1, \mu_2$ are unknown and $\sigma_1^2, \sigma_2^2$ are known, and define the r.v. $T = T_{m,n}(\mu_1 - \mu_2)$ by:
$T_{m,n}(\mu_1 - \mu_2) = \frac{(\bar{X}_m - \bar{Y}_n) - (\mu_1 - \mu_2)}{\sqrt{(\sigma_1^2/m) + (\sigma_2^2/n)}}$.
Then show that:

(i)  A $100(1 - \alpha)\%$ confidence interval for $\mu_1 - \mu_2$, based on $T$, is given by:

$$[(\bar{X}_m - \bar{Y}_n) - b\sqrt{\tfrac{\sigma_1^2}{m} + \tfrac{\sigma_2^2}{n}}, \ (\bar{X}_m - \bar{Y}_n) - a\sqrt{\tfrac{\sigma_1^2}{m} + \tfrac{\sigma_2^2}{n}}]$$ for suitable

constants $a$ and $b$.

(ii)  The confidence interval in part (i) with the shortest length is taken for $b = z_{\alpha/2}$ and $a = -z_{\alpha/2}$.

**1.18**  Refer to Exercise 1.17, and suppose that $\mu_1, \mu_2$ are known and $\sigma_1^2, \sigma_2^2$ are unknown. Then define the r.v. $\bar{T} = \bar{T}_{m,n}(\sigma_1^2/\sigma_2^2) = \frac{\sigma_2^2}{\sigma_1^2} \times \frac{S_Y^2}{S_X^2}$, where $S_X^2 = \frac{1}{m}\sum_{i=1}^m (X_i - \mu_1)^2$ and $S_Y^2 = \frac{1}{n}\sum_{j=1}^n (Y_j - \mu_2)^2$, and show that a $100(1 - \alpha)\%$ confidence interval for $\sigma_1^2/\sigma_2^2$, based on $\bar{T}$, is given by $[a\frac{S_X^2}{S_Y^2}, \ b\frac{S_X^2}{S_Y^2}]$ for $0 < a < b$ with $P(a \le X \le b) = 1 - \alpha, X \sim F_{n,m}$. In particular, we may choose $a = F_{n,m;1-\frac{\alpha}{2}}$ and $b = F_{n,m;\frac{\alpha}{2}}$.

**1.19**  Consider the independent random samples $X_1, \ldots, X_m$ and $Y_1, \ldots, Y_n$ from the Negative Exponential distributions
$$f(x; \theta_1) = \tfrac{1}{\theta_1}e^{-x/\theta_1}, \ x > 0, \ \theta_1 \in \Omega = (0, \infty), \text{ and } f(y; \theta_2) = \tfrac{1}{\theta_2}e^{-y/\theta_2},$$
$y > 0, \ \theta_2 \in \Omega = (0, \infty)$, and set $U = \sum_{i=1}^m X_i, V = \sum_{j=1}^n Y_j$. Then, by Exercise 1.6(ii), $\frac{2U}{\theta_1} \sim \chi_{2m}^2$, $\frac{2V}{\theta_2} \sim \chi_{2n}^2$ and they are independent. It follows that $\frac{\frac{2V}{\theta_2}/2n}{\frac{2U}{\theta_1}/2m} = \frac{\theta_1}{\theta_2} \times \frac{mV}{nU} = \frac{\theta_1}{\theta_2} \times \frac{m\sum_{j=1}^n Y_j}{n\sum_{i=1}^m X_i} \sim F_{2n,2m}$. Use this result in order to construct a $100(1 - \alpha)\%$ confidence interval for $\theta_1/\theta_2$.

**1.20**  In reference to Exercise 4.3 in Chapter 9, construct a $100(1 - \alpha)\%$ (Bayes) confidence interval for $\theta$; that is, determine a set

$$\{\theta \in (0, 1); \ h(\theta|x_1, \ldots, x_n) \ge c(x_1, \ldots, x_n)\},$$

where $c(x_1, \ldots, x_n)$ is determined by the requirement that the $P_\lambda$-probability of this set is equal to $1 - \alpha$. (Notice that this $\alpha$ is unrelated to the parameter $\alpha$ of the Beta distribution.)

**Hint.** Use the graph of the p.d.f. of a suitable Beta distribution.

**1.21**  In reference to Exercise 4.5 in Chapter 9, construct a $100(1 - \alpha)\%$) (Bayes) confidence interval for $\theta$ as in Exercise 1.20 here.

## 10.2 CONFIDENCE INTERVALS IN THE PRESENCE OF NUISANCE PARAMETERS

In Example 1, the position was adopted that only one of the parameters in the $N(\mu, \sigma^2)$ distribution was unknown. This is a rather artificial assumption as, in practice, both $\mu$ and $\sigma^2$ are most often unknown. What was done in that example did, however, pave the way to solving the problem here in its natural setting.

**Example 3.** Let $X_1, \ldots, X_n$ be a random sample from the $N(\mu, \sigma^2)$ distribution, where both $\mu$ and $\sigma^2$ are unknown. Construct confidence intervals for $\mu$ and $\sigma^2$, each with confidence coefficient $1 - \alpha$.

**Discussion.** We have that:

$$\frac{\sqrt{n}(\bar{X} - \mu)}{\sigma} \sim N(0, 1) \quad \text{and} \quad \frac{(n-1)S^2}{\sigma^2} = \sum_{i=1}^{n} \left( \frac{X_i - \bar{X}}{\sigma} \right)^2 \sim \chi_{n-1}^2,$$

where $S^2 = \frac{1}{n-1} \sum_{i=1}^{n} (X_i - \bar{X})^2$, and these two r.v.'s are independent. It follows that
$\frac{\sqrt{n}(\bar{X}-\mu)/\sigma}{\sqrt{(n-1)S^2/\sigma^2(n-1)}} = \frac{\sqrt{n}(\bar{X}-\mu)}{S} \sim t_{n-1}$. From the $t$-tables, determine any pair $(a, b)$
with $a < b$ such that $P(a \leq X \leq b) = 1 - \alpha$, where $X \sim t_{n-1}$. It follows that:

$$P_\theta \left[ a \leq \frac{\sqrt{n}(\bar{X} - \mu)}{S} \leq b \right] = 1 - \alpha, \quad \text{for all } \theta = (\mu, \sigma^2),$$

or

$$P_\theta \left( \bar{X} - b\frac{S}{\sqrt{n}} \leq \mu \leq \bar{X} - a\frac{S}{\sqrt{n}} \right) = 1 - \alpha, \quad \text{for all } \theta.$$

In particular,

$$P_\theta \left( \bar{X} - t_{n-1;\alpha/2}\frac{S}{\sqrt{n}} \leq \mu \leq \bar{X} + t_{n-1;\alpha/2}\frac{S}{\sqrt{n}} \right) = 1 - \alpha, \quad \text{for all } \theta,$$

where $P(X \geq t_{n-1;\alpha/2}) = \frac{\alpha}{2}$ (and $X \sim t_{n-1}$). It follows that the required confidence interval for $\mu$ is:

$$\left[ \bar{X} - t_{n-1;\alpha/2}\frac{S}{\sqrt{n}}, \ \bar{X} + t_{n-1;\alpha/2}\frac{S}{\sqrt{n}} \right] = \bar{X} \pm t_{n-1;\alpha/2}\frac{S}{\sqrt{n}} \quad \text{(for brevity).} \tag{4}$$

The construction of a confidence interval for $\sigma^2$ in the presence of (an unknown) $\mu$ is easier. We have already mentioned that $\frac{(n-1)S^2}{\sigma^2} \sim \chi_{n-1}^2$. Then repeat the process in Example 1(ii), replacing $\chi_n^2$ by $\chi_{n-1}^2$, to obtain the confidence interval.

$$\left[ \frac{(n-1)S^2}{\chi_{n-1;\alpha/2}^2}, \ \frac{(n-1)S^2}{\chi_{n-1;1-\alpha/2}^2} \right], \quad S^2 = \frac{1}{n-1} \sum_{i=1}^{n} (X_i - \bar{X})^2. \tag{5}$$

**Remark 4.** Observe that the confidence interval in (4) differs from that in (2) in that $\sigma$ in (2) is replaced by an estimate $S$, and then the constant $z_{\alpha/2}$ in (2) is adjusted to $t_{n-1;\alpha/2}$. Likewise, the confidence intervals in (3) and (5) are of the same form, with the only difference that (the unknown) $\mu$ in (3) is replaced by its estimate $\bar{X}$ in (5). The constants $n$, $\chi_{n;\alpha/2}^2$, and $\chi_{n;1-\alpha/2}^2$ are also adjusted as indicated in (5).

*Numerical Example.* Let $n = 25$ and $1 - \alpha = 0.95$. Then $t_{n-1;\alpha/2} = t_{24;0.025} = 2.0639$, and the interval in (4) becomes $\bar{X} \pm 0.41278S$. Also, $\chi_{n-1;\alpha/2}^2 = \chi_{24;0.025}^2 = 39.364$, $\chi_{n-1;1-\alpha/2}^2 = \chi_{24;0.975}^2 = 12.401$, so that the interval in (5) is $[\frac{24S^2}{39.364}, \frac{24S^2}{12.401}] \simeq [0.610S^2, 1.935S^2]$.

Actually, a somewhat more important problem from a practical viewpoint is that of constructing confidence intervals for the difference of the means of two Normal populations and the ratio of their variances. This is a way of comparing two Normal populations. The precise formulation of the problem is given below.

**Example 4.** Let $X_1, \ldots, X_m$ and $Y_1, \ldots, Y_n$ be two independent random samples from the $N(\mu_1, \sigma_1^2)$ and $N(\mu_2, \sigma_2^2)$ distributions, respectively, with all $\mu_1, \mu_2, \sigma_1^2$, and $\sigma_2^2$ unknown. We wish to construct confidence intervals for $\mu_1 - \mu_2$ and $\sigma_1^2/\sigma_2^2$.

**Discussion.**

(i) *Confidence interval for* $\mu_1 - \mu_2$. In order to be able to resolve this problem, we have to *assume* that the variances, although unknown, are equal; i.e., $\sigma_1^2 = \sigma_2^2 = \sigma^2$, say.

Let us review briefly some distribution results. Recall that $\bar{X} - \mu_1 \sim N(0, \frac{\sigma^2}{m})$, $\bar{Y} - \mu_2 \sim N(0, \frac{\sigma^2}{n})$, and by independence,

$$[(\bar{X} - \bar{Y}) - (\mu_1 - \mu_2)] / \left(\sigma\sqrt{\frac{1}{m} + \frac{1}{n}}\right) \sim N(0, 1). \tag{6}$$

Also, if

$$S_X^2 = \frac{1}{m-1}\sum_{i=1}^{m}(X_i - \bar{X})^2, \quad S_Y^2 = \frac{1}{n-1}\sum_{j=1}^{n}(Y_j - \bar{Y})^2,$$

then $\frac{(m-1)S_X^2}{\sigma^2} \sim \chi_{m-1}^2$, $\frac{(n-1)S_Y^2}{\sigma^2} \sim \chi_{n-1}^2$, and by independence,

$$\frac{(m-1)S_X^2 + (n-1)S_Y^2}{\sigma^2} \sim \chi_{m+n-2}^2. \tag{7}$$

From (6) and (7), we obtain then:

$$\frac{(\bar{X} - \bar{Y}) - (\mu_1 - \mu_2)}{\sqrt{\frac{(m-1)S_X^2 + (n-1)S_Y^2}{m+n-2}\left(\frac{1}{m} + \frac{1}{n}\right)}} \sim t_{m+n-2}. \tag{8}$$

Then working with (8) as in Example 1(i), we arrive at the following confidence interval

$$\left[(\bar{X} - \bar{Y}) - t_{m+n-2;\alpha/2}\sqrt{\frac{(m-1)S_X^2 + (n-1)S_Y^2}{m+n-2}\left(\frac{1}{m} + \frac{1}{n}\right)},\right.$$

$$\left.(\bar{X} - \bar{Y}) + t_{m+n-2;\alpha/2}\sqrt{\frac{(m-1)S_X^2 + (n-1)S_Y^2}{m+n-2}\left(\frac{1}{m} + \frac{1}{n}\right)}\right]$$

$$= (\bar{X} - \bar{Y}) \pm t_{m+n-2;\alpha/2}\sqrt{\frac{(m-1)S_X^2 + (n-1)S_Y^2}{m+n-2}\left(\frac{1}{m} + \frac{1}{n}\right)}. \tag{9}$$

(ii) *Confidence interval for $\sigma_1^2/\sigma_2^2$.* By the fact that $\frac{(m-1)S_X^2}{\sigma_1^2} \sim \chi_{m-1}^2$, $\frac{(n-1)S_Y^2}{\sigma_2^2} \sim$
$\chi_{n-1}^2$, and independence, we have $\frac{S_Y^2/\sigma_2^2}{S_X^2/\sigma_1^2} = \frac{\sigma_1^2}{\sigma_2^2} \times \frac{S_Y^2}{S_X^2} \sim F_{n-1,m-1}$. From the $F$-
tables, determine any pair $(a, b)$ with $0 < a < b$ such that $P(a \leq X \leq b) =$
$1 - \alpha$, where $X \sim F_{n-1,m-1}$. Then, for all $\theta = (\mu_1, \mu_2, \sigma_1^2, \sigma_2^2)$,

$$P_\theta\left(a \leq \frac{\sigma_1^2}{\sigma_2^2} \times \frac{S_Y^2}{S_X^2} \leq b\right) = 1 - \alpha, \quad \text{or} \quad P_\theta\left(a\frac{S_X^2}{S_Y^2} \leq \frac{\sigma_1^2}{\sigma_2^2} \leq b\frac{S_X^2}{S_Y^2}\right) = 1 - \alpha.$$

In particular, for all $\theta$,

$$P_\theta\left(\frac{S_X^2}{S_Y^2}F_{n-1,m-1;1-\alpha/2} \leq \frac{\sigma_1^2}{\sigma_2^2} \leq \frac{S_X^2}{S_Y^2}F_{n-1,m-1;\alpha/2}\right) = 1 - \alpha,$$

where $P(X \leq F_{n-1,m-1;1-\alpha/2}) = P(X \geq F_{n-1,m-1;\alpha/2}) = \frac{\alpha}{2}$ (and $X \sim F_{n-1,m-1}$).
The required confidence interval is then

$$\left[\frac{S_X^2}{S_Y^2}F_{n-1,m-1;1-\alpha/2}, \quad \frac{S_X^2}{S_Y^2}F_{n-1,m-1;\alpha/2}\right]. \tag{10}$$

*Numerical Example.* Let $m = 13, n = 14$, and $1 - \alpha = 0.95$. Then $t_{m+n-2;\alpha/2} =$
$t_{25;0.025} = 2.0595$, so that the interval in (9) becomes

$$(\bar{X} - \bar{Y}) \pm 2.0595\sqrt{\frac{12S_X^2 + 13S_Y^2}{25}\left(\frac{1}{13} + \frac{1}{14}\right)} \simeq (\bar{X} - \bar{Y}) \pm 0.1586\sqrt{12S_X^2 + 13S_Y^2}.$$

Next, $F_{n-1,m-1;\alpha/2} = F_{13,12;0.025} = 3.2388$, $F_{n-1,m-1;1-\alpha/2} = F_{13,12;0.975} =$
$\frac{1}{F_{12,13;0.025}} = \frac{1}{3.1532} \simeq 0.3171$. Therefore the interval in (10) is $[0.3171\frac{S_X^2}{S_Y^2}, 3.2388\frac{S_X^2}{S_Y^2}]$.

---

## EXERCISES

**2.1** If the independent r.v.'s $X_1, \ldots, X_n$ are $N(\mu, \sigma^2)$ distributed with both $\mu$ and $\sigma^2$ unknown, construct a $100(1 - \alpha)\%$ confidence interval for $\sigma$.

**2.2** Refer to Exercise 1.18 and suppose that all $\mu_1, \mu_2$, and $\sigma_1^2, \sigma_2^2$ are unknown. Then construct a $100(1 - \alpha)\%$ confidence interval for $\sigma_1/\sigma_2$.

---

## 10.3 A CONFIDENCE REGION FOR $(\mu, \sigma^2)$ IN THE $N(\mu, \sigma^2)$ DISTRIBUTION

Refer again to Example 1 and suppose that both $\mu$ and $\sigma^2$ are unknown, as is most often the case. In this section, we wish to construct a *confidence region* for the pair $(\mu, \sigma^2)$; i.e., a subset of the plane determined in terms of statistics and containing $(\mu, \sigma^2)$ with probability $1 - \alpha$. This problem is resolved in the following example.

**Example 5.** On the basis of the random sample $X_1, \ldots, X_n$ from the $N(\mu, \sigma^2)$ distribution, construct a confidence region for the pair $(\mu, \sigma^2)$ with confidence coefficient $1 - \alpha$.

**Discussion.** In solving this problem, we draw heavily on what we have done in the previous example. Let $\bar{X}$ be the sample mean and define $S^2$ by $S^2 = \frac{1}{n-1} \sum_{i=1}^n (X_i - \bar{X})^2$. Then

$$\frac{\sqrt{n}(\bar{X} - \mu)}{\sigma} \sim N(0, 1), \qquad \frac{(n-1)S^2}{\sigma^2} \sim \chi^2_{n-1}, \tag{11}$$

and the two r.v.'s involved here are independent. From the Normal tables, define $c > 0$ so that $P(-c \le Z \le c) = \sqrt{1 - \alpha}$, $Z \sim N(0, 1)$; $c$ is uniquely determined. From the $\chi^2$-tables, determine a pair $(a, b)$ with $0 < a < b$ and $P(a \le X \le b) = \sqrt{1 - \alpha}$, where $X \sim \chi^2_{n-1}$. Then, by means of (11), and with $\theta = (\mu, \sigma^2)$, we have:

$$P_\theta \left[ -c \le \frac{\sqrt{n}(\bar{X} - \mu)}{\sigma} \le c \right] = \sqrt{1 - \alpha},$$

$$\tag{12}$$

$$P_\theta \left[ a \le \frac{(n-1)S^2}{\sigma^2} \le b \right] = \sqrt{1 - \alpha}.$$

These relations are rewritten thus:

$$P_\theta \left[ -c \le \frac{\sqrt{n}(\bar{X} - \mu)}{\sigma} \le c \right] = P_\theta \left[ (\mu - \bar{X})^2 \le \frac{c^2 \sigma^2}{n} \right] (= \sqrt{1 - \alpha}), \tag{13}$$

$$P_\theta \left[ a \le \frac{(n-1)S^2}{\sigma^2} \le b \right] = P_\theta \left[ \frac{(n-1)S^2}{b} \le \sigma^2 \le \frac{(n-1)S^2}{a} \right] (= \sqrt{1 - \alpha}), \tag{14}$$

so that, by means of (12)–(14) and independence, we have:

$$P_\theta \left[ -c \le \frac{\sqrt{n}(\bar{X} - \mu)}{\sigma} \le c, \ a \le \frac{(n-1)S^2}{\sigma^2} \le b \right]$$

$$= P_\theta \left[ -c \le \frac{\sqrt{n}(\bar{X} - \mu)}{\sigma} \le c \right] P_\theta \left[ a \le \frac{(n-1)S^2}{\sigma^2} \le b \right]$$

$$= P_\theta \left[ (\mu - \bar{X})^2 \le \frac{c^2}{n} \sigma^2 \right] P_\theta \left[ \frac{(n-1)S^2}{b} \le \sigma^2 \le \frac{(n-1)S^2}{a} \right] = 1 - \alpha. \tag{15}$$

Let $\bar{x}$ and $s^2$ be the observed values of $\bar{X}$ and $S^2$. Then in a system of orthogonal $(\mu, \sigma^2)$-axis, the equation $(\mu - \bar{x})^2 = \frac{c^2}{n} \sigma^2$ is the equation of a parabola with vertex $V$ located at the point $(\bar{x}, 0)$, with focus $F$ with coordinates $(\bar{x}, \frac{c^2}{4n})$, and with directrix $L$ with equation $\sigma^2 = -\frac{c^2}{4n}$ (see Figure 10.1). Then the part of the plane for which $(\mu - \bar{x})^2 \le \frac{c^2 \sigma^2}{n}$ is the inner part of the parabola along with the points on the parabola. Since

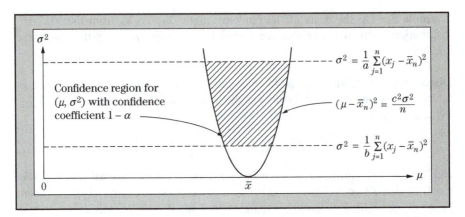

**FIGURE 10.1**

Confidence region for $(\mu, \sigma^2)$ with confidence coefficient $1 - \alpha$.

$$\sigma^2 = \frac{(n-1)s^2}{b} = \frac{1}{b}\sum_{i=1}^{n}(x_i - \bar{x})^2 \quad \text{and} \quad \sigma^2 = \frac{(n-1)s^2}{a} = \frac{1}{a}\sum_{i=1}^{n}(x_i - \bar{x})^2$$

are straight lines parallel to the $\mu$-axis, the set of points $(\mu, \sigma^2)$ in the plane, which satisfy simultaneously all inequalities:

$$(\mu - \bar{x})^2 \leq \frac{c^2\sigma^2}{n}, \quad \frac{1}{b}\sum_{i=1}^{n}(x_i - \bar{x})^2 \leq \sigma^2 \leq \frac{1}{a}\sum_{i=1}^{n}(x_i - \bar{x})^2$$

is the part of the plane between the straight lines mentioned above and the inner part of the parabola (along with the points on the parabola) (see shaded area in Figure 10.1).

From relation (15), it follows then that, when replacing $\bar{x}$ by $\bar{X}$ and $s^2$ by $S^2$, the shaded region with random boundary (determined completely as described above) becomes the required confidence region for $(\mu, \sigma^2)$. What is depicted in Figure 10.1 is a realization of such a confidence region, evaluated for the observed values of the $X_i$'s.

Actually, the point $c$ above is $z_\gamma$, where $\gamma = (1 - \sqrt{1 - \alpha})/2$, and for definiteness, we may choose to split the probability $1 - \sqrt{1 - \alpha}$ equally among the two tails of the Chi-square distribution. Thus, we take $b = \chi^2_{n-1;\gamma}$ and $a = \chi^2_{n-1;1-\gamma}$. Then the confidence region is:

$$(\mu - \bar{X})^2 \leq \frac{z_\gamma^2}{n}\sigma^2, \quad \frac{1}{\chi^2_{n-1;\gamma}}\sum_{i=1}^{n}(X_i - \bar{X})^2 \leq \sigma^2 \leq \frac{1}{\chi^2_{n-1;1-\gamma}}\sum_{i=1}^{n}(X_i - \bar{X})^2,$$

$$\gamma = (1 - \sqrt{1 - \alpha})/2. \tag{16}$$

*Numerical Example.* As a numerical example, take $n = 25$ and $\alpha = 0.05$, so that $\gamma \simeq 0.012661$, and (by linear interpolation) $z_\gamma \simeq 2.236$, $\chi^2_{24;\gamma} \simeq 42.338$, $\chi^2_{24;1-\gamma} \simeq 11.130$, and the confidence region becomes:

$$(\mu - \bar{X})^2 \leq 0.199988\sigma^2, \quad 0.023619 \sum_{i=1}^{25}(X_i - \bar{X})^2 \leq \sigma^2 \leq 0.089847 \sum_{i=1}^{25}(X_i - \bar{X})^2$$

or, approximately,

$$(\mu - \bar{X})^2 \leq 0.2\sigma^2, \quad 0.024 \sum_{i=1}^{25}(X_i - \bar{X})^2 \leq \sigma^2 \leq 0.09 \sum_{i=1}^{25}(X_i - \bar{X})^2.$$

**Remark 5.** A somewhat general theory for constructing confidence regions is discussed in Chapter 12 (see Theorem 4 there and the examples following it).

## 10.4 CONFIDENCE INTERVALS WITH APPROXIMATE CONFIDENCE COEFFICIENT

It is somewhat conspicuous that in this chapter we have not yet dealt with examples, such as the Binomial, the Poisson, and the Negative Exponential. There is a reason, however, behind it, and that is that the expressions which would serve as the basis for constructing confidence intervals do not have a known exact distribution. They do have, however, an approximate Normal distribution, and this fact leads to the construction of confidence intervals with confidence coefficient *approximately* (rather than exact) $1 - \alpha$. The remainder of this section is devoted to constructing such intervals.

**Example 6.** On the basis of the random sample $X_1, \ldots, X_n$ from the $B(1, \theta)$ distribution, construct a confidence interval for $\theta$ with confidence coefficient approximately $1 - \alpha$.

**Discussion.** The tools employed here, as well as in the following two examples, are the CLT and the WLLN in conjunction with either Theorem 7(ii) or Theorem 6(iii) in Chapter 7. It will be assumed throughout that $n$ is large enough, so that these theorems apply.

Recall that $E_\theta X_1 = \theta$ and $\sigma^2_\theta(X_1) = \theta(1 - \theta)$, so that, by the CLT,

$$\frac{\sqrt{n}(\bar{X}_n - \theta)}{\sqrt{\theta(1 - \theta)}} \simeq N(0, 1). \tag{17}$$

In the denominator in (17), replace $\theta(1 - \theta)$ by $S_n^2$, where $S_n^2 = \frac{1}{n}\sum_{i=1}^{n}(X_i - \bar{X}_n)^2 = \frac{1}{n}(\sum_{i=1}^{n}X_i^2 - n\bar{X}_n^2) = \frac{1}{n}(\sum_{i=1}^{n}X_i - n\bar{X}_n^2) = \bar{X} - \bar{X}^2 = \bar{X}(1 - \bar{X})$, in order to obtain (by Theorem 7(ii) in Chapter 7),

$$\frac{\sqrt{n}(\bar{X}_n - \theta)}{\sqrt{\bar{X}_n(1 - \bar{X}_n)}} \simeq N(0, 1). \tag{18}$$

It follows from (18) that

$$P_\theta\left[-z_{\alpha/2} \le \frac{\sqrt{n}(\bar{X}_n - \theta)}{\sqrt{\bar{X}_n(1 - \bar{X}_n)}} \le z_{\alpha/2}\right] \simeq 1 - \alpha, \quad \text{for all } \theta.$$

This expression is equivalent to:

$$P_\theta\left[\bar{X}_n - z_{\alpha/2}\sqrt{\frac{\bar{X}_n(1 - \bar{X}_n)}{n}} \le \theta \le \bar{X}_n + z_{\alpha/2}\sqrt{\frac{\bar{X}_n(1 - \bar{X}_n)}{n}}\right] \simeq 1 - \alpha, \quad \text{for all } \theta,$$

which leads to the confidence interval

$$\left[\bar{X}_n - z_{\alpha/2}\sqrt{\frac{\bar{X}_n(1 - \bar{X}_n)}{n}}, \ \bar{X}_n + z_{\alpha/2}\sqrt{\frac{\bar{X}_n(1 - \bar{X}_n)}{n}}\right] = \bar{X}_n \pm z_{\alpha/2}\sqrt{\frac{\bar{X}_n(1 - \bar{X}_n)}{n}} \tag{19}$$

with confidence coefficient approximately $1 - \alpha$.

*Numerical Example.* For $n = 100$ and $1 - \alpha = 0.95$, the confidence interval in (19) becomes: $\bar{X}_n \pm 1.96\sqrt{\frac{\bar{X}_n(1-\bar{X}_n)}{100}} = \bar{X}_n \pm 0.196\sqrt{\bar{X}_n(1 - \bar{X}_n)}$.

**Example 7.** Construct a confidence interval for $\theta$ with confidence coefficient approximately $1 - \alpha$ on the basis of the random sample $X_1, \ldots, X_n$ from the $P(\theta)$ distribution.

**Discussion.** Here $E_\theta X_1 = \sigma_\theta^2(X_1) = \theta$, so that, working as in the previous example, and employing Theorem 6(iii) in Chapter 7, we have

$$\frac{\sqrt{n}(\bar{X}_n - \theta)}{\sqrt{\theta}} \simeq N(0, 1), \quad \text{or} \quad \frac{\sqrt{n}(\bar{X}_n - \theta)}{\sqrt{\bar{X}_n}} \simeq N(0, 1).$$

Hence $P_\theta[-z_{\alpha/2} \le \frac{\sqrt{n}(\bar{X}_n-\theta)}{\sqrt{\bar{X}_n}} \le z_{\alpha/2}] \simeq 1 - \alpha$, for all $\theta$, which leads to the required confidence interval

$$\left[\bar{X}_n - z_{\alpha/2}\sqrt{\frac{\bar{X}_n}{n}}, \ \bar{X}_n + z_{\alpha/2}\sqrt{\frac{\bar{X}_n}{n}}\right] = \bar{X}_n \pm z_{\alpha/2}\sqrt{\frac{\bar{X}_n}{n}}. \tag{20}$$

*Numerical Example.* For $n = 100$ and $1 - \alpha = 0.95$, the confidence interval in (20) becomes: $\bar{X}_n \pm 0.196\sqrt{\bar{X}_n}$.

**Example 8.** Let $X_1, \ldots, X_n$ be a random sample from the Negative Exponential distribution in the following parameterization: $f(x; \theta) = \frac{1}{\theta}e^{-x/\theta}$, $x > 0$. Construct a confidence interval for $\theta$ with confidence coefficient approximately $1 - \alpha$.

**Discussion.**    In the adopted parameterization above, $E_\theta X_1 = \theta$ and $\sigma_\theta^2(X_1) = \theta^2$. Then working as in the previous example, we have that

$$\frac{\sqrt{n}(\bar{X}_n - \theta)}{\theta} \simeq N(0, 1), \quad \text{or} \quad \frac{\sqrt{n}(\bar{X}_n - \theta)}{\bar{X}_n} \simeq N(0, 1).$$

It follows that the required confidence interval is given by:

$$\left[ \bar{X}_n - z_{\alpha/2}\frac{\bar{X}_n}{\sqrt{n}}, \ \bar{X}_n + z_{\alpha/2}\frac{\bar{X}_n}{\sqrt{n}} \right] = \bar{X}_n \pm z_{\alpha/2}\frac{\bar{X}_n}{\sqrt{n}}. \tag{21}$$

*Numerical Example.*    For $n = 100$ and $1 - \alpha = 0.95$, the confidence interval in (21) becomes: $\bar{X}_n \pm 0.196\bar{X}_n$.

## EXERCISES

**4.1**    Let the independent r.v.'s $X_1, \ldots, X_n$ have unknown (finite) mean $\mu$ and known (finite) variance $\sigma^2$, and suppose that $n$ is large. Then:

(i)   Use the CLT in order to construct a confidence interval for $\mu$ with confidence coefficient approximately $1 - \alpha$.

(ii)  Provide the form of the interval in part (i) for $n = 100$, $\sigma = 1$, and $\alpha = 0.05$.

(iii) Refer to part (i) and suppose that $\sigma = 1$ and $\alpha = 0.05$. Then determine the sample size $n$, so that the length of the confidence interval is 0.1.

(iv)  Observe that the length of the confidence interval in part (i) tends to 0 as $n \to \infty$, for any $\sigma$ and any $\alpha$.

**4.2**    Refer to Exercise 4.1, and suppose that both $\mu$ and $\sigma^2$ are unknown. Then:

(i)   Construct a confidence interval for $\mu$ with confidence coefficient approximately $1 - \alpha$.

(ii)  Provide the form of the interval in part (i) for $n = 100$ and $\alpha = 0.05$.

(iii) Show that the length of the interval in part (i) tends to 0 in probability as $n \to \infty$.

**Hint.** For part (i), refer to Theorem 7(ii) in Chapter 7, and for part (iii), refer to Theorem 7(i) and Theorem 6(ii) in the same chapter.

**4.3**    (i)   Let $X \sim N(\mu, \sigma^2)$, and for $0 < \alpha < 1$, let $x_\alpha$ and $x_{1-\alpha}$ be the $\alpha$th and $(1 - \alpha)$th quantiles, respectively, of $X$; i.e., $P(X \le x_\alpha) = P(X \ge x_{1-\alpha}) = \alpha$, so that $P(x_\alpha \le X \le x_{1-\alpha}) = 2\alpha$. Show that $x_\alpha = \mu + \sigma\Phi^{-1}(\alpha)$, $x_{1-\alpha} = \mu + \sigma\Phi^{-1}(1 - \alpha)$, so that $[x_\alpha, x_{1-\alpha}] = [\mu + \sigma\Phi^{-1}(\alpha), \mu + \sigma\Phi^{-1}(1 - \alpha)]$.

(ii)  Refer to Exercise 4.5(i) of Chapter 9 (see also Exercise 4.4 there), where it is found that the posterior p.d.f. of $\theta$, given $X_1 = x_1, \ldots, X_n = x_n$, $h(\cdot \mid x_1, \ldots, x_n)$, is $N(\frac{n\bar{x}+\mu}{n+1}, \frac{1}{n+1})$. Use part (i) in order to find the expression of the interval $[x_\alpha, x_{1-\alpha}]$ here.

**Remark.** In the present context, the interval $[x_\alpha, x_{1-\alpha}]$ is called *a prediction interval* for $\theta$ with confidence coefficient $1 - 2\alpha$.

(iii) Compute the prediction interval in part (ii) when
$$n = 9, \ \mu = 1, \ \bar{x} = 1.5, \ \text{and} \ \alpha = 0.025.$$

**4.4** Let $X_1, \ldots, X_n$ be independent r.v.'s with strictly increasing d.f. $F$, and let $Y_i$ be the $i$th order statistic of the $X_i$'s, $1 \leq i \leq n$. For $0 < p < 1$, let $x_p$ be the (unique) $p$th quantile of $F$. Then:

(i) Show that for any $i$ and $j$ with $1 \leq i < j \leq n - 1$,

$$P(Y_i \leq x_p \leq Y_j) = \sum_{k=i}^{j-1} \binom{n}{k} p^k q^{n-k} \quad (q = 1 - p).$$

Thus, $[Y_i, Y_j]$ is a confidence interval for $x_p$ with confidence coefficient $\sum_{k=i}^{j-1} \binom{n}{k} p^k q^{n-k}$. This probability is often referred to as *probability of coverage* of $x_p$.

(ii) For $n = 10$ and $p = 0.25$, identify the respective coverage probabilities for the pairs $(Y_1, Y_3)$, $(Y_1, Y_4)$, $(Y_2, Y_4)$, $(Y_2, Y_5)$.

(iii) For $p = 0.50$, do the same as in part (ii) for the pairs $(Y_3, Y_9)$, $(Y_4, Y_7)$, $(Y_4, Y_8)$, $(Y_5, Y_7)$.

(iv) For $p = 0.75$, do the same as in part (ii) for the pairs $(Y_8, Y_{10})$, $(Y_7, Y_{10})$, $(Y_7, Y_9)$, $(Y_6, Y_9)$.

**Hint.** For part (i), observe that: $P(Y_i \leq x_p) = P(\text{at least } i \text{ of } X_1, \ldots, X_n \leq x_p) = \sum_{k=i}^{n} \binom{n}{k} p^k q^{n-k}$, since $P(X_k \leq x_p) = p$ and $q = 1 - p$. Also, $P(Y_i \leq x_p) = P(Y_i \leq x_p, Y_j \geq x_p) + P(Y_i \leq x_p, Y_j < x_p) = P(Y_i \leq x_p \leq Y_j) + P(Y_j < x_p)$, so that $P(Y_i \leq x_p \leq Y_j) = P(Y_i \leq x_p) - P(Y_j \leq x_p)$. For part (iv), observe that $\binom{n}{k} p^k q^{n-k} = \binom{n}{n-r} q^r p^{n-r} = \binom{n}{r} q^r p^{n-r}$ (by setting $n - k = r$ and recalling that $\binom{n}{n-r} = \binom{n}{r}$).

**4.5** Let $X$ be a r.v. with a strictly increasing d.f. $F$ (Figure 10.2), and let $p$ be a number with $0 < p < 1$. Consider the event: $A_p = \{F(X) \leq p\} = \{s \in \mathcal{S};$

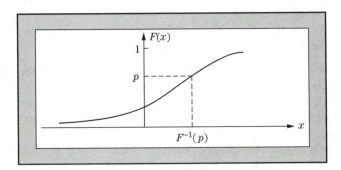

**FIGURE 10.2**

Graph of the d.f. employed in Exercise 4.5.

$F(X(s)) \le p\} = \{s \in \mathcal{S}; X(s) \le F^{-1}(p)\}$. So, $A_p$ is the event in the underlying sample space $\mathcal{S}$ for the sample points $s$ of which $F(X(s)) \le p$. Since for each fixed $x$, $F(x)$ represents the proportion of the (unit) distribution mass of $F$ which is covered (or carried) by the interval $(-\infty, x]$, it follows that the random interval $(-\infty, X]$ covers (carries) the (random) proportion $F(X)$ of the distribution mass of $F$, and on account of the event $A_p$, the random interval $(-\infty, X]$ covers (carries) at most $100p\%$ of the mass of $F$. Equivalently, the random interval $(X, \infty)$ covers (carries) at least $100(1 - p)\%$ of the distribution mass of $F$. Use Theorem 10 in Chapter 6 in order to show that $P(A_p) = p$; i.e., $(-\infty, X]$ covers at most $100p\%$ of the distribution mass of $F$ with probability $p$. Equivalently, the random interval $(X, \infty)$ covers at least $100(1 - p)\%$ of the distribution mass of $F$ with probability $p$.

# Testing hypotheses

In this chapter, the problem of testing hypotheses is considered to some extent. Additional topics are discussed in Chapter 12. The chapter consists of four sections, the first of which is devoted to some general concepts and the formulation of a null hypothesis and its alternative. A number of examples discussed provide sufficient motivation for what is done in this section.

Section 2 is somewhat long and enters into the essence of the testing hypotheses issue. Specifically, the Neyman–Pearson Fundamental Lemma is stated, and the main points of its proof are presented for the case that the underlying r.v.'s are of the continuous type. It is stated that this result by itself is of limited use; nevertheless, it does serve as the stepping stone in establishing other more complicated and truly useful results. This is achieved when the underlying family of distributions is the so-called family of distributions of the Exponential type. Thus, the definition of an Exponential type p.d.f. follows, and it is next illustrated by means of examples that such families occur fairly often. In an Exponential type p.d.f. (in the real-valued parameter $\theta$), Uniformly most powerful (UMP) tests are presented for one-sided and two-sided hypotheses, which arise in practice in a natural way. This is done in Theorems 2 and 3.

In the following section, Theorems 2 and 3 are applied to concrete cases, such as the Binomial distribution, the Poisson distribution, and the Normal distribution. All applications are accompanied by numerical examples.

The last section of this chapter, Section 4, is also rather extensive and deals with likelihood ratio (LR) tests. General concepts, the necessary notation, and some motivation for the tests used are given. The better part of the section is devoted to deriving LR tests in Normal distributions. The problem is divided into two parts. The first part considers the case where we are dealing with one sample from an underlying Normal distribution, and LR tests are derived for the mean and the variance of the distribution. In the second part, two independent random samples are available coming from two underlying Normal populations. Then LR tests are derived in comparing the means and the variances of the distributions. In all cases, the results produced are illustrated by means of numerical examples.

## 11.1 GENERAL CONCEPTS, FORMULATION OF SOME TESTING HYPOTHESES

In order to motivate the formulation of a null hypothesis and its alternative, consider some specific examples. Most of them are taken from Chapter 1.

**Example 1.** In reference to Example 6 in Chapter 1, let $\theta$ be the unknown proportion of unemployed workers, and let $\theta_0$ be an acceptable level of unemployment; e.g., $\theta_0 = 6.25\%$. Then the parameter space is split into the sets $(0.0625, 1)$ and $(0, 0.0625]$, and one of them will be associated with the (null) hypothesis. It is proposed that that set be $(0.0625, 1)$; i.e., $H_0: \theta > 0.0625$ (and therefore $H_A: \theta \leq 0.0625$). The rule of thumb for selecting $H_0$ is this: "Select as null hypothesis that hypothesis whose false rejection has the most serious consequences." Indeed, if $\theta$ is, actually, greater than 6.25% and is (falsely) rejected, then human suffering may occur, due to the fact that the authorities in charge had no incentives to take the necessary measures. On the other hand, if $\theta \leq 0.0625$ was selected as the null hypothesis and was falsely rejected, then the most likely consequence would be for the authorities to undertake some unnecessary measures and, perhaps, waste some money. However, the former consequence is definitely more serious than the latter. Another way of looking at the problem of determining the null hypothesis is to formulate as such a position, which we wish to challenge, and which we are willing to accept only on the face of convincing evidence, provided by the interested party. To summarize then, if $X$ is the r.v. denoting the number of unemployed workers among $n$ sampled, then $X \sim B(n, \theta)$ and the hypothesis to be tested is $H_0: \theta > 0.0625$ against the alternative $H_A: \theta \leq 0.0625$ at (some given) level of significance $\alpha$.

**Example 2.** In reference to Example 8 in Chapter 1, if $X$ is the r.v. denoting those young adults, among the $n$ sampled, who listen to this particular weekend music program, then $X \sim B(n, \theta)$. Then, arguing as in the previous example, we have that the hypothesis to be tested is $H_0: \theta \leq \theta_0 \ (= 100p\%)$ against the alternative $H_A: \theta > \theta_0$ at level of significance $\alpha$.

**Example 3.** Refer to Example 12 of Chapter 1, and let $X$ be the r.v. denoting the mean bacteria count per unit volume of water at a lake beach. Then $X \sim P(\theta)$ and the hypothesis to be tested is $H_0: \theta > 200$ against $H_A: \theta \leq 200$ at level of significance $\alpha$.

**Example 4.** Suppose that the mean $\theta$ of a r.v. $X$ represents the dosage of a drug which is used for the treatment of a certain disease. For this medication to be both safe and effective, $\theta$ must satisfy the requirements $\theta_1 < \theta < \theta_2$, for two specified values $\theta_1$ and $\theta_2$. Then, on the basis of previous discussions, the hypothesis to be tested here is $H_0: \theta \leq \theta_1$ or $\theta \geq \theta_2$ against the alternative $H_A: \theta_1 < \theta < \theta_2$ at the level of significance $\alpha$. Of course, we have to assume a certain distribution for the r.v. $X$, which for good reasons is taken to be $N(\theta, \sigma^2)$, $\sigma$ known.

**Example 5.** Refer to Example 16 in Chapter 1, and suppose that the survival time for a terminal cancer patient treated with the standard treatment is a r.v. $X \sim N(\theta_1, \sigma_1^2)$. Likewise, let the r.v. $Y$ stand for the survival time for such a patient subject to the new treatment, and let $Y \sim N(\theta_2, \sigma_2^2)$. Then the hypothesis to be tested here is $H_0: \theta_2 = \theta_1$ against the alternative $H_A: \theta_2 > \theta_1$ at level of significance $\alpha$; $\sigma_1$ and $\sigma_2$ are assumed to be known.

**Remark 1.** The hypothesis to be tested could also be $\theta_2 \leq \theta_1$, but the possibility that $\theta_2 < \theta_1$ may be excluded; it can be assumed that the new treatment cannot be inferior to the existing one. The supposition that $\theta_2 = \theta_1$, there is no difference between the two treatments, leads to the term "null" for the hypothesis $H_0: \theta_2 = \theta_1$.

Examples 1–4 have the following common characteristics: A r.v. $X$ is distributed according to the p.d.f. $f(\cdot; \theta), \theta \in \Omega \subseteq \Re$, and we are interested in testing one of the following hypotheses, each one at some specified level of significance $\alpha$: $H_0: \theta > \theta_0$ against $H_A: \theta \leq \theta_0$; $H_0: \theta \leq \theta_0$ against $H_A: \theta > \theta_0$; $H_0: \theta \leq \theta_1$ or $\theta \geq \theta_2$ against $H_A: \theta_1 < \theta < \theta_2$. It is understood that in all cases $\theta$ remains in $\Omega$. In Example 5, two Normally distributed populations are compared in terms of their means, and the hypothesis tested is $H_0: \theta_2 = \theta_1$ against $H_A: \theta_2 > \theta_1$. An example of a different nature would lead to testing the hypothesis $H_0: \theta_2 < \theta_1$ against $H_A: \theta_2 = \theta_1$.

In the first four examples, the hypotheses stated are to be tested by means of a random sample $X_1, \ldots, X_n$ from the underlying distribution. In the case of Example 5, the hypothesis is to be tested by utilizing two independent random samples $X_1, \ldots, X_m$ and $Y_1, \ldots, Y_n$ from the underlying distributions.

Observe that in all cases the hypotheses tested are composite, and so are the alternatives. We wish, of course, for the proposed tests to be optimal in some satisfactory sense. If the tests were to be Uniformly Most Powerful (UMP), then they would certainly be highly desirable. In the following section, a somewhat general theory will be provided, which, when applied to the examples under consideration, will produce UMP tests.

## EXERCISES

**1.1** In the following examples, indicate which statements constitute a simple and which a composite hypothesis:

(i) $X$ is a r.v. whose p.d.f. $f$ is given by $f(x) = 2e^{-2x}$, $x > 0$.

(ii) When tossing a coin, let $X$ be the r.v. taking the value 1 if the head appears and 0 if the tail appears. Then the statement is: The coin is biased.

(iii) $X$ is a r.v. whose expectation is equal to 5.

**1.2** Let $X_1, \ldots, X_n$ be i.i.d. r.v.'s with p.d.f. $f$ which may be either Normal, $N(\mu, \sigma^2)$, to be denoted by $f_N$, or Cauchy with parameters $\mu$ and $\sigma^2$, to be denoted by $f_C$, where, we recall that:

$$f_N(x; \mu, \sigma^2) = \frac{1}{\sqrt{2\pi}\sigma} e^{-(x-\mu)^2/2\sigma^2}, \quad x \in \Re, \ \mu \in \Re, \ \sigma > 0,$$

$$f_C(x; \mu, \sigma^2) = \frac{\sigma}{\pi} \times \frac{1}{(x-\mu)^2 + \sigma^2}, \quad x \in \Re, \ \mu \in \Re, \ \sigma > 0.$$

Consider the following null hypotheses and the corresponding alternatives:

  **(i)**   $H_{01}$: $f$ is Normal, $H_{A1}$: $f$ is Cauchy.
 **(ii)**   $H_{02}$: $f$ is Normal with $\mu \le \mu_0$, $H_{A2}$: $f$ is Cauchy with $\mu \le \mu_0$.
**(iii)**   $H_{03}$: $f$ is Normal with $\mu = \mu_0$, $H_{A3}$: $f$ is Cauchy with $\mu = \mu_0$.
 **(iv)**   $H_{04}$: $f$ is Normal with $\mu = \mu_0, \sigma \ge \sigma_0$, $H_{A4}$: $f$ is Cauchy with $\mu = \mu_0, \sigma \ge \sigma_0$.
  **(v)**   $H_{05}$: $f$ is Normal with $\mu = \mu_0, \sigma > \sigma_0$, $H_{A5}$: $f$ is Cauchy with $\mu = \mu_0, \sigma > \sigma_0$.
 **(vi)**   $H_{06}$: $f$ is Normal with $\mu = \mu_0, \sigma = \sigma_0$, $H_{A6}$: $f$ is Cauchy with $\mu = \mu_0, \sigma = \sigma_0$.

State which of the $H_{0i}$, $i = 1, \ldots, 6$ are simple and which are composite, and likewise for $H_{Ai}$, $i = 1, \ldots, 6$.

## 11.2 NEYMAN–PEARSON FUNDAMENTAL LEMMA, EXPONENTIAL TYPE FAMILIES, UMP TESTS FOR SOME COMPOSITE HYPOTHESES

In reference to Example 1, one could certainly consider testing the simple hypothesis $H_0$: $\theta = \theta_0$ (e.g., 0.05) against the simple alternative $H_A$: $\theta = \theta_1$, for some fixed $\theta_1$ either $> \theta_0$ or $< \theta_0$. However, such a testing framework would be highly unrealistic. It is simply not reasonable to isolate two single values from the continuum of values (0, 1) and test one against the other. What is meaningful is the way we actually formulated $H_0$ in this example. Nevertheless, it is still true that a long journey begins with the first step, and this applies here as well. Accordingly, we are going to start out with the problem of testing a simple hypothesis against a simple alternative, which is what the celebrated Neyman–Pearson Fundamental Lemma is all about.

**Theorem 1** (Neyman–Pearson Fundamental Lemma).   *Let $X_1, \ldots, X_n$ be a random sample with p.d.f. $f$ unknown. We are interested in testing the simple hypothesis $H_0$: $f = f_0$ (specified) against the simple alternative $H_A$: $f = f_1$ (specified) at level of significance $\alpha$ $(0 < \alpha < 1)$, on the basis of the observed values $x_i$ of the $X_i$, $i = 1, \ldots, n$. To this end, define the test $\varphi$ as follows:*

$$\varphi(x_1, \ldots, x_n) = \begin{cases} 1 & \text{if} \quad f_1(x_1) \ldots f_1(x_n) > C f_0(x_1) \ldots f_0(x_n) \\ \gamma & \text{if} \quad f_1(x_1) \ldots f_1(x_n) = C f_0(x_1) \ldots f_0(x_n) \\ 0 & \text{if} \quad f_1(x_1) \ldots f_1(x_n) < C f_0(x_1) \ldots f_0(x_n), \end{cases} \tag{1}$$

*where the constants $C$ and $\gamma$ $(C > 0, 0 \le \gamma \le 1)$ are defined through the relationship:*

$$E_{f_0}\varphi(X_1,\ldots,X_n) = P_{f_0}[f_1(X_1)\cdots f_1(X_n) > Cf_0(X_1)\cdots f_0(X_n)]$$
$$+\gamma P_{f_0}[f_1(X_1)\cdots f_1(X_n) = Cf_0(X_1)\cdots f_0(X_n)] = \alpha. \qquad (2)$$

*Then the test $\varphi$ is MP among all tests with level of significance $\leq \alpha$.*

**Remark 2.** The test $\varphi$ is a *randomized* test, if $0 < \gamma < 1$. The necessity for a randomized test stems from relation (2), where the left-hand side has to be equal to $\alpha$. If the $X_i$'s are discrete, the presence of $\gamma$ $(0 < \gamma < 1)$ is indispensable. In case, however, the $X_i$'s are of the continuous type, then $\gamma = 0$ and the test is *nonrandomized*.

    The appearance of $f_0$ as a subscript indicates, of course, that expectations and probabilities are calculated by using the p.d.f. $f_0$ for the $X_i$'s.

*Proof of Theorem 1 (Outline, for $X_i$'s of the Continuous Type).* To simplify the notation, write 0 (or 1) rather than $f_0$ (or $f_1$) when $f_0$ (or $f_1$) occurs as a subscript. Also, it would be convenient to use the vector notation $X = (X_1,\ldots,X_n)$ and $x = (x_1,\ldots,x_n)$. First, we show that the test $\varphi$ is of level $\alpha$. Indeed, let $T = \{x \in \mathfrak{R}^n; L_0(x) > 0\}$, where $L_0(x) = f_0(x_1)\ldots f_0(x_n)$, and likewise $L_1(x) = f_1(x_1)\ldots f_1(x_n)$. Then, if $D = X^{-1}(T)$; i.e., $D = \{s \in S; X(s) \in T\}$, so that $D^c = \{s \in S; X(s) \in T^c\}$, it follows that $P_0(D^c) = P_0(X \in T^c) = \int_{T^c} L_0(x)\, dx = 0$. Therefore, in calculating probabilities by using the p.d.f. $L_0$, it suffices to restrict ourselves to the set $D$. Then, by means of (2),

$$\begin{aligned}
E_0\varphi(X) &= P_0[L_1(X) > CL_0(X)] \\
&= P_0\{[L_1(X) > CL_0(X)] \cap D\} \\
&= P_0\left\{\left[\frac{L_1(X)}{L_0(X)} > C\right] \cap D\right\} \quad \text{(since } L_0(X) > 0 \text{ on } D) \\
&= P_0(Y > C) = 1 - P_0(Y \leq C) = g(C), \text{ say,}
\end{aligned}$$

where $Y = \frac{L_1(X)}{L_0(X)}$ on $D$, and arbitrary on $D^c$. The picture of $1 - P_0(Y \leq C)$ is depicted in Figure 11.1, and it follows that, for each $\alpha$ $(0 < \alpha < 1)$, there is (essentially) a

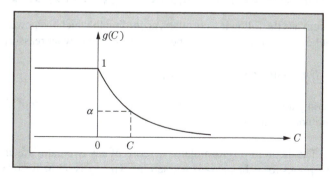

**FIGURE 11.1**

The graph of the function of $(C) = 1 - P_0(\gamma \leq C)$.

unique $C$ such that $1 - P_0(Y \le C) = \alpha$. That is, $E_\theta \varphi(X) = \alpha$, which shows that the test $\varphi$ is of level $\alpha$.

Next, it is shown that $\varphi$ is MP as described by showing that, if $\varphi^*$ is any other test with $E_0 \varphi^*(X) = \alpha^* \le \alpha$, then $\pi_\varphi(1) = E_1 \varphi(X) \ge E_1 \varphi^*(X) = \pi_{\varphi^*}(1)$ (i.e., the power of $\varphi$ is not smaller than the power of any other test $\varphi^*$ of level of significance $\le \alpha$). Indeed, define $B^+$ and $B^-$ by:

$$B^+ = \{x \in \Re^n; \varphi(x) - \varphi^*(x) > 0\} = (\varphi > \varphi^*),$$
$$B^- = \{x \in \Re^n; \varphi(x) - \varphi^*(x) < 0\} = (\varphi < \varphi^*).$$

Then, clearly, $B^+ \cap B^- = \varnothing$, and, by means of (1),

$$B^+ = (\varphi > \varphi^*) \subseteq (\varphi = 1) = (L_1 > CL_0),$$
$$B^- = (\varphi < \varphi^*) \subseteq (\varphi = 0) = (L_1 < CL_0). \tag{3}$$

Therefore

$$\int_{\Re^n} [\varphi(x) - \varphi^*(x)][L_1(x) - CL_0(x)]dx$$

$$= \int_{B^+} [\varphi(x) - \varphi^*(x)][L_1(x) - CL_0(x)]dx$$

$$+ \int_{B^-} [\varphi(x) - \varphi^*(x)][L_1(x) - CL_0(x)]dx \ge 0 \text{ by (3)}.$$

Hence

$$\int_{\Re^n} \varphi(x)L_1(x)dx - \int_{\Re^n} \varphi^*(x)L_1(x)dx$$

$$\ge C\left[\int_{\Re^n} \varphi(x)L_0(x)dx - \int_{\Re^n} \varphi^*(x)L_0(x)dx\right]$$

$$= C(\alpha - \alpha^*) \ge 0 \quad (\text{since } \alpha^* \le \alpha).$$

Hence $\int_{\Re^n} \varphi(x)L_1(x)dx = E_1 \varphi(X) \ge E_1 \varphi^*(X) = \int_{\Re^n} \varphi^*(x)L_1(x)\,dx.$  ∎

This theorem has the following corollary, according to which the power of the MP test $\varphi$ cannot be $< \alpha$; not very much to be sure, but yet somewhat reassuring.

**Corollary.**   *For the MP test* $\varphi$, $\pi_\varphi(1) = E_1 \varphi(x) \ge \alpha$.

*Proof.* Just compare the power of $\varphi$ with that of $\varphi^* \equiv \alpha$ whose level of significance and power are both equal to $\alpha$.  ∎

**Remark 3.**   The theorem was formulated in terms of any two p.d.f.'s $f_0$ and $f_1$ as the two possible options for $f$. In a parametric setting, where $f$ is of the form $f(\cdot; \theta)$, $\theta \in \Omega \subseteq \Re^r$, $r \ge 1$, the p.d.f.'s $f_0$ and $f_1$ will correspond to two specified values of $\theta$; $\theta_0$ and $\theta_1$, say. That is, $f_0 = f(\cdot; \theta_0)$ and $f_1 = f(\cdot; \theta_1)$.

The following examples will help illustrate how the Neyman–Pearson Fundamental Lemma actually applies in concrete cases.

**Example 6.** On the basis of a random sample of size 1 from the p.d.f. $f(x; \theta) = \theta x^{\theta-1}$, $0 < x < 1$ $(\theta > 1)$:

(i) Use the Neyman–Pearson Fundamental Lemma to derive the MP test for testing the hypothesis $H_0$: $\theta = \theta_0$ against the alternative $H_A$: $\theta = \theta_1$ at level of significance $\alpha$.
(ii) Derive the formula for the power $\pi(\theta_1)$.
(iii) Give numerical values for parts (i) and (ii) when $\theta_0 = 4$ and $\theta_1 = 6, \theta_1 = 2$; take $\alpha = 0.05$.

**Discussion.** In the first place, the given function is, indeed, a p.d.f., since $\int_0^1 \theta x^{\theta-1} dx = x^\theta|_0^1 = 1$. Next:

(i) $H_0$ is rejected, if for some positive constant $C^*$:

$$\frac{\theta_1 x^{\theta_1-1}}{\theta_0 x^{\theta_0-1}} > C^*, \quad \text{or} \quad x^{\theta_1-\theta_0} > \frac{\theta_0 C^*}{\theta_1}, \quad \text{or} \quad (\theta_1 - \theta_0)\log x > \log\left(\frac{\theta_0 C^*}{\theta_1}\right).$$

Now, if $\theta_1 > \theta_0$, this last inequality is equivalent to:

$$\log x > \log\left(\frac{\theta_0 C^*}{\theta_1}\right)^{1/(\theta_1-\theta_0)} = \log C \quad \left(C = \left(\frac{\theta_0 C^*}{\theta_1}\right)^{1/(\theta_1-\theta_0)}\right),$$

or $x > C$. If $\theta_1 < \theta_0$, the final form of the inequality becomes $x < C$. For $\theta_1 > \theta_0$, the cutoff point is calculated by:

$$P_{\theta_0}(X > C) = \int_C^1 \theta_0 x^{\theta_0-1} dx = x^{\theta_0}|_C^1 = 1 - C^{\theta_0} = \alpha, \quad \text{or} \quad C = (1-\alpha)^{1/\theta_0}.$$

For $\theta_1 < \theta_0$, we have:

$$P_{\theta_0}(X < C) = x^{\theta_0}|_0^C = C^{\theta_0} = \alpha, \quad \text{or} \quad C = \alpha^{1/\theta_0}.$$

Then, for $\theta_1 > \theta_0$, reject $H_0$ when $x > (1-\alpha)^{1/\theta_0}$; and for $\theta_1 < \theta_0$, reject $H_0$ when $x < \alpha^{1/\theta_0}$.

(ii) For $\theta_1 > \theta_0$, the power of the test is given by:

$$\pi(\theta_1) = P_{\theta_1}(X > C) = \int_C^1 \theta_1 x^{\theta_1-1} dx = x^{\theta_1}|_C^1 = 1 - C^{\theta_1}, \quad \text{or}$$

$\pi(\theta_1) = 1 - (1-\alpha)^{\theta_1/\theta_0}$. For $\theta_1 < \theta_0$, we have:

$$\pi(\theta_1) = P_{\theta_1}(X < C) = \int_0^C \theta_1 x^{\theta_1-1} dx = x^{\theta_1}|_0^C = C^{\theta_1} = \alpha^{\theta_1/\theta_0}.$$

That is,

$$\pi(\theta_1) = 1 - (1-\alpha)^{\theta_1/\theta_0} \quad \text{for} \quad \theta_1 > \theta_0; \pi(\theta_1) = \alpha^{\theta_1/\theta_0} \quad \text{for } \theta_1 < \theta_0.$$

**(iii)** For $\theta_1 = 6$, the cutoff point is:

$$(1 - 0.05)^{1/4} = 0.95^{0.25} \simeq 0.987,$$

and the power is: $\pi(6) = 1 - (0.95)^{1.5} \simeq 1 - 0.926 = 0.074$. For $\theta_1 = 2$, the cutoff point is: $(0.05)^{1/4} = (0.05)^{0.25} \simeq 0.473$, and the power is: $\pi(2) = (0.05)^{1/2} \simeq 0.224$.

**Example 7.**   Refer to Example 6 and:

**(i)** Show that the Neyman–Pearson test which rejects the (simple) hypothesis $H_0$: $\theta = \theta_0$ when tested against the (simple) alternative $H_{A,\theta_1}: \theta = \theta_1$, for some fixed $\theta_1 > \theta_0$, at level of significance $\alpha$, is, actually, UMP for testing $H_0$ against the composite alternative $H_A: \theta > \theta_0$ at level of significance $\alpha$.

**(ii)** Also, show that the Neyman–Pearson test which rejects the hypothesis $H_0: \theta = \theta_0$ when tested against the (simple) alternative $H'_{A,\theta_1}: \theta = \theta_1$, for some fixed $\theta_1 < \theta_0$, at level of significance $\alpha$, is, actually, UMP for testing $H_0$ against the composite alternative $H'_A: \theta < \theta_0$ at level of significance $\alpha$.

**(iii)** Show that there is no UMP test for testing the hypothesis $H_0: \theta = \theta_0$ against the (double-sided) composite alternative $H''_A: \theta \neq \theta_0$ at level of significance $\alpha$.

**Discussion.**

**(i)** Indeed, by part (i) of Example 6, the MP test for testing $H_0: \theta = \theta_0$ against $H_{A,\theta_1}: \theta = \theta_1$ rejects $H_0$ when $x > (1 - \alpha)^{1/\theta_0}$, regardless of the *specific* value of $\theta_1$, provided $\theta_1 > \theta_0$. Thus, this test becomes a UMP test when $H_{A,\theta_1}$ is replaced by $H_A: \theta > \theta_0$.

**(ii)** Likewise, by Example 6(i), the MP test for testing $H_0: \theta = \theta_0$ against $H'_{A,\theta_1}: \theta = \theta_1$ rejects $H_0$ when $x < \alpha^{1/\theta_0}$, regardless of the *specific* value $\theta_1$, provided $\theta_1 < \theta_0$. Thus, this test becomes a UMP test when $H'_{A,\theta_1}$ is replaced by $H'_A: \theta < \theta_0$.

**(iii)** The rejection region for testing the hypotheses $H_0: \theta = \theta_0$ against the alternative $H_A: \theta > \theta_0$ is $R_1 = ((1 - \alpha)^{1/\theta_0}, 1)$, and the rejection region for testing $H_0$ against $H'_A: \theta < \theta_0$ is $R_2 = (0, \alpha^{1/\theta_0})$. Since these MP regions depend on which side of $\theta_0$ lie the alternative $\theta$'s and are different, there cannot exist a UMP test for testing $H_0$ against $H''_A: \theta \neq \theta_0$.

**Example 8.**   On the basis of a random sample of size 1 from the p.d.f. $f(x; \theta) = 1 + \theta^2(\frac{1}{2} - x)$, $0 < x < 1$, $-1 \leq \theta \leq 1$:

**(i)** Use the Neyman–Pearson Fundamental Lemma to derive the MP test for testing the hypothesis $H_0: \theta = 0$ (i.e., the p.d.f. is $U(0, 1)$) against the alternative $H_A: \theta = \theta_1$ at level of significance $\alpha$.

**(ii)** Investigate whether or not the test derived in part (i) is a UMP test for testing $H_0: \theta = 0$ against the alternative $H'_A: \theta \neq 0$.

**(iii)** Determine the test in part (i) for $\alpha = 0.05$.

**(iv)** Determine the power of the test in part (i).

**Discussion.** First, the function given is a p.d.f., because it is nonnegative and $\int_0^1 [1 + \theta^2(\frac{1}{2} - x)]dx = 1 + \theta^2(\frac{1}{2} - \frac{1}{2}) = 1$. Next:

(i)  $H_0$ is rejected whenever $1 + \theta_1^2(\frac{1}{2} - x) > C^*$, or $x < C$, where $C = \frac{1}{2} - (C^* - 1)/\theta_1^2$, and $C$ is determined by $P_0(X < C) = \alpha$, so that $C = \alpha$, since $X \sim U(0, 1)$ under $H_0$. Thus, $H_0$ is rejected when $x < \alpha$.

(ii)  Observe that the test is independent of $\theta_1$, and since it is MP against each fixed $\theta_1$, it follows that it is UMP for testing $H_0$ against $H_A' : \theta \neq 0$.

(iii)  For $\alpha = 0.05$, the test in part (i) rejects $H_0$ whenever $x < 0.05$.

(iv)  For $\theta \neq 0$, the power of the test is:

$$\pi(\theta) = P_\theta(X < \alpha) = \int_0^\alpha \left[1 + \theta^2 \left(\frac{1}{2} - x\right)\right] dx = \frac{1}{2}\alpha(1 - \alpha)\theta^2 + \alpha.$$

Thus, e.g., $\pi(\pm 1) = \frac{1}{2}\alpha(1 - \alpha) + \alpha$, $\pi(\pm\frac{1}{2}) = \frac{1}{8}\alpha(1 - \alpha) + \alpha$, which for $\alpha = 0.05$ become: $\pi(\pm 1) \simeq 0.074$, $\pi(\pm\frac{1}{2}) \simeq 0.056$.

---

## EXERCISES

Exercises 2.1 through 2.8 in this section are meant as application of Theorem 1.

**2.1**  If $X_1, \ldots, X_{16}$ are independent r.v.'s:

    (i)  Construct the MP test of the hypothesis $H_0$: the common distribution of the $X_i$'s is $N(0, 9)$ against the alternative $H_A$: the common distribution of the $X_i$'s is $N(1, 9)$; take $\alpha = 0.05$.

    (ii)  Also, determine the power of the test.

**2.2**  Let $X_1, \ldots, X_n$ be independent r.v.'s distributed as $N(\mu, \sigma^2)$, where $\mu$ is unknown and $\sigma$ is known.

    (i)  For testing the hypothesis $H_0$: $\mu = 0$ against the alternative $H_A$: $\mu = 1$, show that the sample size $n$ can be determined to achieve a given level of significance $\alpha$ and given power $\pi(1)$.

    (ii)  What is the numerical value of $n$ for $\alpha = 0.05$, $\pi(1) = 0.9$ when $\sigma = 1$?

**2.3**  (i)  Let $X_1, \ldots, X_n$ be independent r.v.'s distributed as $N(\mu, \sigma^2)$, where $\mu$ is unknown and $\sigma$ is known. Derive the MP test for testing the hypothesis $H_0$: $\mu = \mu_1$ against the alternative $H_A$: $\mu = \mu_2$ ($\mu_2 > \mu_1$) at level of significance $\alpha$.

    (ii)  Find an expression for computing the power of the test.

    (iii)  Carry out the testing hypothesis and compute the power for $n = 100$, $\sigma^2 = 4$, $\mu_1 = 3$, $\mu_2 = 3.5$, $\bar{x} = 3.2$, and $\alpha = 0.01$.

**2.4**  Let $X_1, \ldots, X_n$ be independent r.v.'s distributed as $N(\mu, \sigma^2)$ with $\mu$ unknown and $\sigma$ known. Suppose we wish to test the hypothesis $H_0$: $\mu = \mu_0$ against the alternative $H_A$: $\mu = \mu_1$ ($\mu_1 > \mu_0$).

    (i)  Derive the MP test for testing $H_0$ against $H_A$.

(ii) For a given level of significance $\alpha_n (< 0.5)$ and given power $\pi_n (> 0.5)$, determine the cutoff point $C_n$ and the sample size for which both $\alpha_n$ and $\pi_n$ are attained.

(iii) Show that $\alpha_n \to 0$ and $\pi_n \to 1$ as $n \to \infty$.

(iv) Determine the sample size $n$ and the cutoff point $C_n$ for $\mu_0 = 0, \mu_1 = 1, \sigma = 1, \alpha_n = 0.001$, and $\pi_n = 0.995$.

**2.5** Let $X_1, \ldots, X_n$ be independent r.v.'s having the Gamma distribution with $\alpha$ known and $\beta$ unknown.

(i) Construct the MP test for testing the hypothesis $H_0: \beta = \beta_1$ against the alternative $H_A: \beta = \beta_2$ $(\beta_2 > \beta_1)$ at level of significance $\alpha$.

(ii) By using the m.g.f. approach, show that, if $X \sim$ Gamma $(\alpha, \beta)$, then $X_1 + \cdots + X_n \sim$ Gamma $(n\alpha, \beta)$, where the $X_i$'s are independent and distributed as $X$.

(iii) Use the CLT to carry out the test when:
$n = 30, \alpha = 10, \beta_1 = 2.5, \beta_2 = 3$, and $\alpha = 0.05$.

(iv) Compute the power of the test, also by using the CLT.

**2.6** The life of an electronic equipment is a r.v. $X$ whose p.d.f. is
$f(x; \theta) = \theta e^{-\theta x}$, $x > 0$, $\theta \in \Omega = (0, \infty)$, and let $\ell$ be its expected lifetime. On the basis of the random sample $X_1, \ldots, X_n$ from this distribution:

(i) Derive the MP test for testing the hypothesis $H_0: \ell = \ell_0$ against the alternative $H_A: \ell = \ell_1$ $(\ell_1 > \ell_0)$ at level of significance $\alpha$, and write down the expression giving the power of the test.

(ii) Use the m.g.f. approach in order to show that the r.v. $Y = 2\theta (\sum_{i=1}^{n} X_i)$ is distributed as $\chi_{2n}^2$.

(iii) Use part (ii) in order to relate the cutoff point and the power of the test to $\chi^2$-percentiles.

(iv) Employ the CLT (assuming that $n$ is sufficiently large) in order to find (approximate) values for the cutoff point and the power of the test.

(v) Use parts (iii) and (iv) in order to carry out the test and also calculate the power when $n = 22, \ell_0 = 10, \ell_1 = 12.5$, and $\alpha = 0.05$.

**2.7** Let $X$ be a r.v. whose p.d.f. $f$ is either the $U(0, 1)$, to be denoted by $f_0$, or the Triangular over the interval $[0, 1]$, to be denoted by $f_1$ (that is, $f_1(x) = 4x$ for $0 \leq x < \frac{1}{2}$; $f_1(x) = 4 - 4x$ for $\frac{1}{2} \leq x \leq 1$, and 0 otherwise).

(i) Test the hypothesis $H_0: f = f_0$ against the alternative $H_A: f = f_1$ at level of significance $\alpha = 0.05$.

(ii) Compute the power of the test.

(iii) Draw the picture of $f_1$ and compute the power by means of Geometric consideration.

**Hint.** Use $P_i$ to denote calculation of probabilities under $f_i$, $i = 0, 1$, and for part (i), write: $P_0[f_1(X) > C] = P_0[f_1(X) > C$ and $0 \leq X < \frac{1}{2}] + P_0[f_1(X) > C$ and $\frac{1}{2} \leq X \leq 1]$. Then use conditional probabilities to conclude that: $P_0(\frac{C}{4} < X < \frac{1}{2}) + P_0(\frac{1}{2} \leq X < 1 - \frac{C}{4}) = 1 - \frac{C}{2} = 0.05$, so that

$C = 1.9$. Finally, for an observed $x$, either $0 \le x < \frac{1}{2}$, in which case $4x > 1.9$, or $\frac{1}{2} \le x \le 1$, in which case $4 - 4x > 1.9$.

**2.8**   Let $X$ be a r.v. with p.d.f. $f$ which is either the $P(1)$ (Poisson with $\lambda = 1$), to be denoted by $f_0$, or the $f_1(x) = 1/2^{x+1}, x = 0, 1, \ldots$. For testing the hypothesis $H_0: f = f_0$ against the alternative $H_A: f = f_1$ on the basis of one observation $x$ on $X$:

   (i)  Show that the rejection region is defined by: $\{x \ge 0$ integer; $1.36 \times \frac{x!}{2^x} \ge C\}$ for some positive number $C$.
   (ii)  Determine the level of significance $\alpha$ of the test when $C = 2$.
   **Hint.** Observe that the function $g(x) = \frac{x!}{2^x}$ is nondecreasing for $x$ integer $\ge 1$.

## 11.2.1 EXPONENTIAL TYPE FAMILIES OF p.d.f.'s

The remarkable thing here is that, if the p.d.f. $f(\cdot; \theta)$ is of a certain general form to be discussed below, then the apparently simple-minded Theorem 1 leads to UMP tests; it is the stepping stone for getting to those tests.

**Definition 1.**   The p.d.f. $f(\cdot; \theta), \theta \in \Omega \subseteq \Re$, is said to be of the *Exponential type*, if

$$f(x; \theta) = C(\theta)e^{Q(\theta)T(x)} \times h(x), \quad x \in \Re, \qquad (4)$$

where $Q$ is strictly monotone and $h$ does not involve $\theta$ in any way; $C(\theta)$ is simply a normalizing constant.

Most of the p.d.f.'s we have encountered so far are of the form (4). Here are some examples.

**Example 9.**   The $B(n, \theta)$ p.d.f. is of the Exponential type.

**Discussion.**   Indeed,

$$f(x; \theta) = \binom{n}{x}\theta^x(1 - \theta)^{n-x}I_A(x), \quad A = \{0, 1, \ldots, n\},$$

where, we recall that $I_A$ is the indicator of $A$; i.e., $I_A(x) = 1$ if $x \in A$, and $I_A(x) = 0$ if $x \in A^c$.
   Hence

$$f(x; \theta) = (1 - \theta)^n \times e^{[\log(\frac{\theta}{1-\theta})]x} \times \binom{n}{x}I_A(x),$$

so that $f(x; \theta)$ is of the form (4) with $C(\theta) = (1 - \theta)^n, Q(\theta) = \log(\frac{\theta}{1-\theta})$ strictly increasing (since $\frac{d}{d\theta}(\frac{\theta}{1-\theta}) = \frac{1}{(1-\theta)^2} > 0$ and $\log(\cdot)$ is strictly increasing), $T(x) = x$, and $h(x) = \binom{n}{x}I_A(x)$.

**Example 10.**   The $P(\theta)$ p.d.f. is of the Exponential type.

**Discussion.** Here

$$f(x; \theta) = \frac{e^{-\theta}\theta^x}{x!}I_A(x), \quad A = \{0, 1, \ldots\}.$$

Hence

$$f(x; \theta) = e^{-\theta} \times e^{(\log \theta)x} \times \frac{1}{x!}I_A(x),$$

so that $f(x; \theta)$ is of the form (4) with $C(\theta) = e^{-\theta}$, $Q(\theta) = \log \theta$ strictly increasing, $T(x) = x$, and $h(x) = \frac{1}{x!}I_A(x)$.

**Example 11.** The $N(\theta, \sigma^2)$ ($\sigma$ known) p.d.f. is of the Exponential type.

**Discussion.** In fact,

$$f(x; \theta) = \frac{1}{\sqrt{2\pi}\sigma}e^{-\frac{(x-\theta)^2}{2\sigma^2}} = \frac{1}{\sqrt{2\pi}\sigma}e^{-\frac{\theta^2}{2\sigma^2}} \times e^{\frac{\theta}{\sigma^2}x} \times e^{-\frac{x^2}{2\sigma^2}},$$

and this is of the form (4) with $C(\theta) = \frac{1}{\sqrt{2\pi}\sigma}e^{-\frac{\theta^2}{2\sigma^2}}$, $Q(\theta) = \frac{\theta}{\sigma^2}$ strictly increasing, $T(x) = x$, and $h(x) = e^{-x^2/2\sigma^2}$.

**Example 12.** The $N(\mu, \theta)$ ($\mu$ known) p.d.f. is of the Exponential type.

**Discussion.** Here

$$f(x; \theta) = \frac{1}{\sqrt{2\pi\theta}}e^{-\frac{1}{2\theta}(x-\mu)^2},$$

and this is of the form (4) with $C(\theta) = \frac{1}{\sqrt{2\pi\theta}}$, $Q(\theta) = -\frac{1}{2\theta}$ strictly increasing (since $\frac{d}{d\theta}(-\frac{1}{2\theta}) = \frac{1}{2\theta^2} > 0$), $T(x) = (x - \mu)^2$, and $h(x) = 1$.

## 11.2.2 UMP TESTS FOR SOME COMPOSITE HYPOTHESES

We may now proceed with the formulation of the following important results. Theorem 2 below gives UMP tests for certain one-sided hypotheses in the context of Exponential families of p.d.f.'s.

**Theorem 2.** *Let $X_1, \ldots, X_n$ be a random sample with Exponential type p.d.f. $f(x; \theta), \theta \in \Omega \subseteq \Re$; i.e.,*

$$f(x; \theta) = C(\theta)e^{Q(\theta)T(x)} \times h(x), \quad x \in \Re,$$

*(with $Q$ strictly monotone and $h$ independent of $\theta$), and set $V(x_1, \ldots, x_n) = \sum_{i=1}^n T(x_i)$. Then each one of the tests defined below is UMP of level $\alpha$ for testing the hypothesis specified against the respective alternative among all tests of level $\leq \alpha$. Specifically:*

**(i)** *Let Q be strictly increasing.*

   *Then for testing $H_0: \theta \leq \theta_0$ against $H_A: \theta > \theta_0$, the UMP test is given by:*

$$\varphi(x_1,\ldots,x_n) = \begin{cases} 1 & \text{if} \quad V(x_1,\ldots,x_n) > C \\ \gamma & \text{if} \quad V(x_1,\ldots,x_n) = C \\ 0 & \text{if} \quad V(x_1,\ldots,x_n) < C, \end{cases} \qquad (5)$$

*where the constants C and $\gamma$ ($C > 0$, $0 \leq \gamma \leq 1$) are determined by:*

$$E_{\theta_0}\varphi(X_1,\ldots,X_n) = P_{\theta_0}[V(X_1,\ldots,X_n) > C]$$
$$+ \gamma P_{\theta_0}[V(X_1,\ldots,X_n) = C] = \alpha. \qquad (6)$$

*The power of the test is given by:*

$$\pi_\varphi(\theta) = P_\theta[V(X_1,\ldots,X_n) > C] + \gamma P_\theta[V(X_1,\ldots,X_n) = C] \quad (\theta > \theta_0). \qquad (7)$$

*If the hypothesis to be tested is $H_0: \theta \geq \theta_0$, so that the alternative is $H_A: \theta < \theta_0$, then the UMP test is given by (5) and (6) with reversed inequalities; i.e.,*

$$\varphi(x_1,\ldots,x_n) = \begin{cases} 1 & \text{if} \quad V(x_1,\ldots,x_n) < C \\ \gamma & \text{if} \quad V(x_1,\ldots,x_n) = C \\ 0 & \text{if} \quad V(x_1,\ldots,x_n) > C, \end{cases} \qquad (8)$$

*where the constants C and $\gamma$ ($C > 0$, $0 \leq \gamma \leq 1$) are determined by:*

$$E_{\theta_0}\varphi(X_1,\ldots,X_n) = P_{\theta_0}[V(X_1,\ldots,X_n) < C]$$
$$+ \gamma P_{\theta_0}[V(X_1,\ldots,X_n) = C] = \alpha. \qquad (9)$$

*The power of the test is given by:*

$$\pi_\varphi(\theta) = P_\theta[V(X_1,\ldots,X_n) < C] + \gamma P_\theta[V(X_1,\ldots,X_n) = C] \quad (\theta < \theta_0).$$
$$(10)$$

**(ii)** *Let Q be strictly decreasing. Then for testing $H_0: \theta \leq \theta_0$ against $H_A: \theta > \theta_0$, the UMP test is given by (8) and (9), and the power is given by (10).*

   *For testing $H_0: \theta \geq \theta_0$ against $H_A: \theta < \theta_0$, the UMP test is given by (5) and (6), and the power is given by (7).*

*Proof (Just Pointing Out the Main Arguments).* The proof of this theorem is based on Theorem 1 and also the specific form assumed for the p.d.f. $f(\cdot; \theta)$. As a rough illustration, consider the case that $Q$ is strictly increasing and the hypothesis to be tested is $H_0: \theta \leq \theta_0$. For an arbitrary $\theta_1 < \theta_0$, it is shown that $E_{\theta_1}\varphi(X_1,\ldots,X_n) < \alpha$. This establishes that $E_\theta\varphi(X_1,\ldots,X_n) \leq \alpha$ for all $\theta \leq \theta_0$, so that $\varphi$ is of level $\alpha$. Next, take an arbitrary $\theta_1 > \theta_0$ and consider the problem of testing the simple hypothesis $H'_0: \theta = \theta_0$ against the simple alternative $H_{A,1}: \theta = \theta_1$. It is shown that the MP test, provided by Theorem 1, actually, coincides with the test $\varphi$ given by (5) and (6). This shows that the test $\varphi$ is UMP. The same reasoning applies for the remaining cases. ■

   Figures 11.2 and 11.3 depict the form of the power of the UMP tests for the one-sided hypotheses $H_0: \theta \leq \theta_0$ and $H_0: \theta \geq \theta_0$.

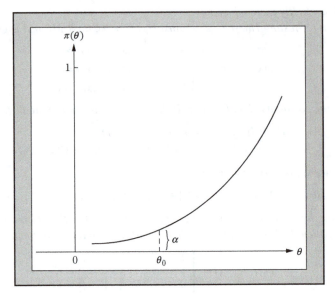

**FIGURE 11.2**

$H_0: \theta \leq \theta_0$, $H_A: \theta > \theta_0$: the power curve.

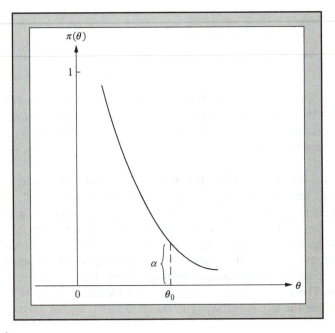

**FIGURE 11.3**

$H_0: \theta \geq \theta_0$, $H_A: \theta < \theta_0$: the power curve.

**Remark 4.**   It is to be pointed out here that the UMP tests, given in Theorem 2, still remain UMP tests of level $\alpha$ if the one-sided hypotheses ($H_0 : \theta \leq \theta_0$, or $H_0 : \theta \geq \theta_0$) are replaced by the simple hypothesis $H_0' : \theta = \theta_0$. This is so, because if $C_0$ is the class of all level $\leq \alpha$ tests for testing $H_0 : \theta \leq \theta_0$ against $H_A : \theta > \theta_0$ and $C_0'$ is the class of all level $\leq \alpha$ tests for testing $H_0' : \theta = \theta_0$ against $H_A$, then, clearly, $C_0 \subseteq C_0'$. Since $\varphi \in C_0$ and is UMP within $C_0'$, it is so within $C_0$. Similarly, if $H_0 : \theta \leq \theta_0$ is replaced by $H_0 : \theta \geq \theta_0$.

## 11.3 SOME APPLICATIONS OF THEOREMS 2

**Application 1** (The Binomial Case).   In reference to Example 9 with $n = 1$, we have $T(x) = x$ where $x = 0, 1$, and then in Theorem 2, $V(x_1, \ldots, x_n) = \sum_{i=1}^{n} x_i$ and $V(X_1, \ldots, X_n) = \sum_{i=1}^{n} X_i \sim B(n, \theta)$. Since $Q(\theta) = \log(\frac{\theta}{1-\theta})$ is strictly increasing, consider relations (5) and (6), which, for testing the hypothesis $H_0: \theta \leq \theta_0$ against the alternative $H_A: \theta > \theta_0$, become here:

$$\varphi(x_1, \ldots, x_n) = \begin{cases} 1 & \text{if } \sum_{i=1}^{n} x_i > C \\ \gamma & \text{if } \sum_{i=1}^{n} x_i = C \\ 0 & \text{if } \sum_{i=1}^{n} x_i < C, \end{cases} \tag{11}$$

$$E_{\theta_0} \varphi(X_1, \ldots, X_n) = P_{\theta_0}(X > C) + \gamma P_{\theta_0}(X = C) = \alpha, \quad X \sim B(n, \theta_0). \tag{12}$$

Relation (12) is rewritten below to allow the usage of the Binomial tables for the determination of $C$ and $\gamma$; namely,

$$P_{\theta_0}(X \leq C) - \gamma P_{\theta_0}(X = C) = 1 - \alpha, \quad X \sim B(n, \theta_0). \tag{13}$$

The power of the test is:

$$\pi_{\varphi}(\theta) = P_{\theta}(X > C) + \gamma P_{\theta}(X = C) = 1 - P_{\theta}(X \leq C) + \gamma P_{\theta}(X = C),$$
$$(\theta > \theta_0), \quad X \sim B(n, \theta). \tag{14}$$

*Numerical Example.*   Refer to Example 2 and suppose that $n = 25, \theta_0 = 100p\% = 0.125$, and $\alpha = 0.05$.

**Discussion.**   Here, for $\theta = \theta_0, X \sim B(25,\ 0.125)$, and (13) becomes

$$P_{0.125}(X \leq C) - \gamma P_{0.125}(X = C) = 0.95.$$

From the Binomial tables, the value of $C$ which renders $P_{0.125}(X \leq C)$ just above 0.95 is 6 and $P_{0.125}(X \leq 6) = 0.9703$. Also, $P_{0.125}(X = 6) = 0.9703 - 0.9169 = 0.0534$, so that $\gamma = \frac{0.9703 - 0.95}{0.0534} = \frac{0.0203}{0.0534} \simeq 0.38$. Thus, the test in (11) is:

$$\varphi(x_1, \ldots, x_n) = \begin{cases} 1 & \text{if } x > 6 \\ 0.38 & \text{if } x = 6 \\ 0 & \text{if } x < 6. \end{cases}$$

Reject outright the hypothesis that $100p\% = 12.5\%$ if the number of listeners among the sample of 25 is 7 or more, reject the hypothesis with probability 0.38 if this number is 6, and accept the hypothesis if this number is 5 or smaller.

The power of the test is calculated to be as follows, by relation (14):

$$\pi_\varphi(0.1875) \simeq 1 - 0.8261 + 0.38 \times 0.1489 \simeq 0.230,$$
$$\pi_\varphi(0.25) \simeq 1 - 0.5611 + 0.38 \times 0.1828 \simeq 0.508,$$
$$\pi_\varphi(0.375) \simeq 1 - 0.1156 + 0.38 \times 0.0652 \simeq 0.909.$$

If we suppose that the observed value of $X$ is 7, then the $P$-value is:

$$1 - P_{0.125}(X \le 7) + 0.38 P_{0.125}(X = 7) = 1 - 0.9910 + 0.38 \times 0.0207 \simeq 0.017,$$

so that the rejection of $H_0$ is strong, or the result is highly statistically significant, as we say.

Next, for testing the hypothesis $H_0 : \theta \ge \theta_0$ against the alternative $H_A : \theta < \theta_0$, relations (8) and (9) become:

$$\varphi(x_1, \ldots, x_n) = \begin{cases} 1 & \text{if } \sum_{i=1}^n x_i < C \\ \gamma & \text{if } \sum_{i=1}^n x_i = C \\ 0 & \text{if } \sum_{i=1}^n x_i > C, \end{cases} \tag{15}$$

and

$$P_{\theta_0}(X \le C - 1) + \gamma P_{\theta_0}(X = C) = \alpha, \quad X = \sum_{i=1}^n X_i \sim B(n, \theta_0). \tag{16}$$

The power of the test is:

$$\pi_\varphi(\theta) = P_\theta(X \le C - 1) + \gamma P_\theta(X = C) \quad (\theta < \theta_0), \quad X \sim B(n, \theta). \tag{17}$$

*Numerical Example.* Refer to Example 1 and suppose that $n = 25, \theta_0 = 0.0625$, and $\alpha = 0.05$.

**Discussion.** Here, under $\theta_0, X \sim B(25, 0.0625)$, and (16) becomes

$$P_{0.0625}(X \le C - 1) + \gamma P_{0.0625}(X = C) = 0.05,$$

so that $C = 0$, and $\gamma P_{0.0625}(X = 0) = 0.1992 \ \gamma = 0.05$. It follows that $\gamma \simeq 0.251$. Therefore, the hypothesis is rejected with probability 0.251, if $x = 0$, and is accepted otherwise.

**Application 2** (The Poisson Case). In reference to Example 10, we have $T(x) = x, x = 0, 1, \ldots$, and then in Theorem 2, $V(x_1, \ldots, x_n) = \sum_{i=1}^n x_i$ and $V(X_1, \ldots, X_n) = \sum_{i=1}^n X_i \sim P(n\theta)$. Since $Q(\theta) = \log \theta$ is strictly increasing, consider relations (8) and (9) for testing $H_0 : \theta \ge \theta_0$ against $H_A : \theta < \theta_0$. They become here:

$$\varphi(x_1, \ldots, x_n) = \begin{cases} 1 & \text{if } \sum_{i=1}^n x_i < C \\ \gamma & \text{if } \sum_{i=1}^n x_i = C \\ 0 & \text{if } \sum_{i=1}^n x_i > C, \end{cases} \tag{18}$$

$$E_{\theta_0}\varphi(X_1, \ldots, X_n) = P_{\theta_0}(X < C) + \gamma P_{\theta_0}(X = C) = \alpha, \quad X = \sum_{i=1}^n X_i \sim P(n\theta_0),$$

or

$$P_{\theta_0}(X \le C - 1) + \gamma P\theta_0(X = C) = \alpha, \quad X \sim P(n\theta_0). \tag{19}$$

The power of the test is:

$$\pi_\varphi(\theta) = P_\theta(X \le C - 1) + \gamma P_\theta(X = C)(\theta < \theta_0), \quad X \sim P(n\theta). \tag{20}$$

Unfortunately, no numerical application for Example 3 can be given as the Poisson tables do not provide entries for $\theta_0 = 200$. In order to be able to apply the test defined by (18) and (19), consider the following example.

**Example 13.** Let $X_1, \ldots, X_{20}$ be i.i.d. r.v.'s denoting the number of typographical errors in 20 pages of a book. We may assume that the $X_i$'s are independently distributed as $P(\theta)$, and let us test the hypothesis $H_0: \theta > 0.5$ (the average number of errors is more than 1 per couple of pages) against the alternative $H_A: \theta \le 0.5$ at level $\alpha = 0.05$.

**Discussion.**   In (19), $X \sim P(10)$, so that

$$P_{0.5}(X \le C - 1) + \gamma P_{0.5}(X = C) = 0.05,$$

and hence $C - 1 = 4$ and $P_{0.5}(X \le 4) = 0.0293, P_{0.5}(X = 5) = 0.0671 - 0.0293 = 0.0378$. It follows that $\gamma = \frac{0.05 - 0.0293}{0.0378} \simeq 0.548$. Therefore by (18), reject the hypothesis outright if $x \le 4$, reject it with probability 0.548 if $x = 5$, and accept it otherwise. The power of the test is: For $\theta = 0.2, X \sim P(4)$ and:

$$\pi_\varphi(0.2) = P_{0.2}(X \le 4) + 0.548P_{0.2}(X = 5) = 0.6288 + 0.548 \times 0.1563 \simeq 0.714.$$

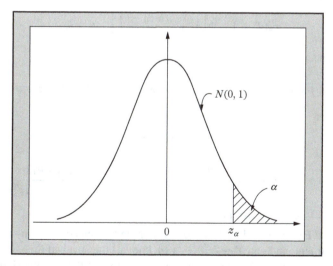

**FIGURE 11.4**

Rejection region of the hypothesis $H_0: \theta \le \theta_0$ (the shaded area) in the form (21′) and the respective probability.

**Application 3** (The Normal Case: Testing Hypotheses About the Mean). Refer to Example 11 and observe that $T(x) = x$ and $Q(\theta)$ is strictly increasing. Therefore, the appropriate test for testing $H_0: \theta \leq \theta_0$ against $H_A: \theta > \theta_0$ at level of significance $\alpha$ is given by (5) and (6) with $\gamma = 0$ (Figure 11.4). That is,

$$\varphi(x_1, \ldots, x_n) = \begin{cases} 1 & \text{if } \sum_{i=1}^n x_i > C \\ 0 & \text{otherwise,} \end{cases} \tag{21}$$

or

$$\varphi(x_1, \ldots, x_n) = \begin{cases} 1 & \text{if } \frac{\sqrt{n}(\bar{x} - \theta_0)}{\sigma} > z_\alpha \\ 0 & \text{otherwise,} \end{cases} \tag{21'}$$

because

$$\alpha = E_{\theta_0} \varphi(X_1, \ldots, X_n) = P_{\theta_0}\left( \sum_{i=1}^n X_i > C \right)$$

$$= P_{\theta_0}\left[ \frac{\sqrt{n}(\bar{X} - \theta_0)}{\sigma} > \frac{C - n\theta_0}{\sigma\sqrt{n}} \right],$$

so that $\frac{C - n\theta_0}{\sigma\sqrt{n}} = z_\alpha$ and therefore

$$C = n\theta_0 + z_\alpha \sigma \sqrt{n}; \tag{22}$$

this is so, because $\frac{\sqrt{n}(\bar{X} - \theta_0)}{\sigma} = Z \sim N(0, 1)$, and recall that $P(Z > z_\alpha) = \alpha$.

The power of the test is given by:

$$\pi_\varphi(\theta) = 1 - \Phi\left[ z_\alpha + \frac{\sqrt{n}(\theta_0 - \theta)}{\sigma} \right], \quad \theta > \theta_0, \tag{23}$$

because, on account of (22):

$$\pi_\varphi(\theta) = P_\theta\left( \sum_{i=1}^n X_i > C \right) = P_\theta\left[ \sum_{i=1}^n X_i - n\theta > n(\theta_0 - \theta) + z_\alpha \sigma \sqrt{n} \right]$$

$$= P_\theta\left[ \frac{\sum_{i=1}^n X_i - n\theta}{\sigma\sqrt{n}} > z_\alpha + \frac{\sqrt{n}(\theta_0 - \theta)}{\sigma} \right]$$

$$= P\left[ Z > z_\alpha + \frac{\sqrt{n}(\theta_0 - \theta)}{\sigma} \right] = 1 - \Phi\left[ z_\alpha + \frac{\sqrt{n}(\theta_0 - \theta)}{\sigma} \right],$$

since $\frac{\sum_{i=1}^n X_i - n\theta}{\sigma\sqrt{n}} = Z \sim N(0, 1)$.

*Numerical Example.* In reference to Example 5, focus on patients treated with the new treatment, and call $Y \sim N(\theta, \sigma^2)$ ($\sigma$ known) the survival time. On the basis of observations on $n = 25$ such patients, we wish to test the hypothesis $H_0: \theta \leq 5$ (in years) against $H_A: \theta > 5$ at level of significance $\alpha = 0.01$. For simplicity, take $\sigma = 1$.

**Discussion.** Here $z_\alpha = z_{0.01} = 2.33$, so that $C = 25 \times 5 + 2.33 \times 1 \times 5 = 136.65$. Thus, reject $H_0$ if the total of survival years is $>136.65$, and accept $H_0$ otherwise.

The power of the test is given by (23) and is:

$$\text{For } \theta = 5.5, \pi_\varphi(5.5) = 1 - \Phi[2.33 + 5(5 - 5.5)] = 1 - \Phi(-0.17)$$

$$= \Phi(0.17) = 0.567495;$$

$$\text{and for } \theta = 6, \pi_\varphi(6) = 1 - \Phi[2.33 + 5(6 - 5.5)] = 1 - \Phi(-2.67)$$

$$= \Phi(2.67) = 0.996207.$$

If we suppose that the observed value of $\sum_{i=1}^{25} x_i$ is equal to 138, then the $P$-value is $P(\sum_{i=1}^{25} X_i > 138) = 1 - \Phi(\frac{138-125}{5}) = 1 - \Phi(2.6) = 1 - 0.995339 = 0.004661$, so that the result is highly statistically significant.

**Application 4** (The Normal Case (continued): Testing Hypotheses About the Variance). Refer to Example 12, where $T(x) = (x - \mu)^2$ ($\mu$ known) and $Q$ is strictly increasing. Then, for testing the hypothesis $H_0: \sigma^2 \geq \sigma_0^2$ against the alternative $H_A$: $\sigma^2 < \sigma_0^2$ (or $\theta \geq \theta_0$ against $\theta < \theta_0$ with $\theta = \sigma^2$ and $\theta_0 = \sigma_0^2$) at level of significance $\alpha$, the appropriate test is given by (8) and (9) (with $\gamma = 0$), and it is here:

$$\varphi(x_1,\ldots,x_n) = \begin{cases} 1 & \text{if } \sum_{i=1}^n (x_i - \mu)^2 < C \\ 0 & \text{otherwise,} \end{cases} \tag{24}$$

or

$$\varphi(x_1,\ldots,x_n) = \begin{cases} 1 & \text{if } \sum_{i=1}^n \left(\frac{x_i - \mu}{\sigma_0}\right)^2 < \chi_{n;1-\alpha}^2 \\ 0 & \text{otherwise,} \end{cases} \tag{24'}$$

because

$$\alpha = E_{\sigma_0^2}\varphi(X_1,\ldots,X_n) = P_{\sigma_0^2}\left[\sum_{i=1}^n (X_i - \mu)^2 < C\right]$$

$$= P_{\sigma_0^2}\left[\sum_{i=1}^n \left(\frac{X_i - \mu}{\sigma_0}\right)^2 < \frac{C}{\sigma_0^2}\right],$$

so that $\frac{C}{\sigma_0^2} = \chi_{n;1-\alpha}^2$ and therefore

$$C = \sigma_0^2 \chi_{n;1-\alpha}^2; \tag{25}$$

this is so, because $\sum_{i=1}^n (\frac{X_i-\mu}{\sigma_0})^2 \sim \chi_n^2$ (Figure 11.5).

By slightly abusing the notation and denoting by $\chi_n^2$ also a r.v. which has the $\chi_n^2$ distribution, the power of the test is given by:

$$\pi_\varphi(\sigma^2) = P\left(\chi_n^2 < \frac{\sigma_0^2}{\sigma^2}\chi_{n;1-\alpha}^2\right), \quad \sigma^2 < \sigma_0^2, \tag{26}$$

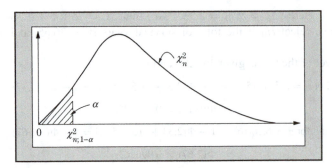

**FIGURE 11.5**

Rejection region of the hypothesis $H_0: \sigma^2 \geq \sigma_0^2$ (the shaded area) in the form $(24')$ and the respective probability.

because, on account of (25):

$$\pi_\varphi(\sigma^2) = P_{\sigma^2}\left[\sum_{i=1}^n (X_i - \mu)^2 < C\right] = P_{\sigma^2}\left[\sum_{i=1}^n \left(\frac{X_i - \mu}{\sigma}\right)^2 < \frac{C}{\sigma^2}\right]$$

$$= P_{\sigma^2}\left[\sum_{i=1}^n \left(\frac{X_i - \mu}{\sigma}\right)^2 < \frac{\sigma_0^2}{\sigma^2} \chi_{n;1-\alpha}^2\right] = P\left(\chi_n^2 < \frac{\sigma_0^2}{\sigma^2} \chi_{n;1-\alpha}^2\right),$$

since $\sum_{i=1}^n (\frac{X_i-\mu}{\sigma})^2 \sim \chi_n^2$.

   *Numerical Example.*    Suppose $n = 40$, $\sigma_0 = 2$, and $\alpha = 0.025$. For simplicity, take $\mu = 0$.

**Discussion.**   Here $\chi_{n;1-\alpha}^2 = \chi_{40;0.975}^2 = 24.433$  and  $C = 4 \times 24.433 = 97.732$. Thus, by means of (24), the hypothesis is rejected if $\sum_{i=1}^{40} x_i^2 < 97.732$, and it is accepted otherwise.

   For $\sigma = 1.25$, for example, the power of the test is, by means of (26),

$$\pi_\varphi(\sigma^2) = \pi_\varphi(2.25) = P\left(\chi_{40}^2 < \frac{97.732}{1.5625}\right)$$

$$= P\left(\chi_{40}^2 < 62.548\right) = 0.986$$

(by linear interpolation).

   If we suppose that the observed value of $\sum_{i=1}^{40} x_i^2$ is 82.828, then the $P$-value is

$$P_4\left(\sum_{i=1}^{40} X_i^2 < 82.828\right) = P_4\left[\sum_{i=1}^{40} \left(\frac{X_i}{2}\right)^2 < 20.707\right] = 0.005,$$

which indicates strong rejection, or that the result is highly statistically significant.

## EXERCISES

**3.1** (i) In reference to Example 8 in Chapter 1, the appropriate model is the Binomial model with $X_i = 1$ if the $i$th young adult listens to the program, and $X_i = 0$ otherwise, where $P(X_i = 1) = p$, and the $X_i$'s are independent, so that $X = \sum_{i=1}^{n} X_i \sim B(n, p)$.

   (ii) The claim is that $p > p_0$ some specified number $0 < p_0 < 1$, and the claim is checked by testing the hypothesis $H_0: p \leq p_0$ against the alternative $H_A: p > p_0$ at level of significance $\alpha$.

   (iii) For $p_0 = 5\%, n = 100$, and $\alpha = 0.02$, use the CLT to carry out the test.

**3.2** (i) In reference to Example 9 in Chapter 1, the appropriate model is the Binomial model with $X_i = 1$ if the $i$th item is defective, and 0 otherwise, where $P(X_i = 1) = p$, and the $X_i$'s are independent, so that $X = \sum_{i=1}^{n} X_i \sim B(n, p)$.

   (ii) The process is under control if $p < p_0$, where $p_0$ is a specified number with $0 < p_0 < 1$, and the hypothesis to be checked is $H_0 : p \geq p_0$ against the alternative $H_A: p < p_0$ at level of significance $\alpha$.

   (iii) For $p_0 = 0.0625, n = 100$, and $\alpha = 0.10$, use the CLT to carry out the test.

**3.3** (i) In reference to Example 10 in Chapter 1, the appropriate model is the Binomial model with $X_i = 1$ if the $i$th flawed specimen is identified as such, and $X_i = 0$, otherwise, where $P(X_i = 1) = p$, and the $X_i$'s are independent, so that $X = \sum_{i=1}^{n} X_i \sim B(n, p)$.

   (ii) The electronic scanner is superior to the mechanical testing if $p > p_0$, some specified $p_0$ with $0 < p_0 < 1$, and this is checked by testing the hypothesis $H_0: p \leq p_0$ against the alternative $H_A: p > p_0$ at level of significance $\alpha$.

   (iii) For $p_0 = 90\%, n = 100$, and $\alpha = 0.05$, use the CLT to carry out the test.

**3.4** (i) In a certain university, 400 students were chosen at random and it was found that 95 of them were women. On the basis of this, test the hypothesis $H_0$: the proportion of women is 25% against the alternative $H_A$: the proportion of women is less than 25% at level of significance $\alpha = 0.05$.

   (ii) Use the CLT in order to determine the cutoff point.

**3.5** Let $X_1, \ldots, X_n$ be independent r.v.'s distributed as $B(1, p)$. For testing the hypothesis $H_0: p \leq \frac{1}{2}$ against the alternative $H_A: p > \frac{1}{2}$, use the CLT in order to determine the sample size $n$ for which the level of significance and power are, respectively, $\alpha = 0.05$ and $\pi(7/8) = 0.95$.

**3.6** Let $X$ be a r.v. distributed as $B(n, \theta), \theta \in \Omega = (0, 1)$.

   (i) Use relations (11) and (13) to set up the UMP test for testing the hypothesis $H_0: \theta \leq \theta_0$ against the alternative $H_A: \theta > \theta_0$ at level of significance $\alpha$.

(ii)  Specify the test in part (i) for $n = 10, \theta_0 = 0.25$, and $\alpha = 0.05$.

(iii)  Compute the power of the test for $\theta_1 = 0.375, 0.500$.

(iv)  For $\theta > 0.5$, show that: $P_\theta(X \leq C) = 1 - P_{1-\theta}(X \leq n - C - 1)$ and hence $P_\theta(X = C) = P_{1-\theta}(X \leq n - C) - P_{1-\theta}(X \leq n - C - 1)$.

(v)  Use part (iv) to compute the power of the test for $\theta_1 = 0.625, 0.875$.

(vi)  Use the CLT in order to determine the sample size $n$ if $\theta_0 = 0.125, \alpha = 0.1$, and $\pi(0.25) = 0.9$.

**3.7**  (i)  In reference to Example 12 in Chapter 1, the appropriate model to be used is the Poisson model; i.e., $X \sim P(\lambda)$.

(ii)  The safety level is specified by $\lambda < 200$, and this is checked by testing the hypothesis $H_0: \lambda \geq 200$ against the alternative $H_A: \lambda < 200$ at level of significance $\alpha$.

(iii)  On the basis of a random sample of size $n = 100$, use the CLT in order to carry out the test for $\alpha = 0.05$.

**3.8**  The number of total traffic accidents in a certain city during a year is a r.v. $X$, which may be assumed to be distributed as $P(\lambda)$. For the last year, the observed value of $X$ was $x = 4$, whereas for the past several years, the average was 10.

(i)  Formulate the hypothesis that the average remains the same against the alternative that there is an improvement.

(ii)  Refer to Application 2 in order to derive the UMP test for testing the hypothesis of part (i) at level $\alpha = 0.01$.

**Hint.** Refer to Remark 4, regarding the null hypothesis.

**3.9**  (i)  In reference to Example 16 in Chapter 1, a suitable model would be to assume that $X \sim N(\mu_1, \sigma^2), Y \sim N(\mu_2, \sigma^2)$ and that they are independent. It is also assumed that $\sigma$ is known.

(ii)  Here $\mu_1$ is the known mean survival period (in years) for the existing treatment. Then the claim is that $\mu_2 > \mu_1$, and this is to be checked by testing the hypothesis $H_0: \mu_2 \leq \mu_1$ against the alternative $H_A: \mu_2 > \mu_1$ at level of significance $\alpha$.

(iii)  Carry out the test if $n = 100, \mu_1 = 5$, and $\alpha = 0.05$.

**3.10**  The lifetime of a 50-watt light bulb of a certain brand is a r.v. $X$, which may be assumed to be distributed as $N(\mu, \sigma^2)$ with unknown $\mu$ and $\sigma$ known. Let $X_1, \ldots, X_n$ be a random sample from this distribution and suppose that we are interested in testing the hypothesis $H_0: \mu = \mu_0$ against the alternative $H_A: \mu < \mu_0$ at level of significance $\alpha$.

(i)  Derive the UMP test.

(ii)  Derive the formula for the power of the test.

(iii)  Carry out the testing hypothesis problem when $n = 25, \mu_0 = 1,800, \sigma = 150$ (in hours), $\alpha = 0.01$, and $\bar{x} = 1,730$. Also, calculate the power at $\mu = 1,700$.

**Hint.** As in Exercise 3.8.

**3.11** The rainfall at a certain station during a year is a r.v. $X$, which may be assumed to be distributed as $N(\mu, \sigma^2)$ with $\mu$ unknown and $\sigma = 3$ inches. For the past 10 years, the record provides the following rainfalls:

$$x_1 = 30.5, \quad x_2 = 34.1, \quad x_3 = 27.9, \quad x_4 = 29.4, \quad x_5 = 35.0,$$
$$x_6 = 26.9, \quad x_7 = 30.2, \quad x_8 = 28.3, \quad x_9 = 31.7, \quad x_{10} = 25.8.$$

Test the hypothesis $H_0$: $\mu = 30$ against the alternative $H_A$: $\mu < 30$ at level of significance $\alpha = 0.05$.
**Hint.** As in Exercise 3.8.

**3.12** **(i)** On the basis of the independent r.v.'s $X_1, \ldots, X_{25}$, distributed as $N(0, \sigma^2)$, test the hypothesis $H_0$: $\sigma \le 2$ against the alternative $H_A$: $\sigma > 2$ at level of significance $\alpha = 0.05$.
**(ii)** Specify the test when the observed values $x_i$'s of the $X_i$'s are such that $\sum_{i=1}^{25} x_i^2 = 120$.

**3.13** The diameters of bolts produced by a certain machine are independent r.v.'s distributed as $N(\mu, \sigma^2)$ with $\mu$ known. In order for the bolts to be usable for the intended purpose, the s.d. $\sigma$ must not exceed 0.04 inch. A random sample of size $n = 16$ is taken and it is found that $s = 0.05$ inch. Formulate the appropriate testing hypothesis problem and carry out the test at level of significance $\alpha = 0.05$.

**3.14** Let $X$ be a r.v. with p.d.f. $f(x; \theta) = \frac{1}{\theta} e^{-x/\theta}$, $x > 0$, $\theta \in \Omega = (0, \infty)$.
**(i)** Refer to Definition 1 in order to show that $f(\cdot; \theta)$ is of the Exponential type.
**(ii)** Use Theorem 2 in order to derive the UMP test for testing the hypothesis $H_0$: $\theta \ge \theta_0$ against the alternative $H_A$: $\theta < \theta_0$ at level of significance $\alpha$, on the basis of the random sample $X_1, \ldots, X_n$ from the above p.d.f.
**(iii)** Use the m.g.f. approach in order to show that the r.v.
$Y = 2 \times (\sum_{i=1}^{n} X_i)/\theta$ is distributed as $\chi_{2n}^2$.
**(iv)** Use parts (ii) and (iii) in order to find an expression for the cutoff point $C$ and the power function of the test.
**(v)** If $\theta_0 = 1,000$ and $\alpha = 0.05$, determine the sample size $n$, so that the power of the test at $\theta_1 = 500$ is at least 0.95.

**3.15** The number of times that an electric light switch can be turned on and off until failure occurs is a r.v. $X$, which may be assumed to have the Geometric p.d.f. with parameter $\theta$; i.e., $f(x; \theta) = \theta(1 - \theta)^{x-1}$, $x = 1, 2, \ldots, \theta \in \Omega = (0, 1)$.
**(i)** Refer to Definition 1 in order to show that $f(\cdot; \theta)$ is of the Exponential type.
**(ii)** Use Theorem 2 in order to derive the UMP test for testing the hypothesis $H_0$: $\theta = \theta_0$ against the alternative $H_A$: $\theta > \theta_0$ at level of significance $\alpha$, on the basis of a random sample of size $n$ from the p.d.f. $f(\cdot; \theta)$.
**(iii)** Use the CLT to find an approximate value for the cutoff point $C$.

**(iv)** Carry out the test if $n = 15$, the observed sample mean $\bar{x} = 15,150$, $\theta_0 = 10^{-4}$, and $\alpha = 0.05$.

**Hint.** As in Exercise 3.8.

## 11.3.1 FURTHER UNIFORMLY MOST POWERFUL TESTS FOR SOME COMPOSITE HYPOTHESES

The result stated below as Theorem 3 gives UMP tests for a certain two-sided hypothesis in the context of Exponential families of p.d.f.'s.

**Theorem 3.** *Let $X_1, \ldots, X_n$ be a random sample with p.d.f. as in Theorem 2, and let $V(x_1, \ldots, x_n)$ be as in the same theorem. Consider the problem of testing the hypothesis $H_0$: $\theta \leq \theta_1$ or $\theta \geq \theta_2$ against the alternative $H_A$: $\theta_1 < \theta < \theta_2$ at level of significance $\alpha$. Then the tests defined below are UMP of level $\alpha$ among all tests of level $\leq \alpha$. Specifically:*

**(i)** *If $Q$ is strictly increasing, the UMP test is given by:*

$$\varphi(x_1, \ldots, x_n) = \begin{cases} 1 & \text{if } C_1 < V(x_1, \ldots, x_n) < C_2 \\ \gamma_1 & \text{if } V(x_1, \ldots, x_n) = C_1 \\ \gamma_2 & \text{if } V(x_1, \ldots, x_n) = C_2 \\ 0 & \text{otherwise,} \end{cases} \tag{27}$$

*where the constants $C_1, C_2$ and $\gamma_1, \gamma_2$ ($C_1 > 0, C_2 > 0$, $0 \leq \gamma_1 \leq 1, 0 \leq \gamma_2 \leq 1$) are determined through the relationships:*

$$E_{\theta_1}\varphi(X_1, \ldots, X_n) = P_{\theta_1}[C_1 < V(X_1, \ldots, X_n) < C_2]$$
$$+ \gamma_1 P_{\theta_1}[V(X_1, \ldots, X_n) = C_1]$$
$$+ \gamma_2 P_{\theta_1}[V(X_1, \ldots, X_n) = C_2] = \alpha, \tag{28}$$

$$E_{\theta_2}\varphi(X_1, \ldots, X_n) = P_{\theta_2}[C_1 < V(X_1, \ldots, X_n) < C_2]$$
$$+ \gamma_1 P_{\theta_2}[V(X_1, \ldots, X_n) = C_1]$$
$$+ \gamma_2 P_{\theta_2}[V(X_1, \ldots, X_n) = C_2] = \alpha. \tag{29}$$

*The power of the test is given by:*

$$\pi_\varphi(\theta) = P_\theta[C_1 < V(X_1, \ldots, X_n) < C_2] + \gamma_1 P_\theta[V(X_1, \ldots, X_n) = C_1]$$
$$+ \gamma_2 P_\theta[V(X_1, \ldots, X_n) = C_2] \quad (\theta_1 < \theta < \theta_2). \tag{30}$$

**(ii)** *If $Q$ is strictly decreasing, then the UMP test is given by (11) and (12)–(13) with reversed inequalities; i.e.,*

$$\varphi(x_1, \ldots, x_n) = \begin{cases} 1 & \text{if } V(x_1, \ldots, x_n) < C_1 \text{ or } V(x_1, \ldots, x_n) > C_2 \\ \gamma_1 & \text{if } V(x_1, \ldots, x_n) = C_1 \\ \gamma_2 & \text{if } V(x_1, \ldots, x_n) = C_2 \\ 0 & \text{otherwise,} \end{cases} \tag{31}$$

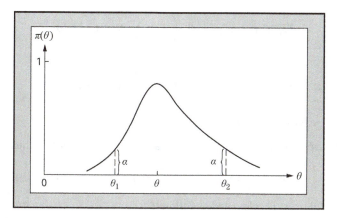

**FIGURE 11.6**

$H_0: \theta \le \theta_1$ or $\theta \ge \theta_2, H_A: \theta_1 < \theta < \theta_2$: the power curve.

*and the constants $C_1$, $C_2$ and $\gamma_1$, $\gamma_2$ are determined by:*

$$E_{\theta_1}\varphi(X_1,\ldots,X_n) = P_{\theta_1}[V(X_1,\ldots,X_n) < C_1 \text{ or } V(X_1,\ldots,X_n) > C_2]$$
$$+\gamma_1 P_{\theta_1}[V(X_1,\ldots,X_n) = C_1]$$
$$+\gamma_2 P_{\theta_1}[V(X_1,\ldots,X_n) = C_2] = \alpha, \tag{32}$$

$$E_{\theta_2}\varphi(X_1,\ldots,X_n) = P_{\theta_2}[V(X_1,\ldots,X_n) < C_1 \text{ or } V(X_1,\ldots,X_n) > C_2]$$
$$+\gamma_1 P_{\theta_2}[V(X_1,\ldots,X_n) = C_1]$$
$$+\gamma_2 P_{\theta_2}[V(X_1,\ldots,X_n) = C_2] = \alpha. \tag{33}$$

*The power of the test is given by:*

$$\pi_\varphi(\theta) = P_\theta[V(X_1,\ldots,X_n) < C_1 \text{ or } V(X_1,\ldots,X_n) > C_2]$$
$$+\gamma_1 P_\theta[V(X_1,\ldots,X_n) = C_1] + \gamma_2 P_\theta[V(X_1,\ldots,X_n) = C_2]$$
$$(\theta_1 < \theta < \theta_2). \tag{34}$$

*The power of the UMP test is depicted in Figure 11.6.*

### 11.3.2 AN APPLICATION OF THEOREM 3

**Application 5** (The Normal Case (continued): Testing Further Hypotheses About the Mean). In reference to Example 11, $T(x) = x$ and $Q(\theta)$ is strictly increasing. Therefore, for testing $H_0: \theta \le \theta_1$ or $\theta \ge \theta_2$ against $H_A: \theta_1 < \theta < \theta_2$ at level of significance $\alpha$, the test to be employed is the one given by (27) and (28)–(29), which here becomes ($\gamma_1 = \gamma_2 = 0$):

$$\varphi(x_1,\ldots,x_n) = \begin{cases} 1 & \text{if } C_1 \le \sum_{i=1}^n x_i \le C_2 \\ 0 & \text{otherwise,} \end{cases} \tag{35}$$

$$E_{\theta_1}\varphi(X_1,\ldots,X_n) = P_{\theta_1}\left(C_1 \le \sum_{i=1}^{n} X_i \le C_2\right) = \alpha,$$

$$\tag{36}$$

$$E_{\theta_2}\varphi(X_1,\ldots,X_n) = P_{\theta_2}\left(C_1 \le \sum_{i=1}^{n} X_i \le C_2\right) = \alpha,$$

and $\sum_{i=1}^{n} X_i \sim N(n\theta_i, n\sigma^2)$, $i = 1, 2$.

For the purpose of utilizing the Normal tables, (36) are rewritten thus:

$$\Phi\left(\frac{C_2 - n\theta_i}{\sigma\sqrt{n}}\right) - \Phi\left(\frac{C_1 - n\theta_i}{\sigma\sqrt{n}}\right) = \alpha, \quad i = 1, 2. \tag{37}$$

The power of the test is calculated as follows:

$$\pi_\varphi(\theta) = \Phi\left(\frac{C_2 - n\theta}{\sigma\sqrt{n}}\right) - \Phi\left(\frac{C_1 - n\theta}{\sigma\sqrt{n}}\right), \quad (\theta_1 < \theta < \theta_2). \tag{38}$$

*Numerical Example.*    In reference to Example 4, suppose $n = 25, \theta_1 = 1, \theta_2 = 3$, and $\alpha = 0.01$. For simplicity, let us take $\sigma = 1$.

**Discussion.**    Here $n\theta_1 = 25, n\theta_2 = 75$, and (37) yields:

$$\Phi\left(\frac{C_2 - 25}{5}\right) - \Phi\left(\frac{C_1 - 25}{5}\right) = \Phi\left(\frac{C_2 - 75}{5}\right) - \Phi\left(\frac{C_1 - 75}{5}\right) = 0.01. \tag{39}$$

Placing the four quantities $\frac{C_1 - 75}{5}, \frac{C_2 - 75}{5}, \frac{C_1 - 25}{5}$, and $\frac{C_2 - 25}{5}$ under the $N(0, 1)$ curve, we observe that relation (39) obtains only for:

$$\frac{C_1 - 25}{5} = -\frac{C_2 - 75}{5} \quad \text{and} \quad \frac{C_2 - 25}{5} = -\frac{C_1 - 75}{5},$$

which imply that $C_1 + C_2 = 100$. Setting $C_1 = C$, we have then that $C_2 = 100 - C$, and (39) gives:

$$\Phi\left(\frac{75 - C}{5}\right) - \Phi\left(\frac{C - 25}{5}\right) = 0.01. \tag{40}$$

From the Normal tables, we find that (40) is closely satisfied for $C = 36.5$. So, $C_1 = 36.5$ and hence $C_2 = 63.5$, and the test rejects the hypothesis $H_0$ whenever $\sum_{i=1}^{25} X_i$ is between 36.5 and 63.5 and accepts it otherwise.

The power of the test, calculated through (38), is, for example, for $\theta = 2.5$ and $\theta = 2$:

$$\pi_\varphi(1.5) = \pi_\varphi(2.5) = 0.57926 \quad \text{and} \quad \pi_\varphi(2) = 0.99307.$$

## EXERCISES

**3.16**  Let $X_i, i = 1,\ldots,4$ and $Y_j, j = 1,\ldots,4$ be two independent random samples from the distributions $N(\mu_1, \sigma_1^2)$ and $N(\mu_2, \sigma_2^2)$, respectively. Suppose that the observed values of the $X_i$'s and the $Y_j$'s are as follows:

$$x_1 = 10.1, \quad x_2 = 8.4, \quad x_3 = 14.3, \quad x_4 = 11.7,$$
$$y_1 = 9.0, \quad y_2 = 8.2, \quad y_3 = 12.1, \quad y_4 = 10.3.$$

Suppose further that $\sigma_1 = 4$ and $\sigma_2 = 3$.

Then test the hypothesis that the two means differ in absolute value by at least 1 unit. That is, if $\theta = \mu_1 - \mu_2$, then the hypothesis to be tested is $H_0: |\theta| \geq 1$, or, equivalently, $H_0: \theta \leq -1$ or $\theta \geq 1$. The alternative is $H_A: -1 < \theta < 1$. Take $\alpha = 0.05$.

**Hint.** Set $Z_i = X_i - Y_i$, so that the $Z_i$'s are independent and distributed as $N(\mu, 25)$. Then use appropriately Theorem 3.

## 11.4 LIKELIHOOD RATIO TESTS

In the previous sections, UMP tests were constructed for several important hypotheses and were illustrated by specific examples. Those tests have the UMP property, provided the underlying p.d.f. is of the Exponential type given in (4). What happens if either the p.d.f. is not of this form and/or the hypotheses to be tested are not of the type for which UMP tests exist? One answer is for sure that the testing activities will not be terminated here; other procedures are to be invented and investigated. Such a procedure is one based on the Likelihood Ratio (LR), which gives rise to the so-called LR tests. The rationale behind this procedure was given in Section 3 of Chapter 8. What we are doing in this section is to apply it to some specific cases and produce the respective LR tests in a usable form.

As already explained, LR tests do have a motivation which is, at least intuitively, satisfactory, although they do not possess, in general, a property such as the UMP property. The LR approach also applies to multidimensional parameters and leads to manageable tests. In addition, much of the work needed to set up a LR test has already been done in Section 1 of Chapter 9 about MLE's. In our discussions below, we restrict ourselves to the Normal case, where exact tests do exist. In the next chapter, we proceed with the Multinomial distribution, where we have to be satisfied with approximations.

The basics here, as we recall from Chapter 8, Section 8.3, are as follows: $X_1, \ldots, X_n$ is a random sample from the p.d.f. $f(\cdot; \boldsymbol{\theta})$, $\boldsymbol{\theta} \in \Omega \subseteq \mathfrak{R}^r$, $r \geq 1$, and $\omega$ is a (proper) subset of $\Omega$. On the basis of this random sample, test the hypothesis $H_0: \boldsymbol{\theta} \in \omega$ at level of significance $\alpha$. (In the present framework, the alternative is $H_A: \boldsymbol{\theta} \notin \omega$, but is not explicitly stated.) Then, by relation (8) in Chapter 8 and the discussion following it, reject $H_0$ whenever

$$\lambda < \lambda_0, \quad \text{where } \lambda_0 \text{ is a constant to be specified,} \tag{41}$$

or

$$g(\lambda) > g(\lambda_0), \text{ or } g(\lambda) < g(\lambda_0), \text{ for some strictly monotone function } g; \tag{42}$$

$g(\lambda) = -2 \log \lambda$ is such a function, and $H_0$ may be rejected whenever

$$-2 \log \lambda > C, \text{ a constant to be determined} \tag{43}$$

(see relation (10) in Chapter 8, and Theorem 1 in Chapter 12).

Recall that

$$\lambda = \lambda(x_1,\ldots,x_n) = \frac{L(\hat{\omega})}{L(\hat{\Omega})}, \tag{44}$$

where, with $x = (x_1,\ldots,x_n)$, the observed value of $X = (X_1,\ldots,X_n)$, $L(\hat{\Omega})$ is the maximum of the likelihood function $L(\theta \mid x)$, which obtains if $\theta$ is replaced by its MLE, and $L(\hat{\omega})$ is again the maximum of the likelihood function under the restriction that $\theta$ lies in $\omega$. Clearly, $L(\hat{\omega}) = L(\hat{\theta}_\omega)$, where $\hat{\theta}_\omega$ is the MLE of $\theta$ under the restriction that $\theta$ lies in $\omega$. Actually, much of the difficulty associated with the present method stems from the fact that, in practice, obtaining $\hat{\theta}_\omega$ is far from a trivial problem.

The following two examples shed some light on how a LR test is actually constructed. These examples are followed by a series of applications to Normal populations.

**Example 14.** Determine the LR test for testing the hypothesis $H_0: \theta = 0$ (against the alternative $H_A: \theta \neq 0$) at level of significance $\alpha$ on the basis of one observation from the p.d.f. $f(x; \theta) = \frac{1}{\pi} \times \frac{1}{1+(x-\theta)^2}, x \in \Re, \theta \in \Omega = \Re$ (the Cauchy p.d.f.).

**Discussion.** First, $f(\cdot; \theta)$ is a p.d.f., since

$$\frac{1}{\pi}\int_{-\infty}^{\infty}\frac{dx}{1+(x-\theta)^2} = \frac{1}{\pi}\int_{-\infty}^{\infty}\frac{dy}{1+y^2} \quad (\text{by setting } x - \theta = y) = \frac{1}{\pi}\int_{-\pi/2}^{\pi/2} dt = 1$$

$$\left(\text{by setting } y = \tan t, \text{ so that } 1+y^2 = 1 + \frac{\sin^2 t}{\cos^2 t} = \frac{1}{\cos^2 t},\right.$$

$$\left.\frac{dy}{dt} = \frac{d}{dt}\left(\frac{\sin t}{\cos t}\right) = \frac{1}{\cos^2 t}, \quad \text{and} \quad -\frac{\pi}{2} < t < \frac{\pi}{2}\right).$$

Next, clearly, $L(\theta \mid x)(= f(x; \theta))$ is maximized for $\theta = x$, so that $\lambda = \frac{1}{\pi} \times \frac{1}{1+x^2}/\frac{1}{\pi} = \frac{1}{1+x^2}$, and $\lambda < \lambda_0$, if and only if $x^2 > \frac{1}{\lambda_0} - 1 = C$, or $x < -C$ or $x > C$, where $C$ is determined through the relation: $P_0(X < -C \text{ or } X > C) = \alpha$, or $P(X > C) = \frac{\alpha}{2}$ due to the symmetry (around 0) of the p.d.f. $f(x; 0)$. But

$$P(X > C) = \int_C^{\infty}\frac{1}{\pi} \times \frac{dx}{1+x^2} = \frac{1}{\pi}\int_{\tan^{-1}C}^{\pi/2} dt = \frac{1}{\pi}\left(\frac{\pi}{2} - \tan^{-1}C\right) = \frac{\alpha}{2},$$

or $\tan^{-1}C = \frac{(1-\alpha)\pi}{2}$, and hence $C = \tan(\frac{(1-\alpha)\pi}{2})$. So, $H_0$ is rejected whenever $x < -\tan(\frac{(1-\alpha)\pi}{2})$ or $x > \tan(\frac{(1-\alpha)\pi}{2})$. For example, for $\alpha = 0.05, C = \tan(0.475\pi) \simeq 12.706$, and $H_0$ is rejected when $x < -12.706$ or $x > 12.706$.

**Example 15.** Let $X_1,\ldots,X_n$ be a random sample of size $n$ from the Negative Exponential p.d.f. $f(x; \theta) = \theta e^{-\theta x}, x > 0 (\theta > 0)$. Derive the LR test for testing the hypothesis $H_0: \theta = \theta_0$ (against the alternative $H_A: \theta \neq \theta_0$) at level of significance $\alpha$.

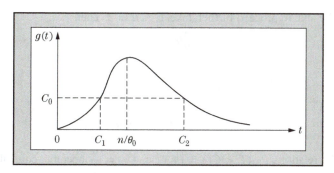

**FIGURE 11.7**

Graphical determination of the rejection region.

**Discussion.**   Here

$$L(\theta \mid \boldsymbol{x}) = \theta^n e^{-\theta t}, \quad \text{where} \quad \boldsymbol{x} = (x_1, \ldots, x_n) \text{ and } t = \sum_{i=1}^{n} x_i.$$

We also know that the MLE of $\theta$ is $\hat{\theta} = 1/\bar{x} = n/t$. Therefore, the LR $\lambda$ is given by:

$$\lambda = \theta_0^n e^{-\theta_0 t} \bigg/ \left(\frac{n}{t}\right)^n e^{-n} = \left(\frac{e\theta_0}{n}\right)^n t^n e^{-\theta_0 t},$$

and hence $\lambda < \lambda_0$, if and only if $te^{-\frac{\theta_0}{n}t} < C_0 (= \frac{n\lambda_0^{1/n}}{e\theta_0})$. We wish to determine the cutoff point $C_0$. To this end, set $g(t) = te^{-dt}$ $(d = \theta_0/n)$ and observe that $g(t)$ is increasing for $0 < t < \frac{1}{d} = \frac{n}{\theta_0}$, decreasing for $t > \frac{n}{\theta_0}$, and attains its maximum at $t = n/\theta_0$ (see Figure 11.7). It follows that $te^{-dt} < C_0$, if and only if $t < C_1$ or $t > C_2$. Therefore, by setting $T = \sum_{i=1}^{n} X_i$, we have: $P_{\theta_0}(Te^{-\frac{\theta_0}{n}T} < C_0) = \alpha$, if and only if $P_{\theta_0}(T < C_1 \text{ or } T > C_2) = \alpha$. For simplicity, let us take the two-tail probabilities equal. Thus,

$$P_{\theta_0}(T < C_1) = P(T > C_2) = \frac{\alpha}{2}.$$

By the fact that the independent $X_i$'s have the $f(x; \theta_0) = \theta_0 e^{-\theta_0 x}$ p.d.f. (under $H_0$), it follows that $T$ is distributed as Gamma with $\alpha = n$ and $\beta = \frac{1}{\theta_0}$. Therefore, its p.d.f. is given by:

$$f_T(t) = \frac{\theta_0^n}{\Gamma(n)} t^{n-1} e^{-\theta_0 t}, \quad t > 0.$$

Then $C_1$ and $C_2$ are determined by:

$$\int_0^{C_1} f_T(t)dt = \int_{C_2}^{\infty} f_T(t)dt = \frac{\alpha}{2}.$$

In order to be able to proceed further, take, e.g., $n = 2$. Then

$$f_T(t) = \theta_0^2 t e^{-\theta_0 t}, \quad t > 0,$$

and

$$\int_0^{C_1} \theta_0^2 t e^{-\theta_0 t} \, dt = 1 - e^{-\theta_0 C_1} - \theta_0 C_1 e^{-\theta_0 C_1},$$

$$\int_{C_2}^{\infty} \theta_0^2 t e^{-\theta_0 t} \, dt = e^{-\theta_0 C_2} + \theta_0 C_2 e^{-\theta_0 C_2}.$$

Thus, the relations $P_{\theta_0}(T < C_1) = P_{\theta_0}(T > C_2) = \dfrac{\alpha}{2}$ become, equivalently, for $\alpha = 0.05$:

$$0.975 e^p - p = 1, \quad \text{with } p = \theta_0 C_1; \qquad 0.025 e^q - q = 1, \quad \text{with } q = \theta_0 C_2.$$

By trial and error, we find: $p = 0.242$ and $q = 5.568$, so that $C_1 = 0.242/\theta_0$ and $C_2 = 5.568/\theta_0$. Thus, for $n = 2$ and by splitting the error $\alpha = 0.05$ equally between the two tails, the LR test rejects $H_0$ when $t(= x_1 + x_2) < 0.242/\theta_0$ or $t > 5.568/\theta_0$. For example, for $\theta_0 = 1$, the test rejects $H_0$ when $t < 0.242$ or $t > 5.568$. (See, however, Exercise 2.7(ii) in this chapter for an alternative approach.)

**APPLICATIONS TO THE NORMAL CASE**     The applications to be discussed here are organized as follows: First, we consider the one-sample case and test a hypothesis about the mean, regardless of whether the variance is known or unknown. Next, a hypothesis is tested about the variance, regardless of whether the mean is known or not. Second, we consider the two-sample problem and make the realistic assumption that all parameters are unknown. Then the hypothesis is tested about the equality of the means, and, finally, the variances are compared through their ratio.

## 11.4.1 TESTING HYPOTHESES FOR THE PARAMETERS IN A SINGLE NORMAL POPULATION

Here $X_1, \ldots, X_n$ is a random sample from the $N(\mu, \sigma^2)$, and we are interested in testing: (i) $H_0$: $\mu = \mu_0$, $\sigma$ known; (ii) $H_0$: $\mu = \mu_0$, $\sigma$ unknown; (iii) $H_0$: $\sigma = \sigma_0$ (or $\sigma^2 = \sigma_0^2$), $\mu$ known; (iv) $H_0$: $\sigma = \sigma_0$ (or $\sigma^2 = \sigma_0^2$), $\mu$ unknown.

**Discussion.**

**(i)** $H_0$: $\mu = \mu_0$, $\sigma$ *known.* Under $H_0$,

$$L(\hat{\omega}) = (2\pi\sigma^2)^{-n/2} \exp\left[ -\frac{1}{2\sigma^2} \sum_{i=1}^{n} (x_i - \mu_0)^2 \right],$$

and

$$L(\hat{\Omega}) = (2\pi\sigma^2)^{-n/2} \exp\left[ -\frac{1}{2\sigma^2} \sum_{i=1}^{n} (x_i - \bar{x})^2 \right], \quad \text{since } \hat{\mu}_\Omega = \bar{x}.$$

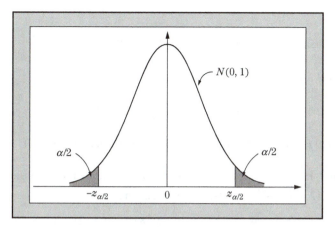

**FIGURE 11.8**

Rejection region of the hypothesis $H_0$ in (i) (the shaded areas), and the respective probabilities of rejection.

Forming the likelihood ratio $\lambda$ and taking $-2 \log \lambda$, we have:

$$-2 \log \lambda = \frac{1}{\sigma^2} \sum_{i=1}^{n} [(x_i - \mu_0)^2 - (x_i - \bar{x})^2] = \left[ \frac{\sqrt{n}(\bar{x} - \mu_0)}{\sigma} \right]^2. \qquad (45)$$

Then

$$-2 \log \lambda > C^2, \quad \text{if and only if} \quad \left[ \frac{\sqrt{n}(\bar{x} - \mu_0)}{\sigma} \right]^2 > C^2 \quad (\text{some } C > 0),$$

and this happens, if and only if

$$\frac{\sqrt{n}(\bar{x} - \mu_0)}{\sigma} < -C, \quad \text{or} \quad \frac{\sqrt{n}(\bar{x} - \mu_0)}{\sigma} > C.$$

Under $H_0$, $\frac{\sqrt{n}(\bar{X} - \mu_0)}{\sigma} \sim N(0, 1)$, so that the relation (Figure 11.8)

$$P_{\mu_0} \left[ \frac{\sqrt{n}(\bar{X} - \mu_0)}{\sigma} < -C, \text{ or } \frac{\sqrt{n}(\bar{X} - \mu_0)}{\sigma} > C \right] = \alpha, \quad \text{gives } C = z_{\alpha/2}.$$

Thus, the likelihood ratio test is:

$$\varphi(x_1, \ldots, x_n) = \begin{cases} 1 & \text{if } \frac{\sqrt{n}(\bar{x} - \mu_0)}{\sigma} < -z_{\alpha/2}, \text{ or } \frac{\sqrt{n}(\bar{x} - \mu_0)}{\sigma} > z_{\alpha/2} \\ 0 & \text{otherwise.} \end{cases} \qquad (46)$$

Since, for any $\mu$,

$$\frac{\sqrt{n}(\bar{x} - \mu_0)}{\sigma} > z_{\alpha/2} \text{ is equivalent to } \frac{\sqrt{n}(\bar{x} - \mu)}{\sigma} > \frac{\sqrt{n}(\mu_0 - \mu)}{\sigma} + z_{\alpha/2}, \qquad (47)$$

and likewise

$$\frac{\sqrt{n}(\bar{x} - \mu_0)}{\sigma} < -z_{\alpha/2} \text{ is equivalent to } \frac{\sqrt{n}(\bar{x} - \mu)}{\sigma} < \frac{\sqrt{n}(\mu_0 - \mu)}{\sigma} - z_{\alpha/2}, \quad (48)$$

it follows that the power of the test is given by:

$$\pi_\varphi(\mu) = 1 - \Phi\left[\frac{\sqrt{n}(\mu_0 - \mu)}{\sigma} + z_{\alpha/2}\right] + \Phi\left[\frac{\sqrt{n}(\mu_0 - \mu)}{\sigma} - z_{\alpha/2}\right]. \quad (49)$$

*Numerical Example.*    Suppose $n = 36$, $\mu_0 = 10$, and let $\alpha = 0.01$.

**Discussion.**    Here $z_{\alpha/2} = z_{0.005} = 2.58$, and, if $\sigma = 4$, the power of the test is:
For $\mu = 12$, $\frac{\sqrt{n}(\mu_0 - \mu)}{\sigma} = \frac{6(10 - 12)}{4} = -3$, so that

$$\pi_\varphi(12) = 1 - \Phi(-0.42) + \Phi(-5.58) \simeq \Phi(0.42) = 0.662757;$$

and for $\mu = 6$, $\pi_\varphi(6) = 1 - \Phi(8.58) + \Phi(3.42) \simeq 0.999687$.
For the computation of $P$-values, take as cutoff point the value of the test statistic for the observed value of $\bar{X}$. Thus, for $\bar{x} = 11.8$, $\frac{\sqrt{n}(\bar{x} - \mu_0)}{\sigma} = 1.5 \times 1.8 = 2.7$, and therefore

$$P\text{-value} = P_{\mu_0}\left[\frac{\sqrt{n}(\bar{X} - \mu_0)}{\sigma} < -2.7\right] + P_{\mu_0}\left[\frac{\sqrt{n}(\bar{X} - \mu_0)}{\sigma} > 2.7\right]$$

$$= \Phi(-2.7) + 1 - \Phi(2.7) = 2[1 - \Phi(2.7)] = 2 \times 0.003467 = 0.006934, \text{ suggest-}$$
ing strong rejection of $H_0$.

**(ii)**   $H_0$: $\mu = \mu_0$, $\sigma$ *unknown*. Under $H_0$,

$$L(\hat{\omega}) = \left(2\pi\hat{\sigma}_\omega^2\right)^{-n/2} \exp\left[-\frac{1}{2\hat{\sigma}_\omega^2}\sum_{i=1}^{n}(x_i - \mu_0)^2\right] = \left(2\pi\hat{\sigma}_\omega^2\right)^{-n/2} \exp\left(-\frac{n}{2}\right),$$

and

$$L(\hat{\Omega}) = \left(2\pi\hat{\sigma}_\Omega^2\right)^{-n/2} \exp\left[-\frac{1}{2\hat{\sigma}_\Omega^2}\sum_{i=1}^{n}(x_i - \bar{x})^2\right] = \left(2\pi\hat{\sigma}_\Omega^2\right)^{-n/2} \exp\left(-\frac{n}{2}\right),$$

since $\hat{\sigma}_\omega^2 = \frac{1}{n}\sum_{i=1}^{n}(x_i - \mu_0)^2$ and $\hat{\sigma}_\Omega^2 = \frac{1}{n}\sum_{i=1}^{n}(x_i - \bar{x})^2$. Then

$$\lambda = \left(\frac{\hat{\sigma}_\Omega^2}{\hat{\sigma}_\omega^2}\right)^{n/2}, \quad \text{or} \quad \lambda^{2/n} = \frac{\sum_{i=1}^{n}(x_i - \bar{x})^2}{\sum_{i=1}^{n}(x_i - \mu_0)^2}.$$

Observe that

$$\sum_{i=1}^{n}(x_i - \mu_0)^2 = \sum_{i=1}^{n}[(x_i - \bar{x}) + (\bar{x} - \mu_0)]^2 = \sum_{i=1}^{n}(x_i - \bar{x})^2 + n(\bar{x} - \mu_0)^2,$$

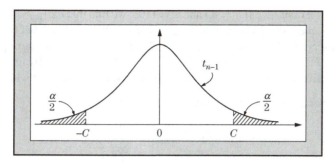

**FIGURE 11.9**

Rejection region of the hypothesis $H_0$ in (ii) (the shaded areas), and the respective probabilities; Here $C = t_{n-1;\alpha/2}$.

and set $t = \sqrt{n}(\bar{x} - \mu_0)\big/\sqrt{\frac{1}{n-1}\sum_{i=1}^{n}(x_i - \bar{x})^2}$. Then

$$\lambda^{2/n} = \frac{\sum_{i=1}^{n}(x_i - \bar{x})^2}{\sum_{i=1}^{n}(x_i - \bar{x})^2 + n(\bar{x} - \mu_0)^2} = \frac{1}{1 + \frac{n(\bar{x}-\mu_0)^2}{\sum_{i=1}^{n}(x_i-\bar{x})^2}}$$

$$= \frac{1}{1 + \frac{1}{n-1} \times \frac{n(\bar{x}-\mu_0)^2}{\frac{1}{n-1}\sum_{i=1}^{n}(x_i-\bar{x})^2}} = \frac{1}{1 + \frac{1}{n-1}\left[\frac{\sqrt{n}(\bar{x}-\mu_0)}{\sqrt{\frac{1}{n-1}\sum_{i=1}^{n}(x_i-\bar{x})^2}}\right]^2}$$

$$= \frac{1}{1 + \frac{t^2}{n-1}}.$$

Since $g(\lambda) = \lambda^{2/n}$ is a strictly increasing function of $\lambda$, the LR test rejects $H_0$ when $\lambda^{2/n} < C_1$ or $\frac{1}{1+\frac{t^2}{n-1}} < C_1$ or $1 + \frac{t^2}{n-1} > C_2$ or $t^2 > C_3$ or, finally, $t < -C$ or $t > C$. Under $H_0$, the distribution of

$$t(X) = \frac{\sqrt{n}(\bar{X} - \mu_0)}{\sqrt{\frac{1}{n-1}\sum_{i=1}^{n}(X_i - \bar{X})^2}}$$

is $t_{n-1}$. Since $P_{\mu_0}[t(X) < -C, \text{ or } t(X) > C] = \alpha$, it follows that $C = t_{n-1;\alpha/2}$ (Figure 11.9). Therefore, the LR test is:

$$\varphi(x_1, \ldots, x_n) = \begin{cases} 1 & \text{if } t < -t_{n-1;\alpha/2}, \text{ or } t > t_{n-1;\alpha/2} \\ 0 & \text{otherwise,} \end{cases} \tag{50}$$

where

$$t = t(x) = \frac{\sqrt{n}(\bar{x} - \mu_0)}{\sqrt{\frac{1}{n-1}\sum_{i=1}^{n}(x_i - \bar{x})^2}}. \tag{51}$$

*Numerical Example.*   If $n = 85$ and $\alpha = 0.05$, we find that $t_{n-1;\alpha/2} = t_{84;0.025} = 1.9886$. Thus, the test rejects $H_0$ whenever $t$ is $< -1.9886$ or $t > 1.9886$.

If the observed value of the test statistic is 2.5 and is taken as the cutoff point, then the $P$-value of the test is: $P\text{-value} = 2[1 - P(t_{84} \leq 2.5)] \simeq 0.015$ (by linear interpolation), a relatively strong rejection of $H_0$; $t_{84}$ is a r.v. distributed as $t$ with 84 d.f..

**(iii)**  $H_0: \sigma = \sigma_0$ (*or* $\sigma^2 = \sigma_0^2$), $\mu$ *known.* Under $H_0$,

$$L(\hat{\omega}) = \left(2\pi\sigma_0^2\right)^{-n/2} \exp\left[-\frac{1}{2\sigma_0^2}\sum_{i=1}^{n}(x_i - \mu)^2\right],$$

and

$$L(\hat{\Omega}) = \left(2\pi\hat{\sigma}_\Omega^2\right)^{-n/2} \exp\left[-\frac{1}{2\hat{\sigma}_\Omega^2}\sum_{i=1}^{n}(x_i - \mu)^2\right] = \left(2\pi\hat{\sigma}_\Omega^2\right)^{-n/2} \exp\left(-\frac{n}{2}\right),$$

since $\hat{\sigma}_\Omega^2 = \frac{1}{n}\sum_{i=1}^{n}(x_i - \mu)^2$. Therefore

$$\lambda = \left(\frac{\hat{\sigma}_\Omega^2}{\sigma_0^2}\right)^{n/2} e^{n/2} \exp\left[-\frac{1}{2\sigma_0^2}\sum_{i=1}^{n}(x_i - \mu)^2\right]$$

$$= e^{n/2}\left[\frac{1}{n}\sum_{i=1}^{n}\left(\frac{x_i - \mu}{\sigma_0}\right)^2\right]^{n/2} \exp\left[-\frac{1}{2}\sum_{i=1}^{n}\left(\frac{x_i - \mu}{\sigma_0}\right)^2\right]$$

$$= e^{n/2}u^{n/2}\exp\left(-\frac{nu}{2}\right), \quad \text{where } u = \frac{1}{n}\sum_{i=1}^{n}\left(\frac{x_i - \mu}{\sigma_0}\right)^2.$$

The function $\lambda = \lambda(u)$, $u \geq 0$, has the following properties (see Exercise 4.16):

$$\left.\begin{array}{l} \lambda(u) \text{ is strictly increasing for } 0 \leq u \leq 1, \\ \lambda(u) \text{ is strictly decreasing for } u > 1, \\ \max\{\lambda(u); 0 \leq u < \infty\} = \lambda(1) = 1, \text{ and} \\ \lambda(u) \to 0, \text{ as } u \to \infty, \text{and, of course,} \\ \lambda(0) = 0. \end{array}\right\} \tag{52}$$

On the basis of these observations, the picture of $\lambda(u)$ is as in Figure 11.10.

Therefore $\lambda(u) \leq \lambda_0$ if and only if $u \leq C_1$, or $u \geq C_2$, where $C_1$ and $C_2$ are determined by the requirement that

$$P_{\sigma_0}(U \leq C_1, \text{ or } U \geq C_2) = \alpha, \quad U = \frac{1}{n}\sum_{i=1}^{n}\left(\frac{X_i - \mu}{\sigma_0}\right)^2.$$

However, under $H_0$, $\sum_{i=1}^{n}(\frac{X_i - \mu}{\sigma_0})^2 \sim \chi_n^2$, so that $U = \frac{X}{n}$ with $X \sim \chi_n^2$. Then $P_{\sigma_0}(U \leq C_1, \text{ or } U \geq C_2) = P_{\sigma_0}(X \leq nC_1, \text{ or } X \geq nC_2) = \alpha$, and, for convenience, we may take the two-tail probabilities equal to $\frac{\alpha}{2}$. Then $nC_1 = \chi_{n;1-\alpha/2}^2, nC_2 = \chi_{n;\alpha/2}^2$ (Figure 11.11). Summarizing what we have done so far, we have:

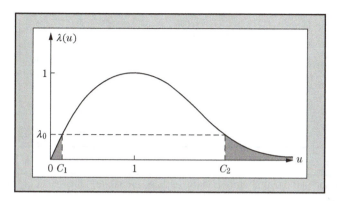

**FIGURE 11.10**

Graphical determination of the rejection region.

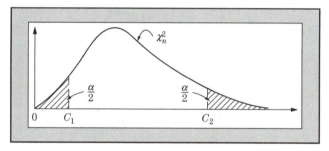

**FIGURE 11.11**

Rejection region of the hypothesis $H_0$ in (iii) (the shaded areas), and the respective probabilities; here $C_1 = \chi^2_{n;1-\frac{\alpha}{2}}$, $C_2 = \chi^2_{n;\frac{\alpha}{2}}$.

$$\varphi(x_1,\ldots,x_n) = \begin{cases} 1 & \text{if } \sum_{i=1}^n \left(\frac{x_i-\mu}{\sigma_0}\right)^2 \leq \chi^2_{n;1-\alpha/2}, \quad \text{or } \sum_{i=1}^n \left(\frac{x_i-\mu}{\sigma_0}\right)^2 \geq \chi^2_{n;\alpha/2} \\ 0 & \text{otherwise.} \end{cases}$$

(53)

(See also Exercise 4.25(i) below.)

*Numerical Example.* For $n = 40$ and $\alpha = 0.01$, we find $\chi^2_{n;1-\alpha/2} = \chi^2_{40;0.995} = 20.707$ and $\chi^2_{n;\alpha/2} = \chi^2_{40;0.005} = 66.766$. Therefore, the test rejects $H_0$ whenever $\sum_{i=1}^{40}\left(\frac{x_i-\mu}{\sigma_0}\right)^2$ is either $\leq 20.707$ or $\geq 66.766$.

**(iv)** $H_0: \sigma = \sigma_0$ (*or* $\sigma^2 = \sigma_0^2$), $\mu$ *unknown.* Under $\omega$,

$$L(\hat{\omega}) = \left(2\pi\sigma_0^2\right)^{-n/2} \exp\left[-\frac{1}{2\sigma_0^2}\sum_{i=1}^n (x_i - \bar{x})^2\right], \quad \text{since } \hat{\mu}_\omega = \bar{x},$$

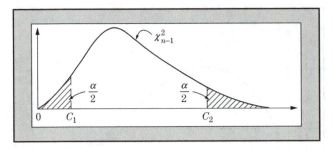

**FIGURE 11.12**

Rejection region of the hypothesis $H_0$ in (iv) (shaded areas), and the respective probabilities; here $C_1 = \chi^2_{n-1;1-\frac{\alpha}{2}}, C_2 = \chi^2_{n-1;\frac{\alpha}{2}}$.

and

$$L(\hat{\Omega}) = \left(2\pi\hat{\sigma}^2_\Omega\right)^{-n/2} \exp\left[-\frac{1}{2\hat{\sigma}^2_\Omega}\sum_{i=1}^n (x_i - \bar{x})^2\right] = \left(2\pi\hat{\sigma}^2_\Omega\right)^{-n/2} \exp\left(-\frac{n}{2}\right),$$

since $\hat{\sigma}^2_\Omega = \frac{1}{n}\sum_{i=1}^n (x_i - \bar{x})^2$. Therefore

$$\lambda = \left(\frac{\hat{\sigma}^2_\Omega}{\sigma^2_0}\right)^{n/2} e^{n/2} \exp\left[-\frac{1}{2\sigma^2_0}\sum_{i=1}^n (x_i - \bar{x})^2\right],$$

and then proceed exactly as in the previous case with $u = \frac{1}{n}\sum_{i=1}^n (\frac{x_i-\bar{x}}{\sigma_0})^2$, in order to arrive at the following modified test; namely,

$$\varphi(x_1,\ldots,x_n) = \begin{cases} 1 & \text{if } \sum_{i=1}^n \left(\frac{x_i-\bar{x}}{\sigma_0}\right)^2 \leq \chi^2_{n-1;1-\alpha/2}, \quad \text{or } \sum_{i=1}^n \left(\frac{x_i-\bar{x}}{\sigma_0}\right)^2 \geq \chi^2_{n-1;\alpha/2} \\ 0 & \text{otherwise.} \end{cases}$$

$$(54)$$

(See also Exercise 4.25(ii) below.)

*Numerical Example.* With the values of $n$ and $\alpha$ as in the previous case ($n = 40$, $\alpha = 0.01$), we find $\chi^2_{n-1;1-\alpha/2} = \chi^2_{39;0.995} = 19.996$ and $\chi^2_{n-1;\alpha/2} = \chi^2_{39;0.005} = 65.476$, so that the test rejects $H_0$ whenever $\sum_{i=1}^n (\frac{x_i-\bar{x}}{\sigma_0})^2$ is $\leq 19.996$ or $\geq 65.476$ (Figure 11.12).

## 11.4.2 COMPARING THE PARAMETERS OF TWO NORMAL POPULATIONS

Here, we have two independent random samples $X_1,\ldots,X_m \sim N(\mu_1,\sigma^2_1)$ and $Y_1,\ldots,Y_n \sim N(\mu_2,\sigma^2_2)$ with all parameters unknown. The two populations are compared, first, by way of their means, and second, through their variances. When comparing these populations through their means, it is necessary from a mathematical viewpoint (i.e., in order to be able to derive an exact distribution for the test statistic) that the variances, although unknown, be equal.

**(i)** $H_0: \mu_1 = \mu_2 = \mu$, say, unknown, $\sigma_1 = \sigma_2 = \sigma$, say, unknown.

The (joint) likelihood function of the $X_i$'s and the $Y_j$'s here is

$$(2\pi\sigma^2)^{-(m+n)/2} \exp\left\{ -\frac{1}{2\sigma^2}\left[ \sum_{i=1}^{m}(x_i - \mu_1)^2 + \sum_{j=1}^{n}(y_j - \mu_2)^2 \right] \right\}. \tag{55}$$

Maximizing (55) with respect to $\mu_1, \mu_2$, and $\sigma^2$, we find for their MLE's:

$$\hat{\mu}_1 = \bar{x}, \quad \hat{\mu}_2 = \bar{y}, \quad \hat{\sigma}_\Omega^2 = \frac{1}{m+n}\left[ \sum_{i=1}^{m}(x_i - \bar{x})^2 + \sum_{j=1}^{n}(y_j - \bar{y})^2 \right]. \tag{56}$$

Hence

$$L(\hat{\Omega}) = \left(2\pi\hat{\sigma}_\Omega^2\right)^{-\frac{m+n}{2}} \exp\left(-\frac{m+n}{2}\right). \tag{57}$$

Next, under $H_0$, the (joint) likelihood function becomes

$$(2\pi\sigma^2)^{-(m+n)/2} \exp\left\{ -\frac{1}{2\sigma^2}\left[ \sum_{i=1}^{m}(x_i - \mu)^2 + \sum_{j=1}^{n}(y_j - \mu)^2 \right] \right\}, \tag{58}$$

from the maximization (with respect to $\mu$ and $\sigma^2$) of which we obtain the MLE's:

$$\hat{\mu}_\omega = \frac{1}{m+n}\left( \sum_{i=1}^{m}x_i + \sum_{j=1}^{n}y_j \right) = \frac{m\bar{x} + n\bar{y}}{m+n},$$

$$\tag{59}$$

$$\hat{\sigma}_\omega^2 = \frac{1}{m+n}\left[ \sum_{i=1}^{m}(x_i - \hat{\mu}_\omega)^2 + \sum_{j=1}^{n}(y_j - \hat{\mu}_\omega)^2 \right].$$

Inserting these expressions in (58), we then have

$$L(\hat{\omega}) = \left(2\pi\hat{\sigma}_\omega^2\right)^{-\frac{m+n}{2}} \exp\left(-\frac{m+n}{2}\right). \tag{60}$$

Thus, the likelihood function becomes, on account of (57) and (60),

$$\lambda = \left(\frac{\hat{\sigma}_\Omega^2}{\hat{\sigma}_\omega^2}\right)^{\frac{m+n}{2}}, \quad \text{or} \quad \lambda^{2/(m+n)} = \frac{\hat{\sigma}_\Omega^2}{\hat{\sigma}_\omega^2}. \tag{61}$$

Next,

$$\sum_{i=1}^{m}(x_i - \hat{\mu}_\omega)^2 = \sum_{i=1}^{m}[(x_i - \bar{x}) + (\bar{x} - \hat{\mu}_\omega)]^2 = \sum_{i=1}^{m}(x_i - \bar{x})^2 + m(\bar{x} - \hat{\mu}_\omega)^2$$

$$= \sum_{i=1}^{m}(x_i - \bar{x})^2 + m\left(\bar{x} - \frac{m\bar{x} + n\bar{y}}{m+n}\right)^2 = \sum_{i=1}^{m}(x_i - \bar{x})^2 + \frac{mn^2(\bar{x} - \bar{y})^2}{(m+n)^2},$$

and likewise,

$$\sum_{j=1}^{n}(y_j - \hat{\mu}_\omega)^2 = \sum_{j=1}^{n}(y_j - \bar{y})^2 + \frac{m^2 n(\bar{x} - \bar{y})^2}{(m+n)^2}.$$

Then, by means of (56) and (59), $\hat{\sigma}_\omega^2$ is written as follows:

$$\hat{\sigma}_\omega^2 = \frac{1}{m+n}\left[\sum_{i=1}^{m}(x_i - \bar{x})^2 + \sum_{j=1}^{n}(y_j - \bar{y})^2\right] + \frac{mn^2(\bar{x} - \bar{y})^2 + m^2 n(\bar{x} - \bar{y})^2}{(m+n)^3}$$

$$= \hat{\sigma}_\Omega^2 + \frac{mn(\bar{x} - \bar{y})^2(m+n)}{(m+n)^3} = \hat{\sigma}_\Omega^2 + \frac{mn(\bar{x} - \bar{y})^2}{(m+n)^2}. \tag{62}$$

Therefore (61) yields, by way of (62) and (56),

$$\lambda^{2/(m+n)} = \frac{\hat{\sigma}_\Omega^2}{\hat{\sigma}_\Omega^2 + \frac{mn(\bar{x}-\bar{y})^2}{(m+n)^2}} = \frac{1}{1 + \frac{mn(\bar{x}-\bar{y})^2}{(m+n)^2} \times \frac{1}{\hat{\sigma}_\Omega^2}}$$

$$= \left\{1 + \left[\frac{mn(\bar{x} - \bar{y})^2}{(m+n)^2}\right]\Big/\hat{\sigma}_\Omega^2\right\}^{-1}$$

$$= \left\{1 + \left[\frac{mn}{m+n}(\bar{x} - \bar{y})^2\right]\Big/\left[\sum_{i=1}^{m}(x_i - \bar{x})^2 + \sum_{j=1}^{n}(y_j - \bar{y})^2\right]\right\}^{-1}$$

$$= \left[1 + \left\{\left[\frac{mn}{m+n}(\bar{x} - \bar{y})^2\right]\Big/(m+n-2)\right\}\Big/\frac{1}{m+n-2}\right.$$

$$\times\left.\left[\sum_{i=1}^{m}(x_i - \bar{x})^2 + \sum_{j=1}^{n}(y_j - \bar{y})^2\right]\right]^{-1}$$

$$= \left[1 + \left\{\frac{\sqrt{\frac{mn}{m+n}}(\bar{x} - \bar{y})}{\sqrt{\frac{1}{m+n-2}\left[\sum_{i=1}^{m}(x_i - \bar{x})^2 + \sum_{j=1}^{n}(y_j - \bar{y})^2\right]}}\right\}^2\Big/(m+n-2)\right]^{-1}$$

$$= \left(1 + \frac{t^2}{m+n-2}\right)^{-1}, \quad \text{where}$$

$$t = t(\boldsymbol{x}, \boldsymbol{y}) = \sqrt{\frac{mn}{m+n}}(\bar{x} - \bar{y})\Big/\sqrt{\frac{1}{m+n-2}\left[\sum_{i=1}^{m}(x_i - \bar{x})^2 + \sum_{j=1}^{n}(y_j - \bar{y})^2\right]}. \tag{63}$$

So, $\lambda^{2/(m+n)} = (1 + \frac{t^2}{m+n-2})^{-1}$ and hence $\lambda = (1 + \frac{t^2}{m+n-2})^{-\frac{m+n}{2}}$. Since $\lambda$ is strictly decreasing in $t^2$ (see Exercise 4.19 below), the LR test rejects $H_0$ whenever $t^2 > C_0$ or, equivalently, $t \leq -C$ or $t \geq C$. The constant $C$ is to be determined by

$$P_{H_0}[t(\boldsymbol{X}, \boldsymbol{Y}) \leq -C, \text{ or } t(\boldsymbol{X}, \boldsymbol{Y}) \geq C] = \alpha, \tag{64}$$

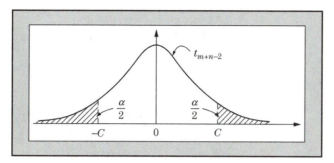

**FIGURE 11.13**

Rejection region of the hypothesis $H_0$ in (i) (shaded areas), and the respective probabilities; here $C = t_{m+n-2;\frac{\alpha}{2}}$.

where $t(X, Y)$ is taken from (63) with the $x_i$'s and the $y_i$'s being replaced by the r.v.'s $X_i$'s and $Y_j$'s. However, under $H_0$,

$$t(X, Y) \sim t_{m+n-2}, \tag{65}$$

so that (64) yields $C = t_{m+n-2;\alpha/2}$. In conclusion, then,

$$\varphi(x, y) = \begin{cases} 1 & \text{if } t(x, y) < -t_{m+n-2;\alpha/2}, \quad \text{or } t(x, y) > t_{m+n-2;\alpha/2} \\ 0 & \text{otherwise,} \end{cases} \tag{66}$$

where $t(x, y)$ is given by (63) (Figure 11.13).

*Numerical Example.* For $m = 40, n = 50$, and $\alpha = 0.05$, we get $t_{m+n-2;\alpha/2} = t_{88;0.025} = 1.9873$, and hence the hypothesis about equality of means is rejected whenever $|t(x, y)| \geq 1.9873$.

When the observed value of the test statistics is 2.52 and is taken as the cutoff point, the $P$-value of the test is $P$-value $= 2[1 - P(t_{88} \leq 1.9873)] \simeq 0.014$ (by linear interpolation), a relatively strong rejection of $H_0$; again, $t_{88}$ is a $t$-distributed r.v. with 88 d.f.

**(ii)** $H_0: \sigma_1 = \sigma_2 = \sigma$, *say (or $\sigma_1^2 = \sigma_2^2 = \sigma^2$, say), $\mu_1, \mu_2$ unknown.*

The (joint) likelihood function of the $X_i$'s and $Y_j$'s is here

$$(2\pi)^{-\frac{m+n}{2}} (\sigma_1^2)^{-m/2} (\sigma_2^2)^{-n/2} \exp\left[ -\frac{1}{2\sigma_1^2} \sum_{i=1}^{m} (x_i - \mu_1)^2 - \frac{1}{2\sigma_2^2} \sum_{j=1}^{n} (y_j - \mu_2)^2 \right]. \tag{67}$$

Maximizing (67) with respect to all four parameters, we find the following MLE's:

$$\hat{\mu}_{1,\Omega} = \bar{x}, \ \hat{\mu}_{2,\Omega} = \bar{y}, \ \hat{\sigma}_{1,\Omega}^2 = \frac{1}{m} \sum_{i=1}^{m} (x_i - \bar{x})^2, \ \hat{\sigma}_{2,\Omega}^2 = \frac{1}{n} \sum_{j=1}^{n} (y_j - \bar{y})^2. \tag{68}$$

Then

$$L(\hat{\Omega}) = (2\pi)^{-\frac{m+n}{2}} (\hat{\sigma}_{1,\Omega}^2)^{-m/2} (\hat{\sigma}_{2,\Omega}^2)^{-n/2} \exp\left(-\frac{m+n}{2}\right). \tag{69}$$

Under $H_0$, the likelihood function has the form (55), and the MLE's are already available and given by (56). That is,

$$\hat{\mu}_{1,\omega} = \bar{x}, \quad \bar{\mu}_{2,\omega} = \bar{y}, \quad \hat{\sigma}_{\omega}^2 = \frac{1}{m+n}\left[\sum_{i=1}^{m}(x_i - \bar{x})^2 + \sum_{j=1}^{n}(y_j - \bar{y})^2\right]. \tag{70}$$

Therefore,

$$L(\hat{\omega}) = (2\pi)^{-\frac{m+n}{2}} (\hat{\sigma}_{\omega}^2)^{-\frac{m+n}{2}} \exp\left(-\frac{m+n}{2}\right). \tag{71}$$

For simplicity, set $\sum_{i=1}^{m}(x_i - \bar{x})^2 = a$, $\sum_{j=1}^{n}(y_j - \bar{y})^2 = b$. Then the LR is, by means of (68) through (71),

$$\lambda = \frac{(\hat{\sigma}_{1,\Omega}^2)^{m/2}(\hat{\sigma}_{2,\Omega}^2)^{n/2}}{(\hat{\sigma}_{\omega}^2)^{(m+n)/2}} = \frac{m^{-m/2}n^{-n/2}a^{m/2}b^{n/2}}{(m+n)^{-(m+n)/2}(a+b)^{(m+n)/2}}$$

$$= \frac{(m+n)^{(m+n)/2}}{m^{m/2}n^{n/2}} \times \frac{(a/b)^{m/2}}{(1+\frac{a}{b})^{(m+n)/2}} \quad \text{(dividing by } b^{(m+n)/2})$$

$$= \frac{(m+n)^{(m+n)/2}}{m^{m/2}n^{n/2}} \times \frac{[(\frac{m-1}{n-1})(\frac{a}{m-1}/\frac{b}{n-1})]^{m/2}}{[1+(\frac{m-1}{n-1})(\frac{a}{m-1}/\frac{b}{n-1})]^{(m+n)/2}}$$

$$= \frac{(m+n)^{(m+n)/2}}{m^{m/2}n^{n/2}} \times \frac{(\frac{m-1}{n-1}u)^{m/2}}{(1+\frac{m-1}{n-1}u)^{(m+n)/2}}, \quad \text{where } u = \frac{a}{m-1}/\frac{b}{n-1}.$$

So

$$\lambda = \lambda(u) = \frac{(m+n)^{(m+n)/2}}{m^{m/2}n^{n/2}} \times \frac{(\frac{m-1}{n-1}u)^{m/2}}{(1+\frac{m-1}{n-1}u)^{(m+n)/2}}, \quad u \geq 0. \tag{72}$$

The function $\lambda(u)$ has the following properties:

$$\left.\begin{array}{l}
\lambda(0) = 0 \text{ and } \lambda(u) \to 0, \text{ as } u \to \infty, \\[4pt]
\frac{d}{du}\lambda(u) = 0 \text{ for } u = u_0 = \frac{m(n-1)}{n(m-1)}, \frac{d}{du}\lambda(u) > 0 \text{ for } u < u_0, \text{ and} \\[4pt]
\frac{d}{du}\lambda(u) < 0 \text{ for } u > u_0, \text{ so that } \lambda(u) \text{ is} \\[4pt]
\text{maximized for } u = u_0, \text{ and } \lambda(u_0) = 1.
\end{array}\right\} \tag{73}$$

On the basis of these properties, the picture of $\lambda(u)$ is as in Figure 11.14.

Therefore $\lambda(u) \leq \lambda_0$, if and only if $u \leq C_1$ or $u > C_2$, where $C_1$ and $C_2$ are determined by the requirement that

$$P_{H_0}(U \leq C_1, \text{ or } U \geq C_2) = \alpha, \quad \text{where } U(X, Y) = \frac{\sum_{i=1}^{m}(X_i - \bar{X})^2/(m-1)}{\sum_{j=1}^{n}(Y_j - \bar{Y})^2/(n-1)}. \tag{74}$$

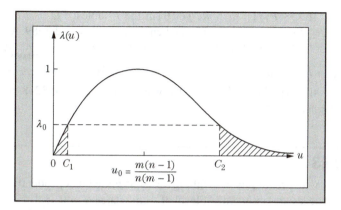

**FIGURE 11.14**

The graph of the function $\lambda = \lambda(u)$ given in relation (72).

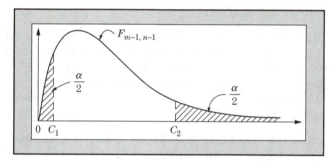

**FIGURE 11.15**

Rejection region of the hypothesis $H_0$ in (ii) (shaded areas), and the respective probabilities; here $C_1 = F_{m-1,n-1;1-\frac{\alpha}{2}}$, $C_2 = F_{m-1,n-1;\frac{\alpha}{2}}$.

Under $H_0$, $U \sim F_{m-1,n-1}$, and this allows the determination of $C_1, C_2$. For simplicity, we may split the probability $\alpha$ equally among the two tails, in which case (Figure 11.15),

$$C_1 = F_{m-1,n-1;1-\alpha/2}, \quad C_2 = F_{m-1,n-1;\alpha/2}.$$

To summarize then, the LR test is as follows:

$$\varphi(\boldsymbol{x},\boldsymbol{y}) = \begin{cases} 1 & \text{if } u(\boldsymbol{x},\boldsymbol{y}) \leq F_{m-1,n-1;1-\alpha/2}, \quad \text{or } u(\boldsymbol{x},\boldsymbol{y}) \geq F_{m-1,n-1;\alpha/2} \\ 0 & \text{otherwise,} \end{cases} \tag{75}$$

where

$$u(\boldsymbol{x},\boldsymbol{y}) = \frac{\sum_{i=1}^{m}(x_i - \bar{x})^2/(m-1)}{\sum_{j=1}^{n}(y_j - \bar{y})^2/(n-1)}. \tag{76}$$

*Numerical Example.* Let $m = 13$, $n = 19$, and take $\alpha = 0.05$.

**Discussion.** If $X \sim F_{12,18}$, then we get from the $F$-tables: $P(X > F_{12,18;0.025}) = 0.025$, or $P(X \leq F_{12,18;0.025}) = 0.975$, and hence $F_{12,18;0.025} = 2.7689$. Also, $P(X \leq F_{12,18;0.975}) = 0.025$, or $P(X > F_{12,18;0.975}) = 0.975$, or $P(\frac{1}{X} < \frac{1}{F_{12,18;0.975}}) = 0.975$. But then $\frac{1}{X} \sim F_{18,12}$, and therefore $\frac{1}{F_{12,18;0.975}} = 3.1076$, and hence $F_{12,18;0.975} \simeq 0.3218$. Thus, the hypothesis $H_0$ is rejected whenever $u(x,y) \leq 0.3218$, or $u(x,y) \geq 2.7689$, and it is accepted otherwise.

## EXERCISES

**4.1**  A coin, with probability $\theta$ of falling heads, is tossed independently 100 times and 60 heads are observed. At level of significance $\alpha = 0.1$:
  **(i)**  Use the LR test in order to test the hypothesis $H_0: \theta = 1/2$ (against the alternative $H_A: \theta \neq 1/2$).
  **(ii)**  Employ the appropriate approximation (see relation (10) in Chapter 8) to determine the cutoff point.

**4.2**  Let $X_1, X_2, X_3$ be independent r.v.'s distributed as $B(1, \theta), \theta \in \Omega = (0, 1)$, and let $t = x_1 + x_2 + x_3$, where the $x_i$'s are the observed values of the $X_i$'s.
  **(i)**  Derive the LR test $\lambda$ for testing the hypothesis $H_0: \theta = 0.25$ (against the alternative $H_A: \theta \neq 0.25$) at level of significance $\alpha = 0.02$.
  **(ii)**  Calculate the distribution of $\lambda(T)$ and carry out the test, where $T = X_1 + X_2 + X_3$.

**4.3**  **(i)**  In reference to Example 15 in Chapter 1, the appropriate model to be employed is the Normal distribution $N(\mu, \sigma^2)$ (with $\mu > 0$, of course). It is assumed, however, that $\sigma$ is unknown.
  **(ii)**  If $\mu_0$ is the stipulated average growth, then this will be checked by testing the hypothesis $H_0: \mu = \mu_0$ (against the alternative $H_A: \mu \neq \mu_0$) at level of significance $\alpha$.
  **(iii)**  On the basis of a random sample of size $n$, use the likelihood ratio test to test $H_0$ when $n = 25, \mu_0 = 6$ inch, and $\alpha = 0.05$.

**4.4**  **(i)**  In reference to Example 17 in Chapter 1, an appropriate model would be the following. Let $X_i$ and $Y_i$ be the blood pressure of the $i$th individual before and after the use of the pill, and set $Z_i = Y_i - X_i, i = 1, \ldots, n$. Furthermore, it is reasonable to assume that the $X_i$'s and the $Y_i$'s are independent and Normally distributed, so that the $Z_i$'s are independently distributed as $N(\mu, \sigma^2)$, but $\sigma$ is unknown.
  **(ii)**  With $\mu$ denoting the difference of blood pressure after the usage of the pill and before it, the claim is that $\mu < 0$. This claim is checked by testing the hypothesis $H_0: \mu = 0$ (against the alternative $H_A: \mu \neq 0$, with the only viable part of it here being $\mu > 0$) at level of significance $\alpha$, by using the likelihood ratio test.
  **(iii)**  Carry out the test if $n = 90$ and $\alpha = 0.05$.

**4.5**   In reference to Example 25 in Chapter 1:

   **(i)**  For $i = 1, \ldots, 15$, let $X_i$ and $Y_i$ be the heights of the cross-fertilized
        plants and self-fertilized plants, respectively. It is reasonable to assume
        that the $X_i$'s and the $Y_i$'s are independent random samples with
        respective distributions $N(\mu_1, \sigma_1^2)$ and $N(\mu_2, \sigma_2^2)$ (the estimates of $\sigma_1^2$
        and $\sigma_2^2$ do not justify the possible assumption of a common variance).
        Setting $Z_i = X_i - Y_i$, we have that the $Z_i$'s are independent and
        distributed as $N(\mu, \sigma^2)$, where $\mu = \mu_1 - \mu_2$, $\sigma^2 = \sigma_1^2 + \sigma_2^2$
        unknown.

   **(ii)**  The claim is that $\mu > 0$, and is to be checked by testing the hypothesis
        $H_0: \mu = 0$ (against the alternative $H_A: \mu \neq 0$, with the only viable part
        of it being that $\mu > 0$) at level of significance $\alpha$, by using the likelihood
        ratio test.

   **(iii)**  Carry out the test when $\alpha = 0.05$ and $\alpha = 0.10$.

**4.6**   The diameters of certain cylindrical items produced by a machine are r.v.'s
     distributed as $N(\mu, \ 0.01)$. A sample of size 16 is taken and it is found that
     $\bar{x} = 2.48$ inches. If the desired value for $\mu$ is 2.5 inches, formulate the
     appropriate testing hypothesis problem and carry out the test if
     $\alpha = 0.05$.

**4.7**   A manufacturer claims that packages of certain goods contain 18 ounces. In
     order to check his claim, 100 packages are chosen at random from a large lot
     and it is found that $\sum_{i=1}^{100} x_i = 1752$ and $\sum_{i=1}^{100} x_i^2 = 31,157$. Assume that the
     observations are Normally distributed, and formulate the manufacturer's
     claim as a testing hypothesis problem. Carry out the test at level of
     significance $\alpha = 0.01$.

**4.8**   The breaking powers of certain steel bars produced by processes A and B are
     independent r.v.'s distributed as Normal with possibly different means but the
     same variance. A random sample of size 25 is taken from bars produced by
     each one of the processes, and it is found that $\bar{x} = 60, s_x = 6, \bar{y} = 65, s_y = 7$.
     Test whether there is a difference between the two processes at the level of
     significance $\alpha = 0.05$.

**4.9**   **(i)**  Let $X_i, i = 1, \ldots, 9$ and $Y_j, j = 1, \ldots, 10$ be independent r.v.'s from the
        distributions $N(\mu_1, \sigma_1^2)$ and $N(\mu_2, \sigma_2^2)$, respectively. Suppose that the
        observed values of the sample s.d.'s are $s_x = 2, \ s_y = 3$. At level of
        significance $\alpha = 0.05$, test the hypothesis $H_0: \sigma_1 = \sigma_2$ (against the
        alternative $H_A: \sigma_1 \neq \sigma_2$.)

   **(ii)**  Find an expression for the computation of the power of the test for
        $\sigma_1 = 2$ and $\sigma_2 = 3$.

**4.10**  Refer to Exercise 3.16, and suppose that the variances $\sigma_1^2$ and $\sigma_2^2$ are
     unknown. Then test the hypothesis $H_0: \sigma_1 = \sigma_2$ (against the alternative
     $H_A: \sigma_1 \neq \sigma_2$) at level of significance $\alpha = 0.05$.

**4.11** The independent random samples $X_i$ and $Y_i$, $i = 1, \ldots, 5$ represent resistance measurements taken on two test pieces, and the observed values (in ohms) are as follows:

$$x_1 = 0.118, \quad x_2 = 0.125, \quad x_3 = 0.121, \quad x_4 = 0.117, \quad x_5 = 0.120,$$

$$y_1 = 0.114, \quad y_2 = 0.115, \quad y_3 = 0.119, \quad y_4 = 0.120, \quad y_5 = 0.110.$$

Assume that the $X_i$'s and the $Y_i$'s are Normally distributed, and test the hypothesis $H_0: \sigma_1 = \sigma_2$ (against the alternative $H_A: \sigma_1 \neq \sigma_2$) at level of signifince $\alpha = 0.05$.

**4.12** Refer to Exercise 4.11, and assume now that $\sigma_1 = \sigma_2 = \sigma$, say, unknown (which is supported by the fact that the hypothesis $H_0: \sigma_1 = \sigma_2$ was not rejected). Then test the hypothesis $H_0: \mu_1 = \mu_2$ (against the alternative $H_A: \mu_1 \neq \mu_2$) at level of significance $\alpha = 0.05$.

**4.13** Consider the independent random samples $X_1, \ldots, X_m$ and $Y_1, \ldots, Y_n$ from the respective distributions $N(\mu_1, \sigma^2)$ and $N(\mu_2, \sigma^2)$ where $\sigma$ is known, and suppose we are interested in testing the hypothesis $H_0: \mu_1 = \mu_2 = \mu$, say, unknown (against the alternative $H_A: \mu_1 \neq \mu_2$) at level of significance $\alpha$, by means of the likelihood ratio test. Set $x = (x_1, \ldots, x_m)$ and $y = (y_1, \ldots, y_n)$ for the observed values of the $X_i$'s and the $Y_i$'s.

(i) Form the joint likelihood function $L(\mu_1, \mu_2 \mid x, y)$ of the $X_i$'s and the $Y_i$'s, as well as the likelihood function $L(\mu \mid x, y)$.

(ii) From part (i), conclude immediately that the MLE's of $\mu_1$ and $\mu_2$ are $\hat{\mu}_1 = \bar{x}$ and $\hat{\mu}_2 = \bar{y}$. Also, show that the MLE of $\mu$ is given by
$\hat{\mu}_\omega = \frac{m\bar{x} + n\bar{y}}{m+n}$.

(iii) Show that $-2 \log \lambda = mn(\bar{x} - \bar{y})^2 / \sigma^2 (m+n)$, where $\lambda = L(\hat{\omega})/L(\hat{\Omega})$.

(iv) From part (iii), conclude that the likelihood ratio test $-2 \log \lambda > C_0$ is equivalent to $|\bar{x} - \bar{y}| > C(= \sigma\sqrt{(m+n)C_0/mn})$.

(v) Show that $C = z_{\alpha/2}\sigma\sqrt{\frac{1}{m} + \frac{1}{n}}$.

(vi) For any $\mu_1$ and $\mu_2$, show that the power of the test depends on $\mu_1$ and $\mu_2$ through their difference $\mu_1 - \mu_2 = \Delta$, say, and is given by the formula:

$$\pi(\Delta) = 2 - \Phi\left(\frac{C - \Delta}{\sigma\sqrt{\frac{1}{m} + \frac{1}{n}}}\right) - \Phi\left(\frac{C + \Delta}{\sigma\sqrt{\frac{1}{m} + \frac{1}{n}}}\right).$$

(vii) Determine the cutoff point when $m = 10, n = 15, \sigma = 1$, and $\alpha = 0.05$.

(viii) Determine the power of the test when $\Delta = 1$ and $\Delta = 2$.

**4.14** In reference to Example 15, verify the results:

$$\int_0^{C_1} \theta_0^2 t e^{-\theta_0 t} dt = 1 - e^{-\theta_0 C_1} - \theta_0 C_1 e^{-\theta_0 C_1},$$

$$\int_{C_2}^{\infty} \theta_0^2 t e^{-\theta_0 t} \, dt = e^{-\theta_0 C_2} + \theta_0 C_2 e^{-\theta_0 C_2}.$$

**4.15**  Verify expression (49) for the power of the test.

**4.16**  Verify the assertions made in expressions (52) about the function
$\lambda = \lambda(u), \ u \geq 0$.

**4.17**  Verify the assertion made in relation (56) that $\hat{\mu}_1, \hat{\mu}_2$, and $\hat{\sigma}_\Omega^2$ are the MLE's
of $\mu_1, \mu_2$, and $\sigma^2$, respectively.

**4.18**  Show that the expressions in relation (59) are, indeed, the MLE's of
$(\mu_1 = \mu_2 =)\mu$ and $\sigma^2$, respectively.

**4.19**  Show that $\lambda = \lambda(t^2) = (1 + \frac{t^2}{m+n-2})^{-\frac{m+n}{2}}$ is, indeed, strictly increasing in $t^2$
as asserted right after relation (63).

**4.20**  Justify the statement made in relation (65) that $t(X, Y) \sim t_{m+n-2}$.

**4.21**  Show that the expressions in relation (68) are, indeed, the MLE's of
$\mu_1, \mu_2, \sigma_1^2$, and $\sigma_2^2$, respectively.

**4.22**  Verify the assertions made in expression (73) about the function
$\lambda = \lambda(u), \ u \geq 0$.

**4.23**  Refer to the Bivariate Normal distribution discussed in Chapter 4, Section 5,
whose p.d.f. is given by:

$$f_{X,Y}(x, y) = \frac{1}{2\pi\sigma_1\sigma_2\sqrt{1 - \rho^2}} e^{-q/2}, \quad x, y \in \Re,$$

where $q = \frac{1}{1-\rho^2}[(\frac{x-\mu_1}{\sigma_1})^2 - 2\rho(\frac{x-\mu_1}{\sigma_1})(\frac{y-\mu_2}{\sigma_2}) + (\frac{y-\mu_2}{\sigma_2})^2], \mu_1, \mu_2 \in \Re, \sigma_1^2,$
$\sigma_2^2 > 0$, and $-1 \leq \rho \leq 1$ are the parameters of the distribution. Also, recall
that independence between $X$ and $Y$ is equivalent to their being uncorrelated;
i.e., $\rho = 0$. In this exercise, a test is derived for testing the hypothesis
$H_0 : \rho = 0$ (against the alternative $H_A : \rho \neq 0$, the $X$ and $Y$ are not
independent). The test statistic is based on the likelihood ratio statistic.

  **(i)**  On the basis of a random sample of size $n$ from a Bivariate Normal
distribution, $(X_i, Y_i), i = 1, \ldots, n$, the MLE's of the parameters involved
are given by:

$$\hat{\mu}_1 = \bar{x}, \quad \hat{\mu}_2 = \bar{y}, \quad \hat{\sigma}_1^2 = S_x, \quad \hat{\sigma}_2^2 = S_y, \quad \hat{\rho} = S_{xy}/\sqrt{S_x S_y},$$

where $S_x = \frac{1}{n} \sum_{i=1}^{n}(x_i - \bar{x})^2, S_y = \frac{1}{n} \sum_{i=1}^{n}(y_i - \bar{y})^2, S_{xy} = \frac{1}{n} \sum_{i=1}^{n}(x_i - \bar{x})(y_i - \bar{y})$, and the $x_i$'s and $y_i$'s are the observed values of
the $X_i$'s and $Y_i$'s. (See Exercise 1.14 (vii) in Chapter 9.)

  **(ii)**  Under the hypothesis $H_0 : \rho = 0$, the MLE's of $\mu_1, \mu_2, \sigma_1^2$, and $\sigma_2^2$ are
the same as in part (i).

**(iii)** When replacing the parameters by their MLE's, the likelihood function, call it $L(x, y)$, is given by:

$$L(x, y) = \left[2\pi \left(S_x S_y - S_{xy}^2\right)\right]^{-\frac{n}{2}} e^{-n},$$

where $x = (x_1, \ldots, x_n)$, $y = (y_1, \ldots, y_n)$.

**(iv)** Under the hypothesis $H_0(\rho = 0)$, when the parameters are replaced by their MLE's, the likelihood function, call it $L_0(x, y)$, is given by:

$$L_0(x, y) = (2\pi S_x S_y)^{-\frac{n}{2}} e^{-n}.$$

**(v)** [rom parts (iii) and (iv), it follows that the likelihood ratio statistic $\lambda$ is given by:

$$\lambda = (1 - \hat{\rho}^2)^{n/2}, \quad \hat{\rho} = S_{xy}/\sqrt{S_x S_y}.$$

**Hint.** Part (ii) follows immediately, because the joint p.d.f. of the pairs factorizes to the joint p.d.f. of the $X_i$'s times the joint p.d.f. of the $Y_i$'s.

**4.24** **(i)** By differentiation, show that the function $f(r) = (1 - r)^{n/2}$ is decreasing in $r$. Therefore, in reference to Exercise 4.23(v), $\lambda < \lambda_0$ is equivalent to $\hat{\rho}^2 > C_1$, some constant $C_1$ (actually, $C_1 = 1 - \lambda_0^{2/n}$); equivalently, $\hat{\rho} < -C_2$ or $\hat{\rho} > C_2$ ($C_2 = \sqrt{C_1}$).

**(ii)** Since the LR test rejects the hypothesis $H_0$ when $\lambda < \lambda_0$, part (i) states that the LR test is equivalent to rejecting $H_0$ wherever $\hat{\rho} < -C_2$ or $\hat{\rho} > C_2$.

**(iii)** In $\hat{\rho}$, replace the $x_i$'s and the $y_i$'s by the respective r.v.'s $X_i$ and $Y_i$, and set $R$ for the resulting r.v. Then, in part (ii), carrying out the test based on $\hat{\rho}$, requires knowledge of the cutoff point $C_2$, which in turn, presupposes knowledge of the distribution of $R$ (under $H_0$). Although the distribution of $R$ can be determined (see, e.g., Corollary to Theorem 7, Section 18.3, in the book *A Course in Mathematical Statistics*, 2nd edition (1997), Academic Press, by G. G. Roussas), it is not of any of the known forms, and hence no tables can be used.

**(iv)** Set $W = W(R) = \frac{\sqrt{n-2}R}{\sqrt{1-R^2}}$, and show that $W$ is an increasing function of $R$ by showing that $\frac{d}{dr} W(r)$ is positive.

**(v)** By parts (ii) and (iv), it follows that the likelihood ratio test is equivalent to rejecting $H_0$ whenever $W(r) < -C$ or $W(r) > C$, where $C$ is determined by the requirement that $P_{H_0}[W(R) < C \text{ or } W(R) > C] = \alpha$ (the given level of significance).

**(vi)** Under $H_0$, it can be shown (see, e.g., Section 18.3, in the book cited in part (iii) above) that $W(R)$ has the $t_{n-2}$ distribution. It follows that $C = t_{n-2;\frac{\alpha}{2}}$.

To summarize then, for testing $H_0 : \rho = 0$ at level of significance $\alpha$, reject $H_0$ whenever $W(r) < -t_{n-2;\frac{\alpha}{2}}$ or $W(r) > t_{n-2;\frac{\alpha}{2}}$, where

$W(r) = \frac{\sqrt{n-2}r}{\sqrt{1-r^2}}, r = \hat{\rho} = S_{xy}/\sqrt{S_xS_y}$; this test is equivalent to the likelihood ratio test.

**4.25** **(i)** In reference to the test given in (53), show that the power at $\sigma$ is given by:

$$P\left(X \le \frac{\sigma_0^2}{\sigma^2}\chi^2_{n;\ 1-\frac{\alpha}{2}}\right) + P\left(X \ge \frac{\sigma_0^2}{\sigma^2}\chi^2_{n;\ \frac{\alpha}{2}}\right),$$

where $X \sim \chi^2_n$.

**(ii)** In reference to the test given in (54), show that the power at $\sigma$ is given by:

$$P\left(Y \le \frac{\sigma_0^2}{\sigma^2}\chi^2_{n-1;\ 1-\frac{\alpha}{2}}\right) + P\left(Y \ge \frac{\sigma_0^2}{\sigma^2}\chi^2_{n-1;\ \frac{\alpha}{2}}\right),$$

where $Y \sim \chi^2_{n-1}$.

# More about testing hypotheses

In this chapter, a few more topics are discussed on testing hypotheses problems. More specifically, likelihood ratio (LR) tests are presented for the Multinomial distribution with further applications to contingency tables. A brief section is devoted to the so-called (Chi-square) goodness-of-fit tests, and another also brief section discusses the decision-theoretic approach to testing hypotheses. The chapter is concluded with a result connecting testing hypotheses and construction of confidence regions.

## 12.1 LIKELIHOOD RATIO TESTS IN THE MULTINOMIAL CASE AND CONTINGENCY TABLES

It was stated in Section 3 of Chapter 8 that the statistic $-2\log\lambda$ is distributed approximately as $\chi_f^2$ with certain degrees of freedom $f$, provided some regularity conditions are met. In this section, this result is stated in a more formal way, although the required conditions will not be spelled out. Also, no attempt will be made towards its justification.

**Theorem 1.** *On the basis of the random sample $X_1, \ldots, X_n$ from the p.d.f. $f(\cdot; \boldsymbol{\theta}), \boldsymbol{\theta} \in \Omega \subseteq \Re^r, r \geq 1$, we wish to test the hypothesis $H_0: \boldsymbol{\theta} \in \omega \subset \Omega$ at level of significance $\alpha$ and on the basis of the Likelihood Ratio statistic $\lambda = \lambda(X_1, \ldots, X_n)$. Then, provided certain conditions are met, it holds that:*

$$-2\log\lambda \simeq \chi_{r-m}^2, \text{ for all sufficiently large } n \text{ and } \boldsymbol{\theta} \in \omega;$$

*more formally,*

$$P_{\boldsymbol{\theta}}(-2\log\lambda \leq x) \to G(x), \quad x \geq 0, \quad \text{as } n \to \infty, \tag{1}$$

*where $G$ is the d.f. of the $\chi_{r-m}^2$ distribution; $r$ is the dimensionality of $\Omega$, $m$ is the dimensionality of $\omega$, and $\boldsymbol{\theta} \in \omega$.*

The practical use of (1) is that (for sufficiently large $n$) we can use the $\chi_{r-m}^2$ distribution in order to determine the cutoff point $C$ of the test, which rejects $H_0$ when $-2\log\lambda \geq C$. Specifically, $C \simeq \chi_{r-m;\alpha}^2$. Thus, for testing the hypothesis $H_0$ at level of significance $\alpha$, $H_0$ is to be rejected whenever $-2\log\lambda$ is $\geq \chi_{r-m;\alpha}^2$ (always provided $n$ is sufficiently large).

**Example 1** (The Multinomial Case). A Multinomial experiment, with $k$ possible outcomes $O_1, \ldots, O_k$ and respective unknown probabilities $p_1, \ldots, p_k$, is carried out independently $n$ times, and let $X_1, \ldots, X_k$ be the r.v.'s denoting the number of times outcomes $O_1, \ldots, O_k$ occur, respectively. Then the joint p.d.f. of the $X_i$'s is:

$$f(x_1, \ldots, x_k; \boldsymbol{\theta}) = \frac{n!}{x_1! \cdots x_k!} p_1^{x_1} \cdots p_k^{x_k}, \tag{2}$$

for $x_1, \ldots, x_k \geq 0$ integers with $x_1 + \cdots + x_k = n$, and $\boldsymbol{\theta} = (p_1, \ldots, p_k)$. The parameter space $\Omega$ is $(k-1)$-dimensional and is defined by:

$$\Omega = \{(p_1, \ldots, p_k) \in \Re^k; \ p_i > 0, \ i = 1, \ldots, k, \ p_1 + \cdots + p_k = 1\}.$$

**Discussion.** Suppose we wish to test the hypothesis $H_0 : p_i = p_{i0}, \ i = 1, \ldots, k$ (specified) (against the alternative $H_A : p_i \neq p_{i0}$ for at least one $i = 1, \ldots, k$) at level of significance $\alpha$. Under $H_0$,

$$L(\hat{\omega}) = \frac{n!}{x_1! \cdots x_k!} p_{10}^{x_1} \cdots p_{k0}^{x_k},$$

and we know that the MLE's of the $p_i$'s are: $\hat{p}_i = \frac{x_i}{n}, \ i = 1, \ldots, k$. Therefore

$$L(\hat{\Omega}) = \frac{n!}{x_1! \cdots x_k!} \hat{p}_1^{x_1} \cdots \hat{p}_k^{x_k} = \frac{n!}{x_1! \cdots x_k!} \left(\frac{x_1}{n}\right)^{x_1} \cdots \left(\frac{x_k}{n}\right)^{x_k}$$

$$= n^{-n} \frac{n!}{x_1! \cdots x_k!} x_1^{x_1} \cdots x_k^{x_k}.$$

Therefore

$$\lambda = n^n \left(\frac{p_{10}}{x_1}\right)^{x_1} \cdots \left(\frac{p_{k0}}{x_k}\right)^{x_k}, \quad \text{or} \quad -2 \log \lambda = 2 \left( \sum_{i=1}^{k} x_i \log x_i \right.$$

$$\left. - \sum_{i=1}^{k} x_i \log p_{i0} - n \log n \right),$$

and $H_0$ is rejected when $-2 \log \lambda \geq \chi^2_{k-1;\alpha}$, since here $r = k - 1$ and $m = 0$.

*Numerical Example.* The fairness of a die is to be tested on the basis of the following outcomes of 30 independent rollings: $x_1 = 4, x_2 = 7, x_3 = 3, x_4 = 8, x_5 = 4, x_6 = 4$. Take $\alpha = 0.05$.

**Discussion.** Here $k = 6$, $H_0: p_{i0} = \frac{1}{6}, \ i = 1, \ldots, 6$ and the $-2 \log \lambda$ is given by:

$$-2 \log \lambda = 2 \left( \sum_{i=1}^{6} x_i \log x_i + 30 \log 6 - 30 \log 30 \right) \simeq 2 \,(50.18827 + 53.75278$$

$$-102.03592) = 3.81026.$$

That is, $-2 \log \lambda \simeq 3.81026$, whereas $\chi^2_{5;0.05} = 11.071$. Thus, the hypothesis $H_0$ is not rejected.

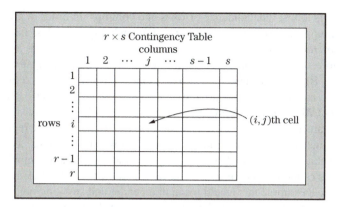

**FIGURE 12.1**

An $r \times s$ contingency table.

*Application to Contingency Tables*  Consider a Multinomial experiment with $r \times s$ possible outcomes arranged in a rectangular array with $r$ rows and $s$ columns. Such a rectangular array is referred to as an $r \times s$ *contingency table*. The $r$ rows and $s$ columns generate $r \times s$ *cells*. (See Figure 12.1.) Denote by $p_{ij}$ the probability that an outcome will fall into the $(i,j)$th cell. Carry out the Multinomial experiment under consideration $n$ independent times, and let $X_{ij}$ be the r.v. denoting the number of outcomes falling into the $(i,j)$th cell. Define $p_{i.}$ and $p_{.j}$ by the formulas:

$$p_{i.} = \sum_{j=1}^{s} p_{ij}, \quad i = 1,\ldots,r, \qquad p_{.j} = \sum_{i=1}^{r} p_{ij}, \quad j = 1,\ldots,s. \tag{3}$$

Then, clearly, $p_{i.}$ is the probability that an outcome falls in the $i$th row regardless of column, and $p_{.j}$ is the probability that an outcome falls in the $j$th column regardless of row. Of course, $\sum_{i=1}^{r} p_{i.} = \sum_{j=1}^{s} p_{.j} = \sum_{i=1}^{r} \sum_{j=1}^{s} p_{ij} = 1$. Also, define the r.v.'s $X_{i.}$ and $X_{.j}$ as follows:

$$X_{i.} = \sum_{j=1}^{s} X_{ij}, \quad i = 1,\ldots,r, \qquad X_{.j} = \sum_{i=1}^{r} X_{ij}, \quad j = 1,\ldots,s. \tag{4}$$

Thus, clearly, $X_{i.}$ denotes the number of outcomes falling in the $i$th row regardless of column, and $X_{.j}$ denotes the number of outcomes falling in the $j$th column regardless of row. It is also clear, that

$$\sum_{i=1}^{r} X_{i.} = \sum_{j=1}^{s} X_{.j} = \sum_{i=1}^{r} \sum_{j=1}^{s} X_{ij} = n.$$

The parameters $p_{ij}$, $i = 1,\ldots,r$, $j = 1,\ldots,s$ are, in practice, unknown and are estimated by the MLE's $\hat{p}_{ij} = \frac{x_{ij}}{n}$. In the testing hypotheses framework, one could test the hypothesis $H_0: p_{ij} = p_{ij0}$, $i = 1,\ldots,r$, $j = 1,\ldots,s$, specified. However, from a practical viewpoint, this is not an interesting hypothesis. What is of true interest here is to test independence of rows and columns. In order to provide some motivation, suppose that some subjects (e.g., human beings) are classified according to two *characteristics* to be denoted by $A$ and $B$ (e.g., human beings are classified according to gender, characteristic $A$, and whether or not they are cigarette smokers, characteristic $B$). Suppose that characteristic $A$ has $r$ *levels* and characteristic $B$ has $s$ *levels*. (In the concrete example at hand, $r = 2$ (Male, Female), and $s = 2$ (Smoker, Nonsmoker).) We agree to have the $r$ rows in an $r \times s$ contingency table represent the $r$ levels of characteristic $A$ and the $s$ columns of the contingency table represent the $s$ levels of characteristic $B$. Then independence of rows and columns, as mentioned earlier, is restated as independence of characteristics $A$ and $B$ or, more precisely, independence of the $r$ levels of characteristic $A$ and the $s$ levels of characteristic $B$. (In the concrete example this would mean that gender and smoking/nonsmoking are independent events.) The probabilistic formulation of the independence stated is as follows:

Observe that $P(A_i \cap B_j) = p_{ij}, P(A_i) = p_{i.}$, and $P(B_j) = p_{.j}$. Independence of $A_i$ and $B_j$ for all $i$ and $j$ means then that

$$P(A_i \cap B_j) = P(A_i)P(B_j), \text{ all } i \text{ and } j, \quad \text{or} \quad p_{ij} = p_{i.}p_{.j}, \text{ all } i \text{ and } j.$$

To put it differently, we wish to test the hypothesis that there exist (probabilities) $p_i > 0, i = 1, \ldots, r, p_1 + \cdots + p_r = 1$ and $q_j > 0, j = 1, \ldots, s, q_1 + \cdots + q_s = 1$, such that

$$H_0 : p_{ij} = p_i q_j, \quad i = 1, \ldots, r, \quad j = 1, \ldots, s. \tag{5}$$

(Of course, then $p_i = p_{i.}$ and $q_j = p_{.j}$, all $i$ and $j$.) The MLE of $p_{ij}$ is $\hat{p}_{ij} = \frac{x_{ij}}{n}$, $i = 1, \ldots, r, \ j = 1, \ldots, s$. Therefore, writing $\prod_{i,j}$ for $\prod_{i=1}^{r} \prod_{j=1}^{s}$ and setting $\boldsymbol{\theta}$ for $(p_{ij}, i = 1, \ldots, r, j = 1, \ldots, s)$, we have, for the likelihood function:

$$L(\boldsymbol{\theta}|x_{ij}, i = 1, \ldots, r, j = 1, \ldots, s) = \frac{n!}{\prod_{i,j} x_{ij}!} \prod_{i,j} p_{ij}^{x_{ij}}, \tag{6}$$

and

$$L(\hat{\Omega}) = \frac{n!}{\prod_{i,j} x_{ij}!} \prod_{i,j} \left(\frac{x_{ij}}{n}\right)^{x_{ij}} = \frac{n!}{n^n \prod_{i,j} x_{ij}!} \prod_{i,j} x_{ij}^{x_{ij}}. \tag{7}$$

Under $H_0$, the likelihood function becomes:

$$\frac{n!}{\prod_{i,j} x_{ij}!} \prod_{i,j} (p_i q_j)^{x_{ij}} = \frac{n!}{\prod_{i,j} x_{ij}!} \prod_i \prod_j p_i^{x_{ij}} q_j^{x_{ij}}$$

$$= \frac{n!}{\prod_{i,j} x_{ij}!} \prod_i p_i^{x_{i.}} q_1^{x_{i1}} \cdots q_s^{x_{is}} = \frac{n!}{\prod_{i,j} x_{ij}!} \left(\prod_i p_i^{x_{i.}}\right)\left(\prod_j q_j^{x_{.j}}\right), \tag{8}$$

because

$$\prod_i p_i^{x_{i.}} q_1^{x_{i1}} \cdots q_s^{x_{is}} = (p_1^{x_{1.}} q_1^{x_{11}} \cdots q_s^{x_{1s}}) \cdots (p_r^{x_{r.}} q_1^{x_{r1}} \cdots q_s^{x_{rs}})$$

$$= \left(p_1^{x_{1.}} \cdots p_r^{x_{r.}}\right)\left(q_1^{x_{11}} \cdots q_1^{x_{r1}}\right) \cdots \left(q_s^{x_{1s}} \cdots q_s^{x_{rs}}\right)$$

$$= \left(\prod_i p_i^{x_{i.}}\right)\left(q_1^{x_{.1}} \cdots q_s^{x_{.s}}\right) = \left(\prod_i p_i^{x_{i.}}\right)\left(\prod_j q_j^{x_{.j}}\right).$$

The MLE's of $p_i$ and $q_j$ are given by

$$\hat{p}_i = \frac{x_{i.}}{n}, \quad i = 1, \ldots, r, \quad \hat{q}_j = \frac{x_{.j}}{n}, \quad j = 1, \ldots, s, \tag{9}$$

so that

$$L(\hat{\omega}) = \frac{n!}{\prod_{i,j} x_{ij}!} \prod_i \left(\frac{x_{i.}}{n}\right)^{x_{i.}} \prod_j \left(\frac{x_{.j}}{n}\right)^{x_{.j}} = \frac{n!}{n^{2n} \prod_{i,j} x_{ij}!} \left(\prod_i x_{i.}^{x_{i.}}\right)\left(\prod_j x_{.j}^{x_{.j}}\right). \tag{10}$$

By (7) and (10), we have then

$$\lambda = \frac{\left(\prod_i x_{i.}^{x_{i.}}\right)\left(\prod_j x_{.j}^{x_{.j}}\right)}{n^n \prod_{i,j} x_{ij}^{x_{ij}}}, \tag{11}$$

and

$$-2\log \lambda = 2\left[\left(n\log n + \sum_{i=1}^r \sum_{j=1}^s x_{ij}\log x_{ij}\right) - \left(\sum_{i=1}^r x_{i.}\log x_{i.} + \sum_{j=1}^s x_{.j}\log x_{.j}\right)\right]. \tag{12}$$

Here the dimension of $\Omega$ is $rs - 1$ because we have $rs$ $p_{ij}$, $i = 1, \ldots, r$, $j = 1, \ldots, s$, which, however, satisfy the relationship $\sum_{i=1}^r \sum_{j=1}^s p_{ij} = 1$. In order to determine the dimension of $\omega$, observe that we have $r + s$ parameters $p_i$, $i = 1, \ldots, r$ and $q_j$, $j = 1, \ldots, s$, which, however, satisfy two relationships; namely, $\sum_{i=1}^r p_i = 1$ and $\sum_{j=1}^s q_j = 1$. Therefore, the dimension of $\omega$ is $r + s - 2$ and

$$\dim \Omega - \dim \omega = (rs - 1) - (r + s - 2) = (r - 1)(s - 1).$$

Furthermore, it so happens that the (unspecified) conditions of Theorem 1 are satisfied here, so that, under $H_0$, $-2\log \lambda$ is distributed approximately (for all sufficiently large $n$) as $\chi^2_{(r-1)(s-1)}$. It follows that the hypothesis (5) about independence is rejected, at level of significance $\alpha$, whenever

$$-2\log \lambda \geq \chi^2_{(r-1)(s-1);\,\alpha}. \tag{13}$$

*Numerical Example.*    A population consisting of $n = 100$ males (M) and females (F) is classified according to their smoking (S) or nonsmoking (NS) cigarettes habit. Suppose the resulting $2 \times 2$ contingency table is as given below. Then test independence of gender and smoking/nonsmoking habit at the level of significance $\alpha = 0.05$.

|   | S | NS |    |
|---|---|----|----|
| M | 20 | 35 | 55 |
| F | 15 | 30 | 45 |
|   | 35 | 65 | 100 |

**Discussion.** The values $x_{ij}$ are shown in the cells and the $x_{i.}, x_{j}$ are shown in the margins, and they are: $x_{11} = 20$, $x_{12} = 35$, $x_{21} = 15$, $x_{22} = 30$, $x_{1.} = 55$, $x_{2.} = 45$, $x_{.1} = 35, x_{.2} = 65$. Replacing these values in the expression of $-2 \log \lambda$ given by (12), we find $-2 \log \lambda \simeq 0.05$. Here $r = s = 2$, so that $\chi^2_{(r-1)(s-1); \alpha} = \chi^2_{1; 0.05} = 3.841$. Therefore, the hypothesis is not rejected.

---

## EXERCISES

**1.1**   (i)  In reference to Example 18 in Chapter 1, the appropriate probability model is the Multinomial distribution with parameters $n$ and $p_A, p_B,$ $p_{AB}, p_O,$ where $p_A$ through $p_O$ are the probabilities that an individual, chosen at random from among the $n$ persons has blood type either $A$ or $B$ or $AB$ or $O$, respectively.

   (ii)  Let $p_{A0}, p_{B0}, p_{AB0},$ and $p_{O0}$ be a priori stipulated numbers. Then, checking agreement of the actual probabilities with the stipulated values amounts to testing the hypothesis

$$H_0: p_A = p_{A0}, \quad p_B = p_{B0}, \quad p_{AB} = p_{AB0}, \quad p_O = p_{O0}.$$

   (iii)  Test the hypothesis $H_0$ by means of the log-LR test at level of significance $\alpha$ (see Example 1 here). Take $\alpha = 0.05$. (See also Exercise 2.1.)

**1.2**   (i)  In reference to Example 19 in Chapter 1, the appropriate probability model is the Multinomial distribution with parameters $n = 41,208$ and $p_i$, $i = 1, \ldots, 12$, where $p_i = P$(a birth chosen at random from among the $n$ births falls in the $i$th month).

   (ii)  Checking Uniform distribution of the $n$ births over the 12 months amounts to testing the hypothesis

$$H_0: p_i = p_{i0} = \frac{1}{12}, \quad i = 1, \ldots, 12.$$

   (iii)  The hypothesis $H_0$, tested by means of the log-LR test at level of significance $\alpha$ (see Example 1 here), is rejected when $-2 \log \lambda > \chi^2_{11; \alpha}$; take $\alpha = 0.01$. (See also Exercise 2.2.)

**1.3**   (i)  In reference to Example 20 in Chapter 1, the appropriate probability model is the $2 \times 3$ contingency table setup in the example.

   (ii)  If $p_{ij}$ is the probability that a randomly chosen subject from among the 150 falls into the $(i, j)$th cell, then independence between the factors health and diet is checked by testing the hypothesis

$$H_0: p_{ij} = p_i q_j, \quad i = 1, 2 \text{ and } j = 1, 2, 3,$$

at level of significance $\alpha$. An appropriate test for testing the hypothesis $H_0$ is the log-LR test (see relation (12)). Use this test by taking $\alpha = 0.05$ and $\alpha = 0.01$. (See also Exercise 2.3.)

**1.4**    **(i)**   In reference to Example 21 in Chapter 1, the appropriate probability model is the $3 \times 3$ contingency table setup in the example.

     **(ii)**   If $p_{ij}$ is the probability that a randomly chosen subject from among the 200 falls into the $(i, j)$th cell, then checking the stipulation that change of bone minerals does not vary for different groups amounts to testing the hypothesis

$$H_0: p_{ij} = p_i q_j, \quad i = 1, 2, 3 \quad \text{and} \quad j = 1, 2, 3,$$

at level of significance $\alpha$. The hypothesis $H_0$ may be checked by means of the log-LR test (see relation (12)). Use this test by taking $\alpha = 0.05$ and $\alpha = 0.01$. (See also Exercise 2.4.)

**1.5**   In reference to Example 1 of Chapter 1, the $n$ landfills are classified according to two levels of concentration (High and Low) and three levels of hazardous chemicals (Arsenic, Barium, and Mercury) to produce the following $2 \times 3$ contingency table:

|  |  | Hazardous Chemicals | | | |
|---|---|---|---|---|---|
|  |  | **Arsenic** | **Barium** | **Mercury** | **Totals** |
| Level of | High | $x_{11}$ | $x_{12}$ | $x_{13}$ | $x_{1.}$ |
| Concentration | Low | $x_{21}$ | $x_{22}$ | $x_{23}$ | $x_{2.}$ |
|  | Totals | $x_{.1}$ | $x_{.2}$ | $x_{.3}$ | $x_{..} = n$ |

Then, if $p_{ij}$ is the probability that a landfill chosen at random from among the $n$ landfills falls into the $(i, j)$th cell, part (ii) of the example becomes that of testing the hypothesis $H_0: p_{ij} = p_{ij0}$ at level of significance $\alpha$, where $p_{ij0}$, $i = 1, 2$ and $j = 1, 2, 3$ are a priori stipulated numbers. The hypothesis $H_0$ is tested by means of the log-LR test. Use this test by taking $\alpha = 0.05$ and $\alpha = 0.01$. (See also Exercise 2.5.)

## 12.2 A GOODNESS-OF-FIT TEST

This test applies primarily to the Multinomial distribution, although other distributions can also be suitably reduced to a Multinomial framework. In the notation of the previous section, we have that, for each fixed $i = 1, \ldots, k$, $X_i \sim B(n, p_i)$, so that $E_\theta X_i = np_i$, $i = 1, \ldots, k$, $\boldsymbol{\theta} = (p_1, \ldots, p_k)$. Thus, the $i$th outcome would be expected to appear $np_i$ times, whereas the actual number of times it appears is $X_i$. It then makes sense to compare what we expect and what we, actually, observe, and do this simultaneously for all $i = 1, \ldots, k$. One way of doing this is to look at the quantity $\sum_{i=1}^{k} (X_i - np_i)^2$. Small values of this quantity would indicate agreement between expected and observed values, and large values would indicate the opposite. For distributional reasons, the above expression is modified as indicated below, and in this form it is denoted by $\chi^2$; namely,

$$\chi^2 = \sum_{i=1}^{k} \frac{(X_i - np_i)^2}{np_i}. \tag{14}$$

Expression (14) is the basis for constructing test statistics for testing various hypotheses. In this setting, we will consider the hypothesis $H_0: p_i = p_{i0}, i = 1, \ldots, k$, specified as we did in the previous section. Under $H_0$, (14) is denoted by $\chi_\omega^2$ and is equal to:

$$\chi_\omega^2 = \sum_{i=1}^{k} \frac{(X_i - np_{i0})^2}{np_{i0}}. \tag{15}$$

This is a statistic and is used for testing $H_0$. Accordingly, $H_0$ is rejected, at level of significance $\alpha$, if $\chi_\omega^2 \geq C$, where $C$ is determined by the requirement $P_{H_0}(\chi_\omega^2 \geq C) = \alpha$. It can be seen (see the statement of Theorem 1' below) that, under $H_0$, $\chi_\omega^2 \simeq \chi_{k-1}^2$ for all sufficiently large $n$. Consequently, $C \simeq \chi_{k-1;\alpha}^2$. The test used here is called a test of *goodness-of-fit* for obvious reasons. It is also referred to as *Chi-square* (or $\chi^2$) *goodness-of-fit* test, because of the symbol used in relation (15), and because its asymptotic distribution (under the null hypothesis) is Chi-square with certain degrees of freedom. Thus, the (Chi-square) goodness-of-fit test rejects $H_0$ whenever $\chi_\omega^2 \geq \chi_{k-1;\alpha}^2$.

For illustrative and also comparative purposes, let us consider the first numerical example in the previous section.

*Numerical Example.* Here $np_{10} = \cdots = np_{60} = \frac{30}{6} = 5$, and then the observed value of $\chi_\omega^2$ is:

$$\chi_\omega^2 = \frac{1}{5}[(4-5)^2 + (7-5)^2 + (3-5)^2 + (8-5)^2 + (4-5)^2 + (4-5)^2] = 4.$$

For $\alpha = 0.05$, $\chi_{k-1;\alpha}^2 = \chi_{5;0.05}^2 = 11.071$, and since $\chi_\omega^2 = 4 < 11.071$, the hypothesis $H_0$ is not rejected, as was also the case with the LR test.

In the framework of a contingency table, expression (14) becomes

$$\chi^2 = \sum_{i=1}^{r} \sum_{j=1}^{s} \frac{(X_{ij} - np_{ij})^2}{np_{ij}}. \tag{16}$$

Under the hypothesis of independence stated in (5), expression (16) takes the form

$$\chi_\omega^2 = \sum_{i=1}^{r} \sum_{j=1}^{s} \frac{(X_{ij} - np_i q_j)^2}{np_i q_j}. \tag{17}$$

From (17), we form the test statistic $\chi_{\hat\omega}^2$ defined below:

$$\chi_{\hat\omega}^2 = \sum_{i=1}^{r} \sum_{j=1}^{s} \frac{(X_{ij} - n\hat p_i \hat q_j)^2}{n\hat p_i \hat q_j}, \tag{18}$$

where $\hat{p}_i$, $i = 1, \ldots, r$ and $\hat{q}_j$, $j = 1, \ldots, s$ are given in (9). Once again, it may be seen that, under $H_0$, $\chi_{\hat{\omega}}^2 \simeq \chi_{(r-1)(s-1)}^2$ for all sufficiently large $n$. Thus, the hypothesis $H_0$ is rejected, at level of significance $\alpha$, whenever $\chi_{\hat{\omega}}^2 \geq \chi_{(r-1)(s-1);\alpha}^2$.

The approximate determination of the cutoff points in testing the hypothesis associated with the test statistics given in (15) and (18) is based on the following result.

**Theorem 1′.**

(i) *Let $X_1, \ldots, X_n$ be a random sample from a Multinomial distribution with parameters $(k; p_1, \ldots, p_k)$. Then under the null hypothesis $H_0 : p_i = p_{i0}$, $i = 1, \ldots, k$, with the $p_{i0}$'s specified, the test statistic $\chi_{\omega}^2$, given in (15), is approximately $\chi_f^2$, $\chi_{\omega}^2 \simeq \chi_f^2$, in the sense that*

$$P_{H_0}(\chi_{\omega}^2 \leq x) \to G(x), \ x > 0, \ \text{as } n \to \infty$$

*where $G$ is the d.f. of the $\chi_f^2$ distribution and $f = k - 1$. Therefore, for level of significance $\alpha$, the approximate value of the cutoff point is $\chi_{k-1;\alpha}^2$.*

(ii) *Let the random sample $X_{ij}$, $i = 1, \ldots, r$, $j = 1, \ldots, s$ be arranged in the form of an $r \times s$ contingency table with cell probabilities $p_{ij}$, $i = 1, \ldots, r$, $j = 1, \ldots, s$. Thus, under the null hypothesis $H_0 : p_{ij} = p_i q_j$, with the $p_i$'s and $q_j$'s unspecified $i = 1, \ldots, r, j = 1, \ldots, s$, the test statistic $\chi_{\hat{\omega}}^2$ given in (18), is approximately $\chi_f^2$, $\chi_{\hat{\omega}}^2 \simeq \chi_f^2$, in the sense that*

$$P_{H_0}(\chi_{\hat{\omega}}^2 \leq x) \to G(x), \ x > 0, \ \text{as } n \to \infty$$

*where $G$ is the d.f. of the $\chi_f^2$ distribution and $f = (r-1)(s-1)$. Therefore, for level of significance $\alpha$, the approximate value of the cutoff point is $\chi_{(r-1)(s-1);\alpha}^2$.*

*The number of degrees of freedom, $f$, of the limiting distribution is determined as follows:*

$$f = (\text{number of independent parameters in the model})$$
$$- (\text{number of independent estimated parameters}).$$

*Observe that in the case of the first hypothesis, the number of independent parameters is $k - 1$, since $p_1 + , \ldots , + p_k = 1$, and the number of (independent) estimated parameter is 0, so that $f = k - 1$. In the case of the second hypothesis, the number of independent parameters is $r \times s - 1$, since $\sum_{i=1}^r \sum_{j=1}^s r_{ij} = 1$, and the number of independent estimated parameters is $(r-1) + (s-1)$, since $\sum_{i=1}^r p_i = 1, \sum_{j=1}^s q_j = 1$. Thus, $f = (r \times s - 1) - [(r-1) + (s-1)] = r \times s - r - s + 1 = (r-1)(s-1)$.*

The contingency table numerical example of the previous section is as below.

*Numerical Example.*    Here $\hat{p}_1 = 0.55$, $\hat{p}_2 = 0.45$, $\hat{q}_1 = 0.35$, $\hat{q}_2 = 0.65$, so that
$$n\hat{p}_1\hat{q}_1 = 19.25, \quad n\hat{p}_1\hat{q}_2 = 35.75, \quad n\hat{p}_2\hat{q}_1 = 15.75, \quad n\hat{p}_2\hat{q}_2 = 29.25.$$

Therefore
$$\chi_\omega^2 = \frac{(20 - 19.25)^2}{19.25} + \frac{(35 - 35.75)^2}{35.75} + \frac{(15 - 15.75)^2}{15.75} + \frac{(30 - 29.25)^2}{29.25} \simeq 0.0998.$$

Since $\chi_{(r-1)(s-1);\alpha}^2 = \chi_{1;0.05}^2 = 3.841$, the hypothesis $H_0$ is not rejected, as was also the case with the LR test.

---

# EXERCISES

In Exercises 2.1-2.5 below, use the $\chi^2$ goodness-of-fit test.

**2.1**    Same as Exercise 1.1.

**2.2**    Same as Exercise 1.2.

**2.3**    Same as Exercise 1.3.

**2.4**    Same as Exercise 1.4.

**2.5**    Same as Exercise 1.5.

**2.6**    A coin, with probability $p$ of falling heads, is tossed independently 100 times, and 60 heads are observed.
    **(i)**    Test the hypothesis $H_0: p = 1/2$ (against the alternative $H_A: p \neq 1/2$) at level of significance $\alpha = 0.1$, by using the appropriate $\chi^2$ goodness-of-fit test.
    **(ii)**    Determine the $P$-value of the test (use linear interpolation).

**2.7**    A die is cast independently 600 times, and the numbers 1 through 6 appear with the frequencies recorded below.

| 1 | 2 | 3 | 4 | 5 | 6 |
|---|---|---|---|---|---|
| 100 | 94 | 103 | 89 | 110 | 104 |

Use the appropriate $\chi^2$ goodness-of-fit test to test fairness for the die at level of significance $\alpha = 0.1$.

**2.8**    In a certain genetic experiment, two different varieties of a certain species are crossed and a specific characteristic of the offspring can occur at only three levels $A$, $B$, and $C$, say. According to a proposed model, the probabilities for $A, B$, and $C$ are $\frac{1}{12}$, $\frac{3}{12}$, and $\frac{8}{12}$, respectively. Out of 60 offspring, 6, 18, and 36 fall into levels $A, B$, and $C$, respectively. Test the validity of the proposed model at the level of significance $\alpha = 0.05$. Use the appropriate $\chi^2$ goodness-of-fit test.

**2.9** Course work grades are often assumed to be Normally distributed. In a certain class, suppose that letter grades are given in the following manner: $A$ for grades in the range from 90 to 100 inclusive, $B$ for grades in the range from 75 to 89 inclusive, $C$ for grades in the range from 60 to 74 inclusive, $D$ for grades in the range from 50 to 59 inclusive, and $F$ for grades in the range from 0 to 49. Use the data given below to check the assumption that the data are coming from an $N(75, 9^2)$ distribution. For this purpose, employ the appropriate $\chi^2$ goodness-of-fit test, and take $\alpha = 0.05$.

| A | B | C | D | F |
|---|---|---|---|---|
| 3 | 12 | 10 | 4 | 1 |

**Hint.** Assuming that the grade of a student chosen at random is a r.v. $X \sim N(75, 81)$, and compute the probabilities of an $A, B, C, D$, and $F$. Then use these probabilities in applying the $\chi^2$ goodness-of-fit test.

**2.10** It is often assumed that the I.Q. scores of human beings are Normally distributed. On the basis of the following data, test this claim at level of significance $\alpha = 0.05$ by using the appropriate $\chi^2$ goodness-of-fit test. Specifically, if $X$ is the r.v. denoting the I.Q. score of an individual chosen at random, then:

   **(i)** Set

$$p_1 = P(X \leq 90), \ p_2 = P(90 < X \leq 100), \ p_3 = P(100 < X \leq 110),$$
$$p_4 = P(110 < X \leq 120), \ p_5 = P(120 < X \leq 130), \ p_6 = P(X > 130).$$

  **(ii)** Calculate the probabilities $p_i, i = 1, \ldots, 6$ under the assumption that $X \sim N(100, 15^2)$ and call them $p_{i0}, i = 1, \ldots, 6$. Then set up the hypothesis $H_0: p_i = p_{i0}, \ i = 1, \ldots, 6$.

 **(iii)** Use the appropriate $\chi^2$ goodness-of-fit test to test the hypothesis at level of significance $\alpha = 0.05$. The available data are given below, where $x$ denotes the observed number of individuals lying in a given interval.

| $x \leq 90$ | $90 < x \leq 100$ | $100 < x \leq 110$ | $110 < x \leq 120$ | $120 < x \leq 130$ | $x > 130$ |
|---|---|---|---|---|---|
| 10 | 18 | 23 | 22 | 18 | 9 |

**2.11** Consider a group of 100 people living and working under very similar conditions. Half of them are given a preventive shot against a certain disease and the other half serve as controls. Of those who received the treatment, 40 did not contract the disease whereas the remaining 10 did so. Of those not treated, 30 did contract the disease and the remaining 20 did not. Test effectiveness of the vaccine at the level of significance $\alpha = 0.05$, by using the appropriate $\chi^2$ goodness-of-fit test.

**Hint.** For an individual chosen at random from the target population of 100 individuals, denote by $T_1, T_2$ and $D_1, D_2$ the following events: $T_1 = $ "treated,"

$T_2 =$ "not treated," $D_1 =$ "diseased," $D_2 =$ "not diseased," and set up the appropriate $2 \times 2$ contingency table.

**2.12** On the basis of the following scores, appropriately taken, test whether there are gender-associated differences in mathematical ability (as is often claimed!). Take $\alpha = 0.05$, and use the appropriate $\chi^2$ goodness-of-fit test.

| Boys: | 80 | 96 | 98 | 87 | 75 | 83 | 70 | 92 | 97 | 82 |
|-------|----|----|----|----|----|----|----|----|----|----|
| Girls: | 82 | 90 | 84 | 70 | 80 | 97 | 76 | 90 | 88 | 86 |

**Hint.** Group the grades into the following intervals: $[70, 75), [75, 80),$ $[80, 85), [85, 90), [90, 95), [95, 100),$ and count the grades of boys and girls falling into each one of these intervals. Then form a $2 \times 6$ contingency table with rows the two levels of gender (Boy, Girl), and columns the six levels of grades. Finally, with $p_{ij}$ standing for the probability that an individual, chosen at random from the target population, falls into the $(i, j)$th cell, stipulate the hypothesis $H_0: p_{ij} = p_i q_j, i = 1, 2$ and $j = 1, \ldots, 6$, and proceed to test it as suggested.

**2.13** From each of four political wards of a city with approximately the same number of voters, 100 voters were chosen at random and their opinions were asked regarding a certain legislative proposal. On the basis of the data given below, test whether the fractions of voters favoring the legislative proposal under consideration differ in the four wards. Take $\alpha = 0.05$, and use the appropriate $\chi^2$ goodness-of-fit test.

|  | Ward | | | | |
|--|------|--|--|--|--|
|  | 1 | 2 | 3 | 4 | Totals |
| Favor proposal | 37 | 29 | 32 | 21 | 119 |
| Do not favor proposal | 63 | 71 | 68 | 79 | 281 |
| Totals | 100 | 100 | 100 | 100 | 400 |

## 12.3 DECISION-THEORETIC APPROACH TO TESTING HYPOTHESES

There are chapters and books written on this subject. What we plan to do in this section is to deal with the simplest possible case of a testing hypothesis problem in order to illustrate the underlying concepts.

To this end, let $X_1, \ldots, X_n$ be a random sample with an unknown p.d.f. $f$. We adopt the (somewhat unrealistic) position that $f$ can be one of two possible specified p.d.f.'s, $f_0$ or $f_1$. On the basis of the observed values $x_1, \ldots, x_n$ of $X_1, \ldots, X_n$, we are invited to decide which is the true p.d.f. This decision will be made on the basis of a (nonrandomized) *decision function* $\delta = \delta(x_1, \ldots, x_n)$ defined on $\Re^n$ into $\Re$. More specifically, let $R$ be a subset of $\Re^n$, and suppose that if $\boldsymbol{x} = (x_1, \ldots, x_n)$ lies in $R$, we decide that $f_1$ is the true p.d.f., and if $x$ lies in $R^c$ (the complement of $R$ with respect to $\Re^n$), we decide in favor of $f_0$. In terms of a decision function, we reach the same conclusion by taking $\delta(x) = I_R(x)$ (the indicator function of R) and deciding in favor of $f_1$ if $\delta(x) = 1$ and in favor of $f_0$ if $\delta(x) = 0$. Or

$$\delta(x) = \begin{cases} 1 & \text{(which happens when } x \in R\text{) leads to selection} \\ & \text{of } f_1, \text{ and hence rejection of } f_0, \\ 0 & \text{(which happens when } x \in R^c\text{) leads to selection} \\ & \text{of } f_0, \text{ and hence rejection of } f_1. \end{cases} \tag{19}$$

At this point, we introduce monetary penalties for making wrong decisions, which are expressed in terms of a *loss function*. Specifically, let $L(f; \delta)$ be a function in two arguments, the p.d.f. $f$ and the decision function $\delta = \delta(x)$. Then it makes sense to define $L(f, \delta)$ in the following way:

$$L(f; \delta) = \begin{cases} 0 & \text{if } f = f_0 \text{ and } \delta(x) = 0 \text{ or } f = f_1 \text{ and } \delta(x) = 1, \\ L_1 & \text{if } f = f_0 \text{ and } \delta(x) = 1, \\ L_2 & \text{if } f = f_1 \text{ and } \delta(x) = 0, \end{cases} \tag{20}$$

where $L_1$ and $L_2$ are positive quantities.

Next, consider the average (expected) loss when the decision function $\delta$ is used, which is denoted by $R(f; \delta)$ and is called the *risk function*. In order to find the expression of $R(f; \delta)$, let us suppose that $P_{f_0}(X \in R) = P_{f_0}[\delta(X) = 1] = \alpha$ and $P_{f_1}(X \in R) = P_{f_1}[\delta(X) = 1] = \pi$. Then $\alpha$ is the probability of deciding in favor of $f_1$ if, actually, $f_0$ is true, and $\pi$ is the probability of deciding in favor of $f_1$ when $f_1$ is, actually, true. Then:

$$R(f; \delta) = \begin{cases} L_1 P_{f_0}(X \in R) = L_1 P_{f_0}[\delta(X) = 1] = L_1\alpha, & \text{if } f = f_0 \\ L_2 P_{f_1}(X \in R^c) = L_2 P_{f_1}[\delta(X) = 0] = L_2(1 - \pi), & \text{if } f = f_1, \end{cases} \tag{21}$$

or,

$$R(f_0; \delta) = L_1 P_{f_0}(X \in R) = L_1\alpha,$$
$$R(f_1; \delta) = L_2 P_{f_1}(X \in R^c) = L_2(1 - \pi). \tag{22}$$

Let us recall that our purpose is to construct an optimal decision function $\delta = \delta(x)$, where optimality is defined below on the basis of two different criteria. From relation (22), we know which is the bigger among the risk values $R(f_0; \delta) = L_1\alpha$ and $R(f_1; \delta) = L_2(1 - \pi)$. That is, we have the quantity $\max\{R(f_0; \delta), R(f_1; \delta)\}$. For any other (nonrandomized) decision function $\delta^*$ the corresponding quantity is $\max\{R(f_0; \delta^*), R(f_1; \delta^*)\}$. Then it makes sense to choose $\delta$ so that

$$\max\{R(f_0; \delta), R(f_1; \delta)\} \le \max\{R(f_0; \delta^*), R(f_1; \delta^*)\} \tag{23}$$

for any other decision function $\delta^*$ as described above. A decision function $\delta$, if it exists, which satisfies inequality (23) is called *minimax* (since it minimizes the maximum risk). The result below, Theorem 2, provides conditions under which the decision function $\delta$ defined by (19) is, actually, minimax. The problem is stated as a testing problem of a simple hypothesis against a simple alternative.

**Theorem 2.** *Let $X_1, \ldots, X_n$ be a random sample with p.d.f. $f$ which is either $f_0$ or $f_1$, both completely specified. For testing the hypothesis $H_0: f = f_0$ against the alternative $H_A: f = f_1$ at level of significance $\alpha$, define the rejection region $R$ by:*

$$R = \{(x_1, \ldots, x_n) \in \mathfrak{R}^n; f_1(x_1) \cdots f_1(x_n) > C f_0(x_1) \cdots f_0(x_n)\},$$

*and let the test function $\delta = \delta(x)$ ($x = (x_1, \ldots, x_n)$) be defined by (19); i.e.,*

$$\delta(x) = \begin{cases} 1 & \text{if } x \in R, \\ 0 & \text{if } x \in R^c. \end{cases}$$

*The constant $C$ is defined by the requirement that $E_{f_0}\delta(X) = P_{f_0}(X \in R) = \alpha$ ($X = (X_1, \ldots, X_n)$), and it is assumed that the level of significance $\alpha$, the power $\pi$ of the test $\delta$, and the quantities $L_1$ and $L_2$ satisfy the relationship*

$$(R(f_0; \delta)=) \quad L_1\alpha = L_2(1 - \pi) \quad (=R(f_1; \delta)). \tag{24}$$

*Then the decision function $\delta = \delta(x)$ is minimax.*

**Remark 1.** In connection with relation (24), observe that, if we determine the level of significance $\alpha$, then the power $\pi$ is also determined, and therefore relation (24) simply specifies a relationship between the losses $L_1$ and $L_2$; they cannot be determined independently but rather one will be a function of the other. In the present context, however, we wish to have the option of specifying the losses $L_1$ and $L_2$, and then see what is a possible determination of the constant $C$, which will produce a test of level of significance $\alpha$ (and of power $\pi$) satisfying relation (24).

*Proof of Theorem 2.* For simplicity, let us write $P_0$ and $P_1$ instead of $P_{f_0}$ and $P_{f_1}$, respectively, and likewise, $R(0; \delta)$ and $R(1; \delta)$ instead of $R(f_0; \delta)$ and $R(f_1; \delta)$, respectively. Then assumption (24) is rewritten thus: $R(0; \delta) = L_1\alpha = L_2(1 - \pi) = R(1; \delta)$. Recall that we are considering only nonrandomized decision functions. With this in mind, let $T$ be any (other than $R$) subset of $\mathfrak{R}^n$, and let $\delta^*$ be its indicator function, $\delta^*(x) = I_T(x)$, so that $\delta^*$ is the decision function associated with $T$. Then, in analogy with (22),

$$R(0; \delta^*) = L_1 P_0(X \in T), \quad R(1; \delta^*) = L_2 P_1(X \in T^c). \tag{25}$$

Look at $R(0; \delta)$ and $R(0; \delta^*)$ and suppose that $R(0; \delta^*) \leq R(0; \delta)$. This is equivalent to $L_1 P_0(X \in T) \leq L_1 P_0(X \in R) = L_1\alpha$, or $P_0(X \in T) \leq \alpha$. So $\delta^*$, being looked upon as a test, is of level of significance $\leq \alpha$. Then by Theorem 1 in Chapter 11, the power of the test $\delta^*$, which is $P_1(X \in T)$, is less than or equal to $P_1(X \in R)$, which is the power of the test $\delta$. This is so because $\delta$ is of level of significance $\alpha$, MP among all tests of level of significance $\leq \alpha$. From $P_1(X \in T) \leq P_1(X \in R)$ we have, equivalently, $P_1(X \in T^c) \geq P_1(X \in R^c)$ or $L_2 P_1(X \in T^c) \geq L_2 P_1(X \in R^c)$ or $R(1; \delta^*) \geq R(1; \delta)$. To summarize, the assumption $R(0; \delta^*) \leq R(0; \delta)$ leads to $R(1; \delta) \leq R(1; \delta^*)$. Hence

$$R(0; \delta^*) \leq R(0; \delta) = R(1; \delta) \leq R(1; \delta^*) \text{(the equality holding by (24)),}$$

and therefore

$$\max\{R(0; \delta^*), R(1; \delta^*)\} = R(1; \delta^*) \geq R(1; \delta) = \max\{R(0; \delta), R(1; \delta)\}, \qquad (26)$$

as desired. Next, the assumption,

$$R(0; \delta) < R(0; \delta^*), \qquad (27)$$

and the fact that

$$\max\{R(0; \delta^*), R(1; \delta^*)\} \geq R(0; \delta^*) \qquad (28)$$

imply

$$\max\{R(0; \delta^*), R(1; \delta^*)\} \geq R(0; \delta^*) > R(0; \delta) = \max\{R(0; \delta), R(1; \delta)\}. \qquad (29)$$

Relations (26) and (29) yield

$$\max\{R(0; \delta), R(1; \delta)\} \leq \max\{R(0; \delta^*), R(1; \delta^*)\},$$

so that $\delta$ is, indeed, minimax. ∎

**Remark 2.** It is to be pointed out that the minimax decision function $\delta = \delta(x)$ above is the MP test of level of significance $P_{f_0}(X \in R)$ for testing the (simple) hypothesis $H_0: f = f_0$ against the (simple) alternative $H_A: f = f_1$.

**Remark 3.** If the underlying p.d.f. $f$ depends on a parameter $\theta \in \Omega$, then the two possible options $f_0$ and $f_1$ for $f$ will correspond to two values of the parameter $\theta$, $\theta_0$, and $\theta_1$, say.

The theorem of this section is illustrated now by two examples.

**Example 2.** On the basis of the random sample $X_1, \ldots, X_n$ from the $N(\theta, 1)$ distribution, determine the minimax decision function $\delta = \delta(x)$ for testing the hypothesis $H_0: \theta = \theta_0$ against the alternative $H_A: \theta = \theta_1$.

**Discussion.** Here the joint p.d.f. of the $X_i$'s is

$$L(x; \theta) = (2\pi)^{-n/2} \exp\left[-\frac{1}{2}\sum_{i=1}^{n}(x_i - \theta)^2\right],$$

so that the rejection region $R$ is defined by $L(x; \theta_1) > CL(x; \theta_0)$ or, equivalently, by

$$\exp[n(\theta_1 - \theta_0)\bar{x}] > C \exp\left[\frac{n}{2}(\theta_1^2 - \theta_0^2)\right],$$

or

$$\bar{x} > C_0 \quad \text{for } \theta_1 > \theta_0, \quad \text{and} \quad \bar{x} < C_0 \quad \text{for } \theta_1 < \theta_0,$$

$$\text{where } C_0 = \frac{1}{2}(\theta_1 + \theta_0) + \frac{\log C}{n(\theta_1 - \theta_0)}. \qquad (30)$$

Then the requirement in (24) becomes, accordingly,

$$L_1 P_{\theta_0}(\bar{X} > C_0) = L_2 P_{\theta_1}(\bar{X} \leq C_0) \quad \text{for } \theta_1 > \theta_0,$$

and

$$L_1 P_{\theta_0}(\bar{X} < C_0) = L_2 P_{\theta_1}(\bar{X} \geq C_0) \quad \text{for } \theta_1 < \theta_0,$$

or

$$L_1\{1 - \Phi[\sqrt{n}(C_0 - \theta_0)]\} = L_2\Phi[\sqrt{n}(C_0 - \theta_1)] \quad \text{for } \theta_1 > \theta_0,$$
$$L_1\Phi[\sqrt{n}(C_0 - \theta_0)] = L_2\{1 - \Phi[\sqrt{n}(C_0 - \theta_1)]\} \quad \text{for } \theta_1 < \theta_0. \tag{31}$$

Consider the following numerical application.

*Numerical Example.* Suppose $n = 25$ and let $\theta_0 = 0$ and $\theta_1 = 1$. In the spirit of Remark 1, take, e.g., $L_1 = 5$ and $L_2 = 2.5$.

**Discussion.** Then the first relation in (31), which is applicable here, becomes

$$\Phi[5(C_0 - 1)] = 2[1 - \Phi(5C_0)] \quad \text{or} \quad 2\Phi(5C_0) - \Phi(5 - 5C_0) = 1.$$

From the Normal tables, we find $C_0 = 0.53$, so that the minimax decision function is given by:

$$\delta(x) = 1 \quad \text{if } \bar{x} > 0.53, \quad \text{and} \quad \delta(x) = 0 \quad \text{if } \bar{x} \leq 0.53.$$

Let us now calculate the level of significance and the power of this test. We have

$$P_0(\bar{X} > 0.53) = 1 - \Phi(5 \times 0.53) = 1 - \Phi(2.65) = 1 - 0.995975 \simeq 0.004,$$

and

$$\pi(1) = P_1(\bar{X} > 0.53) = 1 - \Phi[5(0.53 - 1)] = \Phi(2.35) = 0.990613 \simeq 0.991.$$

**Example 3.** In terms of the random sample $X_1, \ldots, X_n$ from the $B(1, \theta)$ distribution, determine the minimax function $\delta = \delta(x)$ for testing the hypothesis $H_0: \theta = \theta_0$ against the alternative $H_A: \theta = \theta_1$.

**Discussion.** The joint p.d.f. of the $X_i$'s here is

$$L(x; \theta) = \theta^t(1 - \theta)^{n-t}, \quad t = x_1 + \cdots + x_n,$$

so that the rejection region $R$ is determined by

$$L(x; \theta_1) > CL(x; \theta_0) \quad \text{or} \quad [\theta_1(1 - \theta_0)/\theta_0(1 - \theta_1)]^t > C[(1 - \theta_0)/(1 - \theta_1)]^n,$$

or

$$t \log \frac{(1 - \theta_0)\theta_1}{\theta_0(1 - \theta_1)} > C_0' = \log C - n \log \frac{1 - \theta_1}{1 - \theta_0}.$$

This is equivalent to

$$t > C_0 \quad \text{for } \theta_1 > \theta_0, \quad \text{and} \quad t < C_0 \quad \text{for } \theta_1 < \theta_0,$$

where $C_0 = C_0'/\log \frac{(1-\theta_0)\theta_1}{\theta_0(1-\theta_1)}$. With $T = X_1 + \cdots + X_n$, the requirement in (24) becomes here, respectively,

$$L_1 P_{\theta_0}(T > C_0) = L_2 P_{\theta_1}(T \leq C_0) \quad \text{for } \theta_1 > \theta_0,$$

and

$$L_1 P_{\theta_0}(T < C_0) = L_2 P_{\theta_1}(T \geq C_0) \quad \text{for } \theta_1 < \theta_0,$$

or

$$L_1 P_{\theta_0}(T \leq C_0) + L_2 P_{\theta_1}(T \leq C_0) = L_1 \quad \text{for } \theta_1 > \theta_0,$$

and                                                                                                    (32)

$$L_1 P_{\theta_0}(T \leq C_0 - 1) + L_2 P_{\theta_1}(T \leq C_0 - 1) = L_2 \quad \text{for } \theta_1 < \theta_0,$$

where $X \sim B(n, \theta)$.

*Numerical Example.* Let $n = 20$, and suppose $\theta_0 = 0.50$ and $\theta_1 = 0.75$.

**Discussion.** Here, the first relation in (32) is applicable. Since

$$P_{0.75}(T \leq C_0) = P_{0.25}(T \geq 20 - C_0) = 1 - P_{0.25}(T \leq 19 - C_0),$$          (33)

the first relation in (32) becomes

$$L_1 P_{0.50}(T \leq C_0) - L_2 P_{0.25}(T \leq 19 - C_0) = L_1 - L_2,$$

or

$$L_2 = [1 - P_{0.50}(T \leq C_0)]L_1/[1 - P_{0.25}(T \leq 19 - C_0)].$$          (34)

At this point, let us take $L_1 = 1$ and $L_2 = 0.269$. Then the right-hand side of (34) gives, for $C_0 = 13$: $\frac{1-0.9423}{1-0.7858} = \frac{0.0577}{0.2142} \simeq 0.269 = L_2$; i.e., the first relation in (32) obtains. The minimax decision function $\delta = \delta(x)$ is then given by: $\delta(x) = 1$ if $t \geq 14$, and $\delta(x) = 0$ for $t \geq 13$. The level of significance and the power of this test are:

$$P_{0.50}(T \geq 14) = 1 - P_{0.50}(T \leq 13) = 1 - 0.9423 = 0.0577,$$

and, on account of (33),

$$\pi(0.75) = P_{0.75}(T \geq 14) = P_{0.25}(T \leq 6) = 0.7858.$$

Instead of attempting to select $\delta = \delta(x)$ so as to minimize the maximum risk, we may, instead, try to determine $\delta$ so that $\delta$ minimizes the average risk. This approach calls for choosing the p.d.f.'s $f_0$ and $f_1$ according to a probability distribution; choose $f_0$ with probability $p_0$ and choose $f_1$ with probability $p_1$ ($p_0 + p_1 = 1$), and set $\lambda_0 = \{p_0, p_1\}$. If $R_{\lambda_0}(\delta)$ denotes the corresponding average risk, then, on account of (22), this average is given by:

$$R_{\lambda_0}(\delta) = L_1 P_{f_0}(X \in R)p_0 + L_2 P_{f_1}(X \in R^c)p_1$$
$$= p_0 L_1 P_{f_0}(X \in R) + p_1 L_2[1 - P_{f_1}(X \in R)]$$

$$= p_1 L_2 + [p_0 L_1 P_{f_0}(X \in R) - p_1 L_2 P_{f_1}(X \in R)]$$

$$= \begin{cases} p_1 L_2 + \int_R [p_0 L_1 f_0(x_1) \cdots f_0(x_n) - p_1 L_2 f_1(x_1) \cdots f_1(x_n)] \, dx_1 \cdots dx_n \\ p_1 L_2 + \sum_{x \in R} [p_0 L_1 f_0(x_1) \cdots f_0(x_n) - p_1 L_2 f_1(x_1) \cdots f_1(x_n)] \end{cases}$$

for the continuous and the discrete case, respectively. From this last expression, it follows that $R_{\lambda_0}(\delta)$ is minimized, if $p_0 L_1 f_0(x_1) \cdots f_0(x_n) - p_1 L_2 f_1(x_1) \cdots f_1(x_n)$ is $< 0$ on $R$. But $\delta(x) = 1$ on $R$ and $\delta(x) = 0$ on $R^c$. Thus, we may restate these equations as follows:

$$\delta(x) = \begin{cases} 1 & \text{if } f_1(x_1) \cdots f_1(x_n) > \frac{p_0 L_1}{p_1 L_2} f_0(x_1) \cdots f_0(x_n), \\ 0 & \text{otherwise.} \end{cases} \tag{35}$$

Thus, given a probability distribution $\lambda_0 = \{p_0, p_1\}$ on $\{f_0, f_1\}$, there is always a (nonrandomized) decision function $\delta$ which minimizes the average risk $R_{\lambda_0}(\delta)$, and this $\delta$ is given by (35) and is called a *Bayes decision function*.

**Theorem 3.** *The Bayes decision function $\delta_{\lambda_0}(x)$ corresponding to the probability distribution $\lambda_0 = \{p_0, p_1\}$ on $\{f_0, f_1\}$ is given by (35). This decision function is, actually, the MP test for testing the hypothesis $H_0: f = f_0$ against the alternative $H_A: f = f_1$ with cutoff point $C = p_0 L_1 / p_1 L_2$ and level of significance $\alpha$ given by:*

$$P_{f_0}[f_1(X_1) \cdots f_1(X_n) > C f_0(X_1) \cdots f_0(X_n)] = \alpha. \tag{36}$$

**Remark 4.** As mentioned earlier, if the underlying p.d.f. depends on a parameter $\theta \in \Omega$, then the above problem becomes that of testing $H_0: \theta = \theta_0$ against $H_A: \theta = \theta_1$ for some specified $\theta_0$ and $\theta_1$ in $\Omega$.

**Example 4.** Examples 2 and 3 (continued).

**Discussion.** In reference to Example 2 and for the case that $\theta_1 > \theta_0$, $\delta_{\lambda 0}(x) = 1$ if $\bar{x} > C_0$, $C_0 = \frac{1}{2}(\theta_1 + \theta_0) + \frac{\log C}{n(\theta_1 - \theta_0)}$, $C = p_0 L_1 / p_1 L_2$, as follows from relation (30). For the numerical data of the same example, we obtain $C_0 = 0.50 + 0.04 \log \frac{2p_0}{1 - p_0}$. For example, for $p_0 = \frac{1}{2}$, $C_0$ is $\simeq 0.50 + 0.04 \times 0.693 = 0.52772 \simeq 0.53$, whereas for $p_0 = \frac{1}{4}$, $C_0$ is $\simeq 0.50 - 0.04 \times 0.405 = 0.4838 \simeq 0.48$. For $C_0 = 0.53$, the level of significance and the power have already been calculated. For $C_0 = 0.48$, these quantities are, respectively:

$$P_0(\bar{X} > 0.48) = 1 - \Phi(5 \times 0.48) = 1 - \Phi(2.4) = 1 - 0.991802 = 0.008198,$$

$$\pi(1) = P_1(\bar{X} > 0.48) = 1 - \Phi[5(0.48 - 1)] = \Phi(2.6) = 0.995339.$$

In reference to Example 3 and for the case that $\theta_1 > \theta_0$, $\delta_{\lambda_0}(x) = 1$ if $x > C_0$, $C_0 = (\log C - n \log \frac{1 - \theta_1}{1 - \theta_0}) / \log \frac{(1 - \theta_0)\theta_1}{\theta_0(1 - \theta_1)}$, $C = p_0 L_1 / p_1 L_2$. For the numerical data of the same example, we have $C_0 \simeq (15.173 + \log \frac{p_0}{1 - p_0}) / 1.099$. For $p = \frac{1}{2}$, $C_0$ is

13.81, and for $p_0 = \frac{1}{4}$, $C_0$ is 12.81. In the former case, $\delta_{\lambda_0}(x) = 1$ for $x \geq 14$, and in the latter case, $\delta_{\lambda_0}(x) = 0$ for $x \geq 13$. The level of significance and the power have been calculated for the former case. As for the latter case, we have:

$$P_{0.50}(X \geq 13) = 1 - P_{0.50}(X \leq 12) = 1 - 0.8684 = 0.1316,$$

$$\pi(0.75) = P_{0.75}(X \geq 13) = P_{0.25}(X \leq 7) = 0.8982.$$

## 12.4 RELATIONSHIP BETWEEN TESTING HYPOTHESES AND CONFIDENCE REGIONS

In this brief section, we discuss a relationship which connects a testing hypothesis problem and the problem of constructing a confidence region for the underlying parameter. To this effect, suppose $X_1, \ldots, X_n$ is a random sample from the p.d.f. $f(\cdot; \boldsymbol{\theta})$, $\boldsymbol{\theta} \in \Omega \subseteq \mathfrak{R}^r$, $r \geq 1$, and for each $\boldsymbol{\theta}$ in $\Omega$, consider the problem of testing the hypothesis, to be denoted by $H_0(\boldsymbol{\theta})$, that the parameter $\boldsymbol{\theta}^*$, say, in $\Omega$, is actually, equal to the value of $\boldsymbol{\theta}$ considered. That is, $H_0(\boldsymbol{\theta})$: $\boldsymbol{\theta}^* = \boldsymbol{\theta}$ at level of significance $\alpha$. Denote by $A(\boldsymbol{\theta})$ the respective acceptance region in $\mathfrak{R}^n$. As usually, $X = (X_1, \ldots, X_n)$ and $x = (x_1, \ldots, x_n)$ is the observed value of $X$. For each $x \in \mathfrak{R}^n$, define in $\Omega$ the region $T(x)$ as follows:

$$T(x) = \{\boldsymbol{\theta} \in \Omega; \; x \in A(\boldsymbol{\theta})\}. \tag{37}$$

Thus, $T(x)$ consists of all those $\boldsymbol{\theta} \in \Omega$ for which, on the basis of the outcome $x$, the hypothesis $H_0(\boldsymbol{\theta})$ is accepted. On the basis of the definition of $T(x)$ by (37), it is clear that

$$\boldsymbol{\theta} \in T(x) \text{ if and only if } x \in A(\boldsymbol{\theta}).$$

Therefore

$$P_{\boldsymbol{\theta}}[\boldsymbol{\theta} \in T(X)] = P_{\boldsymbol{\theta}}[X \in A(\boldsymbol{\theta})]. \tag{38}$$

But the probability on the right-hand side of (38) is equal to $1 - \alpha$, since the hypothesis $H_0(\boldsymbol{\theta})$ being tested is of level of significance $\alpha$. Thus,

$$P_{\boldsymbol{\theta}}[\boldsymbol{\theta} \in T(X)] = 1 - \alpha,$$

and this means that the region $T(X)$ is a confidence region for $\boldsymbol{\theta}$ with confidence coefficient $1 - \alpha$.

Summarizing what has been discussed so far in the form of a theorem, we have the following result.

**Theorem 4.** *Let $X_1, \ldots, X_n$ be a random sample from the p.d.f. $f(\cdot; \boldsymbol{\theta})$, $\boldsymbol{\theta} \in \Omega \subseteq \mathfrak{R}^r$, $r \geq 1$, and, for each $\boldsymbol{\theta} \in \Omega$, consider the problem of testing the hypothesis $H_0(\boldsymbol{\theta})$: $\boldsymbol{\theta}^* = \boldsymbol{\theta}$ at level of significance $\alpha$. Let $A(\boldsymbol{\theta})$ be the corresponding acceptance region in $\mathfrak{R}^n$, and for each $x \in \mathfrak{R}^n$, define the region $T(x)$ in $\Omega$ as in (37). Then $T(X)$ is a* *confidence region for $\theta$ with confidence coefficient $1 - \alpha$, where $X = (X_1, \ldots, X_n)$ and $x = (x_1, \ldots, x_n)$ is the observed value of $X$.*

This result will now be illustrated below by two examples.

**Example 5.** On the basis of a random sample $X_1, \ldots, X_n$ from the $N(\theta, \sigma^2)$ distribution with $\sigma$ known, construct a confidence interval for $\theta$ with confidence coefficient $1 - \alpha$, by utilizing Theorem 4.

**Discussion.** For each $\theta \in \Omega = \Re$ and for testing the hypothesis $H_0(\theta)$ that the (unknown) parameter $\theta^*$, say, is, actually, equal to $\theta$, it makes sense to reject $H_0(\theta)$ when $\bar{X}$ is either too far to the left or too far to the right of $\theta$. Equivalently, if $\bar{X} - \theta$ is either $< C_1$ or $\bar{X} - \theta$ is $> C_2$ for some constants $C_1, C_2$ (see relation (46) in Chapter 1). If $H_0(\theta)$ is to be of level of significance $\alpha$, we will have $P_\theta(\bar{X} - \theta < C_1 \text{ or } \bar{X} - \theta > C_2) = \alpha$. But under $H_0(\theta)$, the distribution of $\bar{X}$ is symmetric about $\theta$, so that it is reasonable to take $C_1 = -C_2$, and then $C_2 = z_{\alpha/2}\frac{\sigma}{\sqrt{n}}$, $C_1 = -z_{\alpha/2}\frac{\sigma}{\sqrt{n}}$. Thus, $H_0(\theta)$ is accepted whenever $-z_{\alpha/2}\frac{\sigma}{\sqrt{n}} \leq \bar{X} - \theta \leq z_{\alpha/2}\frac{\sigma}{\sqrt{n}}$ or $-z_{\alpha/2} \leq \frac{\sqrt{n}(\bar{X}-\theta)}{\sigma} \leq z_{\alpha/2}$, or the acceptance region (interval here) is given by

$$A(\theta) = \left\{ x \in \Re^n; -z_{\alpha/2} \leq \frac{\sqrt{n}(\bar{x} - \theta)}{\sigma} \leq z_{\alpha/2} \right\},$$

and

$$P_\theta[X \in A(\theta)] = P_\theta\left[ -z_{\alpha/2} \leq \frac{\sqrt{n}(\bar{X} - \theta)}{\sigma} \leq z_{\alpha/2} \right] = 1 - \alpha.$$

Furthermore, by (37),

$$T(x) = \{\theta \in \Re; \ x \in A(\theta)\}$$

$$= \left\{ \theta \in \Re; \ -z_{\alpha/2} \leq \frac{\sqrt{n}(\bar{x} - \theta)}{\sigma} \leq z_{\alpha/2} \right\}$$

$$= \left\{ \theta \in \Re; \ \bar{x} - z_{\alpha/2}\frac{\sigma}{\sqrt{n}} \leq \theta \leq \bar{x} + z_{\alpha/2}\frac{\sigma}{\sqrt{n}} \right\}.$$

In other words, we ended up with the familiar confidence interval for $\theta$, $\bar{x} \pm z_{\alpha/2}\frac{\sigma}{\sqrt{n}}$, we have already constructed in Chapter 10, Example 1(i).

The figure belows (Figure 12.2), is meant to illustrate graphically the interplay between $T(x)$ and $A(\theta)$.

**Example 6.** Let the random sample $X_1, \ldots, X_n$ be from the $N(\mu, \sigma^2)$ distribution, where both $\mu$ and $\sigma^2$ are unknown. Construct a confidence interval for $\sigma^2$ with confidence coefficient $1 - \alpha$, again using Theorem 4.

**Discussion.** Here $S^2 = \frac{1}{n-1}\sum_{i=1}^{n}(X_i - \bar{X})^2$ is an estimate of $\sigma^2$, and, therefore, for testing the hypothesis $H_0(\sigma^2)$: Variance $= \sigma^2$, it is reasonable to reject $H_0(\sigma^2)$ whenever the ratio of $S^2$ over the $\sigma^2$ specified by the hypothesis is either too small or too large (see relation (54) in Chapter 1). That is, reject $H_0(\sigma^2)$ when $\frac{S^2}{\sigma^2} < C_1$ or $\frac{S^2}{\sigma^2} > C_2$ for some ($>0$) constants $C_1, C_2$ to be specified by the requirement that $P_{\sigma^2}(\frac{S^2}{\sigma^2} < C_1$ or $\frac{S^2}{\sigma^2} > C_2) = \alpha$, or

$$P_{\sigma^2}\left[\frac{(n-1)S^2}{\sigma^2} < C_1' \quad \text{or} \quad \frac{(n-1)S^2}{\sigma^2} > C_2'\right] = \alpha,$$
$$C_1' = (n-1)C_1, \quad C_2' = (n-1)C_2.$$

Since under $H_0(\sigma^2)$, $\frac{(n-1)S^2}{\sigma^2} \sim \chi_{n-1}^2$, we may choose to split the probability $\alpha$ equally between the two tails, in which case $C_1' = \chi_{n-1;\,1-\alpha/2}^2$ and $C_2' = \chi_{n-1;\,\alpha/2}^2$, and $H(\sigma^2)$ is accepted whenever

$$\chi_{n-1;\,1-\alpha/2}^2 \leq \frac{(n-1)S^2}{\sigma^2} \leq \chi_{n-1;\,\alpha/2}^2.$$

Of course,

$$P_{\sigma^2}\left[\chi_{n-1;\,1-\alpha/2}^2 \leq \frac{(n-1)S^2}{\sigma^2} \leq \chi_{n-1;\,\alpha/2}^2\right] = 1 - \alpha.$$

Then, with $s^2$ denoting the observed value of $S^2$,

$$A(\sigma^2) = \left\{x \in \mathfrak{R}^n;\; \chi_{n-1;\,1-\alpha/2}^2 \leq \frac{(n-1)s^2}{\sigma^2} \leq \chi_{n-1;\,\alpha/2}^2\right\},$$

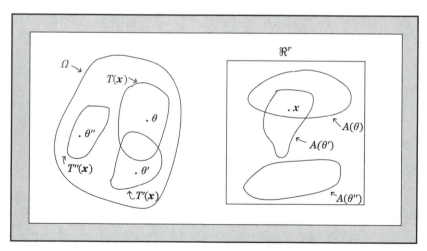

**FIGURE 12.2**

Graphical interplay between acceptance regions and confidence regions.

and therefore (37) becomes here:

$$T(x) = \{\sigma^2 \in (0, \infty); \; x \in A(\sigma^2)\}$$

$$= \left\{\sigma^2 \in (0, \infty); \; \chi^2_{n-1\,;\,1-\alpha/2} \leq \frac{(n-1)s^2}{\sigma^2} \leq \chi^2_{n-1\,;\,\alpha/2}\right\}$$

$$= \left\{\sigma^2 \in (0, \infty); \; \frac{(n-1)s^2}{\chi^2_{n-1;\,\alpha/2}} \leq \sigma^2 \leq \frac{(n-1)s^2}{\chi^2_{n-1;\,1-\alpha/2}}\right\};$$

that is, we have arrived once again at the familiar confidence interval for $\sigma^2$,
$\left[\frac{(n-1)s^2}{\chi^2_{n-1;\alpha/2}}, \frac{(n-1)s^2}{\chi^2_{n-1;1-\alpha/2}}\right]$, (see Example 3 in Chapter 10).

# A simple linear regression model

# 13

This is a rather extensive chapter on an important subject matter with an abundance of diverse applications. The basic idea involved may be described as follows. There is a stimulus, denoted by $x$, and a response to it, denoted by $y$. At different levels of $x$, one observes the respective responses. How are the resulting $(x, y)$ pairs related, if they are related at all? There are all kind of possibilities, and the one discussed in this chapter is the simplest such possibility, namely, the pairs are linearly related.

In reality, what one, actually, observes at $x$, due to errors, is a value of a r.v. Y, and then the question arises as to how we would draw a straight line, which would lie "close" to most of the $(x, y)$ pairs. This leads to the Principle of Least Squares. On the basis of this principle, one is able to draw the so-called fitted linear regression line by computing the Least Squares Estimates (LSE's) of parameters involved. Also, some properties of these estimates are established. These things are done in the first two sections of the chapter.

Up to this point, the errors are not required to have any specific distribution, other than having zero mean and finite variance. However, in order to proceed with further statistical inference about the parameters involved, such as constructing confidence intervals and testing hypotheses, one has to stipulate a distribution for the errors; this distribution, reasonably enough, is assumed to be Normal. As a consequence of it, one is in a position to specify the distribution of all estimates involved and proceed with the inference problems referred to above. These issues are discussed in Sections 13.3 and 13.4.

In the following section, Section 13.5, the problem of predicting the expected value of the observation $Y_0$ at a given point $x_0$ and the problem of predicting a single value of $Y_0$ are discussed. Suitable predictors are provided, and also confidence intervals for them are constructed.

The chapter is concluded with Section 3.7 indicating extensions of the model discussed in this chapter to more general situations covering a much wider class of applications.

## 13.1 SETTING UP THE MODEL—THE PRINCIPLE OF LEAST SQUARES

As has already been mentioned elsewhere, Examples 22 and 23 in Chapter 1 provide motivation for the statistical model to be adopted and studied in this chapter.

Example 22, in particular, will serve throughout the chapter to illustrate the underlying general results. For convenience, the data related to this example are reproduced here in Table 13.1.

**Table 13.1** The Data $x$ = Undergraduate GPA and $y$ = Score in the Graduate Management Aptitude Test (GMAT); There Are 34 $(x, y)$ Pairs Altogether

**Data of Undergraduate GPA ($x$) and GMAT Score ($y$)**

| $x$ | $y$ | $x$ | $y$ | $x$ | $y$ |
|------|-----|------|-----|------|-----|
| 3.63 | 447 | 2.36 | 399 | 2.80 | 444 |
| 3.59 | 588 | 2.36 | 482 | 3.13 | 416 |
| 3.30 | 563 | 2.66 | 420 | 3.01 | 471 |
| 3.40 | 553 | 2.68 | 414 | 2.79 | 490 |
| 3.50 | 572 | 2.48 | 533 | 2.89 | 431 |
| 3.78 | 591 | 2.46 | 509 | 2.91 | 446 |
| 3.44 | 692 | 2.63 | 504 | 2.75 | 546 |
| 3.48 | 528 | 2.44 | 336 | 2.73 | 467 |
| 3.47 | 552 | 2.13 | 408 | 3.12 | 463 |
| 3.35 | 520 | 2.41 | 469 | 3.08 | 440 |
| 3.39 | 543 | 2.55 | 538 | 3.03 | 419 |
|      |     |      |     | 3.00 | 509 |

The first question which arises is whether the pairs $(x, y)$ are related at all and, if they are, how? An indication that those pairs are, indeed, related is borne out by the scatter plot depicted in Figure 13.1. Indeed, taking into consideration that we are operating in a random environment, one sees a conspicuous, albeit somewhat loose, linear relationship between the pairs $(x, y)$.

So, we are not too far off the target by assuming that there is a straight line in the $xy$-plane which is "close" to most of the pairs $(x, y)$. The question now is how to quantify the term "close." The first step toward this end is the adoption of the model described in relation (11) of Chapter 8. Namely, we assume that, for each $i = 1, \ldots, 34$, the respective $y_i$ is the observed value of a r.v. $Y_i$ associated with $x_i$, and if it were not for the random errors involved, the pairs $(x_i, y_i), i = 1, \ldots, 34$ would lie on a straight line $y = \beta_1 + \beta_2 x$; i.e., we would have $y_i = \beta_1 + \beta_2 x_i, i = 1, \ldots, 34$. Thus, the r.v. $Y_i$ itself, whose $y_i$ are simply observed values, would be equal to $\beta_1 + \beta_2 x_i$ except for fluctuations due to a random error $e_i$. In other words, $Y_i = \beta_1 + \beta_2 x_i + e_i$. Next, arguing as in Section 8.4, it is reasonable to assume that the $e_i$'s are independent r.v.'s with $E e_i = 0$ and $Var(e_i) = \sigma^2$ for all $i$'s, so that one arrives at the model described in relation (11) of Chapter 8; namely, $Y_1, \ldots, Y_{34}$ are independent r.v.'s having the structure:

$$Y_i = \beta_1 + \beta_2 x_i + e_i, \quad \text{with } E e_i = 0 \quad \text{and} \quad Var(e_i) = \sigma^2(< \infty), \quad i = 1, \ldots, 34. \quad (1)$$

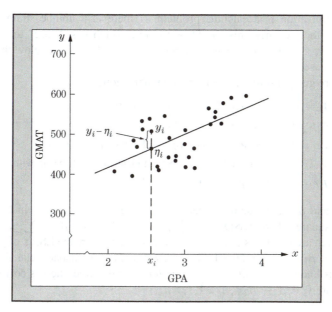

**FIGURE 13.1**

Scatter diagram for Table 13.1.

Set $EY_i = \eta_i$. Then, because of the errors involved, it is, actually, the pairs $(x_i, \eta_i), i = 1, \ldots, 34$ which lie on a straight line $y = \beta_1 + \beta_2 x$; i.e., $\eta_i = \beta_1 + \beta_2 x_i, i = 1, \ldots, 34$. It is in the determination of a particular straight line where the Principle of Least Squares enters the picture. According to this principle, one argues as follows: On the basis of the model described in (1), what we would *expect* to have observed at $x_i$ would be $\eta_i$, whereas what is, *actually*, observed is $y_i$. Thus, there is a deviation measured by $y_i - \eta_i, i = 1, \ldots, 34$ (see Figure 13.1). Some of these deviations are positive, some are negative, and, perhaps, some are zero. In order to deal with non-negative numbers, look at $|y_i - \eta_i|$, which is, actually, the distance between the points $(x_i, y_i)$ and $(x_i, \eta_i)$. Then, draw the line $y = \beta_1 + \beta_2 x$, so that these distances are simultaneously minimized. More formally, first look at the squares of these distances $(y_i - \eta_i)^2$, as it is much easier to work with squares as opposed to absolute values, and in order to account for the simultaneous minimization mentioned earlier, consider the sum $\sum_{i=1}^{34}(y_i - \eta_i)^2$ and seek its minimization. At this point, replace the observed value $y_i$ by the r.v. $Y_i$ itself and set

$$S(Y, \beta) = \sum_{i=1}^{34}(Y_i - \eta_i)^2 = \sum_{i=1}^{34}[Y_i - (\beta_1 + \beta_2 x_i)]^2 \left( = \sum_{i=1}^{34} e_i^2 \right), \qquad (2)$$

where $Y = (Y_1, \ldots, Y_{34})$ and $\beta = (\beta_1, \beta_2)$.

Then the *Principle of Least Squares* calls for the determination of $\beta_1$ and $\beta_2$ which *minimize the sum of squares of errors*; i.e., the quantity $S(Y, \beta)$ in (2). The actual minimization is a calculus problem. If there is a unique straight line so determined, then, clearly, this would be the line which lies "close" to most pairs $(x_i, Y_i), i = 1, \ldots, 34$, in the Least Squares sense. It will be seen below that this is, indeed, the case.

In a more general setting, consider the model below:

$Y_i = \beta_1 + \beta_2 x_i + e_i$, where the random errors

$e_i, i = 1, \ldots, n$ are i.i.d. r.v.'s. with

$Ee_i = 0$, and $Var(e_i) = \sigma^2 (< \infty)$, which imply that the r.v.'s

$Y_i, i = 1, \ldots, n$ are independent, but not identically distributed, with

$$EY_i = \eta_i = \beta_1 + \beta_2 x_i \text{ and } Var(y_i) = \sigma^2. \tag{3}$$

Let $\hat{\beta}_1$ and $\hat{\beta}_2$ be the unique values of $\beta_1$ and $\beta_2$, respectively, which minimize the sum of squares of errors $S(Y, \beta) = \sum_{i=1}^{n}[Y_i - (\beta_1 + \beta_2 x_i)]^2 (= \sum_{i=1}^{n} e_i^2)$. These values, which are functions of the $Y_i$'s as well as the $x_i$'s, are the LSE's of $\beta_1$ and $\beta_2$. Any line $y = \beta_1 + \beta_2 x$ is referred to as a *regression line* and, in particular, the line $\hat{y} = \hat{\beta}_1 + \hat{\beta}_2 x$ is known as the *fitted regression line*. For this line, the $\hat{y}_i$'s corresponding to the $x_i$'s are $\hat{y}_i = \hat{\beta}_1 + \hat{\beta}_2 x_i, i = 1, \ldots, n$.

## 13.2 THE LEAST SQUARES ESTIMATES OF $\beta_1$ AND $\beta_2$ AND SOME OF THEIR PROPERTIES

In this section, the LSE's of $\beta_1$ and $\beta_2$ are derived and some of their properties are obtained. Also, the (unknown) variance $\sigma^2$ is estimated.

**Theorem 1.** *In reference to the model described in (3), the LSE's $\hat{\beta}_1$ and $\hat{\beta}_2$ of $\beta_1$ and $\beta_2$, respectively, are given by the following expressions (which are also appropriate for computational purposes):*

$$\hat{\beta}_1 = \frac{\left(\sum_{i=1}^{n} x_i^2\right)\left(\sum_{i=1}^{n} Y_i\right) - \left(\sum_{i=1}^{n} x_i\right)\left(\sum_{i=1}^{n} x_i Y_i\right)}{n\sum_{i=1}^{n} x_i^2 - \left(\sum_{i=1}^{n} x_i\right)^2}, \tag{4}$$

*and*

$$\hat{\beta}_2 = \frac{n\sum_{i=1}^{n} x_i Y_i - \left(\sum_{i=1}^{n} x_i\right)\left(\sum_{i=1}^{n} Y_i\right)}{n\sum_{i=1}^{n} x_i^2 - \left(\sum_{i=1}^{n} x_i\right)^2}. \tag{5}$$

*Proof.* Consider the partial derivatives:

$$\frac{\partial}{\partial \beta_1} S(Y, \beta) = 2\sum_{i=1}^{n}(Y_i - \beta_1 - \beta_2 x_i)(-1) = -2\left(\sum_{i=1}^{n} Y_i - n\beta_1 - \beta_2 \sum_{i=1}^{n} x_i\right),$$

$$\frac{\partial}{\partial \beta_2} S(Y, \beta) = 2\sum_{i=1}^{n}(Y_i - \beta_1 - \beta_2 x_i)(-x_i) \tag{6}$$

$$= -2\left(\sum_{i=1}^{n} x_i Y_i - \beta_1 \sum_{i=1}^{n} x_i - \beta_2 \sum_{i=1}^{n} x_i^2\right), \tag{7}$$

and solve the so-called *Normal equations:* $\frac{\partial}{\partial \beta_1} S(Y, \beta) = 0$ and $\frac{\partial}{\partial \beta_2} S(Y, \beta) = 0$, or $n\beta_1 + (\sum_{i=1}^{n} x_i)\beta_2 = \sum_{i=1}^{n} Y_i$ and $(\sum_{i=1}^{n} x_i)\beta_1 + (\sum_{i=1}^{n} x_i^2)\beta_2 = \sum_{i=1}^{n} x_i Y_i$ to find:

$$\hat{\beta}_1 = \frac{\begin{vmatrix} \sum_{i=1}^{n} Y_i & \sum_{i=1}^{n} x_i \\ \sum_{i=1}^{n} x_i Y_i & \sum_{i=1}^{n} x_i^2 \end{vmatrix}}{\begin{vmatrix} n & \sum_{i=1}^{n} x_i \\ \sum_{i=1}^{n} x_i & \sum_{i=1}^{n} x_i^2 \end{vmatrix}} = \frac{\left(\sum_{i=1}^{n} x_i^2\right)\left(\sum_{i=1}^{n} Y_i\right) - \left(\sum_{i=1}^{n} x_i\right)\left(\sum_{i=1}^{n} x_i Y_i\right)}{n \sum_{i=1}^{n} x_i^2 - \left(\sum_{i=1}^{n} x_i\right)^2},$$

and

$$\hat{\beta}_2 = \frac{\begin{vmatrix} n & \sum_{i=1}^{n} Y_i \\ \sum_{i=1}^{n} x_i & \sum_{i=1}^{n} x_i Y_i \end{vmatrix}}{n \sum_{i=1}^{n} x_i^2 - \left(\sum_{i=1}^{n} x_i\right)^2} = \frac{n \sum_{i=1}^{n} x_i Y_i - \left(\sum_{i=1}^{n} x_i\right)\left(\sum_{i=1}^{n} Y_i\right)}{n \sum_{i=1}^{n} x_i^2 - \left(\sum_{i=1}^{n} x_i\right)^2}.$$

It remains to show that $\hat{\beta}_1$ and $\hat{\beta}_2$, actually, minimize $S(Y, \beta)$. From (6) and (7), we get:

$$\frac{\partial^2}{\partial \beta_1^2} S(Y, \beta) = 2n, \quad \frac{\partial^2}{\partial \beta_1 \partial \beta_2} S(Y, \beta) = \frac{\partial^2}{\partial \beta_2 \partial \beta_1} S(Y, \beta) = 2\sum_{i=1}^{n} x_i,$$

$$\frac{\partial^2}{\partial \beta_2^2} S(Y, \beta) = 2\sum_{i=1}^{n} x_i^2,$$

and the $2 \times 2$ matrix below is positive semidefinite for all $\beta_1, \beta_2$, since, for all $\lambda_1, \lambda_2$ reals not both 0,

$$(\lambda_1, \lambda_2) \begin{pmatrix} n & \sum_{i=1}^{n} x_i \\ \sum_{i=1}^{n} x_i & \sum_{i=1}^{n} x_i^2 \end{pmatrix} \begin{pmatrix} \lambda_1 \\ \lambda_2 \end{pmatrix}$$

$$= \left(\lambda_1 n + \lambda_2 \sum_{i=1}^{n} x_i \quad \lambda_1 \sum_{i=1}^{n} x_i + \lambda_2 \sum_{i=1}^{n} x_i^2\right) \begin{pmatrix} \lambda_1 \\ \lambda_2 \end{pmatrix}$$

$$= \lambda_1^2 n + 2\lambda_1 \lambda_2 \sum_{i=1}^{n} x_i + \lambda_2^2 \sum_{i=1}^{n} x_i^2$$

$$= \lambda_1^2 n + 2n\lambda_1\lambda_2\bar{x} + \lambda_2^2 \sum_{i=1}^{n} x_i^2 \quad \left(\text{where } \bar{x} = \frac{1}{n}\sum_{i=1}^{n} x_i\right)$$

$$= \lambda_1^2 n + 2n\lambda_1\lambda_2\bar{x} + \lambda_2^2 \left(\sum_{i=1}^{n} x_i^2 - n\bar{x}^2\right) + \lambda_2^2 n\bar{x}^2$$

$$= n\left(\lambda_1^2 + 2\lambda_1\lambda_2\bar{x} + \lambda_2^2\bar{x}^2\right) + \lambda_2^2 \sum_{i=1}^{n}(x_i - \bar{x})^2$$

$$= n(\lambda_1 + \lambda_2\bar{x})^2 + \lambda_2^2 \sum_{i=1}^{n}(x_i - \bar{x})^2 \geq 0.$$

This completes the proof of the theorem. ∎

**Corollary.**    *With* $\bar{x} = (x_1 + \cdots + x_n)/n$ *and* $\bar{Y} = (Y_1 + \cdots + Y_n)/n$, *the LSE's* $\hat{\beta}_1$ *and* $\hat{\beta}_2$ *may also be written as follows (useful expressions for noncomputational purposes):*

$$\hat{\beta}_1 = \bar{Y} - \hat{\beta}_2\bar{x}, \quad \hat{\beta}_2 = \frac{\sum_{i=1}^{n}(x_i - \bar{x})(Y_i - \bar{Y})}{\sum_{i=1}^{n}(x_i - \bar{x})^2} = \frac{1}{\sum_{i=1}^{n}(x_i - \bar{x})^2}\sum_{i=1}^{n}(x_i - \bar{x})Y_i. \quad (8)$$

*Proof.*  In the first place,

$$\sum_{i=1}^{n}(x_i - \bar{x})(Y_i - \bar{Y}) = \sum_{i=1}^{n}(x_i - \bar{x})Y_i - \bar{Y}\sum_{i=1}^{n}(x_i - \bar{x}) = \sum_{i=1}^{n}(x_i - \bar{x})Y_i,$$

which shows that $\hat{\beta}_2$ is as claimed.

Next, replacing $\hat{\beta}_2$ by its expression in (5), we get

$$\bar{Y} - \hat{\beta}_2\bar{x} = \bar{Y} - \frac{n\sum_{i=1}^{n}x_iY_i - \left(\sum_{i=1}^{n}x_i\right)\left(\sum_{i=1}^{n}Y_i\right)}{n\sum_{i=1}^{n}x_i^2 - \left(\sum_{i=1}^{n}x_i\right)^2}\bar{x}$$

$$= \bar{Y} - \frac{n\sum_{i=1}^{n}x_iY_i - n^2\bar{x}\bar{Y}}{n\sum_{i=1}^{n}x_i^2 - n^2\bar{x}^2}\bar{x}$$

$$= \frac{n\bar{Y}\sum_{i=1}^{n}x_i^2 - n^2\bar{x}^2\bar{Y} - n\bar{x}\sum_{i=1}^{n}x_iY_i + n^2\bar{x}^2\bar{Y}}{n\sum_{i=1}^{n}(x_i - \bar{x})^2}$$

$$= \frac{\left(\sum_{i=1}^{n}x_i^2\right)\left(\sum_{i=1}^{n}Y_i\right) - \left(\sum_{i=1}^{n}x_i\right)\left(\sum_{i=1}^{n}x_iY_i\right)}{n\sum_{i=1}^{n}(x_i - \bar{x})^2} = \hat{\beta}_1 \text{ by (4).}$$

∎

The following notation is suggested (at least in part) by the expressions in the LSE's $\hat{\beta}_1$ and $\hat{\beta}_2$, and it will be used extensively and conveniently throughout the rest of this chapter.

Set

$$SS_x = \sum_{i=1}^{n}(x_i - \bar{x})^2 = \sum_{i=1}^{n}x_i^2 - n\bar{x}^2 = \sum_{i=1}^{n}x_i^2 - \frac{1}{n}\left(\sum_{i=1}^{n}x_i\right)^2,$$

and likewise,

$$SS_y = \sum_{i=1}^{n}(Y_i - \bar{Y})^2 = \sum_{i=1}^{n}Y_i^2 - n\bar{Y}^2 = \sum_{i=1}^{n}Y_i^2 - \frac{1}{n}\left(\sum_{i=1}^{n}Y_i\right)^2, \quad (9)$$

and

$$SS_{xy} = \sum_{i=1}^{n}(x_i - \bar{x})(Y_i - \bar{Y}) = \sum_{i=1}^{n}(x_i - \bar{x})Y_i$$

$$= \sum_{i=1}^{n}x_iY_i - \frac{1}{n}\left(\sum_{i=1}^{n}x_i\right)\left(\sum_{i=1}^{n}Y_i\right).$$

Then the LSE's $\hat{\beta}_1$ and $\hat{\beta}_2$ may be rewritten as follows:

$$\hat{\beta}_1 = \frac{1}{n}\sum_{i=1}^{n}Y_i - \hat{\beta}_2\left(\frac{1}{n}\sum_{i=1}^{n}x_i\right), \quad \hat{\beta}_2 = \frac{SS_{xy}}{SS_x}. \quad (10)$$

**Table 13.2** GPA's and the corresponding GMAT scores

| x | y | $x^2$ | xy | x | y | $x^2$ | xy |
|---|---|-------|-----|---|---|-------|-----|
| 3.63 | 447 | 13.1769 | 1622.61 | 3.48 | 528 | 12.1104 | 1837.44 |
| 3.59 | 588 | 12.8881 | 2110.92 | 3.47 | 552 | 12.0409 | 1915.44 |
| 3.30 | 563 | 10.8900 | 1857.90 | 3.35 | 520 | 11.2225 | 1742.00 |
| 3.40 | 553 | 11.5600 | 1880.20 | 3.39 | 543 | 11.4921 | 1840.77 |
| 3.50 | 572 | 12.2500 | 2002.00 | 2.36 | 399 | 5.5696 | 941.64 |
| 3.78 | 591 | 14.2884 | 2233.98 | 2.36 | 482 | 5.5696 | 1137.52 |
| 3.44 | 692 | 11.8336 | 2380.48 | 2.66 | 420 | 7.0756 | 1117.20 |
| 2.68 | 414 | 7.1824 | 1109.52 | 3.01 | 471 | 9.0601 | 1417.71 |
| 2.48 | 533 | 6.1504 | 1321.84 | 2.79 | 490 | 7.7841 | 1367.10 |
| 2.46 | 509 | 6.0516 | 1252.14 | 2.89 | 431 | 8.3521 | 1245.59 |
| 2.63 | 504 | 6.9169 | 1325.52 | 2.91 | 446 | 8.4681 | 1297.86 |
| 2.44 | 336 | 5.9536 | 819.84 | 2.75 | 546 | 7.5625 | 1501.50 |
| 2.13 | 408 | 4.5369 | 869.04 | 2.73 | 467 | 7.4529 | 1274.91 |
| 2.41 | 469 | 5.8081 | 1130.29 | 3.12 | 463 | 9.7344 | 1444.56 |
| 2.55 | 538 | 6.5025 | 1371.90 | 3.08 | 440 | 9.4864 | 1355.20 |
| 2.80 | 444 | 7.8400 | 1243.20 | 3.03 | 419 | 9.1809 | 1269.57 |
| 3.13 | 416 | 9.7969 | 1302.08 | 3.00 | 509 | 9.0000 | 1527.00 |
| Total 50.35 | 8577 | 153.6263 | 25,833.46 | 50.38 | 8126 | 151.1622 | 24,233.01 |

Also, recall that the fitted regression line is given by:

$$\hat{y} = \hat{\beta}_1 + \hat{\beta}_2 x \quad \text{and that} \quad \hat{y}_i = \hat{\beta}_1 + \hat{\beta}_2 x_i, \quad i = 1, \ldots, n. \tag{11}$$

Before we go any further, let us discuss the example below.

**Example 1.** In reference to Table 13.1, compute the LSE's $\hat{\beta}_1$ and $\hat{\beta}_2$ and draw the fitted regression line $\hat{y} = \hat{\beta}_1 + \hat{\beta}_2 x$.

**Discussion.** The application of formula (10) calls for the calculation of $SS_x$ and $SS_{xy}$ given in (9). Table 13.2 facilitates the calculations.

$$\sum_i x_i = 100.73, \quad \sum_i y_i = 16,703, \quad \sum_i x_i^2 = 304.7885, \quad \sum_i x_i y_i = 50,066.47,$$

and then

$$SS_x = 304.7885 - \frac{(100.73)^2}{34} \simeq 304.7885 - 298.4274 \simeq 6.361,$$

$$SS_{xy} = 50,066.47 - \frac{(100.73) \times (16,703)}{34} \simeq 50,066.47 - 49,485.094$$

$$= 581.376.$$

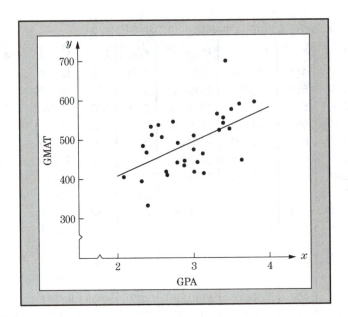

**FIGURE 13.2**

The fitted regression line $\hat{y} = 220.456 + 91.397x$.

Thus,

$$\hat{\beta}_2 = \frac{581.376}{6.361} \simeq 91.397 \quad \text{and} \quad \hat{\beta}_1 = \frac{16,703}{34} - (91.397) \times \frac{100.73}{34}$$
$$\simeq 491.265 - 270.809 = 220.456,$$

and the fitted regression line $\hat{y} = 220.456 + 91.397x$ is depicted in the Figure 13.2.

The LSE's $\hat{\beta}_1$ and $\hat{\beta}_2$ have the desirable property of being unbiased, as shown in the following theorem.

**Theorem 2.** *The LSE's $\hat{\beta}_1$ and $\hat{\beta}_2$ are unbiased; i.e., $E\hat{\beta}_1 = \beta_1$ and $E\hat{\beta}_2 = \beta_2$. Furthermore,*

$$Var(\hat{\beta}_1) = \sigma^2 \left( \frac{1}{n} + \frac{\bar{x}^2}{SS_x} \right) \quad \text{and} \quad Var(\hat{\beta}_2) = \frac{\sigma^2}{SS_x},$$

*where $SS_x$ is given in (9).*

*Proof.* In this proof and also elsewhere, the range of the summation is not explicitly indicated, since it is always from 1 to $n$. Consider $\hat{\beta}_2$ as given in (8). Then: $SS_x \hat{\beta}_2 = \sum_i (x_i - \bar{x}) Y_i$, so that, by taking expectations:

$$SS_x E\hat{\beta}_2 = \sum_i (x_i - \bar{x}) EY_i = \sum_i (x_i - \bar{x})(\beta_1 + \beta_2 x_i)$$

$$= \beta_1 \sum_i (x_i - \bar{x}) + \beta_2 \sum_i x_i(x_i - \bar{x}) = \beta_2 \sum_i x_i(x_i - \bar{x})$$

$$= \beta_2 \left( \sum_i x_i^2 - n\bar{x}^2 \right) = \beta_2 \sum_i (x_i - \bar{x})^2 = SS_x \beta_2.$$

Therefore, dividing through by $SS_x$, we get $E\hat{\beta}_2 = \beta_2$. Next, also from (8),

$$E\hat{\beta}_1 = E(\bar{Y} - \hat{\beta}_2 \bar{x}) = E\bar{Y} - \bar{x}E\hat{\beta}_2 = \frac{1}{n} \sum_i (\beta_1 + \beta_2 x_i) - \bar{x}\beta_2$$

$$= \beta_1 + \beta_2 \bar{x} - \beta_2 \bar{x} = \beta_1.$$

Regarding the variances, we have from (8): $SS_x \hat{\beta}_2 = \sum_i (x_i - \bar{x})Y_i$, so that:

$$SS_x^2 \, Var(\hat{\beta}_2) = Var\left( \sum_i (x_i - \bar{x})Y_i \right) = \sum_i (x_i - \bar{x})^2 Var(Y_i)$$

$$= \sigma^2 \sum_i (x_i - \bar{x})^2 = \sigma^2 SS_x,$$

so that $Var(\hat{\beta}_2) = \sigma^2/SS_x$. Finally, from (8),

$$\hat{\beta}_1 = \bar{Y} - \bar{x}\hat{\beta}_2 = \frac{1}{n} \sum_i Y_i - \frac{\bar{x}}{SS_x} \sum_i (x_i - \bar{x})Y_i = \sum_i \left[ \frac{1}{n} - \frac{\bar{x}(x_i - \bar{x})}{SS_x} \right] Y_i, \qquad (12)$$

so that

$$Var(\hat{\beta}_1) = \sigma^2 \sum_i \left[ \frac{1}{n} - \frac{\bar{x}(x_i - \bar{x})}{SS_x} \right]^2 = \sigma^2 \left( \frac{1}{n} + \frac{\bar{x}^2}{SS_x^2} SS_x \right) = \sigma^2 \left( \frac{1}{n} + \frac{\bar{x}^2}{SS_x} \right). \quad \blacksquare$$

**Example 2.** In reference to Example 1, the variances of the LSE's $\hat{\beta}_1$ and $\hat{\beta}_2$ are given by:

$$Var(\hat{\beta}_1) \simeq \sigma^2 \left( \frac{1}{34} + \frac{8.777}{6.361} \right) \simeq \sigma^2(0.029 + 1.380) = 1.409\sigma^2,$$

and

$$Var(\hat{\beta}_2) = \frac{\sigma^2}{6.361} \simeq 0.157\sigma^2.$$

In fitting a regression line, there are various deviations which occur. At this point, these deviations will be suitably attributed to several sources, certain pieces of terminology will be introduced, and also some formal relations will be established. To this end, look at the observable $Y_i$ and split it as follows: $Y_i = \hat{y}_i + (Y_i - \hat{y}_i)$. The component $\hat{y}_i$ is associated with the point $(x_i, \hat{y}_i)$ which lies on the fitted regression line $\hat{y} = \hat{\beta}_1 + \hat{\beta}_2 x$, and the difference $Y_i - \hat{y}_i$ is the deviation of $Y_i$ from $\hat{y}_i$. We may refer to the component $\hat{y}_i$ as that part of $Y_i$ which is *due to the linear regression*, or it *is explained by the linear regression*, and the component $Y_i - \hat{y}_i$ of $Y_i$ as the *residual*, or the *deviation from the linear regression*, or *variability unexplained by the linear regression.* We can go through the same arguments with reference to the sample mean $\bar{Y}$ of the $Y_i$'s. That is, we consider:

$$Y_i - \bar{Y} = (\hat{y}_i - \bar{Y}) + (Y_i - \hat{y}_i).$$

The interpretation of this decomposition is the same as the one given above, but with reference to $\bar{Y}$. Next, look at the squares of these quantities:

$$(Y_i - \bar{Y})^2, \quad (\hat{y}_i - \bar{Y})^2, \quad (Y_i - \hat{y}_i)^2,$$

and, finally, at their sums:

$$\sum_{i=1}^{n}(Y_i - \bar{Y})^2, \quad \sum_{i=1}^{n}(\hat{y}_i - \bar{Y})^2, \quad \sum_{i=1}^{n}(Y_i - \hat{y}_i)^2.$$

At this point, assume for a moment that:

$$\sum_{i=1}^{n}(Y_i - \bar{Y})^2 = \sum_{i=1}^{n}(\hat{y}_i - \bar{Y})^2 + \sum_{i=1}^{n}(Y_i - \hat{y}_i)^2. \tag{13}$$

Then this relation would state that the *total variability* (of the $Y_i$'s in reference to their mean $\bar{Y}$), $\sum_{i=1}^{n}(Y_i - \bar{Y})^2$, is the sum of the variability $\sum_{i=1}^{n}(\hat{y}_i - \bar{Y})^2$ *due to the linear regression*, or *explained by the linear regression*, and the *sum of residual variability*, $\sum_{i=1}^{n}(Y_i - \hat{y}_i)^2$, or *variability unexplained by the linear regression*.

We proceed in proving relation (13).

**Theorem 3.** *Let* $SS_T(=SS_y$, *see* (9)), $SS_R$, *and* $SS_E$, *respectively, be the total variability, the variability due to the linear regression (or explained by the linear regression), and the residual variability (or variability not explained by the linear regression); i.e.,*

$$SS_T(= SS_y) = \sum_{i=1}^{n}(Y_i - \bar{Y})^2, \quad SS_R = \sum_{i=1}^{n}(\hat{y}_i - \bar{Y})^2, \quad SS_E = \sum_{i=1}^{n}(Y_i - \hat{y}_i)^2, \tag{14}$$

*where* $\hat{y}_i = \hat{\beta}_1 + \hat{\beta}_2 x_i, i = 1, \ldots, n$, *the LSE's* $\hat{\beta}_1$ *and* $\hat{\beta}_2$ *are given by (10) (or (8)), and* $\bar{Y}$ *is the mean of the* $Y_i$'s. *Then:*

**(i)**
$$SS_T = SS_R + SS_E. \tag{15}$$

*Furthermore,*

**(ii)**
$$SS_T = SS_y, \quad SS_R = \frac{SS_{xy}^2}{SS_x}, \quad \text{and hence} \quad SS_E = SS_y - \frac{SS_{xy}^2}{SS_x}, \tag{16}$$

*where* $SS_x$, $SS_y$, *and* $SS_{xy}$ *are given in (9).*

*Proof.*

**(i)** We have:

$$SS_T = \sum_{i}(Y_i - \bar{Y})^2 = \sum_{i}[(\hat{y}_i - \bar{Y}) + (Y_i - \hat{y}_i)]^2$$

$$= \sum_i (\hat{y}_i - \bar{Y})^2 + \sum_i (Y_i - \hat{y}_i)^2 + 2 \sum_i (\hat{y}_i - \bar{Y})(Y_i - \hat{y}_i)$$

$$= SS_R + SS_E + 2 \sum_i (\hat{y}_i - \bar{Y})(Y_i - \hat{y}_i).$$

So, we have to show that the last term on the right-hand side above is equal to 0. To this end, observe that $\hat{y}_i = \hat{\beta}_1 + \hat{\beta}_2 x_i$ and $\hat{\beta}_1 = \bar{Y} - \hat{\beta}_2 \bar{x}$ (by (8)), so that

$$\hat{y}_i - \bar{Y} = \hat{\beta}_1 + \hat{\beta}_2 x_i - \bar{Y} = \bar{Y} - \hat{\beta}_2 \bar{x} + \hat{\beta}_2 x_i - \bar{Y} = \hat{\beta}_2 (x_i - \bar{x}),$$

and

$$Y_i - \hat{y}_i = Y_i - \hat{\beta}_1 - \hat{\beta}_2 x_i = Y_i - \bar{Y} + \hat{\beta}_2 \bar{x} - \hat{\beta}_2 x_i = (Y_i - \bar{Y}) - \hat{\beta}_2 (x_i - \bar{x}),$$

so that

$$(\hat{y}_i - \bar{Y})(Y_i - \hat{y}_i) = \hat{\beta}_2 (x_i - \bar{x})[(Y_i - \bar{Y}) - \hat{\beta}_2 (x_i - \bar{x})]$$

$$= \hat{\beta}_2 (x_i - \bar{x})(Y_i - \bar{Y}) - \hat{\beta}_2^2 (x_i - \bar{x})^2.$$

Therefore, by (9) and (10):

$$\sum_i (\hat{y}_i - \bar{Y})(Y_i - \hat{y}_i) = \frac{SS_{xy}}{SS_x} \times SS_{xy} - \frac{SS_{xy}^2}{SS_x^2} \times SS_x = \frac{SS_{xy}^2}{SS_x} - \frac{SS_{xy}^2}{SS_x} = 0. \qquad (17)$$

Thus, $SS_T = SS_R + SS_E$.

**(ii)** That $SS_T = SS_y$ is immediate from relations (9) and (14). Next, $\hat{y}_i - \bar{Y} = \hat{\beta}_2 (x_i - \bar{x})$ as was seen in the proof of part (i), so that, by (10),

$$SS_R = \sum_i (\hat{y}_i - \bar{Y})^2 = \hat{\beta}_2^2 \sum_i (x_i - \bar{x})^2 = \frac{SS_{xy}^2}{SS_x^2} \times SS_x = \frac{SS_{xy}^2}{SS_x},$$

as was to be seen. Finally, by part (i), $SS_E = SS_T - SS_R = SS_y - \frac{SS_{xy}^2}{SS_x}$, as was to be seen. ∎

This section is closed with some remarks.

**Remark 1.**

(i) The quantities $SS_T, SS_R$, and $SS_E$, given in (14), are computed by way of $SS_x, SS_y$, and $SS_{xy}$ given in (9). This is so because of (16).

(ii) In the next section, an estimate of the (unknown) variance $\sigma^2$ will also be given, based on the residual variability $SS_E$. That this should be the case is intuitively clear by the nature of $SS_E$, and it will be formally justified in the following section.

(iii) From the relation $SS_T = SS_R + SS_E$ given in (15) and the definition of the variability due to regression, $SS_R$, given in (14), it follows that the better the regression fit is, the smaller the value of $SS_R$ is. Then, its ratio to the total variability, $SS_T$, $r = SS_R/SS_T$, can be used as an index of how good the linear regression fit is; the larger $r(\leq 1)$ the better the fit.

## 13.3 NORMALLY DISTRIBUTED ERRORS: MLE'S OF $\beta_1$, $\beta_2$, AND $\sigma^2$, SOME DISTRIBUTIONAL RESULTS

It is to be noticed that in the linear regression model as defined in relation (3), no distribution assumption about the errors $e_i$, and therefore the r.v.'s $Y_i$, was made. Such an assumption was not necessary, neither for the construction of the LSE's of $\beta_1$, $\beta_2$, nor in proving their unbiasedness and in calculating their variances. However, in order to be able to construct confidence intervals for $\beta_1$ and $\beta_2$ and test hypotheses about them, among other things, we have to assume a distribution for the $e_i$'s. The $e_i$'s being errors, it is not unreasonable to assume that they are Normally distributed, and we shall do so. Then the model (3) is supplemented as follows:

$$Y_i = \beta_1 + \beta_2 x_i + e_i, \quad \text{where } e_i, i = 1, \ldots, n \text{ are independent r.v.'s}$$

$$\sim N(0, \sigma^2), \quad \text{which implies that } Y_i, i = 1, \ldots, n \text{ are} \tag{18}$$

$$\text{independent r.v.'s and } Y_i \sim N(\beta_1 + \beta_2 x_i, \sigma^2).$$

We now proceed with the following theorem.

**Theorem 4.**   *Under model (18):*

(i)  *The LSE's $\hat{\beta}_1$ and $\hat{\beta}_2$ of $\beta_1$ and $\beta_2$, respectively, are also MLE's.*
(ii)  *The MLE $\hat{\sigma}^2$ of $\sigma^2$ is given by: $\hat{\sigma}^2 = SS_E/n$.*
(iii)  *The estimates $\hat{\beta}_1$ and $\hat{\beta}_2$ are Normally distributed as follows:*

$$\hat{\beta}_1 \sim N\left(\beta_1, \sigma^2\left(\frac{1}{n} + \frac{\bar{x}^2}{SS_x}\right)\right), \quad \hat{\beta}_2 \sim N\left(\beta_2, \frac{\sigma^2}{SS_x}\right),$$

*where $SS_x$ is given in (9).*

*Proof.*

(i)  The likelihood function of the $Y_i$'s is given by:

$$L(y_1, \ldots, y_n; \beta_1, \beta_2, \sigma^2) = \left(\frac{1}{\sqrt{2\pi\sigma^2}}\right)^n \exp\left[-\frac{1}{2\sigma^2}\sum_i (y_i - \beta_1 - \beta_2 x_i)^2\right].$$

For each fixed $\sigma^2$, maximization of the likelihood function with respect to $\beta_1$ and $\beta_2$, is, clearly, equivalent to minimization of $\sum_i (y_i - \beta_1 - \beta_2 x_i)^2$ with respect to $\beta_1$ and $\beta_2$, which minimization has produced the LSE's $\hat{\beta}_1$ and $\hat{\beta}_2$.

(ii)  The MLE of $\sigma^2$ is to be found by minimizing, with respect to $\sigma^2$, the expression:

$$\log L(y_i, \ldots, y_n; \hat{\beta}_1, \hat{\beta}_2, \sigma^2) = -\frac{n}{2}\log(2\pi) - \frac{n}{2}\log\sigma^2 - \frac{1}{2\sigma^2}SS_E,$$

since, by (14) and (11), and by using the same notation $SS_E$ both for $\sum_{i=1}^{n}(Y_i - \hat{y}_i)^2$ and $\sum_{i=1}^{n}(y_i - \hat{y}_i)^2$, we have $\sum_i (y_i - \hat{\beta}_1 - \hat{\beta}_2 x_i)^2 =$

$\sum_i (y_i - \hat{y}_i)^2 = SS_E$. From this expression, we get:

$$\frac{d}{d\sigma^2} \log L(y_1, \ldots, y_n; \hat{\beta}_1, \hat{\beta}_2, \sigma^2) = -\frac{n}{2} \times \frac{1}{\sigma^2} + \frac{SS_E}{2(\sigma^2)^2} = 0,$$

so that $\sigma^2 = SS_E/n$. Since

$$\frac{d^2}{d(\sigma^2)^2} \log L(y_1, \ldots, y_n; \hat{\beta}_1, \hat{\beta}_2, \sigma^2)\Big|_{\sigma^2 = SS_E/n} = -\frac{n^3}{2SS_E^2} < 0,$$

it follows that $\hat{\sigma}^2 = SS_E/n$ is, indeed, the MLE of $\sigma^2$.

(iii)  From (12), we have: $\hat{\beta}_1 = \sum_i [\frac{1}{n} - \frac{\bar{x}(x_i - \bar{x})}{SS_x}] Y_i$, and we have also seen in Theorem 2 that:

$$E\hat{\beta}_1 = \beta_1, \quad Var(\hat{\beta}_1) = \sigma^2 \left( \frac{1}{n} + \frac{\bar{x}^2}{SS_x} \right).$$

Thus, $\hat{\beta}_1$ is Normally distributed as a linear combination of independent Normally distributed r.v.'s, and its mean and variance must be as stated above. Next, from (8), we have that: $\hat{\beta}_2 = \sum_i (\frac{x_i - \bar{x}}{SS_x}) Y_i$, so that, as above, $\hat{\beta}_2$ is Normally distributed. Its mean and variance have been computed in Theorem 2 and they are $\beta_2$ and $\sigma^2/SS_x$, respectively. ∎

Before proceeding further, we return to Example 1 and compute an estimate for $\sigma^2$. Also, discuss Example 23 in Chapter 1 and, perhaps, an additional example to be introduced here.

**Example 3.**   In reference to Example 1, determine the MLE of $\sigma^2$.

**Discussion.**   By Theorem 4(ii), this estimate is: $\hat{\sigma}^2 = \frac{SS_E}{n}$. For the computation of $SS_E$ by (16), we have to have the quantity $\sum y_i^2$ from Table 13.2, which is calculated to be:

$$\sum_i y_i^2 = 8,373,295. \tag{19}$$

Then, by (9),

$$SS_y = 8,373,295 - \frac{(16,703)^2}{34} \simeq 8,373,295 - 8,205,594.382 = 167,700.618,$$

and therefore

$$SS_E = 167,700.618 - \frac{(581.376)^2}{6.361} \simeq 167,700.618 - 53,135.993 = 114,564.625;$$

i.e.,

$$SS_E = 114,564.625 \quad \text{and then} \quad \hat{\sigma}^2 = \frac{114,564.625}{34} \simeq 3369.548.$$

Since $SS_T = SS_y = 167,700.618$ and $SS_R = 53,135.993$, it follows that only $\frac{53,135.993}{167,700.618} \simeq 31.685\%$ of the variability is explained by linear regression and $\frac{114,564.625}{167,700.618} \simeq 68.315\%$ is not explained by linear regression. The obvious outlier $(3.44, 692)$ may be mainly responsible for it.

**Example 4.**   In reference to Example 23 in Chapter 1, assume a linear relationship between the dose of a compost fertilizer $x$ and the yield of a crop $y$. On the basis of the following summary data recorded:

$$n = 15, \quad \bar{x} = 10.8, \quad \bar{y} = 122.7, \quad SS_x = 70.6, \quad SS_y = 98.5, \quad SS_{xy} = 68.3:$$

(i)  Determine the estimates $\hat{\beta}_1$ and $\hat{\beta}_2$, and draw the fitted regression line.
(ii)  Give the MLE $\hat{\sigma}^2$ of $\sigma^2$.
(iii)  Over the range of $x$ values covered in the study, what would your conjecture be regarding the average increase in yield per unit increase in the compost dose?

**Discussion.**

(i)  By (10),

$$\hat{\beta}_2 = \tfrac{68.3}{70.6} \simeq 0.967, \quad \hat{\beta}_1 = 122.7 - 0.967 \times 10.8 \simeq 112.256,$$
$$\text{and} \quad \hat{y} = 112.256 + 0.967x.$$

(ii)  We have: $\hat{\sigma}^2 = \frac{SS_E}{15}$, where $SS_E = 98.5 - \frac{(68.3)^2}{70.6} \simeq 32.425$, so that $\hat{\sigma}^2 = \frac{32.425}{15} = 2.162$.

(iii)  The conjecture would be a number close to the slope of the fitted regression line, which is 0.967 (Figure 13.3).

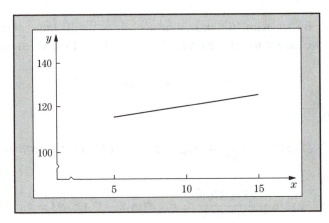

**FIGURE 13.3**

The fitted regression line $\hat{y} = 112.256 + 0.967x$.

**Table 13.3** Dosage ($x$) (in milligrams) and the number of days of relief ($y$) from allergy for 10 patients

|       | $x$ | $y$ | $x^2$ | $y^2$ | $xy$ |
|-------|-----|-----|-------|-------|------|
|       | 3   | 9   | 9     | 81    | 27   |
|       | 3   | 5   | 9     | 25    | 15   |
|       | 4   | 12  | 16    | 144   | 48   |
|       | 5   | 9   | 25    | 81    | 45   |
|       | 6   | 14  | 36    | 196   | 84   |
|       | 6   | 16  | 36    | 256   | 96   |
|       | 7   | 22  | 49    | 484   | 154  |
|       | 8   | 18  | 64    | 324   | 144  |
|       | 8   | 24  | 64    | 576   | 192  |
|       | 9   | 22  | 81    | 484   | 198  |
| Total | 59  | 151 | 389   | 2651  | 1003 |

**Example 5.** In one stage of the development of a new medication for an allergy, an experiment is conducted to study how different dosages of the medication affect the duration of relief from the allergic symptoms. Ten patients are included in the experiment. Each patient receives a specific dosage of the medication and is asked to report back as soon as the protection of the medication seems to wear off. The observations are recorded in Table 13.3, which shows the dosage ($x$) and respective duration of relief ($y$) for the 10 patients.

(i) Draw the scatter diagram of the data in Table 13.3 (which indicate tendency toward linear dependence).

(ii) Compute the estimates $\hat{\beta}_1$ and $\hat{\beta}_2$, and draw the fitted regression line.

(iii) What percentage of the total variability is explained by the linear regression and what percentage remains unexplained?

(iv) Compute the MLE $\hat{\sigma}^2$ of $\sigma^2$.

**Discussion.**

(i),(ii) First, $SS_x = 389 - \frac{59^2}{10} = 40.9$ and $SS_{xy} = 1003 - \frac{59 \times 151}{10} = 112.1$, and hence:

$$\hat{\beta}_2 = \frac{112.1}{40.9} \simeq 2.741 \quad \text{and} \quad \hat{\beta}_1 = \frac{151}{10} - 2.741 \times \frac{59}{10} \simeq -1.072.$$

Then the fitted regression line is $\hat{y} = -1.072 + 2.741x$ (Figure 13.4).

(iii) Since $SS_T = SS_y = 2651 - \frac{151^2}{10} = 370.9$ and $SS_R = \frac{(112.1)^2}{40.9} \simeq 307.247$, it follows that $SS_E = 370.9 - 307.247 = 63.653$. Therefore $\frac{307.247}{370.9} \simeq 82.838\%$

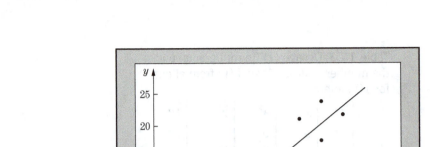

**FIGURE 13.4**

Scatter diagram and the fitted regression line $\hat{y} = -1.072 + 2.741x$.

of the variability is explained by the linear regression and $\frac{63.653}{370.9} \simeq 17.162\%$ remains unexplained.

**(iv)** We have: $\hat{\sigma}^2 = \frac{63.653}{10} = 6.3653 \simeq 6.365$.

For the purpose of constructing confidence intervals for the parameters of the model, and also testing hypotheses about them, we have to know the distribution of $SS_E$ and also establish independence of the statistics $\hat{\beta}_1$ and $SS_E$, as well as independence of the statistics $\hat{\beta}_2$ and $SS_E$. The relevant results are stated in the following theorem, whose proof is deferred to Section 13.6.

**Theorem 5.** *Under model (18):*

**(i)** *The distribution of $SS_E/\sigma^2$ is $\chi^2_{n-2}$.*
**(ii)** *The following statistics are independent:*

      (a) $SS_E$ and $\hat{\beta}_2$;   (b) $\bar{Y}$ and $\hat{\beta}_2$;   (c) $SS_E, \bar{Y}$ and $\hat{\beta}_2$;   (d) $SS_E$ and $\hat{\beta}_1$.

*Proof.* Deferred to Section 13.6. ■

To this theorem, in conjunction with Theorem 4, there is the following corollary.

**Corollary.** *Under model (18):*

**(i)** *The MLE $\hat{\sigma}^2$ of $\sigma^2$ is a biased estimate of $\sigma^2$, but $\frac{n}{n-2}\hat{\sigma}^2 = \frac{SS_E}{n-2}$, call it $S^2$, is an unbiased estimate of $\sigma^2$.*

**(ii)**

$$\frac{\hat{\beta}_1 - \beta_1}{S\sqrt{\frac{1}{n} + \frac{\bar{x}^2}{SS_x}}} \sim t_{n-2}, \quad \text{(iii)} \quad \frac{\hat{\beta}_2 - \beta_2}{S/\sqrt{SS_x}} \sim t_{n-2}, \tag{20}$$

*where*

$$S^2 = SS_E/(n-2). \tag{21}$$

*Proof.*

**(i)** It has been seen in Theorem 4(ii) that $\hat{\sigma}^2 = \frac{SS_E}{n} = \frac{n-2}{n} \times \frac{SS_E}{n-2}$. Since $\frac{SS_E}{\sigma^2} \sim \chi^2_{n-2}$, it follows that $E(\frac{SS_E}{\sigma^2}) = n-2$, or $E(\frac{SS_E}{n-2}) = \sigma^2$, so that $\frac{SS_E}{n-2}$ is an unbiased estimate of $\sigma^2$. Also, $E\hat{\sigma}^2 = \frac{n-2}{n}E(\frac{SS_E}{n-2}) = \frac{n-2}{n}\sigma^2$, so that $\hat{\sigma}^2$ is biased.

**(ii)** By Theorem 4(iii),

$$\frac{\hat{\beta}_1 - \beta_1}{\sigma(\hat{\beta}_1)} = \frac{\hat{\beta}_1 - \beta_1}{\sigma\sqrt{\frac{1}{n} + \frac{\bar{x}^2}{SS_x}}} \sim N(0, 1),$$

and $\frac{SS_E}{\sigma^2} = \frac{(n-2)SS_E}{(n-2)\sigma^2} = \frac{(n-2)S^2}{\sigma^2} \sim \chi^2_{n-2}$. Furthermore, $\frac{\hat{\beta}_1 - \beta_1}{\sigma(\hat{\beta}_1)}$ and $\frac{SS_E}{\sigma^2}$ are independent, since $\hat{\beta}_1$ and $SS_E$ are so. It follows that:

$$\frac{(\hat{\beta}_1 - \beta_1)/\sigma(\hat{\beta}_1)}{\sqrt{\frac{SS_E}{\sigma^2}/(n-2)}} \sim t_{n-2}, \quad \text{or} \quad \frac{(\hat{\beta}_1 - \beta_1)/\sigma\sqrt{\frac{1}{n} + \frac{\bar{x}^2}{SS_x}}}{\sqrt{S^2/\sigma^2}} \sim t_{n-2},$$

or, finally,

$$\frac{\hat{\beta}_1 - \beta_1}{S\sqrt{\frac{1}{n} + \frac{\bar{x}^2}{SS_x}}} \sim t_{n-2}.$$

**(iii)** Again by Theorem 4(iii),

$$\frac{\hat{\beta}_2 - \beta_2}{\sigma(\hat{\beta}_2)} = \frac{\hat{\beta}_2 - \beta_2}{\sigma/\sqrt{SS_x}} \sim N(0, 1),$$

and $\frac{\hat{\beta}_2 - \beta_2}{\sigma(\hat{\beta}_2)}$ and $\frac{SS_E}{\sigma^2}$ are independent, since $\hat{\beta}_2$ and $SS_E$ are so. Then:

$$\frac{(\hat{\beta}_2 - \beta_2)/\sigma(\hat{\beta}_2)}{\sqrt{\frac{SS_E}{\sigma^2}/(n-2)}} \sim t_{n-2}, \quad \text{or} \quad \frac{(\hat{\beta}_2 - \beta_2)/\frac{\sigma}{\sqrt{SS_x}}}{\sqrt{S^2/\sigma^2}} \sim t_{n-2},$$

or, finally, $\frac{\hat{\beta}_2 - \beta_2}{S/\sqrt{SS_x}} \sim t_{n-2}.$ ∎

## EXERCISES

**3.1** Verify the result $\frac{d^2}{dt^2} \log(y_1, \ldots, y_n; \hat{\beta}_1, \hat{\beta}_2, t)\big|_{t=SS_{E/n}} = -\frac{n^3}{2SS_E^2}$ as claimed in the proof of Theorem 4(ii), where $t = \sigma^2$.

**3.2** Consider Table 13.1 and leave out the "outlier" pairs (3.63, 447), (3.44, 692), and (2.44, 336). Then recalculate all quantities below:

$$\sum_i x_i, \quad \sum_i y_i, \quad \sum_i x_i^2, \quad \sum_i x_i y_i, \quad \sum_i y_i^2, \quad SS_x, \quad SS_y, \quad SS_{xy}.$$

**3.3** Use the calculations in Exercise 3.2 to compute the estimates $\hat{\beta}_1, \hat{\beta}_2$ and the fitted regression line.

**3.4** Refer to Exercise 3.2, and compute the variances $Var(\hat{\beta}_1)$, $Var(\hat{\beta}_2)$, and the MLE of $\sigma^2$.

**3.5** By Theorem 5, the r.v. $SS_E/\sigma^2$ is distributed as $\chi^2_{n-2}$. Therefore, in the usual manner,

$$\left[ \frac{SS_E}{\chi^2_{n-2\,;\frac{\alpha}{2}}}, \quad \frac{SS_E}{\chi^2_{n-2\,;1-\frac{\alpha}{2}}} \right]$$

is a confidence interval for $\sigma^2$ with confidence coefficient $1 - \alpha$, where $SS_E$ is given in (16) and (9). That is, $SS_E = SS_y - \frac{SS_{xy}^2}{SS_x}$, where $SS_x$, $SS_y$, and $SS_{xy}$ are given in (9).

  **(i)** Refer to Example 1 (see also Example 3), and construct a 95% confidence interval for $\sigma^2$.
  **(ii)** Refer to Example 4 and do the same as in part (i).
  **(iii)** Refer to Example 5 and do the same as in part (i).
  **(iv)** Refer to Exercise 3.2 and do the same as in part (i).

**3.6** Consider the linear regression model given in relation (18), and let $x_0$ be an unknown point at which observations $Y_{0i}, i = 1, \ldots, m$ are taken. It is assumed that the $Y_{0i}$'s and the $Y_j$'s are independent, and set $Y_0 = \frac{1}{m} \sum_{i=1}^m Y_{0i}$. Set $y = (y_1, \ldots, y_n), y_0 = (y_{01}, \ldots, y_{0m})$ for the observed values of the $Y_j$'s and $Y_{0i}$'s, respectively, and form their joint log-likelihood function:

$$\Lambda = \Lambda(\beta_1, \beta_2, \sigma^2, x_0) = \log L(\beta_1, \beta_2, \sigma^2, x_0 | y, y_0)$$

$$= -\frac{m+n}{2} \log(2\pi) - \frac{m+n}{2} \log \sigma^2$$

$$- \frac{1}{2\sigma^2} \left[ \sum_{j=1}^n (y_j - \beta_1 - \beta_2 x_j)^2 + \sum_{i=1}^m (y_{0i} - \beta_1 - \beta_2 x_0)^2 \right].$$

  **(i)** Show that the log-likelihood equations $\frac{\partial \Lambda}{\partial \beta_1} = 0, \frac{\partial \Lambda}{\partial \beta_2} = 0$, and $\frac{\partial \Lambda}{\partial x_0} = 0$ produce the equations:

$$(m+n)\beta_1 + (mx_0 + n\bar{x})\beta_2 = my_0 + n\bar{y} \tag{a}$$

$$(mx_0 + n\bar{x})\beta_1 + \left(mx_0^2 + \sum_j x_j^2\right)\beta_2 = mx_0y_0 + \sum_j x_jy_j \qquad \text{(b)}$$

$$\beta_1 + x_0\beta_2 = y_0. \qquad \text{(c)}$$

**(ii)** In (c), solve for $\beta_1$, $\beta_1 = y_0 - x_0\beta_2$, replace it in (a) and (b), and solve for $\beta_2$ to obtain, by assuming here and in the sequel that all divisions and cancellations are legitimate,

$$\beta_2 = \frac{\bar{y} - y_0}{\bar{x} - x_0}, \qquad \beta_2 = \frac{\sum_j x_jy_j - n\bar{x}y_0}{\sum_j x_j^2 - nx_0\bar{x}}. \qquad \text{(d)}$$

**(iii)** Equate the $\beta_2$'s in (ii), and solve for $x_0$ to obtain:

$$
\begin{aligned}
x_0 &= \left[ y_0 \sum_j (x_j - \bar{x})^2 + \bar{x}\sum_j x_jy_j - \bar{y}\sum_j x_j^2 \right] \\
&\quad \Big/ \left( \sum_j x_jy_j - n\bar{x}\bar{y} \right) \\
&= \left[ ny_0 \sum_j (x_j - \bar{x})^2 + n\bar{x}\sum_j x_jy_j - n\bar{y}\sum_j x_j^2 \right] \\
&\quad \Big/ \left[ n\sum_j x_jy_j - \left(\sum_j x_j\right)\left(\sum_j y_j\right) \right].
\end{aligned}
\qquad \text{(e)}
$$

**(iv)** Replace $x_0$ in the first expression for $\beta_2$ in (d) in order to get, after some simplifications:

$$\beta_2 = \frac{n\sum_j x_jy_j - \left(\sum_j x_j\right)\left(\sum_j y_j\right)}{n\sum_j x_j^2 - \left(\sum_j x_j\right)^2}, \qquad \text{(f)}$$

and observe that this expression is the MLE (LSE) of $\beta_2$, calculated on the basis of $y_j$ and $x_j, j = 1, \ldots, n$ only (see relation (5) and Theorem 4(i)).

**(v)** Replace $x_0$ and $\beta_2$ in the expression $\beta_1 = y_0 - x_0\beta_2$ in order to arrive at the expression:

$$\beta_1 = \frac{\left(\sum_j x_j^2\right)\left(\sum_j y_j\right) - \left(\sum_j x_j\right)\left(\sum_j x_jy_j\right)}{n\sum_j x_j^2 - \left(\sum_j x_j\right)^2}, \qquad \text{(g)}$$

after some calculations, and observe that this is the MLE (LSE) of $\beta_1$, calculated on the basis of $y_j$ and $x_j, j = 1, \ldots, n$ only (see relation (4) and Theorem 4(i)).

**(vi)** It follows that the MLE's of $\beta_1$, $\beta_2$, and $x_0$, to be denoted by $\hat{\beta}_1$, $\hat{\beta}_2$, and $\hat{x}_0$, respectively, are given by the expressions:

$$\hat{\beta}_1 = \frac{\left(\sum_j x_j^2\right)\left(\sum_j y_j\right) - \left(\sum_j x_j\right)\left(\sum_j x_jy_j\right)}{n\sum_j x_j^2 - \left(\sum_j x_j\right)^2},$$

$$\hat{\beta}_2 = \frac{n \sum_j x_j y_j - (\sum_j x_j)(\sum_j y_j)}{n \sum_j x_j^2 - (\sum_j x_j)^2},$$

and

$$\hat{x}_0 = \frac{y_0 - \hat{\beta}_1}{\hat{\beta}_2}.$$

(vii) Differentiate the log-likelihood function with respect to $\sigma^2$, equate the derivative to zero, and replace $\beta_1, \beta_2$, and $x_0$ by their MLE's in order to obtain the MLE $\hat{\sigma}^2$ of $\sigma^2$, which is given by the expression:

$$\hat{\sigma}^2 = \frac{1}{m+n}(SS_E + SS_{0E}),$$

where

$$SS_E = \sum_{j=1}^{n} (y_j - \hat{y}_j)^2 = \sum_{j=1}^{n} (y_j - \hat{\beta}_1 - \hat{\beta}_2 x_j)^2,$$

and

$$SS_{0E} = \sum_{i=1}^{m} (y_{0i} - \hat{\beta}_1 - \hat{\beta}_2 \hat{x}_0)^2 = \sum_{i=1}^{m} (y_{0i} - y_0)^2.$$

Also, by means of (14) and (16),

$$SS_E = SS_y - \frac{SS_{xy}^2}{SS_x}, \quad \text{where } SS_x = \sum_{j=1}^{n} x_j^2 - \frac{1}{n}\left(\sum_{j=1}^{n} x_j\right)^2,$$

$$SS_y = \sum_{j=1}^{n} y_j^2 - \frac{1}{n}\left(\sum_{j=1}^{n} y_j\right)^2, \quad SS_{xy} = \sum_{j=1}^{n} x_j y_j - \frac{1}{n}\left(\sum_{j=1}^{n} x_j\right)\left(\sum_{j=1}^{n} y_j\right),$$

and

$$SS_{0E} = \sum_{i=1}^{m} y_{0i}^2 - \frac{1}{m}\left(\sum_{i=1}^{m} y_{0i}\right)^2.$$

(viii) Observe that, by Theorem 5(i), $\frac{SS_E}{\sigma^2} \sim \chi_{n-2}^2$, whereas $\frac{SS_{0E}}{\sigma^2} \sim \chi_{m-1}^2$. Then, by independence of the $Y_j$'s and the $Y_{0i}$'s, it follows that $\frac{1}{\sigma^2}(SS_E + SS_{0E}) \sim \chi_{m+n-3}^2$.

(ix) Observe that, by Theorem 8(i), $\hat{y}_0 = \hat{\beta}_1 + \hat{\beta}_2 x_0 \sim$ $N(\beta_1 + \beta_2 x_0, \sigma^2(\frac{1}{n} + \frac{(x_0-\bar{x})^2}{SS_x}))$, whereas $Y_0 \sim N(\beta_1 + \beta_2 x_0, \frac{\sigma^2}{m})$, and $\hat{y}_0$ and $Y_0$ are independent, so that the r.v. $V = Y_0 - \hat{y}_0 \sim N(0, \sigma_V^2)$, where $\sigma_V^2 = \sigma^2(\frac{1}{m} + \frac{1}{n} + \frac{(x_0-\bar{x})^2}{SS_x})$, and $\frac{V}{\sigma_V} \sim N(0, 1)$.

(x) Observe that, by Theorem 5, the r.v.'s $V/\sigma_V$ and $(SS_E + SS_{0E})/\sigma^2$ are independent, so that

$$\frac{V/\sigma_V}{\frac{1}{\sigma}\sqrt{\frac{SS_E+SS_{0E}}{m+n-3}}} = \frac{V/\sigma\sqrt{\frac{1}{m}+\frac{1}{n}+\frac{(x_0-\bar{x})^2}{SS_x}}}{\frac{1}{\sigma}\sqrt{\frac{SS_E+SS_{0E}}{m+n-3}}}$$

$$= \frac{\sqrt{m+n-3}\;V}{\sqrt{\left[\frac{1}{m}+\frac{1}{n}+\frac{(x_0-\bar{x})^2}{SS_x}\right](SS_E+SS_{0E})}} \sim t_{m+n-3}.$$

## 13.4 CONFIDENCE INTERVALS AND HYPOTHESES TESTING PROBLEMS

The results obtained in the corollary to Theorem 5 allow the construction of confidence intervals for the parameters of the model, as well as testing hypotheses about them.

**Theorem 6.** *Under model (18), $100(1-\alpha)\%$ confidence intervals for $\beta_1$ and $\beta_2$ are given, respectively, by:*

$$\left[\hat{\beta}_1 - t_{n-2\,;\alpha/2}S\sqrt{\frac{1}{n}+\frac{\bar{x}^2}{SS_x}},\quad \hat{\beta}_1 + t_{n-2\,;\alpha/2}S\sqrt{\frac{1}{n}+\frac{\bar{x}^2}{SS_x}}\right], \tag{22}$$

*and*

$$\left[\hat{\beta}_2 - t_{n-2\,;\alpha/2}\frac{S}{\sqrt{SS_x}},\quad \hat{\beta}_2 + t_{n-2\,;\alpha/2}\frac{S}{\sqrt{SS_x}}\right], \tag{23}$$

*where $S = \sqrt{SS_E/(n-2)}$, and $SS_E, SS_x$, and $\hat{\beta}_1, \hat{\beta}_2$ are given by (16), (9), and (10).*

*Proof.* The confidence intervals in (22) and (23) follow immediately from results (ii) and (iii), respectively, in the corollary to Theorem 5, and the familiar procedure of constructing confidence intervals. ∎

**Remark 2.** A confidence interval can also be constructed for $\sigma^2$ on the basis of the statistic $SS_E$ and the fact that $SS_E/\sigma^2$ is distributed as $\chi^2_{n-2}$.

Procedures for testing some hypotheses are summarized below in the form of a theorem. The tests proposed here have an obvious intuitive interpretation. However, their justification rests on that they are likelihood ratio tests. For the case of simple hypotheses, this fact can be established directly. For composite hypotheses, it follows as a special case of more general results of testing hypotheses regarding the entire mean $\eta = \beta_1 + \beta_2 x$. See, e.g., Chapter 16 and, in particular, Examples 2 and 3 in the book *A Course in Mathematical Statistics*, 2nd edition (1997), Academic Press, by G. G. Roussas.

**Theorem 7.**    *Under model (18):*

**(i)**  *For testing the hypothesis $H_0$: $\beta_1 = \beta_{10}$ against the alternative $H_A$: $\beta_1 \neq \beta_{10}$ at level of significance $\alpha$, the null hypothesis $H_0$ is rejected whenever*

$$|t| > t_{n-2;\alpha/2}, \quad \text{where } t = (\hat{\beta}_1 - \beta_{10}) \Big/ S\sqrt{\frac{1}{n} + \frac{\bar{x}^2}{SS_x}}. \qquad (24)$$

**(ii)**  *For testing the hypothesis $H_0$: $\beta_2 = \beta_{20}$ against the alternative $H_A$: $\beta_2 \neq \beta_{20}$ at level of significance $\alpha$, the null hypothesis $H_0$ is rejected whenever*

$$|t| > t_{n-2;\alpha/2}, \quad \text{where } t = (\hat{\beta}_2 - \beta_{20}) \Big/ \frac{S}{\sqrt{SS_x}}. \qquad (25)$$

*When the alternative is of the form $H_A$: $\beta_2 > \beta_{20}$, the null hypothesis is rejected whenever $t > t_{n-2;\alpha}$, and it is rejected whenever $t < -t_{n-2;\alpha}$ if the alternative is of the form $H_A$: $\beta_2 < \beta_{20}$.*

**Remark 3.**

**(i)**  In the reference cited above, the test statistic used, actually, has the $F$-distribution under the null hypothesis. It should be recalled, however, that if $t$ has the $t$-distribution with $r$ d.f.; i.e., $t = Z/\sqrt{\chi_r^2/r}$, where $\chi_r^2$ has the $\chi^2$-distribution with $r$ d.f. $Z \sim N(0, 1)$ and $Z$ and $\chi_r^2$ are independent then, $t^2 = \frac{Z^2}{\chi_r^2/r}$ has the $F$-distribution with 1 and $r$ d.f.

**(ii)**  Hypotheses can also be tested about $\sigma^2$ on the basis of the fact that $\frac{SS_E}{\sigma^2} \sim \chi_{n-2}^2$.

**Example 6.**    In reference to Example 1:

**(i)**  Construct 95% confidence intervals for $\beta_1$ and $\beta_2$.

**(ii)**  Test the hypothesis that the GMAT scores increase with increasing GPA scores.

**Discussion.**    (i) The required confidence intervals are given by (22) and (23). In the discussion of Example 1, we have found that: $\bar{x} \simeq 2.963$, $SS_x \simeq 6.361$, $\hat{\beta}_1 \simeq 220.456$, and $\hat{\beta}_2 \simeq 91.397$. Also, in the discussion of Example 3, we saw that $SS_E \simeq 114{,}564.625$, so that, by (21), $S = (\frac{114{,}564.625}{32})^{1/2} \simeq 59.834$. Finally, $t_{32\,;0.025} = 2.0369$. Then

$$\hat{\beta}_1 - t_{n-2\,;\alpha/2} S\sqrt{\frac{1}{n} + \frac{\bar{x}^2}{SS_x}} = 220.456 - 2.0369 \times 59.834 \times \sqrt{\frac{1}{34} + \frac{(2.963)^2}{6.361}}$$

$$\simeq 220.456 - 121.876 \times 1.187 \simeq 220.456 - 144.667 = 75.789,$$

and

$$\hat{\beta}_1 + t_{n-2\,;\alpha/2} S\sqrt{\frac{1}{n} + \frac{\bar{x}^2}{SS_x}} \simeq 220.456 + 144.667 = 365.123.$$

So the observed conference interval is [75.789, 365.123].

Likewise, $t_{n-2\,;\alpha/2}\frac{S}{\sqrt{SS_x}} \simeq 1.6939 \times \frac{59.834}{\sqrt{6.361}} = 1.6939 \times 23.725 \simeq 40.188$, and therefore the observed confidence interval for $\beta_2$ is: [51.209, 131.585].

(ii) Here we are to test $H_0$: $\beta_2 = 0$ against the alternative $H_A$: $\beta_2 > 0$. Let us take $\alpha = 0.05$, so that $t_{32;0.05} = 1.6939$. The observed value of the test statistics is:

$$t = \frac{\hat{\beta}_2 - \beta_{20}}{S/\sqrt{SS_x}} \simeq \frac{91.397}{23.725} = 3.852,$$

and the null hypothesis is rejected; the GMAT scores increase along with increasing GPA scores.

**Example 7.**  In reference to Example 4:

(i) Construct 95% confidence intervals for $\beta_1$ and $\beta_2$.
(ii) Test the hypothesis that crop yield increases with increasing compost fertilizer amounts.

**Discussion.**   (i) In the discussion of Example 4, we have seen that: $n = 15, \bar{x} = 10.8, SS_x = 70.6, SS_y = 98.5, \hat{\beta}_1 \simeq 112.256$, and $\hat{\beta}_2 \simeq 0.967$. It follows that:

$$S = \left(\frac{98.5}{13}\right)^{1/2} \simeq 2.753 \quad \text{and} \quad S\sqrt{\frac{1}{n} + \frac{\bar{x}^2}{SS_x}} = 2.753 \times \sqrt{\frac{1}{15} + \frac{(10.8)^2}{70.6}}$$

$$\simeq 2.753 \times 1.311 \simeq 3.609.$$

Since $t_{13\,;0.025} = 1.1604$, it follows that the required observed confidence interval for $\beta_1$ is: $[112.256 - 1.1604 \times 3.609, 112.256 + 1.1604 \times 3.609]$, or [108.068, 116.444].

Next, $\frac{S}{\sqrt{SS_x}} \simeq \frac{2.753}{\sqrt{70.6}} \simeq 0.328$, and $t_{13\,;0.025}\frac{S}{\sqrt{SS_x}} = 1.1604 \times 0.328 \simeq 0.381$, so that the required observed confidence interval for $\beta_2$ is: [0.586, 1.348].

(ii) The hypothesis to be tested is $H_0$: $\beta_2 = 0$ against the alternative $H_A$: $\beta_2 > 0$. Take $\alpha = 0.05$, so that $t_{13\,;0.05} = 1.7709$. The observed value of the test statistic is:

$$t = \frac{\hat{\beta}_2 - \beta_{20}}{S/\sqrt{SS_x}} \simeq \frac{0.967}{0.328} \simeq 2.948,$$

and therefore the null hypothesis is rejected. Consequently, crop yield increases with increasing amounts of compost fertilizer.

**Example 8.**   In reference to Example 5:

(i) Construct 95% confidence intervals for $\beta_1$ and $\beta_2$.
(ii) Test the hypothesis that the duration of relief increases with higher dosages of the medication.

**Discussion.** (i) From the discussion of Example 5, we have: $n = 10, \bar{x} = 5.9, SS_x = 40.9, SS_E = 63.653, \hat{\beta}_1 = -1.072,$ and $\hat{\beta}_2 \simeq 2.741$. Then $S = (\frac{63.653}{8})^{1/2} \simeq 2.821$. Also, $t_8 ;_{0.025} = 3.3060$. Therefore:

$$t_{n-2} ;_{\alpha/2} S \sqrt{\frac{1}{n} + \frac{\bar{x}^2}{SS_x}} = 3.306 \times 2.821 \times \sqrt{\frac{1}{10} + \frac{(5.9)^2}{40.9}} \simeq 9.326 \times 0.975$$

$$\simeq 9.093.$$

Hence the observed confidence interval for $\beta_1$ is: $[-1.072 - 9.093, -1.072 + 9.093]$, or $[-10.165, 8.021]$. Next,

$$t_{n-2} ;_{\alpha/2} \frac{S}{\sqrt{SS_x}} = 3.306 \times \frac{2.821}{\sqrt{40.9}} \simeq 3.306 \times 0.441 \simeq 1.458,$$

and therefore the observed confidence interval for $\beta_2$ is: $[2.741 - 1.458, 2.741 + 1.458]$, or $[1.283, 4.199]$.

(ii) The hypothesis to be tested is $H_0: \beta_2 = 0$ against $H_A: \beta_2 > 0$, and let us take $\alpha = 0.05$, so that $t_8 ;_{0.05} = 1.8595$. The observed value of the test statistic is:

$$t = \frac{\hat{\beta}_2 - \beta_{20}}{S/\sqrt{SS_x}} \simeq \frac{2.741}{0.441} \simeq 6.215,$$

and the null hypothesis is rejected. Thus, increased dosages of medication provide longer duration of relief.

## EXERCISES

**4.1**   Refer to Exercises 3.2 and 3.4, and compute 95% confidence intervals for $\beta_1$ and $\beta_2$.

**4.2**   Refer to Exercises 3.3 and 4.1, and test the hypotheses $H_0: \beta_1 = 300$ against $H_A: \beta_1 \neq 300$, and $H_0: \beta_2 = 60$ against $H_A: \beta_2 \neq 60$, each at level of significance $\alpha = 0.05$.

**4.3**   Refer to Example 5 and:
   (i)   Derive 95% confidence intervals for $\beta_1$ and $\beta_2$.
   (ii)  Test the hypothesis $H_0: \beta_1 = -1$ against the alternative $H_A: \beta_1 \neq -1$ at level of significance $\alpha = 0.05$.
   (iii) Do the same for the hypothesis $H_0: \beta_2 = 3$ against the alternative $H_A: \beta_2 \neq 3$ at the same level $\alpha = 0.05$.

**4.4**   Suppose the observations $Y_1, \ldots, Y_n$ are of the following structure:
   $Y_i = \beta + \gamma(x_i - \bar{x}) + e_i$, where $\beta$ and $\gamma$ are parameters and the $e_i$'s are independent r.v.'s with mean 0 and unknown variance $\sigma^2$.
   (i)   Set $t_i = x_i - \bar{x}, i = 1, \ldots, n$, and observe that the model
   $Y_i = \beta + \gamma t_i + e_i$ is of the standard form (1) with $\beta_1 = \beta, \beta_2 = \gamma$, and the additional property that $\sum_{i=1}^n t_i = 0$, or $\bar{t} = 0$.

**(ii)** Use expressions (5) and (8) to conclude that the LSE's of $\beta$ and $\gamma$ are given by:

$$\hat{\beta} = \bar{Y}, \quad \hat{\gamma} = \frac{\sum_{i=1}^{n} t_i Y_i}{\sum_{i=1}^{n} t_i^2}.$$

**(iii)** Employ Theorem 4 in order to conclude that:

$$\hat{\beta} \sim N\left(\beta, \frac{\sigma^2}{n}\right) \quad \text{and} \quad \hat{\gamma} \sim N\left(\gamma, \frac{\sigma^2}{SS_t}\right),$$

where (by (9)) $SS_t = \sum_{i=1}^{n} t_i^2$.

**(iv)** Determine the form of the confidence intervals for $\beta$ and $\gamma$ from relations (22) and (23).

**(v)** Determine the expression of the test statistics by means of relations (24) and (25).

**(vi)** How do the confidence intervals in relation (29) and in Theorem 9 (iii) become here?

**4.5** Consider the linear regression models: $Y_i = \beta_1 + \beta_2 x_i + e_i, i = 1, \dots, m$ and $Y_j^* = \beta_1^* + \beta_2^* x_j^* + e_j^*, j = 1, \dots, n$, where the random errors $e_1, \dots, e_m$ and $e_1^*, \dots, e_n^*$ are i.i.d. r.v.'s distributed as $N(0, \sigma^2)$.

**(i)** The independence of $e_1, \dots, e_m$ and $e_1^*, \dots, e_n^*$ implies independence of $Y_1, \dots, Y_m$ and $Y_1^*, \dots, Y_n^*$. Then write down, the joint likelihood of the $Y_i$'s and the $Y_j^*$'s and observe that the MLE's of $\beta_1, \beta_2, \beta_1^*, \beta_2^*,$ and $\sigma^2$, in obvious notation, are given by:

$$\hat{\beta}_1 = \bar{Y} - \hat{\beta}_2 \bar{x}, \quad \hat{\beta}_2 = \frac{m \sum_{i=1}^{m} x_i Y_i - \left(\sum_{i=1}^{m} x_i\right)\left(\sum_{i=1}^{m} Y_i\right)}{m \sum_{i=1}^{m} x_i^2 - \left(\sum_{i=1}^{m} x_i\right)^2},$$

$$\hat{\beta}_1^* = \bar{Y}^* - \hat{\beta}_2^* \bar{x}^*, \quad \hat{\beta}_2^* = \frac{n \sum_{j=1}^{n} x_j^* Y_j^* - \left(\sum_{j=1}^{n} x_j^*\right)\left(\sum_{j=1}^{n} Y_j^*\right)}{n \sum_{j=1}^{n} x_j^{2*} - \left(\sum_{j=1}^{n} x_j\right)^2},$$

$$\hat{\sigma}^2 = (SS_E + SS_E^*)/(m+n),$$

where

$$SS_E = \sum_{i=1}^{m} (Y_i - \hat{\beta}_1 - \hat{\beta}_2 x_i)^2 = SS_y - \frac{SS_{xy}^2}{SS_x},$$

$$SS_x = \sum_{i=1}^{m} x_i^2 - \frac{1}{m}\left(\sum_{i=1}^{m} x_i\right)^2,$$

$$SS_y = \sum_{i=1}^{m} Y_i^2 - \frac{1}{m}\left(\sum_{i=1}^{m} Y_i\right)^2,$$

$$SS_{xy} = \sum_{i=1}^{m} x_i Y_i - \frac{1}{m}\left(\sum_{i=1}^{m} x_i\right)\left(\sum_{i=1}^{m} Y_i\right),$$

and

$$SS_E^* = \sum_{j=1}^{n}(Y_j^* - \hat{\beta}_1^* - \hat{\beta}_2^* x_j^*)^2 = SS_y^* - \frac{SS_{xy}^{*2}}{SS_x^*},$$

$$SS_x^* = \sum_{j=1}^{n}x_j^{2*} - \frac{1}{n}\left(\sum_{j=1}^{n}x_j^*\right)^2,$$

$$SS_y^* = \sum_{j=1}^{n}Y_j^{2*} - \frac{1}{n}\left(\sum_{j=1}^{n}Y_j^*\right)^2,$$

$$SS_{xy}^* = \sum_{j=1}^{n}x_j^* Y_j^* - \frac{1}{n}\left(\sum_{j=1}^{n}x_j^*\right)\left(\sum_{j=1}^{n}Y_j^*\right).$$

**(ii)** In accordance with Theorem 4, observe that

$$\hat{\beta}_1 \sim N\left(\beta_1, \sigma^2\left(\frac{1}{m} + \frac{\bar{x}^2}{SS_x}\right)\right), \quad \hat{\beta}_2 \sim N\left(\beta_2, \frac{\sigma^2}{SS_x}\right),$$

$$\hat{\beta}_1^* \sim N\left(\beta_1^*, \sigma^2\left(\frac{1}{n} + \frac{\bar{x}^{*2}}{SS_x^*}\right)\right), \quad \hat{\beta}_2^* \sim N\left(\beta_2^*, \frac{\sigma^2}{SS_x^*}\right),$$

and

$$\frac{SS_E + SS_E^*}{\sigma^2} \sim \chi^2_{m+n-4}.$$

**(iii)** From part (ii) and Theorem 5, conclude that

$$\frac{\sqrt{m+n-4}\left[(\hat{\beta}_1 - \hat{\beta}_1^*) - (\beta_1 - \beta_1^*)\right]}{\sqrt{(SS_E + SS_E^*)\left(\frac{1}{m} + \frac{1}{n} + \frac{\bar{x}^2}{SS_x} + \frac{\bar{x}^{*2}}{SS_x^*}\right)}} \sim t_{m+n-4},$$

and

$$\frac{\sqrt{m+n-4}\left[(\hat{\beta}_2 - \hat{\beta}_2^*) - (\beta_2 - \beta_2^*)\right]}{\sqrt{(SS_E + SS_E^*)\left(\frac{1}{SS_x} + \frac{1}{SS_x^*}\right)}} \sim t_{m+n-4}.$$

**(iv)** From part (iii), observe that the two regression lines can be compared through the test of the hypotheses $H_0: \beta_1 = \beta_1^*$ against the alternative $H_A: \beta_1 \neq \beta_1^*$, and $H_0': \beta_2 = \beta_2^*$ against the alternative $H_A': \beta_2 \neq \beta_2^*$ by using the respective test statistics:

$$t = \frac{\sqrt{m+n-4}\,(\hat{\beta}_1 - \hat{\beta}_1^*)}{\sqrt{(SS_E + SS_E^*)\left(\frac{1}{m} + \frac{1}{n} + \frac{\bar{x}^2}{SS_x} + \frac{\bar{x}^{*2}}{SS_x^*}\right)}},$$

$$t' = \frac{\sqrt{m+n-4}\,(\hat{\beta}_2 - \hat{\beta}_2^*)}{\sqrt{(SS_E + SS_E^*)\left(\frac{1}{SS_x} + \frac{1}{SS_x^*}\right)}}.$$

At level of significance $\alpha$, the hypothesis $H_0$ is rejected when $|t| > t_{m+n-4;\frac{\alpha}{2}}$, and the hypothesis $H'_0$ is rejected when $|t'| > t_{m+n-4 ;\frac{\alpha}{2}}$.

(v) Again from part (iii), observe that 95% confidence intervals for $\beta_1 - \beta_1^*$ and $\beta_2 - \beta_2^*$ are given by:

$$(\hat{\beta}_1 - \hat{\beta}_1^*) \pm t_{m+n-4;\frac{\alpha}{2}} \sqrt{\frac{SS_E + SS_E^*}{m+n-4} \left( \frac{1}{m} + \frac{1}{n} + \frac{\bar{x}^2}{SS_x} + \frac{\bar{x}^{*2}}{SS_x^*} \right)},$$

and

$$(\hat{\beta}_2 - \hat{\beta}_2^*) \pm t_{m+n-4;\frac{\alpha}{2}} \sqrt{\frac{SS_E + SS_E^*}{m+n-4} \left( \frac{1}{SS_x} + \frac{1}{SS_x^*} \right)},$$

respectively.

(vi) Finally, from part (ii) conclude that a 95% confidence interval for $\sigma^2$ is given by:

$$\left[ \frac{SS_E + SS_E^*}{\chi^2_{m+n-4;\frac{\alpha}{2}}}, \quad \frac{SS_E + SS_E^*}{\chi^2_{m+n-4;1-\frac{\alpha}{2}}} \right].$$

## 13.5 SOME PREDICTION PROBLEMS

According to model (18), the expectation of the observation $Y_i$ at $x_i$ is $EY_i = \beta_1 + \beta_2 x_i$. Now, suppose $x_0$ is a point distinct from all $x_i$'s, but lying in the range that the $x_i$'s span, and we wish to predict the expected value of the observation $Y_0$ at $x_0$; i.e., $EY_0 = \beta_1 + \beta_2 x_0$. An obvious predictor for $EY_0$ is the statistic $\hat{y}_0$ given by the expression below and modified as indicated:

$$\hat{y}_0 = \hat{\beta}_1 + \hat{\beta}_2 x_0 = (\bar{Y} - \hat{\beta}_2 \bar{x}) + \hat{\beta}_2 x_0 = \bar{Y} + (x_0 - \bar{x})\hat{\beta}_2. \tag{26}$$

The result below gives the distribution of $\hat{y}_0$, which also provides for the construction of a confidence interval for $\beta_1 + \beta_2 x_0$.

**Theorem 8.**  *Under model (18) and with $\hat{y}_0$ given by (26), we have:*

(i)

$$\frac{\hat{y}_0 - (\beta_1 + \beta_2 x_0)}{\sigma \sqrt{\frac{1}{n} + \frac{(x_0 - \bar{x})^2}{SS_x}}} \sim N(0, 1). \tag{27}$$

(ii)

$$\frac{\hat{y}_0 - (\beta_1 + \beta_2 x_0)}{S \sqrt{\frac{1}{n} + \frac{(x_0 - \bar{x})^2}{SS_x}}} \sim t_{n-2}. \tag{28}$$

(iii)  *A $100(1 - \alpha)$% confidence interval for $\beta_1 + \beta_2 x_0$ is given by:*

$$\left[ \hat{y}_0 - t_{n-2 ;\alpha/2} S \sqrt{\frac{1}{n} + \frac{(x_0 - \bar{x})^2}{SS_x}}, \quad \hat{y}_0 + t_{n-2 ;\alpha/2} S \sqrt{\frac{1}{n} + \frac{(x_0 - \bar{x})^2}{SS_x}} \right]. \tag{29}$$

*It is recalled that* $S = \sqrt{SS_E/(n-2)}$, $SS_E = SS_y - \frac{SS_{xy}^2}{SS_x}$, *and* $SS_y$ *and* $SS_x$ *are given in (9).*

*Proof.* (i) The assumption that $Y_i \sim N(\beta_1 + \beta_2 x_i, \sigma^2), i = 1, \ldots, n$ independent implies that $\sum_i Y_i \sim N(n\beta_1 + \beta_2 \sum_i x_i, n\sigma^2)$ and hence

$$\bar{Y} \sim N(\beta_1 + \beta_2 \bar{x}, \sigma^2/n). \tag{30}$$

By Theorem 4(iii), $\hat{\beta}_2 \sim N(\beta_2, \sigma^2/SS_x)$, so that

$$(x_0 - \bar{x})\hat{\beta}_2 \sim N\left((x_0 - \bar{x})\beta_2, \frac{\sigma^2(x_0 - \bar{x})^2}{SS_x}\right). \tag{31}$$

Furthermore, by Theorem 5(ii)(b), $\bar{Y}$ and $\hat{\beta}_2$ are independent. Then, relations (26), (30), and (31) yield:

$$\hat{y}_0 = \bar{Y} + (x_0 - \bar{x})\hat{\beta}_2 \sim N\left(\beta_1 + \beta_2 x_0, \ \sigma^2\left(\frac{1}{n} + \frac{(x_0 - \bar{x})^2}{SS_x}\right)\right), \tag{32}$$

and therefore (27) follows by standardization.

(ii) By Theorem 5(ii)(c), $SS_E$ is independent of $\bar{Y}$ and $\hat{\beta}_2$ and hence independent of $\hat{y}_0$ because of (26). Furthermore, by Theorem 5(i),

$$\frac{SS_E}{\sigma^2} = \frac{(n-2)S^2}{\sigma^2} \sim \chi^2_{n-2}. \tag{33}$$

Therefore,

$$\frac{[\hat{y}_0 - (\beta_1 + \beta_2 x_0)]/\sigma\sqrt{\frac{1}{n} + \frac{(x_0-\bar{x})^2}{SS_x}}}{\sqrt{\frac{(n-2)S^2}{\sigma^2}/(n-2)}} = \frac{\hat{y}_0 - (\beta_1 + \beta_2 x_0)}{S\sqrt{\frac{1}{n} + \frac{(x_0-\bar{x})^2}{SS_x}}} \sim t_{n-2},$$

which is relation (28).

(iii) This part follows immediately from part (ii) and the standard procedure of setting up confidence intervals. ∎

Finally, we would like to consider the problem of predicting a *single response* at a given point $x_0$ rather than its *expected value*. Call $Y_0$ the response corresponding to $x_0$ and, reasonably enough, assume that $Y_0$ is independent of the $Y_i$'s. The predictor for $Y_0$ is $\hat{y}_0$, the same as the one given in (26). The objective here is to construct a prediction interval for $Y_0$. This is done indirectly in the following result.

**Theorem 9.** *Under model (18), let $Y_0$ be the (unobserved) observation at $x_0$, and assume that $Y_0$ is independent of the $Y_i$'s. Predict $Y_0$ by $\hat{y}_0 = \hat{\beta}_1 + \hat{\beta}_2 x_0$. Then:*
**(i)**

$$\frac{\hat{y}_0 - Y_0}{\sigma\sqrt{1 + \frac{1}{n} + \frac{(x_0-\bar{x})^2}{SS_x}}} \sim N(0, 1). \tag{34}$$

**(ii)**

$$\frac{\hat{y}_0 - Y_0}{S\sqrt{1 + \frac{1}{n} + \frac{(x_0-\bar{x})^2}{SS_x}}} \sim t_{n-2}. \tag{35}$$

**(iii)** *A* $100(1 - \alpha)\%$ *prediction interval for* $Y_0$ *is given by:*

$$\left[ \hat{y}_0 - t_{n-2\,;\alpha/2}S\sqrt{1 + \frac{1}{n} + \frac{(x_0 - \bar{x})^2}{SS_x}}, \; \hat{y}_0 + t_{n-2\,;\alpha/2}S\sqrt{1 + \frac{1}{n} + \frac{(x_0 - \bar{x})^2}{SS_x}} \right],$$

*where S and* $SS_x$ *are as in Theorem 8(ii), (iii).*

*Proof.* First, we compute the expectation and the variance of $\hat{y}_0 - Y_0$. To this end, we have: $Y_0 = \beta_1 + \beta_2 x_0 + e_0$, predicted by $\hat{y}_0 = \hat{\beta}_1 + \hat{\beta}_2 x_0$. Then $EY_0 = \beta_1 + \beta_2 x_0$ and $E\hat{y}_0 = \beta_1 + \beta_2 x_0$, so that $E(\hat{y}_0 - Y_0) = 0$. In deriving the distribution of $\hat{y}_0 - Y_0$, we need its variance. By (26), we have:

$$Var(\hat{y}_0 - Y_0) = Var(\bar{Y} + (x_0 - \bar{x})\hat{\beta}_2 - Y_0) = Var(\bar{Y}) + (x_0 - \bar{x})^2 Var(\hat{\beta}_2) + Var(Y_0)$$

(since all three r.v.'s, $\bar{Y}, \hat{\beta}_2$, and $Y_0$, are independent)

$$= \frac{\sigma^2}{n} + (x_0 - \bar{x})^2 \times \frac{\sigma^2}{SS_x} + \sigma^2 \quad \text{(by Theorem 2)}$$

$$= \sigma^2 \left\{ 1 + \frac{1}{n} + \frac{(x_0 - \bar{x})^2}{SS_x} \right\}; \quad \text{i.e.,}$$

$$E(\hat{y}_0 - Y_0) = 0 \quad \text{and} \quad Var(\hat{y}_0 - Y_0) = \sigma^2 \left[ 1 + \frac{1}{n} + \frac{(x_0 - \bar{x})^2}{SS_x} \right].$$

We now proceed with parts (i) and (ii).

**(i)** Since $\hat{y}_0$ and $Y_0$ are independent and $Y_0 \sim N(\beta_1 + \beta_2 x_0, \sigma^2)$, then these facts along with (32) yield:

$$\hat{y}_0 - Y_0 \sim N\left( 0, \; \sigma^2 \left[ 1 + \frac{1}{n} + \frac{(x_0 - \bar{x})^2}{SS_x} \right] \right).$$

Relation (34) follows by standardizing $\hat{y}_0 - Y_0$.

**(ii)** It has been argued in the proof of Theorem 8(ii) that $S$ and $\hat{y}_0$ are independent. It follows that $S$ and $\hat{y}_0 - Y_0$ are also independent. Then, dividing the expression on the left-hand side in (34) by $\sqrt{\frac{(n-2)S^2}{\sigma^2}/(n-2)} = \frac{S}{\sigma}$ in (33), we obtain the result in (35), after some simplifications.

**(iii)** This part follows from part (ii) through the usual procedure of setting up confidence intervals. ∎

---

## EXERCISES

**5.1** Refer to Exercises 3.1, 3.3, 4.1, and:
  **(i)** Predict $EY_0$ at $x_0 = 3.25$, and construct a 95% confidence interval of $EY_0$.
  **(ii)** Predict the
      response $Y_0$ at $x_0 = 3.25$, and construct a 95% prediction interval for $Y_0$.

**5.2** In reference to Example 22 in Chapter 1 (see also scatter diagram in Figure 13.1 and Examples 1, 2, 3, and 6 here), do the following:

(i) Predict the $EY_0$, where $Y_0$ is the response at $x_0 = 3.25$.

(ii) Construct a 95% confidence interval for $EY_0 = \beta_1 + \beta_2 x_0 = \beta_1 + 3.25\beta_2$.

(iii) Predict the response $Y_0$ at $x_0 = 2.5$.

(iv) Construct a 90% prediction interval for $Y_0$.

**5.3** In reference to Example 23 in Chapter 1 (see also Examples 4 and 7 here), do the following:

(i) Predict the $EY_0$, where $Y_0$ is the response at $x_0 = 12$.

(ii) Construct a 95% confidence interval for $EY_0 = \beta_1 + \beta_2 x_0 = \beta_1 + 12\beta_2$.

(iii) Predict the response $Y_0$ at $x_0 = 12$.

(iv) Construct a 95% prediction interval for $Y_0$.

**5.4** Refer to Example 5 and:

(i) Predict the $EY_0$ at $x_0 = 6$.

(ii) Construct a 95% confidence interval for $EY_0 = \beta_1 + 6\beta_2$.

(iii) Predict the response $Y_0$ at $x_0 = 6$.

(iv) Construct a 95% prediction interval for $Y_0$.

**5.5** Suppose that the data given in the table below follow model (18).

| $x$ | 5 | 10 | 15 | 20 | 25 | 30 |
|---|---|---|---|---|---|---|
| $y$ | 0.10 | 0.21 | 0.30 | 0.35 | 0.44 | 0.62 |

(i) Determine the MLE's (LSE's) of $\beta_1, \beta_2$, and $\sigma^2$.

(ii) Construct 95% confidence intervals for $\beta_1, \beta_2$, and $\sigma^2$.

(iii) At $x_0 = 17$, predict both $EY_0$ and $Y_0$ (the respective observation at $x_0$), and construct a 95% confidence interval and prediction interval, respectively, for them.

**Hint.** For a confidence interval for $\sigma^2$, see Exercise 3.5.

**5.6** The following table gives the reciprocal temperatures $x$ and the corresponding observed solubilities of a certain chemical substance, and assume that they follow model (18).

| $x$ | 3.80 | 3.72 | 3.67 | 3.60 | 3.54 |
|---|---|---|---|---|---|
|  | 1.27 | 1.20 | 1.10 | 0.82 | 0.65 |
| $y$ | 1.32 | 1.26 | 1.07 | 0.84 | 0.57 |
|  | 1.50 |  |  | 0.80 | 0.62 |

(i) Determine the MLE's (LSE's) of $\beta_1, \beta_2$, and $\sigma^2$.

(ii) Construct 95% confidence intervals for $\beta_1, \beta_2$, and $\sigma^2$.

(iii) At $x_0 = 3.77$, predict both $EY_0$ and $Y_0$ (the respective observation at $x_0$), and construct a 95% confidence interval and prediction interval, respectively, for them.

**Hint.** Here $n = 13$ and $x_1 = x_2 = x_3$, $x_4 = x_5$, $x_6 = x_7, x_8 = x_9 = x_{10}$, and $x_{11} = x_{12} = x_{13}$.

## 13.6 PROOF OF THEOREM 5

This section is solely devoted to justifying Theorem 5. Its proof is presented in considerable detail, and it makes use of some linear algebra results. The sources of those results are cited.

*Proof of Theorem 5.* For later use, let us set

$$U_i = Y_i - \beta_1 - \beta_2 x_i, \quad \text{so that} \quad \bar{U} = \bar{Y} - \beta_1 - \beta_2 \bar{x}, \tag{36}$$

and

$$U_i - \bar{U} = (Y_i - \bar{Y}) - \beta_2(x_i - \bar{x}) \quad \text{and} \quad Y_i - \bar{Y} = (U_i - \bar{U}) + \beta_2(x_i - \bar{x}). \tag{37}$$

Then, by (10),

$$\hat{\beta}_2 SS_x = SS_{xy} = \sum_i (x_i - \bar{x})(Y_i - \bar{Y}), \tag{38}$$

so, that

$$\begin{aligned}
(\hat{\beta}_2 - \beta_2) SS_x &= \sum_i (x_i - \bar{x})(Y_i - \bar{Y}) - \beta_2 SS_x \\
&= \sum_i (x_i - \bar{x})[(U_i - \bar{U}) + \beta_2(x_i - \bar{x})] - \beta_2 SS_x \\
&= \sum_i (x_i - \bar{x})(U_i - \bar{U}) + \beta_2 SS_x - \beta_2 SS_x \\
&= \sum_i (x_i - \bar{x})(U_i - \bar{U}).
\end{aligned} \tag{39}$$

Next,

$$\begin{aligned}
SS_E &= \sum_i (Y_i - \hat{y}_i)^2 = \sum_i (Y_i - \hat{\beta}_1 - \hat{\beta}_2 x_i)^2 \\
&= \sum_i (Y_i - \bar{Y} + \hat{\beta}_2 \bar{x} - \hat{\beta}_2 x_i)^2 \quad (\text{by (8)}) \\
&= \sum_i [(Y_i - \bar{Y}) - \hat{\beta}_2(x_i - \bar{x})]^2 \\
&= \sum_i [(Y_i - \bar{Y}) - \hat{\beta}_2(x_i - \bar{x}) + \beta_2(x_i - \bar{x}) - \beta_2(x_i - \bar{x})]^2 \\
&= \sum_i \{[(Y_i - \bar{Y}) - \beta_2(x_i - \bar{x})] - (\hat{\beta}_2 - \beta_2)(x_i - \bar{x})\}^2 \\
&= \sum_i [(U_i - \bar{U}) - (\hat{\beta}_2 - \beta_2)(x_i - \bar{x})]^2 \quad (\text{by (37)})
\end{aligned}$$

$$= \sum_i (U_i - \bar{U})^2 + (\hat{\beta}_2 - \beta_2)^2 SS_x - 2(\hat{\beta}_2 - \beta_2) \sum_i (x_i - \bar{x})(U_i - \bar{U})$$

$$= \sum_i (U_i - \bar{U})^2 + (\hat{\beta}_2 - \beta_2)^2 SS_x - 2(\hat{\beta}_2 - \beta_2)^2 SS_x \quad \text{(by (39))}$$

$$= \sum_i (U_i - \bar{U})^2 - (\hat{\beta}_2 - \beta_2)^2 SS_x$$

$$= \sum_i U_i^2 - n\bar{U}^2 - (\hat{\beta}_2 - \beta_2)^2 SS_x; \text{ i.e.,}$$

$$SS_E = \sum_i U_i^2 - n\bar{U}^2 - (\hat{\beta}_2 - \beta_2)^2 SS_x. \tag{40}$$

From (18) and (36), we have that the r.v.'s $U_1, \ldots, U_n$ are independent and distributed as $N(0, \sigma^2)$. Transform them into the r.v.'s $V_1, \ldots, V_n$ by means of an orthogonal transformation $\mathbf{C}$ as described below (see also Remark 4):

$$\mathbf{C} = \begin{pmatrix} \frac{x_1 - \bar{x}}{\sqrt{SS_x}} & \frac{x_2 - \bar{x}}{\sqrt{SS_x}} & \cdots & \frac{x_n - \bar{x}}{\sqrt{SS_x}} \\ \frac{1}{\sqrt{n}} & \frac{1}{\sqrt{n}} & \cdots & \frac{1}{\sqrt{n}} \\ \text{(whatever, subject to the res-} \\ \text{triction that } \mathbf{C} \text{ is orthogonal)} \end{pmatrix}$$

That is, with "'" standing for transpose, we have:

$$(V_1, V_2, \ldots, V_n)' = \mathbf{C}(U_1, U_2, \ldots, U_n)'. \tag{41}$$

Then, by Theorem 8 in Chapter 6, the r.v.'s $V_1, \ldots, V_n$ are independent and distributed as $N(0, \sigma^2)$, whereas by relation (21) in the same chapter

$$\sum_i V_i^2 = \sum_i U_i^2. \tag{42}$$

From (41),

$$V_1 = \frac{1}{\sqrt{SS_x}} \sum_i (x_i - \bar{x})U_i, \quad V_2 = \frac{1}{\sqrt{n}} \sum_i U_i = \sqrt{n} \times \frac{1}{n} \sum_i U_i = \sqrt{n}\bar{U}. \tag{43}$$

But

$$\sum_i (x_i - \bar{x})U_i = \sum_i (x_i - \bar{x})(U_i - \bar{U}) = (\hat{\beta}_2 - \beta_2)SS_x, \quad \text{(by (39))},$$

so that

$$V_1 = (\hat{\beta}_2 - \beta_2)\sqrt{SS_x}, \quad V_1^2 = (\hat{\beta}_2 - \beta_2)^2 SS_x, \quad \text{and} \quad V_2^2 = n\bar{U}^2. \tag{44}$$

Then, from relations (40), (42), and (44), it follows that

$$SS_E = \sum_{i=1}^{n} V_i^2 - V_1^2 - V_2^2 = \sum_{i=3}^{n} V_i^2. \tag{45}$$

We now proceed with the justifications of parts (i) and (ii) of the theorem.

**(i)** From (45), $\frac{SS_E}{\sigma^2} = \sum_{i=3}^{n} \left(\frac{V_i}{\sigma}\right)^2 \sim \chi^2_{n-2}$, since $\frac{V_i}{\sigma}, i = 1, \ldots, n$ are independent and distributed as $N(0, 1)$.

**(ii)** **(a)** From (44) and (45), $\hat{\beta}_2$ and $SS_E$ are functions of nonoverlapping $V_i$'s (of $V_1$ the former, and of $V_3, \ldots, V_n$ the latter). Thus, $SS_E$ and $\hat{\beta}_2$ are independent.

    **(b)** By (36) and (43), $\bar{Y} = \bar{U} + (\beta_1 + \beta_2 \bar{x}) = \frac{V_2}{\sqrt{n}} + \beta_1 + \beta_2 \bar{x}$, so that $\bar{Y}$ is a function of $V_2$ and recall that $\hat{\beta}_2$ is a function of $V_1$. Then the independence of $\bar{Y}$ and $\hat{\beta}_2$ follows.

    **(c)** As was seen in (a) and (b), $SS_E$ is a function of $V_3, \ldots, V_n$; $\bar{Y}$ is a function of $V_2$; and $\hat{\beta}_2$ is a function of $V_1$; i.e., they are functions of nonoverlapping $V_i$'s, and therefore independent.

    **(d)** By (8), $\hat{\beta}_1 = \bar{Y} - \hat{\beta}_2 \bar{x}$ and the right-hand side is a function of $V_1$ and $V_2$ alone, by (44) and part (b). Since $SS_E$ is a function of $V_3, \ldots, V_n$, by (45), the independence of $SS_E$ and $\hat{\beta}_1$ follows. ∎

**Remark 4.** There is always an orthogonal matrix $\mathbf{C}$ with the first two rows as given above. Clearly, the vectors $r_1 = (x_1 - \bar{x}, \ldots, x_n - \bar{x})'$ and $r_2 = (\frac{1}{\sqrt{n}}, \ldots, \frac{1}{\sqrt{n}})'$ are linearly independent. Then supplement them with $n - 2$ vectors $r_3, \ldots, r_n$, so that the vectors $r_1, \ldots, r_n$ are linearly independent. Finally, use the Gram-Schmidt orthogonalization process (which leaves $r_1$ and $r_2$ intact) to arrive at an orthogonal matrix $\mathbf{C}$. (See, e.g., Theorem 1.16 and the discussion following it, in pages 33–34, of the book *Linear Algebra for Undergraduates* (1957), John Wiley & Sons, by D. C. Murdoch.)

## 13.7 CONCLUDING REMARKS

In this chapter, we studied the simplest linear regression model, according to which the response $Y$ at a point $x$ is given by $Y = \beta_1 + \beta_2 x + e$. There are extensions of this model in different directions. First, the model may *not* be linear in the parameters involved; i.e., the expectation $\eta = EY$ is not linear. Here are some such examples.

$$\text{(i)}\eta = ae^{bx}; \quad \text{(ii)}\eta = ax^b; \quad \text{(iii)} \eta = \frac{1}{a + bx}; \quad \text{(iv)} \eta = a + b\sqrt{x}.$$

It happens that these particular nonlinear models can be reduced to linear ones by suitable transformations. Thus, in (i), taking the logarithms (always with base e), we have:

$$\log \eta = \log a + bx, \quad \text{or} \quad \eta' = \beta_1 + \beta_2 x',$$

where $\eta' = \log \eta$, $\beta_1 = \log a, \beta_2 = b$, and $x' = x$, and the new model is linear. Likewise, in (ii):

$$\log \eta = \log a + b \log x, \quad \text{or} \quad \eta' = \beta_1 + \beta_2 x',$$

where $\eta' = \log \eta$, $\beta_1 = \log a$, $\beta_2 = b$, and $x' = \log x$, and the transformed model is linear. In (iii), simply set $\eta' = \frac{1}{\eta}$ to get $\eta' = a + bx$, or $\eta' = \beta_1 + \beta_2 x'$, where $\beta_1 = a$, $\beta_2 = b$ and $x' = x$. Finally, in (iv), let $x' = \sqrt{x}$ in order to get the linear model $\eta' = \beta_1 + \beta_2 x'$, with $\eta' = \eta$, $\beta_1 = a$, and $\beta_2 = b$.

Another direction of a generalization is the consideration of the so-called *multiple regression* linear models. In such models, there is more than one input variable $x$ and more than two parameters $\beta_1$ and $\beta_2$. This simply reflects the fact that the response is influenced by more than one factor each time. For example, the observation may be the systolic blood pressure of the individual in a certain group, and the influencing factors may be weight and age. The general form of a multiple regression linear model is as follows:

$$Y_i = x_{1i}\beta_1 + x_{2i}\beta_2 + \cdots + x_{pi}\beta_p + e_i, \quad i = 1, \ldots, n,$$

and the assumptions attached to it are similar to those used in model (18). The analysis of such a model can be done, in principle, along the same lines as those used in analyzing model (18). However, the analysis becomes unwieldy and one has to employ, most efficiently, linear algebra methodology. Such models are referred to as *general linear models* in the statistical literature, and they have proved very useful in a host of applications. The theoretical study of such models can be found, e.g., in Chapter 16 of the book *A Course in Mathematical Statistics*, 2nd edition (1997), Academic Press, by G. G. Roussas.

# Two models of analysis of variance

# 14

This chapter is about statistical analysis in certain statistical models referred to as Analysis of Variance (ANOVA). There is a great variety of such models, and their detailed study constitutes an interesting branch of statistics. What is done presently is to introduce two of the simplest models of ANOVA, underline the basic concepts involved, and proceed with the analysis in the framework of the proposed models.

The first section is devoted to the study of the one-way layout ANOVA with the same number of observations for each combination of the factors involved (cells). The study consists in providing a motivation for the model, in deriving the maximum likelihood estimates (MLE's) of its parameters, and in testing an important hypothesis. In the process of doing so, an explanation is provided for the term ANOVA. Also, several technical results necessary for the analysis are established.

In the second section of the chapter, we construct confidence intervals for all so-called contrasts among the (mean) parameters of the model in Section 14.1.

Section 14.3 is a generalization of the model studied in the first section, in that the outcome of an experiment is due to two factors. Again, a motivation is provided for the model, finally adopted, and then relevant statistical analysis is discussed. This analysis consists in deriving the MLE's of the parameters of the model, and also in testing two hypotheses reflecting the actual influence, or lack thereof, of the factors involved in the outcome of the underlying experiment. Again, in the process of the analysis, an explanation is provided for the term ANOVA. Also, a substantial number of technical results are stated that are necessary for the analysis. Their proofs are deferred to a final subsection of this section in order not to disrupt the continuity of arguments.

In all sections, relevant examples are discussed in detail in order to clarify the underlying ideas and apply the results obtained.

## 14.1 ONE-WAY LAYOUT WITH THE SAME NUMBER OF OBSERVATIONS PER CELL

In this section, we derive the MLE's of the parameters $\mu_i, i = 1, \ldots, I$ and $\sigma^2$ described in relation (13) of Chapter 8. Next, we consider the problem of testing the null hypothesis $H_0: \mu_1 = \cdots = \mu_I = \mu$ (unspecified) for which the MLE's of the parameters $\mu$ and $\sigma^2$ are to be derived under $H_0$. Then we set up the likelihood ratio test, which turns out to be an $F$-test. For the justification of this fact, we have to split sums of squares of variations in a certain way. Actually, it is this splitting from which the name ANOVA derives. Furthermore, the splitting provides insight into what is happening behind the formal analysis.

### 14.1.1 THE MLE's OF THE PARAMETERS OF THE MODEL

First, $Y_{ij} \sim N(\mu_i, \sigma^2), i = 1, \ldots, I, j = 1, \ldots, J$, and all these r.v.'s are independent. Then their likelihood function, to be denoted by $L(y;\mu,\sigma^2)$, is given by the expression below, where $y = (y_1, \ldots, y_j)$ and $\mu = (\mu_1, \ldots, \mu_I)$:

$$
\begin{aligned}
L(y;\mu,\sigma^2) &= \prod_{i,j} \left\{ \frac{1}{\sqrt{2\pi\sigma^2}} \exp\left[ -\frac{1}{2\sigma^2}(y_{ij} - \mu_i)^2 \right] \right\} \\
&= \prod_i \prod_j \left\{ \frac{1}{\sqrt{2\pi\sigma^2}} \exp\left[ -\frac{1}{2\sigma^2}(y_{ij} - \mu_i)^2 \right] \right\} \\
&= \prod_i \left\{ \left( \frac{1}{\sqrt{2\pi\sigma^2}} \right)^J \exp\left[ -\frac{1}{2\sigma^2} \sum_j (y_{ij} - \mu_i)^2 \right] \right\} \\
&= \left( \frac{1}{\sqrt{2\pi\sigma^2}} \right)^{IJ} \prod_i \left\{ \exp\left[ -\frac{1}{2\sigma^2} \sum_j (y_{ij} - \mu_i)^2 \right] \right\} \\
&= \left( \frac{1}{\sqrt{2\pi\sigma^2}} \right)^{IJ} \exp\left[ -\frac{1}{2\sigma^2} \sum_i \sum_j (y_{ij} - \mu_i)^2 \right];
\end{aligned}
$$

following common practice, we do not explicitly indicate the range of $i$ and $j$, since no confusion is possible. Hence

$$
\log L(y;\mu,\sigma^2) = -\frac{IJ}{2} \log(2\pi) - \frac{IJ}{2} \log \sigma^2 - \frac{1}{2\sigma^2} \sum_i \sum_j (y_{ij} - \mu_i)^2. \tag{1}
$$

From (1), we see that, for each fixed $\sigma^2$, the log-likelihood is maximized with respect to $\mu_1, \ldots, \mu_I$, if the exponent

$$
S(\mu_1, \ldots, \mu_I) = \sum_i \sum_j (y_{ij} - \mu_i)^2
$$

is minimized with respect to $\mu_1, \ldots, \mu_I$. By differentiation, we get

$$
\frac{\partial}{\partial \mu_i} S(\mu_1, \ldots, \mu_I) = -2 \sum_j y_{ij} + 2J\mu_i = 0, \quad \text{so that } \mu_i = \frac{1}{J} \sum_j y_{ij}, \tag{2}
$$

$\frac{\partial^2}{\partial \mu_i^2} S(\mu_1, \ldots, \mu_I) = 2J$ and $\frac{\partial^2}{\partial \mu_i \mu_j} S(\mu_1, \ldots, \mu_I) = 0$ for $i \neq j$. The resulting $I \times I$ diagonal matrix is positive definite, since, for all $\lambda_1, \ldots, \lambda_I$ with $\lambda_1^2 + \cdots + \lambda_I^2 > 0$,

$$
(\lambda_1, \ldots, \lambda_I) \begin{pmatrix} 2J & \cdots & 0 \\ \vdots & \ddots & \vdots \\ 0 & \cdots & 2J \end{pmatrix} \begin{pmatrix} \lambda_1 \\ \vdots \\ \lambda_I \end{pmatrix} = (2J\lambda_1, \ldots, 2J\lambda_I) \begin{pmatrix} \lambda_I \\ \vdots \\ \lambda_I \end{pmatrix}
$$

$$
= 2J(\lambda_1^2 + \cdots + \lambda_I^2) > 0.
$$

It follows that the values of $\mu_i$'s given in (2) are, indeed, the MLE's of the $\mu_i$'s. That is,

$$\hat{\mu}_i = y_{i.}, \quad \text{where } y_{i.} = \frac{1}{J} \sum_j y_{ij}, \quad i = 1, \dots, I. \tag{3}$$

Now, in (1), replace the exponent by $\hat{S} = \sum_i \sum_j (y_{ij} - y_{i.})^2$ to obtain in obvious notation

$$\log \hat{L}(y; \hat{\mu}, \sigma^2) = -\frac{IJ}{2} \log(2\pi) - \frac{IJ}{2} \log \sigma^2 - \frac{1}{2\sigma^2} \hat{S}. \tag{4}$$

Differentiating with respect to $\sigma^2$ and equating to 0, we get

$$\frac{d}{d\sigma^2} \log \hat{L}(y; \hat{\mu}, \sigma^2) = -\frac{IJ}{2\sigma^2} + \frac{\hat{S}}{2\sigma^4} = 0, \quad \text{or} \quad \sigma^2 = \frac{\hat{S}}{IJ}. \tag{5}$$

Since $\frac{d^2}{d(\sigma^2)^2} \log \hat{L}(y; \hat{\mu}, \sigma^2) = \frac{IJ}{2(\sigma^2)^2} - \frac{2\hat{S}}{2(\sigma^2)^3}$, which, evaluated at $\sigma^2 = \hat{S}/IJ$, gives: $-\frac{(IJ)^3}{2\hat{S}^2} < 0$, it follows that the value of $\sigma^2$ given in (5) is, indeed, its MLE. That is,

$$\widehat{\sigma^2} = \frac{1}{IJ} SS_e, \quad \text{where } SS_e = \sum_i \sum_j (y_{ij} - y_{i.})^2. \tag{6}$$

The results recorded in (3) and (6) provide the answer to the first objective. That is, we have established the following result.

**Theorem 1.** *Consider the model described in relation (13) of Chapter 8; that is, $Y_{ij} = \mu_i + e_{ij}$ where the $e_{ij}$'s are independently $\sim N(0, \sigma^2)$ r.v.'s, $i = 1, \dots, I(\geq 2), j = 1, \dots, J(\geq 2)$. Then the MLE's of the parameters $\mu_i, i = 1, \dots, I$ and $\sigma^2$ of the model are given by (3) and (6), respectively. Furthermore, the MLE's of $\mu_i$, $i = 1, \dots, I$ are also their least square estimates (LSE's).*

## 14.1.2 TESTING THE HYPOTHESIS OF EQUALITY OF MEANS

Next, consider the problem of testing the null hypothesis

$$H_0: \mu_1 = \cdots = \mu_I = \mu \text{ (unspecified)}. \tag{7}$$

Under $H_0$, the expression in (1) becomes:

$$\log L(y; \mu, \sigma^2) = -\frac{IJ}{2} \log(2\pi) - \frac{IJ}{2} \log \sigma^2 - \frac{1}{2\sigma^2} \sum_i \sum_j (y_{ij} - \mu)^2. \tag{8}$$

Repeating a procedure similar to the one we went through above, we derive the MLE's of $\mu$ and $\sigma^2$ under $H_0$, to be denoted by $\hat{\mu}$ and $\widehat{\sigma^2_{H_0}}$, respectively; i.e.,

$$\hat{\mu} = y_{..}, \quad \text{where } y_{..} = \frac{1}{IJ} \sum_i \sum_j y_{ij}, \quad \widehat{\sigma^2_{H_0}} = \frac{1}{IJ} SS_T,$$

$$\text{where } SS_T = \sum_i \sum_j (y_{ij} - y_{..})^2. \tag{9}$$

We now proceed with the setting up of the likelihood ratio statistic $\lambda = \lambda(y)$ in order to test the hypothesis $H_0$. To this end, first observe that, under $H_0$:

$$\exp\left[-\frac{1}{2\widehat{\sigma_{H_0}^2}}\sum_i\sum_j(y_{ij}-y_{..})^2\right] = \exp\left(-\frac{IJ}{2SS_T}\times SS_T\right) = \exp\left(-\frac{IJ}{2}\right),$$

whereas, under no restrictions imposed,

$$\exp\left[-\frac{1}{2\widehat{\sigma^2}}\sum_i\left(y_{ij}-y_{i.}\right)^2\right] = \exp\left(-\frac{IJ}{2SS_e}\times SS_e\right) = \exp\left(-\frac{IJ}{2}\right).$$

Therefore, after cancellations, the likelihood ratio statistic $\lambda$ is

$$\lambda = \left(\widehat{\sigma^2}/\widehat{\sigma_{H_0}^2}\right)^{IJ/2}.$$

Hence $\lambda < C$, if and only if

$$\left(\frac{\widehat{\sigma^2}}{\widehat{\sigma_{H_0}^2}}\right)^{IJ/2} < C, \quad \text{or} \quad \frac{\widehat{\sigma^2}}{\widehat{\sigma_{H_0}^2}} < C^{2/(IJ)}, \quad \text{or} \quad \frac{\widehat{\sigma_{H_0}^2}}{\widehat{\sigma^2}} > 1/C^{2/(IJ)} = C_0. \tag{10}$$

At this point, we need the following result.

**Lemma 1.**   $SS_T = SS_e + SS_H$, where $SS_e$ and $SS_T$ are given by (6) and (9), respectively, and

$$SS_H = \sum_i\sum_j(y_{i.}-y_{..})^2 = J\sum_i(y_{i.}-y_{..})^2. \tag{11}$$

*Proof.* Deferred to Subsection 14.1.3. ■

According to this lemma, the last expression in (10) becomes:

$$\frac{SS_T}{SS_e} > C_0, \quad \text{or} \quad \frac{SS_e + SS_H}{SS_e} > C_0, \quad \text{or} \quad \frac{SS_H}{SS_e} > C_1 = C_0 - 1.$$

In other words, the likelihood ratio test rejects $H_0$ whenever

$$\frac{SS_H}{SS_e} > C_1, \text{ where } SS_e \text{ and } SS_H \text{ are given by (6) and (11), respectively.} \tag{12}$$

In order to determine the cutoff point $C_1$ in (12), we have to have the distribution, under $H_0$, of the statistic $SS_H/SS_e$, where it is tacitly assumed that the observed values have been replaced by the respective r.v.'s. For this purpose, we need the following result.

**Lemma 2.**   *Consider the model described in Theorem 1, and in the expressions $SS_e, SS_T$, and $SS_H$, defined by (6), (9), and (11), respectively, replace the observed values $y_{ij}, y_{i.}$, and $y_{..}$ by the r.v.'s $Y_{ij}, Y_{i.}$, and $Y_{..}$, respectively, but retain the same notation. Then:*

(i) The r.v. $SS_e/\sigma^2$ is distributed as $\chi^2_{I(J-1)}$.
(ii) The statistics $SS_e$ and $SS_H$ are independent.
Furthermore, if the null hypothesis $H_0$ defined in (7) is true, then:
(iii) The r.v. $SS_H/\sigma^2$ is distributed as $\chi^2_{I-1}$.
(iv) The statistic $\frac{SS_H/(I-1)}{SS_e/I(J-1)} \sim F_{I-1,I(J-1)}$.
(v) The r.v. $SS_T/\sigma^2$ is distributed as $\chi^2_{IJ-1}$.

*Proof.* Deferred to Subsection 14.1.3. ■

To this lemma, there is the following corollary, which also encompasses the $\hat{\mu}_i$'s.

**Corollary.**

(i) *The MLE's $\hat{\mu}_i = Y_{i.}$ are unbiased estimates of $\mu_i, i = 1, \ldots, I$.*
(ii) *The MLE $\widehat{\sigma^2} = SS_e/IJ$ is biased, but the estimate $MS_e = SS_e/I(J-1)$ is unbiased.*

*Proof.* (i) Immediate; (ii) Follow from Lemma 2(i). ■

We may conclude that, on the basis of (12) and Lemma 2(iv), in order to test the hypothesis stated in (7), at level of significance $\alpha$, we reject the null hypothesis $H_0$ whenever

$$\mathcal{F} = \frac{SS_H/(I-1)}{SS_e/I(J-1)} = \frac{MS_H}{MS_e} > F_{I-1,I(J-1);\alpha}. \qquad (13)$$

So, the following result has been established.

**Theorem 2.** *In reference to the model described in Theorem 1, the null hypothesis $H_0$ defined in (7) is rejected whenever the inequality in (13) holds; the quantities $SS_e$ and $SS_H$ are given in relations (6) and (11), respectively, and they can be computed by using the formulas in (14) below.*

**Remark 1.** At this point, it should be recalled that the point $F_{m,n;\alpha}$ is determined, so that $P(X > F_{m,n;\alpha}) = \alpha$, where $X$ is a r.v. distributed as $F_{m,n}$; see Figure 14.1.

**Remark 2.** By rewriting analytically the relation in Lemma 1, we have that:

$$\sum_i \sum_j (y_{ij} - y_{..})^2 = \sum_i \sum_j (y_{ij} - y_{i.})^2 + \sum_i \sum_j (y_{i.} - y_{..})^2.$$

That is, the *total variation* of the $y_{ij}$'s with respect to the grand mean $y_{..}$ is split into two parts: The variation of the $y_{ij}$'s in each $i$th group with respect to their mean $y_{i.}$ in that group (*variation within groups*), and the variation of the $I$ means $y_{i.}, i = 1, \ldots, I$ from the grand mean $y_{..}$ (*variation between groups*). By changing the term "variation" to "variance," we are led to the term ANOVA. The expression $SS_e$ is also referred

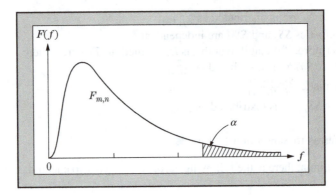

**FIGURE 14.1**

The graph of the p.d.f. of the $F_{m,n}$ distribution along with the rejection region of $H_0$ and the level of significance.

to as the *error sum of squares* for obvious reasons, and the expression $SS_H$ is also referred to as the *treatment sum of squares*, since it reflects variations due to treatment differences. The subscript $H$ accounts for the fact that this statistic is instrumental in testing $H_0$; see also Lemma 3 below. Finally, the expression $SS_T$ is called the *total sum of squares*, again for obvious reasons.

The various quantities employed in carrying out the test described in (13) are usually gathered together in the form of a table, an ANOVA table, as is done in Table 14.1.

**Remark 3.**   For computational purposes, we have:

$$SS_H = J\sum_i Y_{i.}^2 - IJY_{..}^2, \quad SS_e = \sum_i\sum_j Y_{ij}^2 - J\sum_i Y_{i.}^2. \tag{14}$$

**Table 14.1** Analysis of Variance for One-Way Layout

| Source of Variance | Sums of Squares | Degrees of Freedom | Mean Squares |
|---|---|---|---|
| Between groups | $SS_H = J\sum_{i=1}^{I}(Y_{i.} - Y_{..})^2$ | $I-1$ | $MS_H = \frac{SS_H}{I-1}$ |
| Within groups | $SS_e = \sum_{i=1}^{I}\sum_{j=1}^{J}(Y_{ij} - Y_{i.})^2$ | $I(J-1)$ | $MS_e = \frac{SS_e}{I(J-1)}$ |
| Total | $SS_T = \sum_{i=1}^{I}\sum_{j=1}^{J}(Y_{ij} - Y_{..})^2$ | $IJ-1$ | — |

Indeed, first recall the identity

$$\sum_{i=1}^{n}(X_i - \bar{X})^2 = \sum_{i=1}^{n}X_i^2 - n\bar{X}^2 \text{ (Exercise 2.1(i) in Chapter 5),}$$

and then apply it here with $X_i, \bar{X}$ and $n$ replaced by $Y_{i.}, Y_{..}$ and $I$, respectively, to obtain

$$\sum_{i}(Y_{i.} - Y_{..})^2 = \sum_{i}Y_{i.}^2 - IY_{..}^2;$$

then summing over $j$, we get the first result. Next, in the same identity, replace $X_i, \bar{X}$, and $n$ by $Y_{ij}, Y_{i.}$ and $J$, respectively, to obtain

$$\sum_{j}(Y_{ij} - Y_{i.})^2 = \sum_{j}Y_{ij}^2 - JY_{i.}^2.$$

Then summation over $i$ gives the second result.

**Example 1.**   For a numerical example, take $I = 3, J = 5$, and let:

$$\begin{array}{lll}
y_{11} = 82 & y_{21} = 61 & y_{31} = 78 \\
y_{12} = 83 & y_{22} = 62 & y_{32} = 72 \\
y_{13} = 75 & y_{23} = 67 & y_{33} = 74 \\
y_{14} = 79 & y_{24} = 65 & y_{34} = 75 \\
y_{15} = 78 & y_{25} = 64 & y_{35} = 72.
\end{array}$$

(i)   Compute the MLE of $\mu_i, i = 1, 2, 3$.
(ii)   Compute the sum of squares $SS_H$ and $SS_e$, and also the unbiased estimate $MS_e$ of $\sigma^2$.
(iii)   Test the hypothesis $H_0 : \mu_1 = \mu_2 = \mu_3 = \mu$ at level of significance $\alpha = 0.05$.
(iv)   Compute the MLE of $\mu$.

**Discussion.**

(i)   $\hat{\mu}_1 = 79.4, \hat{\mu}_2 = 63.8, \hat{\mu}_3 = 74.2$.
(ii)   Since $y_{i.} = \hat{\mu}_i, i = 1, 2, 3$ and $y_{..} \simeq 72.467$, we get, by (14):

$$SS_H \simeq 79{,}402.2 - 78{,}771.991 = 630.209,$$

$$SS_e = 79{,}491 - 79{,}402.2 = 88.8, \quad \text{and} \quad MS_e = \frac{88.8}{12} = 7.4.$$

So, the unbiased estimate of $\sigma^2$ is 7.4, whereas its MLE is $\frac{88.8}{15} = 5.92$. Since $MS_H \simeq 315.105$, the test statistic is: $\frac{315.105}{7.4} \simeq 42.582$. On the other hand, $F_{2,12;0.05} = 3.8853$, so that the hypothesis $H_0$ is rejected.
(iii)   Finally, $\hat{\mu} = y_{..} = 72.467$.

Here is another example with data from a real experiment.

**Example 2.** In an effort to improve the quality of recording tapes, the effects of four kinds of coatings A, B, C, D on the reproducing quality of sound are compared. Suppose that the measurements of sound distortion given in Table 14.2 are obtained from tapes treated with the four coatings. (This table is a modification of the data given in Example 24 in Chapter 1, in order to be able to apply the theory developed here.) Look at this problem as a one-way layout ANOVA and carry out the analysis; take as level of significance $\alpha = 0.05$.

**Discussion.** Here $I = 4, J = 4$. For the MLE's of the means, we have: $\hat{\mu}_1 = y_1. = 11.25, \hat{\mu}_2 = y_2. = 17.00, \hat{\mu}_3 = y_3. = 15.50, \hat{\mu}_4 = y_4. = 14.75$. Also, $y.. = 14.625$. Next, by (14): $SS_H = 4 \times 873.375 - 16 \times 213.890625 = 3493.50 - 3422.25 = 71.25$, $SS_e = 3568 - 4 \times 873.375 = 3568 - 3493.5 = 74.50$, so that $MS_H = 23.75, MS_e \simeq 6.208$. The observed value of the test statistics is: $\frac{23.75}{6.208} \simeq 3.826$, whereas $F_{3,12;0.05} = 3.4903$. Therefore the null hypothesis about equality of the means is rejected. Finally, $\hat{\mu} = y.. = 14.625$.

The following observations are meant to shed more light on the $F$ test which is used for testing the null hypothesis $H_0$. Recall that $\mathcal{F} = \frac{MS_H}{MS_e}$, where, by (11) and (13),

$$MS_H = \frac{J}{I-1} \sum_i (Y_{i.} - Y_{..})^2,$$

and that $\frac{SS_H}{\sigma^2} \sim \chi^2_{I-1}$, under $H_0$, so that $EMS_H = \sigma^2$. It will be shown below that, regardless whether $H_0$ is true or not,

$$EMS_H = \sigma^2 + \frac{J}{I-1} \sum_i (\mu_i - \mu_.)^2, \quad \text{where } \mu_. = \frac{1}{I} \sum_i \mu_i. \tag{15}$$

Therefore $EMS_H \geq \sigma^2 = E(MS_H \mid H_0)$ and $EMS_H = \sigma^2$ under $H_0$; also, $EMS_e = \sigma^2$. Thus, on the basis of this average criterion, it makes sense to reject $H_0$ when $MS_H$, measured against $MS_e$, takes large values. For reference purposes, relation (15) is stated below as a lemma.

**Table 14.2** Sound Distortion Obtained with Four Types of Coatings

| Coating | Observations | Mean | Grand Mean |
|---------|--------------|------|------------|
| A | 10, 15, 8, 12 | $y_1. = 11.25$ | $y.. = 14.625$ |
| B | 14, 18, 21, 15 | $y_2. = 17.00$ | |
| C | 17, 16, 14, 15 | $y_3. = 15.50$ | |
| D | 12, 15, 17, 15 | $y_4. = 14.75$ | |

**Lemma 3.**  *It holds that:*

$$EMS_H = \frac{1}{I-1}E\sum_i\sum_j(Y_{i.} - Y_{..})^2 = \frac{J}{I-1}E\sum_i(Y_{i.} - Y_{..})^2$$

$$= \sigma^2 + \frac{J}{I-1}\sum_i(\mu_i - \mu_.)^2.$$

*Proof.*  Deferred to Subsection 14.1.3.  ∎

### 14.1.3 PROOF OF LEMMAS IN SECTION 14.1

We now proceed with the justification of Lemmas 1–3 in this section.

*Proof of Lemma 1.*  We have:

$$SS_T = \sum_i\sum_j(y_{ij} - y_{..})^2 = \sum_i\sum_j[(y_{ij} - y_{i.}) + (y_{i.} - y_{..})]^2$$

$$= \sum_i\sum_j(y_{ij} - y_{i.})^2 + \sum_i\sum_j(y_{i.} - y_{..})^2 + 2\sum_i\sum_j(y_{ij} - y_{i.})(y_{i.} - y_{..})$$

$$= SS_e + SS_H, \quad \text{since}$$

$$\sum_i\sum_j(y_{ij} - y_{i.})(y_{i.} - y_{..}) = \sum_i(y_{i.} - y_{..})\sum_j(y_{ij} - y_{i.})$$

$$= \sum_i(y_{i.} - y_{..})(Jy_{i.} - Jy_{i.}) = 0.$$

∎

*Proof of Lemma 2.*  At this point, recall that, if $X_1, \ldots, X_n$ are independent r.v.'s distributed as $N(\mu, \sigma^2)$, then: (a) $\bar{X}$ and $\sum_i(X_i - \bar{X})^2$ are independent; (b) $\frac{1}{\sigma^2}\sum_i (X_i - \bar{X})^2 \sim \chi^2_{n-1}$. Apply these results as follows:

**(i)**  For each $i = 1, \ldots, I$, $\frac{1}{\sigma^2}\sum_j(Y_{ij} - Y_{i.})^2 \sim \chi^2_{J-1}$, by (b) above. Furthermore, for $i' \neq i$, $\sum_j(Y_{i'j} - Y_{i'.})^2$ and $\sum_j(Y_{ij} - Y_{i.})^2$ are independent, since they are defined (separately) on sets of independent r.v.'s. It follows that:

$$\frac{SS_e}{\sigma^2} = \frac{1}{\sigma^2}\sum_i\sum_j(Y_{ij} - Y_{i.})^2 = \sum_i\frac{1}{\sigma^2}\sum_j(Y_{ij} - Y_{i.})^2 \sim \chi^2_{I(J-1)}.$$

**(ii)**  For each $i = 1, \ldots, I$, the independent r.v.'s $Y_{i1}, \ldots, Y_{iJ} \sim N(\mu_i, \sigma^2)$. Hence, by (a) above, $\sum_j(Y_{ij} - Y_{i.})^2$ and $Y_{i.}$ are independent. Furthermore, $\sum_j(Y_{ij} - Y_{i.})^2$ and $Y_{i'.}$ are also independent for $i' \neq i$, because $Y_{i'.}$ is defined on a set of r.v.'s which are independent of $\sum_j(Y_{ij} - Y_{i.})^2$. Thus, each of the statistics $\sum_j(Y_{ij} - Y_{i.})^2, i = 1, \ldots, I$ is independent of the statistics $Y_{1.}, \ldots, Y_{I.}$, and the statistics $\sum_j(Y_{ij} - Y_{i.})^2, i = 1, \ldots, I$ are independent, as was seen in part (i). It follows that the sets $\sum_j(Y_{ij} - Y_{i.})^2, i = 1, \ldots, I$ and $Y_{i.}, i = 1, \ldots, I$ are independent. Then so are functions defined (separately) on them. In particular, the functions $\sum_i \sum_j (Y_{ij} - Y_{i.})^2$ and $\sum_i \sum_j (Y_{i.} - Y_{..})^2$ are independent, or $SS_e$ and $SS_H$ are independent.

(iii) Under $H_0$, the r.v.'s $Y_{1.}, \ldots, Y_{I.}$ are independent and distributed as $N(\mu, \sigma^2/J)$. Therefore $\frac{J}{\sigma^2}\sum_i(Y_{i.} - Y_{..})^2 \sim \chi^2_{I-1}$. Since $\frac{J}{\sigma^2}\sum_i(Y_{i.} - Y_{..})^2 = \frac{1}{\sigma^2}\sum_i\sum_j(Y_{i.} - Y_{..})^2 = \frac{1}{\sigma^2}SS_H$, the result follows.

(iv) It follows from parts (i)–(iii) and the definition of the $F$ distribution.

(v) Under $H_0$, the r.v.'s $Y_{ij}, i = 1, \ldots, I, j = 1, \ldots, J$ are independent and distributed as $N(\mu, \sigma^2)$. Then, by (b) above, $\frac{1}{\sigma^2}\sum_i\sum_j(Y_{ij} - Y_{..})^2$ is distributed as $\chi^2_{IJ-1}$, or $\frac{SS_T}{\sigma^2} \sim \chi^2_{IJ-1}$. ∎

*Proof of Lemma 3.* Before taking expectations, work with $\sum_i(Y_{i.} - Y_{..})^2$ and rewrite it in a convenient form; namely,

$$\sum_i(Y_{i.} - Y_{..})^2 = \sum_i[(Y_{i.} - \mu_.) - (Y_{..} - \mu_.)]^2 = \sum_i(Y_{i.} - \mu_.)^2 - I(Y_{..} - \mu_.)^2,$$

because

$$\sum_i(Y_{i.} - \mu_.) = \sum_i Y_{i.} - I\mu_. = \sum_i \frac{1}{J}\sum_j Y_{ij} - I\mu_.$$

$$= \frac{1}{J}\sum_i\sum_j Y_{ij} - I\mu_. = IY_{..} - I\mu_. = I(Y_{..} - \mu_.),$$

so that

$$-2\sum_i(Y_{i.} - \mu_.)(Y_{..} - \mu_.) = -2(Y_{..} - \mu_.) \times I(Y_{..} - \mu_.) = -2I(Y_{..} - \mu_.)^2.$$

So,

$$E\sum_i(Y_{i.} - Y_{..})^2 = \sum_i E(Y_{i.} - \mu_.)^2 - IE(Y_{..} - \mu_.)^2 = \sum_i E(Y_{i.} - \mu_.)^2 - I\,Var(Y_{..})$$

$$= \sum_i E(Y_{i.} - \mu_.)^2 - I\frac{\sigma^2}{IJ} = \sum_i E(Y_{i.} - \mu_.)^2 - \frac{\sigma^2}{J},$$

and

$$E(Y_{i.} - \mu_.)^2 = E[(Y_{i.} - \mu_i) + (\mu_i - \mu_.)]^2 = E(Y_{i.} - \mu_i)^2 + (\mu_i - \mu_.)^2$$
$$= Var(Y_{i.}) + (\mu_i - \mu_.)^2 = \tfrac{\sigma^2}{J} + (\mu_i - \mu_.)^2.$$

Therefore

$$E\sum_i(Y_{i.} - Y_{..})^2 = \frac{I\sigma^2}{J} + \sum_i(\mu_i - \mu_.)^2 - \frac{\sigma^2}{J} = \frac{I-1}{J}\sigma^2 + \sum_i(\mu_i - \mu_.)^2,$$

and, by (11) and (13),

$$EMS_H = \sigma^2 + \frac{J}{I-1} \sum_i (\mu_i - \mu_.)^2, \quad \text{which is (15).}$$

■

**Remark 4.** In closing this section, it should be pointed out that an obvious generalization of what was done here is to have different $J$'s for each $i = 1, \ldots, I$; i.e., for each $i$, we have $J_i$ observations. The analysis conceptually remains the same, only one would have to carry along the $J_i$'s as opposed to one $J$.

## EXERCISES

**1.1** Apply the one-way layout analysis of variance to the data given in the table below. Take $\alpha = 0.05$.

| A | B | C |
|---|---|---|
| 10.0 | 9.1 | 9.2 |
| 11.5 | 10.3 | 8.4 |
| 11.7 | 9.4 | 9.4 |

**1.2** Consider the log-likelihood function (1) as it becomes under the null hypothesis $H_0$ stated in (7), and show that the MLE's of $\mu$ and $\sigma^2$ are given by the expression in relation (9).

**1.3** In reference to the derivation of the likelihood ratio test $\lambda$ for testing the hypothesis $H_0$ stated in relation (7), show that $\lambda$ is, actually, given by any one of the expressions in relation (10).

**1.4** In reference to the proof of Lemma 3, show that:
   (i)   $E(Y_{..} - \mu_.)^2 = \frac{\sigma^2}{IJ}$.
   (ii)  $E(Y_{i.} - \mu_i)^2 = \frac{\sigma^2}{J}$.
   (iii) $E(Y_{i.} - \mu_.)^2 = E(Y_{i.} - \mu_i)^2 + (\mu_i - \mu_.)^2$.

## 14.2 A MULTICOMPARISON METHOD

Refer to the one-way layout model discussed in the previous section, namely, to the model described in Theorem 1. One of the problems we have studied was that of testing the null hypothesis about equality of the $I$ means; i.e.,

$$H_0 : \mu_1 = \cdots = \mu_I = \mu \text{ unspecified.} \tag{16}$$

Suppose now that the hypothesis $H_0$ is rejected, as, indeed, was the case in Examples 1 and 2. Rejection of $H_0$ simply means that not *all* of the $\mu_i$'s are equal.

Clearly, it would be desirable to know which of the $\mu_i$'s are responsible for the rejection of $H_0$. It is true that we do gain some information about it by looking at the estimates $\hat{\mu}_i$. However, we would like to obtain additional information analogous to that provided by a confidence interval for a real-valued parameter. This is the problem to examine in this section.

In relation (16), the hypothesis $H_0$ compares the parameters involved and, actually, stipulates that they are all equal. This suggests that any attempt to construct a confidence interval should not focus on a single parameter, but rather on two or more parameters simultaneously. For example, we would like to compare all possible pairs $(\mu_i, \mu_j)$ through the differences $\mu_i - \mu_j$. Or, more generally, to compare one subset of these parameters against the complement of this subset. Thus, in Example 1, where we have three parameters $\mu_1, \mu_2$, and $\mu_3$, we may wish, e.g., to compare $\mu_1$ against $(\mu_2, \mu_3)$, or $\mu_2$ against $(\mu_1, \mu_3)$, or $\mu_3$ against $(\mu_1, \mu_2)$. One way of doing it is to look at the respective differences: $\mu_1 - \frac{1}{2}(\mu_1 + \mu_2), \mu_2 - \frac{1}{2}(\mu_1 + \mu_3), \mu_3 - \frac{1}{2}(\mu_1 + \mu_2)$.

At this point, it is to be observed that all expressions we looked at above are of the form $c_1\mu_1 + \cdots + c_I\mu_I$ with $c_1 + \cdots + c_I = 0$. This observation leads to the following definition.

**Definition 1.**    In reference to the model described in Theorem 1, any relation among the parameters $\mu_1, \ldots, \mu_I$ of the form $\Psi = \sum_{i=1}^{I} c_i\mu_i$ with $\sum_{i=1}^{I} c_i = 0$ is called a *contrast* among the $\mu_i$'s.

It follows from the above discussion that what would be really meaningful here would be the construction of confidence intervals for contrasts among the $\mu_i$'s. In particular, it would be clearly, highly desirable to construct confidence intervals for all possible contrast among the $\mu_i$'s, which would *all* have the same confidence coefficient. This is exactly the content of the theorem stated below.

First, let us introduce some pieces of notation needed. To this end, consider the contrast

$$\Psi = \sum_i c_i\mu_i \left( \sum_i c_i = 0 \right), \tag{17}$$

and let us estimate $\Psi$ by $\hat{\Psi}$, where

$$\hat{\Psi} = \sum_i c_i\hat{\mu}_i = \sum_i c_i Y_i. \tag{18}$$

Clearly,

$$E\hat{\Psi} = \Psi \quad \text{and} \quad Var(\hat{\Psi}) = \left( \frac{1}{J}\sum_i c_i^2 \right)\sigma^2, \tag{19}$$

and the variance is estimated, in an obvious manner, by

$$\widehat{Var}(\hat{\Psi}) = \left( \frac{1}{J}\sum_i c_i^2 \right)MS_e, \quad MS_e = SS_e/I(J-1), \tag{20}$$

and $SS_e$ is given in (6) (see also (14)). Finally, define $S^2$ by:

$$S^2 = (I-1)F_{I-1,I(J-1);\alpha.} \tag{21}$$

Then we have the following important result.

**Theorem 3.** *With the notation introduced in (17), (18), (20), and (21), the interval*

$$\left(\hat{\Psi} - S\sqrt{\widehat{Var}(\hat{\Psi})}, \quad \hat{\Psi} + S\sqrt{\widehat{Var}(\hat{\Psi})}\right) \tag{22}$$

*is a confidence interval with confidence coefficient* $1-\alpha$ *simultaneously for all contrasts* $\Psi$.

At this point, it should not come as a surprise that there is an intimate relationship between the null hypothesis $H_0$ and confidence intervals for contrast. The result stated below as a lemma (but not proved!) articulates this relationship. In its statement, we need a concept defined now.

**Definition 2.** Let $\Psi$ and $\hat{\Psi}$ be as in (17) and (18), respectively. Then we say that $\hat{\Psi}$ is *significantly different from zero*, if the interval defined in (22) does not contain zero; equivalently, $|\hat{\Psi}| > S\sqrt{\widehat{Var}(\hat{\Psi})}$.

Then the lemma mentioned above is as follows.

**Lemma 4.** *The null hypothesis $H_0$ stated in (16) is rejected, if and only if there is at least one contrast $\Psi$ for which $\hat{\Psi}$ is significantly different from zero.*

We do not intend to pursue the proof of Theorem 3 here, which can be found in great detail in Section 17.4 of the book *A Course in Mathematical Statistics*, 2nd edition, Academic Press (1997), by G. G. Roussas. Suffice it to say that it follows from the maximization, with respect to $c_1, \ldots, c_I$, subject to the contrast constraint $\sum_i c_i = 0$, of the function

$$f(c_1, \ldots, c_I) = \frac{1}{\sqrt{\frac{1}{J} \sum_i c_i^2}} \sum_i c_i (Y_{i.} - \mu_i),$$

and that this maximization is obtained by means of the so-called Lagrange multipliers. In the process of doing so, we also need the following two facts.

**Lemma 5.**

(i) With $\mu_{.} = \frac{1}{I} \sum_i \mu_i$, the r.v.'s $\sum_j [(Y_{ij} - \mu_i) - (Y_{i.} - \mu_i)]^2, i = 1, \ldots, I$ are independent.

(ii) The r.v.'s $\sum_i [(Y_{i.} - Y_{..}) - (\mu_i - \mu_{.})]^2$ and $SS_e$ are independent.

(iii) Under the null hypothesis $H_0$, $\frac{J}{\sigma^2} \sum_i [(Y_{i.} - Y_{..}) - (\mu_i - \mu_{.})]^2 \sim \chi_{I-1}^2$.

*Proof.* Deferred to the end of the section. ■

We now consider some examples.

**Example 3.** In reference to Example 1, construct a 95% confidence interval for each of the following contrasts:

$$\mu_1 - \mu_2, \quad \mu_1 - \mu_3, \quad \mu_2 - \mu_3, \quad \mu_1 - \tfrac{1}{2}(\mu_2 + \mu_3),$$

$$\mu_2 - \tfrac{1}{2}(\mu_3 + \mu_1), \quad \mu_3 - \tfrac{1}{2}(\mu_1 + \mu_2).$$

**Discussion.** Here $I = 3, J = 5$, and hence $F_{I-1,I(J-1);\alpha} = F_{2,12;0.05} = 3.8853, S^2 = (I-1)F_{I-1,I(J-1);\alpha} = 2 \times 3.8853 = 7.7706$ and $S \simeq 2.788$. Also, $MS_e = 7.4$ from Example 1. From the same example, for $\Psi = \mu_1 - \mu_2$, we have $\hat{\Psi} = y_{1.} - y_{2.} = 79.4 - 63.8 = 15.6, \widehat{Var}(\hat{\Psi}) = \tfrac{2}{5} \times 7.4 = 2.96, \sqrt{\widehat{Var}(\hat{\Psi})} \simeq 1.72$ and $S\sqrt{\widehat{Var}(\hat{\Psi})} = 2.788 \times 1.72 \simeq 4.795$. Therefore, the observed confidence interval for $\mu_1 - \mu_2$ is:

$$[15.6 - 4.795, \quad 15.6 + 4.795] = [10.805, \quad 20.395].$$

Likewise, for $\Psi = \mu_1 - \mu_3$, we have $\hat{\Psi} = y_{1.} - y_{3.} = 79.4 - 74.2 = 5.2, \widehat{Var}(\hat{\Psi}) = 2.96$ the same as before, and hence $S\sqrt{\widehat{Var}(\hat{\Psi})} \simeq 4.795$. Then the observed confidence interval for $\mu_1 - \mu_3$ is:

$$[5.2 - 4.795, \quad 5.2 + 4.795] = [0.405, \quad 9.995].$$

Also, for $\Psi = \mu_2 - \mu_3$, we have $\hat{\Psi} = y_{2.} - y_{3.} = 63.8 - 74.2 = -10.4$. Since the $\widehat{Var}(\hat{\Psi})$ is still 2.96, the observed confidence interval for $\mu_2 - \mu_3$ is:

$$[-10.4 - 4.795, \quad -10.4 + 4.795] = [-15.195, \quad -5.605].$$

Next, let $\Psi = \mu_1 - \tfrac{1}{2}(\mu_2 + \mu_3)$, so that $\hat{\Psi} = 79.4 - \tfrac{1}{2}(63.8 + 74.2) = 79.4 - 69 = 10.4, \widehat{Var}(\hat{\Psi}) = \tfrac{3}{10} \times 7.4 = 2.22, \sqrt{\widehat{Var}(\hat{\Psi})} \simeq 1.49$ and $S\sqrt{\widehat{Var}(\hat{\Psi})} \simeq 4.154$. Therefore, the observed confidence interval is:

$$[10.4 - 4.154, \quad 10.4 + 4.154] = [6.246, \quad 14.554].$$

For $\Psi = \mu_2 - \tfrac{1}{2}(\mu_3 + \mu_1)$, we have $\hat{\Psi} = 63.8 - \tfrac{1}{2}(74.2 + 79.4) = -13$, and therefore the observed confidence interval is:

$$[-13 - 4.154, \quad -13 + 4.154] = [-17.154, \quad -8.846].$$

Finally, for $\Psi = \mu_3 - \tfrac{1}{2}(\mu_1 + \mu_2)$, we have $\hat{\Psi} = 74.2 - \tfrac{1}{2}(79.4 + 63.8) = 2.6$, and the observed confidence interval is:

$$[2.6 - 4.154, \quad 2.6 + 4.154] = [-1.554, \quad 6.754].$$

It is noteworthy that of the six contrasts we have entertained in this example, for only one contrast $\Psi = \mu_3 - \tfrac{1}{2}(\mu_1 + \mu_2)$ the respective quantity $\hat{\Psi} = 2.6$, is *not* significantly different from zero. This is consonant with Lemma 4, since we already know (from Example 1) that $H_0$ is rejected. For example, for the contrast $\Psi = \mu_1 - \mu_2$, we found the confidence interval $[10.805, 20.395]$, which does not contain 0. This simply says that, at the confidence level considered, $\mu_1$ and $\mu_2$ cannot be equal; thus, $H_0$ would have to be rejected. Likewise for the contrasts $\mu_1 - \mu_3$ and $\mu_2 - \mu_3$.

**Example 4.** In reference to Example 2, construct a 95% confidence interval for each of the following contrasts:

$$\mu_1 - \mu_2, \quad \mu_1 - \mu_3, \quad \mu_1 - \mu_4, \quad \mu_2 - \mu_3, \quad \mu_2 - \mu_4, \quad \mu_3 - \mu_4.$$

**Discussion.** Here $I = J = 4, F_{I-1,I(J-1);\alpha} = F_{3,12;0.05} = 3.4903, S^2 = (I-1)$ $F_{I-1,I(J-1);\alpha} = 3 \times 3.4903 = 10.4709$, and $S \simeq 3.236$. For $\Psi = \mu_1 - \mu_2$, we have $\hat{\Psi} = y_{1.} - y_{2.} = 11.25 - 17 = -5.75$, and $\widehat{Var}(\hat{\Psi}) = 0.5 \times 6.208 = 3.104$, $\sqrt{\widehat{Var}(\hat{\Psi})} \simeq 1.762$. Thus, $S\sqrt{\widehat{Var}(\hat{\Psi})} = 3.236 \times 1.762 \simeq 5.702$. Then the observed confidence interval for $\mu_1 - \mu_2$ is:

$$[-5.75 - 5.702, \quad -5.75 + 5.702] = [-11.452, \quad -0.048].$$

For $\Psi = \mu_1 - \mu_3$, we have $\hat{\Psi} = 11.25 - 15.50 = -4.25$, and the observed confidence interval for $\mu_1 - \mu_3$ is:

$$[-4.25 - 5.702, \quad -4.25 + 5.702] = [-9.952, \quad 1.452].$$

For $\Psi = \mu_1 - \mu_4$, we have $\hat{\Psi} = 11.25 - 14.75 = -3.5$, and the observed confidence interval is:

$$[-3.5 - 5.702, \quad -3.5 + 5.702] = [-9.202, \quad 2.202].$$

For $\Psi = \mu_2 - \mu_3$, we have $\hat{\Psi} = 17 - 15.5 = 1.5$, and the observed confidence interval is:

$$[1.5 - 5.702, \quad 1.5 + 5.702] = [-4.202, \quad 7.202].$$

For $\Psi = \mu_2 - \mu_4$, we have $\hat{\Psi} = 17 - 14.75 = 2.25$, and the observed confidence interval is:

$$[2.25 - 5.702, \quad 2.25 + 5.702] = [-3.452, \quad 7.952].$$

Finally, for $\Psi = \mu_3 - \mu_4$, we have $\hat{\Psi} = 15.50 - 14.75 = 0.75$, and the observed confidence interval is:

$$[0.75 - 5.702, \quad 0.75 + 5.702] = [-4.952, \quad 6.452].$$

In this example, we have that for *only one* of the six contrasts considered, $\Psi = \mu_1 - \mu_2$, the respective $\hat{\Psi} = -5.75$, is significantly different from zero. This, of course, suffices for the rejection of the hypothesis $H_0$ (according to Lemma 4), as it actually happened in Lemma 2. So, it appears that the means $\mu_1$ and $\mu_2$ are the culprits here. This fact is also reflected by the estimates of the $\mu_i$'s found in Example 2; namely,

$$\hat{\mu}_1 = 11.25, \quad \hat{\mu}_2 = 17.00, \quad \hat{\mu}_3 = 15.50, \quad \hat{\mu}_4 = 14.75;$$

$\hat{\mu}_1$ and $\hat{\mu}_2$ are the furthest apart.

This section is concluded with the presentation of a justification of Lemma 5.

*Proof of Lemma 5.*

**(i)** Here

$$\sum_j [(Y_{ij} - \mu_i) - (Y_{i.} - \mu_i)]^2 = \sum_j (Y_{ij} - Y_{i.})^2,$$

and the statistics $\sum_j (Y_{ij} - Y_{i.})^2, i = 1, \ldots, I$ are independent, since they are defined (separately) on independent sets (rows) of r.v.'s.

**(ii)** The proof of this part is reminiscent of that of Lemma 2(ii). The independent r.v.'s $Y_{i1} - \mu_i, \ldots, Y_{iJ} - \mu_i$ are distributed as $N(0, \sigma^2)$. Since $\frac{1}{J} \sum_j (Y_{ij} - \mu_i) = Y_{i.} - \mu_i$, it follows, by an application of (a) in the proof of Lemma 2, that, for each $i = 1, \ldots, I, \sum_j [(Y_{ij} - \mu_i) - (Y_{i.} - \mu_i)]^2 = \sum_j (Y_{ij} - Y_{i.})^2$ is independent of $Y_{i.} - \mu_i$. For $i' \neq i$, each of $\sum_j (Y_{ij} - Y_{i.})^2$ is also independent of $Y_{i'.} - \mu_{i'}$. Also, by part (i), $\sum_j (Y_{ij} - Y_{i.})^2, i = 1, \ldots, I$ are independent. It follows that the sets of statistics

$$\sum_j (Y_{ij} - Y_{i.})^2, \quad i = 1, \ldots, I \quad \text{and} \quad Y_{i.} - \mu_i, \quad i = 1, \ldots, I$$

are independent. Then so are functions defined (separately) on them. In particular, this is true for the functions

$$\sum_i \sum_j (Y_{ij} - Y_{i.})^2 = SS_e \quad \text{and}$$

$$\sum_i [(Y_{i.} - \mu_i) - (Y_{..} - \mu_{..})]^2 = \sum_i [(Y_{i.} - Y_{..}) - (\mu_i - \mu_{..})]^2.$$

**(iii)** For $i = 1, \ldots, I$, the r.v.'s $Y_{i.} - \mu_i$ are independent and distributed as $N(0, \sigma^2/J)$, so that the independent r.v.'s $\frac{\sqrt{J}}{\sigma}(Y_{i.} - \mu_i), i = 1, \ldots, I$ are distributed as $N(0, 1)$.

Since $\frac{1}{I} \sum_i \frac{\sqrt{J}}{\sigma}(Y_{i.} - \mu_i) = \frac{\sqrt{J}}{\sigma}(Y_{..} - \mu_{..})$, it follows (by (a) in the proof of Lemma 2) that

$$\sum_i \left[ \frac{\sqrt{J}}{\sigma}(Y_{i.} - \mu_i) - \frac{\sqrt{J}}{\sigma}(Y_{..} - \mu_{..}) \right]^2 \sim \chi_{I-1}^2,$$

or

$$\frac{J}{\sigma^2} \sum_i [(Y_{i.} - Y_{..}) - (\mu_i - \mu_{..})]^2 \sim \chi_{I-1}^2. \qquad \blacksquare$$

## EXERCISES

**2.1**   Refer to Exercise 1.1, and construct 95% confidence intervals for all contrasts of the $\mu_i$'s.

**2.2**   Refer to Example 2, and construct 95% confidence intervals for all contrasts of the $\mu_i$'s.

## 14.3 TWO-WAY LAYOUT WITH ONE OBSERVATION PER CELL

In this section, we pursue the study of the kind of problems considered in Section 14.1, but in a more general framework. Specifically, we consider experiments whose outcomes are influenced by more than one factor. In the model to be analyzed here there will be two such factors, one factor occurring at $I$ levels and the other factor occurring at $J$ levels. The following example will help clarify the underlying ideas and the issues to be resolved.

**Example 5.**   Suppose we are interested in acquiring a fairly large number of equipments from among $I$ brands entertained. The available workforce to use the equipments bought consists of $J$ workers. Before a purchase decision is made, an experiment is carried out whereby each one of the $J$ workers uses each one of the $I$ equipments for one day. It is assumed that the one day's production would be a quantity, denoted by $\mu_{ij}$, depending on the $i$th brand of equipment and the $j$th worker, except for an error $e_{ij}$ associated with the $i$th equipment and the $j$th worker. Thus, the one day's outcome is, actually, an observed value of a r.v. $Y_{ij}$, which has the following structure: $Y_{ij} = \mu_{ij} + e_{ij}$, $i = 1, \ldots, I$, $j = 1, \ldots, J$. For the errors $e_{ij}$ the familiar assumptions are made; namely, the r.v.'s $e_{ij}$, $i = 1, \ldots, I$, $j = 1, \ldots, J$ are independent and distributed as $N(0, \sigma^2)$. It follows that the r.v.'s $Y_{ij}$, $i = 1, \ldots, I$, $j = 1, \ldots, J$ are independent with $Y_{ij} \sim N(\mu_{ij}, \sigma^2)$. At this point, the further reasonable assumption is made that each mean $\mu_{ij}$ consists of three *additive* parts: A quantity $\mu$, the *grand mean*, the same for all $i$ and $j$; an effect due the $i$th equipment, denoted by $\alpha_i$ and usually referred to as the *row effect*; and an effect due to the $j$th worker, denoted by $\beta_j$ and usually referred to as the *column effect*. So, $\mu_{ij} = \mu + \alpha_i + \beta_j$. Now, it is not unreasonable to assume that some of the $\alpha_i$ effects are positive, some are negative, and on the whole their sum is zero. Likewise for the $\beta_j$ effects.

Gathering together the assumptions made so far, we have the following model.

$$
\left.
\begin{array}{l}
Y_{ij} = \mu + \alpha_i + \beta_j + e_{ij}, \sum_{i=1}^{I} \alpha_i = 0 \text{ and } \sum_{j=1}^{J} \beta_j = 0, \text{ the r.v.'s} \\
e_{ij}, i = 1, \ldots, I(\geq 2), j = 1, \ldots, J(\geq 2) \text{ are independent and distributed} \\
\text{as } N(0, \sigma^2). \\
\text{It follows that the r.v.'s } Y_{ij}, i = 1, \ldots, I, j = 1, \ldots, J \\
\text{are independent with } Y_{ij} \sim N(\mu + \alpha_i + \beta_j, \sigma^2).
\end{array}
\right\}
$$

$$(23)$$

Of course, once model (23) is arrived at, it can be detached from the specific example which helped motivate the model.

In reference to model (23), the questions which arise naturally are the following: What are the magnitudes of the grand mean $\mu$, of the row effects $\alpha_i$, of the column effects $\beta_j$, and of the error variance $\sigma^2$? Also, are there, really, any row effects present (does it make a difference, for the output, which equipment is purchased)? Likewise for the column effects. In statistical terminology, the questions posed above translate as follows: Estimate the *parameters* of the model $\mu, \alpha_i, i = 1, \ldots, I, \beta_j, j = 1, \ldots, J,$ and $\sigma^2$. The estimates sought will be the MLE's, which for the parameters $\mu, \alpha_i,$ and $\beta_j$ are also LSE's. Test the null hypothesis of no row effect: $H_{0,A}: \alpha_1 = \cdots = \alpha_I$ (and therefore $= 0$). Test the null hypothesis of no column effect: $H_{0,B}: \beta_1 = \cdots = \beta_J$ (and therefore $= 0$).

## 14.3.1 THE MLE's OF THE PARAMETERS OF THE MODEL

The likelihood function of the $Y_{ij}$'s, to be denoted by $L(y;\mu, \alpha, \beta, \sigma^2)$ in obvious notation, is given by the formula below. In this formula and in the sequel, the precise range of $i$ and $j$ will not be indicated explicitly for notational convenience.

$$L(y;\mu, \alpha, \beta, \sigma^2) = \left(\frac{1}{\sqrt{2\pi\sigma^2}}\right)^{IJ} \exp\left[-\frac{1}{2\sigma^2}\sum_i\sum_j(y_{ij} - \mu - \alpha_i - \beta_j)^2\right].$$

(24)

For each fixed $\sigma^2$, maximization of the likelihood function with respect to $\mu, \alpha_i,$ and $\beta_j$ is equivalent to minimization, with respect to these parameters, of the expression:

$$S(\mu, \alpha_1, \ldots, \alpha_I, \beta_1, \ldots, \beta_J) = S(\mu, \alpha, \beta) = \sum_i\sum_j(y_{ij} - \mu - \alpha_i - \beta_j)^2.$$

(25)

Minimization of $S(\mu, \alpha, \beta)$ with respect to $\mu, \alpha,$ and $\beta$ yields the values given in the following result.

**Lemma 6.** *The unique minimizing values of $\mu, \alpha_i,$ and $\beta_j$ for expression (25) (i.e., the LSE's of $\mu, \alpha_i,$ and $\beta_j$) are given by:*

$$\hat{\mu} = y_{..}, \quad \hat{\alpha}_i = y_{i.} - y_{..}, \quad i = 1, \ldots, I, \quad \hat{\beta}_j = y_{.j} - y_{..}, \quad j = 1, \ldots, J,$$

(26)

*where*

$$y_{i.} = \frac{1}{J}\sum_j y_{ij}, \quad y_{.j} = \frac{1}{I}\sum_i y_{ij}, \quad y_{..} = \frac{1}{IJ}\sum_i\sum_j y_{ij}.$$

(27)

*Proof.* Deferred to Subsection 14.3.3. ∎

For the values in (26), the log-likelihood function becomes, with obvious notation:

$$\log \hat{L}(y;\hat{\mu}, \hat{\alpha}, \hat{\beta}, \sigma^2) = -\frac{IJ}{2}\log(2\pi) - \frac{IJ}{2}\log\sigma^2 - \frac{1}{2\sigma^2}\hat{S},$$

(28)

where $\hat{S} = \sum_i \sum_j (y_{ij} - y_{i.} - y_{.j} + y_{..})^2$. Relation (28) is of exactly the same type as relation (5), maximization of which produced the value

$$\widehat{\sigma^2} = \frac{1}{IJ}\hat{S} = \frac{1}{IJ} \sum_i \sum_j (y_{ij} - y_{i.} - y_{.j} + y_{..})^2. \tag{29}$$

Combining then the results in (26) and (29), we have the following result.

**Theorem 4.**  *Under model (23), the MLE's of the parameters of the model are given by relations (26) and (29). Furthermore, the MLE's of $\mu, \alpha_i$, and $\beta_j$ are also their LSE's.*

## 14.3.2 TESTING THE HYPOTHESIS OF NO ROW OR NO COLUMN EFFECTS

First, consider the null hypothesis of no row effect; namely,

$$H_{0,A} : \alpha_1 = \cdots = \alpha_I = 0. \tag{30}$$

Under $H_{0,A}$, the likelihood function in (24), to be denoted for convenience by $L_A(y; \mu, \boldsymbol{\beta}, \sigma^2)$, becomes:

$$L_A(y; \mu, \boldsymbol{\beta}, \sigma^2) = \left(\frac{1}{\sqrt{2\pi\sigma^2}}\right)^{IJ} \exp\left[-\frac{1}{2\sigma^2} \sum_i \sum_j (y_{ij} - \mu - \beta_j)^2\right]. \tag{31}$$

Maximization of this likelihood with respect to $\beta_j$'s and $\mu$, for each fixed $\sigma^2$, is equivalent to minimization, with respect to $\beta_j$'s and $\mu$ of the expression:

$$S(\mu, \beta_1, \ldots, \beta_J) = S(\mu, \boldsymbol{\beta}) = \sum_i \sum_j (y_{ij} - \mu - \beta_j)^2. \tag{32}$$

Working exactly as in (25), we obtain the following MLE's, under $H_{0,A}$, to be denoted by $\hat{\mu}_A$ and $\hat{\beta}_{j,A}$:

$$\hat{\mu}_A = y_{..} = \hat{\mu}, \quad \hat{\beta}_{j,A} = y_{.j} - y_{..} = \hat{\beta}_j, \quad j = 1, \ldots, J. \tag{33}$$

Then, repeating the steps in relation (28), we obtain the MLE of $\sigma^2$, under $H_{0,A}$:

$$\widehat{\sigma_A^2} = \frac{1}{IJ} \sum_i \sum_j (y_{ij} - y_{.j})^2. \tag{34}$$

The hypothesis $H_{0,A}$ will be tested by means of the likelihood ratio test. First, observe that:

$$\exp\left[-\frac{1}{2\widehat{\sigma_A^2}} \sum_i \sum_j (y_{ij} - y_{.j})^2\right] = \exp\left(-\frac{IJ}{2\widehat{\sigma_A^2}} \times \widehat{\sigma_A^2}\right) = \exp\left(-\frac{IJ}{2}\right),$$

and

$$\exp\left[-\frac{1}{2\widehat{\sigma^2}} \sum_i \sum_j (y_{ij} - y_{i.} - y_{.j} + y_{..})^2\right] = \exp\left(-\frac{IJ}{2\widehat{\sigma^2}} \times \widehat{\sigma^2}\right) = \exp\left(-\frac{IJ}{2}\right).$$

Then the likelihood ratio statistic $\lambda$ is given by:

$$\lambda = \left(\widehat{\sigma^2}/\widehat{\sigma_A^2}\right)^{IJ/2}.$$

Hence $\qquad \left(\dfrac{\widehat{\sigma^2}}{\widehat{\sigma_A^2}}\right)^{IJ/2} < C,\quad$ if and only if $\dfrac{\widehat{\sigma_A^2}}{\widehat{\sigma^2}} > C_0 = 1/C^{2/(IJ)}.$ $\qquad$ (35)

At this point, use the following notation:

$$SS_e = IJ\widehat{\sigma^2} = \sum_i \sum_j (y_{ij} - y_{i.} - y_{.j} + y_{..})^2, \quad SS_A = J\sum_i \hat{\alpha}_i^2 = J\sum_i (y_{i.} - y_{..})^2, \quad (36)$$

by means of which it is shown that:

**Lemma 7.**   With $\widehat{\sigma_A^2}$, $SS_e$ and $SS_A$ defined by (34) and (36), it holds: $IJ\widehat{\sigma_A^2} = IJ\widehat{\sigma^2} + SS_A = SS_e + SS_A$.

*Proof.* Deferred to Subsection 14.3.3. ∎

By means of this lemma, relation (35) becomes:

$$\frac{\widehat{\sigma_A^2}}{\widehat{\sigma^2}} = \frac{IJ\widehat{\sigma_A^2}}{IJ\widehat{\sigma^2}} = \frac{SS_e + SS_A}{SS_e} = 1 + \frac{SS_A}{SS_e} > C_0, \quad \text{or} \quad \frac{SS_A}{SS_e} > C_1 = C_0 - 1.$$

So, the likelihood ratio test rejects $H_{0,A}$ whenever

$$\frac{SS_A}{SS_e} > C_1, \quad \text{where } SS_A \text{ and } SS_e \text{ are given in (36).} \qquad (37)$$

For the determination of the cutoff point $C_1$ in (37), we need the distribution of the statistic $SS_A/SS_e$ under $H_{0,A}$, where it is tacitly assumed that the observed values have been replaced by the respective r.v.'s. For this purpose, we establish the following result.

**Lemma 8.**   *Consider the expressions $SS_e$ and $SS_A$ defined in (36), and replace the observed values $y_{ij}, y_{i.}, y_{.j},$ and $y_{..}$ by the respective r.v.'s $Y_{ij}, Y_{i.}, Y_{.j},$ and $Y_{..}$, but retain the same notation. Then, under model (23):*

**(i)**  *The r.v. $SS_e/\sigma^2$ is distributed as $\chi^2_{(I-1)(J-1)}$.*
**(ii)**  *The statistics $SS_e$ and $SS_A$ are independent.*

*Furthermore, if the null hypothesis $H_{0,A}$ defined in (30) is true, then:*

**(i)**  *The r.v. $SS_A/\sigma^2$ is distributed as $\chi^2_{I-1}$.*
**(ii)**  *The statistic $\dfrac{SS_A/(I-1)}{SS_e/(I-1)(J-1)} \sim F_{I-1,(I-1)(J-1)}$.*

*Proof.* Deferred to Subsection 14.3.3. ∎

To this lemma, there is the following corollary, which also encompasses the estimates $\hat{\mu}, \hat{\alpha}_i$, and $\hat{\beta}_j$.

**Corollary.**
(i) *The MLE's $\hat{\mu} = Y_{..}, \hat{\alpha}_i = Y_{i.} - Y_{..}, i = 1, \ldots, I$, and $\hat{\beta}_j = Y_{.j} - Y_{..}, j = 1, \ldots, J$ are unbiased estimates of the respective parameters $\mu, \alpha_i$, and $\beta_j$.*
(ii) *The MLE $\widehat{\sigma^2} = SS_e/IJ$ of $\sigma^2$ given by (29) and (36) is biased, but the estimate $MS_e = SS_e/(I-1)(J-1)$ is unbiased.*

*Proof.*

(i) It is immediate from the definition of $Y_{i.}, Y_{.j}$, and $Y_{..}$ as (sample) means.
(ii) From the lemma, $\frac{SS_e}{\sigma^2} \sim \chi^2_{(I-1)(J-1)}$, so that

$$E\left(\frac{SS_e}{\sigma^2}\right) = (I-1)(J-1), \quad \text{or} \quad E\left[\frac{SS_e}{(I-1)(J-1)}\right] = \sigma^2,$$

which proves the unbiasedness asserted. Also,

$$E\widehat{\sigma^2} = E\left(\frac{SS_e}{IJ}\right) = \frac{(I-1)(J-1)}{IJ} E\left[\frac{SS_E}{(I-1)(J-1)}\right]$$
$$= \frac{(I-1)(J-1)}{IJ}\sigma^2,$$

which shows that $\widehat{\sigma^2}$ is biased. ∎

By means then of this lemma and relation (37), we reach the following conclusion: The hypothesis $H_{0,A}$ is rejected at level $\alpha$ whenever

$$\mathcal{F}_A = \frac{SS_A/(I-1)}{SS_e/(I-1)(J-1)} = \frac{MS_A}{MS_e} > F_{I-1,(I-1)(J-1);\alpha}. \tag{38}$$

Next, consider the hypothesis of no column effect; i.e.,

$$H_{0,B} : \beta_1 = \cdots = \beta_J = 0. \tag{39}$$

Then, working exactly as in (31) and (32), we obtain:

$$\hat{\mu}_B = y_{..} = \hat{\mu}, \quad \hat{\alpha}_{i,B} = y_{i.} - y_{..} = \hat{\alpha}_i, \quad i = 1, \ldots, I, \tag{40}$$

and

$$\widehat{\sigma^2_B} = \frac{1}{IJ}\sum_i\sum_j(y_{ij} - y_{i.})^2. \tag{41}$$

Thus, as in (35), the hypothesis $H_{0,B}$ is rejected whenever

$$\frac{\widehat{\sigma^2_B}}{\widehat{\sigma^2}} > C_0'. \tag{42}$$

Set

$$SS_B = I\sum_j\hat{\beta}_j^2 = I\sum_j(y_{.j} - y_{..})^2, \tag{43}$$

and consider the following result.

**Lemma 9.** With $SS_e, \widehat{\sigma_B^2}$, and $SS_B$ defined by (36), (41), and (43), it holds: $IJ\widehat{\sigma_B^2} = IJ\widehat{\sigma^2} + SS_B = SS_e + SS_B$.

*Proof.* Deferred to Subsection 14.3.3. ∎

By means of this lemma, relation (42) becomes, as in (37): Reject $H_{0,B}$ whenever

$$\frac{SS_B}{SS_e} > C_1', \quad \text{where } SS_e \text{ and } SS_B \text{ are given in (36) and (43).} \tag{44}$$

Finally, for the determination of the cutoff point $C_1'$ in (44), a certain distribution is needed. In other words, a lemma analogous to Lemma 8 is needed here.

**Lemma 10.** *Consider the expressions $SS_e$ and $SS_B$ defined in (36) and (43), and replace the observed values $y_{ij}, y_{i.}, y_{.j}$, and $y_{..}$ by the respective r.v.'s $Y_{ij}, Y_{i.}, Y_{.j}$, and $Y_{..}$, but retain the same notation. Then, under model (23):*

**(i)** *The r.v. $SS_e/\sigma^2$ is distributed as $\chi^2_{(I-1)(J-1)}$ (restatement of part (i) in Lemma 8).*

**(ii)** *The statistics $SS_e$ and $SS_B$ are independent.*
*Furthermore, if the null hypothesis $H_{0,B}$ defined in (39) is true, then:*

**(iii)** *The r.v. $SS_B/\sigma^2$ is distributed as $\chi^2_{J-1}$.*

**(iv)** *The statistic $\frac{SS_B/(J-1)}{SS_e/(I-1)(J-1)} \sim F_{J-1,(I-1)(J-1)}$.*

*Proof.* Deferred to Subsection 14.3.3. ∎

By means of this lemma and relation (44), we conclude that: The hypothesis $H_{0,B}$ is rejected at level $\alpha$ whenever

$$\mathcal{F}_B = \frac{SS_B/(J-1)}{SS_e/(I-1)(J-1)} = \frac{MS_B}{MS_e} > F_{J-1,(I-1)(J-1);\alpha}. \tag{45}$$

For computational purposes, we need the following result.

**Lemma 11.** *Let $SS_e, SS_A$, and $SS_B$ be given by (36) and (43), and let $SS_T$ be defined by:*

$$SS_T = \sum_i \sum_j (y_{ij} - y_{..})^2. \tag{46}$$

*Then:*

**(i)**

$$SS_A = J\sum_i y_{i.}^2 - IJy_{..}^2, \quad SS_B = I\sum_j y_{.j}^2 - IJy_{..}^2, \quad SS_T = \sum_i \sum_j y_{ij}^2 - IJy_{..}^2. \tag{47}$$

**(ii)**

$$SS_T = SS_e + SS_A + SS_B. \tag{48}$$

*Proof.* Deferred to Subsection 14.3.3. ■

Gathering together the hypotheses testing results obtained, we have the following theorem.

**Theorem 5.** *Under model (23), the hypotheses $H_{0,A}$ and $H_{0,B}$ are rejected at level of significance $\alpha$ whenever inequalities (38) and (45), respectively, hold true. The statistics $SS_A, SS_B$ are computed by means of (47), and the statistic $SS_e$ is computed by means of (47) and (48).*

As in Section 14.1, the various quantities employed in testing the hypotheses $H_{0,A}$ and $H_{0,B}$, and also for estimating the error variance $\sigma^2$, are gathered together in a table, an *ANOVA* table, as in Table 14.3.

**Remark 5.** In the present context, relation (48) is responsible for the term ANOVA. It states that the total variation (variance) $\sum_i \sum_j (Y_{ij} - Y_{..})^2$ (with reference to the grand sample mean $Y_{..}$) is split in three ways: One component $\sum_i \sum_j (Y_{i.} - Y_{..})^2$ *associated with the row effect* (*due to the row effect*, or *explained by the row effect*); one component $\sum_i \sum_j (Y_{.j} - Y_{..})^2$ *associated with the column effect* (*due to the column effect*, or *explained by the column effect*); and the *residual* component $\sum_i \sum_j (Y_{ij} - Y_{i.} - Y_{.j} + Y_{..})^2 = \sum_i \sum_j [(Y_{ij} - Y_{..}) - (Y_{i.} - Y_{..}) - (Y_{.j} - Y_{..})]^2$ (*unexplained by the row and column effects*, the *sum* of *squares of errors*).

Before embarking on the proof of the lemmas stated earlier in this section, let us illustrate the theory developed by a couple of examples. In the first example, we are presented with a set of numbers, not associated with any specific experiment; in the second example, a real-life experiment is considered.

**Example 6.** Apply the two-way layout ANOVA with one observation per cell for the data given in Table 14.4; take $\alpha = 0.05$.

**Table 14.3** Analysis of Variance for Two-Way Layout with One Observation per Cell

| Source of Variance | Sums of Squares | Degrees of Freedom | Mean Squares |
|---|---|---|---|
| Rows | $SS_A = J \sum_{i=1}^{I} \hat{\alpha}_i^2 = J \sum_{i=1}^{I} (Y_{i.} - Y_{..})^2$ | $I - 1$ | $MS_A = \frac{SS_A}{I-1}$ |
| Columns | $SS_B = I \sum_{j=1}^{J} \hat{\beta}_j^2 = I \sum_{j=1}^{J} (Y_{.j} - Y_{..})^2$ | $J - 1$ | $MS_B = \frac{SS_B}{J-1}$ |
| Residual | $SS_e = \sum_{i=1}^{I} \sum_{j=1}^{J} (Y_{ij} - Y_{i.} - Y_{.j} + Y_{..})^2$ | $(I-1) \times (J-1)$ | $MS_e = \frac{SS_e}{(I-1)(J-1)}$ |
| Total | $SS_T = \sum_{i=1}^{I} \sum_{j=1}^{J} (Y_{ij} - Y_{..})^2$ | $IJ - 1$ | — |

**Table 14.4** Data for a Two-Way Layout ANOVA

| | 1 | 2 | 3 | 4 | $y_i.$ |
|---|---|---|---|---|---|
| 1 | 3 | 7 | 5 | 4 | 19/4 |
| 2 | -1 | 2 | 0 | 2 | 3/4 |
| 3 | 1 | 2 | 4 | 0 | 7/4 |
| $y._j$ | 1 | 11/3 | 3 | 2 | $y.. = \frac{29}{12}$ |

Here: $\hat{\mu} = y.. = \frac{29}{12} \simeq 2.417$, and:

$$\hat{\alpha}_1 = y_1. - y.. = \frac{19}{4} - \frac{29}{12} = \frac{7}{3} \simeq 2.333$$
$$\hat{\alpha}_2 = y_2. - y.. = \frac{3}{4} - \frac{29}{12} = -\frac{5}{3} \simeq -1.667 \quad ;$$
$$\hat{\alpha}_3 = y_3. - y.. = \frac{7}{4} - \frac{29}{12} = -\frac{2}{3} \simeq -0.667$$

$$\hat{\beta}_1 = y._1 - y.. = 1 - \frac{29}{12} = -\frac{17}{12} \simeq -1.417$$
$$\hat{\beta}_2 = y._2 - y.. = \frac{11}{3} - \frac{29}{12} = \frac{5}{4} = 1.25$$
$$\hat{\beta}_3 = y._3 - y.. = 3 - \frac{29}{12} = \frac{7}{12} \simeq 0.583$$
$$\hat{\beta}_4 = y._4 - y.. = 2 - \frac{29}{12} = -\frac{5}{12} \simeq -0.417.$$

$$SS_A = 4 \times \left[ \left(\frac{19}{4}\right)^2 + \left(\frac{3}{4}\right)^2 + \left(\frac{7}{4}\right)^2 \right] - 12 \times \left(\frac{29}{12}\right)^2 = \frac{104}{3} \simeq 34.667,$$

$$SS_B = 3 \times \left[ 1^2 + \left(\frac{11}{3}\right)^2 + 3^2 + 2^2 \right] - 12 \times \left(\frac{29}{12}\right)^2 = \frac{147}{12} = 12.25,$$

$$SS_T = [3^2 + 7^2 + 5^2 + 4^2 + (-1)^2 + 2^2 + 0^2 + 2^2 + 1^2 + 2^2 + 4^2 + 0^2]$$
$$- 12 \times \left(\frac{29}{12}\right)^2 = \frac{707}{12} \simeq 58.917,$$

so that

$$SS_e = SS_T - SS_A - SS_B = \frac{707}{12} - \frac{104}{3} - \frac{147}{12} = 12.$$

Hence, the unbiased estimate of $\sigma^2$ is: $\frac{SS_e}{(I-1)(J-1)} = \frac{12}{6} = 2$, whereas its MLE is $\frac{12}{12} = 1$. Furthermore,

$$\mathcal{F}_A = \frac{MS_A}{MS_e} = \frac{104/3 \times 2}{12/6} = \frac{26}{3} \simeq 8.667,$$

$$\mathcal{F}_B = \frac{MS_B}{MS_e} = \frac{147/12 \times 3}{12/6} = \frac{147}{72} \simeq 2.042.$$

Since $F_{I-1,(I-1)(J-1);\alpha} = F_{2,6;0.05} = 5.1433$, $F_{3,6;0.05} = 4.7571$, we see that the hypothesis $H_{0,A}$ is rejected, whereas the hypothesis $H_{0,B}$ is not rejected.

**Example 7.** The cutting speeds of four types of tools are being compared by using five materials of varying degress of hardness. The data pertaining to measurements of cutting time in seconds are given in Table 14.5. Carry out the ANOVA for these data; take $\alpha = 0.05$.

**Table 14.5** Data for a Two-Way Layout ANOVA

|       | 1  | 2  | 3  | 4  | 5  | $y_{i.}$     |
|-------|----|----|----|----|----|--------------|
| 1     | 12 | 2  | 8  | 1  | 7  | 6            |
| 2     | 20 | 14 | 17 | 12 | 17 | 16           |
| 3     | 13 | 7  | 13 | 8  | 14 | 11           |
| 4     | 11 | 5  | 10 | 3  | 6  | 7            |
| $y_{.j}$ | 14 | 7  | 12 | 6  | 11 | $y_{..} = 10$ |

Here $I = 4, J = 5$. From the table: $\hat{\mu} = 10$, and:

$$
\begin{aligned}
\hat{\alpha}_1 &= y_{1.} - y_{..} = 6 - 10 = -4 & \hat{\beta}_1 &= y_{.1} - y_{..} = 14 - 10 = 4 \\
\hat{\alpha}_2 &= y_{2.} - y_{..} = 16 - 10 = 6 & \hat{\beta}_2 &= y_{.2} - y_{..} = 7 - 10 = -3 \\
\hat{\alpha}_3 &= y_{3.} - y_{..} = 11 - 10 = 1 \; ; & \hat{\beta}_3 &= y_{.3} - y_{..} = 12 - 10 = 2 \\
\hat{\alpha}_4 &= y_{4.} - y_{..} = 7 - 10 = -3 & \hat{\beta}_4 &= y_{.4} - y_{..} = 6 - 10 = -4 \\
& & \hat{\beta}_5 &= y_{.5} - y_{..} = 11 - 10 = 1.
\end{aligned}
$$

$$
\begin{aligned}
SS_A &= 5 \times (6^2 + 16^2 + 11^2 + 7^2) - 20 \times 10^2 = 310, \\
SS_B &= 4 \times (14^2 + 7^2 + 12^2 + 6^2 + 11^2) - 20 \times 10^2 = 184, \\
SS_T &= 2158 - 2000 = 518, \\
SS_e &= 518 - 310 - 184 = 24.
\end{aligned}
$$

Hence, the unbiased estimate for $\sigma^2$ is: $\frac{SS_e}{(I-1)(J-1)} = \frac{24}{3 \times 4} = 2$, whereas its MLE is $\frac{24}{4 \times 5} = 1.2$. Furthermore,

$$
\begin{aligned}
\mathcal{F}_A &= \frac{MS_A}{MS_e} = \frac{310/3}{24/12} = \frac{155}{3} \simeq 51.667, \\
\mathcal{F}_B &= \frac{MS_B}{MS_e} = \frac{184/4}{24/12} = 23.
\end{aligned}
$$

Since $F_{I-1,(I-1)(J-1);\alpha} = F_{3,12;.05} = 3.4903$, it follows that both hypotheses $H_{0,A}$ and $H_{0,B}$ are to be rejected. So, the mean cutting times, either for the tools across the material cut, or for the material cut across the tools used, cannot be assumed to be equal (at the $\alpha = 0.05$ level). Actually, this should not come as a surprise when looking at the margin of the table, which provide estimates of these times.

### 14.3.3 PROOF OF LEMMAS IN SECTION 14.3

In this subsection, a justification (or an outline thereof) is provided for the lemmas used in this section.

*Proof of Lemma 6.* Consider the expression $\mathcal{S}(\mu, \boldsymbol{\alpha}, \boldsymbol{\beta}) = \sum_i \sum_j (y_{ij} - \mu - \alpha_i - \beta_j)^2$ and recall that $\sum_i \alpha_i = 0, \sum_j \beta_j = 0$. Following the method of Lagrange multipliers, consider the linear combination

$$S^*(\mu, \boldsymbol{\alpha}, \boldsymbol{\beta}, \lambda_1, \lambda_2) = \sum_i \sum_j (y_{ij} - \mu - \alpha_i - \beta_j)^2 + \lambda_1 \sum_i \alpha_i + \lambda_2 \sum_j \beta_j,$$

where $\lambda_1, \lambda_2$ are constants, determine the partial derivatives of $S^*(\mu, \boldsymbol{\alpha}, \boldsymbol{\beta}, \lambda_1, \lambda_2)$ with respect to $\mu, \alpha_i$, and $\beta_j$, equate them to 0, append to them the side constraints $\sum_i \alpha_i = 0, \sum_j \beta_j = 0$, and solve the resulting system with respect to $\mu$, the $\alpha_i$'s, and the $\beta_j$'s (and also $\lambda_1, \lambda_2$). By implementing these steps, we get:

$$\left.\begin{array}{rcl}
\frac{\partial}{\partial \mu} S^*(\mu, \boldsymbol{\alpha}, \boldsymbol{\beta}, \lambda_1, \lambda_2) = -2 \sum_i \sum_j y_{ij} + 2IJ\mu + J \sum_i \alpha_i + 2I \sum_j \beta_j &=& 0 \\[2mm]
\frac{\partial}{\partial \alpha_i} S^*(\mu, \boldsymbol{\alpha}, \boldsymbol{\beta}, \lambda_1, \lambda_2) = -2 \sum_j y_{ij} + 2J\mu + 2J\alpha_i + 2 \sum_j \beta_j + \lambda_1 &=& 0 \\[2mm]
\frac{\partial}{\partial \beta_j} S^*(\mu, \boldsymbol{\alpha}, \boldsymbol{\beta}, \lambda_1, \lambda_2) = -2 \sum_j y_{ij} + 2I\mu + 2 \sum_i \alpha_i + 2I\beta_j + \lambda_2 &=& 0 \\[2mm]
\sum_i \alpha_i &=& 0 \\[2mm]
\sum_j \beta_j &=& 0
\end{array}\right\},$$

from which we obtain:

$$\mu = \frac{1}{IJ} \sum_i \sum_j y_{ij} = y_{..}, \quad \alpha_i = y_{i.} - y_{..} - \frac{\lambda_1}{2J}, \quad \beta_j = y_{.j} - y_{..} - \frac{\lambda_2}{2I}.$$

But

$$0 = \sum_i \alpha_i = \sum_i y_{i.} - Iy_{..} - \frac{I\lambda_1}{2J} = Iy_{..} - Iy_{..} - \frac{I\lambda_1}{2J} = -\frac{I\lambda_1}{2J}, \quad \text{so that } \lambda_1 = 0,$$

and likewise for $\lambda_2$ by summing up the $\beta_j$'s. Thus,

$$\mu = y_{..}, \quad \alpha_i = y_{i.} - y_{..}, \quad i = 1, \dots, I, \quad \beta_j = y_{.j} - y_{..}, \quad j = 1, \dots, J. \tag{49}$$

Now the parameter $\mu$ is any real number, the $\alpha_i$'s span an $(I-1)$-dimensional hyperplane, and the $\beta_j$'s span a $(J-1)$-dimensional hyperplane. It is then clear that the expression $S(\mu, \boldsymbol{\alpha}, \boldsymbol{\beta})$ (as a function of $\mu$, the $\alpha_i$'s, and the $\beta_j$'s) does not have a maximum. Then the values in (49) are candidates to produce a minimum of $S(\mu, \boldsymbol{\alpha}, \boldsymbol{\beta})$, in which case (26) follows. Again, geometrical considerations suggest that they do produce a minimum, and we will leave it at that presently. ∎

**Remark 6.**   It should be mentioned at this point that ANOVA models are special cases of the so-called *General Linear Models*, and then the above minimization problem is resolved in a general setting by means of linear algebra methodology. For a glimpse at it, one may consult Chapter 17 in the book *A Course in Mathematical Statistics*, 2nd edition, Academic Press (1997), by G. G. Roussas.

*Proof of Lemma 7.* Here, we have to establish the relation:

$$\sum_i \sum_j (y_{ij} - y_{.j})^2 = J \sum_i (y_{i.} - y_{..})^2 + \sum_i \sum_j (y_{ij} - y_{i.} - y_{.j} + y_{..})^2.$$

Indeed,

$$SS_e = \sum_i \sum_j (y_{ij} - y_{i.} - y_{.j} + y_{..})^2 = \sum_i \sum_j [(y_{ij} - y_{.j}) - (y_{i.} - y_{..})]^2$$

$$= \sum_i \sum_j (y_{ij} - y_{.j})^2 + J \sum_i (y_{i.} - y_{..})^2 - 2 \sum_i \sum_j (y_{i.} - y_{..})(y_{ij} - y_{.j})$$

$$= IJ\widehat{\sigma_A^2} + SS_A - 2SS_A = IJ\widehat{\sigma_A^2} - SS_A, \text{ because}$$

$$\sum_i \sum_j (y_{i.} - y_{..})(y_{ij} - y_{.j}) = \sum_i (y_{i.} - y_{..}) \sum_j (y_{ij} - y_{.j})$$

$$= \sum_i (y_{i.} - y_{..})(Jy_{i.} - Jy_{..}) = J \sum_i (y_{i.} - y_{..})^2 = SS_A. \qquad \blacksquare$$

*Proof of Lemma 8.* There are several ways one may attempt to justify the results in this lemma. One would be to refer to Lemma 2 and suggest that a similar approach be used, but that would do no justice. Another approach would be to utilize the theory of quadratic forms, but that would require an extensive introduction to the subject and the statement and/or proof of a substantial number of related results. Finally, the last approach would be to use a Geometric descriptive approach based on fundamental concepts of (finite dimensional) vector spaces. We have chosen to follow this last approach.

All vectors to be used here are column vectors, and the prime notation, "$\prime$", indicates transpose of a vector. Set $Y = (Y_{11}, \ldots, Y_{1J}; Y_{21}, \ldots, Y_{2J}; \ldots; Y_{I1}, \ldots, Y_{IJ})'$, so that $Y$ belongs in an $I \times J$-dimensional vector space to be denoted by $V_{I \times J}$. Also, set

$$\eta = EY = (EY_{11}, \ldots, EY_{1J}; EY_{21}, \ldots, EY_{2J}; \ldots; EY_{I1}, \ldots, EY_{IJ})'$$
$$= (\mu + \alpha_1 + \beta_1, \ldots, \mu + \alpha_1 + \beta_J; \mu + \alpha_2 + \beta_1, \ldots, \mu + \alpha_2 + \beta_J; \ldots;$$
$$\mu + \alpha_I + \beta_1, \ldots, \mu + \alpha_I + \beta_J)'.$$

Although the vector $\eta$ has $I \times J$ coordinates, due to its form and the fact that $\sum_i \alpha_i = \sum_j \beta_j = 0$, it follows that it lies in an $(I + J - 1)$-dimensional space, $V_{I+J-1}$ (see also Exercise 3.4). Finally, if $H_{0,A}: \alpha_1 = \cdots = \alpha_I = 0$ holds, then the respective mean vector, to be denoted by $\eta_A$, is

$$\eta_A = (\mu + \beta_1, \ldots, \mu + \beta_J; \mu + \beta_1, \ldots, \mu + \beta_J; \ldots; \mu + \beta_1, \ldots, \mu + \beta_J)',$$

and reasoning as above, we conclude that $\eta_A \in V_J$. Thus, we have three vector spaces related as follows: $V_J \subset V_{I+J-1} \subset V_{I \times J}$.

It is clear that, if $\mu, \alpha_i$'s, and $\beta_j$'s are replaced by their (least squares) estimates $\hat{\mu}, \hat{\alpha}_i$'s, and $\hat{\beta}_j$'s, the resulting random vector $\hat{\eta}$ still lies in $V_{I+J-1}$, and likewise for the random vector $\hat{\eta}_A$, which we get, if $\mu$ and $\beta_j$'s are replaced by $\hat{\mu}_A = \hat{\mu}$ and $\hat{\beta}_{j,A} = \hat{\beta}_j$; i.e., $\hat{\eta}_A \in V_J$. We now proceed as follows: Let $\alpha_1, \ldots, \alpha_{I+J-1}$ be an orthonormal basis in $V_J$ (i.e., $\alpha_i' \alpha_j = 0$ for $i \neq j$ and $\|\alpha_i\| = 1$), which we extend to an orthonormal basis $\alpha_1, \ldots, \alpha_{I-1}, \alpha_I, \ldots, \alpha_{I+J-1}$ in $V_{I+J-1}$, and then to an orthonormal basis

$$\alpha_1, \ldots, \alpha_{I-1}, \alpha_I, \ldots, \alpha_{I+J-1}, \alpha_{I+J}, \ldots, \alpha_{I \times J}$$

in $V_{I \times J}$. This can be done, as has already been mentioned in a similar context in the proof of Lemma 5 in Chapter 13. Also, see Remark 4 in the same chapter. Since $Y \in V_{I \times J}$, it follows that $Y$ is a linear combination of the $\alpha_i$'s with coefficient some r.v.'s $Z_i$'s. That is, $Y = \sum_{i=1}^{I \times J} Z_i \alpha_i$. Since $\hat{\eta}$ minimizes the quantity $\|Y - \eta\|^2 = \sum_i \sum_j (Y_{ij} - \mu - \alpha_i - \beta_j)^2$, it follows that $\hat{\eta}$ is, actually, the projection of $Y$ into the space $V_{I+J-1}$. It follows then that $\hat{\eta} = \sum_{i=1}^{I+J-1} Z_i \alpha_i$. Under $H_{0,A}$, the vector $\hat{\eta}_A$ minimizes $\|Y - \eta_A\|^2 = \sum_i \sum_j (Y_{ij} - \mu - \beta_j)^2$, and therefore is the projection of $Y$ into the space $V_J$. Thus, $\hat{\eta}_A = \sum_{i=I}^{I+J-1} Z_i \alpha_i$. Then $Y - \hat{\eta} = \sum_{i=I+J}^{I \times J} Z_i \alpha_i$, $Y - \hat{\eta}_A = \sum_{i=1}^{I-1} Z_i \alpha_i + \sum_{i=I+J}^{I \times J} Z_i \alpha_i$, and $\hat{\eta} - \hat{\eta}_A = \sum_{i=1}^{I-1} Z_i \alpha_i$. Because of the orthonormality of the $\alpha_i$'s, it follows that:

$$\|Y - \hat{\eta}\|^2 = \left\| \sum_{i=I+J}^{I \times J} Z_i \alpha_i \right\| = \sum_{i=I+J}^{I \times J} Z_i^2,$$

$$\|Y - \hat{\eta}_A\|^2 = \left\| \sum_{i=1}^{I-1} Z_i \alpha_i + \sum_{i=I+J}^{I \times J} Z_i \alpha_i \right\| = \sum_{i=1}^{I-1} Z_i^2 + \sum_{i=I+J}^{I \times J} Z_i^2,$$

and

$$\|\hat{\eta} - \hat{\eta}_A\|^2 = \left\| \sum_{i=1}^{I-1} Z_i \alpha_i \right\| = \sum_{i=1}^{I-1} Z_i^2.$$

However,

$$\|Y - \hat{\eta}\|^2 = \sum_i \sum_j (Y_{ij} - \hat{\mu} - \hat{\alpha}_i - \hat{\beta}_j)^2 = \sum_i \sum_j (Y_{ij} - Y_{i.} - Y_{.j} + Y_{..})^2 = SS_e,$$

and

$$\|\hat{\eta} - \hat{\eta}_A\|^2 = \sum_i \sum_j \hat{\alpha}_i^2 = \sum_i \sum_j (Y_{i.} - Y_{..})^2 = J \sum_i (Y_{i.} - Y_{..})^2 = SS_A.$$

Therefore

$$SS_A = \sum_{i=1}^{I-1} Z_i^2, \quad SS_e = \sum_{i=I+J}^{I \times J} Z_i^2. \tag{50}$$

Now, observe that the r.v.'s $Z_1, \ldots, Z_{I \times J}$ are the transformation of the r.v.'s $Y_1, \ldots, Y_{I \times J}$ under the orthogonal matrix $P$ whose rows are the vectors $\alpha_i'$'s. This follows immediately from the relation $Y = \sum_{i=1}^{I \times J} Z_i \alpha_i$, if we multiply (in the inner product sense) by $\alpha_j'$. We then get $\alpha_j' Y = \sum_{i=1}^{I \times J} Z_i (\alpha_j' \alpha_i)$, and this is $Z_i$, if $j = i$, and 0 otherwise. So, $Z_i = \alpha_i' Y$, $i = 1, \ldots, I \times J$. Since the $Y_i$'s are independent and Normally distributed with (common) variance $\sigma^2$, it follows that the $Z_i$'s are also independently Normally distributed with specified means and the same variance $\sigma^2$. (See Theorem 8 in Chapter 6.) From the fact that $Y - \hat{\eta}_A = \sum_{i=I+J}^{I \times J} Z_i \alpha_i$, we get $\alpha'_i(Y - \hat{\eta}_A) = Z_i, i = I + J, \ldots, I \times J$, so that $E[\alpha'_i(Y - \hat{\eta}_A)] = EZ_i$. That is, $Z_i, i = 1, \ldots, I \times I$ are independent Normal, $EZ_i = 0$ for the last $(I - 1)(J - 1)$ coordinates, and they all have variance $\sigma^2$. It follows that:

(i) $\frac{SS_e}{\sigma^2} = \frac{1}{\sigma^2} \sum_{i=I+J}^{I \times I} Z_i^2 \sim \chi^2_{(I-1)(J-1)}$.

(ii) The statistics $SS_e$ and $SS_A$ are independent, because they are defined in terms of nonoverlapping sets of the independent r.v.'s $Z_i$'s (see relation (50)).

(iii) The expectations of the coordinates of $\hat{\eta}$ are $\mu + \alpha_i + \beta_j$, and the expectations of the coordinates of $\hat{\eta}_A$ are $\mu + \beta_j$. It follows that the expectations of the coordinates of $\hat{\eta} - \hat{\eta}_A$ are $(\mu + \alpha_i + \beta_j) - (\mu + \beta_j) = \alpha_i$. Therefore, if $H_{0,A}$ is true, these expectations are 0, and then so are the expectations of $Z_i, i = 1, \ldots, I - 1$, since $\hat{\eta} - \hat{\eta}_A = \sum_{i=1}^{I-1} Z_i \alpha_i$. It follows that

$$\frac{SS_A}{\sigma^2} = \frac{1}{\sigma^2} \sum_{i=1}^{I-1} Z_i^2 \sim \chi^2_{I-1}.$$

(iv) Immediate from parts (i)–(iii) and the definition of the $F$ distribution.    ∎

*Proof of Lemma 9.*  We have to show that

$$\sum_i \sum_j (y_{ij} - y_{i.})^2 = I \sum_j (y_{.j} - y_{..})^2 + \sum_i \sum_j (y_{ij} - y_{i.} - y_{.j} + y_{..})^2.$$

∎

As in the proof of Lemma 7,

$$SS_e = \sum_i \sum_j (y_{ij} - y_{i.} - y_{.j} + y_{..})^2 = \sum_i \sum_j [(y_{ij} - y_{i.}) - (y_{.j} - y_{..})]^2$$

$$= \sum_i \sum_j (y_{ij} - y_{i.})^2 + \sum_i \sum_j (y_{.j} - y_{..})^2 - 2 \sum_i \sum_j (y_{.j} - y_{..})(y_{ij} - y_{i.})$$

$$= IJ\widehat{\sigma_B^2} + SS_B - 2SS_B = IJ\widehat{\sigma_B^2} - SS_B, \text{because}$$

$$\sum_i \sum_j (y_{.j} - y_{..})(y_{ij} - y_{..}) = \sum_j (y_{.j} - y_{..}) \sum_i (y_{ij} - y_{..})$$

$$= \sum_j (y_{.j} - y_{..})(Iy_{.j} - Iy_{..}) = I \sum_j (y_{.j} - y_{..})^2 = SS_B.$$

*Proof of Lemma 10.*

(i) It is the same as (i) in Lemma 8.

(ii) It is done as in Lemma 8(ii), where $H_{0,A}$ is replaced by $H_{0,B}$.

(iii) Again, it is a repetition of the arguments in Lemma 8(iii).

(iv) Immediate from parts (i)–(iii).    ∎

*Proof of Lemma 11.*

(i) They are all a direct application of the identity: $\sum_{k=1}^{n}(X_k - \bar{X})^2 = \sum_k X_k^2 - n\bar{X}^2$.

(ii) Clearly,

$$SS_e = \sum_i \sum_j (y_{ij} - y_{i.} - y_{.j} + y_{..})^2$$

$$= \sum_i \sum_j [(y_{ij} - y_{..}) - (y_{i.} - y_{..}) - (y_{.j} - y_{..})]^2$$

$$= \sum_i \sum_j (y_{ij} - y_{..})^2 + J \sum_i (y_{i.} - y_{..})^2 + I \sum_j (y_{.j} - y_{..})^2$$

$$- 2 \sum_i \sum_j (y_{i.} - y_{..})(y_{ij} - y_{..}) - 2 \sum_i \sum_j (y_{.j} - y_{..})(y_{ij} - y_{..})$$

$$+ 2 \sum_i \sum_j (y_{i.} - y_{..})(y_{.j} - y_{..}) = SS_T - SS_A - SS_B,$$

because

$$\sum_i \sum_j (y_{i.} - y_{..})(y_{ij} - y_{..}) = \sum_i (y_{i.} - y_{..})(Jy_{i.} - Jy_{..})$$

$$= J \sum_i (y_{i.} - y_{..})^2 = SS_A,$$

$$\sum_i \sum_j (y_{.j} - y_{..})(y_{ij} - y_{..}) = \sum_j (y_{.j} - y_{..})(Iy_{.j} - Iy_{..})$$

$$= I \sum_j (y_{.j} - y_{..})^2 = SS_B,$$

and

$$\sum_i \sum_j (y_{i.} - y_{..})(y_{.j} - y_{..}) = \sum_i (y_{i.} - y_{..})(Jy_{..} - Jy_{..}) = 0. \qquad \blacksquare$$

**Remark 7.** In a two-way layout of ANOVA, we may have $K(\geq 2)$ observations per cell rather than one. The concepts remain the same, but the analysis is somewhat more complicated. The reader may wish to refer to Section 17.3 in Chapter 17 of the book *A Course in Mathematical Statistics*, 2nd edition, Academic Press (1997), by G. G. Roussas. The more general cases, where there is an unequal number of observations per cell or there are more than two factors influencing the outcome, are the subject matter of the ANOVA branch of statistics and are, usually, not discussed in an introductory course.

## EXERCISES

**13.1** Apply the two-way layout (with one observation per cell) analysis of variance to the data given in the table below. Take $\alpha = 0.05$.

| Levels of Factor A | Levels of Factor B 1 | 2 | 3 | 4 | 5 |
|---|---|---|---|---|---|
| 1 | 110 | 128 | 48 | 123 | 19 |
| 2 | 214 | 183 | 115 | 114 | 129 |
| 3 | 208 | 183 | 130 | 225 | 114 |

**13.2** Under the null hypothesis $H_{0,A}$ stated in relation (30), show that the MLE's of $\mu$ and $\beta_j, j = 1, \ldots, J$ are, indeed, given by the expressions in relation (33).

**13.3** Under the null hypothesis $H_{0,A}$ stated in relation (30), show that the MLE of $\sigma^2$ is, indeed, given by the expression in relation (34).

**13.4** In reference to the proof of Lemma 8, show that $\boldsymbol{\eta}$ is an $(I + J - 1)$-dimensional vector.

**Hint.** This problem may be approached as follows:

$$
X' = \left\{
\begin{array}{c}
\overbrace{\begin{array}{cccc} 1 & 1 & 0 & 0 \\ 1 & 1 & 0 & 0 \\ . & . & . & . \\ 1 & 1 & 0 & 0 \end{array}}^{I} \cdot \cdot \cdot \overbrace{\begin{array}{cccccc} 0 & 1 & 0 & 0 & \cdot \cdot \cdot & 0 \\ 0 & 0 & 1 & 0 & \cdot \cdot \cdot & 0 \\ . & . & . & . & & . \\ 0 & 0 & 0 & 0 & \cdot \cdot & 0 & 1 \end{array}}^{J} \left.\right\} J \\[2em]
\begin{array}{cccc} 1 & 0 & 1 & 0 \\ 1 & 0 & 1 & 0 \\ . & . & . & . \\ 1 & 0 & 1 & 0 \end{array} \cdot \cdot \cdot \begin{array}{cccccc} 0 & 1 & 0 & 0 & \cdot \cdot \cdot & 0 \\ 0 & 0 & 1 & 0 & \cdot \cdot \cdot & 0 \\ . & . & . & . & & . \\ 0 & 0 & 0 & 0 & \cdot \cdot & 0 & 1 \end{array} \left.\right\} J \\[2em]
\cdot \cdot \cdot \\[2em]
\begin{array}{cccc} 1 & 0 & 0 & 0 \\ 1 & 0 & 0 & 0 \\ . & . & . & . \\ 1 & 0 & 0 & 0 \end{array} \cdot \cdot \begin{array}{cccccc} 0 & 1 & 1 & 0 & 0 & \cdot \cdot \cdot & 0 \\ 0 & 1 & 0 & 1 & 0 & \cdot \cdot \cdot & 0 \\ . & . & . & . & . & & . \\ 0 & 1 & 0 & 0 & 0 & \cdot \cdot & 0 & 1 \end{array} \left.\right\} J
\end{array}
\right.
$$

Consider the $IJ \times (I + J + 1)$ matrix $X'$ given above and let the $1 \times (I + J + 1)$ vector $\boldsymbol{\beta}'$ be defined by: $\boldsymbol{\beta}' = (\mu, \alpha_1, \ldots, \alpha_I, \beta_1, \ldots, \beta_J)$. Then do the following:

  **(i)** Observe that $\boldsymbol{\eta} = X'\boldsymbol{\beta}$, so that $\boldsymbol{\eta}$ lies in the vector space generated by the columns of $X'$.
  **(ii)** For $I \geq 2$ and $J \geq \frac{I+1}{I-1}$, observe that rank $X' \leq I + J + 1 = \min\{I + J + 1, IJ\}$.

(iii) Show that rank $X' = I + J - 1$ by showing that: (a) The 1st column of $X'$; is the sum of the subsequent $I$ columns of $X'$; (b) The 2nd column is the ((sum of the last $J$ columns)-(sum of the last $I - 1$ columns in the block of $I$ columns)); (c) The $I + J - 1$ columns, except for the first two, are linearly independent (by showing that any linear combination of them by scalars is the zero vector if and only if all scalars are zero). It will then follow that the dimension of $\eta$ is $I + J - 1$.

**13.5** In reference to the proof of Lemma 8, and under the hypothesis $H_{0,A}$, show that the dimension of the vector $\eta_A$ is $J$.

**Hint.** As in Exercise 3.4, one may use similar steps in showing that $\eta_A$ belongs in a $J$-dimensional vector space and thus is of dimension $J$. To this end, consider the $IJ \times (J + 1)$ matrix $X'_A$ given below, and let $\beta'_A = (\mu, \beta_1, \ldots, \beta_J)$. Then do the following:

$$
X' = \left\{
\begin{array}{c}
\overbrace{\hphantom{xxxxxxxxxxxx}}^{J} \\
\left.
\begin{array}{cccccccc}
1 & 1 & 0 & 0 & \cdot & \cdot & \cdot & 0 \\
1 & 0 & 1 & 0 & \cdot & \cdot & \cdot & 0 \\
\cdot & \cdot & \cdot & \cdot & & & & \cdot \\
1 & 0 & 0 & 0 & \cdot & \cdot & 0 & 1 \\
\end{array}
\right\} J \\
\left.
\begin{array}{cccccccc}
1 & 1 & 0 & 0 & \cdot & \cdot & \cdot & 0 \\
1 & 0 & 1 & 0 & \cdot & \cdot & \cdot & 0 \\
\cdot & \cdot & & & & & & \cdot \\
1 & 0 & 0 & 0 & \cdot & \cdot & 0 & 1 \\
\end{array}
\right\} J \\
\begin{array}{cccccccc}
\cdot & \cdot & \cdot & \cdot & \cdot & \cdot & \cdot & \cdot \\
\cdot & \cdot & \cdot & \cdot & \cdot & \cdot & \cdot & \cdot \\
\end{array} \\
\left.
\begin{array}{cccccccc}
1 & 1 & 0 & 0 & \cdot & \cdot & \cdot & 0 \\
1 & 0 & 1 & 0 & \cdot & \cdot & \cdot & 0 \\
\cdot & \cdot & \cdot & \cdot & & & & \cdot \\
1 & 0 & 0 & 0 & \cdot & \cdot & 0 & 1 \\
\end{array}
\right\} J
\end{array}
\right.
$$

(i) Observe that $\eta_A = X'_A \beta_A$, so that $\eta_A$ lies in the vector space generated by the columns (rows) of $X'_A$.

(ii) For $I \geq 2$, it is always true that $J + 1 \leq IJ$, and therefore rank $X'_A \leq J + 1 = \min\{J + 1, IJ\}$.

(iii) Show that rank $X'_A = J$ by showing that: (a) The 1st column of $X'_A$ is the sum of the subsequent $J$ columns; (b) The $J$ columns, except for the 1st one, are linearly independent. It will then follow that the dimension of $\eta_A$ is $J$.

# Some topics in nonparametric inference

# 15

In Chapters 9, 10, 11, and 12, we concerned ourselves with the question of point estimation, interval estimation, and testing hypotheses about (most of the time) a real-valued parameter $\theta$. This inference was hedged on the basic premise that we were able to stipulate each time a probability model, which was completely known except for a parameter $\theta$ (real-valued or of higher dimension).

The natural question which arises is this: What do we do, if there is no sound basis for the stipulation of a probability model from which the observations are drawn? In such a situation, we don't have parametric inference problems to worry about, because, simply, we don't have a parametric model. In certain situations things may not be as bad as this, but they are nearly so. Namely, we are in a position to assume the existence of a parametric model which governs the observations. However, the number of parameters required to render the model meaningful is exceedingly large, and therefore inference about them is practically precluded.

It is in situations like this, where the so-called nonparametric models and nonparametric inference enter the picture. Accordingly, a nonparametric approach starts out with a bare minimum of assumptions, which certainly do not include the existence of a parametric model, and proceeds to derive inference for a multitude of important quantities. This chapter is devoted to discussing a handful of problems of this variety.

Specifically, in the first section confidence intervals are constructed for the mean $\mu$ of a distribution, and also the value at $x$ of the d.f. $F$, $F(x)$. The confidence coefficients are approximately $1 - \alpha$ for large $n$. Illustrative examples are also provided. In the following section, confidence intervals are constructed for the quantiles of a d.f. $F$. Here the concept of a confidence coefficient is replaced by that of the coverage probability. In the subsequent two sections, two populations are compared by means of the sign test, when the sample sizes are equal, and the rank sum test and the Wilcoxon–Mann–Whitney test in the general case. Some examples are also discussed. The last section consists of two subsections. One is devoted to estimating (nonparametrically) a p.d.f. and the formulation of a number of desirable properties of the proposed estimate. The other subsection addresses very briefly two very important problems; namely, the problem of regression estimation under a fixed design, and the problem of prediction when the design is stochastic.

## 15.1 SOME CONFIDENCE INTERVALS WITH GIVEN APPROXIMATE CONFIDENCE COEFFICIENT

We are in a position to construct a confidence interval for the (unknown) mean $\mu$ of $n$ i.i.d. observations $X_1, \ldots, X_n$ with very little information as to where these observations are coming from. Specifically, all we have to know is that these r.v.'s have finite mean $\mu$ and variance $\sigma^2 \in (0, \infty)$, and nothing else. Then, by the CLT,

$$\frac{\sqrt{n}(\bar{X}_n - \mu)}{\sigma} \xrightarrow[n \to \infty]{d} Z \sim N(0, 1), \quad \bar{X}_n = \frac{1}{n}\sum_{i=1}^{n} X_i. \tag{1}$$

Suppose first that $\sigma$ is known. Then, for all sufficiently large $n$, the Normal approximation in (1) yields:

$$P\left[-z_{\frac{\alpha}{2}} \leq \frac{\sqrt{n}(\bar{X}_n - \mu)}{\sigma} \leq z_{\frac{\alpha}{2}}\right] \simeq 1 - \alpha,$$

or

$$P\left(\bar{X}_n - z_{\frac{\alpha}{2}}\frac{\sigma}{\sqrt{n}} \leq \mu \leq \bar{X}_n + z_{\frac{\alpha}{2}}\frac{\sigma}{\sqrt{n}}\right) \simeq 1 - \alpha.$$

In other words,

$$\left[\bar{X}_n - z_{\frac{\alpha}{2}}\frac{\sigma}{\sqrt{n}}, \quad \bar{X}_n + z_{\frac{\alpha}{2}}\frac{\sigma}{\sqrt{n}}\right] \tag{2}$$

is a confidence interval for $\mu$ with confidence coefficient approximately $1 - \alpha$ ($0 < \alpha < 1$).

Now, if $\mu$ is unknown, it is quite likely that $\sigma$ is also unknown. What we do then is to estimate $\sigma^2$ by

$$S_n^2 = \frac{1}{n}\sum_{i=1}^{n}(X_i - \bar{X}_n)^2 = \frac{1}{n}\sum_{i=1}^{n}X_i^2 - \bar{X}_n^2, \tag{3}$$

and recall that (by Theorem 7(i) in Chapter 7):

$$S_n^2 \xrightarrow[n \to \infty]{P} \sigma^2, \quad \text{or} \quad \frac{S_n^2}{\sigma^2} \xrightarrow[n \to \infty]{P} 1. \tag{4}$$

Then convergences (1) and (4), along with Theorem 6(iii) in Chapter 7, yield:

$$\frac{\sqrt{n}(\bar{X}_n - \mu)/\sigma}{S_n/\sigma} = \frac{\sqrt{n}(\bar{X}_n - \mu)}{S_n} \xrightarrow[n \to \infty]{d} Z \sim N(0, 1).$$

Then, proceeding as before, we obtain that

$$\left[\bar{X}_n - z_{\frac{\alpha}{2}}\frac{S_n}{\sqrt{n}}, \quad \bar{X}_n + z_{\frac{\alpha}{2}}\frac{S_n}{\sqrt{n}}\right] \tag{5}$$

is a confidence interval for $\mu$ with confidence coefficient approximately $1 - \alpha$.

Here is an application of formula (5).

**Example 1.** Refer to the GPA's in Example 22 of Chapter 1, where we assume that the given GPA scores are observed values of r.v.'s $X_i, i = 1, \ldots, 34$ with (unknown) mean $\mu$ and (unknown) variance $\sigma^2$, both finite. Construct a confidence interval for $\mu$ with confidence coefficient approximately 95%.

**Discussion.**   In the discussion of Example 1 in Chapter 13, we saw that: $\sum_i x_i = 100.73$ and $\sum_i x_i^2 = 304.7885$, so that:

$$\bar{x} = \frac{100.73}{34} \simeq 2.963, \quad s_n^2 = \frac{304.7885}{34} - \frac{(100.73)^2}{34^2} \simeq 0.187, \quad \text{and} \quad s_n \simeq 0.432.$$

Since $z_{0.025} = 1.96$, formula (5) gives:

$$\left[ 2.9626 - 1.96 \times \frac{0.432}{5.831}, \; 2.9626 + 1.96 \times \frac{0.432}{5.831} \right] \simeq [2.818, 3.108].$$

Another instance where a nonparametric approach provides a confidence interval is the following. The i.i.d. r.v.'s $X_1, \ldots, X_n$ have (unknown) d.f. $F$, and let $F_n(x)$ be the empirical d.f. based on the $X_i$'s, as defined in Application 5 to the WLLN in Chapter 7. We saw there that

$$F_n(x) = \frac{1}{n} \sum_{i=1}^n Y_i(x), \quad Y_1(x), \ldots, Y_n(x) \text{ independent r.v.'s} \sim B(1, F(x)).$$

Then, by the CLT,

$$\frac{\sqrt{n}[F_n(x) - F(x)]}{\sqrt{F(x)[1 - F(x)]}} \xrightarrow[n \to \infty]{d} Z \sim N(0, 1). \tag{6}$$

Also,

$$F_n(x) \xrightarrow[n \to \infty]{P} F(x), \quad \text{or} \quad \frac{F_n(x)[1 - F_n(x)]}{F(x)[1 - F(x)]} \xrightarrow[n \to \infty]{P} 1. \tag{7}$$

From (6) and (7) and Theorem 6(iii) in Chapter 7, it follows that:

$$\frac{\sqrt{n}[F_n(x) - F(x)]/\sqrt{F(x)[1 - F(x)]}}{\sqrt{F_n(x)[1 - F_n(x)]}/\sqrt{F(x)[1 - F(x)]}} = \frac{\sqrt{n}[F_n(x) - F(x)]}{\sqrt{F_n(x)[1 - F_n(x)]}} \xrightarrow[n \to \infty]{d} Z \sim N(0, 1).$$

It follows that, for all sufficiently large $n$ (depending on $x$), the following interval is a confidence interval for $F(x)$ with confidence coefficient approximately $1 - \alpha$; namely,

$$\left[ F_n(x) - z_{\frac{\alpha}{2}} \sqrt{\frac{F_n(x)[1 - F_n(x)]}{n}}, \; F_n(x) + z_{\frac{\alpha}{2}} \sqrt{\frac{F_n(x)[1 - F_n(x)]}{n}} \right]. \tag{8}$$

As an application of formula (8), consider the following numerical example.

**Example 2.**   Refer again to Example 22 in Chapter 1 (see also Example 1 in this chapter), and construct a confidence interval for $F(3)$ with approximately 95% confidence coefficient, where $F$ is the d.f. of the r.v.'s describing the GPA scores.

**Discussion.**    In this example, $n = 34$ and the number of the observations which are $\leq 3$ are 18 (the following; 2.36, 2.36, 2.66, 2.68, 2.48, 2.46, 2.63, 2.44, 2.13, 2.41, 2.55, 2.80, 2.79, 2.89, 2.91, 2.75, 2.73, and 3.00). Then:

$$F_{34}(3) = \frac{18}{34} = \frac{9}{17} \simeq 0.529, \quad \sqrt{\frac{F_{34}(3)[1 - F_{34}(3)]}{34}} \simeq 0.086,$$

and therefore the observed confidence interval is:

$$[0.529 - 1.96 \times 0.086, 0.529 + 1.96 \times 0.086] \simeq [0.360, 0.698].$$

**Remark 1.**    It should be pointed out that the confidence interval given by (8) is of limited usefulness, because the value of (the large enough) $n$ for which (8) holds depends on $x$.

## 15.2 CONFIDENCE INTERVALS FOR QUANTILES OF A DISTRIBUTION FUNCTION

In the previous section, we constructed a confidence interval for the mean $\mu$ of a distribution, whether its variance is known or not, with confidence coefficient approximately a prescribed number $1 - \alpha$ $(0 < \alpha < 1)$. Also, such an interval was constructed for each value $F(x)$ of a d.f. $F$. Now, we have seen (in Section 3.4 of Chapter 3) that the median, and, more generally, the quantiles of a d.f. $F$ are important quantities through which we gain information about $F$. It would then be worth investigating the possibility of constructing confidence intervals for quantiles of $F$. To simplify matters, it will be assumed that $F$ is continuous, and that for each $p \in (0, 1)$, there is a unique $p$th quantile $x_p$; i.e., $F(x_p) = P(X \leq x_p) = p$. The objective is to construct a confidence interval for $x_p$, and, in particular, for the median $x_{0.50}$. This is done below in a rather neat manner, except that we don't have much control on the confidence coefficient involved. Specifically, the following result is established.

**Theorem 1.**    *Let $X_1, \ldots, X_n$ be i.i.d. r.v.'s with continuous d.f. $F$, and let $Y_1, \ldots, Y_n$ be the order statistics of the $X_i$'s. For $p \in (0, 1)$, let $x_p$ be the unique (by assumption) $p$th quantile of $F$. Then, for any $1 \leq i < j \leq n$, the random interval $[Y_i, Y_j]$ is a confidence interval for $x_p$ with confidence coefficient $\sum_{k=i}^{j-1} \binom{n}{k} p^k (1 - p)^{n-k}$.*

*Proof.* Define the r.v.'s $W_j, j = 1, \ldots, n$ as follows:

$$W_j = \begin{cases} 1 & \text{if} \quad X_j \leq x_p \\ 0 & \text{if} \quad X_j > x_p, \end{cases} \quad j = 1, \ldots, n.$$

Then the r.v.'s $W_1, \ldots, W_n$ are independent and distributed as $B(1, p)$, since $P(W_j = 1) = P(X_j \leq x_p) = F(x_p) = p$, and $W = \sum_{j=1}^{n} W_j \sim B(n, p)$. Therefore,

$$P(\text{at least } i \text{ of } X_1, \ldots, X_n \text{ are } \leq x_p) = \sum_{k=i}^{n} \binom{n}{k} p^k (1-p)^{n-k}.$$

However, $P(\text{at least } i \text{ of } X_1, \ldots, X_n \text{ are } \leq x_p) = P(Y_i \leq x_p)$. Thus,

$$P(Y_i \leq x_p) = \sum_{k=i}^{n} \binom{n}{k} p^k (1-p)^{n-k}. \tag{9}$$

Next, for $1 \leq i < j \leq n$, we, clearly, have;

$$P(Y_i \leq x_p) = P(Y_i \leq x_p, Y_j \geq x_p) + P(Y_i \leq x_p, Y_j < x_p)$$
$$= P(Y_i \leq x_p \leq Y_j) + P(Y_j < x_p)$$
$$= P(Y_i \leq x_p \leq Y_j) + P(Y_j \leq x_p), \tag{10}$$

since $P(Y_i \leq x_p, Y_j < x_p) = P(Y_i \leq x_p, Y_j \leq x_p) = P(Y_j \leq x_p)$ by the fact that $(Y_j \leq x_p) \subseteq (Y_i \leq x_p)$. Then, relations (9) and (10) yield:

$$P(Y_i \leq x_p \leq Y_j) = \sum_{k=i}^{n} \binom{n}{k} p^k (1-p)^{n-k} - P(Y_j \leq x_p)$$

$$= \sum_{k=i}^{n} \binom{n}{k} p^k (1-p)^{n-k} - \sum_{k=j}^{n} \binom{n}{k} p^k (1-p)^{n-k}$$

$$= \sum_{k=i}^{j-1} \binom{n}{k} p^k (1-p)^{n-k}$$

$$= \sum_{k=0}^{j-1} \binom{n}{k} p^k (1-p)^{n-k} - \sum_{k=0}^{i-1} \binom{n}{k} p^k (1-p)^{n-k}. \tag{11}$$

So, the random interval $[Y_i, Y_j]$ contains the point $x_p$ with probability $\sum_{k=i}^{j-1} \binom{n}{k} \times p^k (1-p)^{n-k}$, as was to be seen. ∎

**Remark 2.**

(i) From relation (11), it is clear that, although $p$ is fixed, we can enlarge the confidence coefficient $\sum_{k=i}^{j-1} \binom{n}{k} p^k (1-p)^{n-k}$ by taking a smaller $i$ and/or a larger $j$. The price we pay, however, is that of having a larger confidence interval.

(ii) By the fact that the confidence interval $[Y_i, Y_j]$ does not have a prescribed confidence coefficient $1 - \alpha$, as is the case in the usual construction of confidence intervals, we often refer to the probability $\sum_{k=i}^{j-1} \binom{n}{k} p^k (1-p)^{n-k}$ as the *probability of coverage* of $x_p$ by $[Y_i, Y_j]$.

**Example 3.**    Consider the i.i.d. r.v.'s $X_1, \ldots, X_{20}$ with continuous d.f. $F$, which has unique $x_{0.50}, x_{0.25}$, and $x_{0.75}$, and let $Y_1, \ldots, Y_{20}$ be the corresponding order statistics. Then consider several confidence intervals for $x_{0.50}, x_{0.25}$, and $x_{0.75}$, and calculate the respective coverage probabilities.

**Table 15.1** Confidence intervals for certain quartiles

| Quantile | Confidence Interval | Coverage Probability |
|---|---|---|
| $x_{0.50}$ | $(Y_9, Y_{12})$ | 0.4966 |
| | $(Y_8, Y_{13})$ | 0.7368 |
| | $(Y_7, Y_{14})$ | 0.8846 |
| | $(Y_6, Y_{15})$ | 0.9586 |
| $x_{0.25}$ | $(Y_6, Y_6)$ | 0 |
| | $(Y_3, Y_7)$ | 0.6945 |
| | $(Y_2, Y_8)$ | 0.8739 |
| | $(Y_1, Y_9)$ | 0.9559 |
| $x_{0.75}$ | $(Y_{15}, Y_{17})$ | 0.3920 |
| | $(Y_{14}, Y_{18})$ | 0.6945 |
| | $(Y_{13}, Y_{19})$ | 0.8739 |
| | $(Y_{12}, Y_{20})$ | 0.9559 |

**Discussion.** Using formula (11), we obtain the coverage probabilities listed in Table 15.1 for several confidence intervals for the median $x_{0.50}$ and the first quartile $x_{0.25}$. For the calculation of coverage probabilities for confidence intervals for the third quartile $x_{0.75}$, we employ the following formula, which allows us to use the Binomial tables; namely,

$$\sum_{k=i}^{j-1} \binom{20}{k}(0.75)^k (0.25)^{20-k} = \sum_{r=20-j+1}^{20-i} \binom{20}{r}(0.25)^r (0.75)^{20-r}$$

$$= \sum_{r=0}^{20-i} \binom{20}{r}(0.25)^r (0.75)^{20-r} - \sum_{r=0}^{20-j} \binom{20}{r}(0.25)^r (0.75)^{20-r}.$$

## 15.3 THE TWO-SAMPLE SIGN TEST

In this brief section, we discuss a technique of comparing two populations by means of the so-called *sign test*. The test requires that the two samples available are of the same size, and makes no direct use of the values observed; instead, what is really used is the relative size of the components in the pairs of r.v.'s. Some cases where such a test would be appropriate include those in which one is interested in comparing the effectiveness of two different drugs used for the treatment of the same disease, the efficiency of two manufacturing processes producing the same item, the response of $n$ customers regarding their preferences toward a certain consumer item, etc.

In more precise terms, let $X_1, \ldots, X_n$ be i.i.d. r.v.'s with *continuous* d.f. $F$, and let $Y_1, \ldots, Y_n$ be i.i.d. r.v.'s with *continuous* d.f. $G$; it is assumed that the two-dimensional random vectors $(X_i, Y_i)$, $i = 1, \ldots, n$ are independent with the continuous (joint) d.f. $H$. On the basis of the $X_i$'s and $Y_j$'s, we wish to test the null hypothesis $H_0$: $F = G$ against any one of the alternatives $H_A$: $F > G, H_A'$: $F < G, H_A''$: $F \neq G$. The inequality $F > G$ means that $F(z) \geq G(z)$ for all $z$, and $F(z) > G(z)$ for "sufficiently many $z$'s"; likewise for $F < G$. (See Remark 3 below.) Finally, $F \neq G$ means that either $F > G$ or $F < G$ (excluding the possibility that $F(z) > G(z)$ for some $z$ and $F(z) < G(z)$ for some other $z$).

**Remark 3.**    The rigorous interpretation of "sufficiently many $z$'s" is that the set of such $z$'s has positive (Lebesgue) measure. For example, the strict inequalities occur for sets of $z$'s forming on nondegenerate intervals in the real line.

To this end, set

$$Z_i = \begin{cases} 1 & \text{if } X_i < Y_i \\ 0 & \text{if } X_i \geq Y_i \end{cases}, \quad p = P(X_i < Y_i), \ i = 1, \ldots, n, \quad Z = \sum_{i=1}^{n} Z_i. \tag{12}$$

It is clear that the r.v.'s $Z_1, \ldots, Z_n$ are independent and distributed as $B(1, p)$, so that the r.v. $Z$ is distributed as $B(n, p)$. Under the hypothesis $H_0$, we have, $p = \frac{1}{2}$, whereas under $H_A, H_A'$, and $H_A''$, we have, respectively, $p > \frac{1}{2}$, $p < \frac{1}{2}$, $p \neq \frac{1}{2}$. (See also Exercise 3.2) Thus, the problem of testing $H_0$ becomes, equivalently, that of testing $\bar{H}_0$: $p = \frac{1}{2}$ in the $B(n, p)$ distribution. Formulating the relevant results, and drawing upon Application 1 in Section 11.3 of Chapter 11, we have the following theorem; the test employed here is also referred to as the *sign test*.

**Theorem 2.**    *Consider the i.i.d. r.v.'s $X_1, \ldots, X_n$ and $Y_1, \ldots, Y_n$ with respective continuous d.f.'s $F$ and $G$, and assume that the pair $(X_i, Y_i)$, $i = 1, \ldots, n$ are independent with continuous (joint) d.f. $H$. Then, for testing the null hypothesis $H_0$: $F = G$ against any one of the alternatives $H_A$: $F > G$, or $H_A'$: $F < G$, or $H_A''$: $F \neq G$, at level of significance $\alpha$, the hypothesis $H_0$ is rejected, respectively, whenever*

$$Z > C \quad \text{or} \quad Z < C' \quad \text{or} \quad Z < C_1 \quad \text{or} \quad Z > C_2. \tag{13}$$

*The cutoff points $C, C'$, and $C_1, C_2$ are determined by the relations:*

$$\left. \begin{aligned} &P(Z > C) + \gamma P(Z = C) = \alpha, \ \text{or} \ P(Z \leq C) - \gamma P(Z = C) = 1 - \alpha, \\ &P(Z \leq C' - 1) + \gamma' P(Z = C') = \alpha, \\ &P(Z \leq C_1 - 1) + \gamma_1 P(Z = C_1) = \tfrac{\alpha}{2} \ \text{and} \ P(Z > C_2) + \gamma_2 P(Z = C_2) = \tfrac{\alpha}{2}, \\ &\text{or} \\ &P(Z \leq C_1 - 1) + \gamma_1 P(Z = C_1) = \tfrac{\alpha}{2} \ \text{and} \ P(Z \leq C_2) - \gamma_2 P(Z = C_2) = 1 - \tfrac{\alpha}{2}, \end{aligned} \right\} \tag{14}$$

*and $Z \sim B(n, 1/2)$ under $H_0$.*

*For large values of n, the CLT applies and the cutoff points are given by the relations:*

$$C \simeq \frac{n}{2} + z_\alpha \frac{\sqrt{n}}{2}, \quad C' \simeq \frac{n}{2} - z_\alpha \frac{\sqrt{n}}{2}, \\ C_1 \simeq \frac{n}{2} - z_{\frac{\alpha}{2}} \frac{\sqrt{n}}{2}, \quad C_2 \simeq \frac{n}{2} + z_{\frac{\alpha}{2}} \frac{\sqrt{n}}{2}. \tag{15}$$

**Example 4.** Refer to Example 25 in Chapter 1 regarding the plant height (in $1/8$ inches) of cross-fertilized and self-fertilized plants. Denote by $X_i$'s and $Y_i$'s, respectively, the heights of cross-fertilized and self-fertilized plants. Then the observed values for the 15 pairs are given in Example 25 of Chapter 1, which are reproduced in the Table 15.2 for convenience. At the level of significance $\alpha = 0.05$, test the null hypothesis $H_0: F = G$, where $F$ and $G$ are the d.f.'s of the $X_i$'s and $Y_i$'s, respectively, against the alternatives $H_A'' : F \neq G$ and $H_A' : F < G$.

**Discussion.**   From Table 15.2, we have:

$$Z_1 = 0, \ Z_2 = 1, \ Z_3 = 0, \ Z_4 = 0, \ Z_5 = 0, \ Z_6 = 0, \ Z_7 = 0, \ Z_8 = 0,$$
$$Z_9 = 0, \ Z_{10} = 0, \ Z_{11} = 0, \ Z_{12} = 0, \ Z_{13} = 0, \ Z_{14} = 0, \ Z_{15} = 1,$$

so that $Z = 2$. Suppose first that the alternative is $H_A'': F \neq G$ (equivalently, $p \neq \frac{1}{2}$). Then (by relations (13) and (14)) $H_0$ is rejected in favor of $H_A''$ whenever $Z < C_1$ or $Z > C_2$, where: $P(Z \leq C_1 - 1) + \gamma_1 P(Z = C_1) = 0.025$ and $P(Z \leq C_2) - \gamma_2 P(Z = C_2) = 0.975$, and $Z \sim B(15, 1/2)$.

From the Binomial tables, we find $C_1 = 4, C_2 = 11$, and $\gamma_1 = \gamma_2 \simeq 0.178$. Since $Z = 2 < C_1(= 4)$, the null hypothesis is rejected. Next, test $H_0$ against the alternative $H_A': F < G$ (equivalently, $p < \frac{1}{2}$) again at level $\alpha = 0.05$. Then (by relations (13) and (14)) $H_0$ is rejected in favor of $H_A'$ whenever $Z \leq C' - 1$, where $C'$ is determined by:

$$P(Z \leq C' - 1) + \gamma' P(Z = C') = 0.05, \quad Z \sim B(15, \ 1/2).$$

**Table 15.2** Heights of cross-fertilized and self-fertilized plants

| Pair | X Cross- | Y Self- | Pair | X Cross- | Y Self- |
|------|----------|---------|------|----------|---------|
| 1 | 188 | 139 | 9 | 146 | 132 |
| 2 | 96 | 163 | 10 | 173 | 144 |
| 3 | 168 | 160 | 11 | 186 | 130 |
| 4 | 176 | 160 | 12 | 168 | 144 |
| 5 | 153 | 147 | 13 | 177 | 102 |
| 6 | 172 | 149 | 14 | 184 | 124 |
| 7 | 177 | 149 | 15 | 96 | 144 |
| 8 | 163 | 122 | | | |

From the Binomial tables, we find $C' = 4$ and $\gamma' \simeq 0.779$. Since $Z = 2 < C'$ $(= 4)$, $H_0$ is rejected in favor of $H'_A$.

For the Normal approximation, we get from (15): $z_{0.025} = 1.96$, so that $C_1 \simeq 3.704$, $C_2 \simeq 11.296$, and $H_0$ is rejected again, since $Z = 2 < C_1(\simeq 3.704)$. Also, $z_{0.05} = 1.645$, and hence $C' \simeq 4.314$. Again, $H_0$ is rejected in favor of $H'_A$, since $Z = 2 < C'(\simeq 4.314)$.

**Example 5.** Ten patients suffering from a certain disease were treated by a medication with a possible undesirable side effect. For the detection of this side effect, measurements of a blood substance are taken before the medication is taken and also a specified time after the medication is administered. The resulting measurements are listed below.

| Patient | 1 | 2 | 3 | 4 | 5 | 6 | 7 | 8 | 9 | 10 |
|---|---|---|---|---|---|---|---|---|---|---|
| Before medication | 5.04 | 5.16 | 4.75 | 5.25 | 4.80 | 5.10 | 6.05 | 5.27 | 4.77 | 4.86 |
| After medication | 4.82 | 5.20 | 4.30 | 5.06 | 5.38 | 4.89 | 5.22 | 4.69 | 4.52 | 4.72 |

Under the usual assumptions (see Theorem 2), use the sign test to test the hypothesis $H_0 : F = G$ (there is no medication side effect) against the alternative $H''_A : F \neq G$ (there is a side effect) at level of significance $\alpha = 0.05$. Use both the exact test and the Normal approximation for the determination of the cutoff points $C_1$ and $C_2$ (see relations (14) and (15)).

**Discussion.** Here $n = 10$ and for the observed values of the $X_i$'s and $Y_i$'s, we have:

$$z_1 = 0, \ z_2 = 1, \ z_3 = 0, \ z_4 = 0, \ z_5 = 1,$$

$$z_6 = 0, \ z_7 = 0, \ z_8 = 0, \ z_9 = 0, \ z_{10} = 0,$$

so that $z = \sum_{i=1}^{10} z_i = 2$.

Since under $H_0$, $Z = \sum_{i=1}^{10} Z_i \sim B(10, 0.5)$, relations (14) give:

$$P_{0.5} (Z \leq C_1 - 1) + \gamma_1 P_{0.5}(Z = C_1) = 0.025,$$
$$P_{0.5}(Z \leq C_2) - \gamma_2 P_{0.5}(Z = C_2) = 0.975,$$

from which we find $C_1 = 2$, $C_2 = 8$, and $\gamma_1 = \gamma_2 = 0.325$. Since $z = 2$ is neither $\leq C_1 - 1 = 1$ nor $> C_2 = 8$, $H_0$ is not rejected.

For the Normal approximation, since $z_{\alpha/2} = z_{0.025} = 1.96$, relations (15) give: $C_1 \simeq 5 - 3.099 = 1.901$, $C_2 \simeq 5 + 3.099 = 8.099$. Here $H_0$ is not rejected again.

**Example 6.** Twenty eight students in a statistics course were paired out at random and were assigned to each of two sections, one taught by instructor $X$ and the other taught by instructed $Y$. Instructor $X$ is a senior faculty member, whereas instructor $Y$ is a junior faculty member. The final numerical grades for each pair are recorded below.

| Pairs | 1 | 2 | 3 | 4 | 5 | 6 | 7 | 8 | 9 | 10 | 11 | 12 | 13 | 14 |
|---|---|---|---|---|---|---|---|---|---|---|---|---|---|---|
| Instructor $X$ | 83 | 75 | 75 | 60 | 72 | 55 | 94 | 85 | 78 | 96 | 80 | 75 | 66 | 55 |
| Instructor $Y$ | 88 | 91 | 72 | 70 | 80 | 65 | 90 | 89 | 85 | 93 | 86 | 79 | 64 | 68 |

Under the usual assumptions (see Theorem 2), use the sign test to test the hypothesis $H_0 : F = G$ against the alternative $H'' : F \neq G$ at level of significance $\alpha = 0.05$. Use both the exact test and the Normal approximation for the determination of the cutoff points $C_1$ and $C_2$ (see relations (14) and (15)).

**Discussion.** Here $n = 14$ and for the observed values of the $X_i$'s and $Y_i$, we have:

$$z_1 = 1, \; z_2 = 1, \; z_3 = 0, \; z_4 = 1, \; z_5 = 1, \; z_6 = 1, \; z_7 = 0,$$
$$z_8 = 1, \; z_9 = 1, \; z_{10} = 0, \; z_{11} = 1, \; z_{12} = 1, \; z_{13} = 0, \; z_{14} = 1,$$

so that $z = \sum_{i=1}^{14} z_i = 10$.

Since under $H_0$, $Z = \sum_{i=1}^{14} Z_i \sim B(14, 0.5)$, relations (14) give:
$$P_{0.5}(Z \leq C_1 - 1) + \gamma_1 P_{0.5}(Z = C_1) = 0.025,$$
$$P_{0.5}(Z \leq C_2) - \gamma_2 P_{0.5}(Z = C_2) = 0.975,$$

from which we find $C_1 = 3$, $C_2 = 11$, and $\gamma_1 = \gamma_2 \simeq 0.833$. Since $z = 10$, $H_0$ is not rejected.

For the Normal approximation, since $z_{\alpha/2} = z_{0.025} = 1.96$, relations (15) give: $C_1 \simeq 7 - 3.667 = 3.333$, $C_2 \simeq 7 + 3.667 = 10.667$. Again, 10 is neither $< 3.333$ nor $> 10.667$, and therefore $H_0$ is not rejected. Accordingly, the distributions of grades assigned by the two faculty members do not differ at the assumed level of significance.

---

# EXERCISES

**3.1** Let $X$ and $Y$ be r.v.'s with respective d.f.'s $F$ and $G$.

(i) If the r.v.'s are of the discrete type and $Y$ takes on values $y_j$ with respectively probability $P(Y = y_j) = f_Y(y_j)$, $j \geq 1$, then

$$P(X < Y) = \sum_j F(y_j) f_Y(y_j).$$

(ii) If the r.v.'s have continuous d.f.'s, then

$$P(X < Y) = \int_{-\infty}^{\infty} F(y) f_Y(y) dy = \int_{-\infty}^{\infty} F(y) dG(y).$$

**Remark.** Observe that in case (ii) the summation $\sum_j F(y_j) f_Y(y_j)$ is replaced by the integral $\int_{-\infty}^{\infty} F(y) f_Y(y) dy$, (since $y$ is a continuous variable here), and then $f_Y(y)$ is replaced by $dG(y)$ (since $dG(y) = \frac{dG(y)}{dy} dy = f_Y(y) dy$).

**3.2**  Let $X$ and $Y$ be r.v.'s with respective continuous d.f.'s $F$ and $G$, and recall that $F > G$ means that $F(z) \geq G(z)$ for all $z$, and $F(z) > G(z)$ for "sufficiently many $z$'s" (see also Remark 3); likewise $F < G$ means that $F(z) \leq G(z)$ for all $z$, and $F(z) < G(z)$ for "sufficiently many $z$'s"; whereas, $F \neq G$ means that $F < G$ or $F > G$ (excluding the possibility that $F(z) > G(z)$ for some $z$ and $F(z) < G(z)$ for some other $z$). Set $p = P(X < Y)$, and show that:

   **(i)**  $F = G$ implies $p = 1/2$.
   **(ii)**  $F > G$ implies $p > 1/2$.
   **(iii)**  $F < G$ implies $p < 1/2$.
   **(iv)**  Combine parts (ii) and (iii) to conclude that $F > G$ if and only if $p > 1/2$, and $F < G$ if and only if $p < 1/2$.
   **(v)**  Combine parts (i)-(iii) to conclude that $F = G$ if and only if $p = 1/2$.
   **Hint.** Use Exercise 3.1.

**3.3**  Let $F$ be a continuous d.f. and let $G(x) = F(x - \Delta)$, $x \in \Re$, for some $\Delta \in \Re$. The relations $F > G$, $F < G$, and $F \neq G$ are to be understood the way they were defined in Exercise 3.2. Then show that:

   **(i)**  $\Delta > 0$ if and only if $F > G$.
   **(ii)**  $\Delta < 0$ if and only if $F < G$.
   **(iii)**  $\Delta = 0$ if and only if $F = G$.

**3.4**  In reference to Theorem 2, consider the problem of testing the null hypothesis $H_0 : F = G$ against any one of the alternatives stated there. Then, for $n = 100$ and $\alpha = 0.05$, determine the cutoff points $C$, $C'$, $C_1$ and $C_2$.

## 15.4 THE RANK SUM AND THE WILCOXON–MANN–WHITNEY TWO-SAMPLE TESTS

The purpose of this section is the same as that of the previous section; namely, the comparison of the d.f.'s of two sets of i.i.d. r.v.'s. However, the technique used in the last section may not apply here, as the two samples may be of different size, and therefore no pairwise comparison is possible (without discarding observations!). It is assumed, however, that all the r.v.'s involved are independent.

So, what we have here is two independent samples consisting of the i.i.d. r.v.'s $X_1, \ldots, X_m$ with *continuous* d.f. $F$, and the i.i.d. r.v.'s $Y_1, \ldots, Y_n$ with *continuous* d.f. $G$. The problem is that of testing the null hypothesis $H_0$: $F = G$ against any one of the alternatives $H_A$: $F > G$, or $H'_A$: $F < G$, or $H''_A$: $F \neq G$. The test statistic to be used makes no use of the actual values of the $X_i$'s and the $Y_i$'s, but rather of their ranks in the combined sample, which are defined as follows. Consider the combined sample of $X_1, \ldots, X_m$ and $Y_1, \ldots, Y_n$, and order them in ascending order. Because of the assumption of continuity of $F$ and $G$, we are going to have strict inequalities with probability one. Then the *rank* of $X_i$, to be denoted by $R(X_i)$, is that integer among the integers $1, 2, \ldots, m + n$, which corresponds to the position of $X_i$. The *rank* $R(Y_j)$ of $Y_j$ is defined similarly. Next, consider the *rank sums* $R_X$ and $R_Y$ defined by:

$$R_X = \sum_{i=1}^{m} R(X_i), \quad R_Y = \sum_{j=1}^{n} R(Y_j). \tag{16}$$

Then

$$R_X + R_Y = \frac{(m+n)(m+n+1)}{2}, \tag{17}$$

because $R_X + R_Y = \sum_{i=1}^{m} R(X_i) + \sum_{j=1}^{n} R(Y_j) = 1 + 2 + \cdots + m + n$
$= \frac{(m+n)(m+n+1)}{2}$. Before we go further, let us illustrate the concepts introduced so far by a numerical example.

**Example 7.** Let $m = 5, n = 4$, and suppose that:

$$X_1 = 78, X_2 = 65, X_3 = 74, X_4 = 45, X_5 = 82,$$
$$Y_1 = 110, Y_2 = 71, Y_3 = 53, Y_4 = 50.$$

Combining the $X_i$'s and the $Y_j$'s and arranging them in ascending order, we get:

| 45 | 50 | 53 | 65 | 71 | 74 | 78 | 82 | 110 |
|----|----|----|----|----|----|----|----|-----|
| (X) | (Y) | (Y) | (X) | (Y) | (X) | (X) | (X) | (Y). |

Then: $R(X_1) = 7, R(X_2) = 4, R(X_3) = 6, R(X_4) = 1, R(X_5) = 8,$
$R(Y_1) = 9, R(Y_2) = 5, R(Y_3) = 3, R(Y_4) = 2.$

It follows that: $R_X = 26, R_Y = 19$, and, of course, $R_X + R_Y = 45 = \frac{9 \times 10}{2}$. The $m$ ranks $(R(X_1), \ldots, R(X_m))$ can be placed in $m$ positions out of $m + n$ possible in $\binom{m+n}{m}$ different ways (the remaining $n$ positions will be taken up by the $n$ ranks $(R(Y_1), \ldots, R(Y_n))$), and under the null hypothesis $H_0$, each one of them is equally likely to occur. So, each one of the $\binom{m+n}{m}$ positions of $(R(X_1), \ldots, R(X_m))$ has probability $1/\binom{m+n}{m}$. The alternative $H_A$ stipulates that $F > G$; i.e., $F(z) \geq G(z)$ for all $z$, or $P(X \leq z) \geq P(Y \leq z)$ for all $z$, with the inequalities strict for "sufficiently many $z$'s" (see Remark 3), where the r.v.'s $X$ and $Y$ are distributed as the $X_i$'s and the $Y_j$'s, respectively. That is, under $H_A$, the $X_i$'s tend to be smaller than any $z$ with higher probability than the $Y_j$'s are smaller than any $z$. Consequently, since $R_X + R_Y$ is fixed $= (m + n) \times (m + n + 1)/2$, this suggests that the rank sum $R_X$ would tend to take small values. Therefore, $H_0$ should be rejected in favor of $H_A$ whenever $R_X < C$. The rejection region is determined as follows: Consider the $\binom{m+n}{m}$ positions of the ranks $(R(X_1), \ldots, R(X_m))$, and for each one of them, form the respective sum $R_X$. We start with the smallest value of $R_X$ and proceed with the next smallest, etc., until we get to the $k$th smallest, where $k$ is determined by: $k/\binom{m+n}{m} = \alpha$. (In the present setting, the level of significance $\alpha$ is taken to be an integer multiple of $1/\binom{m+n}{m}$, if we wish to have an exact level.) So, the rejection region consists of the $k$ smallest values of $R_X$, where $k/\binom{m+n}{m}$ equals $\alpha$.

Likewise, the hypothesis $H_0$ is rejected in favor of the alternative $H'_A: F < G$ whenever $R_X > C'$, and the rejection region consists of the $k$ largest values of $R_X$, where $k/\binom{m+n}{m}$ equals $\alpha$. Also, $H_0$ is rejected in favor of $H''_A: F \neq G$ whenever

$R_X < C_1$ or $R_X > C_2$, and the rejection region consists of the smallest $r$ values of $R_X$ and the largest $r$ values of $R_X$, where $r$ satisfies the requirement $r/\binom{m+n}{m}$ equals $\alpha/2$.

Summarize these results in the following theorem.

**Theorem 3.** *Consider the independent samples of the i.i.d. r.v.'s $X_1, \ldots, X_m$ and $Y_1, \ldots, Y_n$ with respective continuous d.f.'s $F$ and $G$. Then, for testing the null hypothesis $H_0: F = G$ against any one of the alternatives $H_A: F > G$, or $H'_A: F < G$, or $H''_A: F \neq G$, at level of significance $\alpha$ (so that $\alpha$ or $\frac{\alpha}{2}$ are integer multiples of $1/\binom{m+n}{m}$), the respective rejection regions of the rank sum tests consist of: The $k$ smallest values of the rank sum $R_X$, where $k/\binom{m+n}{m} = \alpha$; the $k$ largest values of the rank sum $R_X$, where $k$ is as above; the $r$ smallest and the $r$ largest values of the rank sum $R_X$, where $r/\binom{m+n}{m} = \frac{\alpha}{2}$.*

*Or, for any $\alpha$ $(0 < \alpha < 1)$, reject $H_0$ (in favor of $H_A$) when $R_X \leq C$, where $C$ is determined by $P(R_X \leq C|H_0) = \alpha$. Likewise, reject $H_0$ (in favor of $H'_A$) when $R_X \geq C'$, where $C'$ is determined by $P(R_X \geq C'|H_0) = \alpha$; and reject $H_0$ (in favor of $H''_A$) when $R_X \leq C_1$ or $R_X \geq C_2$, where $C_1$ and $C_2$, are determined by $P(R_X \leq C_1|H_0) = P(R_X \geq C_2) = \alpha/2$. (Here, typically, randomization will be required).*

**Remark 4.** In theory, carrying out the test procedures described in Theorem 3 is straightforward and neat. Their practical implementation, however, is another matter. To illustrate the difficulties involved, consider Example 5, where $m = 5$ and $n = 4$, so that $\binom{m+n}{m} = \binom{9}{5} = 126$. Thus, one would have to consider the 126 possible arrangements of the ranks $(R(X_1), \ldots, R(X_5))$, form the respective rank sums, and see which ones of their values are to be included in the rejection regions. Clearly, this is not an easy task even for such small sample sizes.

A special interesting case where the rank sum test is appropriate is that where the d.f. $G$ of the $Y_i$'s is assumed to be of the form:

$$G(x) = F(x - \Delta), x \in \mathfrak{R}, \text{ for some unknown } \Delta \in \mathfrak{R}.$$

In such a case, we say that $G$ is a *shift* of $F$ (to the right, if $\Delta > 0$, and to the left, if $\Delta < 0$). Then the hypothesis $H_0: F = G$ is equivalent to testing $\Delta = 0$, and the alternatives $H_A: F > G$, $H'_A: F < G$, $H''_A: F \neq G$ are equivalent to: $\Delta > 0$, $\Delta < 0$, $\Delta \neq 0$.

Because of the difficulties associated with the implementation of the rank sum tests, there is an alternative closely related to it, for which a Normal approximation may be used. This is the *Wilcoxon–Mann–Whitney two-sample test*. For the construction of the relevant test statistic, consider all $mn$ pairs $(X_i, Y_j)$ (with the first coordinate being an X and the second coordinate being a Y), and among them, count those for which $X_i > Y_j$. The resulting r.v. is denoted by $U$ (and is referred to as the *U-statistic*) and is the statistic to be employed. More formally, let the function $u$ be defined by:

$$u(z) = \begin{cases} 1 & \text{if } z > 0 \\ 0 & \text{if } z \leq 0. \end{cases} \tag{18}$$

Then, clearly, the statistic $U$ may be written thus:

$$U = \sum_{i=1}^{m} \sum_{j=1}^{n} u(X_i - Y_j). \tag{19}$$

The statistics $U, R_X,$ and $R_Y$ are related as follows.

**Lemma 1.**   *Let $R_X$, $R_Y$, and $U$ be defined, respectively, by (16) and (19). Then:*

$$U = R_X - \frac{m(m+1)}{2} = mn + \frac{n(n+1)}{2} - R_Y. \tag{20}$$

*Proof.* Deferred to Subsection 15.4.1.  ∎

On the basis of (20), Theorem 3 may be rephrased as follows in terms of the $U$-statistic for any level of significance $\alpha$; randomization may be required.

**Theorem 4.**   *In the notation of Theorem 3 and for testing the null hypothesis $H_0$ against any one of the alternatives $H_A$, or $H_A'$, or $H_A''$ as described there, at level of significance $\alpha$, the Wilcoxon–Mann–Whitney test rejects $H_0$, respectively:*

$$\left.\begin{array}{ll}
\text{For} & U \leq C, \text{ where } C \text{ is determined by } P(U \leq C \mid H_0) = \alpha; \\
\text{for} & U \geq C', \text{ where } C' \text{ is determined by } P(U \geq C' \mid H_0) = \alpha; \\
\text{for} & U \leq C_1 \text{ or } U \geq C_2, \text{ where } C_1 \text{ and } C_2 \text{ are determined} \\
& \text{by } P(U \leq C_1 \mid H_0) = P(U \geq C_2 \mid H_0) = \frac{\alpha}{2}.
\end{array}\right\} \tag{21}$$

In determining the cutoff points $C$, $C'$, $C_1$, and $C_2$ above, we are faced with the same difficulty as that in the implementation of the rank sum tests. However, presently, there are two ways out of it. First, tables are available for small values of $m$ and $n$ ($n \leq m \leq 10$) (see page 341 in the book *Handbook of Statistical Tables*, Addison-Wesley (1962), by D. B. Owen), and second, for large values of $m$ and $n$, and under $H_0$,

$$\frac{U - EU}{\sigma(U)} \text{ is approximately distributed as } Z \sim N(0, 1), \tag{22}$$

where, under $H_0$,

$$EU = \frac{mn}{2}, \quad Var(U) = \frac{mn(m+n+1)}{12}. \tag{23}$$

**Example 8.**   Refer to Example 7 and test the hypothesis $H_0: F = G$ against the alternatives $H_A'': F \neq G$ and $H_A: F > G$.

**Discussion.**   In Example 5 we saw that $R_X = 26$ (and $R_Y = 19$). Since $m = 5$ and $n = 4$, relation (20) gives: $U = 11$. From the tables cited above, we have: $P(U \leq 2) = P(U \geq 17) = 0.032$. So, $C_1 = 2$, $C_2 = 17$, and $H_0$ is rejected in favor of $H_A''$ at the level of significance 0.064. From the same tables, we have that $P(U \leq 3) = 0.056$, so that $C = 3$, and $H_0$ is rejected in favor of $H_A$ at level of significance 0.056.

The results stated in (23) are formulated as a lemma below.

**Lemma 2.** *With U defined by (19), the relations in (23) hold true under $H_0$.*

*Proof.* Deferred to Subsection 15.4.1. ■

For the statistic $U$, the CLT holds. Namely,

**Lemma 3.** *With U, EU, and Var(U) defined, respectively, by (19) and (23), and under $H_0$,*

$$\frac{U - EU}{\sigma(U)} \xrightarrow{d} Z \sim N(0,1), \tag{24}$$

*provided $m, n \to \infty$, so that*

$$\frac{m}{n} \to \lambda \ (0 < \lambda < \infty). \tag{24'}$$

*Proof.* It is omitted. ■

By means of the result in (24), the cutoff points in (21) may be determined approximately, by means of the Normal tables. That is, we have the following corollary.

**Corollary** (to Theorem 4 and Lemma 3). *For sufficiently large m and n (subject to condition (24')), the cutoff points in (21) are given by the following approximate quantities:*

$$\left.\begin{array}{cc} C \simeq \frac{mn}{2} - z_\alpha \sqrt{\frac{mn(m+n+1)}{12}}, & C' \simeq \frac{mn}{2} + z_\alpha \sqrt{\frac{mn(m+n+1)}{12}}, \\ C_1 \simeq \frac{mn}{2} - z_{\frac{\alpha}{2}} \sqrt{\frac{mn(m+n+1)}{12}}, & C_2 \simeq \frac{mn}{2} + z_{\frac{\alpha}{2}} \sqrt{\frac{mn(m+n+1)}{12}}. \end{array}\right\} \tag{25}$$

*Proof.* Follows immediately, from (21), (23), and (24). ■

**Example 9.** Refer to Example 25 in Chapter 1 (see also Example 4 here), and test the null hypothesis $H_0$: $F = G$ against the alternatives $H_A''$ : $F \neq G$, $H_A'$ : $F < G$, and $H_A$ : $F > G$ at level of significance $\alpha = 0.05$ by using Theorem 4 and the above corollary.

**Discussion.** Here $m = n = 15$, $z_{0.05} = 1.645$, and $z_{0.025=1.96}$. Then:

$$C_1 \simeq 112.50 - 47.254 \simeq 65.246,$$
$$C_2 \simeq 112.50 + 47.254 \simeq 159.754,$$
$$C \simeq 112.50 - 39.659 = 72.841,$$
$$C' \simeq 112.50 + 39.659 = 152.159.$$

**Table 15.3** Heights of cross-fertilized (X) and self-fertilized (Y) plants

|   | 1 | 2 | 3 | 4 | 5 | 6 | 7 | 8 | 9 | 10 | 11 | 12 | 13 | 14 | 15 |
|---|---|---|---|---|---|---|---|---|---|----|----|----|----|----|----|
| X | 188 | 96 | 168 | 176 | 153 | 172 | 177 | 163 | 146 | 173 | 186 | 168 | 177 | 184 | 96 |
| Y | 139 | 163 | 160 | 160 | 147 | 149 | 149 | 122 | 132 | 144 | 130 | 144 | 102 | 124 | 144 |

Next, comparing all $15 \times 15$ pairs in Table 15.3, we get the observed value of $U = 185$.

Therefore, the hypothesis $H_0$ is rejected in favor of $H''_A$, since $U = 185 > 159.746 = C_2$; the null hypothesis is also rejected in favor of $H'_A$, since $U = 185 > 152.153 = C'$; but the null hypothesis is not rejected when the alternative is $H_A$, because $U = 185 \not< 72.847 = C$.

The null hypothesis $H_0$ can be tested against any one of the alternatives $H_A$, $H'_A$, $H''_A$ by using the test statistics $R_X$ or $R_Y$ instead of $U$, and determine the approximate cutoff points. This is done as follows. By relations (20) and (21),

$$U < C, \text{ if and only if } R_X - \frac{m(m+1)}{2} < C \text{ or } R_X < C + \frac{m(m+1)}{2},$$

where (by (25)) $C \simeq \frac{mn}{2} - z_\alpha \sqrt{\frac{mn(m+n+1)}{12}}$, so that $R_X < \frac{m(m+n+1)}{2} - z_\alpha \sqrt{\frac{mn(m+n+1)}{12}}$.
So, for testing $H_0$ against $H_A$ at level of significance $\alpha$ by means of $R_X$, $H_0$ is rejected whenever $R_X < C^*$, where $C^* \simeq \frac{m(m+n+1)}{2} - z_\alpha \sqrt{\frac{mn(m+n+1)}{12}}$. Likewise, $H_0$ is rejected in favor of $H'_A$ whenever $R_X > C^{*\prime}$, where $C^{*\prime} \simeq \frac{m(m+n+1)}{2} + z_\alpha \sqrt{\frac{mn(m+n+1)}{12}}$, and $H_0$ is rejected in favor of $H''_A$ whenever $R_X < C_1^*$ or $R_X > C_2^*$, where $C_1^* \simeq \frac{m(m+n+1)}{2} - z_{\alpha/2} \sqrt{\frac{mn(m+n+1)}{12}}$, $C_2^* \simeq \frac{m(m+n+1)}{2} + z_{\alpha/2} \sqrt{\frac{mn(m+n+1)}{12}}$.

**Remark 5.**    From relations (20) and (23), we have that $R_X = U + \frac{m(m+1)}{2}$, so that $ER_X = \frac{mn}{2} + \frac{m(m+1)}{2} = \frac{m(m+n+1)}{2}$, and $\sigma^2(R_X) = \sigma^2(U) = \frac{mn(m+n+1)}{12}$. It follows that the tests based on $R_X$ are of the from $R_X < ER_X - z_\alpha \sigma(R_X)$; $R_X > ER_X + z_\alpha \sigma(R_X)$; $R_X < ER_X - z_{\alpha/2}\sigma(R_X)$ or $R_X > ER_X + z_{\alpha/2}\sigma(R_X)$.

If the test statistics $R_X$ were to be replaced by $R_Y$, one would proceed as follows. By (20) and (21) again,

$$U < C \text{ if and only if } mn + \frac{n(n+1)}{2} - R_Y < C \text{ or } R_Y > mn + \frac{n(n+1)}{2} - C,$$

where (by (25)) $C \simeq \frac{mn}{2} - z_\alpha \sqrt{\frac{mn(m+n+1)}{12}}$, so that $R_Y > \frac{n(m+n+1)}{2} + z_\alpha \sqrt{\frac{mn(m+n+1)}{12}}$.
So, for testing $H_0$ against $H_A$ at level of significance $\alpha$ by means of $R_Y$, $H_0$ is rejected whenever $R_Y > \bar{C}'$, where $\bar{C}' \simeq \frac{n(m+n+1)}{2} + z_\alpha \sqrt{\frac{mn(m+n+1)}{12}}$. Likewise, $H_0$ is rejected in favor of $H'_A$ whenever $R_Y < \bar{C}$, where $\bar{C} \simeq \frac{n(m+n+1)}{2} - z_\alpha \sqrt{\frac{mn(m+n+1)}{12}}$, and $H_0$

is rejected in favor of $H_A''$ whenever $R_Y > \bar{C}_2$ or $R_Y < \bar{C}_1$, where $\bar{C}_1 \simeq \frac{n(m+n+1)}{2} -$ $z_{\alpha/2}\sqrt{\frac{mn(m+n+1)}{12}}$, $\bar{C}_2 \simeq \frac{n(m+n+1)}{2} + z_{\alpha/2}\sqrt{\frac{mn(m+n+1)}{12}}$.

**Remark 6.** Once again, from relations (20) and (23), it follows that $R_Y = mn + \frac{n(n+1)}{2} - U$, so that $ER_Y = mn + \frac{n(n+1)}{2} - \frac{mn}{2} = \frac{n(m+n+1)}{2}$, and $\sigma^2(R_Y) = \sigma^2(U) = \frac{mn(m+n+1)}{12}$. Therefore, the tests based on $R_Y$ are of the from $R_Y < ER_Y - z_\alpha \sigma(R_Y)$; $R_Y > ER_Y + z_\alpha \sigma(R_Y)$; $R_Y < ER_Y - z_{\alpha/2}\sigma(R_Y)$ or $R_Y > ER_Y + z_{\alpha/2}\sigma(R_Y)$. The results obtained above are gathered together in the following proposition.

**Proposition 1.** *In the notation of Theorem 3, consider the problem of testing the null hypothesis $H_0$ against any one of the alternatives $H_A$, or $H_A'$, or $H_A''$ as described there, at level of significance $\alpha$, by using either one of the rank sum test statistics $R_X$ or $R_Y$. Suppose $m$ and $n$ are sufficiently large (subject to relation (24')). Then, on the basis of $R_X$, $H_0$ is rejected, respectively, when:*

$$R_X < C^*; \ R_X > C^{*\prime}; \ R_X < C_1^*; \ or \ R_X > C_2^*,$$

*where*

$$C^* \simeq ER_X - z_\alpha \sigma(R_X), \ C^{*\prime} \simeq ER_X + z_\alpha \sigma(R_X),$$
$$C_1^* \simeq ER_X - z_{\alpha/2}\sigma(R_X), C_2^* \simeq ER_X + z_{\alpha/2}\sigma(R_X);$$

*here*

$$ER_X = \frac{m(m+n+1)}{2}, \ \sigma(R_X) = \sqrt{\frac{mn(m+n+1)}{12}}.$$

*On the basis of $R_Y$, $H_0$ is rejected, respectively, when:*

$$R_Y > \bar{C}'; \ R_Y < \bar{C}; \ R_Y < \bar{C}_1; \ or \ R_Y > \bar{C}_2,$$

*where*

$$\bar{C} \simeq ER_Y - z_\alpha \sigma(R_Y), \ \bar{C}' \simeq ER_Y + z_\alpha \sigma(R_Y),$$
$$\bar{C}_1 \simeq ER_Y - z_{\alpha/2}\sigma(R_Y), \bar{C}_2 \simeq ER_Y + z_{\alpha/2}\sigma(R_Y);$$

*here*

$$ER_Y = \frac{n(m+n+1)}{2}, \ \sigma(R_Y) = \sqrt{\frac{mn(m+n+1)}{12}}.$$

**Example 10.** In reference to Example 5, use the rank sum test based on $R_X$ in order to carry out the testing problem described there. Employ the Normal approximation for the determination of the cutoff points given in Proposition 1.

**Discussion.** Here $m = n = 10$ and $z_{\alpha/2} = z_{0.025} = 1.96$, so that: $\bar{C}_1 \simeq 105 - 25.928 = 79.072$, $\bar{C}_2 \simeq 105 + 25.928 = 130.928$. Next, arranging the combined observed values of the $X_i$'s and the $Y_i$'s, accompanied by their ranks, we have the following display, where the x-values are underlined.

| Values | Ranks | Values | Ranks | Values | Ranks | Values | Ranks |
|--------|-------|--------|-------|--------|-------|--------|-------|
| 4.30   | 1     | 4.77   | 6     | 5.04   | 11    | 5.22   | 16    |
| 4.52   | 2     | 4.80   | 7     | 5.06   | 12    | 5.25   | 17    |
| 4.69   | 3     | 4.82   | 8     | 5.10   | 13    | 5.27   | 18    |
| 4.72   | 4     | 4.86   | 9     | 5.16   | 14    | 5.38   | 19    |
| 4.75   | 5     | 4.89   | 10    | 5.20   | 15    | 6.05   | 20    |

Therefore the observed rank sum of the $X_i$'s is $R_X = 120$. Since 120 is between 79.072 and 130.928, $H_0$ is not rejected. This is the same conclusion reached in Example 5 when the Normal approximation was used for the determination of the cutoff points.

**Example 11.** Refer to Example 6 and use the rank sum test based on $R_X$ in order to carry out the testing problem described there. Employ the Normal approximation for the determination of the cutoff points given in Proposition 1.

**Discussion.** Here $m = n = 14$ and $z_{\alpha/2} = z_{0.025} = 1.96$, so that: $\bar{C}_1 = 203 - 42.657 = 160.347$ and $\bar{C}_2 = 203 + 42.657 = 245.657$. Next, arranging the combined observed values of the $X_i$'s and $Y_i$'s in ascending order, we have:

| $\underline{55}$ | $\underline{55}$ | $\underline{60}$ | 64 | 65 | $\underline{66}$ | 68 | 70 | $\underline{72}$ | 72 |
|------|------|------|----|----|------|----|----|------|----|
| $\underline{75}$ | $\underline{75}$ | $\underline{75}$ | $\underline{78}$ | 79 | $\underline{80}$ | 80 | $\underline{83}$ | 85 | 85 |
| 86 | 88 | 89 | 90 | 91 | 93 | 94 | 96, | | |

where the observed x-values are underlined. In an attempt to rank these values, we face the problem of ties; the value of 55 appears twice, and so do the values 72, 80, and 85, whereas the value 75 appears three times. One way of resolving this issue is the following: Look at the above arrangement, where 55 appears in the first and the second place, and assign to each one of them the rank $(1 + 2)/2 = 1.5$. Next, 72 appears in the 9-th and the 10-th place, so assign to each one of them the rank $(9 + 10)/2 = 9.5$. Likewise, 75 appears in the 11-th, 12-th, and 13-th place, so assign to each one of them the rank $(11 + 12 + 13)/3 = 12$. In the same manner, assign the rank $(16 + 17)/2 = 16.5$ to each one of the 80's, and the rank $(19 + 20)/2 = 19.5$ to each one of the 85's. In so doing, we have then the following arrangement of the combined observed values of the $X_i$'s and the $Y_i$'s, accompanied by the respective ranks; we recall at this point that the x-values are underlined.

| Values | Ranks | Values | Ranks | Values | Ranks | Values | Ranks |
|--------|-------|--------|-------|--------|-------|--------|-------|
| 55 | 1.5 | 70 | 8 | 79 | 15 | 88 | 22 |
| 55 | 1.5 | 72 | 9.5 | 80 | 16.5 | 89 | 23 |
| 60 | 3 | 72 | 9.5 | 80 | 16.5 | 90 | 24 |
| 64 | 4 | 75 | 12 | 83 | 18 | 91 | 25 |
| 65 | 5 | 75 | 12 | 85 | 19.5 | 93 | 26 |
| 66 | 6 | 75 | 12 | 85 | 19.5 | 94 | 27 |
| 68 | 7 | 78 | 14 | 86 | 21 | 96 | 28 |

Therefore the observed rank sum of the $X_i$'s is $R_X = 180.5$. Since 180.5 is neither smaller than 160.447 nor larger than 245.657, $H_0$ is not rejected. This conclusion is consonant with the one reached in Example 6 when the Normal approximation was used.

**Remark.** The same ranking issue appears also in Example 9 if the testing hypotheses are to be tested by means of the rank sames $R_X$ or $R_Y$.

### 15.4.1 PROOFS OF LEMMAS 1 AND 2

*Proof of Lemma 1.* Let $X_{(1)}, \ldots, X_{(m)}$ be the order statistics of the r.v.'s $X_1, \ldots, X_m$, and look at the rank $R(X_{(i)})$ in the combined (ordered) sample of the $X_i$'s and the $Y_i$'s. For each $R(X_{(i)})$, there are $R(X_{(i)}) - 1$ $X_i$'s and $Y_j$'s preceding $X_{(i)}$. Of these, $i - 1$ are $X_i$'s and hence $R(X_{(i)}) - 1 - (i - 1) = R(X_{(i)}) - i$ are $Y_j$'s. Therefore:

$$
\begin{aligned}
U &= \left[R\left(X_{(1)}\right) - 1\right] + \cdots + \left[R\left(X_{(m)}\right) - m\right] \\
&= \left[R\left(X_{(1)}\right) + \cdots + R\left(X_{(m)}\right)\right] - (1 + \cdots + m) \\
&= [R(X_1) + \cdots + R(X_m)] - \frac{m(m+1)}{2} = R_X - \frac{m(m+1)}{2},
\end{aligned}
$$

since $R(X_{(1)}) + \cdots + R(X_{(m)})$ is simply a rearrangement of the terms in the rank sum $R(X_1) + \cdots + R(X_m) = R_X$. Next, from the result just obtained and (17), we have:

$$
\begin{aligned}
U &= \frac{(m+n)(m+n+1)}{2} - R_Y - \frac{m(m+1)}{2} \\
&= \frac{(m+n)(m+n+1) - m(m+1)}{2} - R_Y = mn + \frac{n(n+1)}{2} - R_Y. \quad \blacksquare
\end{aligned}
$$

*Proof of Lemma 2.* Recall that all derivations below are carried out under the assumption that $H_0(F = G)$ holds. Next, for any $i$ and $j$:

$$
Eu(X_i - Y_j) = 1 \times P(X_i > Y_j) = \frac{1}{2}, \quad Eu^2(X_i - Y_j) = 1^2 \times P(X_i > Y_j) = \frac{1}{2},
$$

so that

$$
Var(u(X_i - Y_j)) = \frac{1}{2} - \frac{1}{4} = \frac{1}{4}.
$$

Therefore $EU = \sum_{i=1}^{m} \sum_{j=1}^{n} \frac{1}{2} = \frac{mn}{2}$, and

$$Var(U) = \sum_{i=1}^{m} \sum_{j=1}^{n} Var(u(X_i - Y_j)) + \sum_{i=1}^{m} \sum_{j=1}^{n} \sum_{k=1}^{m} \sum_{l=1}^{n} Cov(u(X_i - Y_j), u(X_k - Y_l))$$

$$= \frac{mn}{4} + \text{sum of the covariances on the right-hand side above.} \tag{26}$$

Regarding the covariances, we consider the following cases. First, let $i \neq k$ and $j \neq l$. Then $Cov(u(X_i - Y_j), u(X_k - Y_l)) = 0$ by independence. Thus, it suffices to restrict ourselves to pairs $(X_i, Y_j)$ and $(X_k, Y_l)$ for which $i = k$ and $j \neq l$, and $i \neq k$ and $j = l$. In order to see how many such pairs there are, consider the following array:

$$(X_1, Y_1,), \quad (X_1, Y_2), \ldots, (X_1, Y_n)$$
$$(X_2, Y_1), \quad (X_2, Y_2), \ldots, (X_2, Y_n)$$
$$\vdots \qquad \vdots \qquad \vdots$$
$$(X_m, Y_1), \quad (X_m, Y_2), \ldots, (X_m, Y_n).$$

From each one of the $m$ rows, we obtain $\binom{n}{2} \times 2 = n(n-1)$ terms of the form: $Cov(u(X - Y), u(X - Z))$, where $X, Y, Z$ are independent r.v.'s with d.f. $F = G$. Since $Cov(u(X - Y), u(X - Z)) = P(X > Y \text{ and } X > Z) - \frac{1}{4}$, we have then $n(n-1) \times P(X > Y \text{ and } X > Z) - \frac{n(n-1)}{4}$ as a contribution to the sum of the covariances on the right-hand side in relation (26) from each row, and therefore from the $m$ rows, the contribution to the sum of the covariances is:

$$mn(n-1)P(X > Y \text{ and } X > Z) - \frac{mn(n-1)}{4}. \tag{27}$$

Next, from each one of the $n$ columns, we obtain $\binom{m}{2} \times 2 = m(m-1)$ terms of the form: $Cov(u(X - Z), u(Y - Z)) = P(X > Z \text{ and } Y > Z) - \frac{1}{4}$. Therefore, the contribution from the $n$ columns to the sum of the covariances on the right-hand side in relation (26) is:

$$mn(m-1)P(X > Z \text{ and } Y > Z) - \frac{mn(m-1)}{4}. \tag{28}$$

Now,

$$(X > Y \text{ and } X > Z) = (X > Y, Y > Z, X > Z) \cup (X > Y, Y \leq Z, X > Z)$$
$$= (X > Y, Y > Z) \cup (X > Z, Z \geq Y) = (X > Y > Z) \cup (X > Z \geq Y),$$

since

$$(X > Y, Y > Z) \subseteq (X > Z), \quad \text{and} \quad (X > Z, Z \geq Y) \subseteq (X > Y).$$

Thus,

$$P(X > Y \text{ and } X > Z) = P(X > Y > Z) + P(X > Z > Y). \tag{29}$$

Likewise,

$$(X > Z \quad \text{and} \quad Y > Z) = (X > Y, Y > Z, X > Z) \cup (X \le Y, Y > Z, X > Z)$$
$$= (X > Y, Y > Z) \cup (Y \ge X, X > Z) = (X > Y > Z) \cup (Y \ge X > Z),$$

since

$$(X > Y, Y > Z) \subseteq (X > Z), \quad \text{and} \quad (Y \ge X, X > Z) \subseteq (Y > Z).$$

Thus,

$$P(X > Z \text{ and } Y > Z) = P(X > Y > Z) + P(Y > X > Z). \tag{30}$$

The r.v.'s $X$, $Y$, and $Z$ satisfy, with probability one, exactly one of the inequalities:

$$X > Y > Z, \quad X > Z > Y, \quad Y > X > Z,$$
$$Y > Z > X, \quad Z > X > Y, \quad Z > Y > X,$$

and each one of these inequalities has probability $1/6$. Then, the expressions in (27) and (28) become, by means of (29) and (30), respectively:

$$\frac{mn(n-1)}{3} - \frac{mn(n-1)}{4} = \frac{mn(n-1)}{12}, \quad \frac{mn(m-1)}{3} - \frac{mn(m-1)}{4} = \frac{mn(m-1)}{12}.$$

Then, formula (26) yields:

$$Var(U) = \frac{mn}{4} + \frac{mn(n-1)}{12} + \frac{mn(m-1)}{12} = \frac{mn(m+n+1)}{12}. \qquad \blacksquare$$

## 15.5 NONPARAMETRIC CURVE ESTIMATION

For quite a few years now, work on nonparametric methodology has switched decisively in what is referred to as nonparametric *curve estimation*. Such estimation includes estimation of d.f.'s, of p.d.f.'s or functions thereof, regression functions, etc. The empirical d.f. is a case of nonparametric estimation of a d.f., although there are others as well. In this section, we are going to describe briefly a way of estimating nonparametrically a p.d.f., and record some of the (asymptotic) desirable properties of the proposed estimate. Also, the problem of estimating, again nonparametrically, a regression function will be discussed very briefly. There is already a huge statistical literature in this area, and research is currently very active.

### 15.5.1 NONPARAMETRIC ESTIMATION OF A PROBABILITY DENSITY FUNCTION

The problem we are faced with here is the following: We are given $n$ i.i.d. r.v.'s $X_1, \ldots, X_n$ with p.d.f. $f$ (of the continuous type), for which very little is known, and we are asked to construct a nonparametric estimate $\hat{f}_n(x)$ of $f(x)$, for each $x \in \Re$, based on the random sample $X_1, \ldots, X_n$. The approach to be used here is the so-called *kernel-estimation* approach. According to this method, we select a (known) p.d.f. (of the continuous type) to be denoted by $K$ and to be termed a *kernel*, subject to some rather minor requirements. Also, we choose a sequence of positive numbers, denoted by $\{h_n\}$, which has the property that $h_n \to 0$ as $n \to \infty$ and also satisfies some additional requirements. The numbers $h_n, n \geq 1$, are referred to as *bandwidth* for a reason to be seen below (see Example 12). Then, on the basis of the random sample $X_1, \ldots, X_n$, the kernel $K$, and the bandwidths $h_n, n \geq 1$, the proposed estimate of $f(x)$ is $\hat{f}_n(x)$ given by:

$$\hat{f}_n(x) = \frac{1}{nh_n} \sum_{i=1}^{n} K\left(\frac{x - X_i}{h_n}\right). \tag{31}$$

**Remark 7.** In the spirit of motivation for using the estimate in (31), observe first that $\frac{1}{h}K\left(\frac{x-y}{h_n}\right)$ is a p.d.f. as a function of $y$ for fixed $x$. Indeed,

$$\int_{-\infty}^{\infty} \frac{1}{h_n} K\left(\frac{x - y}{h_n}\right) dy = \int_{\infty}^{-\infty} \frac{1}{h_n} K(t)(-h_n) dt \quad \text{(by setting } \frac{x - y}{h_n} = t\text{)}$$

$$= \int_{-\infty}^{\infty} K(t) dt = 1.$$

Next, evaluate $\frac{1}{h_n} K\left(\frac{x-y}{h_n}\right)$ at $y = X_i$, $i = 1, \ldots, n$, and then form the average of these values to produce $\hat{f}_n(x)$.

A further motivation for the proposed estimate is the following. Let $F$ be the d.f. of the $X_i$'s, and let $F_n$ be the empirical d.f. based on $X_1, \ldots, X_n$, so that, for fixed $x$, $F_n(y)$ takes on the value $\frac{1}{n}$ at each one of the points $y = X_i$, $i = 1, \ldots, n$. Then weigh $\frac{1}{h_n} K\left(\frac{x-y}{h_n}\right)$ by $\frac{1}{n}$ and sum up from 1 to $n$ to obtain $\hat{f}_n(x)$ again.

**Example 12.** Construct the kernel estimate of $f(x)$, for each $x \in \mathcal{R}$, by using the $U(-1, 1)$ kernel; i.e., by taking

$$K(x) = \frac{1}{2}, \quad \text{for } -1 \leq x \leq 1, \text{ and } 0, \text{ otherwise.}$$

**Discussion.** Here, it is convenient to use the indicator notation; namely, $K(x) = I_{[-1,1]}(x)$ (where, it is recalled, $I_A(x) = 1$ if $x \in A$, and $0$ if $x \in A^c$). Then the estimate (31) becomes as follows:

$$\hat{f}_n(x) = \frac{1}{nh_n} \sum_{i=1}^{n} I_{[-1,1]}\left(\frac{x - X_i}{h_n}\right), \quad x \in R. \tag{32}$$

So, $I_{[-1,1]}(\frac{x-X_i}{h_n}) = 1$, if and only if $x - h_n \leq X_i \leq x + h_n$; in other words, in forming $\hat{f}_n(x)$, we use only those observations $X_i$ which lie in the window $[x - h_n, x + h_n]$. The breadth of this window is, clearly, determined by $h_n$, and this is the reason that $h_n$ is referred to as the bandwidth.

Usually, the minimum of assumptions required of the kernel $K$ and the bandwidth $h_n$, in order for us to be able to establish some desirable properties of the estimate $\hat{f}_n(x)$ given in (31), are the following:

$$\left.\begin{array}{l} K \text{ is bounded; i.e., } \sup\{K(x); x \in \Re\} < \infty. \\ xK(x) \text{ tends to } 0 \text{ as } x \to \pm\infty; \text{ i.e., } |xK(x)| \underset{|x| \to \infty}{\longrightarrow} 0. \\ K \text{ is symmetric about } 0; \text{ i.e., } K(-x) = K(x), \ x \in R. \end{array}\right\} \quad (33)$$

$$\left.\begin{array}{lll} \text{As } n \to \infty: & \text{(i)} & (0 <)h_n \to 0 \\ & \text{(ii)} & nh_n \to \infty \end{array}\right\} \quad (34)$$

**Remark 8.**   Observe that requirements (33) are met for the kernel used in (32). Furthermore, the convergences in (34) are satisfied if one takes, e.g., $h_n = n^{-\alpha}$ with $0 < \alpha < 1/2$. Below, we record three (asymptotic) results regarding the estimate $\hat{f}_n(x)$ given in (31).

**Theorem 5.**   *Under assumptions (33) and (34)(i), the estimate $\hat{f}_n(x)$ given in (31) is an asymptotically unbiased estimate of $f(x)$ for every $x \in \Re$ at which $f$ is continuous; i.e.,*

$$E\hat{f}_n(x) \to f(x) \text{ as } n \to \infty.$$

**Theorem 6.**   *Under assumptions (33) and (34)(i)–(ii), the estimate $\hat{f}_n(x)$ given in (31) is a consistent in quadratic mean estimate of $f(x)$ for every $x \in \Re$ at which $f$ is continuous; i.e.,*

$$E[\hat{f}_n(x) - f(x)]^2 \to 0 \text{ as } n \to \infty.$$

**Theorem 7.**   *Under assumptions (33) and (34)(i)–(ii), the estimate $\hat{f}_n(x)$ given in (31) is asymptotically Normal, when properly normalized, for every $x \in \Re$ at which $f$ is continuous; i.e.,*

$$\frac{\hat{f}_n(x) - E\hat{f}_n(x)}{\sigma[\hat{f}_n(x)]} \xrightarrow[n \to \infty]{d} Z \sim N(0, 1).$$

**Corollary.**   *For any continuity point $x$ of $f$, $\hat{f}_n(x)$ is a consistent estimate of $f(x)$ (in the probability sense); i.e., $\hat{f}_n(x) \xrightarrow[n \to \infty]{P} f(x)$.*
*(For its justification, see Exercise 5.1(iv)).*

At this point, it is only fitting to mention that the concept of kernel estimation of a p.d.f. was introduced by Murray Rosenblatt in 1956, and it was popularized in a fundamental paper by E. Parzen in 1962. There, one can find the proofs of the above theorems, along with other results. The relevant references are as follows:

"Remarks on some nonparametric estimates of a density function", by M. Rosenblatt, in the *Annals of Mathematical Statistics*, Vol. 27 (1956), pages 823–835. "On estimation of a probability density function and mode", by E. Parzen, in the *Annals of Mathematical Statistics*, Vol. 33 (1962), pages 1065–1076.

This section is concluded with the formulation of a result (Theorem A (Bochner)) which plays a central role in the proofs of asymptotic properties of the estimate $\hat{f}_n(x)$, including Theorem 5 and 6 stated above; its proof can be found in Parzen's paper just cited.

**Theorem A** (Bochner).   *Let $K^*$ be a real-valued function (not necessarily a p.d.f.), $K^* : \Re \rightarrow \Re$, which satisfies the following properties:*

(i)  $|K^*(y)| \leq M(< \infty)$ *for all y.*
(ii)  $\int_{-\infty}^{\infty} |K^*(y)|dy \leq \infty.$
(iii)  $|yK^*(y)| \rightarrow 0$ *as* $|y| \rightarrow \infty.$

*Let $g : \Re \rightarrow \Re$ be such that $\int_{-\infty}^{\infty} |g(y)|dy < \infty$; and let $\{h_n, n \geq 1\}$, be a sequence of (real) numbers such that $0 < h_n \rightarrow 0$ as $n \rightarrow \infty$.*
*For $x \in \Re$, set*

$$g_n(x) = \int_{-\infty}^{\infty} K^*(y)g(x - h_n y)dy.$$

*Then, for every continuity point x of g, it holds*

$$g_n(x) \xrightarrow[n \to \infty]{} g(x) \int_{-\infty}^{\infty} K^*(y)dy.$$

As an application of Bochner's theorem, we derive the proofs of Theorems 5 and 6 here.

*Proof of Theorem 5.* Let $X$ be a r.v. distributed as the $X_i$'s. Then, from $\hat{f}_n(x) = \frac{1}{nh_n} \sum_{i=1}^{n} K\left(\frac{x-X_i}{h_n}\right)$, we get

$$E\hat{f}_n(x) = \frac{1}{h_n}EK\left(\frac{x - X}{h_n}\right) = \frac{1}{h_n}\int_{-\infty}^{\infty} K\left(\frac{x - y}{h_n}\right)f(y)dy$$

$$= \frac{1}{h_n}\int_{\infty}^{-\infty} K(u)f(x - h_n u)(-h_n)du \quad \text{(by setting } \frac{x - y}{h_n} = u)$$

$$= \int_{-\infty}^{\infty} K(u)f(x - h_n u)du.$$

At this point, in Bochner's theorem, replace $K^*$ by $K$, and $g$ by $f$, and observe that the assumptions of the theorem are satisfied. Therefore, for any continuity point $x$ of $f$, it holds

$$\int_{-\infty}^{\infty} K(u)f(x - h_n u)du \xrightarrow[n \to \infty]{} f(x) \int_{-\infty}^{\infty} K(u)du = f(x),$$

as was to be seen.  ∎

*Proof of Theorem 6.*  In the first place, for any r.v. $Y$ with $EY^2 < \infty$ and any constant $c \in \Re$, it holds

$$E(Y - c)^2 = \sigma^2(Y) + (EY - c)^2.$$

Indeed,

$$E(Y - c)^2 = E[(Y - EY) + (EY - c)]^2$$

$$= E(Y - EY)^2 + (EY - c)^2 + 2E[(Y - EY)(EY - c)]$$

$$= E(Y - EY)^2 + (EY - c)^2 = \sigma^2(Y) + (EY - c)^2.$$

In this identity, replace $Y$ by $\hat{f}_n(x)$ and $c$ by $f(x)$ to obtain

$$E\left[\hat{f}_n(x) - f(x)\right]^2 = Var\left[\hat{f}_n(x)\right] + \left[E\hat{f}_n(x) - f(x)\right]^2.$$

Since (for any continuity point $x$ of $f$) $E\hat{f}_n(x) - f(x) \xrightarrow[n \to \infty]{} 0$ (by Theorem 5), it suffices to show that $Var\left[\hat{f}_n(x)\right] \xrightarrow[n \to \infty]{} 0$ (for any continuity point $x$ of $f$).

By Exercise 5.1 (i), (iii) (and for a r.v. $X$ distributed as the $X_i$'s),

$$Var\left[\hat{f}_n(x)\right] = \frac{1}{nh_n^2}Var\left[K\left(\frac{x - X}{h_n}\right)\right]$$

$$= \frac{1}{nh_n} \times \frac{1}{h_n}Var\left[K\left(\frac{x - X}{h_n}\right)\right]$$

$$= \frac{1}{nh_n}\left\{\int_{-\infty}^{\infty} K^2(u)f(x - h_n u)du - h_n\left[\int_{-\infty}^{\infty} K(u)f(x - h_n u)du\right]^2\right\},$$

whereas, by Bochner's theorem, $\int_{-\infty}^{\infty} K(u)f(x - h_n u)du \xrightarrow[n \to \infty]{} f(x)$ (as was seen in the proof of Theorem 5), so that

$$h_n\int_{-\infty}^{\infty} K(u)f(x - h_n u) \xrightarrow[n \to \infty]{} 0.$$

Next, $K^2(y) \le M^2(< \infty)$ for all $y$, $\int_{-\infty}^{\infty} K^2(y)dy \le M\int_{-\infty}^{\infty} K(y)dy = M(< \infty)$ and $|yK^2(y)| \le M|y|K(y) \to 0$ as $|y| \to \infty$. So, $K^2$ satisfies the same condition as $K^*$ in Bochner's theorem. Therefore, for any continuity point $x$ of $f$,

$$\int_{-\infty}^{\infty} K^2(u)f(x - h_n u)du \xrightarrow[n \to \infty]{} f(x) \int_{-\infty}^{\infty} K^2(u)du.$$

Thus, the quantity in the squared bracket above converges to $f(x) \int_{-\infty}^{\infty} K^2(u)du$ (as $n \to \infty$ and for any continuity point $x$ of $f$). Since $nh_n \to \infty$ as $n \to \infty$, it follows that $Var[\hat{f}_n(x)] \underset{n \to \infty}{\longrightarrow} 0$, which completes the proof of the theorem. ∎

**Remark 9.** A comment is appropriate here, regarding Theorem 7. This result is in essence, a CLT. However, the framework here is different from the standard setting, as we are faced with a triangular array of r.v.'s. Even an outline of the proof of this theorem would take us far afield, so not such an attempt is made.

## 15.5.2 NONPARAMETRIC REGRESSION ESTIMATION

In Chapter 13, a simple linear regression model was studied and its usefulness was demonstrated by means of specific examples. It was also stated that there is a definite need for more general regression models, where the linearity is retained, or it is discarded altogether. This issue is addressed, to a considerable extent, in this section.

Specifically, the model considered here is the following: For each $n = 1, 2, \ldots$, consider points $x_{n1}, \ldots, x_{nn}$ in a bounded subset $S$ of $\Re$, and at each one of them, an observation is taken, to be denoted by $Y_{ni}, i = 1, \ldots, n$. It is assumed that $Y_{ni}$ is equal to some unknown function $g$ evaluated at $x_{ni}$ except for an error $e_{ni}$; i.e.,

$$Y_{ni} = g(x_{ni}) + e_{ni}, \quad i = 1, \ldots, n. \tag{35}$$

On the errors $e_{ni}, i = 1, \ldots, n$, we make the usual assumptions that they are i.i.d. r.v.'s with $Ee_{ni} = 0$ and $Var(e_{ni}) = \sigma^2 < \infty$.

The model in (1) of Chapter 13 is a very special case of the model just described. In the first place, the points where observations are taken are allowed here to depend on $n$, and second, the regression function in (1) of Chapter 13 is taken from here by setting $g(x) = \beta_1 + \beta_2 x$, so that $\beta_1 + \beta_2 x_i = g(x_i), i = 1, \ldots, n$. The function $g$ in (35) is subject only to the requirement that it is defined on a bounded subset $S$ of $\Re$ and that it is continuous.

The objective here is to (nonparametrically) estimate the function $g(x)$, for each $x \in S$, by means of the observations $Y_{n1}, \ldots, Y_{nn}$. The proposed estimate is the statistic $\hat{g}_n(x; \mathbf{x}_n)$ defined as follows:

$$\hat{g}_n(x; \mathbf{x}_n) = \sum_{i=1}^{n} w_{ni}(x; \mathbf{x}_n) Y_{ni}, \tag{36}$$

where $\mathbf{x}_n = (x_{n1}, \ldots, x_{nn})$ and $w_{ni}, i = 1, \ldots, n$ are *weights*, properly chosen, which depend on the particular point $x$ in $S$ and also the points $x_{n1}, \ldots, x_{nn}$, where observations are taken. The weights are required to satisfy certain conditions, and there is considerable flexibility in choosing them. We do not intend to enter here into this kind of detail. Instead, we restrict ourselves to stating three basic properties that the estimate defined in (36) satisfies.

**Theorem 8.**    *Under suitable regularity conditions, the estimate $\hat{g}_n(x; \mathbf{x}_n)$ defined in (36) is an asymptotically unbiased estimate of $g(x)$; i.e.,*

$$E\hat{g}_n(x; \mathbf{x}_n) \xrightarrow[n \to \infty]{} g(x), \quad \text{for every } x \in S. \tag{37}$$

**Theorem 9.**    *Under suitable regularity conditions, the estimate $\hat{g}_n(x; \mathbf{x}_n)$ is a consistent in quadratic mean estimate of $g(x)$; i.e.,*

$$E[\hat{g}_n(x; \mathbf{x}_n) - g(x)]^2 \xrightarrow[n \to \infty]{} 0, \quad \text{for every } x \in S. \tag{38}$$

**Theorem 10.**    *Under suitable regularity conditions, the estimate $\hat{g}_n(x; \mathbf{x}_n)$, properly normalized, is asymptotically Normal; i.e.,*

$$\frac{\hat{g}_n(x; \mathbf{x}_n) - E\hat{g}_n(x; \mathbf{x}_n)}{\sigma[\hat{g}_n(x; \mathbf{x}_n)]} \xrightarrow[n \to \infty]{d} Z \sim N(0, 1), \quad \text{for every } x \in S.$$

*Also,*

$$\frac{\hat{g}_n(x; \mathbf{x}_n) - g(x)}{\sigma[\hat{g}_n(x; \mathbf{x}_n)]} \xrightarrow[n \to \infty]{d} Z \sim N(0, 1), \quad \text{for every } x \in S. \tag{39}$$

Convergences (37) and (38) provide asymptotic optimal properties for the estimate proposed in (36). If it happens that the error variance $\sigma^2$ is known, or an estimate of it is available, then convergence (39) provides a way of constructing confidence interval for $g(x)$ with confidence coefficient approximately equal to $1 - \alpha$ for large $n$. (See also Exercise 5.2.)

In the regression model considered in Chapter 13, one of the basic tenets was that the point $x$ at which an observation $Y$ is to be made can be chosen, more or less, at will. This, however, need not always be the case. Instead, it may happen that the point $x$ itself is the observed value of a r.v. $X$. Thus, the setup here is as follows: A r.v. $X$ is observed, and if $X = x$, then an observation is taken at the point $x$.

In this framework, several questions may be posed. One of the most important is this: Given that $X = x$, construct a *predictor* of $Y$ corresponding to $x$. The proposed predictor is the conditional expectation of $Y$, given $X = x$, call it $m(x)$; i.e.,

$$m(x) = E(Y \mid X = x). \tag{40}$$

The quantity defined in (40) is an *unknown* function of $x$, since the conditional p.d.f. $Y$, given $X = x$, is unknown. The problem which then arises is that of estimating $m(x)$. The discussion in the remainder of this section revolves around this question.

Clearly, the estimation of $m(x)$ must be made on the basis of available data. To this effect, we assume that we have at our disposal $n$ pairs of r.v.'s $(X_i, Y_i), i = 1, \ldots, n$ which are independent and distributed as the pair $(X, Y)$. Then the proposed estimate of $m(x)$, call it $\hat{m}_n(x)$, is the following:

$$\hat{m}_n(x) = \frac{\hat{w}_n(x)}{\hat{f}_n(x)}, \quad \text{where } \hat{w}_n(x) = \frac{1}{nh_n} \sum_{i=1}^{n} Y_i K\left(\frac{x - X_i}{h_n}\right), \tag{41}$$

and

$$\hat{f}_n(x) \text{ is given in (31); i.e., } \hat{f}_n(x) = \frac{1}{nh_n} \sum_{i=1}^{n} K\left(\frac{x - X_i}{h_n}\right).$$

The estimated predictor $\hat{m}_n(x)$ has several asymptotic optimal properties of which we single out only one here; namely, asymptotic normality.

**Theorem 11.** *Let $\hat{m}_n(x)$ be the estimate of the predictor $m(x)$ given by (41) and (40), respectively, and let $\sigma^2(x)$ be defined by:*

$$\sigma^2(x) = \frac{\sigma_0^2(x)}{f(x)} \int_{-\infty}^{\infty} K^2(t)\, dt \quad (\text{for } f(x) > 0), \tag{42}$$

*where $\sigma_0^2(x)$ is the conditional variance of $Y$, given $X = x$. Then, under suitable regularity conditions, the estimated predictor $\hat{m}_n(x)$, properly normalized, is asymptotically Normal; i.e.,*

$$\sqrt{nh_n}[\hat{m}_n(x) - m(x)] \xrightarrow[n \to \infty]{d} Y \sim N(0, \sigma^2(x)). \tag{43}$$

The variance $\sigma^2(x)$ of the limiting Normal distribution is unknown, but an estimate of it may be constructed. Then the convergence (43) may be used to set up a confidence interval for $m(x)$ with confidence coefficient approximately equal to $1 - \alpha$ for large $n$. (See also Exercise 5.3.)

In closing this section, it should be mentioned that its purpose has been not to list detailed assumptions and present proofs (many of which are beyond the assumed level of this book, anyway), but rather to point out that there are regression results available in the literature, way beyond the simple linear model studied in Chapter 13. Finally, let us mention a piece a terminology used in the literature; namely, the regression model defined by (35) is referred to as a *fixed design* regression model, whereas the one defined by (40) is called a *stochastic design* regression model. The reasons for this are obvious. In the former case, the points where observations are taken are *fixed*, whereas in the latter case they are *values of a r.v. X.*

## EXERCISES

**5.1**  **Note:**  All convergences in this exercise hold for continuity points $x$ of $f(x)$.

In Theorem 7, it is stated that

$$\frac{\hat{f}_n(x) - E\hat{f}_n(x)}{\sigma[\hat{f}_n(x)]} \xrightarrow[n \to \infty]{d} Z \sim N(0, 1).$$

By this fact, Theorem 5, and some additional assumptions, it is also shown that

$$\frac{\hat{f}_n(x) - f(x)}{\sigma[\hat{f}_n(x)]} \xrightarrow[n \to \infty]{d} Z \sim N(0, 1). \tag{44}$$

(i)  Use expression (31) in order to show that

$$Var[\hat{f}_n(x)] = \frac{1}{nh_n^2} Var\left[K\left(\frac{x - X_1}{h_n}\right)\right], \quad \text{so that}$$

$$(nh_n)Var[\hat{f}_n(x)] = \frac{1}{h_n} Var\left[K\left(\frac{x - X_1}{h_n}\right)\right].$$

(ii)  Use the formula $Var(X) = EX^2 - (EX)^2$, and the transformation $\frac{x-y}{h_n} = u$ with $-\infty < u < \infty$, in order to show that

$$(nh_n)Var(\hat{f}_n(x)) = \int_{-\infty}^{\infty} K^2(u)f(x - h_n u)\, du$$

$$- h_n \left[\int_{-\infty}^{\infty} K(u)f(x - h_n u)\, du\right]^2.$$

Now, by Bochner's theorem (see also the proof of Theorem 6),

$$\int_{-\infty}^{\infty} K(u)f(x - h_n u)\, du \xrightarrow[n \to \infty]{} f(x) \int_{-\infty}^{\infty} K(u)du = f(x),$$

and

$$\int_{-\infty}^{\infty} K^2(u)f(x - h_n u)du \xrightarrow[n \to \infty]{} f(x) \int_{-\infty}^{\infty} K^2(u)\, du.$$

From these results, assumption (34)(i), and part (ii), it follows then that

$$\sigma_n^2(x) \stackrel{\text{def}}{=} (nh_n)Var[\hat{f}_n(x)] \xrightarrow[n \to \infty]{} f(x) \int_{-\infty}^{\infty} K^2(u)\, du \stackrel{\text{def}}{=} \sigma^2(x). \tag{45}$$

(iii)  From convergence (45) and Theorem 7, conclude (by means of the Corollary to Theorem 5 in Chapter 7) that

$$\hat{f}_n(x) - E\hat{f}_n(x) \xrightarrow[n \to \infty]{d} 0, \quad \text{and hence } \hat{f}_n(x) - E\hat{f}_n(x) \xrightarrow[n \to \infty]{P} 0. \tag{46}$$

(iv)  Use convergence (46) and Theorem 5 in order to conclude that $\hat{f}_n(x)$ is a *consistent* estimate of $f(x)$ (in the probability sense); i.e., $\hat{f}_n(x) \xrightarrow[n \to \infty]{P} f(x)$.

Set

$$\hat{\sigma}_n^2(x) = \hat{f}_n(x) \int_{-\infty}^{\infty} K^2(u)\, du. \tag{47}$$

**(v)** Use relations (45) and (47) to conclude that

$$\frac{\hat{\sigma}_n^2(x)}{\sigma_n^2(x)} \xrightarrow[n\to\infty]{P} 1, \quad \text{or} \quad \frac{\hat{\sigma}_n(x)}{\sigma_n(x)} \xrightarrow[n\to\infty]{P} 1. \tag{48}$$

Since, by (44) and (45),

$$\frac{\hat{f}_n(x) - f(x)}{\sigma[\hat{f}_n(x)]} = \frac{\sqrt{nh_n}[\hat{f}_n(x) - f(x)]}{\sqrt{nh_n Var[\hat{f}_n(x)]}}$$

$$= \frac{\sqrt{nh_n}[\hat{f}_n(x) - f(x)]}{\sigma_n(x)} \xrightarrow[n\to\infty]{d} Z \sim N(0,1),$$

it follows from this and (48) (by means of Theorem 6 in Chapter 7) that

$$\frac{\sqrt{nh_n}[\hat{f}_n(x) - f(x)]/\sigma_n(x)}{\hat{\sigma}_n(x)/\sigma_n(x)} = \frac{\sqrt{nh_n}[\hat{f}_n(x) - f(x)]}{\hat{\sigma}_n(x)} \xrightarrow[n\to\infty]{d} Z \sim N(0,1). \tag{49}$$

**(vi)** Use convergence (49) in order to conclude that, for all sufficiently large $n$,

$$P\left[\hat{f}_n(x) - \frac{\hat{\sigma}_n(x)}{\sqrt{nh_n}} z_{\alpha/2} \le f(x) \le \hat{f}_n(x) + \frac{\hat{\sigma}_n(x)}{\sqrt{nh_n}} z_{\alpha/2}\right] \simeq 1 - \alpha;$$

i.e., the interval $[\hat{f}_n(x) - \frac{\hat{\sigma}_n(x)}{\sqrt{nh_n}} z_{\alpha/2}, \hat{f}_n(x) + \frac{\hat{\sigma}_n(x)}{\sqrt{nh_n}} z_{\alpha/2}]$ is a confidence interval for $f(x)$ with confidence coefficient approximately $1 - \alpha$, for all sufficiently large $n$.

**5.2**   Refer to convergence (39), and set $\sigma_n(x) = \sigma[\hat{g}_n(x; x_n)]$. Use relation (39) in order to conclude that, for all sufficiently, large $n$,

$$P[\hat{g}_n(x; x_n) - z_{\alpha/2}\sigma_n(x) \le g(x) \le \hat{g}_n(x; x_n) + z_{\alpha/2}\sigma_n(x)] \simeq 1 - \alpha. \tag{50}$$

Thus, if $\sigma_n(x)$ is known, then expression (50) states that the interval $[\hat{g}_n(x; x_n) - z_{\alpha/2}\sigma_n(x), \hat{g}_n(x; x_n) + z_{\alpha/2}\sigma_n(x)]$ is a confidence interval for $g(x)$ with confidence coefficient approximately $1 - \alpha$, for all sufficiently large $n$. If $\sigma_n(x)$ is not known, but a suitable estimate of it, $\hat{\sigma}_n(x)$, can be constructed, then the interval $[\hat{g}_n(x; x_n) - z_{\alpha/2}\hat{\sigma}_n(x), \hat{g}_n(x; x_n) + z_{\alpha/2}\hat{\sigma}_n(x)]$ is a confidence interval for $g(x)$ with confidence coefficient approximately $1 - \alpha$, for all sufficiently large $n$. One arrives at this conclusion working as in Exercise 5.1.

**5.3**   Refer to convergence (43), and go through the usual manipulations to conclude that, for all sufficiently large $n$,

$$P\left[\hat{m}_n(x) - z_{\alpha/2}\frac{\sigma(x)}{\sqrt{nh_n}} \le m(x) \le \hat{m}_n(x) + z_{\alpha/2}\frac{\sigma(x)}{\sqrt{nh_n}}\right] \simeq 1 - \alpha. \tag{51}$$

Thus, if $\sigma(x)$ is known, then expression (51) states that the interval $[\hat{m}_n(x) - \frac{\sigma(x)}{\sqrt{nh_n}}z_{\alpha/2}, \hat{m}_n(x) + \frac{\sigma(x)}{\sqrt{nh_n}}z_{\alpha/2}]$ is a confidence interval for $m(x)$ with confidence coefficient approximately $1 - \alpha$, for all sufficiently large $n$. If $\sigma(x)$ is not known, but a suitable estimate of it, $\hat{\sigma}_n(x)$, can be constructed, then the interval $[\hat{m}_n(x) - \frac{\hat{\sigma}_n(x)}{\sqrt{nh_n}}z_{\alpha/2}, \hat{m}_n(x) + \frac{\hat{\sigma}_n(x)}{\sqrt{nh_n}}z_{\alpha/2}]$ is a confidence interval for $m(x)$ with confidence coefficient approximately $1 - \alpha$, for all sufficiently large $n$.

# Tables

## Table 1 The Cumulative Binomial Distribution

The tabulated quantity is

$$\sum_{j=0}^{k} \binom{n}{j} p^j (1-p)^{n-j}.$$

| n | k | 1/16 | 2/16 | 3/16 | 4/16 | 5/16 | 6/16 | 7/16 | 8/16 |
|---|---|------|------|------|------|------|------|------|------|
| 2 | 0 | 0.8789 | 0.7656 | 0.6602 | 0.5625 | 0.4727 | 0.3906 | 0.3164 | 0.2500 |
|   | 1 | 0.9961 | 0.9844 | 0.9648 | 0.9375 | 0.9023 | 0.8594 | 0.8086 | 0.7500 |
|   | 2 | 1.0000 | 1.0000 | 1.0000 | 1.0000 | 1.0000 | 1.0000 | 1.0000 | 1.0000 |
| 3 | 0 | 0.8240 | 0.6699 | 0.5364 | 0.4219 | 0.3250 | 0.2441 | 0.1780 | 0.1250 |
|   | 1 | 0.9888 | 0.9570 | 0.9077 | 0.8437 | 0.7681 | 0.6836 | 0.5933 | 0.5000 |
|   | 2 | 0.9998 | 0.9980 | 0.9934 | 0.9844 | 0.9695 | 0.9473 | 0.9163 | 0.8750 |
|   | 3 | 1.0000 | 1.0000 | 1.0000 | 1.0000 | 1.0000 | 1.0000 | 1.0000 | 1.0000 |
| 4 | 0 | 0.7725 | 0.5862 | 0.4358 | 0.3164 | 0.2234 | 0.1526 | 0.1001 | 0.0625 |
|   | 1 | 0.9785 | 0.9211 | 0.8381 | 0.7383 | 0.6296 | 0.5188 | 0.4116 | 0.3125 |
|   | 2 | 0.9991 | 0.9929 | 0.9773 | 0.9492 | 0.9065 | 0.8484 | 0.7749 | 0.6875 |
|   | 3 | 1.0000 | 0.9998 | 0.9988 | 0.9961 | 0.9905 | 0.9802 | 0.9634 | 0.9375 |
|   | 4 | 1.0000 | 1.0000 | 1.0000 | 1.0000 | 1.0000 | 1.0000 | 1.0000 | 1.0000 |
| 5 | 0 | 0.7242 | 0.5129 | 0.3541 | 0.2373 | 0.1536 | 0.0954 | 0.0563 | 0.0312 |
|   | 1 | 0.9656 | 0.8793 | 0.7627 | 0.6328 | 0.5027 | 0.3815 | 0.2753 | 0.1875 |
|   | 2 | 0.9978 | 0.9839 | 0.9512 | 0.8965 | 0.8200 | 0.7248 | 0.6160 | 0.5000 |
|   | 3 | 0.9999 | 0.9989 | 0.9947 | 0.9844 | 0.9642 | 0.9308 | 0.8809 | 0.8125 |
|   | 4 | 1.0000 | 1.0000 | 0.9998 | 0.9990 | 0.9970 | 0.9926 | 0.9840 | 0.9687 |
|   | 5 | 1.0000 | 1.0000 | 1.0000 | 1.0000 | 1.0000 | 1.0000 | 1.0000 | 1.0000 |
| 6 | 0 | 0.6789 | 0.4488 | 0.2877 | 0.1780 | 0.1056 | 0.0596 | 0.0317 | 0.0156 |
|   | 1 | 0.9505 | 0.8335 | 0.6861 | 0.5339 | 0.3936 | 0.2742 | 0.1795 | 0.1094 |
|   | 2 | 0.9958 | 0.9709 | 0.9159 | 0.8306 | 0.7208 | 0.5960 | 0.4669 | 0.3437 |
|   | 3 | 0.9998 | 0.9970 | 0.9866 | 0.9624 | 0.9192 | 0.8535 | 0.7650 | 0.6562 |
|   | 4 | 1.0000 | 0.9998 | 0.9988 | 0.9954 | 0.9868 | 0.9694 | 0.9389 | 0.8906 |
|   | 5 | 1.0000 | 1.0000 | 1.0000 | 0.9998 | 0.9991 | 0.9972 | 0.9930 | 0.9844 |
|   | 6 | 1.0000 | 1.0000 | 1.0000 | 1.0000 | 1.0000 | 1.0000 | 1.0000 | 1.0000 |

*Continued*

**Table 1** The Cumulative Binomial Distribution *Continued*

| | | | | | p | | | | |
|---|---|---|---|---|---|---|---|---|---|
| n | k | 1/16 | 2/16 | 3/16 | 4/16 | 5/16 | 6/16 | 7/16 | 8/16 |
| 7 | 0 | 0.6365 | 0.3927 | 0.2338 | 0.1335 | 0.0726 | 0.0373 | 0.0178 | 0.0078 |
| | 1 | 0.9335 | 0.7854 | 0.6114 | 0.4449 | 0.3036 | 0.1937 | 0.1148 | 0.0625 |
| | 2 | 0.9929 | 0.9537 | 0.8728 | 0.7564 | 0.6186 | 0.4753 | 0.3412 | 0.2266 |
| | 3 | 0.9995 | 0.9938 | 0.9733 | 0.9294 | 0.8572 | 0.7570 | 0.6346 | 0.5000 |
| | 4 | 1.0000 | 0.9995 | 0.9965 | 0.9871 | 0.9656 | 0.9260 | 0.8628 | 0.7734 |
| | 5 | 1.0000 | 1.0000 | 0.9997 | 0.9987 | 0.9952 | 0.9868 | 0.9693 | 0.9375 |
| | 6 | 1.0000 | 1.0000 | 1.0000 | 0.9999 | 0.9997 | 0.9990 | 0.9969 | 0.9922 |
| | 7 | 1.0000 | 1.0000 | 1.0000 | 1.0000 | 1.0000 | 1.0000 | 1.0000 | 1.0000 |
| 8 | 0 | 0.5967 | 0.3436 | 0.1899 | 0.1001 | 0.0499 | 0.0233 | 0.0100 | 0.0039 |
| | 1 | 0.9150 | 0.7363 | 0.5406 | 0.3671 | 0.2314 | 0.1350 | 0.0724 | 0.0352 |
| | 2 | 0.9892 | 0.9327 | 0.8238 | 0.6785 | 0.5201 | 0.3697 | 0.2422 | 0.1445 |
| | 3 | 0.9991 | 0.9888 | 0.9545 | 0.8862 | 0.7826 | 0.6514 | 0.5062 | 0.3633 |
| | 4 | 1.0000 | 0.9988 | 0.9922 | 0.9727 | 0.9318 | 0.8626 | 0.7630 | 0.6367 |
| | 5 | 1.0000 | 0.9999 | 0.9991 | 0.9958 | 0.9860 | 0.9640 | 0.9227 | 0.8555 |
| | 6 | 1.0000 | 1.0000 | 0.9999 | 0.9996 | 0.9983 | 0.9944 | 0.9849 | 0.9648 |
| | 7 | 1.0000 | 1.0000 | 1.0000 | 1.0000 | 0.9999 | 0.9996 | 0.9987 | 0.9961 |
| | 8 | 1.0000 | 1.0000 | 1.0000 | 1.0000 | 1.0000 | 1.0000 | 1.0000 | 1.0000 |
| 9 | 0 | 0.5594 | 0.3007 | 0.1543 | 0.0751 | 0.0343 | 0.0146 | 0.0056 | 0.0020 |
| | 1 | 0.8951 | 0.6872 | 0.4748 | 0.3003 | 0.1747 | 0.0931 | 0.0451 | 0.0195 |
| | 2 | 0.9846 | 0.9081 | 0.7707 | 0.6007 | 0.4299 | 0.2817 | 0.1679 | 0.0898 |
| | 3 | 0.9985 | 0.9817 | 0.9300 | 0.8343 | 0.7006 | 0.5458 | 0.3907 | 0.2539 |
| | 4 | 0.9999 | 0.9975 | 0.9851 | 0.9511 | 0.8851 | 0.7834 | 0.6506 | 0.5000 |
| | 5 | 1.0000 | 0.9998 | 0.9978 | 0.9900 | 0.9690 | 0.9260 | 0.8528 | 0.7461 |
| | 6 | 1.0000 | 1.0000 | 0.9998 | 0.9987 | 0.9945 | 0.9830 | 0.9577 | 0.9102 |
| | 7 | 1.0000 | 1.0000 | 1.0000 | 0.9999 | 0.9994 | 0.9977 | 0.9926 | 0.9805 |
| | 8 | 1.0000 | 1.0000 | 1.0000 | 1.0000 | 1.0000 | 0.9999 | 0.9994 | 0.9980 |
| | 9 | 1.0000 | 1.0000 | 1.0000 | 1.0000 | 1.0000 | 1.0000 | 1.0000 | 1.0000 |
| 10 | 0 | 0.5245 | 0.2631 | 0.1254 | 0.0563 | 0.0236 | 0.0091 | 0.0032 | 0.0010 |
| | 1 | 0.8741 | 0.6389 | 0.4147 | 0.2440 | 0.1308 | 0.0637 | 0.0278 | 0.0107 |
| | 2 | 0.9790 | 0.8805 | 0.7152 | 0.5256 | 0.3501 | 0.2110 | 0.1142 | 0.0547 |
| | 3 | 0.9976 | 0.9725 | 0.9001 | 0.7759 | 0.6160 | 0.4467 | 0.2932 | 0.1719 |

*Continued*

**Table 1**  The Cumulative Binomial Distribution *Continued*

| | | | | | p | | | | |
|---|---|---|---|---|---|---|---|---|---|
| **n** | **k** | **1/16** | **2/16** | **3/16** | **4/16** | **5/16** | **6/16** | **7/16** | **8/16** |
| 10 | 4 | 0.9998 | 0.9955 | 0.9748 | 0.9219 | 0.8275 | 0.6943 | 0.5369 | 0.3770 |
| | 5 | 1.0000 | 0.9995 | 0.9955 | 0.9803 | 0.9428 | 0.8725 | 0.7644 | 0.6230 |
| | 6 | 1.0000 | 1.0000 | 0.9994 | 0.9965 | 0.9865 | 0.9616 | 0.9118 | 0.8281 |
| | 7 | 1.0000 | 1.0000 | 1.0000 | 0.9996 | 0.9979 | 0.9922 | 0.9773 | 0.9453 |
| | 8 | 1.0000 | 1.0000 | 1.0000 | 1.0000 | 0.9998 | 0.9990 | 0.9964 | 0.9893 |
| | 9 | 1.0000 | 1.0000 | 1.0000 | 1.0000 | 1.0000 | 0.9999 | 0.9997 | 0.9990 |
| | 10 | 1.0000 | 1.0000 | 1.0000 | 1.0000 | 1.0000 | 1.0000 | 1.0000 | 1.0000 |
| 11 | 0 | 0.4917 | 0.2302 | 0.1019 | 0.0422 | 0.0162 | 0.0057 | 0.0018 | 0.0005 |
| | 1 | 0.8522 | 0.5919 | 0.3605 | 0.1971 | 0.0973 | 0.0432 | 0.0170 | 0.0059 |
| | 2 | 0.9724 | 0.8503 | 0.6589 | 0.4552 | 0.2816 | 0.1558 | 0.0764 | 0.0327 |
| | 3 | 0.9965 | 0.9610 | 0.8654 | 0.7133 | 0.5329 | 0.3583 | 0.2149 | 0.1133 |
| | 4 | 0.9997 | 0.9927 | 0.9608 | 0.8854 | 0.7614 | 0.6014 | 0.4303 | 0.2744 |
| | 5 | 1.0000 | 0.9990 | 0.9916 | 0.9657 | 0.9068 | 0.8057 | 0.6649 | 0.5000 |
| | 6 | 1.0000 | 0.9999 | 0.9987 | 0.9924 | 0.9729 | 0.9282 | 0.8473 | 0.7256 |
| | 7 | 1.0000 | 1.0000 | 0.9999 | 0.9988 | 0.9943 | 0.9807 | 0.9487 | 0.8867 |
| | 8 | 1.0000 | 1.0000 | 1.0000 | 0.9999 | 0.9992 | 0.9965 | 0.9881 | 0.9673 |
| | 9 | 1.0000 | 1.0000 | 1.0000 | 1.0000 | 0.9999 | 0.9996 | 0.9983 | 0.9941 |
| | 10 | 1.0000 | 1.0000 | 1.0000 | 1.0000 | 1.0000 | 1.0000 | 0.9999 | 0.9995 |
| | 11 | 1.0000 | 1.0000 | 1.0000 | 1.0000 | 1.0000 | 1.0000 | 1.0000 | 1.0000 |
| 12 | 0 | 0.4610 | 0.2014 | 0.0828 | 0.0317 | 0.0111 | 0.0036 | 0.0010 | 0.0002 |
| | 1 | 0.8297 | 0.5467 | 0.3120 | 0.1584 | 0.0720 | 0.0291 | 0.0104 | 0.0032 |
| | 2 | 0.9649 | 0.8180 | 0.6029 | 0.3907 | 0.2240 | 0.1135 | 0.0504 | 0.0193 |
| | 3 | 0.9950 | 0.9472 | 0.8267 | 0.6488 | 0.4544 | 0.2824 | 0.1543 | 0.0730 |
| | 4 | 0.9995 | 0.9887 | 0.9429 | 0.8424 | 0.6900 | 0.5103 | 0.3361 | 0.1938 |
| | 5 | 1.0000 | 0.9982 | 0.9858 | 0.9456 | 0.8613 | 0.7291 | 0.5622 | 0.3872 |
| | 6 | 1.0000 | 0.9998 | 0.9973 | 0.9857 | 0.9522 | 0.8822 | 0.7675 | 0.6128 |
| | 7 | 1.0000 | 1.0000 | 0.9996 | 0.9972 | 0.9876 | 0.9610 | 0.9043 | 0.8062 |
| | 8 | 1.0000 | 1.0000 | 1.0000 | 0.9996 | 0.9977 | 0.9905 | 0.9708 | 0.9270 |
| | 9 | 1.0000 | 1.0000 | 1.0000 | 1.0000 | 0.9997 | 0.9984 | 0.9938 | 0.9807 |
| | 10 | 1.0000 | 1.0000 | 1.0000 | 1.0000 | 1.0000 | 0.9998 | 0.9992 | 0.9968 |
| | 11 | 1.0000 | 1.0000 | 1.0000 | 1.0000 | 1.0000 | 1.0000 | 1.0000 | 0.9998 |
| | 12 | 1.0000 | 1.0000 | 1.0000 | 1.0000 | 1.0000 | 1.0000 | 1.0000 | 1.0000 |

*Continued*

**Table 1** The Cumulative Binomial Distribution *Continued*

| n | k | 1/16 | 2/16 | 3/16 | 4/16 | 5/16 | 6/16 | 7/16 | 8/16 |
|---|---|------|------|------|------|------|------|------|------|
| | | | | | | *p* | | | |
| 13 | 0 | 0.4321 | 0.1762 | 0.0673 | 0.0238 | 0.0077 | 0.0022 | 0.0006 | 0.0001 |
| | 1 | 0.8067 | 0.5035 | 0.2690 | 0.1267 | 0.0530 | 0.0195 | 0.0063 | 0.0017 |
| | 2 | 0.9565 | 0.7841 | 0.5484 | 0.3326 | 0.1765 | 0.0819 | 0.0329 | 0.0112 |
| | 3 | 0.9931 | 0.9310 | 0.7847 | 0.5843 | 0.3824 | 0.2191 | 0.1089 | 0.0461 |
| | 4 | 0.9992 | 0.9835 | 0.9211 | 0.7940 | 0.6164 | 0.4248 | 0.2565 | 0.1334 |
| | 5 | 0.9999 | 0.9970 | 0.9778 | 0.9198 | 0.8078 | 0.6470 | 0.4633 | 0.2905 |
| | 6 | 1.0000 | 0.9996 | 0.9952 | 0.9757 | 0.9238 | 0.8248 | 0.6777 | 0.5000 |
| | 7 | 1.0000 | 1.0000 | 0.9992 | 0.9944 | 0.9765 | 0.9315 | 0.8445 | 0.7095 |
| | 8 | 1.0000 | 1.0000 | 0.9999 | 0.9990 | 0.9945 | 0.9795 | 0.9417 | 0.8666 |
| | 9 | 1.0000 | 1.0000 | 1.0000 | 0.9999 | 0.9991 | 0.9955 | 0.9838 | 0.9539 |
| | 10 | 1.0000 | 1.0000 | 1.0000 | 1.0000 | 0.9999 | 0.9993 | 0.9968 | 0.9888 |
| | 11 | 1.0000 | 1.0000 | 1.0000 | 1.0000 | 1.0000 | 0.9999 | 0.9996 | 0.9983 |
| | 12 | 1.0000 | 1.0000 | 1.0000 | 1.0000 | 1.0000 | 1.0000 | 1.0000 | 0.9999 |
| | 13 | 1.0000 | 1.0000 | 1.0000 | 1.0000 | 1.0000 | 1.0000 | 1.0000 | 1.0000 |
| 14 | 0 | 0.4051 | 0.1542 | 0.0546 | 0.0178 | 0.0053 | 0.0014 | 0.0003 | 0.0001 |
| | 1 | 0.7833 | 0.4626 | 0.2312 | 0.1010 | 0.0388 | 0.0130 | 0.0038 | 0.0009 |
| | 2 | 0.9471 | 0.7490 | 0.4960 | 0.2811 | 0.1379 | 0.0585 | 0.0213 | 0.0065 |
| | 3 | 0.9908 | 0.9127 | 0.7404 | 0.5213 | 0.3181 | 0.1676 | 0.0756 | 0.0287 |
| | 4 | 0.9988 | 0.9970 | 0.8955 | 0.7415 | 0.5432 | 0.3477 | 0.1919 | 0.0898 |
| | 5 | 0.9999 | 0.9953 | 0.9671 | 0.8883 | 0.7480 | 0.5637 | 0.3728 | 0.2120 |
| | 6 | 1.0000 | 0.9993 | 0.9919 | 0.9167 | 0.8876 | 0.7581 | 0.5839 | 0.3953 |
| | 7 | 1.0000 | 0.9999 | 0.9985 | 0.9897 | 0.9601 | 0.8915 | 0.7715 | 0.6047 |
| | 8 | 1.0000 | 1.0000 | 0.9998 | 0.9978 | 0.9889 | 0.9615 | 0.8992 | 0.7880 |
| | 9 | 1.0000 | 1.0000 | 1.0000 | 0.9997 | 0.9976 | 0.9895 | 0.9654 | 0.9102 |
| | 10 | 1.0000 | 1.0000 | 1.0000 | 1.0000 | 0.9996 | 0.9979 | 0.9911 | 0.9713 |
| | 11 | 1.0000 | 1.0000 | 1.0000 | 1.0000 | 1.0000 | 0.9997 | 0.9984 | 0.9935 |
| | 12 | 1.0000 | 1.0000 | 1.0000 | 1.0000 | 1.0000 | 1.0000 | 0.9998 | 0.9991 |
| | 13 | 1.0000 | 1.0000 | 1.0000 | 1.0000 | 1.0000 | 1.0000 | 1.0000 | 0.9999 |
| | 14 | 1.0000 | 1.0000 | 1.0000 | 1.0000 | 1.0000 | 1.0000 | 1.0000 | 1.0000 |
| 15 | 0 | 0.3798 | 0.1349 | 0.0444 | 0.0134 | 0.0036 | 0.0009 | 0.0002 | 0.0000 |
| | 1 | 0.7596 | 0.4241 | 0.1981 | 0.0802 | 0.0283 | 0.0087 | 0.0023 | 0.0005 |
| | 2 | 0.9369 | 0.7132 | 0.4463 | 0.2361 | 0.1069 | 0.0415 | 0.0136 | 0.0037 |

*Continued*

**Table 1** The Cumulative Binomial Distribution *Continued*

| n | k | 1/16 | 2/16 | 3/16 | 4/16 | 5/16 | 6/16 | 7/16 | 8/16 |
|---|---|------|------|------|------|------|------|------|------|
| 15 | 3 | 0.9881 | 0.8922 | 0.6946 | 0.4613 | 0.2618 | 0.1267 | 0.0518 | 0.0176 |
|    | 4 | 0.9983 | 0.9689 | 0.8665 | 0.6865 | 0.4729 | 0.2801 | 0.1410 | 0.0592 |
|    | 5 | 0.9998 | 0.9930 | 0.9537 | 0.8516 | 0.6840 | 0.4827 | 0.2937 | 0.1509 |
|    | 6 | 1.0000 | 0.9988 | 0.9873 | 0.9434 | 0.8435 | 0.6852 | 0.4916 | 0.3036 |
|    | 7 | 1.0000 | 0.9998 | 0.9972 | 0.9827 | 0.9374 | 0.8415 | 0.6894 | 0.5000 |
|    | 8 | 1.0000 | 1.0000 | 0.9995 | 0.9958 | 0.9799 | 0.9352 | 0.8433 | 0.6964 |
|    | 9 | 1.0000 | 1.0000 | 0.9999 | 0.9992 | 0.9949 | 0.9790 | 0.9364 | 0.8491 |
|    | 10 | 1.0000 | 1.0000 | 1.0000 | 0.9999 | 0.9990 | 0.9947 | 0.9799 | 0.9408 |
|    | 11 | 1.0000 | 1.0000 | 1.0000 | 1.0000 | 0.9999 | 0.9990 | 0.9952 | 0.9824 |
|    | 12 | 1.0000 | 1.0000 | 1.0000 | 1.0000 | 1.0000 | 0.9999 | 0.9992 | 0.9963 |
|    | 13 | 1.0000 | 1.0000 | 1.0000 | 1.0000 | 1.0000 | 1.0000 | 0.9999 | 0.9995 |
|    | 14 | 1.0000 | 1.0000 | 1.0000 | 1.0000 | 1.0000 | 1.0000 | 1.0000 | 1.0000 |
|    | 15 | 1.0000 | 1.0000 | 1.0000 | 1.0000 | 1.0000 | 1.0000 | 1.0000 | 1.0000 |
| 16 | 0 | 0.3561 | 0.1181 | 0.0361 | 0.0100 | 0.0025 | 0.0005 | 0.0001 | 0.0000 |
|    | 1 | 0.7359 | 0.3879 | 0.1693 | 0.0635 | 0.0206 | 0.0057 | 0.0014 | 0.0003 |
|    | 2 | 0.9258 | 0.6771 | 0.3998 | 0.1971 | 0.0824 | 0.0292 | 0.0086 | 0.0021 |
|    | 3 | 0.9849 | 0.8698 | 0.6480 | 0.4050 | 0.2134 | 0.0947 | 0.0351 | 0.0106 |
|    | 4 | 0.9977 | 0.9593 | 0.8342 | 0.6302 | 0.4069 | 0.2226 | 0.1020 | 0.0384 |
|    | 5 | 0.9997 | 0.9900 | 0.9373 | 0.8103 | 0.6180 | 0.4067 | 0.2269 | 0.1051 |
|    | 6 | 1.0000 | 0.9981 | 0.9810 | 0.9204 | 0.7940 | 0.6093 | 0.4050 | 0.2272 |
|    | 7 | 1.0000 | 0.9997 | 0.9954 | 0.9729 | 0.9082 | 0.7829 | 0.6029 | 0.4018 |
|    | 8 | 1.0000 | 1.0000 | 0.9991 | 0.9925 | 0.9666 | 0.9001 | 0.7760 | 0.5982 |
|    | 9 | 1.0000 | 1.0000 | 0.9999 | 0.9984 | 0.9902 | 0.9626 | 0.8957 | 0.7728 |
|    | 10 | 1.0000 | 1.0000 | 1.0000 | 0.9997 | 0.9977 | 0.9888 | 0.9609 | 0.8949 |
|    | 11 | 1.0000 | 1.0000 | 1.0000 | 1.0000 | 0.9996 | 0.9974 | 0.9885 | 0.9616 |
|    | 12 | 1.0000 | 1.0000 | 1.0000 | 1.0000 | 0.9999 | 0.9995 | 0.9975 | 0.9894 |
|    | 13 | 1.0000 | 1.0000 | 1.0000 | 1.0000 | 1.0000 | 0.9999 | 0.9996 | 0.9979 |
|    | 14 | 1.0000 | 1.0000 | 1.0000 | 1.0000 | 1.0000 | 1.0000 | 1.0000 | 0.9997 |
|    | 15 | 1.0000 | 1.0000 | 1.0000 | 1.0000 | 1.0000 | 1.0000 | 1.0000 | 1.0000 |
|    | 16 | 1.0000 | 1.0000 | 1.0000 | 1.0000 | 1.0000 | 1.0000 | 1.0000 | 1.0000 |
| 17 | 0 | 0.3338 | 0.1033 | 0.0293 | 0.0075 | 0.0017 | 0.0003 | 0.0001 | 0.0000 |
|    | 1 | 0.7121 | 0.3542 | 0.1443 | 0.0501 | 0.0149 | 0.0038 | 0.0008 | 0.0001 |

*Continued*

**Table 1** The Cumulative Binomial Distribution *Continued*

| n | k | 1/16 | 2/16 | 3/16 | 4/16 | 5/16 | 6/16 | 7/16 | 8/16 |
|---|---|------|------|------|------|------|------|------|------|
| | | | | | | *p* | | | |
| 17 | 2 | 0.9139 | 0.6409 | 0.3566 | 0.1637 | 0.0631 | 0.0204 | 0.0055 | 0.0012 |
| | 3 | 0.9812 | 0.8457 | 0.6015 | 0.3530 | 0.1724 | 0.0701 | 0.0235 | 0.0064 |
| | 4 | 0.9969 | 0.9482 | 0.7993 | 0.5739 | 0.3464 | 0.1747 | 0.0727 | 0.0245 |
| | 5 | 0.9996 | 0.9862 | 0.9180 | 0.7653 | 0.5520 | 0.3377 | 0.1723 | 0.0717 |
| | 6 | 1.0000 | 0.9971 | 0.9728 | 0.8929 | 0.7390 | 0.5333 | 0.3271 | 0.1662 |
| | 7 | 1.0000 | 0.9995 | 0.9927 | 0.9598 | 0.8725 | 0.7178 | 0.5163 | 0.3145 |
| | 8 | 1.0000 | 0.9999 | 0.9984 | 0.9876 | 0.9484 | 0.8561 | 0.7002 | 0.5000 |
| | 9 | 1.0000 | 1.0000 | 0.9997 | 0.9969 | 0.9828 | 0.9391 | 0.8433 | 0.6855 |
| | 10 | 1.0000 | 1.0000 | 1.0000 | 0.9994 | 0.9954 | 0.9790 | 0.9323 | 0.8338 |
| | 11 | 1.0000 | 1.0000 | 1.0000 | 0.9999 | 0.9990 | 0.9942 | 0.9764 | 0.9283 |
| | 12 | 1.0000 | 1.0000 | 1.0000 | 1.0000 | 0.9998 | 0.9987 | 0.9935 | 0.9755 |
| | 13 | 1.0000 | 1.0000 | 1.0000 | 1.0000 | 1.0000 | 0.9998 | 0.9987 | 0.9936 |
| | 14 | 1.0000 | 1.0000 | 1.0000 | 1.0000 | 1.0000 | 1.0000 | 0.9998 | 0.9988 |
| | 15 | 1.0000 | 1.0000 | 1.0000 | 1.0000 | 1.0000 | 1.0000 | 1.0000 | 0.9999 |
| | 16 | 1.0000 | 1.0000 | 1.0000 | 1.0000 | 1.0000 | 1.0000 | 1.0000 | 1.0000 |
| 18 | 0 | 0.3130 | 0.0904 | 0.0238 | 0.0056 | 0.0012 | 0.0002 | 0.0000 | 0.0000 |
| | 1 | 0.6885 | 0.3228 | 0.1227 | 0.0395 | 0.0108 | 0.0025 | 0.0005 | 0.0001 |
| | 2 | 0.9013 | 0.6051 | 0.3168 | 0.1353 | 0.0480 | 0.0142 | 0.0034 | 0.0007 |
| | 3 | 0.9770 | 0.8201 | 0.5556 | 0.3057 | 0.1383 | 0.0515 | 0.0156 | 0.0038 |
| | 4 | 0.9959 | 0.9354 | 0.7622 | 0.5187 | 0.2920 | 0.1355 | 0.0512 | 0.0154 |
| | 5 | 0.9994 | 0.9814 | 0.8958 | 0.7175 | 0.4878 | 0.2765 | 0.1287 | 0.0481 |
| | 6 | 0.9999 | 0.9957 | 0.9625 | 0.8610 | 0.6806 | 0.4600 | 0.2593 | 0.1189 |
| | 7 | 1.0000 | 0.9992 | 0.9889 | 0.9431 | 0.8308 | 0.6486 | 0.4335 | 0.2403 |
| | 8 | 1.0000 | 0.9999 | 0.9973 | 0.9807 | 0.9247 | 0.8042 | 0.6198 | 0.4073 |
| | 9 | 1.0000 | 1.0000 | 0.9995 | 0.9946 | 0.9721 | 0.9080 | 0.7807 | 0.5927 |
| | 10 | 1.0000 | 1.0000 | 0.9999 | 0.9988 | 0.9915 | 0.9640 | 0.8934 | 0.7597 |
| | 11 | 1.0000 | 1.0000 | 1.0000 | 0.9998 | 0.9979 | 0.9885 | 0.9571 | 0.8811 |
| | 12 | 1.0000 | 1.0000 | 1.0000 | 1.0000 | 0.9996 | 0.9970 | 0.9860 | 0.9519 |
| | 13 | 1.0000 | 1.0000 | 1.0000 | 1.0000 | 0.9999 | 0.9994 | 0.9964 | 0.9846 |
| | 14 | 1.0000 | 1.0000 | 1.0000 | 1.0000 | 1.0000 | 0.9999 | 0.9993 | 0.9962 |
| | 15 | 1.0000 | 1.0000 | 1.0000 | 1.0000 | 1.0000 | 1.0000 | 0.9999 | 0.9993 |
| | 16 | 1.0000 | 1.0000 | 1.0000 | 1.0000 | 1.0000 | 1.0000 | 1.0000 | 0.9999 |
| | 17 | 1.0000 | 1.0000 | 1.0000 | 1.0000 | 1.0000 | 1.0000 | 1.0000 | 1.0000 |

*Continued*

**Table 1** The Cumulative Binomial Distribution *Continued*

| n | k | 1/16 | 2/16 | 3/16 | 4/16 | 5/16 | 6/16 | 7/16 | 8/16 |
|---|---|------|------|------|------|------|------|------|------|
| 19 | 0 | 0.2934 | 0.0791 | 0.0193 | 0.0042 | 0.0008 | 0.0001 | 0.0000 | 0.0000 |
| | 1 | 0.6650 | 0.2938 | 0.1042 | 0.0310 | 0.0078 | 0.0016 | 0.0003 | 0.0000 |
| | 2 | 0.8880 | 0.5698 | 0.2804 | 0.1113 | 0.0364 | 0.0098 | 0.0021 | 0.0004 |
| | 3 | 0.9722 | 0.7933 | 0.5108 | 0.2631 | 0.1101 | 0.0375 | 0.0103 | 0.0022 |
| | 4 | 0.9947 | 0.9209 | 0.7235 | 0.4654 | 0.2440 | 0.1040 | 0.0356 | 0.0096 |
| | 5 | 0.9992 | 0.9757 | 0.8707 | 0.6678 | 0.4266 | 0.2236 | 0.0948 | 0.0318 |
| | 6 | 0.9999 | 0.9939 | 0.9500 | 0.8251 | 0.6203 | 0.3912 | 0.2022 | 0.0835 |
| | 7 | 1.0000 | 0.9988 | 0.9840 | 0.9225 | 0.7838 | 0.5779 | 0.3573 | 0.1796 |
| | 8 | 1.0000 | 0.9998 | 0.9957 | 0.9713 | 0.8953 | 0.7459 | 0.5383 | 0.3238 |
| | 9 | 1.0000 | 1.0000 | 0.9991 | 0.9911 | 0.9573 | 0.8691 | 0.7103 | 0.5000 |
| | 10 | 1.0000 | 1.0000 | 0.9998 | 0.9977 | 0.9854 | 0.9430 | 0.8441 | 0.0672 |
| | 11 | 1.0000 | 1.0000 | 1.0000 | 0.9995 | 0.9959 | 0.9793 | 0.9292 | 0.8204 |
| | 12 | 1.0000 | 1.0000 | 1.0000 | 0.9999 | 0.9990 | 0.9938 | 0.9734 | 0.9165 |
| | 13 | 1.0000 | 1.0000 | 1.0000 | 1.0000 | 0.9998 | 0.9985 | 0.9919 | 0.9682 |
| | 14 | 1.0000 | 1.0000 | 1.0000 | 1.0000 | 1.0000 | 0.9997 | 0.9980 | 0.9904 |
| | 15 | 1.0000 | 1.0000 | 1.0000 | 1.0000 | 1.0000 | 1.0000 | 0.9996 | 0.9978 |
| | 16 | 1.0000 | 1.0000 | 1.0000 | 1.0000 | 1.0000 | 1.0000 | 1.0000 | 0.9996 |
| | 17 | 1.0000 | 1.0000 | 1.0000 | 1.0000 | 1.0000 | 1.0000 | 1.0000 | 1.0000 |
| | 18 | 1.0000 | 1.0000 | 1.0000 | 1.0000 | 1.0000 | 1.0000 | 1.0000 | 1.0000 |
| 20 | 0 | 0.2751 | 0.0692 | 0.0157 | 0.0032 | 0.0006 | 0.0001 | 0.0000 | 0.0000 |
| | 1 | 0.6148 | 0.2669 | 0.0883 | 0.0243 | 0.0056 | 0.0011 | 0.0002 | 0.0000 |
| | 2 | 0.8741 | 0.5353 | 0.2473 | 0.0913 | 0.0275 | 0.0067 | 0.0013 | 0.0002 |
| | 3 | 0.9670 | 0.7653 | 0.4676 | 0.2252 | 0.0870 | 0.0271 | 0.0067 | 0.0013 |
| | 4 | 0.9933 | 0.9050 | 0.6836 | 0.4148 | 0.2021 | 0.0790 | 0.0245 | 0.0059 |
| | 5 | 0.9989 | 0.9688 | 0.8431 | 0.6172 | 0.3695 | 0.1788 | 0.0689 | 0.0207 |
| | 6 | 0.9999 | 0.9916 | 0.9351 | 0.7858 | 0.5598 | 0.3284 | 0.1552 | 0.0577 |
| | 7 | 1.0000 | 0.9981 | 0.9776 | 0.8982 | 0.7327 | 0.5079 | 0.2894 | 0.1316 |
| | 8 | 1.0000 | 0.9997 | 0.9935 | 0.9591 | 0.8605 | 0.6829 | 0.4591 | 0.2517 |
| | 9 | 1.0000 | 0.9999 | 0.9984 | 0.9861 | 0.9379 | 0.8229 | 0.6350 | 0.4119 |
| | 10 | 1.0000 | 1.0000 | 0.9997 | 0.9961 | 0.9766 | 0.9153 | 0.7856 | 0.5881 |
| | 11 | 1.0000 | 1.0000 | 0.9999 | 0.9991 | 0.9926 | 0.9657 | 0.8920 | 0.7483 |
| | 12 | 1.0000 | 1.0000 | 1.0000 | 0.9998 | 0.9981 | 0.9884 | 0.9541 | 0.8684 |
| | 13 | 1.0000 | 1.0000 | 1.0000 | 1.0000 | 0.9996 | 0.9968 | 0.9838 | 0.9423 |

*Continued*

**Table 1** The Cumulative Binomial Distribution *Continued*

| n | k | 1/16 | 2/16 | 3/16 | 4/16 | 5/16 | 6/16 | 7/16 | 8/16 |
|---|---|------|------|------|------|------|------|------|------|
| 20 | 14 | 1.0000 | 1.0000 | 1.0000 | 1.0000 | 0.9999 | 0.9993 | 0.9953 | 0.9793 |
| | 15 | 1.0000 | 1.0000 | 1.0000 | 1.0000 | 1.0000 | 0.9999 | 0.9989 | 0.9941 |
| | 16 | 1.0000 | 1.0000 | 1.0000 | 1.0000 | 1.0000 | 1.0000 | 0.9998 | 0.9987 |
| | 17 | 1.0000 | 1.0000 | 1.0000 | 1.0000 | 1.0000 | 1.0000 | 1.0000 | 0.9998 |
| | 18 | 1.0000 | 1.0000 | 1.0000 | 1.0000 | 1.0000 | 1.0000 | 1.0000 | 1.0000 |
| | 19 | 1.0000 | 1.0000 | 1.0000 | 1.0000 | 1.0000 | 1.0000 | 1.0000 | 1.0000 |
| 21 | 0 | 0.2579 | 0.0606 | 0.0128 | 0.0024 | 0.0004 | 0.0001 | 0.0000 | 0.0000 |
| | 1 | 0.6189 | 0.2422 | 0.0747 | 0.0190 | 0.0040 | 0.0007 | 0.0001 | 0.0000 |
| | 2 | 0.8596 | 0.5018 | 0.2175 | 0.0745 | 0.0206 | 0.0046 | 0.0008 | 0.0001 |
| | 3 | 0.9612 | 0.7366 | 0.4263 | 0.1917 | 0.0684 | 0.0195 | 0.0044 | 0.0007 |
| | 4 | 0.9917 | 0.8875 | 0.6431 | 0.3674 | 0.1662 | 0.0596 | 0.0167 | 0.0036 |
| | 5 | 0.9986 | 0.9609 | 0.8132 | 0.5666 | 0.3172 | 0.1414 | 0.0495 | 0.0133 |
| | 6 | 0.9998 | 0.9888 | 0.9179 | 0.7436 | 0.5003 | 0.2723 | 0.1175 | 0.0392 |
| | 7 | 1.0000 | 0.9973 | 0.9696 | 0.8701 | 0.6787 | 0.4405 | 0.2307 | 0.0946 |
| | 8 | 1.0000 | 0.9995 | 0.9906 | 0.9439 | 0.8206 | 0.6172 | 0.3849 | 0.1917 |
| | 9 | 1.0000 | 0.9999 | 0.9975 | 0.9794 | 0.9137 | 0.7704 | 0.5581 | 0.3318 |
| | 10 | 1.0000 | 1.0000 | 0.9995 | 0.9936 | 0.9645 | 0.8806 | 0.7197 | 0.5000 |
| | 11 | 1.0000 | 1.0000 | 0.9999 | 0.9983 | 0.9876 | 0.9468 | 0.8454 | 0.6682 |
| | 12 | 1.0000 | 1.0000 | 1.0000 | 0.9996 | 0.9964 | 0.9799 | 0.9269 | 0.8083 |
| | 13 | 1.0000 | 1.0000 | 1.0000 | 0.9999 | 0.9991 | 0.9936 | 0.9708 | 0.9054 |
| | 14 | 1.0000 | 1.0000 | 1.0000 | 1.0000 | 0.9998 | 0.9983 | 0.9903 | 0.9605 |
| | 15 | 1.0000 | 1.0000 | 1.0000 | 1.0000 | 1.0000 | 0.9996 | 0.9974 | 0.9867 |
| | 16 | 1.0000 | 1.0000 | 1.0000 | 1.0000 | 1.0000 | 0.9999 | 0.9994 | 0.9964 |
| | 17 | 1.0000 | 1.0000 | 1.0000 | 1.0000 | 1.0000 | 1.0000 | 0.9999 | 0.9993 |
| | 18 | 1.0000 | 1.0000 | 1.0000 | 1.0000 | 1.0000 | 1.0000 | 1.0000 | 0.9999 |
| | 19 | 1.0000 | 1.0000 | 1.0000 | 1.0000 | 1.0000 | 1.0000 | 1.0000 | 1.0000 |
| | 20 | 1.0000 | 1.0000 | 1.0000 | 1.0000 | 1.0000 | 1.0000 | 1.0000 | 1.0000 |
| 22 | 0 | 0.2418 | 0.0530 | 0.0104 | 0.0018 | 0.0003 | 0.0000 | 0.0000 | 0.0000 |
| | 1 | 0.5963 | 0.2195 | 0.0631 | 0.0149 | 0.0029 | 0.0005 | 0.0001 | 0.0000 |
| | 2 | 0.8445 | 0.4693 | 0.1907 | 0.0606 | 0.0154 | 0.0031 | 0.0005 | 0.0001 |
| | 3 | 0.9548 | 0.7072 | 0.3871 | 0.1624 | 0.0535 | 0.0139 | 0.0028 | 0.0004 |

*Continued*

**Table 1** The Cumulative Binomial Distribution *Continued*

| n | k | 1/16 | 2/16 | 3/16 | 4/16 | 5/16 | 6/16 | 7/16 | 8/16 |
|---|---|------|------|------|------|------|------|------|------|
| 22 | 4 | 0.9898 | 0.8687 | 0.6024 | 0.3235 | 0.1356 | 0.0445 | 0.0133 | 0.0022 |
| | 5 | 0.9981 | 0.9517 | 0.7813 | 0.5168 | 0.2700 | 0.1107 | 0.0352 | 0.0085 |
| | 6 | 0.9997 | 0.9853 | 0.8983 | 0.6994 | 0.4431 | 0.2232 | 0.0877 | 0.0267 |
| | 7 | 1.0000 | 0.9963 | 0.9599 | 0.8385 | 0.6230 | 0.3774 | 0.1812 | 0.0669 |
| | 8 | 1.0000 | 0.9992 | 0.9866 | 0.9254 | 0.7762 | 0.5510 | 0.3174 | 0.1431 |
| | 9 | 1.0000 | 0.9999 | 0.9962 | 0.9705 | 0.8846 | 0.7130 | 0.4823 | 0.2617 |
| | 10 | 1.0000 | 1.0000 | 0.9991 | 0.9900 | 0.9486 | 0.8393 | 0.6490 | 0.4159 |
| | 11 | 1.0000 | 1.0000 | 0.9998 | 0.9971 | 0.9804 | 0.9220 | 0.7904 | 0.5841 |
| | 12 | 1.0000 | 1.0000 | 1.0000 | 0.9993 | 0.9936 | 0.9675 | 0.8913 | 0.7383 |
| | 13 | 1.0000 | 1.0000 | 1.0000 | 0.9999 | 0.9982 | 0.9885 | 0.9516 | 0.8569 |
| | 14 | 1.0000 | 1.0000 | 1.0000 | 1.0000 | 0.9996 | 0.9966 | 0.9818 | 0.9331 |
| | 15 | 1.0000 | 1.0000 | 1.0000 | 1.0000 | 0.9999 | 0.9991 | 0.9943 | 0.9739 |
| | 16 | 1.0000 | 1.0000 | 1.0000 | 1.0000 | 1.0000 | 0.9998 | 0.9985 | 0.9915 |
| | 17 | 1.0000 | 1.0000 | 1.0000 | 1.0000 | 1.0000 | 1.0000 | 0.9997 | 0.9978 |
| | 18 | 1.0000 | 1.0000 | 1.0000 | 1.0000 | 1.0000 | 1.0000 | 1.0000 | 0.9995 |
| | 19 | 1.0000 | 1.0000 | 1.0000 | 1.0000 | 1.0000 | 1.0000 | 1.0000 | 0.9999 |
| | 20 | 1.0000 | 1.0000 | 1.0000 | 1.0000 | 1.0000 | 1.0000 | 1.0000 | 1.0000 |
| 23 | 0 | 0.2266 | 0.0464 | 0.0084 | 0.0013 | 0.0002 | 0.0000 | 0.0000 | 0.0000 |
| | 1 | 0.5742 | 0.1987 | 0.0532 | 0.0116 | 0.0021 | 0.0003 | 0.0000 | 0.0000 |
| | 2 | 0.8290 | 0.4381 | 0.1668 | 0.0492 | 0.0115 | 0.0021 | 0.0003 | 0.0000 |
| | 3 | 0.9479 | 0.6775 | 0.3503 | 0.1370 | 0.0416 | 0.0099 | 0.0018 | 0.0002 |
| | 4 | 0.9876 | 0.8485 | 0.5621 | 0.2832 | 0.1100 | 0.0330 | 0.0076 | 0.0013 |
| | 5 | 0.9976 | 0.9413 | 0.7478 | 0.4685 | 0.2280 | 0.0859 | 0.0247 | 0.0053 |
| | 6 | 0.9996 | 0.9811 | 0.8763 | 0.6537 | 0.3890 | 0.1810 | 0.0647 | 0.0173 |
| | 7 | 1.0000 | 0.9949 | 0.9484 | 0.8037 | 0.5668 | 0.3196 | 0.1403 | 0.0466 |
| | 8 | 1.0000 | 0.9988 | 0.9816 | 0.9037 | 0.7283 | 0.4859 | 0.2578 | 0.1050 |
| | 9 | 1.0000 | 0.9998 | 0.9944 | 0.9592 | 0.8507 | 0.6522 | 0.4102 | 0.2024 |
| | 10 | 1.0000 | 1.0000 | 0.9986 | 0.9851 | 0.9286 | 0.7919 | 0.5761 | 0.3388 |
| | 11 | 1.0000 | 1.0000 | 0.9997 | 0.9954 | 0.9705 | 0.8910 | 0.7285 | 0.5000 |
| | 12 | 1.0000 | 1.0000 | 0.9999 | 0.9988 | 0.9895 | 0.9504 | 0.8471 | 0.6612 |
| | 13 | 1.0000 | 1.0000 | 1.0000 | 0.9997 | 0.9968 | 0.9806 | 0.9252 | 0.7976 |
| | 14 | 1.0000 | 1.0000 | 1.0000 | 0.9999 | 0.9992 | 0.9935 | 0.9686 | 0.8950 |
| | 15 | 1.0000 | 1.0000 | 1.0000 | 1.0000 | 0.9998 | 0.9982 | 0.9888 | 0.9534 |

*Continued*

**Table 1** The Cumulative Binomial Distribution *Continued*

| n | k | 1/16 | 2/16 | 3/16 | 4/16 | 5/16 | 6/16 | 7/16 | 8/16 |
|---|---|------|------|------|------|------|------|------|------|
| 23 | 16 | 1.0000 | 1.0000 | 1.0000 | 1.0000 | 1.0000 | 0.9996 | 0.9967 | 0.9827 |
| | 17 | 1.0000 | 1.0000 | 1.0000 | 1.0000 | 1.0000 | 0.9999 | 0.9992 | 0.9947 |
| | 18 | 1.0000 | 1.0000 | 1.0000 | 1.0000 | 1.0000 | 1.0000 | 0.9998 | 0.9987 |
| | 19 | 1.0000 | 1.0000 | 1.0000 | 1.0000 | 1.0000 | 1.0000 | 1.0000 | 0.9998 |
| | 20 | 1.0000 | 1.0000 | 1.0000 | 1.0000 | 1.0000 | 1.0000 | 1.0000 | 1.0000 |
| | 21 | 1.0000 | 1.0000 | 1.0000 | 1.0000 | 1.0000 | 1.0000 | 1.0000 | 1.0000 |
| 24 | 0 | 0.2125 | 0.0406 | 0.0069 | 0.0010 | 0.0001 | 0.0000 | 0.0000 | 0.0000 |
| | 1 | 0.5524 | 0.1797 | 0.0448 | 0.0090 | 0.0015 | 0.0002 | 0.0000 | 0.0000 |
| | 2 | 0.8131 | 0.4082 | 0.1455 | 0.0398 | 0.0086 | 0.0014 | 0.0002 | 0.0000 |
| | 3 | 0.9405 | 0.6476 | 0.3159 | 0.1150 | 0.0322 | 0.0070 | 0.0011 | 0.0001 |
| | 4 | 0.9851 | 0.8271 | 0.5224 | 0.2466 | 0.0886 | 0.0243 | 0.0051 | 0.0008 |
| | 5 | 0.9970 | 0.9297 | 0.7130 | 0.4222 | 0.1911 | 0.0661 | 0.0172 | 0.0033 |
| | 6 | 0.9995 | 0.9761 | 0.8522 | 0.6074 | 0.3387 | 0.1453 | 0.0472 | 0.0113 |
| | 7 | 0.9999 | 0.9932 | 0.9349 | 0.7662 | 0.5112 | 0.2676 | 0.1072 | 0.0320 |
| | 8 | 1.0000 | 0.9983 | 0.9754 | 0.8787 | 0.6778 | 0.4235 | 0.2064 | 0.0758 |
| | 9 | 1.0000 | 0.9997 | 0.9920 | 0.9453 | 0.8125 | 0.5898 | 0.3435 | 0.1537 |
| | 10 | 1.0000 | 0.9999 | 0.9978 | 0.9787 | 0.9043 | 0.7395 | 0.5035 | 0.2706 |
| | 11 | 1.0000 | 1.0000 | 0.9995 | 0.9928 | 0.9574 | 0.8538 | 0.6618 | 0.4194 |
| | 12 | 1.0000 | 1.0000 | 0.9999 | 0.9979 | 0.9835 | 0.9281 | 0.7953 | 0.5806 |
| | 13 | 1.0000 | 1.0000 | 1.0000 | 0.9995 | 0.9945 | 0.9693 | 0.8911 | 0.7294 |
| | 14 | 1.0000 | 1.0000 | 1.0000 | 0.9999 | 0.9984 | 0.9887 | 0.9496 | 0.8463 |
| | 15 | 1.0000 | 1.0000 | 1.0000 | 1.0000 | 0.9996 | 0.9964 | 0.9799 | 0.9242 |
| | 16 | 1.0000 | 1.0000 | 1.0000 | 1.0000 | 0.9999 | 0.9990 | 0.9932 | 0.9680 |
| | 17 | 1.0000 | 1.0000 | 1.0000 | 1.0000 | 1.0000 | 0.9998 | 0.9981 | 0.9887 |
| | 18 | 1.0000 | 1.0000 | 1.0000 | 1.0000 | 1.0000 | 1.0000 | 0.9996 | 0.9967 |
| | 19 | 1.0000 | 1.0000 | 1.0000 | 1.0000 | 1.0000 | 1.0000 | 0.9999 | 0.9992 |
| | 20 | 1.0000 | 1.0000 | 1.0000 | 1.0000 | 1.0000 | 1.0000 | 1.0000 | 0.9999 |
| | 21 | 1.0000 | 1.0000 | 1.0000 | 1.0000 | 1.0000 | 1.0000 | 1.0000 | 1.0000 |
| | 22 | 1.0000 | 1.0000 | 1.0000 | 1.0000 | 1.0000 | 1.0000 | 1.0000 | 1.0000 |
| 25 | 0 | 0.1992 | 0.0355 | 0.0056 | 0.0008 | 0.0001 | 0.0000 | 0.0000 | 0.0000 |
| | 1 | 0.5132 | 0.1623 | 0.0377 | 0.0070 | 0.0011 | 0.0001 | 0.0000 | 0.0000 |
| | 2 | 0.7968 | 0.3796 | 0.1266 | 0.0321 | 0.0064 | 0.0010 | 0.0001 | 0.0000 |
| | 3 | 0.9325 | 0.6176 | 0.2840 | 0.0962 | 0.0248 | 0.0049 | 0.0007 | 0.0001 |

*Continued*

**Table 1** The Cumulative Binomial Distribution *Continued*

| | | | | | p | | | | |
|---|---|---|---|---|---|---|---|---|---|
| n | k | 1/16 | 2/16 | 3/16 | 4/16 | 5/16 | 6/16 | 7/16 | 8/16 |
| 25 | 4 | 0.9823 | 0.8047 | 0.4837 | 0.2137 | 0.0710 | 0.0178 | 0.0033 | 0.0005 |
| | 5 | 0.9962 | 0.9169 | 0.6772 | 0.3783 | 0.1591 | 0.0504 | 0.0119 | 0.0028 |
| | 6 | 0.9993 | 0.9703 | 0.8261 | 0.5611 | 0.2926 | 0.1156 | 0.0341 | 0.0073 |
| | 7 | 0.9999 | 0.9910 | 0.9194 | 0.7265 | 0.4573 | 0.2218 | 0.0810 | 0.0216 |
| | 8 | 1.0000 | 0.9977 | 0.9678 | 0.8506 | 0.6258 | 0.3651 | 0.1630 | 0.0539 |
| | 9 | 1.0000 | 0.9995 | 0.9889 | 0.9287 | 0.7704 | 0.5275 | 0.2835 | 0.1148 |
| | 10 | 1.0000 | 0.9999 | 0.9967 | 0.9703 | 0.8756 | 0.6834 | 0.4335 | 0.2122 |
| | 11 | 1.0000 | 1.0000 | 0.9992 | 0.9893 | 0.9408 | 0.8110 | 0.5926 | 0.3450 |
| | 12 | 1.0000 | 1.0000 | 0.9998 | 0.9966 | 0.9754 | 0.9003 | 0.7369 | 0.5000 |
| | 13 | 1.0000 | 1.0000 | 1.0000 | 0.9991 | 0.9911 | 0.9538 | 0.8491 | 0.6550 |
| | 14 | 1.0000 | 1.0000 | 1.0000 | 0.9998 | 0.9972 | 0.9814 | 0.9240 | 0.7878 |
| | 15 | 1.0000 | 1.0000 | 1.0000 | 1.0000 | 0.9992 | 0.9935 | 0.9667 | 0.8852 |
| | 16 | 1.0000 | 1.0000 | 1.0000 | 1.0000 | 0.9998 | 0.9981 | 0.9874 | 0.9462 |
| | 17 | 1.0000 | 1.0000 | 1.0000 | 1.0000 | 1.0000 | 0.9995 | 0.9960 | 0.9784 |
| | 18 | 1.0000 | 1.0000 | 1.0000 | 1.0000 | 1.0000 | 0.9999 | 0.9989 | 0.9927 |
| | 19 | 1.0000 | 1.0000 | 1.0000 | 1.0000 | 1.0000 | 1.0000 | 0.9998 | 0.9980 |
| | 20 | 1.0000 | 1.0000 | 1.0000 | 1.0000 | 1.0000 | 1.0000 | 1.0000 | 0.9995 |
| | 21 | 1.0000 | 1.0000 | 1.0000 | 1.0000 | 1.0000 | 1.0000 | 1.0000 | 0.9999 |
| | 22 | 1.0000 | 1.0000 | 1.0000 | 1.0000 | 1.0000 | 1.0000 | 1.0000 | 1.0000 |

**Table 2** The Cumulative Poisson Distribution

The tabulated quantity is

$$\sum_{j=0}^{k} e^{-\lambda} \frac{\lambda^j}{j!}.$$

| | | | | λ | | |
|---|---|---|---|---|---|---|
| k | 0.001 | 0.005 | 0.010 | 0.015 | 0.020 | 0.025 |
| 0 | 0.9990 0050 | 0.9950 1248 | 0.9900 4983 | 0.9851 1194 | 0.9801 9867 | 0.9753 099 |
| 1 | 0.9999 9950 | 0.9999 8754 | 0.9999 5033 | 0.9998 8862 | 0.9998 0264 | 0.9996 927 |
| 2 | 1.0000 0000 | 0.9999 9998 | 0.9999 9983 | 0.9999 9945 | 0.9999 9868 | 0.9999 974 |
| 3 | | 1.0000 0000 | 1.0000 0000 | 1.0000 0000 | 0.9999 9999 | 1.0000 000 |
| 4 | | | | | 1.0000 0000 | 1.0000 000 |

*Continued*

**Table 2** The Cumulative Poisson Distribution *Continued*

| | | | | λ | | |
|---|---|---|---|---|---|---|
| *k* | 0.030 | 0.035 | 0.040 | 0.045 | 0.050 | 0.055 |
| 0 | 0.970 446 | 0.965 605 | 0.960 789 | 0.955 997 | 0.951 229 | 0.946 485 |
| 1 | 0.999 559 | 0.999 402 | 0.999 221 | 0.999 017 | 0.998 791 | 0.998 542 |
| 2 | 0.999 996 | 0.999 993 | 0.999 990 | 0.999 985 | 0.999 980 | 0.999 973 |
| 3 | 1.000 000 | 1.000 000 | 1.000 000 | 1.000 000 | 1.000 000 | 1.000 000 |

| | | | | λ | | |
|---|---|---|---|---|---|---|
| *k* | 0.060 | 0.065 | 0.070 | 0.075 | 0.080 | 0.085 |
| 0 | 0.941 765 | 0.937 067 | 0.932 394 | 0.927 743 | 0.923 116 | 0.918 512 |
| 1 | 0.998 270 | 0.997 977 | 0.997 661 | 0.997 324 | 0.996 966 | 0.996 586 |
| 2 | 0.999 966 | 0.999 956 | 0.999 946 | 0.999 934 | 0.999 920 | 0.999 904 |
| 3 | 0.999 999 | 0.999 999 | 0.999 999 | 0.999 999 | 0.999 998 | 0.999 998 |
| 4 | 1.000 000 | 1.000 000 | 1.000 000 | 1.000 000 | 1.000 000 | 1.000 000 |

| | | | | λ | | |
|---|---|---|---|---|---|---|
| *k* | 0.090 | 0.095 | 0.100 | 0.200 | 0.300 | 0.400 |
| 0 | 0.913 931 | 0.909 373 | 0.904 837 | 0.818 731 | 0.740 818 | 0.670 320 |
| 1 | 0.996 185 | 0.995 763 | 0.995 321 | 0.982 477 | 0.963 064 | 0.938 448 |
| 2 | 0.999 886 | 0.999 867 | 0.999 845 | 0.998 852 | 0.996 401 | 0.992 074 |
| 3 | 0.999 997 | 0.999 997 | 0.999 996 | 0.999 943 | 0.999 734 | 0.999 224 |
| 4 | 1.000 000 | 1.000 000 | 1.000 000 | 0.999 998 | 0.999 984 | 0.999 939 |
| 5 | | | | 1.000 000 | 0.999 999 | 0.999 996 |
| 6 | | | | | 1.000 000 | 1.000 000 |

| | | | | λ | | |
|---|---|---|---|---|---|---|
| *k* | 0.500 | 0.600 | 0.700 | 0.800 | 0.900 | 1.000 |
| 0 | 0.606 531 | 0.548 812 | 0.496 585 | 0.449 329 | 0.406 329 | 0.367 879 |
| 1 | 0.909 796 | 0.878 099 | 0.844 195 | 0.808 792 | 0.772 482 | 0.735 759 |
| 2 | 0.985 612 | 0.976 885 | 0.965 858 | 0.952 577 | 0.937 143 | 0.919 699 |
| 3 | 0.998 248 | 0.996 642 | 0.994 247 | 0.990 920 | 0.986 541 | 0.981 012 |
| 4 | 0.999 828 | 0.999 606 | 0.999 214 | 0.998 589 | 0.997 656 | 0.996 340 |
| 5 | 0.999 986 | 0.999 961 | 0.999 910 | 0.999 816 | 0.999 657 | 0.999 406 |
| 6 | 0.999 999 | 0.999 997 | 0.999 991 | 0.999 979 | 0.999 957 | 0.999 917 |
| 7 | 1.000 000 | 1.000 000 | 0.999 999 | 0.999 998 | 0.999 995 | 0.999 990 |
| 8 | | | 1.000 000 | 1.000 000 | 1.000 000 | 0.999 999 |
| 9 | | | | | | 1.000 000 |

*Continued*

**Table 2** The Cumulative Poisson Distribution *Continued*

| k | 1.20 | 1.40 | 1.60 | 1.80 | 2.00 | 2.50 | 3.00 | 3.50 |
|---|------|------|------|------|------|------|------|------|
| 0 | 0.3012 | 0.2466 | 0.2019 | 0.1653 | 0.1353 | 0.0821 | 0.0498 | 0.0302 |
| 1 | 0.6626 | 0.5918 | 0.5249 | 0.4628 | 0.4060 | 0.2873 | 0.1991 | 0.1359 |
| 2 | 0.8795 | 0.8335 | 0.7834 | 0.7306 | 0.6767 | 0.5438 | 0.4232 | 0.3208 |
| 3 | 0.9662 | 0.9463 | 0.9212 | 0.8913 | 0.8571 | 0.7576 | 0.6472 | 0.5366 |
| 4 | 0.9923 | 0.9857 | 0.9763 | 0.9636 | 0.9473 | 0.8912 | 0.8153 | 0.7254 |
| 5 | 0.9985 | 0.9968 | 0.9940 | 0.9896 | 0.9834 | 0.9580 | 0.9161 | 0.8576 |
| 6 | 0.9997 | 0.9994 | 0.9987 | 0.9974 | 0.9955 | 0.9858 | 0.9665 | 0.9347 |
| 7 | 1.0000 | 0.9999 | 0.9997 | 0.9994 | 0.9989 | 0.9958 | 0.9881 | 0.9733 |
| 8 | | 1.0000 | 1.0000 | 0.9999 | 0.9998 | 0.9989 | 0.9962 | 0.9901 |
| 9 | | | | 1.0000 | 1.0000 | 0.9997 | 0.9989 | 0.9967 |
| 10 | | | | | | 0.9999 | 0.9997 | 0.9990 |
| 11 | | | | | | 1.0000 | 0.9999 | 0.9997 |
| 12 | | | | | | | 1.0000 | 0.9999 |
| 13 | | | | | | | | 1.0000 |

| k | 4.00 | 4.50 | 5.00 | 6.00 | 7.00 | 8.00 | 9.00 | 10.00 |
|---|------|------|------|------|------|------|------|-------|
| 0 | 0.0183 | 0.0111 | 0.0067 | 0.0025 | 0.0009 | 0.0003 | 0.0001 | 0.0000 |
| 1 | 0.0916 | 0.0611 | 0.0404 | 0.0174 | 0.0073 | 0.0030 | 0.0012 | 0.0005 |
| 2 | 0.2381 | 0.1736 | 0.1247 | 0.0620 | 0.0296 | 0.0138 | 0.0062 | 0.0028 |
| 3 | 0.4335 | 0.3423 | 0.2650 | 0.1512 | 0.0818 | 0.0424 | 0.0212 | 0.0103 |
| 4 | 0.6288 | 0.5321 | 0.4405 | 0.2851 | 0.1730 | 0.0996 | 0.0550 | 0.0293 |
| 5 | 0.7851 | 0.7029 | 0.6160 | 0.4457 | 0.3007 | 0.1912 | 0.1157 | 0.0671 |
| 6 | 0.8893 | 0.8311 | 0.7622 | 0.6063 | 0.4497 | 0.3134 | 0.2068 | 0.1301 |
| 7 | 0.9489 | 0.9134 | 0.8666 | 0.7440 | 0.5987 | 0.4530 | 0.3239 | 0.2202 |
| 8 | 0.9786 | 0.9597 | 0.9319 | 0.8472 | 0.7291 | 0.5925 | 0.4577 | 0.3328 |
| 9 | 0.9919 | 0.9829 | 0.9682 | 0.9161 | 0.8305 | 0.7166 | 0.5874 | 0.4579 |
| 10 | 0.9972 | 0.9933 | 0.9863 | 0.9574 | 0.9015 | 0.8159 | 0.7060 | 0.5830 |
| 11 | 0.9991 | 0.9976 | 0.9945 | 0.9799 | 0.9467 | 0.8881 | 0.8030 | 0.6968 |
| 12 | 0.9997 | 0.9992 | 0.9980 | 0.9912 | 0.9730 | 0.9362 | 0.8758 | 0.7916 |
| 13 | 0.9999 | 0.9997 | 0.9993 | 0.9964 | 0.9872 | 0.9658 | 0.9261 | 0.8645 |
| 14 | 1.0000 | 0.9999 | 0.9998 | 0.9986 | 0.9943 | 0.9827 | 0.9585 | 0.9165 |
| 15 | | 1.0000 | 0.9999 | 0.9995 | 0.9976 | 0.9918 | 0.9780 | 0.9513 |
| 16 | | | 1.0000 | 0.9998 | 0.9990 | 0.9963 | 0.9889 | 0.9730 |
| 17 | | | | 0.9999 | 0.9996 | 0.9984 | 0.9947 | 0.9857 |
| 18 | | | | 1.0000 | 0.9999 | 0.9993 | 0.9976 | 0.9928 |
| 19 | | | | | | 0.9997 | 0.9989 | 0.9965 |
| 20 | | | | | 1.0000 | 0.9999 | 0.9996 | 0.9984 |

*Continued*

**Table 2** The Cumulative Poisson Distribution *Continued*

| k | 4.00 | 4.50 | 5.00 | 6.00 | 7.00 | 8.00 | 9.00 | 10.00 |
|---|------|------|------|------|------|------|------|-------|
| | | | | | | | λ | |
| 21 | | | | | | 1.0000 | 0.9998 | 0.9993 |
| 22 | | | | | | | 0.9999 | 0.9997 |
| 23 | | | | | | | 1.0000 | 0.9999 |
| 24 | | | | | | | | 1.0000 |

**Table 3** The Normal Distribution

The tabulated quantity is

$$\Phi(x) = \frac{1}{\sqrt{2\pi}} \int_{-\infty}^{x} e^{-t^2/2} dt,$$
$$[\Phi(-x) = 1 - \Phi(x)].$$

| x | Φ(x) | x | Φ(x) | x | Φ(x) | x | Φ(x) |
|---|------|---|------|---|------|---|------|
| 0.00 | 0.500000 | 0.45 | 0.673645 | 0.90 | 0.815940 | 1.35 | 0.911492 |
| 0.01 | 0.503989 | 0.46 | 0.677242 | 0.91 | 0.818589 | 1.36 | 0.913085 |
| 0.02 | 0.507978 | 0.47 | 0.680822 | 0.92 | 0.821214 | 1.37 | 0.914657 |
| 0.03 | 0.511966 | 0.48 | 0.684386 | 0.93 | 0.823814 | 1.38 | 0.916207 |
| 0.04 | 0.515953 | 0.49 | 0.687933 | 0.94 | 0.826391 | 1.39 | 0.917736 |
| 0.05 | 0.519939 | 0.50 | 0.691462 | 0.95 | 0.828944 | 1.40 | 0.919243 |
| 0.06 | 0.523922 | 0.51 | 0.694974 | 0.96 | 0.831472 | 1.41 | 0.920730 |
| 0.07 | 0.527903 | 0.52 | 0.698468 | 0.97 | 0.833977 | 1.42 | 0.922196 |
| 0.08 | 0.531881 | 0.53 | 0.701944 | 0.98 | 0.836457 | 1.43 | 0.923641 |
| 0.09 | 0.535856 | 0.54 | 0.705401 | 0.99 | 0.838913 | 1.44 | 0.925066 |
| 0.10 | 0.539828 | 0.55 | 0.708840 | 1.00 | 0.841345 | 1.45 | 0.926471 |
| 0.11 | 0.543795 | 0.56 | 0.712260 | 1.01 | 0.843752 | 1.46 | 0.927855 |
| 0.12 | 0.547758 | 0.57 | 0.715661 | 1.02 | 0.846136 | 1.47 | 0.929219 |
| 0.13 | 0.551717 | 0.58 | 0.719043 | 1.03 | 0.848495 | 1.48 | 0.930563 |
| 0.14 | 0.555670 | 0.59 | 0.722405 | 1.04 | 0.850830 | 1.49 | 0.931888 |
| 0.15 | 0.559618 | 0.60 | 0.725747 | 1.05 | 0.853141 | 1.50 | 0.933193 |
| 0.16 | 0.563559 | 0.61 | 0.729069 | 1.06 | 0.855428 | 1.51 | 0.934478 |
| 0.17 | 0.567495 | 0.62 | 0.732371 | 1.07 | 0.857690 | 1.52 | 0.935745 |
| 0.18 | 0.571424 | 0.63 | 0.735653 | 1.08 | 0.859929 | 1.53 | 0.936992 |
| 0.19 | 0.575345 | 0.64 | 0.738914 | 1.09 | 0.862143 | 1.54 | 0.938220 |
| 0.20 | 0.579260 | 0.65 | 0.742154 | 1.10 | 0.864334 | 1.55 | 0.939429 |
| 0.21 | 0.583166 | 0.66 | 0.745373 | 1.11 | 0.866500 | 1.56 | 0.940620 |
| 0.22 | 0.587064 | 0.67 | 0.748571 | 1.12 | 0.868643 | 1.57 | 0.941792 |
| 0.23 | 0.590954 | 0.68 | 0.751748 | 1.13 | 0.870762 | 1.58 | 0.942947 |

*Continued*

**Table 3** The Normal Distribution *Continued*

| x | Φ(x) | x | Φ(x) | x | Φ(x) | x | Φ(x) |
|---|---|---|---|---|---|---|---|
| 0.24 | 0.594835 | 0.69 | 0.754903 | 1.14 | 0.872857 | 1.59 | 0.944083 |
| 0.25 | 0.598706 | 0.70 | 0.758036 | 1.15 | 0.874928 | 1.60 | 0.945201 |
| 0.26 | 0.602568 | 0.71 | 0.761148 | 1.16 | 0.876976 | 1.61 | 0.946301 |
| 0.27 | 0.606420 | 0.72 | 0.764238 | 1.17 | 0.879000 | 1.62 | 0.947384 |
| 0.28 | 0.610261 | 0.73 | 0.767305 | 1.18 | 0.881000 | 1.63 | 0.948449 |
| 0.29 | 0.614092 | 0.74 | 0.770350 | 1.19 | 0.882977 | 1.64 | 0.949497 |
| 0.30 | 0.617911 | 0.75 | 0.773373 | 1.20 | 0.884930 | 1.65 | 0.950529 |
| 0.31 | 0.621720 | 0.76 | 0.776373 | 1.21 | 0.886861 | 1.66 | 0.951543 |
| 0.32 | 0.625516 | 0.77 | 0.779350 | 1.22 | 0.888768 | 1.67 | 0.952540 |
| 0.33 | 0.629300 | 0.78 | 0.782305 | 1.23 | 0.890651 | 1.68 | 0.953521 |
| 0.34 | 0.633072 | 0.79 | 0.785236 | 1.24 | 0.892512 | 1.69 | 0.954486 |
| 0.35 | 0.636831 | 0.80 | 0.788145 | 1.25 | 0.894350 | 1.70 | 0.955435 |
| 0.36 | 0.640576 | 0.81 | 0.791030 | 1.26 | 0.896165 | 1.71 | 0.956367 |
| 0.37 | 0.644309 | 0.82 | 0.793892 | 1.27 | 0.897958 | 1.72 | 0.957284 |
| 0.38 | 0.648027 | 0.83 | 0.796731 | 1.28 | 0.899727 | 1.73 | 0.958185 |
| 0.39 | 0.651732 | 0.84 | 0.799546 | 1.29 | 0.901475 | 1.74 | 0.959070 |
| 0.40 | 0.655422 | 0.85 | 0.802337 | 1.30 | 0.903200 | 1.75 | 0.959941 |
| 0.41 | 0.659097 | 0.86 | 0.805105 | 1.31 | 0.904902 | 1.76 | 0.960796 |
| 0.42 | 0.662757 | 0.87 | 0.807850 | 1.32 | 0.906582 | 1.77 | 0.961636 |
| 0.43 | 0.666402 | 0.88 | 0.810570 | 1.33 | 0.908241 | 1.78 | 0.962462 |
| 0.44 | 0.670031 | 0.89 | 0.813267 | 1.34 | 0.909877 | 1.79 | 0.963273 |
| 1.80 | 0.964070 | 2.30 | 0.989276 | 2.80 | 0.997445 | 3.30 | 0.999517 |
| 1.81 | 0.964852 | 2.31 | 0.989556 | 2.81 | 0.997523 | 3.31 | 0.999534 |
| 1.82 | 0.965620 | 2.32 | 0.989830 | 2.82 | 0.997599 | 3.32 | 0.999550 |
| 1.83 | 0.966375 | 2.33 | 0.990097 | 2.83 | 0.997673 | 3.33 | 0.999566 |
| 1.84 | 0.967116 | 2.34 | 0.990358 | 2.84 | 0.997744 | 3.34 | 0.999581 |
| 1.85 | 0.967843 | 2.35 | 0.990613 | 2.85 | 0.997814 | 3.35 | 0.999596 |
| 1.86 | 0.968557 | 2.36 | 0.990863 | 2.86 | 0.997882 | 3.36 | 0.999610 |
| 1.87 | 0.969258 | 2.37 | 0.991106 | 2.87 | 0.997948 | 3.37 | 0.999624 |
| 1.88 | 0.969946 | 2.38 | 0.991344 | 2.88 | 0.998012 | 3.38 | 0.999638 |
| 1.89 | 0.970621 | 2.39 | 0.991576 | 2.89 | 0.998074 | 3.39 | 0.999651 |
| 1.90 | 0.971283 | 2.40 | 0.991802 | 2.90 | 0.998134 | 3.40 | 0.999663 |
| 1.91 | 0.971933 | 2.41 | 0.992024 | 2.91 | 0.998193 | 3.41 | 0.999675 |
| 1.92 | 0.972571 | 2.42 | 0.992240 | 2.92 | 0.998250 | 3.42 | 0.999687 |
| 1.93 | 0.973197 | 2.43 | 0.992451 | 2.93 | 0.998305 | 3.43 | 0.999698 |
| 1.94 | 0.973810 | 2.44 | 0.992656 | 2.94 | 0.998359 | 3.44 | 0.999709 |
| 1.95 | 0.974412 | 2.45 | 0.992857 | 2.95 | 0.998411 | 3.45 | 0.999720 |
| 1.96 | 0.975002 | 2.46 | 0.993053 | 2.96 | 0.998462 | 3.46 | 0.999730 |

*Continued*

**Table 3** The Normal Distribution *Continued*

| x | Φ(x) | x | Φ(x) | x | Φ(x) | x | Φ(x) |
|------|----------|------|----------|------|----------|------|----------|
| 1.97 | 0.975581 | 2.47 | 0.993244 | 2.97 | 0.998511 | 3.47 | 0.999740 |
| 1.98 | 0.976148 | 2.48 | 0.993431 | 2.98 | 0.998559 | 3.48 | 0.999749 |
| 1.99 | 0.976705 | 2.49 | 0.993613 | 2.99 | 0.998605 | 3.49 | 0.999758 |
| 2.00 | 0.977250 | 2.50 | 0.993790 | 3.00 | 0.998650 | 3.50 | 0.999767 |
| 2.01 | 0.977784 | 2.51 | 0.993963 | 3.01 | 0.998694 | 3.51 | 0.999776 |
| 2.02 | 0.978308 | 2.52 | 0.994132 | 3.02 | 0.998736 | 3.52 | 0.999784 |
| 2.03 | 0.978822 | 2.53 | 0.994297 | 3.03 | 0.998777 | 3.53 | 0.999792 |
| 2.04 | 0.979325 | 2.54 | 0.994457 | 3.04 | 0.998817 | 3.54 | 0.999800 |
| 2.05 | 0.979818 | 2.55 | 0.994614 | 3.05 | 0.998856 | 3.55 | 0.999807 |
| 2.06 | 0.980301 | 2.56 | 0.994766 | 3.06 | 0.998893 | 3.56 | 0.999815 |
| 2.07 | 0.980774 | 2.57 | 0.994915 | 3.07 | 0.998930 | 3.57 | 0.999822 |
| 2.08 | 0.981237 | 2.58 | 0.995060 | 3.08 | 0.998965 | 3.58 | 0.999828 |
| 2.09 | 0.981691 | 2.59 | 0.995201 | 3.09 | 0.998999 | 3.59 | 0.999835 |
| 2.10 | 0.982136 | 2.60 | 0.995339 | 3.10 | 0.999032 | 3.60 | 0.999841 |
| 2.11 | 0.982571 | 2.61 | 0.995473 | 3.11 | 0.999065 | 3.61 | 0.999847 |
| 2.12 | 0.982997 | 2.62 | 0.995604 | 3.12 | 0.999096 | 3.62 | 0.999853 |
| 2.13 | 0.983414 | 2.63 | 0.995731 | 3.13 | 0.999126 | 3.63 | 0.999858 |
| 2.14 | 0.983823 | 2.64 | 0.995855 | 3.14 | 0.999155 | 3.64 | 0.999864 |
| 2.15 | 0.984222 | 2.65 | 0.995975 | 3.15 | 0.999184 | 3.65 | 0.999869 |
| 2.16 | 0.984614 | 2.66 | 0.996093 | 3.16 | 0.999211 | 3.66 | 0.999874 |
| 2.17 | 0.984997 | 2.67 | 0.996207 | 3.17 | 0.999238 | 3.67 | 0.999879 |
| 2.18 | 0.985371 | 2.68 | 0.996319 | 3.18 | 0.999264 | 3.68 | 0.999883 |
| 2.19 | 0.985738 | 2.69 | 0.996427 | 3.19 | 0.999289 | 3.69 | 0.999888 |
| 2.20 | 0.986097 | 2.70 | 0.996533 | 3.20 | 0.999313 | 3.70 | 0.999892 |
| 2.21 | 0.986447 | 2.71 | 0.996636 | 3.21 | 0.999336 | 3.71 | 0.999896 |
| 2.22 | 0.986791 | 2.72 | 0.996736 | 3.22 | 0.999359 | 3.72 | 0.999900 |
| 2.23 | 0.987126 | 2.73 | 0.996833 | 3.23 | 0.999381 | 3.73 | 0.999904 |
| 2.24 | 0.987455 | 2.74 | 0.996928 | 3.24 | 0.999402 | 3.74 | 0.999908 |
| 2.25 | 0.987776 | 2.75 | 0.997020 | 3.25 | 0.999423 | 3.75 | 0.999912 |
| 2.26 | 0.988089 | 2.76 | 0.997110 | 3.26 | 0.999443 | 3.76 | 0.999915 |
| 2.27 | 0.988396 | 2.77 | 0.997197 | 3.27 | 0.999462 | 3.77 | 0.999918 |
| 2.28 | 0.988696 | 2.78 | 0.997282 | 3.28 | 0.999481 | 3.78 | 0.999922 |
| 2.29 | 0.988989 | 2.79 | 0.997365 | 3.29 | 0.999499 | 3.79 | 0.999925 |
| 3.80 | 0.999928 | 3.85 | 0.999941 | 3.90 | 0.999952 | 3.95 | 0.999961 |
| 3.81 | 0.999931 | 3.86 | 0.999943 | 3.91 | 0.999954 | 3.96 | 0.999963 |
| 3.82 | 0.999933 | 3.87 | 0.999946 | 3.92 | 0.999956 | 3.97 | 0.999964 |
| 3.83 | 0.999936 | 3.88 | 0.999948 | 3.93 | 0.999958 | 3.98 | 0.999966 |
| 3.84 | 0.999938 | 3.89 | 0.999950 | 3.94 | 0.999959 | 3.99 | 0.999967 |

**Table 4** Critical Values for Student's $t$-Distribution

Let $t_r$ be a random variable having the Student's $t$-distribution with $r$ degrees of freedom. Then the tabulated quantities are the numbers $x$ for which

$$P(t_r \leq x) = \gamma.$$

| $r$ | $\gamma$ 0.75 | 0.90 | 0.95 | 0.975 | 0.99 | 0.995 |
|---|---|---|---|---|---|---|
| 1 | 1.0000 | 3.0777 | 6.3138 | 12.7062 | 31.8207 | 63.6574 |
| 2 | 0.8165 | 1.8856 | 2.9200 | 4.3027 | 6.9646 | 9.9248 |
| 3 | 0.7649 | 1.6377 | 2.3534 | 3.1824 | 4.5407 | 5.8409 |
| 4 | 0.7407 | 1.5332 | 2.1318 | 2.7764 | 3.7649 | 4.6041 |
| 5 | 0.7267 | 1.4759 | 2.0150 | 2.5706 | 3.3649 | 4.0322 |
| 6 | 0.7176 | 1.4398 | 1.9432 | 2.4469 | 3.1427 | 3.7074 |
| 7 | 0.7111 | 1.4149 | 1.8946 | 2.3646 | 2.9980 | 3.4995 |
| 8 | 0.7064 | 1.3968 | 1.8595 | 2.3060 | 2.8965 | 3.3554 |
| 9 | 0.7027 | 1.3830 | 1.8331 | 2.2622 | 2.8214 | 3.2498 |
| 10 | 0.6998 | 1.3722 | 1.8125 | 2.2281 | 2.7638 | 3.1693 |
| 11 | 0.6974 | 1.3634 | 1.7959 | 2.2010 | 2.7181 | 3.1058 |
| 12 | 0.6955 | 1.3562 | 1.7823 | 2.1788 | 2.6810 | 3.0545 |
| 13 | 0.6938 | 1.3502 | 1.7709 | 2.1604 | 2.6503 | 3.0123 |
| 14 | 0.6924 | 1.3450 | 1.7613 | 2.1448 | 2.6245 | 2.9768 |
| 15 | 0.6912 | 1.3406 | 1.7531 | 2.1315 | 2.6025 | 2.9467 |
| 16 | 0.6901 | 1.3368 | 1.7459 | 2.1199 | 2.5835 | 2.9208 |
| 17 | 0.6892 | 1.3334 | 1.7396 | 2.1098 | 2.5669 | 2.8982 |
| 18 | 0.6884 | 1.3304 | 1.7341 | 2.1009 | 2.5524 | 2.8784 |
| 19 | 0.6876 | 1.3277 | 1.7291 | 2.0930 | 2.5395 | 2.8609 |
| 20 | 0.6870 | 1.3253 | 1.7247 | 2.0860 | 2.5280 | 2.8453 |
| 21 | 0.6864 | 1.3232 | 1.7207 | 2.0796 | 2.5177 | 2.8314 |
| 22 | 0.6858 | 1.3212 | 1.7171 | 2.0739 | 2.5083 | 2.8188 |
| 23 | 0.6853 | 1.3195 | 1.7139 | 2.0687 | 2.4999 | 2.8073 |
| 24 | 0.6848 | 1.3178 | 1.7109 | 2.0639 | 2.4922 | 2.7969 |
| 25 | 0.6844 | 1.3163 | 1.7081 | 2.0595 | 2.4851 | 2.7874 |
| 26 | 0.6840 | 1.3150 | 1.7056 | 2.0555 | 2.4786 | 2.7787 |
| 27 | 0.6837 | 1.3137 | 1.7033 | 2.0518 | 2.4727 | 2.7707 |
| 28 | 0.6834 | 1.3125 | 1.7011 | 2.0484 | 2.4671 | 2.7633 |
| 29 | 0.6830 | 1.3114 | 1.6991 | 2.0452 | 2.4620 | 2.7564 |
| 30 | 0.6828 | 1.3104 | 1.6973 | 2.0423 | 2.4573 | 2.7500 |
| 31 | 0.6825 | 1.3095 | 1.6955 | 2.0395 | 2.4528 | 2.7440 |
| 32 | 0.6822 | 1.3086 | 1.6939 | 2.0369 | 2.4487 | 2.7385 |

*Continued*

**Table 4** Critical Values for Student's *t*-Distribution
*Continued*

| r | 0.75 | 0.90 | 0.95 | 0.975 | 0.99 | 0.995 |
|---|------|------|------|-------|------|-------|
| | | | | γ | | |
| 33 | 0.6820 | 1.3077 | 1.6924 | 2.0345 | 2.4448 | 2.7333 |
| 34 | 0.6818 | 1.3070 | 1.6909 | 2.0322 | 2.4411 | 2.7284 |
| 35 | 0.6816 | 1.3062 | 1.6896 | 2.0301 | 2.4377 | 2.7238 |
| 36 | 0.6814 | 1.3055 | 1.6883 | 2.0281 | 2.4345 | 2.7195 |
| 37 | 0.6812 | 1.3049 | 1.6871 | 2.0262 | 2.4314 | 1.7154 |
| 38 | 0.6810 | 1.3042 | 1.6860 | 2.0244 | 2.4286 | 2.7116 |
| 39 | 0.6808 | 1.3036 | 1.6849 | 2.0227 | 2.4258 | 2.7079 |
| 40 | 0.6807 | 1.3031 | 1.6839 | 2.0211 | 2.4233 | 2.7045 |
| 41 | 0.6805 | 1.3025 | 1.6829 | 2.0195 | 2.4208 | 2.7012 |
| 42 | 0.6804 | 1.3020 | 1.6820 | 2.0181 | 2.4185 | 2.6981 |
| 43 | 0.6802 | 1.3016 | 1.6811 | 2.0167 | 2.4163 | 2.6951 |
| 44 | 0.6801 | 1.3011 | 1.6802 | 2.0154 | 2.4141 | 2.6923 |
| 45 | 0.6800 | 1.3006 | 1.6794 | 2.0141 | 2.4121 | 2.6896 |
| 46 | 0.6799 | 1.3002 | 1.6787 | 2.0129 | 2.4102 | 2.6870 |
| 47 | 0.6797 | 1.2998 | 1.6779 | 2.0117 | 2.4083 | 2.6846 |
| 48 | 0.6796 | 1.2994 | 1.6772 | 2.0106 | 2.4066 | 2.6822 |
| 49 | 0.6795 | 1.2991 | 1.6766 | 2.0096 | 2.4069 | 2.6800 |
| 50 | 0.6794 | 1.2987 | 1.6759 | 2.0086 | 2.4033 | 2.6778 |
| 51 | 0.6793 | 1.2984 | 1.6753 | 2.0076 | 2.4017 | 2.6757 |
| 52 | 0.6792 | 1.2980 | 1.6747 | 2.0066 | 2.4002 | 2.6737 |
| 53 | 0.6791 | 1.2977 | 1.6741 | 2.0057 | 2.3988 | 2.6718 |
| 54 | 0.6791 | 1.2974 | 1.6736 | 2.0049 | 2.3974 | 2.6700 |
| 55 | 0.6790 | 1.2971 | 1.6730 | 2.0040 | 2.3961 | 2.6682 |
| 56 | 0.6789 | 1.2969 | 1.6725 | 2.0032 | 2.3948 | 2.6665 |
| 57 | 0.6788 | 1.2966 | 1.6720 | 2.0025 | 2.3936 | 2.6649 |
| 58 | 0.6787 | 1.2963 | 1.6716 | 2.0017 | 2.3924 | 2.6633 |
| 59 | 0.6787 | 1.2961 | 1.6711 | 2.0010 | 2.3912 | 2.6618 |
| 60 | 0.6786 | 1.2958 | 1.6706 | 2.0003 | 2.3901 | 2.6603 |
| 61 | 0.6785 | 1.2956 | 1.6702 | 1.9996 | 2.3890 | 2.6589 |
| 62 | 0.6785 | 1.2954 | 1.6698 | 1.9990 | 2.3880 | 2.6575 |
| 63 | 0.6784 | 1.2951 | 1.6694 | 1.9983 | 2.3870 | 2.6561 |
| 64 | 0.6783 | 1.2949 | 1.6690 | 1.9977 | 2.3860 | 2.6549 |
| 65 | 0.6783 | 1.2947 | 1.6686 | 1.9971 | 2.3851 | 2.6536 |
| 66 | 0.6782 | 1.2945 | 1.6683 | 1.9966 | 2.3842 | 2.6524 |
| 67 | 0.6782 | 1.2943 | 1.6679 | 1.9960 | 2.3833 | 2.6512 |
| 68 | 0.6781 | 1.2941 | 1.6676 | 1.9955 | 2.3824 | 2.6501 |

*Continued*

**Table 4** Critical Values for Student's *t*-Distribution
*Continued*

| r | γ | | | | | |
|---|---|---|---|---|---|---|
| | 0.75 | 0.90 | 0.95 | 0.975 | 0.99 | 0.995 |
| 69 | 0.6781 | 1.2939 | 1.6672 | 1.9949 | 2.3816 | 2.6490 |
| 70 | 0.6780 | 1.2938 | 1.6669 | 1.9944 | 2.3808 | 2.6479 |
| 71 | 0.6780 | 1.2936 | 1.6666 | 1.9939 | 2.3800 | 2.6469 |
| 72 | 0.6779 | 1.2934 | 1.6663 | 1.9935 | 2.3793 | 2.6459 |
| 73 | 0.6779 | 1.2933 | 1.6660 | 1.9930 | 2.3785 | 2.6449 |
| 74 | 0.6778 | 1.2931 | 1.6657 | 1.9925 | 2.3778 | 2.6439 |
| 75 | 0.6778 | 1.2929 | 1.6654 | 1.9921 | 2.3771 | 2.6430 |
| 76 | 0.6777 | 1.2928 | 1.6652 | 1.9917 | 2.3764 | 2.6421 |
| 77 | 0.6777 | 1.2926 | 1.6649 | 1.9913 | 2.3758 | 2.6412 |
| 78 | 0.6776 | 1.2925 | 1.6646 | 1.9908 | 2.3751 | 2.6403 |
| 79 | 0.6776 | 1.2924 | 1.6644 | 1.9905 | 2.3745 | 2.6395 |
| 80 | 0.6776 | 1.2922 | 1.6641 | 1.9901 | 2.3739 | 2.6387 |
| 81 | 0.6775 | 1.2921 | 1.6639 | 1.9897 | 2.3733 | 2.6379 |
| 82 | 0.6775 | 1.2920 | 1.6636 | 1.9893 | 2.3727 | 2.6371 |
| 83 | 0.6775 | 1.2918 | 1.6634 | 1.9890 | 2.3721 | 2.6364 |
| 84 | 0.6774 | 1.2917 | 1.6632 | 1.9886 | 2.3716 | 2.6356 |
| 85 | 0.6774 | 1.2916 | 1.6630 | 1.9883 | 2.3710 | 2.6349 |
| 86 | 0.6774 | 1.2915 | 1.6628 | 1.9879 | 2.3705 | 2.6342 |
| 87 | 0.6773 | 1.2914 | 1.6626 | 1.9876 | 2.3700 | 2.6335 |
| 88 | 0.6773 | 1.2912 | 1.6624 | 1.9873 | 2.3695 | 2.6329 |
| 89 | 0.6773 | 1.2911 | 1.6622 | 1.9870 | 2.3690 | 2.6322 |
| 90 | 0.6772 | 1.2910 | 1.6620 | 1.9867 | 2.3685 | 2.6316 |

**Table 5** Critical Values for the Chi-square Distribution

Let $\chi_r^2$ be a random variable having the Chi-square distribution
with $r$ degrees of freedom. Then the tabulated quantities are the
numbers $x$ for which

$$P(\chi_r^2 \le x) = \gamma.$$

| r | γ | | | | | |
|---|---|---|---|---|---|---|
| | 0.005 | 0.01 | 0.025 | 0.05 | 0.10 | 0.25 |
| 1 | — | — | 0.001 | 0.004 | 0.016 | 0.102 |
| 2 | 0.010 | 0.020 | 0.051 | 0.103 | 0.211 | 0.575 |
| 3 | 0.072 | 0.115 | 0.216 | 0.352 | 0.584 | 1.213 |
| 4 | 0.207 | 0.297 | 0.484 | 0.711 | 1.064 | 1.923 |
| 5 | 0.412 | 0.554 | 0.831 | 1.145 | 1.610 | 2.675 |
| 6 | 0.676 | 0.872 | 1.237 | 1.635 | 2.204 | 3.455 |

*Continued*

**Table 5** Critical Values for the Chi-square Distribution
*Continued*

| | | | | $\gamma$ | | | |
|---|---|---|---|---|---|---|
| r | 0.005 | 0.01 | 0.025 | 0.05 | 0.10 | 0.25 |
| 7 | 0.989 | 1.239 | 1.690 | 2.167 | 2.833 | 4.255 |
| 8 | 1.344 | 1.646 | 2.180 | 2.733 | 3.490 | 5.071 |
| 9 | 1.735 | 2.088 | 2.700 | 2.325 | 4.168 | 5.899 |
| 10 | 2.156 | 2.558 | 3.247 | 3.940 | 4.865 | 6.737 |
| 11 | 2.603 | 3.053 | 3.816 | 4.575 | 5.578 | 7.584 |
| 12 | 3.074 | 3.571 | 4.404 | 5.226 | 6.304 | 9.438 |
| 13 | 3.565 | 4.107 | 5.009 | 5.892 | 7.042 | 9.299 |
| 14 | 4.075 | 4.660 | 5.629 | 6.571 | 7.790 | 10.165 |
| 15 | 4.601 | 5.229 | 6.262 | 7.261 | 8.547 | 11.037 |
| 16 | 5.142 | 5.812 | 6.908 | 7.962 | 9.312 | 11.912 |
| 17 | 5.697 | 6.408 | 7.564 | 8.672 | 10.085 | 12.792 |
| 18 | 6.265 | 7.015 | 8.231 | 8.390 | 10.865 | 13.675 |
| 19 | 6.844 | 7.633 | 8.907 | 10.117 | 11.651 | 14.562 |
| 20 | 7.434 | 8.260 | 9.591 | 10.851 | 12.443 | 15.452 |
| 21 | 8.034 | 8.897 | 10.283 | 11.591 | 13.240 | 16.344 |
| 22 | 8.643 | 9.542 | 10.982 | 12.338 | 14.042 | 17.240 |
| 23 | 9.260 | 10.196 | 11.689 | 13.091 | 14.848 | 18.137 |
| 24 | 9.886 | 10.856 | 12.401 | 13.848 | 15.659 | 19.037 |
| 25 | 10.520 | 11.524 | 13.120 | 14.611 | 16.473 | 19.939 |
| 26 | 11.160 | 12.198 | 13.844 | 13.379 | 17.292 | 20.843 |
| 27 | 11.808 | 12.879 | 14.573 | 16.151 | 18.114 | 21.749 |
| 28 | 12.461 | 13.565 | 15.308 | 16.928 | 18.939 | 22.657 |
| 29 | 13.121 | 14.257 | 16.047 | 17.708 | 19.768 | 23.567 |
| 30 | 13.787 | 14.954 | 16.791 | 18.493 | 20.599 | 24.478 |
| 31 | 14.458 | 15.655 | 17.539 | 19.281 | 21.434 | 25.390 |
| 32 | 15.134 | 16.362 | 18.291 | 20.072 | 22.271 | 26.304 |
| 33 | 15.815 | 17.074 | 19.047 | 20.867 | 23.110 | 27.219 |
| 34 | 16.501 | 17.789 | 19.806 | 21.664 | 23.952 | 28.136 |
| 35 | 17.192 | 18.509 | 20.569 | 22.465 | 24.797 | 29.054 |
| 36 | 17.887 | 19.233 | 21.336 | 23.269 | 25.643 | 29.973 |
| 37 | 18.586 | 19.960 | 22.106 | 24.075 | 26.492 | 30.893 |
| 38 | 19.289 | 20.691 | 22.878 | 24.884 | 27.343 | 31.815 |
| 39 | 19.996 | 21.426 | 23.654 | 25.695 | 28.196 | 32.737 |
| 40 | 20.707 | 22.164 | 24.433 | 26.509 | 29.051 | 33.660 |
| 41 | 21.421 | 22.906 | 25.215 | 27.326 | 29.907 | 34.585 |
| 42 | 22.138 | 23.650 | 25.999 | 28.144 | 30.765 | 35.510 |
| 43 | 22.859 | 24.398 | 26.785 | 28.965 | 31.625 | 36.436 |
| 44 | 23.584 | 25.148 | 27.575 | 29.787 | 32.487 | 37.363 |
| 45 | 24.311 | 25.901 | 28.366 | 30.612 | 33.350 | 38.291 |

*Continued*

**Table 5** Critical Values for the Chi-square Distribution
*Continued*

| | | | | $\gamma$ | | | |
|---|---|---|---|---|---|---|
| $r$ | 0.75 | 0.90 | 0.95 | 0.975 | 0.99 | 0.995 |
| 1 | 1.323 | 2.706 | 3.841 | 5.024 | 6.635 | 7.879 |
| 2 | 2.773 | 4.605 | 5.991 | 7.378 | 9.210 | 10.597 |
| 3 | 4.108 | 6.251 | 7.815 | 9.348 | 11.345 | 12.838 |
| 4 | 5.385 | 7.779 | 9.488 | 11.143 | 13.277 | 14.860 |
| 5 | 6.626 | 9.236 | 11.071 | 12.833 | 15.086 | 16.750 |
| 6 | 7.841 | 10.645 | 12.592 | 14.449 | 16.812 | 18.548 |
| 7 | 9.037 | 12.017 | 14.067 | 16.013 | 18.475 | 20.278 |
| 8 | 10.219 | 13.362 | 15.507 | 17.535 | 20.090 | 21.955 |
| 9 | 11.389 | 14.684 | 16.919 | 19.023 | 21.666 | 23.589 |
| 10 | 12.549 | 15.987 | 18.307 | 20.483 | 23.209 | 25.188 |
| 11 | 13.701 | 17.275 | 19.675 | 21.920 | 24.725 | 26.757 |
| 12 | 14.845 | 18.549 | 21.026 | 23.337 | 26.217 | 28.299 |
| 13 | 15.984 | 19.812 | 23.362 | 24.736 | 27.688 | 29.819 |
| 14 | 17.117 | 21.064 | 23.685 | 26.119 | 29.141 | 31.319 |
| 15 | 18.245 | 22.307 | 24.996 | 27.488 | 30.578 | 32.801 |
| 16 | 19.369 | 23.542 | 26.296 | 28.845 | 32.000 | 34.267 |
| 17 | 20.489 | 24.769 | 27.587 | 30.191 | 33.409 | 35.718 |
| 18 | 21.605 | 25.989 | 28.869 | 31.526 | 34.805 | 37.156 |
| 19 | 22.718 | 27.204 | 30.144 | 32.852 | 36.191 | 38.582 |
| 20 | 23.828 | 28.412 | 31.410 | 34.170 | 37.566 | 39.997 |
| 21 | 24.935 | 29.615 | 32.671 | 35.479 | 38.932 | 41.401 |
| 22 | 26.039 | 30.813 | 33.924 | 36.781 | 40.289 | 42.796 |
| 23 | 27.141 | 32.007 | 35.172 | 38.076 | 41.638 | 44.181 |
| 24 | 28.241 | 33.196 | 36.415 | 39.364 | 42.980 | 45.559 |
| 25 | 29.339 | 34.382 | 37.652 | 40.646 | 44.314 | 46.928 |
| 26 | 30.435 | 35.563 | 38.885 | 41.923 | 45.642 | 48.290 |
| 27 | 31.528 | 36.741 | 40.113 | 43.194 | 46.963 | 49.645 |
| 28 | 32.620 | 37.916 | 41.337 | 44.641 | 48.278 | 50.993 |
| 29 | 33.711 | 39.087 | 42.557 | 45.722 | 49.588 | 52.336 |
| 30 | 34.800 | 40.256 | 43.773 | 46.979 | 50.892 | 53.672 |
| 31 | 35.887 | 41.422 | 44.985 | 48.232 | 51.191 | 55.003 |
| 32 | 36.973 | 42.585 | 46.194 | 49.480 | 53.486 | 56.328 |
| 33 | 38.058 | 43.745 | 47.400 | 50.725 | 54.776 | 57.648 |
| 34 | 39.141 | 44.903 | 48.602 | 51.966 | 56.061 | 58.964 |
| 35 | 40.223 | 46.059 | 49.802 | 53.203 | 57.342 | 60.275 |
| 36 | 41.304 | 47.212 | 50.998 | 54.437 | 58.619 | 61.581 |
| 37 | 42.383 | 48.363 | 52.192 | 55.668 | 59.892 | 62.883 |
| 38 | 43.462 | 49.513 | 53.384 | 56.896 | 61.162 | 64.181 |
| 39 | 44.539 | 50.660 | 54.572 | 58.120 | 62.428 | 65.476 |
| 40 | 45.616 | 51.805 | 55.758 | 59.342 | 63.691 | 66.766 |
| 41 | 46.692 | 52.949 | 56.942 | 60.561 | 64.950 | 68.053 |
| 42 | 47.766 | 54.090 | 58.124 | 61.777 | 66.206 | 69.336 |
| 43 | 48.840 | 55.230 | 59.304 | 62.990 | 67.459 | 70.616 |
| 44 | 49.913 | 56.369 | 60.481 | 64.201 | 68.710 | 71.893 |
| 45 | 50.985 | 57.505 | 61.656 | 65.410 | 69.957 | 73.166 |

## Table 6 Critical Values for the $F$-Distribution

Let $F_{r_1, r_2}$ be a random variable having the $F$-distribution with $r_1$ and $r_2$ degrees of freedom. Then the tabulated quantities are the numbers $x$ for which

$$P(F_{r_1, r_2} \le x) = \gamma.$$

| | $\gamma$ | 1 | 2 | 3 | 4 | 5 | 6 | $\gamma$ | |
|---|---|---|---|---|---|---|---|---|---|
| | 0.500 | 1.0000 | 1.5000 | 1.7092 | 1.8227 | 1.8937 | 1.9422 | 0.500 | |
| | 0.750 | 5.8285 | 7.5000 | 8.1999 | 8.5810 | 8.8198 | 8.9833 | 0.750 | |
| | 0.900 | 39.864 | 49.500 | 53.593 | 55.833 | 57.241 | 58.204 | 0.900 | |
| 1 | 0.950 | 161.45 | 199.50 | 215.71 | 224.58 | 230.16 | 233.99 | 0.950 | 1 |
| | 0.975 | 647.79 | 799.50 | 864.16 | 899.58 | 921.85 | 937.11 | 0.975 | |
| | 0.990 | 4052.2 | 4999.5 | 5403.3 | 5624.6 | 5763.7 | 5859.0 | 0.990 | |
| | 0.995 | 16,211 | 20,000 | 21,615 | 22,500 | 23,056 | 23,437 | 0.995 | |
| | 0.500 | 0.66667 | 1.0000 | 1.1349 | 1.2071 | 1.2519 | 1.2824 | 0.500 | |
| | 0.750 | 2.5714 | 3.0000 | 3.1534 | 3.2320 | 3.2799 | 3.3121 | 0.750 | |
| | 0.900 | 8.5623 | 9.0000 | 9.1618 | 9.2434 | 9.2926 | 9.3255 | 0.900 | |
| 2 | 0.950 | 18.513 | 19.000 | 19.164 | 19.247 | 19.296 | 19.330 | 0.950 | 2 |
| | 0.975 | 38.506 | 39.000 | 39.165 | 39.248 | 39.298 | 39.331 | 0.975 | |
| | 0.990 | 98.503 | 99.000 | 99.166 | 99.249 | 99.299 | 99.332 | 0.990 | |
| | 0.995 | 198.50 | 199.00 | 199.17 | 199.25 | 199.30 | 199.33 | 0.995 | |
| | 0.500 | 0.58506 | 0.88110 | 1.0000 | 1.0632 | 1.1024 | 1.1289 | 0.500 | |
| | 0.750 | 2.0239 | 2.2798 | 2.3555 | 2.3901 | 2.4095 | 2.4218 | 0.750 | |
| | 0.900 | 5.5383 | 5.4624 | 5.3908 | 5.3427 | 5.3092 | 5.2847 | 0.900 | |
| 3 | 0.950 | 10.128 | 9.5521 | 9.2766 | 9.1172 | 9.0135 | 8.9406 | 0.950 | 3 |
| | 0.975 | 17.443 | 16.044 | 15.439 | 15.101 | 14.885 | 14.735 | 0.975 | |
| | 0.990 | 34.116 | 30.817 | 29.457 | 28.710 | 28.237 | 27.911 | 0.990 | |
| | 0.995 | 55.552 | 49.799 | 47.467 | 46.195 | 45.392 | 44.838 | 0.995 | |
| | 0.500 | 0.54863 | 0.82843 | 0.94054 | 1.0000 | 1.0367 | 1.0617 | 0.500 | |
| | 0.750 | 1.8074 | 2.0000 | 2.0467 | 2.0642 | 2.0723 | 2.0766 | 0.750 | |
| | 0.900 | 4.5448 | 4.3246 | 4.1908 | 4.1073 | 4.0506 | 4.0098 | 0.900 | |
| 4 | 0.950 | 7.7086 | 6.9443 | 6.5914 | 6.3883 | 6.2560 | 6.1631 | 0.950 | 4 |
| | 0.975 | 12.218 | 10.649 | 9.9792 | 9.6045 | 9.3645 | 9.1973 | 0.975 | |
| | 0.990 | 21.198 | 18.000 | 16.694 | 15.977 | 15.522 | 15.207 | 0.990 | |
| | 0.995 | 31.333 | 26.284 | 24.259 | 23.155 | 22.456 | 21.975 | 0.995 | |
| | 0.500 | 0.52807 | 0.79877 | 0.90715 | 0.96456 | 1.0000 | 1.0240 | 0.500 | |
| | 0.750 | 1.6925 | 1.8528 | 1.8843 | 1.8927 | 1.8947 | 1.8945 | 0.750 | |
| | 0.900 | 4.0604 | 3.7797 | 3.6195 | 3.5202 | 3.4530 | 3.4045 | 0.900 | |
| 5 | 0.950 | 6.6079 | 5.7861 | 5.4095 | 5.1922 | 5.0503 | 4.9503 | 0.950 | 5 |
| | 0.975 | 10.007 | 8.4336 | 7.7636 | 7.3879 | 7.1464 | 6.9777 | 0.975 | |
| | 0.990 | 16.258 | 13.274 | 12.060 | 11.392 | 10.967 | 10.672 | 0.990 | |
| | 0.995 | 22.785 | 18.314 | 16.530 | 15.556 | 14.940 | 14.513 | 0.995 | |
| | 0.500 | 0.51489 | 0.77976 | 0.88578 | 0.94191 | 0.97654 | 1.0000 | 0.500 | |
| | 0.750 | 1.6214 | 1.7622 | 1.7844 | 1.7872 | 1.7852 | 1.7821 | 0.750 | |
| | 0.900 | 3.7760 | 3.4633 | 3.2888 | 3.1808 | 3.1075 | 3.0546 | 0.900 | |
| 6 | 0.950 | 5.9874 | 5.1433 | 4.7571 | 4.5337 | 4.3874 | 4.2839 | 0.950 | 6 |
| | 0.975 | 8.8131 | 7.2598 | 6.5988 | 6.2272 | 5.9876 | 5.8197 | 0.975 | |
| | 0.990 | 13.745 | 10.925 | 9.7795 | 9.1483 | 8.7459 | 8.4661 | 0.990 | |
| | 0.995 | 18.635 | 14.544 | 12.917 | 12.028 | 11.464 | 11.073 | 0.995 | |

*Continued*

**Table 6** Critical Values for the *F*-Distribution *Continued*

| | | | 7 | 8 | 9 | 10 | 11 | 12 | | |
|---|---|---|---|---|---|---|---|---|---|---|
| | | $\gamma$ | | | | | | | $\gamma$ | |
| $r_2$ | 1 | 0.500 | 1.9774 | 2.0041 | 2.0250 | 2.0419 | 2.0558 | 2.0674 | 0.500 | |
| | | 0.750 | 9.1021 | 9.1922 | 9.2631 | 9.3202 | 9.3672 | 9.4064 | 0.750 | |
| | | 0.900 | 58.906 | 59.439 | 59.858 | 60.195 | 60.473 | 60.705 | 0.900 | |
| | | 0.950 | 236.77 | 238.88 | 240.54 | 241.88 | 242.99 | 243.91 | 0.950 | 1 |
| | | 0.975 | 948.22 | 956.66 | 963.28 | 968.63 | 973.04 | 976.71 | 0.975 | |
| | | 0.990 | 5928.3 | 5981.1 | 6022.5 | 6055.8 | 6083.3 | 6106.3 | 0.990 | |
| | | 0.995 | 23,715 | 23,925 | 24,091 | 24,224 | 24,334 | 24,426 | 0.995 | |
| | 2 | 0.500 | 1.3045 | 1.3213 | 1.3344 | 1.3450 | 1.3537 | 1.3610 | 0.500 | |
| | | 0.750 | 3.3352 | 3.3526 | 3.3661 | 3.3770 | 3.3859 | 3.3934 | 0.750 | |
| | | 0.900 | 9.3491 | 9.3668 | 9.3805 | 9.3916 | 9.4006 | 9.4081 | 0.900 | |
| | | 0.950 | 19.353 | 19.371 | 19.385 | 19.396 | 19.405 | 19.413 | 0.950 | 2 |
| | | 0.975 | 39.355 | 39.373 | 39.387 | 39.398 | 39.407 | 39.415 | 0.975 | |
| | | 0.990 | 99.356 | 99.374 | 99.388 | 99.399 | 99.408 | 99.416 | 0.990 | |
| | | 0.995 | 199.36 | 199.37 | 199.39 | 199.40 | 199.41 | 199.42 | 0.995 | |
| | 3 | 0.500 | 1.1482 | 1.1627 | 1.1741 | 1.1833 | 1.1909 | 1.1972 | 0.500 | |
| | | 0.750 | 2.4302 | 2.4364 | 2.4410 | 2.4447 | 2.4476 | 2.4500 | 0.750 | |
| | | 0.900 | 5.2662 | 5.2517 | 5.2400 | 5.2304 | 5.2223 | 5.2156 | 0.900 | |
| | | 0.950 | 8.8868 | 8.8452 | 8.8123 | 8.7855 | 8.7632 | 8.7446 | 0.950 | 3 |
| | | 0.975 | 14.624 | 14.540 | 14.473 | 14.419 | 14.374 | 14.337 | 0.975 | |
| | | 0.990 | 27.672 | 27.489 | 27.345 | 27.229 | 27.132 | 27.052 | 0.990 | |
| | | 0.995 | 44.434 | 44.126 | 43.882 | 43.686 | 43.523 | 43.387 | 0.995 | |
| | 4 | 0.500 | 1.0797 | 1.0933 | 1.1040 | 1.1126 | 1.1196 | 1.1255 | 0.500 | |
| | | 0.750 | 2.0790 | 2.0805 | 2.0814 | 2.0820 | 2.0823 | 2.0826 | 0.750 | |
| | | 0.900 | 3.9790 | 3.9549 | 3.9357 | 3.9199 | 3.9066 | 3.8955 | 0.900 | |
| | | 0.950 | 6.0942 | 6.0410 | 5.9988 | 5.9644 | 5.9357 | 5.9117 | 0.950 | 4 |
| | | 0.975 | 9.0741 | 8.9796 | 8.9047 | 8.8439 | 8.7933 | 8.7512 | 0.975 | |
| | | 0.990 | 14.976 | 14.799 | 14.659 | 14.546 | 14.452 | 14.374 | 0.990 | |
| | | 0.995 | 21.622 | 21.352 | 21.139 | 20.967 | 20.824 | 20.705 | 0.995 | |
| | 5 | 0.500 | 1.0414 | 1.0545 | 1.0648 | 1.0730 | 1.0798 | 1.0855 | 0.500 | |
| | | 0.750 | 1.8935 | 1.8923 | 1.8911 | 1.8899 | 1.8887 | 1.8877 | 0.750 | |
| | | 0.900 | 3.3679 | 3.3393 | 3.3163 | 3.2974 | 3.2815 | 3.2682 | 0.900 | |
| | | 0.950 | 4.8759 | 4.8183 | 4.7725 | 4.7351 | 4.7038 | 4.6777 | 0.950 | 5 |
| | | 0.975 | 6.8531 | 6.7572 | 6.6810 | 6.6192 | 6.5676 | 6.5246 | 0.975 | |
| | | 0.990 | 10.456 | 10.289 | 10.158 | 10.051 | 9.9623 | 9.8883 | 0.990 | |
| | | 0.995 | 14.200 | 13.961 | 13.772 | 13.618 | 13.490 | 13.384 | 0.995 | |
| | 6 | 0.500 | 1.0169 | 1.0298 | 1.0398 | 1.0478 | 1.0545 | 1.0600 | 0.500 | |
| | | 0.750 | 1.7789 | 1.7760 | 1.7733 | 1.7708 | 1.7686 | 1.7668 | 0.750 | |
| | | 0.900 | 3.0145 | 2.9830 | 2.9577 | 2.9369 | 2.9193 | 2.9047 | 0.900 | |
| | | 0.950 | 4.2066 | 4.1468 | 4.0990 | 4.0600 | 4.0272 | 3.9999 | 0.950 | 6 |
| | | 0.975 | 5.6955 | 5.5996 | 5.5234 | 5.4613 | 5.4094 | 5.3662 | 0.975 | |
| | | 0.990 | 8.2600 | 8.1016 | 7.9761 | 7.8741 | 7.7891 | 7.7183 | 0.990 | |
| | | 0.995 | 10.786 | 10.566 | 10.391 | 10.250 | 10.132 | 10.034 | 0.995 | |

*Continued*

**Table 6** Critical Values for the *F*-Distribution *Continued*

| $r_2$ | $\gamma$ | $r_1$ = 13 | 14 | 15 | 18 | 20 | 24 | $\gamma$ | $r_2$ |
|---|---|---|---|---|---|---|---|---|---|
| | 0.500 | 2.0773 | 2.0858 | 2.0931 | 2.1104 | 2.1190 | 2.1321 | 0.500 | |
| | 0.750 | 9.4399 | 9.4685 | 9.4934 | 9.5520 | 9.5813 | 9.6255 | 0.750 | |
| | 0.900 | 60.903 | 61.073 | 61.220 | 61.567 | 61.740 | 62.002 | 0.900 | |
| 1 | 0.950 | 244.69 | 245.37 | 245.95 | 247.32 | 248.01 | 249.05 | 0.950 | 1 |
| | 0.975 | 979.85 | 982.54 | 984.87 | 990.36 | 993.10 | 997.25 | 0.975 | |
| | 0.990 | 6125.9 | 6142.7 | 6157.3 | 6191.6 | 6208.7 | 6234.6 | 0.990 | |
| | 0.995 | 24,504 | 24,572 | 24,630 | 24,767 | 24,836 | 24,940 | 0.995 | |
| | 0.500 | 1.3672 | 1.3725 | 1.3771 | 1.3879 | 1.3933 | 1.4014 | 0.500 | |
| | 0.750 | 3.3997 | 3.4051 | 3.4098 | 3.4208 | 3.4263 | 3.4345 | 0.750 | |
| | 0.900 | 9.4145 | 9.4200 | 9.4247 | 9.4358 | 9.4413 | 9.4496 | 0.900 | |
| 2 | 0.950 | 19.419 | 19.424 | 19.429 | 19.440 | 19.446 | 19.454 | 0.950 | 2 |
| | 0.975 | 39.421 | 39.426 | 39.431 | 39.442 | 39.448 | 39.456 | 0.975 | |
| | 0.990 | 99.422 | 99.427 | 99.432 | 99.443 | 99.449 | 99.458 | 0.990 | |
| | 0.995 | 199.42 | 199.43 | 199.43 | 199.44 | 199.45 | 199.46 | 0.995 | |
| | 0.500 | 1.2025 | 1.2071 | 1.2111 | 1.2205 | 1.2252 | 1.2322 | 0.500 | |
| | 0.750 | 2.4520 | 2.4537 | 2.4552 | 2.4585 | 2.4602 | 2.4626 | 0.750 | |
| | 0.900 | 5.2097 | 5.2047 | 5.2003 | 5.1898 | 5.1845 | 5.1764 | 0.900 | |
| 3 | 0.950 | 8.7286 | 8.7148 | 8.7029 | 8.6744 | 8.6602 | 8.6385 | 0.950 | 3 |
| | 0.975 | 14.305 | 14.277 | 14.253 | 14.196 | 14.167 | 14.124 | 0.975 | |
| | 0.990 | 26.983 | 26.923 | 26.872 | 26.751 | 26.690 | 26.598 | 0.990 | |
| | 0.995 | 43.271 | 43.171 | 43.085 | 42.880 | 42.778 | 42.622 | 0.955 | |
| | 0.500 | 1.1305 | 1.1349 | 1.1386 | 1.1473 | 1.1517 | 1.1583 | 0.500 | |
| | 0.750 | 2.0827 | 2.0828 | 2.0829 | 2.0828 | 2.0828 | 2.0827 | 0.750 | |
| | 0.900 | 3.8853 | 3.8765 | 3.8689 | 3.8525 | 3.8443 | 3.8310 | 0.900 | |
| 4 | 0.950 | 5.8910 | 5.8732 | 5.8578 | 5.8209 | 5.8025 | 5.7744 | 0.950 | 4 |
| | 0.975 | 8.7148 | 8.6836 | 8.6565 | 8.5921 | 8.5599 | 8.5109 | 0.975 | |
| | 0.990 | 14.306 | 14.248 | 14.198 | 14.079 | 14.020 | 13.929 | 0.990 | |
| | 0.995 | 20.602 | 20.514 | 20.438 | 20.257 | 20.167 | 20.030 | 0.995 | |
| | 0.500 | 1.0903 | 1.0944 | 1.0980 | 1.1064 | 1.1106 | 1.1170 | 0.500 | |
| | 0.750 | 1.8867 | 1.8858 | 1.8851 | 1.8830 | 1.8820 | 1.8802 | 0.750 | |
| | 0.900 | 3.2566 | 3.2466 | 3.2380 | 3.2171 | 3.2067 | 3.1905 | 0.900 | |
| 5 | 0.950 | 4.6550 | 4.6356 | 4.6188 | 4.5783 | 4.5581 | 4.5272 | 0.950 | 5 |
| | 0.975 | 6.4873 | 6.4554 | 6.4277 | 6.3616 | 6.3285 | 6.2780 | 0.975 | |
| | 0.990 | 9.8244 | 9.7697 | 9.7222 | 9.6092 | 9.5527 | 9.4665 | 0.990 | |
| | 0.995 | 13.292 | 13.214 | 13.146 | 12.984 | 12.903 | 12.780 | 0.995 | |
| | 0.500 | 1.0647 | 1.0687 | 1.0722 | 1.0804 | 1.0845 | 1.0907 | 0.500 | |
| | 0.750 | 1.7650 | 1.7634 | 1.7621 | 1.7586 | 1.7569 | 1.7540 | 0.750 | |
| | 0.900 | 2.8918 | 2.8808 | 2.8712 | 2.8479 | 2.8363 | 2.8183 | 0.900 | |
| 6 | 0.950 | 3.9761 | 3.9558 | 3.9381 | 3.8955 | 3.8742 | 3.8415 | 0.950 | 6 |
| | 0.975 | 5.3287 | 5.2966 | 5.2687 | 5.2018 | 5.1684 | 5.1172 | 0.975 | |
| | 0.990 | 7.6570 | 7.6045 | 7.5590 | 7.4502 | 7.3958 | 7.3127 | 0.990 | |
| | 0.995 | 9.9494 | 9.8769 | 9.8140 | 9.6639 | 9.5888 | 9.4741 | 0.995 | |

*Continued*

**Table 6** Critical Values for the *F*-Distribution *Continued*

| | | | | | $r_1$ | | | | | | |
|---|---|---|---|---|---|---|---|---|---|---|---|
| | | $\gamma$ | 30 | 40 | 48 | 60 | 120 | $\infty$ | $\gamma$ | | |
| | | 0.500 | 2.1452 | 2.1584 | 2.1650 | 2.1716 | 2.1848 | 2.1981 | 0.500 | | |
| | | 0.750 | 9.6698 | 9.7144 | 9.7368 | 9.7591 | 9.8041 | 9.8492 | 0.750 | | |
| | | 0.900 | 62.265 | 62.529 | 62.662 | 62.794 | 63.061 | 63.328 | 0.990 | | |
| | 1 | 0.950 | 250.09 | 251.14 | 251.67 | 252.20 | 253.25 | 254.32 | 0.950 | 1 | |
| | | 0.975 | 1001.4 | 1005.6 | 1007.7 | 1009.8 | 1014.0 | 1018.3 | 0.975 | | |
| | | 0.990 | 6260.7 | 6286.8 | 6299.9 | 6313.0 | 6339.4 | 6366.0 | 0.990 | | |
| | | 0.995 | 25,044 | 25,148 | 25,201 | 25,253 | 25,359 | 25,465 | 0.995 | | |
| | | 0.500 | 1.4096 | 1.4178 | 1.4220 | 1.4261 | 1.4344 | 1.4427 | 0.500 | | |
| | | 0.750 | 3.4428 | 3.4511 | 3.4553 | 3.4594 | 3.4677 | 3.4761 | 0.750 | | |
| | | 0.900 | 9.4579 | 9.4663 | 9.4705 | 9.4746 | 9.4829 | 9.4913 | 0.900 | | |
| | 2 | 0.950 | 19.462 | 19.471 | 19.475 | 19.479 | 19.487 | 19.496 | 0.950 | 2 | |
| | | 0.975 | 39.465 | 39.473 | 39.477 | 39.481 | 39.490 | 39.498 | 0.975 | | |
| | | 0.990 | 99.466 | 99.474 | 99.478 | 99.483 | 99.491 | 99.499 | 0.990 | | |
| | | 0.995 | 199.47 | 199.47 | 199.47 | 199.48 | 199.49 | 199.51 | 0.995 | | |
| | | 0.500 | 1.2393 | 1.2464 | 1.2500 | 1.2536 | 1.2608 | 1.2680 | 0.500 | | |
| | | 0.750 | 2.4650 | 2.4674 | 2.4686 | 2.4697 | 2.4720 | 2.4742 | 0.750 | | |
| | | 0.900 | 5.1681 | 5.1597 | 5.1555 | 5.1512 | 5.1425 | 5.1337 | 0.900 | | |
| | 3 | 0.950 | 8.6166 | 8.5944 | 8.5832 | 8.5720 | 8.5494 | 8.5265 | 0.950 | 3 | |
| | | 0.975 | 14.081 | 14.037 | 14.015 | 13.992 | 13.947 | 13.902 | 0.975 | | |
| | | 0.990 | 26.505 | 26.411 | 26.364 | 26.316 | 26.221 | 26.125 | 0.990 | | |
| $r_2$ | | 0.995 | 42.466 | 42.308 | 42.229 | 42.149 | 41.989 | 41.829 | 0.995 | | $r_2$ |
| | | 0.500 | 1.1649 | 1.1716 | 1.1749 | 1.1782 | 1.1849 | 1.1916 | 0.500 | | |
| | | 0.750 | 2.0825 | 2.0821 | 2.0819 | 2.0817 | 2.0812 | 2.0806 | 0.750 | | |
| | | 0.900 | 3.8174 | 3.8036 | 3.7966 | 3.7896 | 3.7753 | 3.7607 | 0.900 | | |
| | 4 | 0.950 | 5.7459 | 5.7170 | 5.7024 | 5.6878 | 5.6581 | 5.6281 | 0.950 | 4 | |
| | | 0.975 | 8.4613 | 8.4111 | 8.3858 | 8.3604 | 8.3092 | 8.2573 | 0.975 | | |
| | | 0.990 | 13.838 | 13.745 | 13.699 | 13.652 | 13.558 | 13.463 | 0.990 | | |
| | | 0.995 | 19.892 | 19.752 | 19.682 | 19.611 | 19.468 | 19.325 | 0.995 | | |
| | | 0.500 | 1.1234 | 1.1297 | 1.1329 | 1.1361 | 1.1426 | 1.1490 | 0.500 | | |
| | | 0.750 | 1.8784 | 1.8763 | 1.8753 | 1.8742 | 1.8719 | 1.8694 | 0.750 | | |
| | | 0.900 | 3.1741 | 3.1573 | 3.1488 | 1.1402 | 3.1228 | 3.1050 | 0.900 | | |
| | 5 | 0.950 | 4.4957 | 4.4638 | 4.4476 | 4.4314 | 4.3984 | 4.3650 | 0.950 | 5 | |
| | | 0.975 | 6.2269 | 6.1751 | 6.1488 | 6.1225 | 6.0693 | 6.0153 | 0.975 | | |
| | | 0.990 | 9.3793 | 9.2912 | 9.2466 | 9.2020 | 9.1118 | 0.0204 | 0.990 | | |
| | | 0.995 | 12.656 | 12.530 | 12.466 | 12.402 | 12.274 | 12.144 | 0.995 | | |
| | | 0.500 | 1.0969 | 1.1031 | 1.1062 | 1.1093 | 1.1156 | 1.1219 | 0.500 | | |
| | | 0.750 | 1.7510 | 1.7477 | 1.7460 | 1.7443 | 1.7407 | 1.7368 | 0.750 | | |
| | | 0.900 | 2.8000 | 2.7812 | 2.7716 | 2.7620 | 2.7423 | 2.7222 | 0.900 | | |
| | 6 | 0.950 | 3.8082 | 3.7743 | 3.7571 | 3.7398 | 3.7047 | 3.6688 | 0.950 | 6 | |
| | | 0.975 | 5.0652 | 5.0125 | 4.9857 | 4.9589 | 4.9045 | 4.9491 | 0.975 | | |
| | | 0.990 | 7.2285 | 7.1432 | 7.1000 | 7.0568 | 6.9690 | 6.8801 | 0.990 | | |
| | | 0.995 | 9.3583 | 9.2408 | 9.1814 | 9.1219 | 9.0015 | 8.8793 | 0.995 | | |

*Continued*

**Table 6** Critical Values for the *F*-Distribution *Continued*

| | | | | | $r_1$ | | | | | | |
|---|---|---|---|---|---|---|---|---|---|---|---|
| | | $\gamma$ | 1 | 2 | 3 | 4 | 5 | 6 | $\gamma$ | | |
| | | 0.500 | 0.50572 | 0.76655 | 0.87095 | 0.92619 | 0.96026 | 0.98334 | 0.500 | | |
| | | 0.750 | 1.5732 | 1.7010 | 1.7169 | 1.7157 | 1.7111 | 1.7059 | 0.750 | | |
| | | 0.900 | 3.5894 | 3.2574 | 3.0741 | 2.9605 | 2.8833 | 2.8274 | 0.900 | | |
| | 7 | 0.950 | 5.5914 | 4.7374 | 4.3468 | 4.1203 | 3.9715 | 3.8660 | 0.950 | 7 | |
| | | 0.975 | 8.0727 | 6.5415 | 5.8898 | 5.5226 | 5.2852 | 5.1186 | 0.975 | | |
| | | 0.990 | 12.246 | 9.5466 | 8.4513 | 7.8467 | 7.4604 | 7.1914 | 0.990 | | |
| | | 0.995 | 16.236 | 12.404 | 10.882 | 10.050 | 9.5221 | 9.1554 | 0.995 | | |
| | | 0.500 | 0.49898 | 0.75683 | 0.86004 | 0.91464 | 0.94831 | 0.97111 | 0.500 | | |
| | | 0.750 | 1.5384 | 1.6569 | 1.6683 | 1.6642 | 1.6575 | 1.6508 | 0.750 | | |
| | | 0.900 | 3.4579 | 3.1131 | 2.9238 | 2.8064 | 2.7265 | 2.6683 | 0.900 | | |
| | 8 | 0.950 | 5.3177 | 4.4590 | 4.0662 | 3.8378 | 3.6875 | 3.5806 | 0.950 | 8 | |
| | | 0.975 | 7.5709 | 6.0595 | 5.4160 | 5.0526 | 4.8173 | 4.6517 | 0.975 | | |
| | | 0.990 | 11.259 | 8.6491 | 7.5910 | 7.0060 | 6.6318 | 6.3707 | 0.990 | | |
| | | 0.995 | 14.688 | 11.042 | 9.5965 | 8.8051 | 8.3018 | 7.9520 | 0.995 | | |
| | | 0.500 | 0.49382 | 0.74938 | 0.85168 | 0.90580 | 0.93916 | 0.96175 | 0.500 | | |
| | | 0.750 | 1.5121 | 1.6236 | 1.6315 | 1.6253 | 1.6170 | 1.6091 | 0.750 | | |
| | | 0.900 | 3.3603 | 3.0065 | 2.8129 | 2.6927 | 2.6106 | 2.5509 | 0.900 | | |
| | 9 | 0.950 | 5.1174 | 4.2565 | 3.8626 | 3.6331 | 3.4817 | 3.3738 | 0.950 | 9 | |
| | | 0.975 | 7.2093 | 5.7147 | 5.0781 | 4.7181 | 4.4844 | 4.3197 | 0.975 | | |
| | | 0.990 | 10.561 | 8.0215 | 6.9919 | 6.4221 | 6.0569 | 5.8018 | 0.990 | | |
| $r_2$ | | 0.995 | 13.614 | 10.107 | 8.7171 | 7.9559 | 7.4711 | 7.1338 | 0.995 | | $r_2$ |
| | | 0.500 | 0.48973 | 0.74349 | 0.84508 | 0.89882 | 0.93193 | 0.95436 | 0.500 | | |
| | | 0.750 | 1.4915 | 1.5975 | 1.6028 | 1.5949 | 1.5853 | 1.5765 | 0.750 | | |
| | | 0.900 | 3.2850 | 2.9245 | 2.7277 | 2.6053 | 2.5216 | 2.4606 | 0.900 | | |
| | 10 | 0.950 | 4.9646 | 4.1028 | 3.7083 | 3.4780 | 3.3258 | 3.2172 | 0.950 | 10 | |
| | | 0.975 | 6.9367 | 5.4564 | 4.8256 | 4.4683 | 4.2361 | 4.0721 | 0.975 | | |
| | | 0.990 | 10.044 | 7.5594 | 6.5523 | 5.9943 | 5.6363 | 5.3858 | 0.990 | | |
| | | 0.995 | 12.826 | 9.4270 | 8.0807 | 7.3428 | 6.8723 | 6.5446 | 0.995 | | |
| | | 0.500 | 0.48644 | 0.73872 | 0.83973 | 0.89316 | 0.92608 | 0.94837 | 0.500 | | |
| | | 0.750 | 1.4749 | 1.5767 | 1.5798 | 1.5704 | 1.5598 | 1.5502 | 0.750 | | |
| | | 0.900 | 3.2252 | 2.8595 | 2.6602 | 2.5362 | 2.4512 | 2.3891 | 0.900 | | |
| | 11 | 0.950 | 4.8443 | 3.9823 | 3.5874 | 3.3567 | 3.2039 | 3.0946 | 0.950 | 11 | |
| | | 0.975 | 6.7241 | 5.2559 | 4.6300 | 4.2751 | 4.0440 | 3.8807 | 0.975 | | |
| | | 0.990 | 9.6460 | 7.2057 | 6.2167 | 5.6683 | 5.3160 | 5.0692 | 0.990 | | |
| | | 0.995 | 12.226 | 8.9122 | 7.6004 | 6.8809 | 6.4217 | 6.1015 | 0.995 | | |
| | | 0.500 | 0.48369 | 0.73477 | 0.83530 | 0.88848 | 0.92124 | 0.94342 | 0.500 | | |
| | | 0.750 | 1.4613 | 1.5595 | 1.5609 | 1.5503 | 1.5389 | 1.5286 | 0.750 | | |
| | | 0.900 | 3.1765 | 2.8068 | 2.6055 | 2.4801 | 2.3940 | 2.3310 | 0.900 | | |
| | 12 | 0.950 | 4.7472 | 3.8853 | 3.4903 | 3.2592 | 3.1059 | 2.9961 | 0.950 | 12 | |
| | | 0.975 | 6.5538 | 5.0959 | 4.4742 | 4.1212 | 3.8911 | 3.7283 | 0.975 | | |
| | | 0.990 | 9.3302 | 6.9266 | 5.9526 | 5.4119 | 5.0643 | 4.8206 | 0.990 | | |
| | | 0.995 | 11.754 | 8.5096 | 7.2258 | 6.5211 | 6.0711 | 5.7570 | 0.995 | | |

*Continued*

**Table 6** Critical Values for the *F*-Distribution *Continued*

| | $\gamma$ | 7 | 8 | 9 | 10 | 11 | 12 | $\gamma$ | |
|---|---|---|---|---|---|---|---|---|---|
| | | | | | $r_1$ | | | | |
| | 0.500 | 1.0000 | 1.0216 | 1.0224 | 1.0304 | 1.0369 | 1.0423 | 0.500 | |
| | 0.750 | 1.7011 | 1.6969 | 1.6931 | 1.6898 | 1.6868 | 1.6843 | 0.750 | |
| | 0.900 | 2.7849 | 2.7516 | 2.7247 | 2.7025 | 2.6837 | 2.6681 | 0.900 | |
| 7 | 0.950 | 3.7870 | 3.7257 | 3.6767 | 3.6365 | 3.6028 | 3.5747 | 0.950 | 7 |
| | 0.975 | 4.9949 | 4.8994 | 4.8232 | 4.7611 | 4.7091 | 4.6658 | 0.975 | |
| | 0.990 | 6.9928 | 6.8401 | 6.7188 | 6.6201 | 6.5377 | 6.4691 | 0.990 | |
| | 0.995 | 8.8854 | 8.6781 | 8.5138 | 8.3803 | 8.2691 | 8.1764 | 0.995 | |
| | 0.500 | 0.98757 | 1.0000 | 1.0097 | 1.0175 | 1.0239 | 1.0293 | 0.500 | |
| | 0.750 | 1.6448 | 1.6396 | 1.6350 | 1.6310 | 1.6274 | 1.6244 | 0.750 | |
| | 0.900 | 2.6241 | 2.5893 | 2.5612 | 2.5380 | 2.5184 | 2.5020 | 0.900 | |
| 8 | 0.950 | 3.5005 | 3.4381 | 3.3881 | 3.3472 | 3.3127 | 3.2840 | 0.950 | 8 |
| | 0.975 | 4.5286 | 4.4332 | 4.3572 | 4.2951 | 4.2431 | 4.1997 | 0.975 | |
| | 0.990 | 6.1776 | 6.0289 | 5.9106 | 5.8143 | 5.7338 | 5.6668 | 0.990 | |
| | 0.995 | 7.6942 | 7.4960 | 7.3386 | 7.2107 | 7.1039 | 7.0149 | 0.995 | |
| | 0.500 | 0.97805 | 0.99037 | 1.0000 | 1.0077 | 1.0141 | 1.0194 | 0.500 | |
| | 0.750 | 1.6022 | 1.5961 | 1.5909 | 1.5863 | 1.5822 | 1.5788 | 0.750 | |
| | 0.900 | 2.5053 | 2.4694 | 2.4403 | 2.4163 | 2.3959 | 2.3789 | 0.900 | |
| 9 | 0.950 | 3.2927 | 3.2296 | 3.1789 | 3.1373 | 3.1022 | 3.0729 | 0.950 | 9 |
| | 0.975 | 4.1971 | 4.1020 | 4.0260 | 3.9639 | 3.9117 | 3.8682 | 0.975 | |
| | 0.990 | 5.6129 | 5.4671 | 5.3511 | 5.2565 | 5.1774 | 5.1114 | 0.990 | |
| | 0.995 | 6.8849 | 6.6933 | 6.5411 | 6.4171 | 6.3136 | 6.2274 | 0.995 | |
| | 0.500 | 0.97054 | 0.98276 | 0.99232 | 1.0000 | 1.0063 | 1.0166 | 0.500 | |
| | 0.750 | 1.5688 | 1.5621 | 1.5563 | 1.5513 | 1.5468 | 1.5430 | 0.750 | |
| | 0.900 | 2.4140 | 2.3772 | 2.3473 | 2.3226 | 2.3016 | 2.2841 | 0.900 | |
| 10 | 0.950 | 3.1355 | 3.0717 | 3.0204 | 2.9782 | 2.9426 | 2.9130 | 0.950 | 10 |
| | 0.975 | 3.9498 | 3.8549 | 3.7790 | 3.7168 | 3.6645 | 3.6209 | 0.975 | |
| | 0.990 | 5.2001 | 5.0567 | 4.9424 | 4.8492 | 4.7710 | 4.7059 | 0.990 | |
| | 0.995 | 6.3025 | 6.1159 | 5.9676 | 5.8467 | 5.7456 | 5.6613 | 0.995 | |
| | 0.500 | 0.96445 | 0.97661 | 0.98610 | 0.99373 | 0.99999 | 1.0052 | 0.500 | |
| | 0.750 | 1.5418 | 1.5346 | 1.5284 | 1.5230 | 1.5181 | 1.5140 | 0.750 | |
| | 0.900 | 2.3416 | 2.3040 | 2.2735 | 2.2482 | 2.2267 | 2.2087 | 0.900 | |
| 11 | 0.950 | 3.0123 | 2.9480 | 2.8962 | 2.8536 | 2.8176 | 2.7876 | 0.950 | 11 |
| | 0.975 | 3.7586 | 3.6638 | 3.5879 | 3.5257 | 3.4733 | 3.4296 | 0.975 | |
| | 0.990 | 4.8861 | 4.7445 | 4.6315 | 4.5393 | 4.4619 | 4.3974 | 0.990 | |
| | 0.995 | 5.8648 | 5.6821 | 5.5368 | 5.4182 | 5.3190 | 5.2363 | 0.995 | |
| | 0.500 | 0.95943 | 0.97152 | 0.98097 | 0.98856 | 0.99480 | 1.0000 | 0.500 | |
| | 0.750 | 1.5197 | 1.5120 | 1.5054 | 1.4996 | 1.4945 | 1.4902 | 0.750 | |
| | 0.900 | 2.2828 | 2.2446 | 2.2135 | 2.1878 | 1.1658 | 1.1474 | 0.900 | |
| 12 | 0.950 | 2.9134 | 2.8486 | 2.7964 | 2.7534 | 2.7170 | 2.6866 | 0.950 | 12 |
| | 0.975 | 3.6065 | 3.5118 | 3.4358 | 3.3736 | 3.3211 | 3.2773 | 0.975 | |
| | 0.990 | 4.6395 | 4.4994 | 4.3875 | 4.2961 | 4.2193 | 4.1553 | 0.990 | |
| | 0.995 | 5.5245 | 5.3451 | 5.2021 | 5.0855 | 4.9878 | 4.9063 | 0.995 | |

$r_2$ (left margin) $\qquad$ $r_2$ (right margin)

*Continued*

**Table 6** Critical Values for the *F*-Distribution *Continued*

| | | | | | | | | | |
|---|---|---|---|---|---|---|---|---|---|
| | | | | | $r_1$ | | | | |
| | $\gamma$ | 13 | 14 | 15 | 18 | 20 | 24 | $\gamma$ | |
| | 0.500 | 1.0469 | 1.0509 | 1.0543 | 1.0624 | 1.0664 | 1.0724 | 0.500 | |
| | 0.750 | 1.6819 | 1.6799 | 1.6781 | 1.6735 | 1.6712 | 1.6675 | 0.750 | |
| | 0.900 | 2.6543 | 2.6425 | 2.6322 | 2.6072 | 2.5947 | 2.5753 | 0.900 | |
| 7 | 0.950 | 3.5501 | 3.5291 | 3.5108 | 3.4666 | 3.4445 | 3.4105 | 0.950 | 7 |
| | 0.975 | 4.6281 | 4.5958 | 4.5678 | 4.5004 | 4.4667 | 4.4150 | 0.975 | |
| | 0.990 | 6.4096 | 6.3585 | 6.3143 | 6.2084 | 6.1554 | 6.0743 | 0.990 | |
| | 0.995 | 8.0962 | 8.0274 | 7.9678 | 7.8253 | 7.7540 | 7.6450 | 0.995 | |
| | 0.500 | 1.0339 | 1.0378 | 1.0412 | 1.0491 | 1.0531 | 1.0591 | 0.500 | |
| | 0.750 | 1.6216 | 1.6191 | 1.6170 | 1.6115 | 1.6088 | 1.6043 | 0.750 | |
| | 0.900 | 2.4875 | 2.4750 | 2.4642 | 2.4378 | 2.4246 | 2.4041 | 0.900 | |
| 8 | 0.950 | 3.2588 | 3.2371 | 3.2184 | 3.1730 | 3.1503 | 3.1152 | 0.950 | 8 |
| | 0.975 | 4.1618 | 4.1293 | 4.1012 | 4.0334 | 3.9995 | 3.9472 | 0.975 | |
| | 0.990 | 5.6085 | 5.5584 | 5.5151 | 5.4111 | 5.3591 | 5.2793 | 0.990 | |
| | 0.995 | 6.9377 | 6.8716 | 6.8143 | 6.6769 | 6.6082 | 6.5029 | 0.995 | |
| | 0.500 | 1.0239 | 1.0278 | 1.0311 | 1.0390 | 1.0429 | 1.0489 | 0.500 | |
| | 0.750 | 1.5756 | 1.5729 | 1.5705 | 1.5642 | 1.5611 | 1.5560 | 0.750 | |
| | 0.900 | 2.3638 | 2.3508 | 2.3396 | 2.3121 | 2.9893 | 2.2768 | 0.900 | |
| 9 | 0.950 | 3.0472 | 3.0252 | 3.0061 | 2.9597 | 2.9365 | 2.9005 | 0.950 | 9 |
| | 0.975 | 3.8302 | 3.7976 | 3.7694 | 3.7011 | 3.6669 | 3.6142 | 0.975 | |
| | 0.990 | 5.0540 | 5.0048 | 4.9621 | 4.8594 | 4.8080 | 4.7290 | 0.990 | |
| | 0.995 | 6.1524 | 6.0882 | 6.0325 | 5.8987 | 5.8318 | 5.7292 | 0.995 | |
| | 0.500 | 1.0161 | 1.0199 | 1.0232 | 1.0310 | 1.0349 | 1.0408 | 0.500 | |
| | 0.750 | 1.5395 | 1.5364 | 1.5338 | 1.5269 | 1.5235 | 1.5179 | 0.750 | |
| | 0.900 | 2.2685 | 2.2551 | 2.2435 | 2.2150 | 2.2007 | 2.1784 | 0.900 | |
| 10 | 0.950 | 2.8868 | 2.8644 | 2.8450 | 2.7977 | 2.7740 | 2.7372 | 0.950 | 10 |
| | 0.975 | 3.5827 | 3.5500 | 3.5217 | 3.4530 | 3.4186 | 3.3654 | 0.975 | |
| | 0.990 | 4.6491 | 4.6004 | 4.5582 | 4.4563 | 3.4054 | 3.3269 | 0.990 | |
| | 0.995 | 5.5880 | 5.5252 | 5.4707 | 5.3396 | 5.2740 | 5.1732 | 0.995 | |
| | 0.500 | 1.0097 | 1.0135 | 1.0168 | 1.0245 | 1.0284 | 1.0343 | 0.500 | |
| | 0.750 | 1.5102 | 1.5069 | 1.5041 | 1.4967 | 1.4930 | 1.4869 | 0.750 | |
| | 0.900 | 2.1927 | 2.1790 | 2.1671 | 2.1377 | 2.1230 | 2.1000 | 0.900 | |
| 11 | 0.950 | 2.7611 | 2.7383 | 2.7186 | 2.6705 | 2.6464 | 2.6090 | 0.950 | 11 |
| | 0.975 | 3.3913 | 3.3584 | 3.3299 | 3.2607 | 3.2261 | 3.1725 | 0.975 | |
| | 0.990 | 4.3411 | 4.2928 | 4.2509 | 4.1496 | 4.0990 | 4.0209 | 0.990 | |
| | 0.995 | 5.1642 | 5.1024 | 5.0489 | 4.9198 | 4.8552 | 4.7557 | 0.995 | |
| | 0.500 | 1.0044 | 1.0082 | 1.0115 | 1.0192 | 1.0231 | 1.0289 | 0.500 | |
| | 0.750 | 1.4861 | 1.4826 | 1.4796 | 1.4717 | 1.4678 | 1.4613 | 0.750 | |
| | 0.900 | 2.1311 | 2.1170 | 2.1049 | 2.0748 | 2.0597 | 2.0360 | 0.900 | |
| 12 | 0.950 | 2.6598 | 2.6368 | 2.6169 | 2.5680 | 2.5436 | 2.5055 | 0.950 | 12 |
| | 0.975 | 3.2388 | 3.2058 | 3.1772 | 3.1076 | 3.0728 | 3.0187 | 0.975 | |
| | 0.990 | 4.0993 | 4.0512 | 4.0096 | 3.9088 | 3.8584 | 3.7805 | 0.990 | |
| | 0.995 | 4.8352 | 4.7742 | 4.7214 | 4.5937 | 4.5299 | 4.4315 | 0.995 | |

*Continued*

**Table 6** Critical Values for the *F*-Distribution *Continued*

| | $r_2$ | $\gamma$ | 30 | 40 | 48 | 60 | 120 | $\infty$ | $\gamma$ | $r_2$ |
|---|---|---|---|---|---|---|---|---|---|---|
| | | 0.500 | 1.0785 | 1.0846 | 1.0877 | 1.0908 | 1.0969 | 1.1031 | 0.500 | |
| | | 0.750 | 1.6635 | 1.6593 | 1.6571 | 1.6548 | 1.6502 | 1.6452 | 0.750 | |
| | | 0.900 | 2.5555 | 2.5351 | 2.5427 | 2.5142 | 2.4928 | 2.4708 | 0.900 | |
| | 7 | 0.950 | 3.3758 | 3.3404 | 3.3224 | 3.3043 | 3.2674 | 3.2298 | 0.950 | 7 |
| | | 0.975 | 4.3624 | 4.3089 | 4.2817 | 4.2544 | 4.1989 | 4.1423 | 0.975 | |
| | | 0.990 | 5.9921 | 5.9084 | 5.8660 | 5.8236 | 5.7372 | 5.6495 | 0.990 | |
| | | 0.995 | 7.5345 | 7.4225 | 7.3657 | 7.3088 | 7.1933 | 7.0760 | 0.995 | |
| | | 0.500 | 1.0651 | 1.0711 | 1.0741 | 1.0771 | 1.0832 | 1.0893 | 0.500 | |
| | | 0.750 | 1.5996 | 1.5945 | 1.5919 | 1.5892 | 1.5836 | 1.5777 | 0.750 | |
| | | 0.900 | 2.3830 | 2.3614 | 2.3503 | 2.3391 | 2.3162 | 2.2926 | 0.900 | |
| | 8 | 0.950 | 3.0794 | 3.0428 | 3.0241 | 3.0053 | 2.9669 | 2.9276 | 0.950 | 8 |
| | | 0.975 | 3.8940 | 3.8398 | 3.8121 | 3.7844 | 3.7279 | 3.6702 | 0.975 | |
| | | 0.990 | 5.1981 | 5.1156 | 5.0736 | 5.0316 | 4.9460 | 4.8588 | 0.990 | |
| | | 0.995 | 6.3961 | 6.2875 | 6.2324 | 6.1772 | 6.0649 | 5.9505 | 0.995 | |
| | | 0.500 | 1.0548 | 1.0608 | 1.0638 | 1.0667 | 1.0727 | 1.0788 | 0.500 | |
| | | 0.750 | 1.5506 | 1.5450 | 1.5420 | 1.5389 | 1.5325 | 1.5257 | 0.750 | |
| | | 0.900 | 2.2547 | 2.2320 | 2.2203 | 2.2085 | 2.1843 | 2.1592 | 0.900 | |
| | 9 | 0.950 | 2.8637 | 2.8259 | 2.8066 | 2.7872 | 2.7475 | 2.7067 | 0.950 | 9 |
| | | 0.975 | 3.5604 | 3.5055 | 3.4774 | 3.4493 | 3.3918 | 3.3329 | 0.975 | |
| | | 0.990 | 4.6486 | 4.5667 | 4.5249 | 4.4831 | 4.3978 | 4.3105 | 0.990 | |
| $r_2$ | | 0.995 | 5.6248 | 5.5186 | 5.4645 | 5.4104 | 5.3001 | 5.1875 | 0.995 | $r_2$ |
| | | 0.500 | 1.0467 | 1.0526 | 1.0556 | 1.0585 | 1.0645 | 1.0705 | 0.500 | |
| | | 0.750 | 1.5119 | 1.5056 | 1.5023 | 1.4990 | 1.4919 | 1.4843 | 0.750 | |
| | | 0.900 | 2.1554 | 1.1317 | 2.1195 | 2.1072 | 2.0818 | 2.0554 | 0.900 | |
| | 10 | 0.950 | 2.6996 | 2.6609 | 2.6410 | 2.6211 | 2.5801 | 2.5379 | 0.950 | 10 |
| | | 0.975 | 3.3110 | 3.2554 | 3.2269 | 3.1984 | 3.1399 | 3.0798 | 0.975 | |
| | | 0.990 | 4.2469 | 4.1653 | 4.1236 | 4.0819 | 3.9965 | 3.9090 | 0.990 | |
| | | 0.995 | 5.0705 | 4.9659 | 4.9126 | 4.8592 | 4.7501 | 4.6385 | 0.995 | |
| | | 0.500 | 1.0401 | 1.0460 | 1.0490 | 1.0519 | 1.0578 | 1.0637 | 0.500 | |
| | | 0.750 | 1.4805 | 1.4737 | 1.4701 | 1.4664 | 1.4587 | 1.4504 | 0.750 | |
| | | 0.900 | 2.0762 | 2.0516 | 2.0389 | 2.0261 | 1.9997 | 1.9721 | 0.900 | |
| | 11 | 0.950 | 2.5705 | 2.5309 | 2.5105 | 2.4901 | 2.4480 | 2.4045 | 0.950 | 11 |
| | | 0.975 | 3.1176 | 3.0613 | 3.0324 | 3.0035 | 2.9441 | 2.8828 | 0.975 | |
| | | 0.990 | 3.9411 | 3.8596 | 3.8179 | 3.7761 | 3.6904 | 3.6025 | 0.990 | |
| | | 0.995 | 4.6543 | 4.5508 | 4.4979 | 4.4450 | 4.3367 | 4.2256 | 0.995 | |
| | | 0.500 | 1.0347 | 1.0405 | 1.0435 | 1.0464 | 1.0523 | 1.0582 | 0.500 | |
| | | 0.750 | 1.4544 | 1.4471 | 1.4432 | 1.4393 | 1.4310 | 1.4221 | 0.750 | |
| | | 0.900 | 2.0115 | 1.9861 | 1.9729 | 1.9597 | 1.9323 | 1.9036 | 0.900 | |
| | 12 | 0.950 | 2.4663 | 2.4259 | 2.4051 | 2.3842 | 2.3410 | 2.2962 | 0.950 | 12 |
| | | 0.975 | 2.9633 | 2.9063 | 2.8771 | 2.8478 | 2.7874 | 2.7249 | 0.975 | |
| | | 0.990 | 3.7008 | 3.6192 | 3.5774 | 3.5355 | 3.4494 | 3.3608 | 0.990 | |
| | | 0.995 | 4.3309 | 4.2282 | 4.1756 | 4.1229 | 4.0149 | 3.9039 | 0.995 | |

*Continued*

**Table 6** Critical Values for the *F*-Distribution *Continued*

| | $\gamma$ | 1 | 2 | 3 | 4 | 5 | 6 | $\gamma$ | |
|---|---|---|---|---|---|---|---|---|---|
| | | | | | $r_1$ | | | | |
| | 0.500 | 0.48141 | 0.73145 | 0.83159 | 0.88454 | 0.91718 | 0.93926 | 0.500 | |
| | 0.750 | 1.4500 | 1.5452 | 1.5451 | 1.5336 | 1.5214 | 1.5105 | 0.750 | |
| | 0.900 | 3.1362 | 2.7632 | 2.5603 | 2.4337 | 2.3467 | 2.2830 | 0.900 | |
| 13 | 0.950 | 4.6672 | 3.8056 | 3.4105 | 3.1791 | 3.0254 | 2.9153 | 0.950 | 13 |
| | 0.975 | 6.4143 | 4.9653 | 4.3472 | 3.9959 | 3.7667 | 3.6043 | 0.975 | |
| | 0.990 | 9.0738 | 6.7010 | 5.7394 | 5.2053 | 4.8616 | 4.6204 | 0.990 | |
| | 0.995 | 11.374 | 8.1865 | 6.9257 | 6.2335 | 5.7910 | 5.4819 | 0.995 | |
| | 0.500 | 0.47944 | 0.72862 | 0.82842 | 0.88119 | 0.91371 | 0.93573 | 0.500 | |
| | 0.750 | 1.4403 | 1.5331 | 1.5317 | 1.5194 | 1.5066 | 1.4952 | 0.750 | |
| | 0.900 | 3.1022 | 2.7265 | 2.5222 | 2.3947 | 2.3069 | 2.2426 | 0.900 | |
| 14 | 0.950 | 4.6001 | 3.7389 | 3.3439 | 3.1122 | 2.9582 | 2.8477 | 0.950 | 14 |
| | 0.975 | 6.2979 | 4.8567 | 4.2417 | 3.8919 | 3.6634 | 3.5014 | 0.975 | |
| | 0.990 | 8.8616 | 6.5149 | 5.5639 | 5.0354 | 4.6950 | 4.4558 | 0.990 | |
| | 0.995 | 11.060 | 7.9216 | 6.6803 | 5.9984 | 5.5623 | 5.2574 | 0.995 | |
| | 0.500 | 0.47775 | 0.72619 | 0.82569 | 0.87830 | 0.91073 | 0.93267 | 0.500 | |
| | 0.750 | 1.4321 | 1.5227 | 1.5202 | 1.5071 | 1.4938 | 1.4820 | 0.750 | |
| | 0.900 | 3.0732 | 2.6952 | 2.4898 | 2.3614 | 2.2730 | 2.2081 | 0.900 | |
| 15 | 0.950 | 4.5431 | 3.6823 | 3.2874 | 3.0556 | 2.9013 | 2.7905 | 0.950 | 15 |
| | 0.975 | 6.1995 | 4.7650 | 4.1528 | 3.8043 | 3.5764 | 3.4147 | 0.975 | |
| | 0.990 | 8.6831 | 6.3589 | 5.4170 | 4.8932 | 4.5556 | 4.3183 | 0.990 | |
| $r_2$ | 0.995 | 10.798 | 7.7008 | 6.4760 | 5.8029 | 5.3721 | 5.0708 | 0.995 | $r_2$ |
| | 0.500 | 0.47628 | 0.72406 | 0.82330 | 0.87578 | 0.90812 | 0.93001 | 0.500 | |
| | 0.750 | 1.4249 | 1.5137 | 1.5103 | 1.4965 | 1.4827 | 1.4705 | 0.750 | |
| | 0.900 | 3.0481 | 2.6682 | 2.4618 | 2.3327 | 2.2438 | 2.1783 | 0.900 | |
| 16 | 0.950 | 4.4940 | 3.6337 | 3.2389 | 3.0069 | 2.8524 | 2.7413 | 0.950 | 16 |
| | 0.975 | 6.1151 | 4.6867 | 4.0768 | 3.7294 | 3.5021 | 3.3406 | 0.975 | |
| | 0.990 | 8.5310 | 6.2262 | 5.2922 | 4.7726 | 4.4374 | 4.2016 | 0.990 | |
| | 0.995 | 10.575 | 7.5138 | 6.3034 | 5.6378 | 5.2117 | 4.9134 | 0.995 | |
| | 0.500 | 0.47499 | 0.72219 | 0.82121 | 0.87357 | 0.90584 | 0.92767 | 0.500 | |
| | 0.750 | 1.4186 | 1.5057 | 1.5015 | 1.4873 | 1.4730 | 1.4605 | 0.750 | |
| | 0.900 | 3.0262 | 2.6446 | 2.4374 | 2.3077 | 2.2183 | 2.1524 | 0.900 | |
| 17 | 0.950 | 4.4513 | 3.5915 | 3.1968 | 2.9647 | 2.8100 | 2.6987 | 0.950 | 17 |
| | 0.975 | 6.0420 | 4.6189 | 4.0112 | 3.6648 | 3.4379 | 3.2767 | 0.975 | |
| | 0.990 | 8.3997 | 6.1121 | 5.1850 | 4.6690 | 4.3359 | 4.1015 | 0.990 | |
| | 0.995 | 10.384 | 7.3536 | 6.1556 | 5.4967 | 5.0746 | 5.7789 | 0.995 | |
| | 0.500 | 0.47385 | 0.72053 | 0.81936 | 0.87161 | 0.90381 | 0.92560 | 0.500 | |
| | 0.750 | 1.4130 | 1.4988 | 1.4938 | 1.4790 | 1.4644 | 1.4516 | 0.750 | |
| | 0.900 | 3.0070 | 2.6239 | 2.4160 | 2.2858 | 2.1958 | 1.1296 | 0.900 | |
| 18 | 0.950 | 4.4139 | 3.5546 | 3.1599 | 2.9277 | 2.7729 | 2.6613 | 0.950 | 18 |
| | 0.975 | 5.9781 | 4.5597 | 3.9539 | 3.6083 | 3.3820 | 3.2209 | 0.975 | |
| | 0.990 | 8.2854 | 6.0129 | 5.0919 | 4.5790 | 4.2479 | 4.0146 | 0.990 | |
| | 0.995 | 10.218 | 7.2148 | 6.0277 | 5.3746 | 4.9560 | 4.6627 | 0.995 | |

*Continued*

**Table 6** Critical Values for the *F*-Distribution *Continued*

| | $\gamma$ | 7 | 8 | 9 | 10 | 11 | 12 | $\gamma$ | |
|---|---|---|---|---|---|---|---|---|---|
| | | | | $r_1$ | | | | | |
| | 0.500 | 0.95520 | 0.96724 | 0.97665 | 0.98421 | 0.99042 | 0.99560 | 0.500 | |
| | 0.750 | 1.5011 | 1.4931 | 1.4861 | 1.4801 | 1.4746 | 1.4701 | 0.750 | |
| | 0.900 | 2.2341 | 2.1953 | 2.1638 | 1.1376 | 1.1152 | 2.0966 | 0.900 | |
| 13 | 0.950 | 2.8321 | 2.7669 | 2.7144 | 2.6710 | 2.6343 | 2.6037 | 0.950 | 13 |
| | 0.975 | 3.4827 | 3.3880 | 3.3120 | 3.2497 | 3.1971 | 3.1532 | 0.975 | |
| | 0.990 | 4.4410 | 4.3021 | 4.1911 | 4.1003 | 4.0239 | 3.9603 | 0.990 | |
| | 0.995 | 5.2529 | 5.0761 | 4.9351 | 4.8199 | 4.7234 | 4.6429 | 0.995 | |
| | 0.500 | 0.95161 | 0.96360 | 0.97298 | 0.98051 | 0.98670 | 0.99186 | 0.500 | |
| | 0.750 | 1.4854 | 1.4770 | 1.4697 | 1.4634 | 1.4577 | 1.4530 | 0.750 | |
| | 0.900 | 2.1931 | 2.1539 | 2.1220 | 2.0954 | 2.0727 | 2.0537 | 0.900 | |
| 14 | 0.950 | 2.7642 | 2.6987 | 2.6548 | 2.6021 | 2.5651 | 2.5342 | 0.950 | 14 |
| | 0.975 | 3.3799 | 2.2853 | 3.2093 | 3.1469 | 3.0941 | 3.0501 | 0.975 | |
| | 0.990 | 4.2779 | 4.1399 | 4.0297 | 3.9394 | 3.8634 | 3.8001 | 0.990 | |
| | 0.995 | 5.0313 | 4.8566 | 4.7173 | 4.6034 | 4.5078 | 4.4281 | 0.995 | |
| | 0.500 | 0.94850 | 0.96046 | 0.96981 | 0.97732 | 0.98349 | 0.98863 | 0.500 | |
| | 0.750 | 1.4718 | 1.4631 | 1.4556 | 1.4491 | 1.4432 | 1.4383 | 0.750 | |
| | 0.900 | 2.1582 | 2.1185 | 2.0862 | 2.0593 | 2.0363 | 2.0171 | 0.900 | |
| 15 | 0.950 | 2.7066 | 2.6408 | 2.5876 | 2.5437 | 2.5064 | 2.4753 | 0.950 | 15 |
| | 0.975 | 3.2934 | 3.1987 | 3.1227 | 3.0602 | 3.0073 | 2.9633 | 0.975 | |
| | 0.990 | 4.1415 | 4.0045 | 3.8948 | 3.8049 | 3.7292 | 3.6662 | 0.990 | |
| $r_2$ | 0.995 | 4.8473 | 4.6743 | 4.5364 | 4.4236 | 4.3288 | 4.2498 | 0.995 | $r_2$ |
| | 0.500 | 0.94580 | 0.95773 | 0.96705 | 0.97454 | 0.98069 | 0.98582 | 0.500 | |
| | 0.750 | 1.4601 | 1.4511 | 1.4433 | 1.4366 | 1.4305 | 1.4255 | 0.750 | |
| | 0.900 | 2.1280 | 2.0880 | 2.0553 | 2.0281 | 2.0048 | 1.9854 | 0.900 | |
| 16 | 0.950 | 2.6572 | 2.5911 | 2.5377 | 2.4935 | 2.4560 | 2.4247 | 0.950 | 16 |
| | 0.975 | 3.2194 | 3.1248 | 3.0488 | 2.9862 | 2.9332 | 2.8890 | 0.975 | |
| | 0.990 | 4.0259 | 3.8896 | 3.7804 | 3.6909 | 3.6155 | 3.5527 | 0.990 | |
| | 0.995 | 4.6920 | 4.5207 | 4.3838 | 4.2719 | 4.1778 | 4.0994 | 0.995 | |
| | 0.500 | 0.94342 | 0.95532 | 0.96462 | 0.97209 | 0.97823 | 0.98334 | 0.500 | |
| | 0.750 | 1.4497 | 1.4405 | 1.4325 | 1.4256 | 1.4194 | 1.4142 | 0.750 | |
| | 0.900 | 2.1017 | 2.0613 | 2.0284 | 2.0009 | 1.9773 | 1.9577 | 0.900 | |
| 17 | 0.950 | 2.6143 | 2.5480 | 2.4943 | 2.4499 | 2.4122 | 2.3807 | 0.950 | 17 |
| | 0.975 | 3.1556 | 3.0610 | 2.9849 | 2.9222 | 2.8691 | 2.8249 | 0.975 | |
| | 0.990 | 3.9267 | 3.7910 | 3.6822 | 3.5931 | 3.5179 | 3.4552 | 0.990 | |
| | 0.995 | 4.5594 | 4.3893 | 4.2535 | 4.1423 | 4.0488 | 3.9709 | 0.995 | |
| | 0.500 | 0.94132 | 0.95319 | 0.96247 | 0.96993 | 0.97606 | 0.98116 | 0.500 | |
| | 0.750 | 1.4406 | 1.4312 | 1.4320 | 1.4159 | 1.4095 | 1.4042 | 0.750 | |
| | 0.900 | 2.0785 | 2.0379 | 2.0047 | 1.9770 | 1.9532 | 1.9333 | 0.900 | |
| 18 | 0.950 | 2.5767 | 2.5102 | 2.4563 | 2.4117 | 2.3737 | 2.3421 | 0.950 | 18 |
| | 0.975 | 3.0999 | 3.0053 | 2.9291 | 2.8664 | 2.8132 | 2.7689 | 0.975 | |
| | 0.990 | 3.8406 | 3.7054 | 3.5971 | 3.5082 | 3.4331 | 3.3706 | 0.990 | |
| | 0.995 | 4.4448 | 4.2759 | 4.1410 | 4.0305 | 3.9374 | 3.8599 | 0.995 | |

*These tables have been adapted from Donald B. Owen's* Handbook of Statistical Tables, *published by Addison-Wesley, by permission of the publishers.*
*For more extensive F-tables look at the source just cited, pages 63–87.*

**Table 7** Table of  Selected Discrete and Continuous Distributions and Some of Their Characteristics

| Distribution | Probability Density Functions in one Variable | | |
| --- | --- | --- | --- |
| | **Probability Density Function** | **Mean** | **Variance** |
| **Binomial, $B(n, p)$** | $f(x) = \binom{n}{x}p^x q^{n-x}, x = 0, 1, \ldots, n;$ $0 < p < 1, q = 1 - p$ | $np$ | $npq$ |
| **Bernoulli, $B(1, p)$** | $f(x) = p^x q^{1-x}, x = 0, 1$ | $p$ | $pq$ |
| **Geometric** | $f(x) = pq^{x-1}, x = 1, 2, \ldots;$ $0 < p < 1, q = 1 - p$ | $\frac{1}{p}$ | $\frac{q}{p^2}$ |
| **Poisson, $P(\lambda)$** | $f(x) = e^{-\lambda}\frac{\lambda^x}{x!}, x = 0, 1, \ldots; \lambda > 0$ | $\lambda$ | $\lambda$ |
| **Hypergeometric** | $f(x) = \dfrac{\binom{m}{x}\binom{n}{r-x}}{\binom{m+n}{r}},$ where $x = 0, 1, \ldots, r\left(\binom{m}{r} = 0, r > m\right)$ | $\frac{mr}{m+n}$ | $\frac{mnr(m+n-r)}{(m+n)^2(m+n-1)}$ |
| **Gamma** | $f(x) = \frac{1}{\Gamma(\alpha)\beta^\alpha}x^{\alpha-1}\exp\left(-\frac{x}{\beta}\right), x > 0;$ $\alpha, \beta > 0$ | $\alpha\beta$ | $\alpha\beta^2$ |
| **Negative Exponential** | $f(x) = \lambda\exp(-\lambda x), x > 0; \lambda > 0;$ or $f(x) = \frac{1}{\mu}e^{-x/\mu}, x > 0; \mu > 0$ | $\frac{1}{\lambda}$ $\mu$ | $\frac{1}{\lambda^2}$ $\mu^2$ |
| **Chi-square** | $f(x) = \frac{1}{\Gamma(\frac{r}{2})2^{r/2}}x^{\frac{r}{2}-1}\exp\left(-\frac{x}{2}\right), x > 0;$ $r > 0$ integer | $r$ | $2r$ |
| **Normal, $N(\mu, \sigma^2)$** | $f(x) = \frac{1}{\sqrt{2\pi}\sigma}\exp\left[-\frac{(x-\mu)^2}{2\sigma^2}\right],$ $x \in \Re; \ \mu \in \Re, \sigma > 0$ | $\mu$ | $\sigma^2$ |
| **Standard Normal, $N(0, 1)$** | $f(x) = \frac{1}{\sqrt{2\pi}}\exp\left(-\frac{x^2}{2}\right), x \in \Re$ | $0$ | $1$ |
| **Uniform, $U(\alpha, \beta)$** | $f(x) = \frac{1}{\beta-\alpha}, \alpha \le x \le \beta;$ $-\infty < \alpha < \beta < \infty$ | $\frac{\alpha+\beta}{2}$ | $\frac{(\alpha-\beta)^2}{12}$ |
| **Multinomial** | $f(x_1, \ldots, x_k) = \frac{n!}{x_1!x_2!\cdots x_k!} \times$ $p_1^{x_1}p_2^{x_2}\cdots p_k^{x_k}, x_i \ge 0$ integers, $x_1 + x_2 + \cdots + x_k = n; p_j > 0, j = 1,$ $2, \ldots, k, p_1 + p_2 + \cdots + p_k = 1$ | $np_1, \ldots, np_k$ $= 1, \ldots, k$ | $np_1q_1, \ldots, np_kq_k.$ $q_i = 1 - p_i, j$ |
| **Bivariate Normal** | $f(x_1, x_2) = \frac{1}{2\pi\sigma_1\sigma_2\sqrt{1-\rho^2}}\exp\left(-\frac{q}{2}\right),$ $q = \frac{1}{1-\rho^2}\left[\left(\frac{x_1-\mu_1}{\sigma_1}\right)^2 - 2\rho\left(\frac{x_1-\mu_1}{\sigma_1}\right)\right.$ $\left.\times\left(\frac{x_2-\mu_2}{\sigma_2}\right) + \left(\frac{x_2-\mu_2}{\sigma_2}\right)^2\right],$ $x_1, x_2 \in \Re; \mu_1, \mu_2 \in \Re, \sigma_1, \sigma_2 > 0, -1 \le \rho \le 1, \rho = $ correlation coefficient | $\mu_1, \mu_2$ | $\sigma_1^2, \sigma_2^2$ |

*Continued*

**Table 7** Table of Selected Discrete and Continuous Distributions and Some of Their Characteristics *Continued*

| Distribution | Probability Density Functions in one Variable | | |
| --- | --- | --- | --- |
| | **Probability Density Function** | **Mean** | **Variance** |
| *k*-Variate Normal, $N(\boldsymbol{\mu}, \boldsymbol{\Sigma})$ | $f(\mathbf{x}) = (2\pi)^{-k/2}\lvert \boldsymbol{\Sigma} \rvert^{-1/2} \times$ $\exp\left[-\frac{1}{2}(\mathbf{x}-\boldsymbol{\mu})'\boldsymbol{\Sigma}^{-1}(\mathbf{x}-\boldsymbol{\mu})\right],$ $\mathbf{x} \in \Re^k; \boldsymbol{\mu} \in \Re^k, \boldsymbol{\Sigma}: k \times k$ nonsingular symmetric matrix | $\mu_1, \dots, \mu_k$ | Covariance matrix: $\boldsymbol{\Sigma}$ |

| Distribution | Moment Generating Function |
| --- | --- |
| **Binomial,** $B(n,p)$ | $M(t) = (pe^t + q)^n, t \in \Re$ |
| **Bernoulli,** $B(1,p)$ | $M(t) = pe^t + q, t \in \Re$ |
| **Geometric** | $M(t) = \frac{pe^t}{1-qe^t}, t < -\log q$ |
| **Poisson,** $P(\lambda)$ | $M(t) = \exp(\lambda e^t - \lambda), t \in \Re$ |
| **Hypergeometric** | — |
| **Gamma** | $M(t) = \frac{1}{(1-\beta t)^\alpha}, t < \frac{1}{\beta}$ |
| **Negative Exponential** | $M(t) = \frac{\lambda}{\lambda - t}, t < \lambda; \text{ or } M(t) = \frac{1}{1-\mu t}, t < \frac{1}{\mu}$ |
| **Chi-square** | $M(t) = \frac{1}{(1-2t)^{r/2}}, t < \frac{1}{2}$ |
| **Normal,** $N(\mu, \sigma^2)$ | $M(t) = \exp\left(\mu t + \frac{\sigma^2 t^2}{2}\right), t \in \Re$ |
| **Standard Normal,** $N(0,1)$ | $M(t) = \exp\left(\frac{t^2}{2}\right), t \in \Re$ |
| **Uniform,** $U(\alpha, \beta)$ | $M(t) = \frac{e^{t\beta} - e^{t\alpha}}{t(\beta - \alpha)}, t \in \Re$ |
| **Multinomial** | $M(t_1, \dots, t_k) = (p_1 e^{t_1} + \cdots + p_k e^{t_k})^n,$ $t_1, \dots, t_k \in \Re$ |
| **Bivariate Normal** | $M(t_1, t_2) = \exp\Big[\mu_1 t_1 + \mu_2 t_2$ $+ \frac{1}{2}(\sigma_1^2 t_1^2 + 2\rho\sigma_1\sigma_2 t_1 t_2 + \sigma_2^2 t_2^2)\Big],$ $t_1, t_2 \in \Re$ |
| **$k$-Variate Normal,** $N(\boldsymbol{\mu}, \boldsymbol{\Sigma})$ | $M(\mathbf{t}) = \exp\left(\mathbf{t}'\boldsymbol{\mu} + \frac{1}{2}\mathbf{t}'\boldsymbol{\Sigma}\mathbf{t}\right),$ $\mathbf{t} \in \Re^k$ |

**Table 8** Handy Reference to Some Formulas Used in the Text

| | |
|---|---|
| 1 | $\sum_{k=1}^{n} k = \frac{n(n+1)}{2}$, $\sum_{k=1}^{n} k^2 = \frac{n(n+1)(2n+1)}{6}$, $\sum_{k=1}^{n} k^3 = \left(\frac{n(n+1)}{2}\right)^2$ |
| 2 | $(a+b)^n = \sum_{x=0}^{n} \binom{n}{x} a^x b^{n-x}$ |
| 3 | $(a_1 + \cdots + a_k)^n = \sum \frac{n!}{x_1! \cdots x_k!} a_1^{x_1} \cdots a_k^{x_k}$ where the summation is over all non-Negative integers $x_1, \ldots, x_k$ with $x_1 + \cdots + x_k = n$ |
| 4 | $\sum_{n=k}^{\infty} r^n = \frac{r^k}{1-r}$, $k = 0, 1, \ldots$, $|r| < 1$ |
| 5 | $\sum_{n=1}^{\infty} n r^n = \frac{r}{(1-r)^2}$, $\sum_{n=2}^{\infty} n(n-1) r^n = \frac{2r^2}{(1-r)^3}$, $|r| < 1$ |
| 6 | $e^x = \lim_{n \to \infty} \left(1 + \frac{x}{n}\right)^n = \lim_{n \to \infty} \left(1 + \frac{x_n}{n}\right)^n$, $x_n \to x$, $e^x = \sum_{n=0}^{\infty} \frac{x^n}{n!}$, $x \in \Re$ |
| 7 | $\{a\,u(x) + b\,v(x)\}' = a\,u'(x) + b\,v'(x)$, $\{u(x)\,v(x)\}' = u'(x)\,v(x) + u(x)\,v'(x)$ |
| | $\left(\frac{u(x)}{v(x)}\right)' = \frac{u'(x)\,v(x) - u(x)\,v'(x)}{v^2(x)}$, $\frac{d}{dx} u(v(x)) = \{\frac{d}{dv(x)} u(v(x))\} \frac{d}{dx} v(x)$ |
| 8 | $\frac{\partial^2}{\partial x \partial y} w(x,y) = \frac{\partial^2}{\partial y \partial x} w(x,y)$ (under certain conditions) |
| 9 | $\frac{\partial}{\partial t} \sum_{n=1}^{\infty} w(n,t) = \sum_{n=1}^{\infty} \frac{\partial}{\partial t} w(n,t)$ (under certain conditions) |
| 10 | $\frac{\partial}{\partial t} \int_a^b w(x,t) dx = \int_a^b (\frac{\partial}{\partial t} w(x,t)) dx$ ($-\infty \le a < b \le \infty$) (under certain conditions) |
| 11 | $\sum_{n=1}^{\infty} (c\,x_n + d\,y_n) = c \sum_{n=1}^{\infty} x_n + d \sum_{n=1}^{\infty} y_n$ |
| 12 | $\int_a^b (c\,u(x) + dv(x)) dx = c \int_a^b u(x) dx + d \int_a^b v(x) dx$ ($-\infty \le a < b \le \infty$) |
| 13 | $\int_a^b u(x) dv(x) = u(x)\,v(x)\|_a^b - \int_a^b v(x) du(x)$ ($-\infty \le a < b \le \infty$) |
| | In particular $\int_a^b x^n dx = \frac{x^{n+1}}{n+1}\|_a^b = \frac{b^{n+1} - a^{n+1}}{n+1}$, ($n \ne -1$), |
| | $\int_a^b \frac{dx}{x} = \log x\|_a^b = \log b - \log a$ ($0 < a < b$, $\log x$ is the natural algorithm of $x$), |
| | $\int_a^b e^x dx = e^x\|_a^b = e^b - e^a$ |
| 14 | If $u'(x_0) = 0$, then $x_0$ maximizes $u(x)$ if $u''(x_0) < 0$, and $x_0$ minimizes $u(x)$ if $u''(x_0) > 0$ |
| 15 | Let $\frac{\partial}{\partial x} w(x,y)\|_{\substack{x=x_0 \\ y=y_0}} = 0$, $\frac{\partial}{\partial y} w(x,y)\|_{\substack{x=x_0 \\ y=y_0}} = 0$, and set $c_{11} = \frac{\partial^2}{\partial x^2} w(x,y)\|_{\substack{x=x_0 \\ y=y_0}}$, $c_{12} = \frac{\partial^2}{\partial x \partial y} w(x,y)\|_{\substack{x=x_0 \\ y=y_0}} = c_{21} = \frac{\partial^2}{\partial y \partial x} w(x,y)\|_{\substack{x=x_0 \\ y=y_0}}$ $c_{22} = \frac{\partial^2}{\partial y^2} w(x,y)\|_{\substack{x=x_0 \\ y=y_0}}$, $C = \begin{pmatrix} c_{11} & c_{12} \\ c_{21} & c_{22} \end{pmatrix}$ Then $(x_0, y_0)$ maximizes $w(x,y)$ if the matrix $C$ is Negative definite; that is, for all real $\lambda_1, \lambda_2$ not both 0, it holds: $(\lambda_1, \lambda_2) \begin{pmatrix} c_{11} & c_{12} \\ c_{21} & c_{22} \end{pmatrix} \begin{pmatrix} \lambda_1 \\ \lambda_2 \end{pmatrix} = \lambda_1^2 c_{11} + 2\lambda_1 \lambda_2 c_{12} + \lambda_2^2 c_{22} < 0$; $(x_0, y_0$ minimizes $w(x,y)$ if the matrix $C$ is positive definite; that is, $\lambda_1^2 c_{11} + 2\lambda_1 \lambda_2 c_{12} + \lambda_2^2 c_{22} > 0$ with $\lambda_1, \lambda_2$ as above |
| 16 | Criteria analogous to those in #15 hold for a function in $k$ variables $w(x_1, \ldots, x_k)$ |

# Some notation and abbreviations

| | |
|---|---|
| $\Re$ | real line |
| $\Re^k, k \geq 1$ | $k$-dimensional Euclidean space |
| $\uparrow, \downarrow$ | increasing (nondecreasing) and decreasing (nonincreasing), respectively |
| $\mathcal{S}$ | sample space; also, sure (or certain) event |
| $\emptyset$ | empty set; also, impossible event |
| $A \subseteq B$ | event $A$ is contained in event $B$ (event $A$ implies event $B$) |
| $A^c$ | complement of event $A$ |
| $A \cup B$ | union of events $A$ and $B$ |
| $A \cap B$ | intersection of events $A$ and $B$ |
| $A - B$ | difference of events $A$ and $B$ (in this order) |
| r.v. | random variable |
| $I_A$ | indicator of the set $A$: $I_A(x) = 1$ if $x \in A$, $I_A(x) = 0$ if $x \notin A$ |
| $(X \in B) = X^{-1}(B)$ | inverse image of the set $B$ under $X$: $X^{-1}(B) = \{s \in \mathcal{S}; X(s) \in B\}$ |
| $X(\mathcal{S})$ | range of $X$ |
| $P$ | probability function (measure) |
| $P(A)$ | probability of the event $A$ |
| $P_X$ | probability distribution function of $X$ (or just distribution of $X$) |
| $F_X$ | distribution function (d.f.) of $X$ |
| $f_X$ | probability density function (p.d.f.) of $X$ |
| $P(A|B)$ | conditional probability of $A$, given $B$ |
| $\binom{n}{k}$ | combinations of $n$ objects taken $k$ at a time |
| $P_{n,k}$ | permutations of $n$ objects taken $k$ at a time |
| $n!$ | $n$ factorial |
| $EX$ or $\mu(X)$ or $\mu_X$ or just $\mu$ | expectation (expected value, mean value, mean) of $X$ |
| $\text{Var}(X)$ or $\sigma^2(X)$ or $\sigma_X^2$ or just $\sigma^2$ | variance of $X$ |
| $\sqrt{\text{Var}(X)}$ or $\sigma(X)$ or $\sigma_X$ or just $\sigma$ | standard deviation (s.d.) of $X$ |
| $M_X$ or just $M$ | moment generating function (m.g.f.) of $X$ |
| $B(n, p)$ | Binomial distribution with parameters $n$ and $p$ |
| $P(\lambda)$ | Poisson distribution with parameter $\lambda$ |
| $\chi_r^2$ | Chi-square distribution with $r$ degrees of freedom (d.f.) |
| $N(\mu, \sigma^2)$ | Normal distribution with parameters $\mu$ and $\sigma^2$ |
| $\Phi$ | distribution function (d.f.) of the standard $N(0, 1)$ distribution |

*(Continued)*

| | |
|---|---|
| $U(\alpha, \beta)$ or $R(\alpha, \beta)$ | Uniform (or Rectangular) distribution with parameters $\alpha$ and $\beta$ |
| $X \sim B(n, p)$ etc. | the r.v. $X$ has the distribution indicated |
| $\chi^2_{r;\alpha}$ | the point for which $P(X > \chi^2_{r;\alpha}) = \alpha$, $X \sim \chi^2_r$ |
| $z_\alpha$ | the point for which $P(Z > z_\alpha) = \alpha$, where $Z \sim N(0, 1)$ |
| $P_{X_1,\dots,X_n}$ or $P_X$ | joint probability distribution function of the r.v.'s $X_1, \dots, X_n$ or probability distribution function of the random vector $X$ |
| $F_{X_1,\dots,X_n}$ or $F_X$ | joint d.f. of the r.v.'s $X_1, \dots, X_n$ or d.f. of the random vector $X$ |
| $f_{X_1,\dots,X_n}$ or $f_X$ | joint p.d.f. of the r.v.'s $X_1, \dots, X_n$ or p.d.f. of the random vector $X$ |
| $M_{X_1,\dots,X_n}$ or $M_X$ | joint m.g.f. of the r.v.'s $X_1, \dots, X_n$ or m.g.f. of the random vector $X$ |
| i.i.d. (r.v.'s) | independent identically distributed (r.v.'s) |
| $f_{X|Y}(\cdot|Y = y)$ or $f_{X|Y}(\cdot|y)$ | conditional p.d.f. of $X$, given $Y = y$ |
| $E(X|Y = y)$ | conditional expectation of $X$, given $Y = y$ |
| $Var(X|Y = y)$ or $\sigma^2(X|Y = y)$ | conditional variance of $X$, given $Y = y$ |
| $Cov(X, Y)$ | covariance of $X$ and $Y$ |
| $\rho(X, Y)$ or $\rho_{X,Y}$ | correlation coefficient of $X$ and $Y$ |
| $t_r$ | (Student's) $t$-distribution with $r$ degrees of freedom (d.f.) |
| $t_{r;\alpha}$ | the point for which $P(X > t_{r;\alpha}) = \alpha$, $X \sim t_r$ |
| $F_{r_1,r_2}$ | $F$ distribution with $r_1$ and $r_2$ degrees of freedom (d.f.) |
| $F_{r_1,r_2;\alpha}$ | the point for which $P(X > F_{r_1,r_2;\alpha}) = \alpha$, $X \sim F_{r_1,r_2}$ |
| $X_{(j)}$ or $Y_j$ | $j$th order statistic of $X_1, \dots, X_n$ |
| $\xrightarrow{P}, \xrightarrow{d}, \xrightarrow{q.m.}$ | convergence in probability, distribution, quadratic mean, respectively |
| WLLN | Weak Law of Large Numbers |
| CLT | Central Limit Theorem |
| $\theta$ | letter used for a one-dimensional parameter |
| $\boldsymbol{\theta}$ | symbol used for a multidimensional parameter |
| $\Omega$ | letter used for a parameter space |
| ML | maximum likelihood |
| MLE | Maximum Likelihood Estimate |
| UMV | uniformly minimum variance |
| UMVU | Uniformly Minimum Variance Unbiased |
| LS | least squares |
| LSE | Least Squares Estimate |
| $H_0$ | null hypothesis |
| $H_A$ | alternative hypothesis |
| $\varphi$ | letter used for a test function |

| | |
|---|---|
| $\alpha$ | letter used for level of significance |
| $\beta(\theta)$ or $\beta(\boldsymbol{\theta})$ | probability of type II error at $\theta(\boldsymbol{\theta})$ |
| $\pi(\theta)$ or $\pi(\boldsymbol{\theta})$ | power of a test at $\theta(\boldsymbol{\theta})$ |
| MP | Most Powerful (test) |
| UMP | Uniformly Most Powerful (test) |
| LR | likelihood ratio |
| $\lambda = \lambda(x_1, \ldots, x_n)$ | likelihood ratio test function |
| $\log x$ | the logarithm of $x(>0)$ with base always $e$ whether it is so explicitly stated or not |

# Answers to even-numbered exercises

---

## CHAPTER 1
### Section 1.2

**2.2**   **(i)**   $S = \{(r,r,r),(r,r,b),(r,r,g),(r,b,r),(r,b,b),(r,b,g),(r,g,r),$
           $(r,g,b),(r,g,g),(b,r,r),(b,r,b),(b,r,g),(b,b,r),(b,b,b),$
           $(b,b,g),(b,g,r),(b,g,b),(b,g,g),(g,r,r),(g,r,b),(g,r,g),$
           $(g,b,r),(g,b,b),(g,b,g),(g,g,r),(g,g,b),(g,g,g)\}.$

     **(ii)**   $A = \{(r,b,g),(r,g,b),(b,r,g),(b,g,r),(g,r,b),(g,b,r)\},$
         $B = \{(r,r,b),(r,r,g),(r,b,r),(r,b,b),(r,g,r),(r,g,g),(b,r,r),$
           $(b,r,b),(b,b,r),(b,b,g),(b,g,b),(b,g,g),(g,r,r),$
           $(g,r,g),(g,b,b),(g,b,g),(g,g,r),(g,g,b)\},$
         $C = A \cup B = S - \{(r,r,r),(b,b,b),(g,g,g)\}.$ ∎

**2.4**   **(i)**   Denoting by $(x_1,x_2)$ the cars sold in the first and the second sales,
        we have:
        $S = \{(a_1,a_1),(a_1,a_2),(a_1,a_3),(a_2,a_1),(a_2,a_2),(a_2,a_3),(a_3,a_1),$
           $(a_3,a_2),(a_3,a_3),(a_1,b_1),(a_1,b_2),(a_2,b_1),(a_2,b_2),(a_3,b_1),$
           $(a_3,b_2),(a_1,c),(a_2,c),(a_3,c),(b_1,a_1),(b_1,a_2),(b_1,a_3),(b_2,a_1),$
           $(b_2,a_2),(b_2,a_3),(b_1,b_1),(b_1,b_2),(b_2,b_1),(b_2,b_2),(b_1,c),(b_2,c),$
           $(c,a_1),(c,a_2),(c,a_3),(c,b_1),(c,b_2),(c,c)\}.$

     **(ii)**   $A = \{(a_1,a_1),(a_1,a_2),(a_1,a_3),(a_2,a_1),(a_2,a_2),(a_2,a_3),(a_3,a_1),$
           $(a_3,a_2),(a_3,a_3)\},$
         $B = \{(a_1,b_1),(a_1,b_2),(a_2,b_1),(a_2,b_2),(a_3,b_1),(a_3,b_2)\},$
         $C = B \cup \{(b_1,a_1),(b_1,a_2),(b_1,a_3),(b_2,a_1),(b_2,a_2),(b_2,a_3)\},$
         $D = \{(c,b_1),(c,b_2),(b_1,c),(b_2,c)\}.$ ∎

**2.6**   $E = A^c, F = C - D = C \cap D^c, G = B - C = B \cap C^c,$
     $H = A^c - B = A^c \cap B^c = (A \cup B)^c, I = B^c.$ ∎

**2.8**   **(i)**   $B_0 = A_1^c \cap A_2^c \cap A_3^c.$
     **(ii)**   $B_1 = (A_1 \cap A_2^c \cap A_3^c) \cup (A_1^c \cap A_2 \cap A_3^c) \cup (A_1^c \cap A_2^c \cap A_3).$
     **(iii)**   $B_2 = (A_1 \cap A_2 \cap A_3^c) \cup (A_1 \cap A_2^c \cap A_3) \cup (A_1^c \cap A_2 \cap A_3).$
     **(iv)**   $B_3 = A_1 \cap A_2 \cap A_3.$
     **(v)**   $C = B_0 \cup B_1 \cup B_2.$
     **(vi)**   $D = B_1 \cup B_2 \cup B_3 = A_1 \cup A_2 \cup A_3.$ ∎

**2.10**  If $A = \emptyset$, then $A \cap B^c = \emptyset$, $A^c \cap B = S \cap B = B$, so that
$(A \cap B^c) \cup (A^c \cap B) = B$ for every $B$. Next, let $(A \cap B^c) \cup (A^c \cap B) = B$ and
take $B = \emptyset$ to obtain $A \cap B^c = A$, $A^c \cap B = \emptyset$, so that $A = \emptyset$. ∎

**2.12**  $A \subseteq B$ implies that, for every $s \in A$, we have $s \in B$, whereas $B \subseteq C$ implies
that, for every $s \in B$, we have $s \in C$. Thus, for every $s \in A$, we have $s \in C$, so
that $A \subseteq C$. ∎

**2.14**  For $s \in \cup_j A_j$, let $j_0 \geq 1$ be the first $j$ for which $s \in A_{j_0}$. Then, if $j_0 = 1$, it
follows that $s \in A_1$ and therefore $s$ belongs in the right-hand side of the
relation. If $j_0 > 1$, then $s \notin A_j, j = 1, \ldots, j_0 - 1$, but $s \in A_{j_0}$, so that
$s \in A_1^c \cap \cdots \cap A_{j_0-1}^c \cap A_{j_0}$ and hence $s$ belongs to the right-hand side of the
relation. Next, let $s$ belongs to the right-hand side event. Then, if $s \in A_1$, it
follows that $s \in \cup_j A_j$. If $s \notin A_j$ for $j = 1, \ldots, j_0 - 1$, but $s \in A_{j_0}$, it follows
that $s \in \cup_j A_j$. The identity is established. ∎

**2.16**  **(i)**  Since $-5 + \frac{1}{n+1} < -5 + \frac{1}{n}$ and $20 - \frac{1}{n} < 20 - \frac{1}{n+1}$, it follows that
$(-5 + \frac{1}{n}, 20 - \frac{1}{n}) \subset (-5 + \frac{1}{n+1}, 20 - \frac{1}{n+1})$, or $A_n \subset A_{n+1}$, so that $\{A_n\}$
is increasing.
Likewise, $7 + \frac{3}{n+1} < 7 + \frac{3}{n}$, so that $(0, 7 + \frac{3}{n+1}) \subset (0, 7 + \frac{3}{n})$, or
$B_{n+1} \subset B_n$; thus, $\{B_n\}$ is decreasing.
**(ii)**  $\cup_{n=1}^{\infty} A_n = \cup_{n=1}^{\infty}(-5 + \frac{1}{n}, 20 - \frac{1}{n}) = (-5, 20)$, and $\cap_{n=1}^{\infty} B_n = $
$\cap_{n=1}^{\infty}(0, 7 + \frac{3}{n}) = (0, 7]$. ∎

## Section 1.3

**3.2**  Each one of the r.v.'s $X_i, i = 1, 2, 3$, takes on the values: $0, 1, 2, 3$ and
$X_1 + X_2 + X_3 = 3$. ∎

**3.4**  $X$ takes on the values: $-3, -2, -1, 0, 1, 2, 3, 4, 5, 6, 7$,
$(X \leq 2) = \{(-3, 0), (-3, 1), (-3, 2), (-3, 3), (-3, 4), (-2, 0), (-2, 1),$
$(-2, 2), (-2, 3), (-2, 4), (-1, 0), (-1, 1), (-1, 2), (-1, 3),$
$(0, 0), (0, 1), (0, 2), (1, 0), (1, 1), (2, 0)\}$,
$(3 < X \leq 5) = (4 \leq X \leq 5) = (X = 4 \text{ or } X = 5)$
$= \{(0, 4), (1, 3), (1, 4), (2, 2), (2, 3), (3, 1), (3, 2)\}$,
$(X > 6) = (X \geq 7) = \{(3, 4)\}$. ∎

**3.6**  **(i)**  $S = \{(1, 1), (1, 2), (1, 3), (1, 4), (2, 1), (2, 2), (2, 3), (2, 4), (3, 1), (3, 2),$
$(3, 3), (3, 4), (4, 1), (4, 2), (4, 3), (4, 4)\}$.
**(ii)**  The values of $X$ are $2, 3, 4, 5, 6, 7, 8$.
**(iii)**  $(X \leq 3) = (X = 2 \text{ or } X = 3) = \{(1, 1), (1, 2), (2, 1)\}$,
$(2 \leq X < 5) = (2 \leq X \leq 4) = (X = 2 \text{ or } X = 3 \text{ or } X = 4) =$
$\{(1, 1), (1, 2), (2, 1), (1, 3), (2, 2), (3, 1)\}$, $(X > 8) = \emptyset$. ∎

**3.8**  **(i)**  $S = [8{:}00, \ 8{:}15]$.
**(ii)**  The values of $X$ consist of the interval $[8{:}00, \ 8{:}15]$.
**(iii)**  The event described is the interval $[8{:}10, \ 8{:}15]$. ∎

# CHAPTER 2

## Section 2.1

**1.2**  Since $A \cup B \supseteq A$, we have $P(A \cup B) \geq P(A) = \frac{3}{4}$. Also, $A \cap B \subseteq B$ implies
$P(A \cap B) \leq P(B) = \frac{3}{8}$. Finally, $P(A \cap B) = P(A) + P(B) - P(A \cup B) =$
$\frac{3}{4} + \frac{3}{8} - P(A \cup B) = \frac{9}{8} - P(A \cup B) \geq \frac{9}{8} - 1 = \frac{1}{8}$. ■

**1.4**  We have $A^c \cap B = B \cap A^c = B - A$ and $A \subset B$. Therefore, $P(A^c \cap B) =$
$P(B - A) = P(B) - P(A) = \frac{5}{12} - \frac{1}{4} = \frac{1}{6} \simeq 0.167$. Likewise,
$A^c \cap C = C - A$ with $A \subset C$, so that $P(A^c \cap C) = P(C - A) =$
$P(C) - P(A) = \frac{7}{12} - \frac{1}{4} = \frac{1}{3} \simeq 0.333$, $B^c \cap C = C - B$ with $B \subset C$, so that
$P(B^c \cap C) = P(C - B) = P(C) - P(B) = \frac{7}{12} - \frac{5}{12} = \frac{1}{6} \simeq 0.167$. Next,
$A \cap B^c \cap C^c = A \cap (B^c \cap C^c) = A \cap (B \cup C)^c = A \cap C^c = A - C = \emptyset$, so
that $P(A \cap B^c \cap C^c) = 0$, and $A^c \cap B^c \cap C^c = (A \cup B \cup C)^c = C^c$, so that
$P(A^c \cap B^c \cap C^c) = P(C^c) = 1 - P(C) = 1 - \frac{7}{12} = \frac{5}{12} \simeq 0.417$. ■

**1.6**  The event $A$ is defined as follows: $A =$ "$x = 7n, n = 1, \ldots, 28$," so that
$P(A) = \frac{28}{200} = \frac{7}{50} = 0.14$. Likewise, $B =$ "$x = 3n + 10, \ n = 1, \ldots, 63$," so
that $P(B) = \frac{63}{200} = 0.315$, and $C =$ "$x^2 + 1 \leq 375$" = "$x^2 \leq 374$" =
"$x \leq \sqrt{374}$" = "$x \leq 19$," and then $P(C) = \frac{19}{200} = 0.095$. ■

**1.8**  Denote by $A, B$, and $C$ the events that a student reads news magazines $A, B$,
and $C$, respectively. Then the required probability is $P(A^c \cap B^c \cap C^c)$.
However,

$$P(A^c \cap B^c \cap C^c) = P((A \cup B \cup C)^c) = 1 - P(A \cup B \cup C)$$
$$= 1 - [P(A) + P(B) + P(C) - P(A \cap B) - P(A \cap C)$$
$$- P(B \cap C) + P(A \cap B \cap C)]$$
$$= 1 - (0.20 + 0.15 + 0.10 - 0.05 - 0.04 - 0.03 + 0.02)$$
$$= 1 - 0.35 = 0.65. ■$$

**1.10**  From the definition of $A, B$, and $C$, we have:
$A = \{(0, 4), (0, 6), (1, 3), (1, 5), (1, 9), (2, 2), (2, 4), (2, 8), (3, 1), (3, 3),$

$\quad (3, 7), (4, 0), (4, 2), (4, 6), (5, 1), (5, 5), (6, 0), (6, 4)\},$
$B = \{(0, 0), (1, 2), (2, 4), (3, 6), (4, 8)\},$
$C = \{(0, 1), (0, 2), (0, 3), (0, 4), (0, 5), (0, 6), (0, 7), (0, 8), (0, 9), (1, 0),$

$\quad (1, 2), (1, 3), (1, 4), (1, 5), (1, 6), (1, 7), (1, 8), (1, 9), (2, 0), (2, 1),$

$\quad (2, 3), (2, 4), (2, 5), (2, 6), (2, 7), (2, 8), (2, 9), (3, 0), (3, 1), (3, 2),$

$\quad (3, 4), (3, 5), (3, 6), (3, 7), (3, 8), (3, 9), (4, 0), (4, 1), (4, 2), (4, 3),$

$\quad (4, 5), (4, 6), (4, 7), (4, 8), (4, 9), (5, 0), (5, 1), (5, 2), (5, 3), (5, 4),$

$\quad (5, 6), (5, 7), (5, 8), (5, 9), (6, 0), (6, 1), (6, 2), (6, 3), (6, 4), (6, 5),$

$\quad (6, 7), (6, 8), (6, 9)\}$

or

$$C^c = \{(0, 0), (1, 1), (2, 2), (3, 3), (4, 4), (5, 5), (6, 6)\}.$$

Therefore, since the number of points in $\mathcal{S}$ is $7 \times 10 = 70$, we have

$$P(A) = \frac{18}{70} = \frac{9}{35} \simeq 0.257, \qquad P(B) = \frac{5}{70} = \frac{1}{14} \simeq 0.071,$$

$$P(C) = \frac{63}{70} = \frac{9}{10} = 0.9, \quad \text{or} \quad P(C) = 1 - P(C^c) = 1 - \frac{7}{70} = \frac{63}{70} = 0.9. \blacksquare$$

**1.12** $P(\bigcap_{i=1}^{n} A_i^c) = P[(\bigcup_{i=1}^{n} A_i)^c] = 1 - P(\bigcup_{i=1}^{n} A_i) \geq 1 - \sum_{i=1}^{n} P(A_i).$

**1.14** (i) $P(C_1) = \frac{4}{15} \simeq 0.267, P(C_2) = \frac{5}{15} \simeq 0.333, P(C_3) = \frac{3}{15} = 0.2,$
$P(C_4) = \frac{2}{15} \simeq 0.133, P(C_5) = \frac{1}{15} \simeq 0.067.$

(ii) $P(F_1) = \frac{4}{36} \simeq 0.111, P(F_2) = \frac{10}{36} \simeq 0.278, P(F_3) = \frac{9}{36} = 0.25,$
$P(F_4) = \frac{8}{36} \simeq 0.222, P(F_5) = \frac{5}{36} \simeq 0.139. \blacksquare$

## Section 2.2

**2.2** (i) For $0 < x \leq 2, f(x) = \frac{d}{dx}(2c(x^2 - \frac{1}{3}x^3)) = 2c(2x - x^2)$. Thus,
$f(x) = 2c(2x - x^2), \quad 0 < x \leq 2$ (and 0 elsewhere).

(ii) From $\int_0^2 2c(2x - x^2)\, dx = 1$, we get $\frac{8c}{3} = 1$, so that $c = 3/8. \blacksquare$

**2.4** (i)

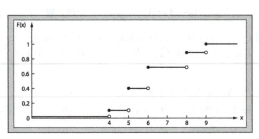

(ii) $P(X \leq 6.5) = 0.7, P(X > 8.1) = 1 - P(X \leq 8.1) = 1 - 0.9 = 0.1, P(5 < X < 8) = P(X < 8) - P(X \leq 5) = 0.7 - 0.4 = 0.3. \blacksquare$

**2.6** (i) We need two relations which are provided by $\int_0^1 (cx + d)\, dx = 1$ and $\int_{1/2}^1 (cx + d)\, dx = 1/3$, or: $c + 2d = 2$ and $9c + 12d = 8$, and hence $c = -\frac{4}{3}, d = \frac{5}{3}.$

(ii) For $0 \leq x \leq 1$, $F(x) = \int_0^x (-\frac{4}{3}t + \frac{5}{3})\, dt = -\frac{2x^2}{3} + \frac{5x}{3}$. Thus,

$$F(x) = \begin{cases} 0, & x < 0, \\ -\frac{2x^2}{3} + \frac{5x}{3}, & 0 \leq x \leq 1, \\ 1, & x > 1. \end{cases} \qquad \blacksquare$$

**2.8** From $\sum_{x=0}^{\infty} c\alpha^x = c\sum_{x=0}^{\infty} \alpha^x = c \times \frac{1}{1-\alpha} = 1$, we get $c = 1 - \alpha. \blacksquare$

**2.10** (i) $\sum_{x=0}^{\infty} c(\frac{1}{3})^x = c[1 + \frac{1}{3} + (\frac{1}{3})^2 + \cdots] = \frac{c}{1-\frac{1}{3}} = \frac{3c}{2} = 1$ and $c = \frac{2}{3}.$

(ii) $P(X \geq 3) = \frac{2}{3}\sum_{x=3}^{\infty}(\frac{1}{3})^x = \frac{2}{3} \times \frac{1/3^3}{2/3} = \frac{1}{27} \simeq 0.037. \blacksquare$

**2.12**  **(i)** $\int_0^\infty ce^{-cx}dx = -\int_0^\infty de^{-cx} = -e^{-cx}|_0^\infty = -(0-1) = 1$ for every $c > 0$.

**(ii)** $P(X \geq 10) = \int_{10}^\infty ce^{-cx}dx = -e^{-cx}|_{10}^\infty = -(0 - e^{-10c}) = e^{-10c}$.

**(iii)** $P(X \geq 10) = 0.5$ implies $e^{-10c} = \frac{1}{2}$, so that $-10c = -\log 2$ and $c = \frac{1}{10}\log 2 \simeq \frac{0.693}{10} \simeq 0.069$. ∎

**2.14**  **(i)** From $\sum_{j=0}^\infty \frac{c}{3^j} = c\sum_{j=0}^\infty \frac{1}{3^j} = c \times \frac{1}{1-\frac{1}{3}} = \frac{3c}{2} = 1$, we get $c = \frac{2}{3}$.

**(ii)** $P(X \geq 3) = c\sum_{j\geq 3}^\infty \frac{1}{3^j} = c \times \frac{1/3^3}{1-\frac{1}{3}} = c \times \frac{1}{2\times 3^2} = \frac{2}{3} \times \frac{1}{2\times 3^2} = \frac{1}{3^3} = \frac{1}{27} \simeq 0.037$.

**(iii)** $P(X = 2k+1, k = 0,1,\ldots) = c\sum_{k=0}^\infty \frac{1}{3^{2k+1}} = c(\frac{1}{3} + \frac{1}{3^3} + \frac{1}{3^5} + \cdots) = c \times \frac{1/3}{1-\frac{1}{9}} = c \times \frac{3}{8} = \frac{2}{3} \times \frac{3}{8} = 0.25$.

**(iv)** $P(X = 3k+1, k = 0,1,\ldots) = c\sum_{k=0}^\infty \frac{1}{3^{3k+1}} = c(\frac{1}{3} + \frac{1}{3^4} + \frac{1}{3^7} + \cdots) = c \times \frac{1/3}{1-\frac{1}{27}} = c \times \frac{9}{26} = \frac{2}{3} \times \frac{9}{26} = \frac{3}{13} \simeq 0.231$. ∎

**2.16**  **(i)** $P(\text{no items are sold}) = f(0) = \frac{1}{2} = 0.5$.

**(ii)** $P(\text{more than three items are sold})$
$= \sum_{x=4}^\infty (\frac{1}{2})^{x+1} = (\frac{1}{2})^5 \times \frac{1}{1-\frac{1}{2}} = \frac{1}{16} = 0.0625$.

**(iii)** $P(\text{an odd number of items are sold})$
$= (\frac{1}{2})^2 + (\frac{1}{2})^4 + (\frac{1}{2})^6 + \cdots = (\frac{1}{2})^2 \times \frac{1}{1-\frac{1}{4}} = \frac{1}{3} \simeq 0.333$. ∎

**2.18**  **(i)** Since $\int_0^\infty c^2xe^{-cx}dx = -cxe^{-cx}|_0^\infty - e^{-cx}|_0^\infty = 1$ for all $c > 0$, the given function is a p.d.f. for all $c > 0$.

**(ii)** From part (i),
$$P(X \geq t) = -cxe^{-cx}|_t^\infty - e^{-cx}|_t^\infty = cte^{-ct} + e^{-ct} = \frac{ct+1}{e^{ct}}.$$

**(iii)** Here $ct + 1 = 2 + 1 = 3$, so that $\frac{ct+1}{e^{ct}} = \frac{3}{e^2} \simeq 0.405$. ∎

**2.20**  We have:
$P(X > x_0) = \int_{x_0}^1 n(1-x)^{n-1}dx = -\int_{x_0}^1 d(1-x)^n = -(1-x)^n|_{x_0}^1 = (1-x_0)^n$, and it is given that this probability is $1/10^{2n}$. Thus, $(1-x_0)^n = \frac{1}{10^{2n}}$, or $1 - x_0 = \frac{1}{100}$ and $x_0 = 0.99$. ∎

**2.22**  For $y > 0$, $F_Y(y) = F_X(y) - F_X(-y)$, and $f_Y(y) = f_X(y) + f_X(-y)$.

**2.24**  **(i)** By setting $\log x = y$, we have
$$\int_0^\infty \frac{1}{x\beta\sqrt{2\pi}}e^{-(\log x - \log \alpha)^2/2\beta^2}dx = \int_{-\infty}^\infty \frac{1}{\beta\sqrt{2\pi}}e^{-(y-\log \alpha)^2/2\beta^2}dy = 1.$$

**(ii)** $F_Y(y) = F_X(e^y)$, and $f_Y(y) = \frac{1}{\beta\sqrt{2\pi}}e^{-(y-\log\alpha)^2/2\beta^2}$.

## Section 2.3

**3.2** We have: $P(A|A \cup B) = \frac{P(A \cap (A \cup B))}{P(A \cup B)} = \frac{P(A)}{P(A \cup B)} = \frac{P(A)}{P(A)+P(B)}$ (since

$A \cap B = \emptyset$), and likewise, $P(B|A \cup B) = \frac{P(B \cap (A \cup B))}{P(A \cup B)} = \frac{P(B)}{P(A)+P(B)}$. ∎

**3.4** (i) $P(b_2|b_1) = 15/26 \simeq 0.577$; (ii) $P(g_2|g_1) = 13/24 \simeq 0.542$;
(iii) $P(b_2) = 0.52$; (iv) $P(b_1 \cap g_2) = 0.22$. ∎

**3.6** Parts (i) and (ii) follow without any calculations by using the fact that $P(\cdot|B)$ and $P(\cdot|C)$ are probability functions, or directly as follows:

(i) $P(A^c|B) = \frac{P(A^c \cap B)}{P(B)} = \frac{P(B-A \cap B)}{P(B)} = \frac{P(B)-P(A \cap B)}{P(B)} = 1 - \frac{P(A \cap B)}{P(B)}$
$= 1 - P(A|B)$.

(ii) $P(A \cup B|C) = \frac{P((A \cup B) \cap C)}{P(C)} = \frac{P((A \cap C) \cup (B \cap C))}{P(C)}$
$= \frac{P(A \cap C) + P(B \cap C) - P(A \cap B \cap C)}{P(C)} = \frac{P(A \cap C)}{P(C)} + \frac{P(B \cap C)}{P(C)} - \frac{P((A \cap B) \cap C)}{P(C)}$
$= P(A|C) + P(B|C) - P(A \cap B|C)$.

(iii) In the sample space $\mathcal{S} = \{HHH, HHT, HTH, THH, HTT, THT, TTH, TTT\}$ with all outcomes being equally likely, define the events:

$$A = \text{"the \# of } H\text{'s is } \leq 2\text{"} = \{TTT, TTH, THT, HTT, THH,$$
$$HTH, HHT\},$$
$$B = \text{"the \# of } H\text{'s is } > 1\text{"} = \{HHT, HTH, THH, HHH\}.$$

Then $B^c = \{HTT, THT, TTH, TTT\}$, $A \cap B^c = B^c$, $A \cap B = \{HHT, HTH, THH\}$, so that $P(A|B^c) = \frac{P(A \cap B^c)}{P(B^c)} = \frac{P(B^c)}{P(B^c)} = 1$ and
$1 - P(A|B) = 1 - \frac{P(A \cap B)}{P(B)} = 1 - \frac{3/8}{4/8} = 1 - \frac{3}{4} = \frac{1}{4}$. Thus,
$P(A|B^c) \neq 1 - P(A|B)$.

(iv) In the sample space $\mathcal{S} = \{1, 2, 3, 4, 5\}$ with all outcomes being equally likely, consider the events $A = \{1, 2\}$, $B = \{3, 4\}$, and $C = \{2, 3\}$, so that $A \cap B = \emptyset$ and $A \cup B = \{1, 2, 3, 4\}$, Then
$P(C|A \cup B) = \frac{P(C \cap (A \cup B))}{P(A \cup B)} = \frac{2/5}{4/5} = \frac{2}{4} = \frac{1}{2}$, whereas
$P(C|A) = \frac{P(A \cap C)}{P(A)} = \frac{1/5}{2/5} = \frac{1}{2}$, $P(C|B) = \frac{P(B \cap C)}{P(B)} = \frac{1/5}{2/5} = \frac{1}{2}$, so that
$P(C|A \cup B) \neq P(C|A) + P(C|B)$. ∎

**3.8** For $n = 2$, the theorem is true since $P(A_2|A_1) = \frac{P(A_1 \cap A_2)}{P(A_1)}$ yields
$P(A_1 \cap A_2) = P(A_2|A_1)P(A_1)$. Next, assume
$P(A_1 \cap \cdots \cap A_k) = P(A_k|A_1 \cap \cdots \cap A_{k-1}) \cdots P(A_2|A_1)P(A_1)$ and show that
$P(A_1 \cap \cdots \cap A_{k+1}) =$
$P(A_{k+1}|A_1 \cap \cdots \cap A_k)P(A_k|A_1 \cap \cdots \cap A_{k-1}) \cdots P(A_2|A_1)P(A_1)$. Indeed,
$P(A_1 \cap \cdots \cap A_{k+1}) = P((A_1 \cap \cdots \cap A_k) \cap A_{k+1}) =$
$P(A_{k+1}|A_1 \cap \cdots \cap A_k)P(A_1 \cap \cdots \cap A_k)$ (by applying the theorem for two
events $A_1 \cap \cdots \cap A_k$ and $A_{k+1}$)
$= P(A_{k+1}|A_1 \cap \cdots \cap A_k)P(A_k|A_1 \cap \cdots \cap A_{k-1}) \cdots P(A_2|A_1)P(A_1)$ (by the induction hypothesis). ∎

**3.10** With obvious notation, we have $P$ (1st white and 4th white)

$= P(W_1 \cap W_2 \cap W_3 \cap W_4) + P(W_1 \cap W_2 \cap B_3 \cap W_4) + P(W_1 \cap B_2 \cap W_3 \cap W_4) + P(W_1 \cap B_1 \cap B_2 \cap W_4)$

$= P(W_4|W_1 \cap W_2 \cap W_3)P(W_3|W_1 \cap W_2)P(W_2|W_1)P(W_1)$

$+ P(W_4|W_1 \cap W_2 \cap B_3)P(B_3|W_1 \cap W_2)P(W_2|W_1)P(W_1)$

$+ P(W_4|W_1 \cap B_2 \cap W_3)P(W_3|W_1 \cap B_2)P(B_2|W_1)P(W_1)$

$+ P(W_4|W_1 \cap B_1 \cap B_2)P(B_2|W_1 \cap B_1)P(B_1|W_1) \times P(W_1)$

$= \frac{7}{12} \times \frac{8}{13} \times \frac{9}{14} \times \frac{10}{15} + \frac{8}{12} \times \frac{5}{13} \times \frac{9}{14} \times \frac{10}{15} + \frac{8}{12}$

$\times \frac{9}{13} \times \frac{5}{14} \times \frac{10}{15} + \frac{9}{12} \times \frac{4}{13} \times \frac{5}{14} \times \frac{10}{15}$

$= \frac{1}{12 \times 13 \times 14 \times 15}(7 \times 8 \times 9 \times 10 + 5 \times 8 \times 9 \times 10 \times 2 + 4 \times 5 \times 9 \times 10)$

$= \frac{9 \times 10 \times 156}{12 \times 13 \times 14 \times 15} = \frac{3}{7} \simeq 0.429.$ ∎

**3.12** (i) $P(+) = 0.01188$; (ii) $P(D|+) = \frac{190}{1188} \simeq 0.16.$ ∎

**3.14** Let I = "switch I is open," II = "switch II is open," S = "signal goes through." Then (i) $P(S) = 0.48$; (ii) $P(I|S^c) = \frac{5}{13} \simeq 0.385$; (iii) $P(II|S^c) = \frac{10}{13} \simeq 0.769.$ ∎

**3.16** With F="an individual is female," M="an individual is male," and C="an individual is color-blind," we have:
$P(F) = 0.52, P(M) = 0.48, P(C|F) = 0.25, P(C|M) = 0.05$, and therefore $P(C) = 0.154, P(M|C) = \frac{12}{77} \simeq 0.156.$ ∎

**3.18** With obvious notation, we have:
   (i) $P(D) = 0.029$; (ii) $P(I|D) = \frac{12}{29} \simeq 0.414$; (iii) $P(II|D) = \frac{9}{29} \simeq 0.310$, and $P(III|D) = \frac{8}{29} \simeq 0.276.$ ∎

**3.20** (i) $P(X > t) = \int_t^\infty \lambda e^{-\lambda x} dx = -\int_t^\infty de^{-\lambda x} = -e^{-\lambda x}|_t^\infty = e^{-\lambda t}.$

   (ii) $P(X > s + t|X > s) = \frac{P(X > s+t, X > s)}{P(X > s)} = \frac{P(X > s+t)}{P(X > s)}$

   $= \frac{e^{-\lambda(s+t)}}{e^{-st}}$   (by part (i))

   $= e^{-\lambda t}.$

   (iii) The conditional probability that $X$ is greater than $t$ units beyond $s$, given that it has been greater than $s$, does not depend on $s$ and is the same as the (unconditional) probability that $X$ is greater than $t$. That is, this distribution has some sort of "memoryless" property. ∎

**3.22** (i) $P(V) = 0.10, P(V^c) = 0.90.$
   (ii) (a)  $P(G|V^c) = 0.08$
      (b)  $P(I|V) = 0.15.$
   (iii) $P(G) = 0.157.$
   (iv) $P(V|G) = \frac{85}{157} \simeq 0.541.$

**3.24** (i) $P(+) = 0.0545$, (ii) $P(D|+) = 475/5450 \simeq 0.087.$

**3.26** (i) $P(G_i) = 0.59$, $i = 1, 2, 3$. (ii) $P(G|G_i) = 56/59 \simeq 0.949$, $i = 1, 2, 3$.
(iii) Required probability $= 0.633542$.

**3.30** $P(\text{ball is black}) = \frac{n}{m+n}$.

**3.32** (i) $\frac{r(r-1)}{r(r-1)+b(b-1)}$. (ii) $\frac{2r(r-1)b}{(r+b)(r+b-1)(r+b-2)}$. (iii) $2/7 \simeq 0.286$, $0.2$.

**3.34** (i) $P(C_2) = P(D_2) = P(H_2) = P(S_2) = 0.25$. (ii)
$P(C_2) = 0.25 = P(D_2) = P(H_2) = P(S_2)$.

**3.36** Required probability $= 43/144 \simeq 0.299$.

**3.38** (i)

$$
\text{Required probability} = \frac{1}{4}\left[ \frac{m_1(m_2 + 1)}{(m_1 + n_1)(m_2 + n_2 + 1)} + \frac{n_1(n_1 - 1)}{(m_1 + n_1)(m_1 + n_1 - 1)} \right.
$$
$$
+ \frac{n_1(n_2 + 1)}{(m_1 + n_1)(m_2 + n_2 + 1)} + \frac{m_2(m_1 + 1)}{(m_2 + n_2)(m_1 + n_1 + 1)}
$$
$$
\left. + \frac{n_1(n_1 + 1)}{(m_1 + n_1)(m_1 + n_1 + 1)} + \frac{n_1(n_2 - 1)}{(m_1 + n_1)(m_2 + n_2 - 1)} \right].
$$

**(ii)** $= 1773/5280 \simeq 0.336$.

## Section 2.4

**4.2** Here $P(A) = P(A \cap A) = P(A)P(A) = [P(A)]^2$, and this happens if $P(A) = 0$, whereas, if $P(A) \neq 0$, it happens only if $P(A) = 1$. ∎

**4.4** Since $P(A_1 \cap A_2) = P(A_1)P(A_2)$, we have to show that

$$
P(A_1 \cap (B_1 \cup B_2)) = P(A_1)P(B_1 \cup B_2), P(A_2 \cap (B_1 \cup B_2))
$$
$$
= P(A_2)P(B_1 \cup B_2), P(A_1 \cap A_2 \cap (B_1 \cup B_2))
$$
$$
= P(A_1)P(A_2)P(B_1 \cup B_2).
$$

Indeed,

$$
P(A_1 \cap (B_1 \cup B_2)) = P((A_1 \cap B_1) \cup (A_1 \cap B_2))
$$
$$
= P(A_1 \cap B_1) + P(A_1 \cap B_2) = P(A_1)P(B_1) + P(A_1)P(B_2)
$$
$$
= P(A_1)P(B_1 \cup B_2), \quad \text{and similarly for } P(A_2 \cap (B_1 \cup B_2)).
$$

Finally,

$$
P(A_1 \cap A_2 \cap (B_1 \cup B_2)) = P((A_1 \cap A_2 \cap B_1) \cup (A_1 \cap A_2 \cap B_2))
$$
$$
= P(A_1 \cap A_2 \cap B_1) + P(A_1 \cap A_2 \cap B_2)
$$
$$
= P(A_1)P(A_2)P(B_1) + P(A_1)P(A_2)P(B_2)
$$
$$
= P(A_1)P(A_2)P(B_1 \cup B_2). \quad ∎
$$

**4.6** (i) Clearly, $A = (A \cap B \cap C) \cup (A \cap B^c \cap C) \cup (A \cap B \cap C^c) \cup (A \cap B^c \cap C^c)$ and hence $P(A) = 0.6875$. Likewise, $P(B) = 0.4375$, $P(C) = 0.5625$.

    (ii) $A, B$, and $C$ are not independent.

    (iii) $P(A \cap B) = \frac{4}{16}$, and then $P(A|B) = \frac{4}{7} \simeq 0.571$.

    (iv) $A$ and $B$ are not independent. ∎

**4.8** (i) $\mathcal{S} = \{HHH, HHT, HTH, THH, HTT, THT, TTH, TTT\}, A = \{HHH, TTT\}$
with $P(A) = p^3 + q^3$ $(q = 1 - p)$.

    (ii) $P(A) = 0.28$. ∎

**4.10** (i) $c = 1/25$.

    (ii) See the below figure.

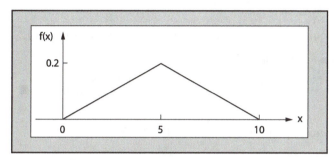

    (iii) $P(A) = P(X > 5) = 0.50, P(B) = P(5 < X < 7.5) = 0.375$.

    (iv) $P(B|A) = 0.75$; (v) $A$ and $B$ are not independent. ∎

**4.12** (i) $P((W_1 \cap R_1^c \cap \cdots \cap R_{n-2}^c \cap R_{n-1}) \cup (W_1^c \cap W_2 \cap R_1^c \cap \cdots \cap R_{n-3}^c \cap R_{n-2}) \cup \cdots \cup (W_1^c \cap \cdots \cap W_{n-2}^c \cap W_{n-1} \cap R_n))$
$= 0.54 \sum_{i=1}^{n-1} (0.1)^{i-1} (0.4)^{n-i-1}$.

    (ii) For $n = 5$, the probability in part (i) is 0.0459. ∎

**4.14** (i) $P(\text{no circuit is closed}) = (1 - p_1) \ldots (1 - p_n)$.

    (ii) $P(\text{at least 1 circuit is closed}) = 1 - (1 - p_1) \ldots (1 - p_n)$.

    (iii) $P(\text{exactly 1 circuit is closed}) = p_1(1 - p_2) \ldots (1 - p_n) + (1 - p_1)p_2 \times (1 - p_3) \ldots (1 - p_n) + \cdots + (1 - p_1) \ldots (1 - p_{n-1})p_n$.

    (iv) The answers above are: $(1 - p)^n, 1 - (1 - p)^n, np(1 - p)^{n-1}$.

    (v) The numerical values are: 0.01024, 0.98976, 0.0768.

**4.16** (i) $p_1 + p_2 - 2p_1p_2$. (ii) 0.172.

**4.18** (i) $D = (A \cap B \cap C^c) \cup (A \cap B^c \cap C) \cup (A^c \cap B \cap C)$. (ii)
$P(D) = P(A \cap B) + P(A \cap C) + P(B \cap C) - 3P(A \cap B \cap C)$. (iii)
$P(D) = 0.332$.

## Section 2.5

**5.2** (i) $3 \times 4 \times 5 = 60$; (ii) $1 \times 2 \times 5 = 10$; (iii) $3 \times 4 \times 1 = 12$. ∎

**5.4** (i) $3 \times 2 \times 3 \times 2 \times 3 = 108$; (ii) $3 \times 2 \times 2 \times 1 \times 1 = 12$. ∎

**5.6**  $2^n$; $2^5 = 32$, $2^{10} = 1{,}024$, $2^{15} = 32{,}768$, $2^{20} = 1{,}048{,}576$,
$2^{25} = 33{,}554{,}432$.  ∎

**5.8**  The required probability is $\frac{1}{360} \simeq 0.003$.  ∎

**5.10**  Start with $\binom{n+1}{m+1} / \binom{n}{m}$, expand in terms of factorial, do the cancellations, and you end up with $(n+1)/(m+1)$.  ∎

**5.12**  Selecting $r$ out of $m+n$ in $\binom{m+n}{r}$ ways is equivalent to selecting $x$ out of $m$ in $\binom{m}{x}$ ways and $r - x$ out of $n$ in $\binom{n}{r-x}$ ways where $x = 0, 1, \ldots, r$. Then $\binom{m+n}{r} = \sum_{x=0}^{r} \binom{m}{x}\binom{n}{r-x}$.  ∎

**5.14**  The required number is $\binom{n}{3}$, which for $n = 10$ becomes $\binom{10}{3} = 120$.  ∎

**5.16**  The required  probability is
$$\frac{\binom{10}{2} \times \binom{15}{3} \times \binom{30}{4} \times \binom{5}{1}}{\binom{60}{10}} = \frac{2{,}480{,}625}{66{,}661{,}386} \simeq 0.037.  ∎$$

**5.18**  The required probability is $\dfrac{\binom{n-1}{m}}{\binom{n}{m}} = 1 - \dfrac{m}{n}$.  ∎

**5.20**  The required probability is $(0.5)^{2n} \sum_{m=0}^{n} \binom{n}{m}^2$, which for $n = 5$, becomes $252 \times (0.5)^{10} \simeq 0.246$.  ∎

**5.22**  The required probability is $\sum_{x=5}^{10} \binom{10}{x}(0.2)^x(0.8)^{10-x} = 0.03279$.  ∎

**5.24**  **(a)**  (i) $\left(\frac{n_R}{n}\right)^3$; (ii) $1 - \left(\frac{n_B + n_W}{n}\right)^3$; (iii) $\frac{6 n_R n_B n_W}{n^3}$.
  **(b)**  (i) $\frac{\binom{n_R}{3}}{\binom{n}{3}}$; (ii) $1 - \frac{\binom{n_B + n_W}{3}}{\binom{n}{3}}$; (iii) $\frac{\binom{n_R}{1}\binom{n_B}{1}\binom{n_W}{1}}{\binom{n}{3}}$.  ∎

**5.26**  360 ways.

**5.28**  (i) 0.3. (ii) 0.3.

**5.30**  (i) 540. (ii) 60.

**5.32**  $(2n)!$.

**5.34**  (i) $1 - (5/6)^4 \simeq 0.518$. (ii) $1 - (35/36)^{24} \simeq 0.491$.

**5.36**  (i) Choose $k$ out of $\{1, 2, \ldots, n - 1\}$ in $\binom{n-1}{k}$ ways, next choose $k - 1$ out of the same set in $\binom{n-1}{k-1}$ ways and then attach $n$. (ii) Algebra.

**5.38**  Write $\binom{i}{k} = \binom{i-1}{k} + \binom{i-1}{k-1}$, and then sum over $i = 1, \ldots, n$.

**5.40**  (i) $\frac{2^k(n-1)\ldots(n-k+1)}{2n(2n-1)\ldots(2n-k+1)}$. (ii) $1 -$ probability in part (i). (iii) For (i): $8/95 \simeq 0.084$, for (ii): $87/95 \simeq 0.916$.

**5.42**  (i) $5/6 \simeq 0.833$. (ii) $5/12 \simeq 0.417$. (iii) $5/9 \simeq 0.054$. (iv) $5/54 \simeq 0.093$. (v) $1/9 \simeq 0.111$.

# CHAPTER 3
## Section 3.1

**1.2** (i) $EX = 0, EX^2 = c^2$, and $\text{Var}(X) = c^2$.

(ii) $P(|X - EX| \leq c) = P(-c \leq X \leq c) = P(X = -c, X = c) = 1 = \frac{c^2}{c^2} = \frac{\text{Var}(X)}{c^2}$. ∎

**1.4** If $Y$ is the net loss to the company, then $EY = \$200$, and if $P$ is the premium to be charged, then the expected profit is $P - EY = P - 200 = 100$, so that $P = \$300$. ∎

**1.6** $\text{Var}(X) = EX^2 - (EX)^2$ by expanding and taking expectations. Also, $E[X(X - 1)] = \text{Var}(X) + (EX)^2 - EX$ by expanding, taking expectations, and using the first result. That $\text{Var}(X) = E[X(X - 1)] + EX - (EX)^2$ follows from the first two results. ∎

**1.8** (i) $EX = 2$, $E[X(X - 1)] = 4$; (ii) $\text{Var}(X) = 2$. ∎

**1.10** $EX = \frac{4}{3} \simeq 1.333$, $EX^2 = 2$, so that $\text{Var}(X) = \frac{2}{9}$ and s.d. of $X = \frac{\sqrt{2}}{3} \simeq 0.471$. ∎

**1.12** $c_1 = -1/12$, $c_2 = 5/3$. ∎

**1.14** (i) By adding and subtracting $\mu$, we get: $E(X - c)^2 = \text{Var}(X) + (\mu - c)^2$.

(ii) Immediate from part (i). ∎

**1.16** (i) $\int_{-\infty}^{\infty} \frac{dx}{1+x^2} = \arctan x \big|_{-\infty}^{\infty} = \arctan(\infty) - \arctan(-\infty) = \pi$, so that $\int_{-\infty}^{\infty} \frac{1}{\pi} \times \frac{dx}{1+x^2} = 1$.

(ii) $\frac{1}{\pi} \int_{-\infty}^{\infty} x \times \frac{dx}{1+x^2} = \frac{1}{2\pi} \int_{-\infty}^{\infty} \frac{d(1+x^2)}{1+x^2} = \frac{1}{2\pi} \times \log(1 + x^2) \big|_{-\infty}^{\infty} = \frac{1}{2\pi}(\infty - \infty)$. ∎

**1.18** For the discrete case, $X \geq c$ means $x_i \geq c$ for all values $x_i$ of $X$. Then $x_i f_X(x_i) \geq c f_X(x_i)$ and hence $\sum_{x_i} x_i f_X(x_i) \geq \sum_{x_i} c f_X(x_i)$. But $\sum_{x_i} x_i f_X(x_i) = EX$ and $\sum_{x_i} c f_X(x_i) = c \sum_{x_i} f_X(x_i) = c$. Thus, $EX \geq c$. The particular case follows, of course, by taking $c = 0$. In the continuous case, summation signs are replaced by integrals. ∎

**1.20** (i) $P(40 \leq X \leq 60) = 0.32$

(ii) $\mu = EX = 10 \times 0.13 + 20 \times 0.12 + 30 \times 0.10 + 40 \times 0.13 + 50 \times 0.10 + 60 \times 0.09 + 70 \times 0.11 + 80 \times 0.08 + 90 \times 0.08 + 100 \times 0.06 = 49.6$, $EX^2 = 3{,}232$, so that $\text{Var}(X) = 3{,}232 - (49.6)^2 = 3{,}232 - 2{,}460.16 = 771.84$, and $\sigma \simeq 27.782$.

(iii) $[\mu - \sigma, \ \mu + \sigma] = [49.6 - 27.782, \ 49.6 + 27.782] = [21.818, 77.382]$, $[\mu - 1.5\sigma, \ \mu + 1.5\sigma] = [49.6 - 41.673, \ 49.6 + 41.673] = [7.927, 91.273]$,

$[\mu - 2\sigma, \ \mu + 2\sigma] = [49.6 - 55.564, \ 49.6 + 55.564] = [-5.964, 105.164]$.

It follows that the respective proportions of claims are 53%, 94%, and 100%. ∎

**1.22** 90.

**1.24** $a = 1.2, b = 0.4$.

**1.26** (i) $\sigma_Y = |a|\sigma$. (ii) 12.

**1.28** Take $X = -1I_{A_1} + 0I_{A_2} + 3I_{A_3}$, $Y = -4I_{A_1} + 0I_{A_2} + 2I_{A_3}$ with $P(A_1) = P(A_2) = P(A_3) = 1/3$.

**1.30** $EX = P(A) = EX^2$, so that $\sigma^2(X) = P(A) - [P(A)]^2 = P(A)P(A^c)$.

**1.32** $EX = \frac{1}{n}\sum_{k=1}^{n} k = \frac{n+1}{2}$, $EX^2 = \frac{1}{n}\sum_{k=1}^{n} k^2 = \frac{(n+1)(2n+1)}{6}$, so that $\sigma^2(X) = (n+1)(n-1)/12$.

**1.34** (i) $h(x) = h(x_0) + (x - x_0)h'(x_0) + \frac{(x-x_0)^2}{2!}h''(x^*)$, for some $x^*$ between $x$ and $x_0$. Then $h(x) \geq h(x_0) + (x - x_0)h'(x_0)$. Replace $x$ by $X$, $x_0$ by $EX$, and take expectations to obtain the result. (ii) Immediate. (iii) Since $h''(x) = \frac{n(n+1)}{x^{n+2}} \geq 0$, part (i) gives $E(1/X)^n \geq (1/EX)^n$.

**1.36** For the continuous case, split the integral from $-\infty$ to 0 and 0 to $\infty$, set $(x - \mu)/\sigma = y$, and arrive at
$$-\sigma \int_{-\mu/\sigma}^{\infty} y^3 f(\mu + \sigma y) + \sigma \int_{-\mu/\sigma}^{\infty} y^3 f(\mu + \sigma y)dy = 0.$$

**1.38** (i) $EX = 4$, $EX^2 = 64/3$, $\sigma^2(X) = 16/3$. (ii) $P(|X - 4| > 3) = 0.25$. (iii) $P(|X - 4| > 3) \leq 16/27 \simeq 0.593$.

**1.40** (i) For $0 < y < 1$: $F_Y(y) = 1 - F_X(\sqrt{1-y})$ and $f_Y(y) = 3\sqrt{1-y}/2$. (ii) $P(a < Y < b) = (1 - a)^{3/2} - (1 - b)^{3/2}$ and $P(1/3 < Y < 2/3) = (2/3)^{3/2} - (1/3)^{3/2} \simeq 0.352$. (iii) $EY = 2/5$, $\sigma^2(Y) = 12/175$.

**1.42** (i) $\sum_{x=1}^{\infty} \left(\frac{1}{2}\right)^x = 1$. (ii) $EX = 2, E[X(X - 1)] = 4$. (iii) $\sigma^2(X) = EX^2 - (EX)^2 = E[X(X - 1)] + EX - (EX)^2 = 2$.

**1.44** (i) In the integral, use the transformation $\log x = y$ to arrive at $\int_{-\infty}^{\infty} e^y \times \frac{1}{\beta\sqrt{2\pi}}e^{-(y-\log\alpha)^2/2\beta^2}dy$ which is the m.g.f. of $Y \sim N(\log\alpha, \beta^2)$ evaluated at 1. Thus, $M_Y(1) = e^{\log\alpha+\beta^2/2} = \alpha e^{\beta^2/2}$. (ii) Likewise, $M_Y(2) = e^{2\log\alpha+2\beta^2} = e^{\log\alpha^2+2\beta^2} = \alpha^2 e^{2\beta^2}$. (iii) Follows from (i) and (ii).

**1.46** (i) Split the integral from $-\infty$ to 0 and 0 to $\infty$, use the transformation $x - \mu = -y$ in the first integral and $x - \mu = y$ in the second integral to arrive at $\int_0^{\infty} e^{-y}dy = 1$. (ii) Work as in part (i). (iii) Again, work as in part (i) to compute $EX^2$. The $\sigma^2(X)$ follows immediately.

**1.48** (i) $EX = \sum_{j=1}^{\infty} jP(X = j) = \lim_{n\to\infty} \sum_{j=1}^{n} jP(X = j) = \lim_{n\to\infty} \sum_{j=1}^{n} P(X \geq j) = \sum_{j=1}^{\infty} P(X \geq j)$. (ii) and (iii) follow by simple manipulations.

**1.50** (i) Split the integral from $-\infty$ to $0$ and from $0$ to $\infty$, set first $x - c = y$ and then $y = -t$ to arrive at the conclusion. (ii) Follows from part (i).

**1.52** Split the integral from $-\infty$ to $0$ and from $0$ to $\infty$, set $(x - \mu)/\sigma = -y$ in the first integral and $(x - \mu)/\sigma = y$ in the second integral to arrive (essentially) at $\log(1 + y^2)\big|_{-\mu/\sigma}^{\infty} = \infty$, $\log(1 + y^2)\big|_{-\infty}^{-\mu/\sigma} = \infty$, so that $E|X| = \infty$.

## Section 3.2

**2.2** (i) $c = \sigma/\sqrt{1 - \alpha}$; (ii) $c = \frac{1}{\sqrt{0.05}} \simeq 4.472$. ■

**2.4** (i) By the Tchebichev inequality, $P(|X - \mu| \geq c) = 0$ for all $c > 0$.
(ii) Consider a sequence $0 < c_n \downarrow 0$ as $n \to \infty$. Then $P(|X - \mu| \geq c_n) = 0$ for all $n$, or equivalently, $P(|X - \mu| < c_n) = 1$ for all $n$, whereas, clearly, $\{(|X - \mu| < c_n)\}$ is a nonincreasing sequence of events and its limits is $\cap_{n=1}^{\infty}(|X - \mu| < c_n)$. Then, by Theorem 2 in Chapter 2, $1 = \lim_{n\to\infty} P(|X - \mu| < c_n) = P(\cap_{n=1}^{\infty}(|X - \mu| < c_n))$. However, it is clear that $\cap_{n=1}^{\infty}(|X - \mu| < c_n) = (|X - \mu| \leq 0) = (X = \mu)$. Thus, $P(X = \mu) = 1$, as was to be seen. ■

## Section 3.3

**3.2** (i) It follows by using the identity $\binom{n+1}{x} = \binom{n}{x} + \binom{n}{x-1}$.
(ii) $B(26, 0.25; 10) = 0.050725$. ■

**3.4** If $X$ is the number of those favoring the proposal, then $X \sim B(15, 0.4375)$. Therefore: (i) $P(X \geq 5) = 0.859$; (ii) $P(X \geq 8) = 0.3106$. ■

**3.6** If $X$ is the number of times the bull's eye is hit, then $X \sim B(100, p)$. Therefore
(i) $P(X \geq 40) = \sum_{x=40}^{100} \binom{100}{x} p^x q^{100-x}$  $(q = 1 - p)$.
(ii) $P(X \geq 40) = \sum_{x=40}^{100} \binom{100}{x} (0.25)^x (0.75)^{100-x}$.
(iii) $EX = np = 100p$, $Var(X) = npq = 100pq$, and for $p = 0.25$, $EX = 25$, $Var(X) = 18.75$, s.d. of $X = \sqrt{18.75} \simeq 4.33$. ■

**3.8** From the Tchebichev inequality $n = 8,000$. ■

**3.10** (i) Writing $\binom{n}{x}$ in terms of factorials, and after cancellations, we get
$EX = np \sum_{y=0}^{n-1} \binom{n-1}{y} p^y q^{(n-1)-y} = np \times 1 = np$. Likewise,
$E[X(X - 1)] = n(n - 1)p^2 \sum_{y=0}^{n-2} \binom{n-2}{y} p^y q^{(n-2)-y} = n(n - 1)p^2 \times 1 = n(n - 1)p^2$.
(ii) From Exercise 1.6, $Var(X) = n(n - 1)p^2 + np - (np)^2 = npq$. ■

**3.12** (i) $P(X \le 10) = 1 - q^{10}$; (ii) $1 - (0.8)^{10} \simeq 0.893$. ■

**3.14** If $X$ is the number of tosses to the first success, then $X$ has the Geometric distribution with $p = 1/6$. Then:
(i) $P(X = 3) = \frac{25}{216} \simeq 0.116$.
(ii) $P(X \ge 5) = \left(\frac{5}{6}\right)^4 \simeq 0.482$. ■

**3.16** (i) $EX = \frac{1}{p}$, $E[X(X-1)] = \frac{2q}{p^2}$; (ii) $\text{Var}(X) = \frac{q}{p^2}$. ■

**3.18** $\lambda = 2$. ■

**3.20** $f(x+1) = e^{-\lambda} \frac{\lambda^{x+1}}{(x+1)!} = \frac{\lambda}{x+1} \times e^{-\lambda} \frac{\lambda^x}{x!} = \frac{\lambda}{x+1} f(x)$. ■

**3.22** (i) $M_X(t) = e^{\lambda(e^t - 1)}$, $t \in \Re$, by applying the definition of $M_X$.
(ii) $EX = \frac{d}{dt} M_X(t)|_{t=0} = \lambda$, $EX^2 = \frac{d^2}{dt^2} M_X(t)|_{t=0} = \lambda(\lambda + 1)$, so that $\text{Var}(X) = \lambda$. ■

**3.24** (i) $\frac{\binom{70}{5}\binom{10}{0}}{\binom{80}{5}} \simeq 0.718$; (ii) $\frac{1}{\binom{80}{5}}\left[\binom{70}{3}\binom{10}{2} + \binom{70}{4}\binom{10}{1} + \binom{70}{5}\binom{10}{0}\right] \simeq 0.987$. ■

**3.26** Writing the combinations in terms of factorials, and recombining the terms, we get the result. ■

**3.28**   a. By integration, using the definition of $\Gamma(\alpha)$ and the recursive relation for $\Gamma(\alpha + 1)$, we get $EX = \frac{\beta}{\Gamma(\alpha)} \times \Gamma(\alpha + 1) = \alpha\beta$. Likewise,
$EX^2 = \frac{\beta^2}{\Gamma(\alpha)} \times \Gamma(\alpha + 2) = \alpha(\alpha + 1)\beta^2$, so that $\text{Var}(X) = \alpha\beta^2$.
(ii) $EX = 1/\lambda$, $\text{Var}(X) = 1/\lambda^2$ from part (i).
(iii) $EX = r$, $\text{Var}(X) = 2r$ from part (i). ■

**3.30** (i) (a) With $g(X) = cX$, we have $Eg(X) = c/\lambda$.
(b) With $g(X) = c(1 - 0.5e^{-\alpha X})$, we have $Eg(X) = \frac{(\alpha + 0.5\lambda)c}{\alpha + \lambda}$.
(ii) (a) 10; (b) 1.5. ■

**3.32** Indeed, $P(T > t) = P$ (0 events occurred in the time interval $(0, t)) = e^{-\lambda t} \frac{(\lambda t)^0}{0!} = e^{-\lambda t}$. So, $1 - F_T(t) = e^{-\lambda t}$, $t > 0$, and hence $f_T(t) = \lambda e^{-\lambda t}$, $t > 0$, and $T$ is as described. ■

**3.34** (i) $\int_0^\infty \alpha\beta x^{\beta-1} e^{-\alpha x^\beta} dx = -\int_0^\infty de^{-\alpha x^\beta} = -e^{-\alpha x^\beta}|_0^\infty = 1$.
(ii) $\beta = 1$ and any $\alpha > 0$.
(iii) For $n = 1, 2, \ldots, EX^n = \Gamma(\frac{n}{\beta} + 1)/\alpha^{n/\beta}$. Then $EX = \Gamma(\frac{1}{\beta} + 1)/\alpha^{1/\beta}$, $EX^2 = \Gamma(\frac{2}{\beta} + 1)/\alpha^{2/\beta}$, and hence $\text{Var}(X) = \{\Gamma(\frac{2}{\beta} + 1) - [\Gamma(\frac{1}{\beta} + 1)]^2\}/\alpha^{2/\beta}$. ■

**3.36** All parts (i)-(iv) are immediate. ■

**3.38** (i) $P(X \le c) = 0.875$, and $c = \mu + 1.15\sigma$; (ii) $c = 7.30$. ■

**3.40** (i) 0.997020; (ii) 0.10565; (iii) 0.532807. ∎

**3.42** Let $X$ be the diameter of a ball bearing, and let $p$ be the probability that a ball bearing is defective. Then $p = P(X > 0.5 + 0.0006 \text{ or } X < 0.5 - 0.0006)$, and by splitting the probabilities and normalizing, we have that it is 0.583921. Thus, if $Y$ is the r.v. denoting the number of defective balls from among those in a day's production, then $Y \sim B(n, p)$, so that $EY = np$ and $E\left(\frac{Y}{n}\right) = p \simeq 0.583921$. ∎

**3.44** **(i)** By the hint,

$$I^2 = \frac{1}{2\pi} \int_{-\infty}^{\infty} \int_{-\infty}^{\infty} e^{-\frac{x^2+y^2}{2}} \, dx \, dy = \frac{1}{2\pi} \left( \int_0^{2\pi} d\theta \right) \left( \int_0^{\infty} r e^{-r^2/2} \, dr \right)$$

$$= \frac{1}{2\pi} \times 2\pi \left( -e^{-r^2/2} \big|_0^{\infty} \right) = 1.$$

**(ii)** $\frac{1}{\sqrt{2\pi}\sigma} \int_{-\infty}^{\infty} e^{-\frac{(x-\mu)^2}{2\sigma^2}} \, dx = \frac{1}{\sqrt{2\pi}\sigma} \int_{-\infty}^{\infty} e^{-y^2/2} \sigma \, dy = I = 1.$ ∎

**3.46** **(i)** $M_X(t) = e^{t^2/2} \times \frac{1}{\sqrt{2\pi}} \int_{-\infty}^{\infty} e^{-\frac{(x-t)^2}{2}} \, dt = e^{t^2/2} \times 1 = e^{t^2/2}, \ t \in \mathfrak{R}.$

**(ii)** With $Z = \frac{X-\mu}{\sigma}$, $e^{t^2/2} = M_Z(t) = M_{\frac{1}{\sigma}X + \frac{-\mu}{\sigma}}(t) = e^{-\frac{\mu t}{\sigma}} M_X\left(\frac{t}{\sigma}\right)$,

so that $M_X\left(\frac{t}{\sigma}\right) = e^{\frac{\mu t}{\sigma} + \frac{t^2}{2}}$, and $M_X(t) = e^{\mu t + \frac{\sigma^2 t^2}{2}}$ by replacing $\frac{t}{\sigma}$ by $t \in \mathfrak{R}$.

**(iii)** By differentiation of $M_X(t)$ and evaluating at 0, we get $EX = \mu, EX^2 = \mu^2 + \sigma^2$, so that $\text{Var}(X) = \sigma^2$. ∎

**3.48** **(i)** $EX^{2n+1} = 0$, and by the hint $EX^{2n} = (2n-1)(2n-3)\cdots 1$
$= \frac{1 \times 2 \times \cdots \times (2n-1) \times (2n)}{(2 \times 1) \times \cdots \times [2 \times (n-1)] \times (2 \times n)} = \frac{(2n)!}{2^n[1 \cdots (n-1)n]} = \frac{(2n)!}{2^n(n!)}.$
**(ii)** $EX = 0, EX^2 = 1$, so that $\text{Var}(X) = 1.$
**(iii)** With $Z = \frac{X-\mu}{\sigma}, 0 = EZ = \frac{1}{\sigma}(EX - \mu)$, so that $EX = \mu$, and $1 = \text{Var}(Z) = \frac{1}{\sigma^2}\text{Var}(X)$, so that $\text{Var}(X) = \sigma^2$. ∎

**3.50** **(i)** $P(-1 < X < 2) = \frac{3}{2\alpha} = 0.75$ and $\alpha = 2.$
**(ii)** $P(|X| < 1) = P(|X| > 2)$ is equivalent to $\frac{1}{\alpha} = 1 - \frac{2}{\alpha}$ from which $\alpha = 3.$ ∎

**3.52** $EX = \frac{1}{2}, EX^2 = \frac{1}{3}$, so that: (i) $-0.5$ and (ii) $2(e-1) \simeq 3.44$. ∎

**3.54** **(i)** $EX_1 = 2np_1, EX_2 = 2np_2.$
**(ii)** $P(X_1 \leq n) = \sum_{x=0}^{n} \binom{2n}{x} p_1^x (1 - p_1)^{2n-x}.$
**(iii)** $P(X_2 \geq n) = 1 - \sum_{x=0}^{n-1} \binom{2n}{x} p_2^x (1 - p_2)^{2n-x}.$
**(iv)** $EX_1 = 7.5, EX_2 = 10.5, P(X_1 \leq 12) = 0.9835, P(X_2 \geq 12) = 0.3382.$
∎

**3.56** **(i)** $P(X = 2) = 3P(X = 4)$ gives $\lambda = 2$. Hence $EX = \text{Var}(X) = 2.$
**(ii)** $P(2 \leq X \leq 4) = 0.5413, P(X \geq 5) = 0.0527.$ ∎

**3.58**

$$\frac{\binom{m}{x}\binom{n}{r-x}}{\binom{m+n}{r}} = \binom{r}{x}\left(\frac{m}{m+n}\right)\left(\frac{m}{m+n} - \frac{1}{m+n}\right)\cdots\left(\frac{m}{m+n} - \frac{x-1}{m+n}\right)$$

$$\times \left(\frac{n}{m+n}\right)\left(\frac{n}{m+n} - \frac{1}{m+n}\right)\cdots\left(\frac{n}{m+n} - \frac{r-x-1}{m+n}\right)$$

$$\times \left[1 \times \left(1 - \frac{1}{m+n}\right)\cdots\left(1 - \frac{r-1}{m+n}\right)\right]^{-1}$$

$$\xrightarrow[m,n\to\infty]{} \binom{r}{x}p^x q^{r-x}, \quad \text{since } \frac{m}{m+n} \to p, \frac{n}{m+n} \to 1 - p = q. \quad \blacksquare$$

**3.60** (i) $EX = 3.125, \sigma^2(X) \simeq 2.669, \sigma(X) \simeq 1.634$.

(ii) $P(3.125 - 1.634 \leq X \leq 3.125 + 1.634) =$

$\sum_{x=0}^{4}\binom{125}{x}\binom{875}{25-x}/\binom{1,000}{25} \overset{\text{def}}{=} \alpha$.

(iii) Since $\frac{m}{m+n} = \frac{2}{16}, \alpha = 0.6424$. Next, $25 \times 0.125 = 3.125$, and the Poisson tables give $\alpha \simeq 0.6095$ (by linear interpolation). $\blacksquare$

**3.62** $EX^n = \int_\alpha^\beta x^n \frac{dx}{\beta-\alpha} = \frac{\beta^{n+1}-\alpha^{n+1}}{(n+1)(\beta-\alpha)}$.

**3.64** (i) $P(X = x) = \frac{\binom{w}{x}\binom{m}{r-x}}{\binom{m+w}{r}}, \quad x = 0, 1, \ldots, r, EX = \frac{wr}{m+w}$. (ii) $EX = 1.2$,

$P(X = x) = \frac{\binom{4}{x}\binom{6}{3-x}}{\binom{10}{3}}, \quad x = 0, 1, 2, 3. P(X = 0) \simeq 0.167, P(X = 1) = 0.5$,

$P(X = 2) = 0.3, P(X = 3) \simeq 0.033$.

**3.66** (i) $r = (m+n)/2$. (ii) $EX = m/2, \sigma^2(X) = \frac{mn}{4(m+m-1)}$.

**3.68** The relevant r.v. has the Negative Binomial distribution with $p = 1/2$ and $r = 10$. Then the answers are: (i) 20, and $2\sqrt{5} \simeq 4.472$, respectively. (ii) $P(X > 15) \simeq 0.849$.

**3.70** (i) $\lambda$ and $\lambda/n$, respectively. (ii) $n$ is the smallest integer $\geq \lambda/(1-p)c^2$. (iii) $n = 320$.

**3.72** (i) $a = \int_a^b af(x)dx \leq \int_a^b xf(x)dx = EX \leq \int_a^b bf(x)dx = b\int_a^b f(x)dx = b$ for the continuous case. (ii) Since (by Exercise 1.14(ii)) $\sigma^2(X) \leq E(X-c)^2$ for any $c$, we have $\sigma^2(X) \leq E\left(X - \frac{a+b}{2}\right)^2 \leq \frac{(b-a)^2}{4}$.

**3.74** (i) $x = 26$ min and 34.8 sec. (ii) Depart on 9 hour 33 min and 25.2 sec.

**3.76** (i) No. (ii) Yes. (iii) Yes. (iv) $p_0 = 1/18, p_1 = 16/18, p_2 = 1/18$.

**3.78** By simple manipulations.

**3.80** (i) $E[X(X-1)(X-2)] = \lambda^3 e^{-\lambda}e^\lambda = \lambda^3$, or $E(X^3 - 3X^2 + 2X) = \lambda^3$, and hence $EX^3 = \lambda^3 + 3\lambda^2 + \lambda$, since $EX = \sigma^2(X) = \lambda$, so that $EX^2 = \lambda + \lambda^2$. Then $E(X-\lambda)^3 = \lambda$ follows by expanding $(X-\lambda)^3$ and taking expectations. (ii) Immediate from part (i).

**3.82** (i) $P(X > Y) = \sum_{y=0}^{n} P(X > y)P(Y = y) = 1 - \sum_{y=0}^{n} P(X \leq y)P(Y = y)$.
(ii) $P(X > Y) \simeq 0.529$.

**3.84** (i) For $\alpha = \beta = 1, f(x) = 1, 0 < x < 1$. (ii) By differentiation,
$f(x) = 6x(1 - x), 0 < x < 1$, is maximized at $x = 1/2$. (iii) For $\alpha, \beta > 1$, the
$f(x)$ is maximized at $x = \frac{\alpha-1}{\alpha+\beta-2}$.

**3.86** It follows by simple manipulations.

**3.88** $P(X > \lambda) = \frac{1}{\Gamma(n)} \int_{\lambda}^{\infty} x^{n-1} e^{-x} dx = \frac{e^{-\lambda}}{\Gamma(n)} \sum_{x=0}^{n-1} \lambda^x \binom{n-1}{x} \Gamma(n - x) = $
$\frac{e^{-\lambda}}{\Gamma(n)} \sum_{x=0}^{n-1} \lambda^x \frac{(n-1)!}{x!(n-x-1)!} \times (n - x - 1)! = \frac{(n-1)!}{\Gamma(n)} \sum_{x=0}^{n-1} e^{-\lambda} \frac{\lambda^x}{x!} = \sum_{x=0}^{n-1} e^{-\lambda} \frac{\lambda^x}{x!}$.

**3.90** (i) $EX^n = \int_0^{\infty} x^n \times \lambda e^{-\lambda x} dx = \frac{1}{\lambda^n} \int_0^{\infty} y^n e^{-y} dy = \frac{1}{\lambda^n} \int_0^{\infty} y^{(n+1)-1} e^{-y} dy = $
$\frac{\Gamma(n+1)}{\lambda^n} = \frac{n!}{\lambda^n}$. (ii) Follow from part (i). (iii) Expanding $(X - \mu)^3$, taking
expectations and replacing the values from part (i), we get
$E(X - \mu)^3 = 2/\lambda^3$. This together with $\sigma^2(X) = 1/\lambda^2$, give $\gamma_1 = 2$.

**3.92** (i) It follows by a series of manipulations. (ii) The $E[X(X - 1)]$ also follows
by a series of manipulations, and the $EX^2$ is computed from
$EX^2 = E[X(X - 1)] + EX]$. (iii) It follows from part (i) and (ii).

## Section 3.4

**4.2** (i) $x_p = [(n + 1)p]^{1/(n+1)}$; (ii) For $p = 0.5$ and $n = 3$: $x_{0.5} = 2^{1/4} \simeq 1.189$. ■

**4.4** (i) $c_1 = c_2 = 1$; (ii) $x_{1/3} = 0$. ■

**4.6** (i)

| $x$ | 2 | 3 | 4 | 5 | 6 | 7 | 8 | 9 | 10 | 11 | 12 |
|---|---|---|---|---|---|---|---|---|---|---|---|
| $f(x)$ | 1/36 | 2/36 | 3/36 | 4/36 | 5/36 | 6/36 | 5/36 | 4/36 | 3/36 | 2/36 | 1/36 |

(ii) $EX = 7$; (iii) median = mode = mean = 7. ■

**4.8** Mode = 25 and $f(25) = \binom{100}{25}(\frac{1}{4})^{25}(\frac{3}{4})^{75}$; one would bet on $X = 25$. ■

**4.10** By the hint, $P(X \leq c) = \int_{-\infty}^{c} f(x) dx = \int_0^{\infty} f(c - y) dy$, and
$P(X \geq c) = \int_c^{\infty} f(x) dx = \int_0^{\infty} f(c + y) dy$. Since $f(c - y) = f(c + y)$, it
follows that $P(X \leq c) = P(X \geq c)$, and hence $c$ is the median. ■

**4.12** (i) $p = P(Y \leq y_p) = P[g(X) \leq y_p] = P[X \leq g^{-1}(y_p)]$, so that
$g^{-1}(y_p) = x_p$ and $y_p = g(x_p)$.
(ii) $x_p = -\log(1 - p)$.
(iii) $y_p = 1/(1 - p)$.
(iv) $x_{0.5} = -\log(0.5) \simeq 0.693$, and $y_{0.5} = 2$. ■

**4.14** (i) $c = (n + 1)^{1/(n+1)}$. (ii) $x_p = [(n + 1)p]^{1/(n+1)}$. (iii) $c = 4^{0.25} \simeq 1.414$,
$x_{0.50} = 2^{0.25} \simeq 1.189$.

## CHAPTER 4

### Section 4.1

**1.2**   $P(X = 0, Y = 1) = P(X = 0, Y = 2) = P(X = 1, Y = 2) = 0,$
$P(X = 0, Y = 0) = 0.3, P(X = 1, Y = 0) = 0.2, P(X = 1, Y = 1) = 0.2,$
$P(X = 2, Y = 0) = 0.075, P(X = 2, Y = 1) = 0.15, P(X = 2, Y = 2) = 0.075.$ ■

**1.4**   (i) $\int_0^2 \int_0^1 (x^2 + \frac{xy}{2}) \, dx \, dy = \frac{6}{7} \times \frac{7}{6} = 1$; (ii) $P(X > Y) = \frac{15}{56} \simeq 0.268.$ ■

**1.6**   (i) $P(X \leq x) = 1 - e^{-x}, x > 0$; (ii) $P(Y \leq y) = 1 - e^{-y}, y > 0$;
(iii) $P(X < Y) = 0.5$; (iv) $P(X + Y < 3) = 1 - 4e^{-3} \simeq 0.801.$ ■

**1.8**   $c = 1/\sqrt{2\pi}.$ ■

**1.10**   $c = 6/7.$ ■

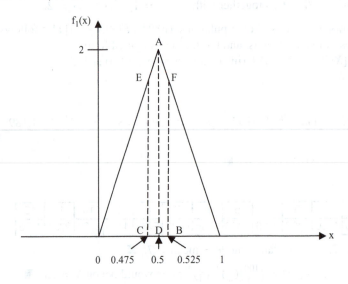

**1.12**   From the picture and symmetry the required area is $\pi/4$. Alternatively, one can integrate with respective to $y$ from 0 to $\sqrt{1 - x^2}$ and with respect to $x$ from 0 to 1.

### Section 4.2

**2.2**   $f_X(0) = 0.3, f_X(1) = 0.4, f_X(2) = 0.3$;
$f_Y(0) = 0.575, f_Y(1) = 0.35, f_Y(2) = 0.075.$ ■

**2.4**   (i) $f_X(1) = 7/36, f_X(2) = 17/36, f_X(3) = 12/36$;
$f_Y(1) = 7/36, f_Y(2) = 14/36, f_Y(3) = 15/36.$

(ii) $f_{X|Y}(1|1) = 2/7, f_{X|Y}(2|1) = 2/7, f_{X|Y}(3|1) = 3/7;$

$\quad f_{X|Y}(1|2) = 1/14, f_{X|Y}(2|2) = 10/14, f_{X|Y}(3|2) = 3/14;$

$\quad f_{X|Y}(1|3) = 4/15, f_{X|Y}(2|3) = 5/15, f_{X|Y}(3|3) = 6/15;$

$\quad f_{Y|X}(1|1) = 2/7, f_{Y|X}(2|1) = 2/7, f_{Y|X}(3|1) = 3/7;$

$\quad f_{Y|X}(1|2) = 2/17, f_{Y|X}(2|2) = 10/17, f_{Y|X}(3|2) = 5/17;$

$\quad f_{Y|X}(1|3) = 3/12, f_{Y|X}(2|3) = 3/12, f_{Y|X}(3|3) = 6/12.$ ■

**2.6** (i) $f_X(x) = \frac{2x}{n(n+1)}, \quad x = 1, \ldots, n; f_Y(y) = \frac{2(n-y+1)}{n(n+1)}, \ y = 1, \ldots, n.$

(ii) $f_{X|Y}(x|y) = \frac{1}{n-y+1}, \quad x = 1, \ldots, n; f_{Y|X}(y|x) = \frac{1}{x}, \quad y = 1, \ldots, x;$
$\qquad\qquad\qquad\qquad\qquad y = 1, \ldots, x \qquad\qquad\qquad\qquad x = 1, \ldots, n.$

(iii) $E(X|Y = y) = \frac{n(n+1)-(y-1)y}{2(n-y+1)}, y = 1, \ldots, n;$

$\qquad E(Y|X = x) = \frac{x+1}{2}, x = 1, \ldots, n.$ ■

**2.8** $f_X(x) = \frac{2}{5}(3x + 1), 0 \leq x \leq 1; f_Y(y) = \frac{3}{5}(2y^2 + 1), 0 \leq y \leq 1.$ ■

**2.10** (i) $f_X(x) = xe^{-x}, x > 0; f_Y(y) = e^{-y}, y > 0.$
(ii) $f_{Y|X}(y|x) = e^{-y}, x > 0, y > 0.$
(iii) $P(X > \log 4) = \frac{1+\log 4}{4} \simeq 0.597.$ ■

**2.12** (i) $f_X(x) = \frac{6x}{7} \times (2x + 1), 0 < x \leq 1; f_Y(y) = \frac{3y}{14} + \frac{2}{7}, 0 \leq y \leq 2;$

$\qquad f_{Y|X}(y|x) = \frac{2x+y}{4x+2}, 0 < x \leq 1, 0 \leq y \leq 2.$

(ii) $EY = \frac{8}{7}; E(Y|X = x) = \frac{2}{3} \times \frac{3x+2}{2x+1}, 0 < x \leq 1.$

(iii) It follows by a direct integration.

(iv) $P(Y > \frac{1}{2}|X < \frac{1}{2}) = \frac{207}{280} \simeq 0.739.$ ■

**2.14** $f_{X|Y}(x|y) = \frac{1}{2}ye^{\frac{y^2}{2}}e^{-\frac{y}{2}x}, 0 < y < x.$ ■

**2.16** (i)

$$f_X(x) = \begin{cases} 6x/7, & 0 < x \leq 1, \\ 6x(2-x)/7, & 1 < x < 2, \\ 0, & \text{elsewhere.} \end{cases}$$

(ii) $f_{Y|X}(y|x)$ is 1 for $0 < x \leq 1$, and is $1/(2 - x)$ for $1 < x < 2$ (and 0 otherwise), whereas $1 \leq x + y < 2.$ ■

**2.18** (i) $f_{X|Y}(\cdot|y)$ is the Poisson p.d.f. with parameter $y$.
(ii) $f_{X,Y}(x, y) = e^{-2y}\frac{y^x}{x!}, x = 0, 1, \ldots.$
(iii) $f_X(x) = \frac{1}{2^{x+1}}, x = 0, 1, \ldots.$ ■

**2.20** By applying the definitions, (i) and (ii) follow. ∎

**2.22** For the continuous case, $E[g(X)h(Y)|X = x] = \int_{-\infty}^{\infty} g(x)h(y)f_{Y|X}(y|x)dy = g(x)\int_{-\infty}^{\infty} h(y)f_{Y|X}(y|x)dy = g(x)E[g(Y)|X = x]$.

**2.24** (i) $f_X(x) = 4x(1 - x)^2$. (ii) $f_{Y|X}(y|x) = \frac{2y}{1-x^2}, 0 < x < y < 1$.
(iii) $E(Y|X = x) = \frac{2(1-x^3)}{3(1-x^2)}, 0 < x < 1$. (iv) $E(Y|X = 1/2) = \frac{7}{9}$.

**2.28** (i) $c = 10$. (ii) $f_X(x) = \frac{10}{3}x(1 - x^3), 0 < x < 1; f_Y(y) = 5y^4, 0 < y < 1$.
(iii) $f_{X|Y}(x|y) = \frac{2x}{y^2}, 0 < x < y < 1; f_{Y|X}(y|x) = \frac{3y^2}{1-x^3}, 0 < x < y < 1$.
(iv) $f_{X|Y}(x|y = 1/2) = 8x, 0 < x < \frac{1}{2}; f_{Y|X}(y|x) = \frac{81}{26}y^2, 1/3 < y < 1$.

**2.30** (i) $c = 9$. (ii) $f_X(x) = -9x^2 \log x, 0 < x < 1; f_Y(y) = 3y^2, 0 < y < 1$.
(iii) $f_{X|Y}(x|y) = \frac{3x^2}{y^3}, 0 < x < y < 1; f_{Y|X}(y|x) = \frac{-1}{y \log x}, 0 < x < y < 1$.

**2.32** (i) $c = 18$. (ii) $f_X(x) = 6x(1 - x), 0 < x < 1; f_Y(y) = 3y^2, 0 < y < 1$.
(iii) $EX = 1/2, EY = 3/4$. (iv) $P(X < Y) = 0.8$.

**2.34** $P(X = x) = \sum_{y=0}^{\infty} P(X = x|Y = y)P(Y = y) = \sum_{y=0}^{\infty} \binom{y}{x}p^x q^{y-x} \times e^{-\lambda} \frac{\lambda^y}{y!} = \frac{e^{-\lambda}(\lambda p)^x}{x!} \sum_{k=0}^{\infty} \frac{(\lambda q)^k}{k!} = e^{-\lambda p} \frac{(\lambda p)^x}{x!}$.

**2.36** (i) $EX = 5/9, \sigma^2(X) = 55/1,134$. (ii) $EY = 5/6, \sigma^2(Y) = 5/252$.
(iii) $E(X|Y = y) = 2y/3, 0 < y < 1; E(Y|X = x) = 3(1 - x^4)/4(1 - x^3)$,
$0 < x < 1; E(Y|X = x) = \frac{3(1-x^4)}{4(1-x^3)}, 0 < x < 1; E(X|Y = 1/3) = 2/9$;
$E(Y|X = 1/2) = 45/56$.
(iv) $E[E(X|Y)] = E\left(\frac{2Y}{3}\right) = \frac{2}{3}\int_0^1 y \times 5y^4 dy = \frac{5}{9} = EX$;
$E[E(Y|X)] = E\left[\frac{3(1-X^4)}{4(1-X^3)}\right] = \frac{3}{4}\int_0^1 \frac{1-x^4}{1-x^3} \times \frac{10}{3}x(1 - x^3)dx = \frac{5}{6} = EY$.

**2.38** (i) $c = 90$. (ii) $f_X(x) = 30x(1 - x)^4, 0 < x < 1; f_Y(y) = 15y^2(1 - 3y^2 + 2y^3)$,
$0 < y < 1$. (iii) $f_{X|Y}(x|y) = \frac{6x(1-x)}{1-3y^2+2y^3}, x > 0, y > 0, (0 <)x + y < 1$;
$f_{Y|X}(y|x) = \frac{3y^2}{(1-x)^3}, x > 0, y > 0, (0 <)x + y < 1$. (iv) $EX = 2/7$,
$EY = 15/28$. (v) $E(X|Y = y) = \frac{(1-y)(1+3y)}{1+2y}, x > 0, y > 0, (0 <)x + y < 1$;
$E(X|Y = 1/2) = 5/8$.

## Section 4.3

**3.2** It follows by an application of the definition and properties of a m.g.f. ∎

**3.4** Apply the exercise cited in the hint with $Z = X - Y$ and $Z = X + Y$. ∎

**3.6** (i) $EX = 1, EY = 0.5, EX^2 = 1.6, EY^2 = 0.65$, so that $\text{Var}(X) = 0.6$ and $\text{Var}(Y) = 0.4$.

(ii)  $E(XY) = 0.8$, so that $\text{Cov}(X, Y) = 0.3$, and
$\rho(X, Y) = 1.25\sqrt{0.24} \simeq 0.612$.

(iii)  The r.v.'s $X$ and $Y$ are positively correlated.  ∎

**3.8**  (i)  $EX = \frac{77}{36}, EY = \frac{20}{9}, EX^2 = \frac{183}{36}, EY^2 = \frac{99}{18}$, so that $\text{Var}(X) = 659/36^2$
and $\text{Var}(Y) = 728/36^2$.

(ii)  $E(XY) = \frac{171}{36}$, so that $\text{Cov}(X, Y) = -\frac{4}{36^2}$, and
$\rho(X, Y) = -\frac{2}{\sqrt{182 \times 659}} \simeq -0.006$.  ∎

**3.10**  $EX = 0$, $\text{Var}(X) = 10/4$, $EY = 5/2, EY^2 = 34/4$, $\text{Var}(Y) = 9/4$, $E(XY) = 0$,
so that $\text{Cov}(X, Y) = 0$, and $\rho(X, Y) = 0$.  ∎

**3.12**  (i)  $EX = EY = 7/12$.

(ii)  $EX^2 = EY^2 = 5/12$, so that $\text{Var}(X) = \text{Var}(Y) = 11/144$.

(iii)  $E(XY) = \frac{1}{3}$, so that $\text{Cov}(X, Y) = -\frac{1}{144}$, and $\rho(X, Y) = -\frac{1}{11} \simeq -0.091$.

(iv)  $X$ and $Y$ are Negatively correlated.  ∎

**3.14**  With $\text{Var}(X) = \sigma^2$, we get $\text{Cov}(X, Y) = a\sigma^2$, and $\rho(X, Y) = \frac{a}{|a|}$.
Thus, $|\rho(X, Y)| = 1$, and $\rho(X, Y) = 1$ if and only if $a > 0$, and $\rho(X, Y) = -1$
if and only if $a < 0$.  ∎

**3.16**  By differentiation, with respect to $\alpha$ and $\beta$, of the function $g(\alpha, \beta) = E[Y - (\alpha X + \beta)]^2$, and by equating the derivatives to 0, we find:
$\hat{\alpha} = \frac{\sigma_Y}{\sigma_X}\rho(X, Y), \hat{\beta} = EY - \hat{\alpha}EX$. The $2 \times 2$ matrix $M$ of the second-order

derivatives is given by $M = 4\begin{pmatrix} EX^2 & EX \\ EX & 1 \end{pmatrix}$, which is positive definite. Then

$\hat{\alpha}$ and $\hat{\beta}$ are minimizing values.  ∎

**3.18**  (i) $EX = 0$. (ii) Values: $-8, -1, 1, 8$, each taken with probability $1/4$.
(iii) $\text{Cov}(X, Y) = 0$. (v) No! they are not *linearly* related.

**3.20**  (i) $f_X(100) = f_X(250) = 0.50; f_Y(0) = f_Y(100) = 0.25, f_Y(200) = 0.50$.
(ii) $EX = 175, EX^2 = 36250, \sigma^2(X) = 5625$. (iii) $EY = 125, EY^2 = 22500$,
$\sigma^2(Y) = 6875$. (iv) $E(XY) = 23750, \text{Cov}(X, Y) = 1875$,
$\rho(X, Y) = \sqrt{11}/11 \simeq 0.302$.

**3.22**  (i) $EXY = \frac{3n^2 + 7n + 2}{12}$. (ii) $\text{Cov}(X, Y) = \frac{n^2 + n - 2}{36}, \rho(X, Y) = 1/2$.

**3.24**  (i) $E(XY) = 10/21$. (ii) $\text{Cov}(X, Y) = 5/378$. (iii) $\rho(X, Y) = \frac{\sqrt{99,792}}{189\sqrt{11}} \simeq 0.504$.
(iv) $E(X/Y) = 2/3, E(X/Y^2) = 5/6, E(Y/X) = 2$.

**3.26**  (iii) $\sigma^2(X) = 11/225, \sigma^2(Y) = 6/225$. (iv) $E(XY) = 4/9$.
(v) $\text{Cov}(X, Y) = 4/225$. (vi)

$\rho(X, Y) = 2\sqrt{66}/33 \simeq 0.492$.

**3.28**  (i) $0.3024$. (ii) $P(|X - Y| < 2) = 4,267/11,875 \simeq 0.359$.

## Section 4.4

**4.2**  $M_{X_1, X_2, X_3}(t_1, t_2, t_3) = c^3/(c - t_1)(c - t_2)(c - t_2)$, provided $t_1, t_2, t_3$ are $<c$. ∎

**4.4**  Follows by applying properties of expectations. ∎

## Section 4.5

**5.2**  If $X_1, X_2$, and $X_3$ are the numbers of customers buying brand $A$, brand $B$, or just browsing, then $X_1, X_2, X_3$ have the Multinomial distribution with parameters $n = 10, p_1 = 0.25, p_2 = 0.40$, and $p_3 = 0.35$. Therefore

(i)  $P(X_1 = 2, X_2 = 3, X_3 = 5) = \frac{10!}{2!3!5!}(0.25)^2 \times (0.40)^3 \times (0.35)^5 \simeq 0.053$.

(ii)  $P(X_1 = 1, X_2 = 3 | X_3 = 6) = \frac{4!}{1!3!}(\frac{5}{13})^1(\frac{8}{13})^3 \simeq 0.359$. ∎

**5.4**  They follow by taking the appropriate derivatives and evaluating them at 0. ∎

**5.6**  The second line in (51) follows from the first line by adding and subtracting the quantity $\rho^2(\frac{x_1 - \mu_1}{\sigma_1})^2$. The expression in the following line follows by the fact that the first three terms on the previous line form a perfect square. What follows is obvious and results from a suitable regrouping of the entities involved. ∎

**5.8**  In Exercise 5.7, it was found that $E(XY) = \mu_1 \mu_2 + \rho \sigma_1 \sigma_2$, where $\mu_1 = EX, \mu_2 = EY$, $\sigma_1 = $ s.d. of $X$ and $\sigma_2 = $ s.d. of $Y$. Then

$$\rho(X, Y) = \frac{\text{Cov}(X, Y)}{\sigma_1 \sigma_2} = \frac{(\mu_1 \mu_2 + \rho \sigma_1 \sigma_2) - \mu_1 \mu_2}{\sigma_1 \sigma_2} = \rho.$$ ∎

**5.10**  They follow by differentiating the m.g.f. and evaluating the derivatives at 0.
∎

**5.12**  (i)  It is not a Bivariate Normal p.d.f., because, for $x$ and $y$ outside the interval $[-1, 1]$, the given p.d.f. becomes $f(x, y) = \frac{1}{2\pi}\exp[-(x^2 + y^2)/2]$, which is the Bivariate Normal with $\mu_1 = \mu_2 = 0, \sigma_1 = \sigma_2 = 1$, and $\rho = 0$, whereas $f(-1, -1) = \frac{1}{\pi e} \neq \frac{1}{2\pi e}$, the value of the Bivariate Normal just mentioned evaluated at $x = y = -1$. The function given integrates to 1 on $\Re^2$, as is seen by splitting the integration over $I = [-1, 1] \times [-1, 1]$ and its complements $I^c$, so that it is a p.d.f.

(ii)  $f_2(y) = \frac{1}{\sqrt{2\pi}}e^{-y^2/2}$, which is the p.d.f. of the $N(0, 1)$ distribution. Similarly, $f_1(x) = \frac{1}{\sqrt{2\pi}}e^{-x^2/2}$. ∎

**5.14**  $P(Y > 92 | X = 84) = 0.456205$; $P(X > Y) = 0.119$; $P(X + Y > 180) = 0.326355$. ∎

**5.16**  $E(Y | X = 69) = 69.4$; conditional s.d. of $Y | X = 69$ is 0.6; $P(Y > 70 | X = 69) = 0.158655$; $P(Y > X) = 0.773373$. ∎

**5.18**  Setting $u = \frac{x-\mu_1}{\sigma_1}$, $v = \frac{y-\mu_2}{\sigma_2}$, the function $f(x, y)$ is maximized if and only if $g(u, v) = u^2 - 2\rho uv + v^2$ is minimized. The partial derivatives with respect to $u$ and $v$, when equated to 0, give $u = v = 0$, and the determinant of the second order derivatives is, essentially, $\begin{pmatrix} 1 & -\rho \\ -\rho & 1 \end{pmatrix}$, which is positive definite.  ∎

**5.20**  Setting $f(x, y) = c$, some constant, or
$\frac{1}{\sigma_1^2}(x - \mu_1)^2 - 2\frac{\rho}{\sigma_1\sigma_2}(x - \mu_1)(y - \mu_2) + \frac{1}{\sigma_2^2}(y - \mu_2)^2 = c_0$, some other constant, the equation represents an ellipse if and only if $\rho^2 < 1$. It represents a circle if and only if $\rho = 0$.  ∎

**5.22**  $E(L - S) = 3.2$, $\sigma^2(L - S) = 2.56$, $\sigma^2(L - S) = 1.6$, and $P(S > L) = 0.02275$.  ∎

# CHAPTER 5
## Section 5.1

**1.2**  The relation $f_{X,Y}(x, y) = f_X(x)f_Y(y)$ holds true for all values of $x$ and $y$, and therefore $X$ and $Y$ are independent.  ∎

**1.4**  The r.v.'s $X$ and $Y$ are not independent, since, for example, $f_{X,Y}(0.1, 0.1) = 0.132 \neq 0.31824 = 0.52 \times 0.612 = f_X(0.1)f_Y(0.1)$.  ∎

**1.6**  (i)  $f_X(x) = \frac{6}{5}(x^2 + \frac{1}{2}), 0 \le x \le 1$; $f_Y(y) = \frac{6}{5}(y + \frac{1}{3}), 0 \le y \le 1$.

 (ii)  The r.v.'s are not independent, since, for example,
$f_{X,Y}(\frac{1}{2}, \frac{1}{4}) = \frac{3}{5} \neq \frac{9}{10} \times \frac{7}{10} = f_X(\frac{1}{2})f_Y(\frac{1}{4})$.  ∎

**1.8**  (i)  $f_X(x) = 2x, 0 < x < 1$; $f_Y(y) = 2y, 0 < y < 1$; $f_Z(z) = 2z, 0 < z < 1$.

 (ii)  The r.v.'s are independent because, clearly,

$$f_{X,Y,Z}(x, y, z) = f_X(x)f_Y(y)f_Z(z),$$

 (iii)  $P(X < Y < Z) = 1/6$.  ∎

**1.10**  (i)  $c$ can be any positive constant.
 (ii)  $f_{X,Y}(x, y) = c^2e^{-cx-cy}, x > 0, y > 0$, and likewise for $f_{X,Z}$ and $f_{Y,Z}$.
 (iii)  $f_X(x) = ce^{-cx}, x > 0$, and likewise for $f_Y$ and $f_Z$.
 (iv)  The r.v.'s $X$ and $Y$ are independent, and likewise for the r.v.'s $X, Z$ and $Y, Z$. Finally, from part (iii), it follows that the r.v.'s $X, Y$, and $Z$ are also independent.  ∎

**1.12**  (i)  $EX = 200$ days; (ii) $M_{X+Y}(t) = 1/(1 - 200t)^2, t < 0.005$, and $f_{X+Y}(t) = (0.005)^2te^{-0.005t}, t > 0$.

 (iii)  $P(X + Y \ge 500) = 2.5e^{-2.5} + e^{-2.5} \simeq 0.287$.  ∎

**1.14** (i) $M_U(t) = \exp[(a\mu_1 + b)t + \frac{(a\sigma_1)^2 t^2}{2}]$ which is the m.g.f. of the
$N(a\mu_1 + b, (a\sigma_1)^2)$ distribution. Likewise for $V$.

(ii) $M_{U,V}(t_1, t_2) = \exp[(a\mu_1 + b)t_1 + \frac{(a\sigma_1)^2 t_1^2}{2} + (c\mu_2 + d)t_2 + \frac{(c\sigma_2)^2 t_2^2}{2}]$.

(iii) Follows from parts (i) and (ii), since $M_U(t_1)M_V(t_2) = M_{U,V}(t_1, t_2)$ for all $t_1, t_2$. ∎

**1.16** $M_{\bar{X}}(t) = [M(\frac{t}{n})]^n$, ∎

**1.18** (i) $E\bar{X} = p$ and $\text{Var}(\bar{X}) = pq/n$; (ii) $n = 10,000$. ∎

**1.20** (i) $f_X(-1) = 2\alpha + \beta, f_X(0) = 2\beta, f_X(1) = 2\alpha + \beta$;
$f_Y(-1) = 2\alpha + \beta, f_Y(0) = 2\beta, f_Y(1) = 2\alpha + \beta$.

(ii) $EX = EY = 0$, and $E(XY) = 0$; (iii) $\text{Cov}(X, Y) = 0$.

(iv) The r.v.'s are not independent, since, for example,
$f(0,0) = 0 \neq (2\beta) \times (2\beta) = f_X(0)f_Y(0)$. ∎

**1.22** (i) $E\bar{X} = \mu$ and $\text{Var}(\bar{X}) = \sigma^2/n$.

(ii) For $k = 1, n = 100$; for $k = 2, n = 25$; and for $k = 3, n = 12$. ∎

**1.24** (i) $E\bar{X} = \mu$ and $\text{Var}(\bar{X}) = \sigma^2/n$.

(ii) The smallest $n$ that is $\geq 1/(1 - \alpha)c^2$.

(iii) For $c = 0.1$, the required $n$ is its smallest value $\geq 100/(1 - \alpha)$.
For $\alpha = 0.90, n = 1000$; for $\alpha = 0.95, n = 2000$; for $\alpha = 0.99, n = 10,000$. ∎

**1.26** Theorem 1(ii) implies that $M_{Y_1,\ldots,Y_k}(t_1,\ldots,t_k) = M_{Y_1}(t_1)\ldots M_{Y_k}(t_k)$, so that, by Theorem 1(iii), $Y_1, \cdots, Y_k$ are independent. ∎

**1.28** They are not independent, since e.g., $f_{X,Y}(1/3, 1/4) = 7/24 \neq \frac{245}{3888} = f_X(1/3)f_Y(1/4)$.

**1.30** $f_{X,Y}(x, y) == \frac{1}{2\pi\sigma_1\sigma_2}e^{-\frac{1}{2}\left[\left(\frac{x-\mu_1}{\sigma_1}\right)^2 + \left(\frac{y-\mu_2}{\sigma_2}\right)^2\right]}$; $\mu_1, \mu_2, \sigma_1^2, \sigma_2^2, \rho = 0$.

**1.32** (i) $f_X(x) = \frac{2\sqrt{1-x^2}}{\pi}, -1 \leq x \leq 1; f_Y(y) = \frac{2\sqrt{1-y^2}}{\pi}, -1 \leq y \leq 1$. (ii) No, since $f_X(x)f_Y(y) = \frac{4}{\pi^2}\sqrt{(1-x^2)(1-y^2)} \neq \frac{1}{\pi}$.

**1.34** (ii) No, since, e.g., $f_{X,Y}(1/2, 1/2) = (1.5)^2 \neq 6 = f_X(1/2)f_Y(1/2)$.

(iii) $f_{X|Y}(x|y) = \frac{2x}{(1-y)^2}, 0 \leq x < 1, 0 \leq y < 1, 0 \leq x + y < 1$.

(iv) $E(X|Y = y) = \frac{2}{3}(1 - y), 0 \leq y < 1$, and $E(X|Y = 1/2) = 1/3$.

**1.36** (i) $P(39 \leq X \leq 42) =\simeq 0.658448$. (ii) $P$(at least one of $X_1, X_2, X_3, X_4 \geq 42) = 1 - P$(all are $< 42) \simeq 1 - 0.685 = 0.315$.

**1.38** (i) $X + Y \sim P(\lambda_1 + \lambda_2)$. (ii) $E(X + Y) = \lambda_1 + \lambda_2, \sigma^2(X + Y) = \lambda_1 + \lambda_2$,
$\sigma(X + Y) = (\lambda_1 + \lambda_2)^{1/2}$. (iii) $E(X + Y) = 18$,
$\sigma(X + Y) = \sqrt{18} = 3\sqrt{2} \simeq 4.243$.

## Section 5.2

**2.2**  $X + Y \sim B(30, 1/6)$ and $P(X + Y \le 10) = \sum_{t=0}^{10} \binom{30}{t} (\frac{1}{6})^t (\frac{5}{6})^{30-t}$.  ∎

**2.4**  (i)  $S_n \sim B(n, p)$.
    (ii)  $EX_i = p$, $\mathrm{Var}(X_i) = pq$ ($q = 1 - p$).
    (iii)  $ES_n = np$, $\mathrm{Var}(S_n) = npq$.  ∎

**2.6**  Let $X$ be the r.v. denoting the breakdown voltage, then $X \sim N(40, 1.5^2)$, and therefore:
    (i)  $P(39 < X < 42) = 0.656812$; (ii) $= 0.382$.  ∎

**2.8**  (i)  $X_1 + \cdots + X_n \sim P(\lambda_1 + \cdots + \lambda_n)$.
    (ii)  $E\bar{X} = (\lambda_1 + \cdots + \lambda_n)/n$, $\mathrm{Var}(\bar{X}) = (\lambda_1 + \cdots + \lambda_n)/n^2$.
    (iii)  $E\bar{X} = \lambda$, $\mathrm{Var}(\bar{X}) = \lambda/n$.  ∎

**2.10**  (i)  $P(X_1 = x_1 | T = t) = \binom{t}{x_1}(\frac{\lambda_1}{\lambda})^{x_1} (1 - \frac{\lambda_1}{\lambda})^{t-x_1}$, so the
        $X_1 | T = t \sim B(t, \frac{\lambda_1}{\lambda})$, and likewise for the other r.v.'s.
    (ii)  Here $\lambda = nc$, and therefore $X_1 | T = t \sim B(t, \frac{1}{n})$.  ∎

**2.12**  (i)  $P(\bar{X} > \bar{Y}) = 1 - \Phi\left( - \dfrac{\mu_1 - \mu_2}{\sqrt{\frac{\sigma_1^2}{m} + \frac{\sigma_2^2}{n}}} \right)$;

    (ii)  $P(\bar{X} > \bar{Y}) = 0.5$.  ∎

**2.14**  (i)  For $r > 0$, $P(R \le r) = P(U \le r^2/\sigma^2)$, $U \sim \chi_2^2$.
    (ii)  For $\sigma = 1$ and the given values of $r$, the respective probabilities are
        $0.75, 0.90, 0.95, 0.975, 0.99$, and $0.995$.  ∎

**2.16**  Here $\frac{X+Y}{2} \sim N(\mu, \frac{3\sigma^2}{4})$, so that $P(|\frac{X+Y}{2}| \le 1.5\sigma) = 2\Phi(\sqrt{3}) - 1 = 0.91637$.
    ∎

---

# CHAPTER 6

## Section 6.1

**1.2**  (i)  $X \sim N(\frac{5\mu - 160}{9}, \frac{25\sigma^2}{81})$; (ii) $a \simeq 32.222, b = 35$.
    (iii)  $a_k = \frac{5\mu - 160}{9} - k\frac{5\sigma}{9}, b_k = \frac{5\mu - 160}{9} + k\frac{5\sigma}{9}$.  ∎

**1.4**  $f_Y(y) = \lambda y^{-(\lambda+1)}, y > 1; f_Z(z) = \lambda e^{z - \lambda e^z}, z \in \mathcal{R}$.  ∎

**1.6**  (i)  $f_Y(y) = \frac{1}{2}e^{-y/2}, y > 0$, which is the p.d.f. of a $\chi_2^2$.
    (ii)  $\sum_{i=1}^{n} Y_i \sim \chi_{2n}^2$, since $Y_i \sim \chi_2^2, i = 1, \ldots, n$ independent.  ∎

**1.8**  $f_Y(y) = \frac{1}{\Gamma(\frac{3}{2})m^{3/2}} y^{\frac{3}{2}-1} e^{-y/m}, y > 0$.  ∎

**1.10**  $y = \sqrt{x}, y^2 = x, \frac{dx}{dy} = 2y, y > 0. f_Y(y) = 2ye^{-y^2}, y > 0$.

## Section 6.2

**2.2** (i) $f_{U,V}(u,v) = \frac{u}{(1+v)^2}e^{-u}, u > 0, v > 0.$

(ii) $f_U(u) = u e^{-u}, u > 0; f_V(v) = 1/(1+v)^2, v > 0.$

(iii) $U$ and $V$ are independent. ∎

**2.4** (i) $f_{U,V}(u,v) = \frac{1}{|ac|}f_X(\frac{u-b}{a})f_Y(\frac{v-d}{c}), (u,v) \in T.$

(ii) $f_{U,V}(u,v) =$

$\frac{1}{\sqrt{2\pi}|a|\sigma_1}\exp\{-\frac{[u-(a\mu_1+b)]^2}{2(a\sigma_1)^2}\} \times \frac{1}{\sqrt{2\pi}|c|\sigma_2}\exp\{-\frac{[v-(c\mu_2+d)]^2}{2(c\sigma_2)^2}\}$, and

therefore $U$ and $V$ are independently distributed as $N(a\mu_1 + b, (a\sigma_1)^2)$ and $N(c\mu_2 + d, (c\sigma_2)^2)$, respectively. ∎

**2.6** (i) $f_{U,V}(u,v) = \frac{1}{\sqrt{2\pi}}e^{-u^2/2} \times \frac{1}{\sqrt{2\pi}}e^{-v^2/2}, u, v \in \Re.$

(ii) $U \sim N(0,1), V \sim N(0,1).$

(iii) $U$ and $V$ are independent.

(iv) By parts (ii) and (iii), $X + Y \sim N(0,2)$ and $X - Y \sim N(0,2).$ ∎

**2.8** $f_U(u) = 1, 0 \leq u \leq 1.$ ∎

**2.10** We have $f_{X_r}(t) = \frac{\Gamma\left(\frac{r+1}{2}\right)}{\sqrt{r}\Gamma\left(\frac{r}{2}\right)} \times \frac{1}{\sqrt{\pi}}\left(1 + \frac{t^2}{r}\right)^{-\frac{r+1}{2}}$ with the second factor on the

right-hand side $\xrightarrow[r\to\infty]{} e^{-t^2/2}/\sqrt{\pi}$. Next,

$$\frac{\Gamma\left(\frac{r+1}{2}\right)}{\sqrt{r}\Gamma\left(\frac{r}{2}\right)} = \frac{\Gamma\left(\frac{r+1}{2}\right)/\sqrt{2\pi}\left(\frac{r+1}{2}\right)^{r/2}e^{-\frac{r+1}{2}}}{\Gamma\left(\frac{r}{2}\right)/\sqrt{2\pi}\left(\frac{r}{2}\right)^{\frac{r-1}{2}}e^{-r/2}} \times \left[\left(1 + \frac{1}{r}\right)^r\right]^{1/2} \times \frac{1}{\sqrt{2}e^{1/2}}$$

$$\xrightarrow[r\to\infty]{} 1 \times e^{1/2}\frac{1}{\sqrt{2}e^{1/2}} = 1/\sqrt{2},$$

so that $f_{X_r}(t) \xrightarrow[r\to\infty]{} e^{-t^2/2}/\sqrt{2\pi}$. ∎

**2.12** $X \sim \chi_2^2, Y \sim \chi_2^2$ independent, so that $\frac{Y/2}{X/2} = W \sim F_{2,2}, U = \frac{1}{1+W}$. From

$u = \frac{1}{1+w}$, we get $w = \frac{1-u}{u}, \frac{dw}{du} = -\frac{1}{u^2}$, and since $f_W(w) = \frac{1}{(1+w)^2}, w > 0$, we

have $f_U(u) = u^2 \times \frac{1}{u^2} = 1, 0 < u < 1.$

**2.14** (i) From $u = x, v = xy, w = xyz$, we get: $x = u, y = v/u, z = w/v, J = 1/uv$,

$0 < w < v < u < 1$. Then, $f_{U,V,W}(u,v,w) = 8 \times u \times \frac{v}{u} \times \frac{w}{v} \times \frac{1}{uv} = \frac{8w}{uv}$,

$0 < w < v < u < 1$. (ii) $\int_0^1 \int_0^u \int_0^v \frac{8w}{uv}dwdvdu = 1.$

**2.16** (i) $F_X(x) = 1 - \frac{1}{1+x}, x > 0; F_Y(y) = 1 - \frac{1}{1+y}, y > 0; F_Z(z) = 1 - \frac{1}{1+z}$,

$z > 0; f_X(x) = \frac{1}{(1+x)^2}, x > 0; f_Y(y) = \frac{1}{(1+y)^2}, y > 0; f_Z(z) = \frac{1}{(1+z)^2}, z > 0.$

(ii) $F_{X,Y}(x,y) = \left(1 - \frac{1}{1+x}\right) \times \left(1 - \frac{1}{1+y}\right), x > 0, y > 0;$

$F_{X,Z}(x,z) = \left(1 - \frac{1}{1+x}\right) \times \left(1 - \frac{1}{1+z}\right), x > 0, z > 0;$

$$F_{Y,Z}(y,z) = \left(1 - \tfrac{1}{1+y}\right) \times \left(1 - \tfrac{1}{1+z}\right), y > 0, z > 0;$$

$$f_{X,Y}(x,y) = \tfrac{1}{(1+x)^2(1+y)^2}, x > 0, y > 0; f_{X,Z}(x,z) = \tfrac{1}{(1+x)^2(1+z)^2}, x > 0,$$

$$z > 0; f_{Y,Z}(y,z) = \tfrac{1}{(1+y)^2(1+z)^2}, y > 0, z > 0.$$

**(iii)** Clearly, $f_{X,Y}(x,y) = f_X(x)f_Y(x), f_{X,Z}(x,z) = f_X(x)f_Z(z),$
$f_{Y,Z}(y,z) = f_Y(y)f_Z(z)$, but $F(x,y,z) = F_X(x)F_Y(y)F_Z(z)$ if and only if
$x + y + z = \infty$ which is not true.

**(iv)** (a) 0.5; (b) 0.25; (c) 0.09375; (d) 0.5; (e) 0.25, 0.25; (f) 0.25; (g) 0.15625,
0.15625, 0.15625; (h) 0.09375, 0.09375, $5/48 \simeq 0.104$;
(i) $7/48 \simeq 0.146$.

# Section 6.3

**3.2**  It follows by forming the inner products of the row vectors. ■

**3.4**  It follows from the joint p.d.f. $f_{X,Y}$, the transformations $u = \tfrac{x-\mu_1}{\sigma_1}, v = \tfrac{y-\mu_2}{\sigma_2}$,
and the fact that the Jacobian $J = \sigma_1\sigma_2$. ■

**3.6**
  **(i)**  It follows from the joint p.d.f. $f_{X,Y}$, the transformations
  $u = x + y, v = x - y$, and the fact that the Jacobian $J = -1/2$.
  **(ii)**  $U$ and $V$ are independent by the fact that they have the Bivariate Normal
  distribution and their correlation coefficient is 0.
  **(iii)**  It follows from part (i) as marginals of the Bivariate Normal. ■

**3.8**  **(i)**  $P(a\mu < \bar{X} < b\mu, 0 < S^2 < c\sigma^2) = \{\Phi[k(b-1)\sqrt{n}] - \Phi[k(a-1)\sqrt{n}]\}$
  $\times P[\chi^2_{n-1} < c(n-1)]$.
  **(ii)**  The probability is 0.89757. ■

# Section 6.5

**5.2**  $EY_1 = \tfrac{1}{n+1}, EY_n = \tfrac{n}{n+1}$, and $EY_n \to 1$ as $n \to \infty$. ■

**5.4**  $E(Y_1 Y_n) = \tfrac{1}{n+2}$. Therefore, by Exercise 5.2, $\mathrm{Cov}(Y_1, Y_n) = \tfrac{1}{(n+1)^2(n+2)}$. ■

**5.6**  $f_Z(z) = \lambda e^{-\lambda z}, z > 0$. ■

**5.8**
  **(i)**  $g_n(y_n) = n\lambda e^{-\lambda y_n}(1 - e^{-\lambda y_n})^{n-1}, y_n > 0$.
  **(ii)**  For $n = 2, EY_2 = 3/2\lambda$, and for $n = 3, EY_3 = 11/6\lambda$. ■

**5.10**  $g_{1n}(y_1, y_n) = n(n-1)[F(y_n) - F(y_1)]^{n-2}f(y_1)f(y_n), a < y_1 < y_n < b.$ ■

**5.12**  (i) $EY_n = \tfrac{n}{n+1}, \sigma^2(Y_n) = \tfrac{n}{(n+1)^2(n+2)}.$

**5.14**  (i) $f_Y(y) = (n\alpha)\beta y^{\beta-1}e^{-(n\alpha)y^\beta}, y > 0$, Weibull with parameter $n\alpha$ and $\beta$.

(ii) $EY = \tfrac{\Gamma\left(\frac{1}{\beta}+1\right)}{(n\alpha)^{1/\beta}}, \sigma^2(Y) = \tfrac{\Gamma\left(\frac{2}{\beta}+1\right)-\left[\Gamma\left(\frac{1}{\beta}+1\right)\right]^2}{(n\alpha)^{2/\beta}}.$

**5.16** (i) $g(y_1, y_5) = 120 \int_{y_1}^{y_5} \int_{y_1}^{y_4} \int_{y_1}^{y_3} e^{-(y_1 + \ldots + y_5)} dy_2 dy_3 dy_4 = 20e^{-(y_1 + y_5)}(e^{-y_1} - e^{-y_5})^3, 0 < y_1 < y_5$. (ii) From $R = Y_5 - Y_1, S = Y_1$, we get $r = y_5 - y_1, s = y_1$, so that $y_1 = s, y_5 = r + s, r > 0, s > 0, |J| = 1$. Then, $f_{R,S}(r, s) = 20(e^{-r} - 3e^{-2r} + 3e^{-3r} - e^{-4r})e^{-5s}, r > 0, s > 0$. (iii) $f_R(r) = 20(e^{-r} - 3e^{-2r} + 3e^{-3r} - e^{-4r}) \int_0^\infty e^{-5s} ds = 4(e^{-r} - 3e^{-2r} + 3e^{-3r} - e^{-4r}), r > 0$. (iv) $ER = 25/12$.

# CHAPTER 7

## Section 7.1

**1.2** For every $\varepsilon > 0, P(|X_n| > \varepsilon) = P(X_n = 1) = p_n$, and therefore $X_n \overset{P}{\to} 0$ if and only if $p_n \to 0$ as $n \to \infty$. ∎

**1.4** (i) $P(|Y_{1,n}| > \varepsilon) = (1 - \varepsilon)^n \to 0$, as $n \to \infty$.

(ii) $P(|Y_{n,n} - 1| > \varepsilon) = 1 - P(|Y_{n,n} - 1| \leq \varepsilon)$ and $P(|Y_{n,n} - 1| \leq \varepsilon) = 1 - (1 - \varepsilon)^n \to 1$, so that $P(|Y_{n,n} - 1| > \varepsilon) \to 0$, as $n \to \infty$. ∎

**1.6** $E\bar{X}_n = \mu$ and $E(\bar{X}_n - \mu)^2 = \text{Var}(\bar{X}_n) = \frac{\sigma^2}{n} \to 0$, as $n \to \infty$. ∎

**1.8** $E(Y_n - X)^2 = E(Y_n - X_n)^2 + E(X_n - X)^2 + 2E[(Y_n - X_n)(X_n - X)] \to 0$, as $n \to \infty$, by the assumptions made and the fact that $|E[(Y_n - X_n)(X_n - X)]| \leq E^{1/2}|X_n - Y_n|^2 \times E^{1/2}|X_n - X|^2$. ∎

**1.10** $\sqrt{n} \sum_{i=1}^n X_i \overset{d}{\longrightarrow} Z \sim N(0, 1)$ by the CLT, and $\frac{1}{n} \sum_{i=1}^n X_i^2 \overset{P}{\longrightarrow} EX_i^2 = \sigma^2(X_i) = 1$ by the WLLN. Then their ratio tends in distribution (as $n \to \infty$) to $\frac{Z}{1} \sim N(0, 1)$. Similarly, $\sqrt{n} \sum_{i=1}^n X_i \overset{d}{\longrightarrow} Z \sim N(0, 1)$, and $\frac{1}{n} \sum_{i=1}^n X_i^2 \overset{P}{\longrightarrow} 1$, so that their ratio tends in distribution (as $n \to \infty$) to $Z \sim N(0, 1)$.

## Section 7.2

**2.2** (i) $M_X(t) = (1 - \alpha)/(1 - \alpha e^t), \quad t < -\log \alpha$.

(ii) $EX = \alpha/(1 - \alpha)$.

(iii) $M_{\bar{X}_n}(t) = \left(\frac{1-\alpha}{1-\alpha e^{t/n}}\right)^n = \{1 - \frac{\alpha t/(1-\alpha) + [\alpha/(1-\alpha)]nR(\frac{t}{n})}{n}\}^{-n} \underset{n \to \infty}{\longrightarrow} e^{\alpha t/(1-\alpha)}$, since $\frac{n}{t} R(\frac{t}{n}) \underset{n \to \infty}{\longrightarrow} 0$ for fixed $t$, and $e^{\alpha t/(1-\alpha)}$ is the m.g.f. of $\frac{\alpha}{1-\alpha}$. ∎

**2.4** Since $X \sim B(1000, p)$, we have

(i) $P(1000p - 50 \leq X \leq 1000p + 50) = \sum_{x=1000p-50}^{1000p+50} \binom{1000}{x} p^x q^{1000-x}$, $q = 1 - p$. For $p = \frac{1}{2}$ and $p = \frac{1}{4}$:

$$P(450 \le X \le 550) = \sum_{x=450}^{550} \binom{1000}{x} (0.5)^{1000},$$

$$P(200 \le X \le 300) = \sum_{x=200}^{300} \binom{1000}{x} (0.25)^x \times (0.75)^{1000-x}.$$

(ii)  For $p = \frac{1}{2}$ and $p = \frac{1}{4}$, the approximate probabilities are
$\Phi(3.16) + \Phi(3.23) - 1 = 0.99892$, and
$\Phi(3.65) + \Phi(3.72) - 1 = 0.999769$. ∎

**2.6**  $EX_i = \frac{7}{2}$, $EX_i^2 = \frac{91}{6}$, so that $\text{Var}(X_i) = \frac{35}{12}$. Therefore
$P(150 \le X \le 200) \simeq \Phi(3.65) + \Phi(3.72) - 1 = 0.999769$. ∎

**2.8**  Since $X \sim B(1000, 0.03)$, the required approximate probability is
$P(X \le 50) \simeq \Phi(3.71) = 0.999896$. ∎

**2.10**  $P(|\frac{X}{n} - 0.53| \le 0.02) \simeq 2\Phi(\frac{0.02\sqrt{n}}{\sqrt{0.2491}}) - 1 = 0.99$, so that $n = 4146$. ∎

**2.12**  With $S_n = \sum_{i=1}^{n} X_i$, we have:
(i)

$$P(S_n \le \lambda n) = P(-0.5 < S_n \le \lambda n) \simeq \Phi(0) - \Phi\left(-\frac{0.5 + \lambda n}{\sqrt{\lambda n}}\right)$$

$$= \Phi\left(\frac{0.5 + \lambda n}{\sqrt{\lambda n}}\right) - 0.50.$$

(ii)  If $\lambda n$ is not an integer, and $[\lambda n]$ is its integer part,

$$P(S_n \ge \lambda n) \simeq 1 + \Phi\left(\frac{\lambda n - [\lambda n]}{\sqrt{\lambda n}}\right) - \Phi\left(\frac{0.5 + \lambda n}{\sqrt{\lambda n}}\right).$$

If $\lambda n$ is an integer,

$$P(S_n \ge \lambda n) \simeq 1 + \Phi\left(\frac{1}{\sqrt{\lambda n}}\right) - \Phi\left(\frac{0.5 + \lambda n}{\sqrt{\lambda n}}\right).$$

(iii)  If $\lambda n/2$ is not an integer,

$$P\left(\frac{\lambda n}{2} \le S_n \le \frac{3\lambda n}{4}\right) \simeq \Phi\left(\frac{\lambda n - [\lambda n/2]}{\sqrt{\lambda n}}\right) - \Phi\left(\frac{\sqrt{\lambda n}}{4}\right).$$

If $\lambda n/2$ is an integer,

$$P\left(\frac{\lambda n}{2} \le S_n \le \frac{3\lambda n}{4}\right) \simeq \Phi\left(\frac{2 + \lambda n}{\sqrt{\lambda n}}\right) - \Phi\left(\frac{\lambda n}{4}\right).$$

(iv)  For (i): $P(S_n \le 100) \simeq 0.5$.
For (ii): $P(S_n > 100) \simeq 0.539828$.
For (iii): $P(50 < S_n \le 75) \simeq 0.00621$. ∎

**2.14** The total life time is $X = \sum_{i=1}^{50} X_i$, where $X_i$'s are independently distributed as Negative Exponential with $\lambda = 1/1{,}500$. Then $P(X \geq 80{,}000) \simeq 1 - \Phi(0.47) = 0.319178.$ ∎

**2.16** (i) $P(a \leq \bar{X} \leq b) \simeq \Phi[(2b-1)\sqrt{3n}] - \Phi[(2a-1)\sqrt{3n}]$.

(ii) Here $(2b-1)\sqrt{3n} = 0.75$, $(2a-1)\sqrt{3n} = -0.75$, and the above probability is: $2\Phi(0.75) - 1 = 0.546746.$ ∎

**2.18** $P(|\bar{X} - \mu| \leq 0.0001) \simeq 2\Phi(0.2\sqrt{n}) - 1 = 0.99$, and then $n = 167.$ ∎

**2.20** (i) $P(|\bar{X}_n - \mu| < k\sigma) \simeq 2\Phi(k\sqrt{n}) - 1 = p$, so that $n$ is the smallest integer $\geq [\frac{1}{k}\Phi^{-1}(\frac{1+p}{2})]^2$.

(ii) Here $n$ is the smallest integer $\geq 1/(1-p)k^2$.

(iii) For $p = 0.90$, $p = 0.95$, and $p = 0.99$, and the respective values of $k$, we determine the values of $n$ by means of the CLT (by linear interpolation for $p = 0.90$ and $p = 0.99$), and the Tchebichev inequality.

Then, for the various values of $k$, the respective values of $n$ are given in the following table for part (i):

| $k \backslash p$ | 0.90 | 0.95 | 0.99 |
|---|---|---|---|
| 0.50 | 11 | 16 | 27 |
| 0.25 | 44 | 62 | 107 |
| 0.10 | 271 | 385 | 664 |

For the Tchebichev inequality, the values of $n$ are given in the following table:

| $k \backslash p$ | 0.90 | 0.95 | 0.99 |
|---|---|---|---|
| 0.50 | 40 | 80 | 400 |
| 0.25 | 160 | 320 | 1,600 |
| 0.10 | 1,000 | 2,000 | 10,000 |

∎

**2.22** (i) $P(|\bar{X} - \bar{Y}| \leq 0.25\sigma) = P(|\bar{Z}| \leq 0.25\sigma) \simeq 2\Phi(\frac{0.25\sqrt{n}}{\sqrt{2}}) - 1 = 0.95$ and then $n = 123$.

(ii) From $1 - \frac{2}{0.0625n} \geq 0.95$, we find $n = 640.$ ∎

**2.24** We have $EX_i = \frac{1}{\lambda}$, $\text{Var}(X_i) = \frac{1}{\lambda^2}$, so that $ES_n = \frac{n}{\lambda}$, $\text{Var}(S_n) = \frac{n}{\lambda^2}$, $\sigma(S_n) = \frac{\sqrt{n}}{\lambda}$. Then:

(i) $P(S_n < nP) = P(0 \leq S_n \leq nP) \simeq \Phi[\sqrt{n}(\lambda P - 1)]$ (since $n$ is expected to be large), and $P = \frac{1}{\lambda}[1 + \frac{1}{\sqrt{n}}\Phi^{-1}(p)]$.

(ii) For the given values, $P = 1000(1 + \frac{1}{100} \times 2.33) = 1023.3.$ ∎

**2.26** Since $EX = 0$ and $\text{Var}(X) = 1/12$, we have: $P(|\bar{X}_{100}| \leq 0.1) \simeq 2\Phi(\sqrt{12}) - 1 = 0.99946.$ ∎

**2.28** (i) $S_{100} \sim B(1000.1)$. (ii) $ES_{100} = 10$, $\text{Var}(S_{100}) = 9$.

(iii) $P(S_{100} = 16) = \binom{100}{16}(0.1)^{16}(0.9)^{84} \simeq 0.019$.

(iv) $P(S_{100} = 16) = P(15 < S_{100} \leq 16) \simeq 0.025$.

**2.30** (i) $P(S_n > k \times n/\lambda) \simeq 1 - \Phi[(k-1)\sqrt{n}]$.

(ii)

$$k = 1.2, \quad 1 - \Phi(1.26) = 1 - 0.896165 = 0.103835$$
$$n = 40, \quad k = 1.4, \quad 1 - \Phi(2.53) = 1 - 0.994297 = 0.005703$$
$$k = 1.5, \quad 1 - \Phi(3.16) = 1 - 0.999211 = 0.000789.$$

**2.32** (i) $P[S_n > (c+1)n\mu] \simeq 1 - \Phi\left(\frac{c\mu\sqrt{n}}{\sigma}\right)$.

(ii) $P(S_n > 505\sqrt{n}) \simeq 1 - \Phi(0.05\sqrt{n})$. (iii) For $n = 1000$:
$P(S_{1000} > 505,000) \simeq 0.057063$. For $n = 5000$: $P(S_{5000} > 2,525,000) \simeq 0$.

## Section 7.3

**3.2** By Exercise 3.1(i) in Chapter 10, $\sum_{i=1}^{n}(X_i - \bar{X}_n)^2 = \sum_{i=1}^{n} X_i^2 - n\bar{X}^2$, so that
$\frac{1}{n-1}\sum_{i=1}^{n}(X_i - \bar{X})^2 = \frac{n}{n-1} \times \frac{1}{n}\sum_{i=1}^{n} X_i^2 - \frac{n}{n-1}\bar{X}_n^2$. However, by the WLLN,
$\frac{1}{n}\sum_{i=1}^{n} X_i^2 \xrightarrow[n\to\infty]{P} EX_1^2 = \sigma^2 + \mu^2$, $\bar{X}_n \xrightarrow[n\to\infty]{P} \mu$, so that by Theorem 5(i),
$\bar{X}_n^2 \xrightarrow[n\to\infty]{P} \mu^2$, and by Theorem 6(ii) (and Theorem 1),
$\frac{n}{n-1} \times \frac{1}{n}\sum_{i=1}^{n} X_i^2 \xrightarrow[n\to\infty]{P}$
$\sigma^2 + \mu^2$, $\frac{n}{n-1}\bar{X}_n^2 \xrightarrow[n\to\infty]{P} \mu^2$. Finally, by Theorem 5(ii),

$$\frac{n}{n-1} \times \frac{1}{n}\sum_{i=1}^{n} X_i^2 - \frac{n}{n-1}\bar{X}_n^2 \xrightarrow[n\to\infty]{P} \sigma^2 + \mu^2 - \mu^2 = \sigma^2. \blacksquare$$

---

# CHAPTER 9
## Section 9.1

**1.2** The matrix is Negative definite, because for $\lambda_1, \lambda_2$ with $\lambda_1^2 + \lambda_2^2 \neq 0$, we have:

$$(\lambda_1, \lambda_2)\begin{pmatrix} -n/s^2 & 0 \\ 0 & -n/2s^4 \end{pmatrix}\begin{pmatrix} \lambda_1 \\ \lambda_2 \end{pmatrix} = -\lambda_1^2\frac{n}{s^2} - \lambda_2^2\frac{n}{2s^4} < 0. \blacksquare$$

**1.4** With $Y = \sum_{i=1}^{n}\left(\frac{X_i - \bar{X}}{\sigma}\right)^2 \sim \chi_{n-1}^2$ and $S^2 = \sigma^2 Y/(n-1)$, we have
$\text{Var}_{\sigma^2}(S^2) = 2\sigma^4/(n-1)$. $\blacksquare$

**1.6** With $L(\theta \mid \boldsymbol{x}) = \theta^n(1-\theta)^{n(\bar{x}-1)}$, $\boldsymbol{x} = (x_1, \ldots, x_n)$, we have
$\frac{\partial}{\partial\theta}\log L(\theta \mid \boldsymbol{x}) = 0$ produces $\theta = 1/\bar{x}$, and
$\frac{\partial^2}{\partial\theta^2}\log L(\theta \mid \boldsymbol{x}) = -\frac{n}{\theta^2} - \frac{n(\bar{x}-1)}{(1-\theta)^2} < 0$, so that $\hat{\theta} = 1/\bar{x}$ is the MLE of $\theta$. $\blacksquare$

**1.8** The MLE of $\theta$ is $\hat{\theta} = -n/\sum_{i=1}^{n}\log x_i$. $\blacksquare$

**1.10** (i) It follows from what is given.
(ii) $\hat{\theta} = 2/\bar{x}$. $\blacksquare$

**1.12**   **(i)** The first two expressions are immediate. The third follows from $\frac{d}{d\sigma_1^2}\left(\frac{1}{\sigma_1^2}\right) = -\frac{1}{2\sigma_1^3}$, and similarly for the fourth. The fifth follows from

$$\frac{d}{d\rho}\left(\frac{1}{1-\rho^2}\right) = \frac{2\rho}{(1-\rho^2)^2} \text{ and } \frac{d}{d\rho}\left(\frac{\rho}{1-\rho^2}\right) = \frac{1+\rho^2}{(1-\rho^2)^2}.$$

  **(ii)** Immediate from part (i).

  **(iii)** $\tilde{\mu}_1 = \bar{x}$ and $\tilde{\mu}_2 = \bar{y}$ follow by solving the equations:

$$\sigma_2\mu_1 - \sigma_1\rho\mu_2 = \sigma_2\bar{x} - \sigma_1\rho\bar{y}, \qquad \sigma_2\rho\mu_1 - \sigma_1\mu_2 = \sigma_2\rho\bar{x} - \sigma_1\bar{y}$$

following from the first two likelihood equations.

  **(iv)** They follow from part (ii).

  **(v)** They follow from part (iv) by solving for $\sigma_1^2, \sigma_2^2$, and $\rho$. ∎

**1.14**   **(i)** Immediate by the fact that $\tilde{d}_{13}(=\tilde{d}_{31}) = \tilde{d}_{14}(=\tilde{d}_{41}) = \tilde{d}_{15}(=\tilde{d}_{51})=\tilde{d}_{23}(=\tilde{d}_{32}) = \tilde{d}_{24}(=\tilde{d}_{42}) = \tilde{d}_{25}(=\tilde{d}_{52}) = 0.$

  **(ii)** Immediate by the fact that $\tilde{d}_{12} = \tilde{d}_{21}$, and $\tilde{d}_{34} = \tilde{d}_{43}$.

  **(iii)** Immediate by the fact that $\tilde{d}_{21} = \tilde{d}_{12}, \tilde{d}_{43} = \tilde{d}_{34}, \tilde{d}_{54} = \tilde{d}_{45}$, and $\tilde{d}_{53} = \tilde{d}_{35}.$

  **(iv)** $\tilde{D}_1 = -\frac{n\beta}{\delta}, \tilde{D}_2 = \frac{n^2}{\delta}, \tilde{D}_3 = -\frac{n^3(\alpha\beta+\delta)}{4\alpha^2\delta^2}, \tilde{D}_4 = \frac{n^4}{4\alpha\beta\delta}.$

  **(v)** $A = \frac{\alpha^3\beta n^2}{2\delta^3}, B = -\frac{\alpha\beta\gamma^2 n^2}{2\delta^3}, C = \frac{\alpha^{1/2}\gamma n^2}{4\beta^{1/2}\delta^2}.$

  **(vi)** $\tilde{D}_5 = -\frac{\alpha\beta n^5}{4\delta^4}.$

  **(vii)** $\tilde{D}_0 = 1 > 0, \tilde{D}_1 = -\frac{n\beta}{\delta} < 0, \tilde{D}_2 = \frac{n^2}{\delta} > 0, \tilde{D}_3 = -\frac{n^3(\alpha\beta+\delta)}{4\alpha^2\delta^2} < 0,$
$\tilde{D}_4 = \frac{n^4}{4\alpha\beta\delta^2} > 0$, and $\tilde{D}_5 = -\frac{\alpha\beta n^5}{4\delta^4} < 0.$ ∎

## Section 9.2

**2.2**   **(i)** $R(x;\theta) = e^{-x/\theta}$; **(ii)** The MLE of $R(x;\theta)$ is $e^{-x/\bar{x}}$. ∎

**2.4**   **(i)** $\hat{\theta} = \frac{1}{n}\sum_{i=1}^{n} x_i^\gamma$; **(ii)** $\hat{\theta} = \bar{x}$. ∎

**2.6**   **(i)** It follows by integration.

  **(ii)** $\prod_{i=1}^{n} X_i$ is a sufficient statistic for $\theta$, and so is $\sum_{i=1}^{n} \log X_i$. ∎

**2.8**   $\sum_{i=1}^{n} |X_i|$ is a sufficient statistic for $\theta$. ∎

**2.10**  Here $(X_1, \ldots, X_r)$ is a set of statistics sufficient for $(p_1, \ldots, p_r)$, or $(X_1, \ldots, X_{r-1})$ is a set of statistics sufficient for $(p_1, \ldots, p_{r-1})$. Furthermore, $(X_{(1)}, X_{(n)})$ is a set of statistics sufficient for $(\alpha, \beta)$. ∎

**2.12**  They follow by integration. ∎

## Section 9.3

**3.2**   **(i)** $E_\theta\bar{X} = \frac{1}{\theta}, E_\theta(nY_1) = \frac{1}{\theta}.$

  **(ii)** $\text{Var}_\theta(\bar{X}) = \frac{1}{n\theta^2} \leq \frac{1}{\theta^2} = \text{Var}_\theta(nY_1)$. ∎

**3.4**    **(i)**  $g_1(y) = \frac{n}{\theta_2-\theta_1}(\frac{\theta_2-y}{\theta_2-\theta_1})^{n-1}, \quad \theta_1 \leq y \leq \theta_2,$

          $g_n(y) = \frac{n}{\theta_2-\theta_1}(\frac{y-\theta_1}{\theta_2-\theta_1})^{n-1}, \quad \theta_1 \leq y \leq \theta_2.$

          Then, with $\boldsymbol{\theta} = (\theta_1, \theta_2)$,

          $E_{\boldsymbol{\theta}} Y_1 = \frac{n\theta_1+\theta_2}{n+1}, \quad E_{\boldsymbol{\theta}} Y_n = \frac{\theta_1+n\theta_2}{n+1}$, and

          $E_{\boldsymbol{\theta}}(\frac{Y_1+Y_2}{2}) = \frac{\theta_1+\theta_2}{2}, \quad E_{\boldsymbol{\theta}}(Y_n - Y_1) = \frac{n-1}{n+1}(\theta_2 - \theta_1),$

          $E_{\boldsymbol{\theta}}[\frac{n+1}{n-1}(Y_n - Y_1)] = \theta_2 - \theta_1.$

    **(ii)**  The pair of statistics $(Y_1, Y_n)$ is sufficient for the pair of the parameters $(\theta_1, \theta_2)$. ∎

**3.6**    They follow by integration and by using part (ii) of Exercise 3.5. ∎

**3.8**    **(i)**  The $\text{Var}_{\theta}(U_1)$ and $\text{Var}_{\theta}(U_2)$ follow from the $\text{Var}_{\theta}(Y_1)$ and $\text{Var}_{\theta}(Y_n)$ in Exercise 3.6, and the $\text{Cov}_{\theta}(Y_1, Y_n)$ follows from Exercise 3.7(ii).

    **(ii)**  Immediate by comparing variances. ∎

**3.10**  **(i)**  The required condition is $\sum_{i=1}^{n} c_i = 1$.

     **(ii)**  $c_i = \frac{1}{n}, i = 1, \ldots, n$. ∎

**3.12**  **(i)**  $E_{\theta}\bar{X} = \theta$, and $\text{Var}_{\theta}(\bar{X}) = \frac{1}{n \times \frac{1}{\sigma^2}} = \frac{1}{nI(\theta)}$.

     **(ii)**  $E_{\theta}S^2 = \theta$, and $\text{Var}_{\theta}(S^2) = \frac{1}{n \times \frac{1}{2\theta^2}} = \frac{1}{nI(\theta)}$. ∎

**3.14**  It follows by the hint and Exercise 3.13. ∎

**3.16**  **(i)**  See application to Theorem 3.

     **(ii)**  If $h : \{0, 1, \ldots, n\} \to \Re$, then $E_{\theta}h(X) = 0$ is equivalent to $\sum_{x=0}^{n} h(x) \times \binom{n}{x}t^x = 0 \, (t = \frac{\theta}{1-\theta})$ from which it follows $h(x) = 0, x = 0, 1, \ldots, n$.

     **(iii)**  It follows by the fact that $\bar{X} = \frac{T}{n}$ and $T \sim B(n, \theta)$ is sufficient and complete. ∎

**3.18**  **(i)**  $X$ is sufficient by the factorization theorem, and also complete, because, if $h : \{0, 1, \ldots, n\} \to \Re$,

$$E_{\theta}h(X) = \frac{1-t}{t} \sum_{x=1}^{\infty} h(x)t^x = 0 \, (t = 1 - \theta)$$

        implies $h(x) = 0, x = 1, 2, \ldots$.

     **(ii)**  $U$ is unbiased, because:

        $E_{\theta}U = 1 \times P_{\theta}(U = 1) = P_{\theta}(U = 1) = P_{\theta}(X = 1) = \theta$.

     **(iii)**  $U$ is UMVU, because it is unbiased and depends only on the sufficient statistic $X$.

     **(iv)**  $\text{Var}_{\theta}(U) = \theta(1 - \theta) > \theta^2(1 - \theta) = \frac{1}{I(\theta)} \, (0 < \theta < 1)$. ∎

**3.20** (i) Sufficiency of $Y_n$ follows from Exercise 2.9(iii) (with $\alpha = 0$); completeness cannot be established here.

(ii) It follows from part (i) and Exercise 3.3(ii).

(iii) Because the function $L(\theta \mid x)$ $(=f(x; \theta))$ is not differentiable at $\theta = x(> 0)$. ∎

## Section 9.4

**4.2** It follows by the fact that $\int_0^1 \frac{\Gamma(\alpha+\beta)}{\Gamma(\alpha)\Gamma(\beta)} x^{\alpha-1}(1-x)^{\beta-1} dx = 1$ and the recursive relation of the Gamma function. ∎

**4.4** Immediate from the hint. ∎

**4.6** $R(\theta;d) = \frac{1}{n}$, independent of $\theta$. However, Theorem 9 need not apply. For this to be the case, $\bar{x}$ must be the Bayes estimate with respect to some prior p.d.f. ∎

## Section 9.5

**5.2** (i) $\tilde{\theta} = (1 - 2\bar{X})/(\bar{X} - 1)$; (ii) $\tilde{\theta} = 3$ and $\hat{\theta} \simeq 3.116$. ∎

**5.4** Follows from Exercise 3.15(i). ∎

**5.6** (i) The $EX$ and $EX^2$ follow by integration, and $\text{Var}(X) = \beta^2$.

(ii) $\tilde{\alpha} = \bar{X} - S$, $\tilde{\beta} = S$, where $S^2 = \frac{1}{n}\sum_{i=1}^n (X_i - \bar{X})^2$. ∎

**5.8** By equating the second-order sample moment to $\frac{(2\theta)^2}{12}$, we get $\tilde{\theta} = (\frac{3}{n}\sum_{i=1}^n X_i^2)^{1/2}$. ∎

**5.10** (i) It follows by integration.

(ii) They also follow by integration.

(iii) $\tilde{\theta} = 3\bar{X}$ and hence $E_\theta \tilde{\theta} = \theta$, and $\text{Var}_\theta(\tilde{\theta}) = \theta^2/2n$. ∎

**5.12** (i) and (ii) follow by straightforward calculations.

(iii) It follows by the expression of $\hat{\rho}_n(X, Y)$ and the WLLN applied to $n^{-1}\sum_{i=1}^n (X_iY_i)$, $\bar{X}$, $\bar{Y}$, $n^{-1}\sum_{i=1}^n (X_i - \bar{X})^2$, and $n^{-1}\sum_{i=1}^n (Y_i - \bar{Y})^2$. ∎

**5.14** $\bar{x} = \frac{2,423}{15} \simeq 161.533$, $s_x = \frac{\sqrt{175,826}}{15} \simeq 27.954$, $\bar{y} = 140.6$, and $s_y = \frac{\sqrt{56,574}}{15} \simeq 15.857$. ∎

## CHAPTER 10
## Section 10.1

**1.2** $m = 4n$. ∎

**1.4** (i) $n = (2z_{\alpha/2}\sigma/l)^2$; (ii) $n = 1537$. ∎

**1.6**    (i)   $M_U(t) = 1/(1 - \theta t)^n, t < \frac{1}{\theta}$, which is the m.g.f. of the Gamma distribution with $\alpha = n$ and $\beta = \theta$.

      (ii)   $M_V(t) = 1/(1 - 2t)^{2n/2}, t < \frac{1}{2}$, which is the m.g.f. of the $\chi^2_{2n}$ distribution, so that $V \sim \chi^2_{2n}$.

      (iii)   $[2b^{-1} \sum_{i=1}^{n} X_i, \ \ 2a^{-1} \sum_{i=1}^{n} X_i]$.   ∎

**1.8**    (i)   For $x > \theta$, $F(x; \theta) = 1 - e^{-(x-\theta)}$ and hence $g(y; \theta) = ne^{-n(y-\theta)}, y > \theta$.

      (ii)   By setting $t = 2n(y - \theta)$, we get $f_T(t; \theta) = \frac{1}{2} e^{-t/2}, t > 0$, which is the p.d.f. of the $\chi^2_2$ distribution.

      (iii)   It follows by the usual arguments.   ∎

**1.10**    (i)   The transformation $y = |x|$ yields $x = y$ for $x > 0, x = -y$ for $x < 0$, and $\left| \frac{dx}{dy} \right| = 1$. Then, by Theorem 6 (applied with $k = 2$),

         $g(y; \theta) = \frac{1}{\theta} e^{-y/\theta}, y > 0$.

      (ii)   and (iii) are as in Exercise 1.9(ii) and (iii), respectively.   ∎

**1.12**    (i)   $e^{-a} - e^{-b} = 1 - \alpha$.

      (ii)   Immediate by part (i) and the fact that $T$ has the p.d.f. $e^{-t}, t > 0$.

      (iii)   Follows from the hint.   ∎

**1.14**   From the transformations $r = y_n - y_1$, and $s = y_1$, we get $y_1 = s, y_n = r + s$, and $|J| = 1$. Then $f_{R,S}(r, s; \theta) = \frac{n(n-1)}{\theta^n} r^{n-2}, 0 < r < \theta, 0 < s < \theta - r$, and then $f_R(r; \theta) = \frac{n(n-1)}{\theta^n} r^{n-2}(\theta - r), 0 < r < \theta$.   ∎

**1.16**    (i)   It follows by using the transformation $t = r/\theta$.

      (ii)   The required confidence interval is $[R, \frac{R}{C}]$. The relation $c^{n-1}[n - (n - 1)c] = \alpha$ follows from

$$1 - \alpha = P_\theta(c \leq T \leq 1) = \int_c^1 n(n - 1)t^{n-2}(1 - t) \, dt$$

$$= n - nc^{n-1} - (n - 1) + (n - 1)c^n. \ ∎$$

**1.18**   Follows by the usual procedure and the fact that $\frac{mS_X^2}{\sigma_1^2} \sim \chi^2_m, \frac{nS_Y^2}{\sigma_2^2} \sim \chi^2_n$ independent, so that $\frac{S_Y^2/\sigma_2^2}{S_X^2/\sigma_1^2} \sim F_{n,m}$.   ∎

**1.20**   Use the graph of the p.d.f. of the Beta distribution with parameters $\alpha + t$ and $\beta + n - t$.   ∎

## Section 10.2

**2.2**   The required interval is $\left[ \sqrt{a} \frac{S_X^*}{S_Y^*}, \sqrt{b} \frac{S_X^*}{S_Y^*} \right]$, where $S_X^{*2} = \frac{1}{m-1} \sum_{i=1}^{m} (X_i - \bar{X})^2$, $S_Y^{*2} = \frac{1}{n-1} \sum_{j=1}^{n} (Y_j - \bar{Y})^2, 0 < a < b$ with $P(a \leq X \leq b) = 1 - \alpha, X \sim F_{n-1,m-1}$. In particular, $a = F_{n-1,m-1;1-\frac{\alpha}{2}}, b = F_{n-1,m-1;\frac{\alpha}{2}}$.   ∎

## Section 10.4

**4.2**   (i) The required confidence interval is $\bar{X}_n \pm z_{\frac{\alpha}{2}} \frac{S_n}{\sqrt{n}}$.

(ii) Here $\bar{X}_{100} \pm 0.196 S_{100}$.

(iii) The length is $2z_{\frac{\alpha}{2}} \frac{S_n}{\sqrt{n}}$, which converges in probability to 0 since

$$S_n \xrightarrow[n \to \infty]{P} \sigma. \ \blacksquare$$

**4.4**   (i) $P(Y_i \le x_p) = P(\text{at least } i \text{ of } X_1, \ldots, X_n \le x_p) = \sum_{k=i}^{n} \binom{n}{k} p^k q^{n-k}$, and
also $P(Y_i \le x_p) = P(Y_i \le x_p \le Y_j) + P(Y_j \le x_p)$, so that

$$P(Y_i \le x_p \le Y_j) = \sum_{k=i}^{n} \binom{n}{k} p^k q^{n-k} - \sum_{k=j}^{n} \binom{n}{k} p^k q^{n-k} = \sum_{k=i}^{j-1} \binom{n}{k} p^k q^{n-k}.$$

(ii) and (iii) follow from part (i) and the Binomial tables.

(iv) It follows from the hint and the Binomial tables. $\blacksquare$

# CHAPTER 11

## Section 11.1

**1.2**   $H_{0i}$ and $H_{Ai}, i = 1, \ldots, 5$ are all composite; $H_{06}$ and $H_{A6}$ are both simple. $\blacksquare$

## Section 11.2

**2.2**   (i) The required $n$ is determined by solving for $n$ (and $C$) the two
equations: $\frac{C\sqrt{n}}{\sigma} = \Phi^{-1}(1-\alpha)$ and $\frac{\sqrt{n}(C-1)}{\sigma} = \Phi^{-1}(1-\pi(1))$.

(ii) $n = 9$ (and $C \simeq 0.562$). $\blacksquare$

**2.4**   (i) The MP test rejects $H_0$ when $\frac{L_1(x)}{L_0(x)} > C_0$, equivalently, when $\bar{x} > C_n$,
where $C_n = [2\sigma^2 \log C_0 + n(\mu_1^2 - \mu_0^2)]/2n(\mu_1 - \mu_0)$.

(ii) The required $C_n$ and $n$ are given by

$$C_n = \mu_0 + \frac{(\mu_1 - \mu_0)\Phi^{-1}(1 - \alpha_n)}{\Phi^{-1}(1 - \alpha_n) - \Phi^{-1}(1 - \pi_n)},$$

$$n \text{ is the smallest integer} \ge \left\{ \frac{\sigma}{\mu_1 - \mu_0} \left[ \Phi^{-1}(1 - \alpha_n) - \Phi^{-1}(1 - \pi_n) \right] \right\}^2.$$

(iii) That $\alpha_n \xrightarrow[n \to \infty]{} 0$ follows from $1 - \Phi\left[ \frac{\sqrt{n}(C_n - \mu_0)}{\sigma} \right] = \alpha_n$ and the fact
that $\mu_0 < C_n$; that $\pi_n \xrightarrow[n \to \infty]{} 1$ follows from

$$1 - \Phi^{-1}\left[ \frac{\sqrt{n}(C_n - \mu_1)}{\sigma} \right] = \pi_n, \text{ and the fact that } C_n < \mu_1.$$

(iv) $n = 33$ and $C_{33} \simeq 0.546$. $\blacksquare$

**2.6**  **(i)** Reject $H_0$ when the observed value $t$ of $T = X_1 + \cdots + X_n$ is $> C$,
where $P_{\theta_0}(T > C) = \alpha$; $\pi = P_{\theta_1}(T > C)$.

**(ii)** $M_{X_i}(t) = 1/\left(1 - \frac{t}{\theta}\right)$, $t < \theta$, so that $M_{2\theta X_i}(t) = \frac{1}{1-2t}$, $t < \frac{1}{2}$ and hence
$2\theta T \sim \chi^2_{2n}$.

**(iii)** $P_{\theta_0}(T > C) = P(\chi^2_{2n} > 2\theta_0 C)$, so that $C = \chi^2_{2n;\alpha}/2\theta_0$, and
$\pi = P(\chi^2_{2n} > 2\theta_1 C)$.

**(iv)** $C \simeq \left[n + \sqrt{n}\Phi^{-1}(1 - \alpha)\right]/\theta_0$, and $\pi \simeq 1 - \Phi\left(\frac{\theta_1 C - n}{\sqrt{n}}\right)$.

**(v)** $C \simeq 302.405$, and $\pi \simeq 0.31$ (by linear interpolation). ∎

**2.8**  **(i)** The MP test rejects $H_0$ when $x \in \{x \geq 0 \text{ integer}; 1.36 \times \frac{x!}{2^x} \geq C\}$, where
$C$ is determined by $P_{H_0}(\{1.36 \times \frac{x!}{2^x} \geq C\}) = \alpha$.

**(ii)** For $C = 2$, if follows that $\alpha = 0.018988 \simeq 0.02$. ∎

# Section 11.3

**3.2**  **(ii)** The UMP test is given by

$$\varphi(x_1, \ldots, x_n) = \begin{cases} 1, & \text{if } \sum_{i=1}^n x_i < C \\ \gamma, & \text{if } \sum_{i=1}^n x_i = C \\ 0, & \text{if } \sum_{i=1}^n x_i > C, \end{cases}$$

where $C$ and $\gamma$ are determined by

$$P_{p_0}(X < C) + \gamma P_{p_0}(X = C) = \alpha, \quad X \sim B(n, p_0).$$

**(iii)** Here $C \simeq 3.139015$, and $H_0$ is rejected when $X \leq 3$. ∎

**3.4**  **(i)** With each student associate a $B(1, p)$ r.v. $X_i, i = 1, \ldots, 400$, so that
$X = \sum_{i=1}^{400} X_i \sim B(400, p)$. Then the UMP test is given by:

$$\varphi(x_1, \ldots, x_{400}) = \begin{cases} 1, & \text{if } \sum_{i=1}^{400} x_i < C \\ \gamma, & \text{if } \sum_{i=1}^{400} x_i = C \\ 0, & \text{if } \sum_{i=1}^{400} x_i > C, \end{cases}$$

where $C$ and $\gamma$ are determined by

$$P_{0.25}(X < C) + \gamma P_{0.25}(X = C) = 0.05, \quad X \sim B(400, 0.25).$$

**(ii)** Here $C \simeq 85.754$, and $H_0$ is rejected when $\sum_{i=1}^{400} x_i \leq 85$. ∎

**3.6**  **(i)** The UMP test is given by relation (11) with $C$ and $\gamma$ defined by relation
(12) (or (13)).

**(ii)** Here $C = 5$ and $\gamma \simeq 0.519$.

**(iii)** $\pi(0.375) \simeq 0.220$ and $\pi(0.500) \simeq 0.505$.

**(iv)** For $\theta > 0.5$, $\pi(\theta) = P_{1-\theta}(X \leq n - C - 1) + \gamma P_{1-\theta}(X = n - C)$.

**(v)** $\pi(0.625) \simeq 0.787$, and $\pi(0.875) \simeq 0.998$.

**(vi)** $n = 62$. ∎

**3.8**   (i) $H_0 : \lambda = 10, H_A : \lambda < 10$.

(ii) The UMP test is given by

$$\varphi(x) = \begin{cases} 1, & \text{if} \quad x < C \\ \gamma, & \text{if} \quad x = C \\ 0, & \text{if} \quad x > C, \end{cases}$$

$$P_{10}(X \le C - 1) + \gamma P_{10}(X = C),$$

$$= 0.01, \ X \sim P(10).$$

From the Poisson tables, $C = 3$ and $\gamma = 0.96$, and since $x = 4, H_0$ is not rejected. ■

**3.10**   (i) $H_0$ is rejected when $\sum_{i=1}^{n} x_i < C, C = n\mu_0 - z_\alpha \sigma \sqrt{n}$.

(ii) $\pi(\mu) = \Phi(\frac{C - n\mu}{\sigma\sqrt{n}})$.

(iii) $C = 43252.5, H_0$ is rejected, and $\pi(1,700) = 0.841345$. ■

**3.12**   (i) In Application 4, the roles of $H_0$ and $H_A$ are reversed. Therefore, by (32), $H_0$ is rejected when $\sum_{i=1}^{n}(x_i - \mu)^2 > C$, where $C$ is determined by

$$P_{\sigma_0}\left(\sum_{i=1}^{n}(X_i - \mu)^2 > C\right) = P_{\sigma_0}\left[\sum_{i=1}^{n}\left(\frac{X_i - \mu}{\sigma_0}\right)^2 > \frac{C}{\sigma_0^2}\right] = \alpha,$$

so that $\frac{C}{\sigma_0^2} = \chi_{n;\alpha}^2$ and $C = \sigma_0^2 \chi_{n;\alpha}^2$.

(ii) Here $\chi_{n;\alpha}^2 = \chi_{25;0.05}^2 = 37.652$, so that $C = 4 \times 37.652 = 150.608$. Since $\sum_{i=1}^{25}(x_i - \mu)^2 = \sum_{i=1}^{25} x_i^2 = 120 < 150.608$, the hypothesis $H_0$ is not rejected. ■

**3.14**   (i) $C(\theta) = \frac{1}{\theta}, Q(\theta) = -\frac{1}{\theta}$ strictly increasing, $T(x) = x$, and $h(x) = I_{(0,\infty)}(x)$.

(ii) The UMP test rejects $H_0$ when $\sum_{i=1}^{n} x_i < C$, where $C$ is determined by $P_{\theta_0}(\sum_{i=1}^{n} X_i < C) = \alpha$.

(iii) $M_{\sum_{i=1}^{n} X_i}(t) = \frac{1}{(1-\theta t)^n}$ $(t < \frac{1}{\theta})$, and $M_{2(\sum_{i=1}^{n} X_i)/\theta}(t) = M_{\sum_{i=1}^{n} X_i}(\frac{2t}{\theta}) = \frac{1}{(1-\theta \times \frac{2t}{\theta})^n} = \frac{1}{(1-2t)^{2n/2}}$, which is the m.g.f. of the $\chi_{2n}^2$ distribution.

(iv) $C = \frac{\theta_0}{2} \chi_{2n;1-\alpha}^2, \pi(\theta_1) = P(X < \frac{2C}{\theta_1}), X \sim \chi_{2n}^2$.

(v) The closest value we can get by means of the Chi-square-tables is $n = 23$. ■

**3.16** $H_0$ is rejected when $C_1 < \sum_{i=1}^{4} z_i < C_2$, where $C_1$ and $C_2$ are determined by $\Phi(0.4 + x) - \Phi(0.4 - x) = 0.05, x = \frac{C}{10}$. From the Normal tables, $x \simeq 0.07$, so that $C=0.7$, and $H_0$ is rejected when $-0.7 < z_1 + z_2 + z_3 + z_4 < 0.7$. Here $z_1 + z_2 + z_3 + z_4 = 4.9$, and therefore $H_0$ is not rejected. ■

## Section 11.4

**4.2**   (i)  $\lambda = (0.25)^t (0.75)^{3-t} / (\frac{t}{3})^t (1 - \frac{t}{3})^{3-t}, t = 0, 1, 2, 3$, and $H_0$ is rejected when $\lambda < C$, where $C$ is determined by $P_{0.25}(\lambda < C) = 0.02$.

   (ii)  At level $\alpha = 0.02$, $H_0$ is outright rejected when $\lambda = 0.015625$ (which is equivalent to $t = 3$), and is rejected with probability $(0.02 - 0.0156)/0.1407 \simeq 0.031$ when $\lambda = 0.31640625$ (which is equivalent to $t = 2$). ∎

**4.4**   (ii)  With $t(z) = \sqrt{n}\bar{z}/\sqrt{\frac{1}{n-1} \sum_{i=1}^{n} (z_i - \bar{z})^2}, z = (z_1, \ldots, z_n)$, $H_0$ is rejected when $t(z) < -t_{n-1;\frac{\alpha}{2}}$ or $t(z) > t_{n-1;\frac{\alpha}{2}}$.

   (iii)  Here $t_{89;0.025} = 1.9870$, and therefore $H_0$ is rejected when
   $$|3\sqrt{10}\bar{z}|/\sqrt{\frac{1}{89} \sum_{i=1}^{90} (z_i - \bar{z})^2} > 1.9870. \quad ∎$$

**4.6**   Here $H_0 : \mu = 2.5$ and $H_A : \mu \neq 2.5$, and $H_0$ is rejected when $\frac{\sqrt{n}(\bar{x}-\mu_0)}{\sigma} < -z_{\alpha/2}$ or $\frac{\sqrt{n}(\bar{x}-\mu_0)}{\sigma} > z_{\alpha/2}$. Since $z_{0.025} = 1.96$ and $\frac{\sqrt{n}(\bar{x}-\mu_0)}{\sigma} = -0.8$, $H_0$ is not rejected. ∎

**4.8**   Here $H_0 : \mu_1 = \mu_2, H_A : \mu_1 \neq \mu_2$, and $H_0$ is rejected when $t(x, y) < -t_{m+n-2 ;\alpha/2}$ or $t(x, y) > t_{m+n-2 ;\alpha/2}$, where
   $$t(x, y) = \sqrt{\frac{mn}{m+n}}(\bar{x} - \bar{y})/\sqrt{\frac{1}{m+n-2} \left[ \sum_{i=1}^{m} (x_i - \bar{x})^2 + \sum_{j=1}^{n} (y_j - \bar{y})^2 \right]},$$
   $x = (x_1, \ldots, x_m), y = (y_1, \ldots, y_n)$. Since $t_{48;0.025} = 2.0106$ and $t(x, y) = -2.712$, the hypothesis $H_0$ is rejected. ∎

**4.10**  $H_0$ is rejected when $u(x, y) \leq F_{m-1, n-1; 1-\frac{\alpha}{2}}$ or $u(x, y) \geq F_{m-1, n-1; \frac{\alpha}{2}}$, where $u(x, y) = \frac{1}{m-1} \sum_{i=1}^{m} (x_i - \bar{x})^2 / \frac{1}{n-1} \sum_{j=1}^{n} (y_j - \bar{y})^2$,
   $x = (x_1, \ldots, x_m), y = (y_1, \ldots, y_n)$. Here $F_{3,3;0.975} \simeq 0.0648$, $F_{3,3;0.025} = 15.439$, and $u(x, y) \simeq 2.1685$. Therefore $H_0$ is not rejected. ∎

**4.12**  The LR test is the same as that given in Exercise 4.8. For the given numerical data, $t_{8;0.025} = 3.3060, t(x, y) \simeq 2.014$, and therefore $H_0$ is not rejected. ∎

**4.14**  They follow by integrating by parts. ∎

**4.16**  With $\lambda(u) = e^{n/2} u^{n/2} e^{-nu/2}, u \geq 0$, we have $\lambda'(u) = \frac{d\lambda(u)}{du} = \frac{n}{2} e^{n/2} u^{\frac{n}{2}-1} e^{-nu/2} \times (1 - u)$, so that $\lambda'(u) \geq 0$ if $(0 \leq) u \leq 1$ and $\lambda'(u) \leq 0$ if $u > 1$. It follows that $\lambda(u)$ is strictly increasing for $u \leq 1$ and strictly decreasing for $u > 1$. Since $\lambda'(u) = 0$ gives $u = 1$, and $\frac{d^2}{du^2} \lambda(u)|_{u=1} = -\frac{n}{2} < 0$, it follows that $\lambda(u)$ is maximized for $u = 1$. That $\lambda(0) = 0$ is immediate, and that $\lambda(u) \to 0$ as $u \to \infty$ follows by taking the limit of the ratio of derivatives of sufficiently high order. ∎

**4.18**  Maximization of (58) is equivalent to maximization of
   $$g = g(\mu, \tau) = -\frac{m+n}{2} \log \tau - \frac{1}{2\tau} \left[ \sum_{i=1}^{m} (x_i - \mu)^2 + \sum_{j=1}^{n} (y_j - \mu)^2 \right], \text{where}$$

$\tau = \sigma^2$. From $\frac{\partial g}{\partial \mu} = 0$, $\frac{\partial g}{\partial \tau} = 0$, we find $\hat{\mu}_\omega = \frac{n\bar{x}+n\bar{y}}{m+n}$,

$\hat{\tau} = \frac{1}{m+n}\left[\sum_{i=1}^{m}(x_i - \hat{\mu}_\omega)^2 + \sum_{j=1}^{n}(y_j - \hat{\mu}_\omega)^2\right]$. Next, $\frac{\partial^2 g}{\partial \mu^2}$, $\frac{\partial^2 g}{\partial \mu \partial \tau} = \frac{\partial^2 g}{\partial \tau \partial \mu}$,

and $\frac{\partial^2 g}{\partial \tau^2}$ evaluated of $\mu = \hat{\mu}_\omega$ and $\tau = \hat{\tau}$, yield, respectively; $-\frac{m+n}{\hat{\tau}}, 0$, and

$-\frac{m+n}{2\hat{\tau}^2}$. Setting $C$ for the $2 \times 2$ matrix of the second-order derivatives of $g$, we have, for $\lambda_1, \lambda_2$ with $\lambda_1^2 + \lambda_2^2 \neq 0$:

$$(\lambda_1, \lambda_2)C\begin{pmatrix}\lambda_1 \\ \lambda_2\end{pmatrix} = -\frac{m+n}{\hat{\tau}}\left(\lambda_1^2 + \frac{\lambda_2^2}{2\hat{\tau}}\right) < 0,$$

so that $C$ is Negative definite, and hence $\hat{\mu}_\omega$ and $\hat{\tau}$ are the MLE of $\mu$ and $\tau$, respectively. ∎

**4.20** From the assumptions made, it follows that

$$\frac{\bar{X} - \bar{Y}}{\sigma\sqrt{\frac{1}{m} + \frac{1}{n}}} \sim N(0, 1); \quad \sum_{i=1}^{m}\left(\frac{X_i - \bar{X}}{\sigma}\right)^2 \sim \chi^2_{m-1}, \quad \sum_{j=1}^{n}\left(\frac{Y_j - \bar{Y}}{\sigma}\right)^2 \sim \chi^2_{n-1}$$

independent, so that their sum is $\sim \chi^2_{m+n-2}$. This sum is also independent of $\bar{X} - \bar{Y}$. It follows that $\frac{\bar{X}-\bar{Y}}{\sigma\sqrt{\frac{1}{m}+\frac{1}{n}}}$ divided by $\left[\sum_{i=1}^{m}(\frac{X_i-\bar{X}}{\sigma})^2 + \sum_{j=1}^{n}(\frac{Y_j-\bar{Y}}{\sigma})^2\right] \big/$

$(m + n - 2)$ is distributed as $t_{m+n-2}$. The cancellation of $\sigma$ leads to the assertion made. ∎

**4.22** Set $c = (m + n)^{\frac{m+n}{2}}/m^{\frac{m}{2}}n^{\frac{n}{2}}$ and $d = \frac{m-1}{n-1}$, so that $\lambda = \lambda(u)$
$= c(du)^{m/2}/(1 + du)^{(m+n)/2}$. That $\lambda(0) = 0$ is immediate. Next, $\lambda \to 0$ as $u \to$
$\infty$ is also clear. Furthermore, $\frac{d\lambda(u)}{du} = \frac{cd}{2} \times \frac{(du)^{\frac{m}{2}-1}}{(1+du)^{(m+n+2)/2}} \times (m - n\,du) = 0$
yields $u = \frac{m}{nd} = \frac{m(n-1)}{n(m-1)}$, call it $u_0$. Also, $\frac{d\lambda(u)}{du} > 0$ for $u < u_0$,
and $\frac{d\lambda(u)}{du} < 0$ for $u > u_0$, so that $\lambda(u)$ increases for $u < u_0$ and decreases
for $u > u_0$. It follows that $\lambda(u)$ attains its maximum for $u = u_0$. This
maximum is 1. ∎

**4.24** (i) Since $\frac{df(r)}{dr} = -\frac{n}{2}(1 - r)^{\frac{n}{2}-1} < 0, f(r)$ is decreasing in $r$.

(iv) Since $\frac{dw(r)}{dr} = \frac{\sqrt{n-2}}{(1-r^2)^{3/2}} > 0, w(r)$ is increasing in $r$. ∎

# CHAPTER 12
## Section 12.1

**1.2** (ii) and (iii) $-2\log\lambda \simeq 2(\sum_{i=1}^{12} x_i \log x_i - 335, 494.1508) =$
$2(335, 530.1368 - 335, 494.1508)$, where $x_i$ is the number of births falling
into the $i$th month. Finally, $-2\log\lambda = 71.972$ and $\chi^2_{11;0.01} = 24.725$. The
hypothesis $H_0$ is rejected. ∎

**1.4** **(iii)** Here $-2\log\lambda \simeq 27.952$, and $\chi^2_{4;0.05} = 9.488$. The hypothesis $H_0$ is rejected. Also, $\chi^2_{4;0.01} = 13.277$ and the hypothesis $H_0$ is still rejected. ■

## Section 12.2

**2.2** Here $\chi^2_\omega \simeq 72.455 > 24.725 = \chi^2_{11;0.01}$, and $H_0$ is rejected. ■

**2.4** Here $\chi^2_{\hat\omega} \simeq 28.161 > 9.488 = \chi^2_{4;0.05}$, and $H_0$ is rejected. Also, $\chi^2_{\hat\omega} \simeq 28.161 > 13.277 = \chi^2_{4;0.01}$, and $H_0$ is still rejected. ■

**2.6** **(i)** Here $\chi^2_\omega = 4 > 2.706 = \chi^2_{1;0.1}$, and $H_0$ is rejected.
**(ii)** The $P$-value is approximately 0.047. ■

**2.8** Here $\chi^2_\omega = 1.2 < 5.991 = \chi^2_{2;0.05}$, and $H_0$ is not rejected. ■

**2.10** **(ii)** $p_{10} = 0.251429, p_{20} = 0.248571, p_{30} = 0.248571,$
$p_{40} = 0.15967, p_{50} = 0.069009, p_{60} = 0.02275.$
**(iii)** $\chi^2_\omega \simeq 51.1607 > 11.071 = \chi^2_{5;0.05}$, and $H_0$ is rejected. ■

**2.12** Here $\chi^2_{\hat\omega} \simeq 1.668 < 11.071 = \chi^2_{5;0.05}$, and $H_0$ is not rejected. ■

---

# CHAPTER 13
## Section 13.3

**3.2** $\sum_i x_i = 91.22, \sum_i y_i = 15228, \sum_i x_i^2 = 273.8244, \sum_i x_i y_i = 45243.54$. Also, $SS_x \simeq 5.4022, SS_y \simeq 101,339.419$, and $SS_{xy} \simeq 433.92$. ■

**3.4** $\text{Var}(\hat\beta_1) \simeq 1.635\sigma^2, \text{Var}(\hat\beta_2) \simeq 0.185\sigma^2$, and $\hat\sigma^2 \simeq 2144.701$. ■

**3.6** **(i)** It follows by differentiating and equating the derivatives to 0.
**(ii)** $-(\mathbf{x})$ as indicated. ■

## Section 13.4

**4.2** From relations (24) and (25), we find $t \simeq -0.737$ and $t \simeq 0.987$. Since $t_{29;0.025} = 2.0452$, none of the hypotheses is rejected. ■

**4.4** **(ii)** Replacing the $x_i$'s by the $t_i$'s and taking into consideration that $\bar{t} = 0$, we get from (5) and (8) the values specified for $\hat\gamma$ and $\hat\beta$.
**(iii)** Immediate by the fact that $\bar{t} = 0$.
**(iv)** $\left[\hat\beta \pm t_{n-2;\frac{\alpha}{2}}\frac{S}{\sqrt{n}}\right]$ and $\left[\hat\gamma \pm t_{n-2;\frac{\alpha}{2}}\frac{S}{\sqrt{SS_t}}\right]$, where $S = \sqrt{SS_E/(n-2)}$,
$SS_E = SS_y - \frac{SS_{xy}^2}{SS_x}, SS_y = \sum_i Y_i^2 - \left(\frac{1}{n}\sum_i Y_i\right)^2,$
$SS_{xy} = \sum_i t_i Y_i, SS_x = SS_t.$
**(v)** $t = \frac{\hat\beta - \beta_0}{S/\sqrt{n}}, t = \frac{\hat\gamma - \gamma_0}{S/\sqrt{SS_t}}.$

(vi) $\left[\hat{y}_0 \pm t_{n-2;\frac{\alpha}{2}} S\sqrt{\frac{1}{n} + \frac{t_0^2}{SS_t}}, \; \hat{y}_0 \pm t_{n-2;\frac{\alpha}{2}} S\sqrt{1 + \frac{1}{n} + \frac{t_0^2}{SS_t}}\right].$ ■

## Section 13.5

**5.2**   (i) $\hat{y}_0 = 517.496$.

  (ii) $S \simeq 59.834, \sqrt{\frac{1}{n} + \frac{(x_0 - \bar{x})^2}{SS_x}} \simeq 0.205$, and $t_{32;0.025} = 2.0369$. The observed confidence interval is $[492.511, 542.481]$.

  (iii) $\hat{y}_0 = 448.949$.

  (iv) The observed confidence interval is $[344.454, 553.444]$. ■

**5.4**   (i) $\hat{y}_0 = 15.374$.

  (ii) The observed confidence interval is $[13.315, 17.433]$.

  (iii) $\hat{y}_0 = 15.374$.

  (iv) The observed confidence interval is $[14.278, 16.470]$. ■

**5.6**   (i) $\hat{\beta}_1 \simeq -9.768, \hat{\beta}_2 \simeq 2.941, \hat{\sigma}^2 \simeq 0.004$.

  (ii) $t_{11;0.025} = 2.201, S \simeq 0.071, \sqrt{\frac{1}{n} + \frac{\bar{x}^2}{SS_x}} \simeq 10.565$, so that the observed confidence intervals for $\beta_1$ and $\beta_2$ are $[-11.457, -8.080]$ and $[2.481, 3.401]$, respectively. Since $\chi^2_{11;0.025} = 21.92$ and $\chi^2_{11;0.975} = 3.816$, the observed confidence interval for $\sigma^2$ is $[0.003, 0.015]$.

  (iii) Both $EY_0$ and $Y_0$ are predicted by 1.32. The respective observed confidence intervals are $[1.253, 1.387]$ and $[1.15, 1.49]$. ■

---

# CHAPTER 14
## Section 14.1

**1.2**   In (1), set $\mu_1 = \cdots = \mu_I = \mu$ to obtain:

$$\log L(\mathbf{y}; \mu, \sigma^2) = -\frac{IJ}{2}\log(2\pi) - \frac{IJ}{2}\log\sigma^2 - \frac{1}{2\sigma^2}\sum_i\sum_j(y_{ij} - \mu)^2.$$

Set $S(\mu) = \sum_i\sum_j(y_{ij} - \mu)^2$, and observe that
$\frac{d}{d\mu}S(\mu) = -2\sum_i\sum_j(y_{ij} - \mu) = 0$ gives $\mu = \frac{1}{IJ}\sum_i\sum_j y_{ij} = y_{..}$, and
$\frac{d^2}{d\mu^2}S(\mu) = 2IJ > 0$ for all values of $\mu$, so that $\hat{\mu} = y_{..}$ minimizes $S(\mu)$.
Replacing $\mu$ by $\hat{\mu}$ in the above expressions, and setting

$$\hat{S} = S(\hat{\mu}), \quad \log \hat{L}(\mathbf{y}; \hat{\mu}, \sigma^2) = -\frac{IJ}{2}\log(2\pi) - \frac{IJ}{2}\log\sigma^2 - \frac{1}{2\sigma^2}\hat{S},$$

we obtain $\sigma^2_{H_0} = \frac{\hat{S}}{IJ}$ from $\frac{d}{d\sigma^2}\log\hat{L}(\mathbf{y}; \hat{\mu}, \sigma^2) = 0$. Also,

$$\frac{d^2}{d(\sigma^2)^2}\log\hat{L}(\mathbf{y}; \hat{\mu}, \sigma^2)|_{\sigma^2 = \sigma^2_{H_0}} = -\frac{IJ}{(\sigma^2_{H_0})^2} < 0,$$

so that $\widehat{\sigma_{H_0}^2} = \frac{\hat{S}}{IJ} = \frac{1}{IJ}\sum_i\sum_j(y_{ij} - y_{..})^2 = \frac{SS_T}{IJ}$ is the MLE of $\sigma^2$ under $H_0$. ■

**1.4** Recall that for $i = 1,\ldots,I$ and $j = 1,\ldots,J$, the r.v.'s $Y_{ij}$ are independent with $EY_{ij} = \mu_i$ and Var $(Y_{ij}) = \sigma^2$.

(i) From $Y_{..} = \frac{1}{IJ}\sum_i\sum_j Y_{ij}$, we have then

$EY_{..} = \frac{1}{IJ}\sum_i\sum_j EY_{ij} = \frac{1}{J}\times j \times \frac{1}{I}\sum_i\mu_i = \mu_{.}$, and therefore

$E(Y_{..} - \mu_{.})^2 = Var(Y_{..}) = Var(\frac{1}{IJ}\sum_i\sum_j Y_{ij}) = \frac{1}{(IJ)^2}IJ\sigma^2 = \frac{\sigma^2}{IJ}$.

(ii) From $Y_{i.} = \frac{1}{J}\sum_j Y_{ij}$, we have $EY_{i.} = \frac{1}{J}\sum_j\mu_i = \frac{1}{J}\times J\mu_i = \mu_i$, so that

$E(Y_{i.} - \mu_i)^2 = Var(Y_{i.}) = Var(\frac{1}{J}\sum_j Y_{ij}) = \frac{1}{J^2}\sum_j\sigma^2 = \frac{\sigma^2}{J}$.

(iii) $E(Y_{i.} - \mu_{.})^2 = E[(Y_{i.} - \mu_i) + (\mu_i - \mu_{.})]^2 =$
$E(Y_{i.} - \mu_i)^2 + (\mu_i - \mu_{.})^2$, because
$E[(Y_{i.} - \mu_i)(\mu_i - \mu_{.})] = (\mu_i - \mu_{.})E(Y_{i.} - \mu_i) = (\mu_i - \mu_{.})\times 0 = 0$.
■

## Section 14.2

**2.2** We have: $\hat{\mu}_1 = 11.25$, $\hat{\mu}_2 = 17.00$, $\hat{\mu}_3 = 15.50$, $\hat{\mu}_4 = 14.75$, so that
$\hat{\psi} = 11.25c_1 + 17.00c_2 + 15.50c_3 + 14.75c_4$. Also,
$\widehat{Var}(\hat{\psi}) \simeq 1.552(\sum_{i=1}^4 c_i^2)$, and $S^2 = 10.471$, so that
$S\sqrt{\widehat{Var}(\hat{\psi})} \simeq 4.031\sqrt{\sum_{i=1}^4 c_i^2}$. Therefore the required observed confidence interval is:

$$\left[11.25c_1 + 17.00c_2 + 15.50c_3 + 14.75c_4 \pm 4.031\sqrt{\sum_{i=1}^4 c_i^2}\right]. ■$$

## Section 14.3

**3.2** By (14.31) and (14.32), $L_A(y;\mu,\boldsymbol{\beta},\sigma^2) = (\frac{1}{\sqrt{2\pi\sigma^2}})^{IJ}\exp[-\mathcal{S}(\mu,\boldsymbol{\beta})]$, where
$\mathcal{S}(\mu,\boldsymbol{\beta}) = \sum_i\sum_j(y_{ij} - \mu - \beta_j)^2$. For each fixed $\sigma^2$, minimize $\mathcal{S}(\mu,\boldsymbol{\beta})$ with respect to $\mu$ and the $\beta_j$'s subject to the restriction $\sum_j\beta_j = 0$. Doing this minimization by using Langrange multipliers, we find the required MLE's; namely, $\hat{\mu}_A = y_{..}$, $\hat{\beta}_{j,A} = y_{.j} - y_{..}$, $j = 1,\ldots,J$. ■

**3.4** (i) That $\boldsymbol{\eta} = X'\boldsymbol{\beta}$ is immediate, and from this it follows that $\boldsymbol{\eta}$ lies in the vector space generated by the columns (rows) of $X'$.

(ii) Here $I + J + 1 \leq IJ$, or $J \geq \frac{I+1}{I-1}$, provided $I \geq 2$. Thus, for $I \geq 2$ and $J \geq (I+1)/(I-1)$, it follows that $\min\{I+J+1, IJ\} = I+J+1$, and hence rank $X' \leq I+J+1$.

(iii) Parts (a) and (b) are immediate. It then follows that rank $X' \leq I + J - 1$. To see part (c), multiply the columns specified by the respective scalars $a_1, a_2, \ldots, a_{I-1}, b_1, \ldots, b_J$ and add them up to obtain

$$(b_1, b_2, \ldots, b_J, a_1 + b_1, a_1 + b_2, \ldots, a_1 + b_J, \ldots, a_{I-1} + b_1, a_{I-1}$$
$$+ b_2, \ldots, a_{I-1} + b_J),$$

and this vector is zero if and only if
$b_1 = \cdots = b_J = 0 = a_1 = \cdots = a_{I-1}$. The conclusion of independence follows.

So, $\eta$, although it has $IJ$ coordinates, belongs in an $(I + J - 1)$-dimensional space $(I + J - 1 \leq IJ)$, and therefore the dimension of $\eta$ is $I + J - 1$. ∎

## CHAPTER 15

### Section 15.3

**3.2**    **(i)** $F = G$ implies $p = 1/2$ by integration.

    **(ii)** $F > G$ implies $p = 1 - \int_{-\infty}^{\infty} G(y) dF(y)$ (by integration by parts)
$> 1 - \int_{-\infty}^{\infty} F(y) dF(y) = 1 - 1/2 = 1/2$.

    **(iii)** $F < G$ implies $p = 1 - \int_{-\infty}^{\infty} G(y) dF(y)$ (as in part (ii))
$< 1 - \int_{-\infty}^{\infty} F(y) dF(y) = 1 - 1/2 = 1/2$.

    **(iv)** $p > 1/2$ implies $F > G$ by contradiction, and $p < 1/2$ implies $F < G$
likewise. Thus, $F > G$ if and on if $p > \frac{1}{2}$, and $F < G$ if and only if
$p < 1/2$.

    **(v)** $p = 1/2$ if and only if $F = G$ by contradiction. ∎

**3.4**    $C \simeq 58.225$, $C' \simeq 41.775$, $C_1 \simeq 40.2$, $C_2 \simeq 59.8$. ∎

### Section 15.5

**5.2**    As indicated. ∎

# Index

Note: Page numbers followed by *f* indicate figures and *t* indicate tables.

| Distribution | Moment Generating Function |
|---|---|
| Binomial, $B(n, p)$ | $M(t) = (pe^t + q)^n, t \in \Re$ |
| Bernoulli, $B(1, p)$ | $M(t) = pe^t + q, t \in \Re$ |
| Geometric | $M(t) = \frac{pe^t}{1-qe^t}, t < -\log q$ |
| Poisson, $P(\lambda)$ | $M(t) = \exp(\lambda e^t - \lambda), t \in \Re$ |
| Hypergeometric | — |
| Gamma | $M(t) = \frac{1}{(1-\beta t)^\alpha}, t < \frac{1}{\beta}$ |
| Negative Exponential | $M(t) = \frac{\lambda}{\lambda - t}, t < \lambda;$ or $M(t) = \frac{1}{1-\mu t}, t < \frac{1}{\mu}$ |
| Chi-square | $M(t) = \frac{1}{(1-2t)^{r/2}}, t < \frac{1}{2}$ |
| Normal, $N(\mu, \sigma^2)$ | $M(t) = \exp\left(\mu t + \frac{\sigma^2 t^2}{2}\right), t \in \Re$ |
| Standard Normal, $N(0, 1)$ | $M(t) = \exp\left(\frac{t^2}{2}\right), t \in \Re$ |
| Uniform, $U(\alpha, \beta)$ | $M(t) = \frac{e^{t\beta} - e^{t\alpha}}{t(\beta - \alpha)}, t \in \Re$ |
| Multinomial | $M(t_1, \ldots, t_k) = (p_1 e^{t_1} + \cdots + p_k e^{t_k})^n,$ $t_1, \ldots, t_k \in \Re$ |
| Bivariate Normal | $M(t_1, t_2) = \exp\left[\mu_1 t_1 + \mu_2 t_2 + \frac{1}{2}\left(\sigma_1^2 t_1^2 + 2\rho\sigma_1\sigma_2 t_1 t_2 + \sigma_2^2 t_2^2\right)\right],$ $t_1, t_2 \in \Re$ |
| $k$-Variate Normal, $N(\mu, \Sigma)$ | $M(\mathbf{t}) = \exp\left(\mathbf{t}'\mu + \frac{1}{2}\mathbf{t}'\Sigma\mathbf{t}\right),$ $\mathbf{t} \in \Re^k$ |

CPSIA information can be obtained at www.ICGtesting.com
Printed in the USA
LVOW05*1624050215

425797LV00002B/4/P